The Scientific Method Proves that the Theory of Evolution Is False

Book Two of the Philosophy of Science Series

By: Mark My Words

Copyright: July 2016 Onward
Original Release: 12JUL2016
Final Version: 25FEB2020
Current Version: 24FEB2020

Documenting my current understanding of the Scientific Method and demonstrating how I use the Scientific Method in my pursuit of THE TRUTH.

I love studying methodology, or the different versions of the Scientific Method. There are many different Scientific Methods. Observation is my favorite scientific method. Through observation and experience, we can identify and falsify the lies in science. Through observation, we can go directly to "knowing the truth". One of my specialties is using the different Scientific Methods to falsify the deceptions and the lies in science, starting with the Theory of Evolution. If you are intelligent enough and creative enough, all of the different Scientific Methods can be used to falsify the Theory of Evolution. You just have to be willing to give it some study and thought.

The Scientific Methods can be used to prove theories false! It's called "falsifying" a theory or "negating the consequent". Our scientists haven't yet discovered the most powerful and most reliable version of the Scientific Method – the negating the consequent version of the Scientific Method. They don't know what it is or how to use it. The version of the scientific method that our scientists currently use has a couple of serious flaws or logic fallacies built into it; and, our scientists don't even know it because they have never studied the Philosophy of Science or the scientific methods.

In this book, I used Science and the Scientific Method to falsify Darwinism, Materialism, Naturalism, and the Theory of Evolution – in a number of different ways. It's easy to do once you know how, and a great deal of fun too. I have never seen anybody do this before; and, I wish that I would have known how to falsify the Theory of Evolution forty years ago. It's so obvious, I wonder why nobody ever thought of it before now.

The purpose of the Scientific Method is to help us to find THE TRUTH, through a preponderance of the evidence, and a process of elimination. We find the Truth in Science by identifying and eliminating everything that is false. If you successfully identify and eliminate everything that is false, then ONLY the truth will remain. This is Logic 101. The truths that remain after you have eliminated everything that is false ends up being the things that have been experienced and observed by someone, somewhere, sometime. If a phenomenon or theory has never been experienced nor observed by anyone, not even God, then it can't possibly be real or true.

The Scientific Method has no value to us if we use it to convince ourselves that a LIE is TRUE, as the Materialists, Atheists, and Darwinists always seem to do.

The Scientific Methods prove that the Theory of Evolution and the Second Law of Thermodynamics are false. That's what I discovered during my Pursuit of the True Reality of All Things, and during my study and usage of the Scientific Method. It's not a gimmick. I actually use the Scientific Method and negating the consequent to falsify Materialism, Naturalism, Darwinism, Nihilism, Atheism, and their derivatives. It's easy to do once you know how. I'm no longer afraid to have this conversation because I'm no longer a materialist, naturalist, nihilist, or atheist.

In this book, I skipped past all the lies and went directly to the truth. In this book, I also used observed truths to falsify the obvious lies. I'm looking for the Truths in Science. The lies have no value to us. This book was years in the making, and by now it has become a pretty solid piece of information.

This is one of the most powerful and comprehensive books about the Scientific Method and the Philosophy of Science that has ever been designed and made. Buy it, read it, and see if you agree.

—

Collectively, my books stand as a powerful witness that by 2020, the Truths in Science can now be found and known if we are willing to give up the deceptions and the lies. I now know what everything is and how it works. I couldn't say that with complete confidence before 2020, but now I can. I'm running out of questions to ask that need to be answered. I'm no longer fretting about any of this science and philosophy stuff like I used to do when I was a materialist, naturalist, nihilist, and atheist. By allowing quantum mechanics and conserved quanta in to play, we can literally explain everything that comes our way. We no longer have to be enslaved to the deceptions and the lies if we don't want to be.

My Amazon Author Page: https://amazon.com/author/science
The Associated Facebook Page: https://www.facebook.com/MarkMyScience/
The Associated Twitter Page: https://twitter.com/Mark_Me_Words

The Website: https://markme.website/scientific-method/
Associated Website: https://evolution-is-entropy.com/
Philosophy of Science: https://philosophy-of-science.com/
The Associated Forum: https://markme.us/forums/forum/scientific-method/
—

The Scientific Method Proves that the Theory of Evolution Is False
https://evolution-is-entropy.com/
https://philosophy-of-science.com/
https://markme.us/forums/forum/scientific-method-proves-evolution-false/
https://markme.us/forums/forum/scientific-method/
https://markme.website/scientific-method-proves-evolution-false/
https://markme.website/scientific-method/
https://www.amazon.com/gp/product/B01IAAIRT2
—

YOUTUBE VIDEOS ON THIS TOPIC:
Origin: Probability of a Single Protein Forming by Chance
https://www.youtube.com/watch?v=W1_KEVaCyaA
Origin of Life - the Probability of Making a Protein
https://www.youtube.com/watch?v=cQoQgTqj3pU
Evidence for Creation by Outside Intervention
https://www.youtube.com/watch?v=t_IT6cazmuI

Mark My Words

Table of Contents

Introduction 13
Dedication and Acknowledgements.. 13
   "Why the Universe Is the Way It Is" 13
   "Introduction to Psychology" 14
   "Genetic Entropy"... 15
PART 0 – PROOF OF CONCEPT 16
   Philosophies of Science Are Interpretations of Science 17
   Introducing the Denialistic Philosophies of Science 19
   Conserved Means Entropyless... 29
      True Definitions for Entropy... 31
      The Mass-Based Definition for Entropy 32
   Transforming the Physical into the Non-Physical 33
   The Theory of Evolution Proves that God Exists 36
   The Second Law of Thermodynamics Is Obviously False 37
   Evolution Is Entropy or Creation by Random Disorder.................... 40
   Science and the Scientific Method Falsify Naturalism and Darwinism 43
      Source 46
   The Scientific Methods... 47
   Truth in Science 50
      Scientists Lie... 55
      Correlation Is Not Causation 61
      The Null Hypothesis and the Alternative Hypothesis.................. 67
      Truth in Science – Proof of Concept 74
      Using the Scientific Method to Find the Truth.............................. 81
   Using the Scientific Method to Eliminate the Usual Suspects and to Prove the Truth 85
   The Probability of Protein Synthesis by Chance Alone 96
   A Binomial Distribution Table Falsifies Creation by Chance............100
      Chance Cannot Do Causation! 104
   The Perpetual Motion Cycle 109
   Where's the Science? 113
   Summarizing the Different Types of Syntropy...................................123
   The Ultimate LAW of Thermodynamics 127

The Quantum LAW of Thermodynamics	132
What Do Genes Code For?	133
Protein Synthesis	136
Discussing Protein Synthesis or Gene Expression	136
Could You Do the Logistics for Protein Synthesis?	143
Source	146
Entropy Falsifies Evolution	147
Genetic Entropy	148
Can Natural Selection Create?	152
Evolution Is Creation by Death	158
Source	160
Reference Materials	161
Kin Selection	161
Source	164
Reference	164
Proof of God's Existence	164
Source	166
Convincing Proof of God's Existence	167
Shifting One's Perspective	167
Source	169
Defining Syntropy	169
Entropy of Any Kind Cannot Design and Create	174
Summarizing Syntropy	176
Source	179
References with a Similar Theme	179
Syntropy Is Order, Power, Organization, Creation, and Life	180
Syntropy versus Entropy	182
The Fruits of Entropy	182
Source Material	185
Physical Matter and Entropy Were Made	186
PART I — GENETIC ENTROPY PUT AN END TO THE THEORY OF EVOLUTION FOR ME	199
Mutation/Selection Cannot Design and Create Anything!	199
That's Powerful Science, Isn't It?	209
I Enter into Evidence	210

- A Free Tip Which Will Make Your Purchase of This Book Worth Every Cent You Paid 211
- Abductive Reasoning and a Process of Elimination 213
- Now Let's Look at What the Opposition Has to Offer 215
- Waiting for Nothing to Do Something 217

PART II — MATERIALISM IS USELESS AND FALSE 219
- Materialism Is a Denial of Reality 219
  - Spirit Is Light, and Light Is Spirit 221
  - Physical Matter Is Enlivened by Light 223
  - Dark Energy 224
  - Laughing at What They Know Not 225
  - Exotic Dark Matter 227
  - Ordinary Matter 229
  - A Multitude of Different Kinds of Light or Luminosity 230
  - Speculating about the Purest Light 231
  - Materialism Is the Study of Reactions 234
  - Providing a Capstone to Spiritualism and a Gravestone for Materialism 237
  - Light, Truth, and Intelligence 240
  - Intelligence 245
  - The Light of Christ or Dark Energy 246
    - The Light of Christ or the Quantum Fields 247
  - Our Spirits or Our Living Lights Seek Nirvana, Bliss, and Peace 248
  - Matching with Reality Makes It Scientifically Accurate 249
  - Materialism Is THE LIE That Everyone Wants to Believe In 250

PART III — THE SCIENTIFIC METHOD 252
- The Scientific Method Proves That the Theory of Evolution Is False 253
  - Macro-Evolution IS Creation by Evolution 253
  - Creation by Evolution and Creation by Chance Have Been FALSIFIED 254
  - Follow the Logic and the Evidence Wherever They Might Lead You 256
  - The Scientific Method 257
  - APPLYING the Scientific Method TO CREATION BY EVOLUTION 259
  - the Scientific Method DEMANDS THAT WE ELIMINATE THE FALSE AND THE IMPOSSIBLE 260
  - SUMMARIZING WHAT I HAVE LEARNED FROM the Scientific Method 262
  - Eliminate the Impossible Premises 263
  - The Existence of God IS Demonstrable 264

- Removing Accountability and Responsibility Is Sinful and Wrong 265
- Seek the Most Parsimonious Interpretation or Explanation 266
- The Theory of Evolution Proves that God Exists 267
- Show NO Mercy towards Creation by Evolution 268
- The Scientific Method 268
- The True Nature of Proof 275
- Talking About What Chance Can Do for You 277
- The Theory of Evolution Violates the Law of NonContradiction 278
- Each Scientific Discipline IS Proof of God's Existence 281
- My Scientific Theory 284

PART IV — USING THE SCIENTIFIC METHOD TO "PROVE" THAT GOD EXISTS 286
- My Method for Detecting THE TRUTH 288
- Proof of God by the Impossibility of Creation by Evolution 289
  - Introducing the Scientific Method 289
  - A Definition of Terms 290
  - Examining These Scientific Hypotheses 291
  - Running a Few Science Experiments 293
  - Premises, Observations, Comparisons, and Scientific Evidence 293
  - THE CONCLUSION 299
  - Reviewing Our CONSLUSION 299
- "Adventures Beyond the Body: How to Experience Out-of-Body Travel" by William Buhlman. 302
- My Favorite Near-Death and Out-of-Body Testimonies 309
- The Inadequacy of Evolution Proves that God Must Exist 311
  - The Battlefield 311
  - Applying the Scientific Method 313

PART V — THE ULTIMATE MODEL OF REALITY 315
- Abstract – Psyche Is the Ultimate Cause 317
- Aristotle's Four Causes and a Fifth 318
  - Psyche Is the Ultimate Cause 321
- Introduction to Ultimate Cause 323
- The Best Example Which I Currently Have of This Ultimate Reality 326
  - William Buhlman Wrote 327
- Proof of Concept 331
  - Verification and Validation 332

- Falsification 339
- The Ultimate Alternative for Knowing the Truth 342
- Definitions Explaining Ultimate Cause 345
- A Lesson in Quantum Mechanics or Spiritual Mechanics 364
- The Philosophical, Scientific, and Empirical Foundation Supporting Ultimate Cause 374
- Every Spiritual Experience Supports Psyche as the Ultimate Cause 386
- The Ramifications of Psyche and Ultimate Cause 388
  - What is truth, and how do we know it? 388
  - Ramifications of this Ultimate Cause Personality Theory 390
  - Conclusions Regarding this Ultimate Model of Reality 395
- Psyche Is the Ultimate Cause 395
  - Core Set of References 398
- PART VI — SCIENTISM 402
- The Scientific Method Is Based Upon Affirming the Consequent 403
- Scientism 409
  - Defining Scientism 409
  - The Weaknesses of Naturalism 411
  - Science and Human Behavior 412
  - Science Is Based Upon Philosophical Assumptions and Philosophical Interpretations 413
  - Defining Science 414
  - The Purpose of Science 415
  - Behavioral Sciences 416
  - Knowledge 419
  - Positivism and Realism 420
  - Observation 421
  - Definition of Behaviorism 421
  - Some Limitations and Weaknesses of Scientific Methods 422
  - The Affirming the Consequent Logic Fallacy 423
  - Science, Logic, and Reason 425
  - Psyche and Lived Experiences Are Axiomatic 427
  - Unidentified Variables and Unknown Gods 427
  - The Problem of Operationalization 428
  - Objectivity 429
  - Alternative Epistemologies 430

    Qualitative Methods 431
    Lived Experiences 431
    Conclusion and Solution 432
PART VII — VERIFICATION VERSUS FALSIFICATION 434
  Verificationism 434
  Falsificationism 435
  Applying Falsification to Materialism 437
  Reference Articles 438
PART VIII — LIVED EXPERIENCES 441
PART IX — MY SCIENTIFIC DISCOVERIES 459
  A Philosophy of Science for Personality Theory or Psyche Theory 470
  A Psyche Ontology – The Grand Unified Theory of Everything 474
  Evolution Is Entropy 477
    The Fruits of Evolution or Entropy 479
    Scientific Inference 480
    What is the difference between a hypothesis and a theory? 484
    ORIGIN: Probability of a Single Protein Forming by Chance 486
    Quantum Mechanics and Quantum Neuroscience 486
    Evolution Is Entropy 488
    Evolution or Entropy Puts on the Brakes 489
    The Summary 491
    Source Material 494
    References 495
  Using the Scientific Method to Falsify Theories 496
    Begging the Question 496
    Affirming the Consequent 497
    Proving the Theory of Evolution True 500
    Falsifying Theories by Negating the Consequent 504
    Falsifying the Theory of Evolution 506
    Science 2.0 Is the Way that Science Should Have Been Done 511
    Analysis of Falsification 514
    Begging the Question 517
    Mission Statement 521
    The Fruits of These Truths 521
    Source 522

- Reference Materials ... 522
- Obviously Made ... 524
  - Natural Selection Is a Blind Watchmaker ... 525
  - It's Obvious that Genomes Were Made ... 528
  - Belief in Darwinism Requires an Infinite Amount of Blind Faith ... 530
  - The Philosophy of Science Falsifies Darwinism ... 531
  - Computer Science Falsifies Darwinism ... 534
  - Scientific Proof of God's Existence ... 534
    - The Probability of a Protein Forming by Chance ... 536
    - The Probability of Making a Protein ... 536
    - Evidence for Creation by Outside Intervention ... 536
  - Genomes Were Obviously Made ... 538
  - Source ... 539
- My Philosophical Proof of God's Existence ... 540
  - Kalam Cosmological Argument ... 540
  - Proofs Demand Perfectly Sound Premises ... 541
  - My Adjustments to the Kalam Cosmological Argument ... 542
  - My Philosophical Proof of God's Existence ... 545
  - Commentary on My Philosophical Proof of God ... 547
  - Introducing Science 2.0 ... 553
  - Go Out of Your Way to Get the Right Interpretation ... 555
  - A Proof or Argument Has to Be Logically Sound ... 555
  - Observational Reality and Common-Sense Logic ... 559
  - Critiquing Craig's Cosmological Argument ... 559
  - The Appearance of Design ... 562
  - Conclusions ... 563
- Quantum Mechanics Is Spiritual Mechanics ... 567
  - Interpretations of Quantum Mechanics and Science ... 567
  - Source Material ... 572
  - Entropy or the Passage of Time ... 572
  - Quantum Mechanics Is Mind Over Matter ... 573
  - The Self Does Not Die ... 577
  - Orthodox Interpretation of Quantum Mechanics ... 580
  - The Mindful Universe ... 592
  - On the Nature of Things ... 599

    Quantum Physics in Neuroscience and Psychology 599
    The Grand Designer 600
    Synaptogenesis Is Applied Quantum Mechanics 603
    Source 607
    Reference Material 608
PART X – EVOLUTION IS ENTROPY 609
    The Dedicated Website 612
PART XI — FIXING OR REPAIRING THE SECOND LAW OF THERMODYNAMICS..... 616
    The Key to Entropy 621
PART XII – THE OBVIOUS LIES AND HOLES IN OUR SCIENCE 626
    SOCIOLOGY IN THE NEWS 632
    SIGNS OF PSYCHE 634
    FALSIFIED PHILOSOPHIES OF SCIENCE 640
    NEGATING THE CONSEQUENT 658
    NEGATING THE CONSEQUENT 2.0 663
        MATERIALISM 664
        SOLIPSISM 665
        FALSIFYING THE MAJOR DENIALISTIC PHILOSOPHIES 666
        NATURALISM.. 667
        ATHEISM 668
        NIHILISM 669
        INTRODUCING VERIFICATION 671
        BEHAVIORISM OR HARD DETERMINISM 673
        CREATION BY CHANCE 674
        RELATIVISM... 677
        SCIENTISM 678
        PHYSICAL REDUCTIONISM 679
        CREATION EX NIHILO 681
        THE THEORY OF EVOLUTION 681
        THE SECOND LAW OF THERMODYNAMICS 683
        CONCLUSION 685
    COMPARATIVE SCIENCE 688
        WHY OUR SCIENTISTS GET THINGS WRONG 689
        CHANCE CANNOT DO CAUSATION 690

- WHO MAKES THE QUANTUM WAVES AND COLLAPSES THE WAVE FUNCTION? 692
- SYNTROPY MUST EXIST OR WE WOULDN'T EXIST 695
- SOME OF THE THINGS OUR SCIENTISTS HAVEN'T DISCOVERED YET 698
- SOME OF THE FALSEHOODS OUR SCIENTISTS HAVE EMBRACED 709
- TRUTH IS REPEATEDLY EXPERIENCED AND OBSERVED 713
- THE OBVIOUS LIES IN OUR SCIENCE 715
  - DENYING THE PERPETUAL MOTION CYCLE 719
  - CHANCE CANNOT DO CAUSATION 725
  - THE ULTIMATE TEST OF TRUTH IN SCIENCE AND STATISTICS 727
  - THE PRIMARY AXIOM OF SCIENCE AND STATISTICS 731
  - NEGATING THE CONSEQUENT 736
  - OBSERVATION AS A SCIENTIFIC METHOD 743
  - EXPANDING CLOUDS OF HYDROGEN GAS 747
  - INFINITE MEMORY STORAGE CAPACITY 749
  - QUANTUM FIELDS WERE OBVIOUSLY MADE 753
- HOW THINGS REALLY WORK 756
- APPENDIX 773
  - Replacing the Lies with the Truth 773
    - Affirming the Consequent 775
    - How They Trick Us and Deceive Us 775
  - Einstein's Biggest Blunders 776
  - ACKNOWLEDGEMENT OF MENTORS AND TEACHERS 794
  - Peer-Reviewed Science Articles Supporting Intelligent Design! 799
  - NOT A CHANCE 800
  - The Associated Websites 803
  - Other Books by This Author, Mark My Words! 805
  - Salutations! 818

Introduction

During the past year or so, I have had Atheists, Materialists, and Darwinists mock me and ridicule me online, telling me that I don't understand the Scientific Method, that I don't understand anything about THE THEORY OF EVOLUTION, and that I don't understand how SCIENCE really works.

They don't even know me; but, based upon my interaction with them, they have concluded that I don't know anything about Science or the Scientific Method. It was their psychological defense mechanism coming into play. I got them stirred up a bit too much.

These essays in this book were produced and presented to document my current understanding of the Scientific Method, and to demonstrate how I currently use the Scientific Method, the Rules of Science, and SCIENCE itself in my personal Pursuit of THE TRUTH.

The reason we use SCIENCE and the Scientific Method is to help us get at THE TRUTH. These things are of no value to us unless they, in reality, help us to ascertain THE TRUTH. It's BAD SCIENCE, if it is in fact being used to cover up THE TRUTH.

After you have read these essays, you will have to decide for yourself if I know anything about SCIENCE, the Theory of Evolution, or the Scientific Method — or not.

Meanwhile, in the following essays, I will APPLY the Scientific Method to "Creation by Evolution", which is in fact The Theory of Evolution, and see what we end up with.

Dedication and Acknowledgements

The following are the individuals and the books which I referenced most while producing this book

—

"Why the Universe Is the Way It Is"

I am thankful for the book, "Why the Universe Is the Way It Is", by Hugh Ross PhD. Within that particular book of his, Hugh Ross provided some convincing and useful SCIENCE that I had never thought about before, which I can now use to defend against Materialism and to debunk Materialism. My own personal Theory of Reality — both the Physical Reality and the Metaphysical Reality — is based in large part upon what I learned from Hugh Ross and his books. Hugh Ross takes a Scientific Approach to everything, including the Bible; and, I like that a great deal. Thank you!

"Introduction to Psychology"

I am extremely grateful to James W. Kalat PhD for Module 2.1 entitled, "Thinking Critically and Evaluating Evidence", as found in the 9th Edition of his book "Introduction to Psychology" on pages 28 through 36. His Module 2.1 provides THE BEST introduction, summary, and explanation of the Scientific Method that I have encountered to-date. Everything I have done in relationship to SCIENCE during the past year has been based upon what I learned in that particular Module about the Scientific Method and about how the critical evaluation of evidence should be done.

His Module about the Scientific Method is some of my most favorite SCIENCE and reading material. That Module made it clear to me what SCIENCE is all about. That Module explained how the Burden of Proof Methodology should be implemented and employed when dealing with the Non-physical Sciences, the Non-replicable and Non-falsifiable Sciences, the NON-demonstrable Sciences, the Hands-off Sciences, the Historical Sciences, the Immaterial Sciences, the Spiritual Sciences, the Philosophical Sciences, and the Metaphysical Sciences such as Forensics, Archeology, the Theory of Evolution, Paleontology, Materialism, Spiritualism, Philosophy, Theology, God, and Psychology!

Sciences that cannot be REPLICATED ON DEMAND must meet their Burden of Proof through a Preponderance of the Evidence or be dismissed for a lack of evidence.

Kalat's emphasis on Critical Thinking and the Evaluation of Evidence sunk into me so deeply that it has informed every essay, book, and thought that I have had about the Scientific Method and about SCIENCE, ever since I read that Module.

I periodically go back to that Module, re-read it, while thinking about it critically and evaluating each and every sentence. I have written large books based upon what I learned from that Module; and, my essay within this book entitled, "the Scientific Method Proves That the Theory of Evolution is False" was inspired, directed, and motivated by Kalat's Module 2.1 in the 9th Edition of his College Textbook entitled, "Introduction to Psychology".

This is one student who was ready for that particular Teacher and Lesson to appear. I have made good use of that Module ever since. It was in fact my most favorite part of the whole book: and, it motivated me to really dig in and think about the Theory of Evolution critically, logically, and scientifically. Thanks to the information and lessons in that Module, I was able to use the Scientific Method to PROVE to myself how and why The Theory of Evolution or Creation by Evolution must be FALSE.

While "thinking critically" about the Theory of Evolution, I discovered and learned that there NO evidence to consider or "evaluate" when it comes to the Theory of Evolution or Creation by Evolution, because there is NO evidence and can be NO evidence that supports Creation by Evolution or the Theory of Evolution. Study these essays and see if you agree.

"Genetic Entropy"

I express gratitude to John C. Sanford for his written permission to quote from any of his books as much as I want, so long as I cite the source. His books are in fact the top three that I recommend to my readers and critics as the first books about Evolution that they should buy and read. It also helps that a couple of them can be found for free online. John C. Sanford is a wonderful individual, which I can't say about many if not most of the Darwinists whom I have encountered online. When it comes to the Theory of Evolution, the Darwinists themselves are often its worst enemy. If you are trying to take down the Theory of Evolution, all you really need is John C. Sanford's scientific research and his books.

I archived his Free Book which he permitted me to quote at the following link:

http://www.markme.us/wp-content/uploads/2016/06/Biological-Information-Synopsis.pdf

I downloaded and zipped up their much larger FREE 584-page science book For Personal Use at this link:

http://www.markme.us/wp-content/uploads/2016/06/Biological-Information-New-Perspectives.zip

The original sources are listed at this link:

http://www.markme.us/id/

You should have noticed that I quoted John Sanford's science book "Genetic Entropy" 4th Edition THE MOST in this book, because that book was the book which did me the most good when it came to the Theory of Evolution. Go with the best! That book, "Genetic Entropy", became my "bible" when it comes to the Theory of Evolution or Creation by Evolution.

PART 0 – PROOF OF CONCEPT

Science is observation and experience, NOT Materialism and Naturalism.

I present some of the science up-front in this edition of this book because people were complaining that there wasn't any science in the free sample that they were reading online.

If they can't see the science now, it's not my problem.

Philosophies of Science Are Interpretations of Science

Materialism, Naturalism, Darwinism, Nihilism, Atheism, Creation by Chance, Causation by Chance, and even the Theory of Evolution and the Second Law of Thermodynamics ARE philosophies of science. They are different ways that science gets interpreted.

Technically, these things are not Science or Knowledge – they are philosophies of science or different interpretations of science. In fact, Materialism, Naturalism, Darwinism, Nihilism, Atheism, and their derivatives are faulty and falsified interpretations of science. They have been FALSIFIED by Science, the Scientific Methods, Science Experiments, as well as the Observations and the Lived Experiences of the human race as a whole.

For example, everything that exists falsifies the second law of thermodynamics which says that it shouldn't exist. The very fact that you exist and are reading this right now is scientific proof that the second law of thermodynamics is false. The second law of thermodynamics states that the total amount of entropy or disorder in our universe is constantly increasing and that it can never decrease and go to zero. This is NOT what we experience and observe! Instead, we observe constantly conserved Order and Organization all throughout our universe. We don't observe any of the things that the second law predicts, which is why it is false.

Both the Theory of Evolution and the Second Law of Thermodynamics ARE Creation by Chance, Creation by Random Disorder, or Chance Causality. Given the proven and verified fact that chance cannot design and create anything, and that it has NEVER been observed doing so, this reality serves as scientific proof that the Theory of Evolution and the Second Law of Thermodynamics as well as Materialism, Naturalism, and Atheism ARE false because these things are in fact Creation by Chance or the Null Hypothesis and were produced or "caused" by chance alone.

Once we accept the fact that CHANCE cannot design and create anything, then we automatically KNOW that Materialism, Naturalism, Darwinism, Nihilism, Atheism, the Theory of Evolution, the Second Law of Thermodynamics, and every other form of Creation by Chance or Chance Causality is FALSE because creation by chance, chance causality, spontaneous generation, abiogenesis, or creation ex nihilo has NEVER been experienced NOR observed. The false is falsified by the truth, and the truth is repeatedly experienced and observed. If something has never been experienced nor observed, then we KNOW that it is in fact false and that it doesn't exist.

Most of my books are about the Philosophy of Science. I've been trying to find a better, more realistic, and more parsimonious Philosophy of Science that is actually true and that actually matches with the observations and lived experiences of the human race. Such a thing doesn't exist, so I had to make one which I called Science 2.0. I even have a book on the subject which I use to introduce the Philosophy of Science to the world.

I have a half a dozen different books on the Philosophy of Science or the interpretation of science, in which I falsify the faulty and false philosophies of science and then try to develop a better Philosophy of Science or Interpretation of Science that actually matches with Reality, which consists of the Observations and the Lived Experiences of the human race as a whole.

You see, we will NEVER know the truth as long as we keep using faulty and falsified "philosophies of science" to interpret our scientific evidence and scientific data. A faulty, falsified, or false interpretation of the data or evidence doesn't magically provide us with a True Science or a True Philosophy of Science. Lies produce additional lies, which is why I have gone in search of Truth in Science; and, that often starts by getting a Philosophy of Science that is actually true and that has been experienced and observed.

Materialism, Naturalism, Darwinism, Nihilism, and Atheism make the CLAIM that the non-physical does not exist. In other words, these falsified religions or falsified philosophies make the claim that the non-physical, massless, chargeless, and entropyless quantum waves or photons DO NOT EXIST. Darwinism and Naturalism also make the erroneous claim that space and time do not exist. Space and time are obviously massless or non-physical; and, Naturalism or Atheism of any kind makes the claim that the non-physical DOES NOT EXIST. The Quantum Fields are obviously massless, heatless, and entropyless non-physical perpetual motion machines. Materialism, Naturalism, Darwinism, Nihilism, and Atheism make the claim that the non-physical Quantum Fields DO NOT EXIST.

We don't observe the non-existence of these things!

In fact, we don't observe anything that Materialism, Naturalism, Darwinism, Nihilism, or Atheism predicts. We don't observe anything that the Second Law of Thermodynamics predicts either, except for the inability of physical matter to function as a massless and entropyless perpetual motion machine, quantum wave, photon, or quantum field. Likewise, we don't observe anything that the Theory of Evolution predicts, except for the extinction of species. The Theory of Evolution predicts spontaneous generation, abiogenesis, chemical evolution, protein synthesis by chance alone, creation by death or natural selection, creation by chance or random disorder, and macro-evolution. We don't observe any of these things, which means that the Theory of Evolution is false.

Faulty and falsified philosophies of science will NEVER provide us with a True Interpretation of our scientific data or scientific evidence. If something hasn't been experienced nor observed, then it isn't real, and it doesn't exist. A phenomenon has to be experienced and observed in order for it to be real and true. That's just the way it is. This is logical common sense; and, it falsifies the false philosophies of science or the false interpretations of science, because they have never been experienced nor observed.

I've been searching for a Philosophy of Science that is true for five years now. I'm getting a bit closer to finding what I am searching for, as I become more aware of what's involved in my quest.

Mark My Words

Introducing the Denialistic Philosophies of Science

In science and statistics, our theories, hypotheses, philosophies, beliefs, and ideas are considered to be FALSE, unless we can find some way to observe them, experience them, replicate them, predict them, control them, or verify them. This PRESUMPTION OF FALSEHOOD is the main hallmark of Good Science and Real Science, and it is intended to lead us towards True Science or Truth in Science. Our scientific theories should always be considered false, pending some type of verification.

According to this model or methodology, the false is falsified by the truth; and, the truth ends up being everything that has ever been experienced, observed, replicated, or verified by Someone Psyche or Someone Intelligent somewhere, somehow, sometime.

The presumption of falsehood is the way that Science should be done; but, it isn't the way that "science" or philosophy is typically done. There have been some *special pleadings* or *special exemptions* granted that allow a certain group of "scientists" and philosophers to get around this presumption of falsehood, and I will try to explain in this essay how this deceptive process works. By using *special pleadings* or *special exemptions*, this group of "scientists" or philosophers trick us and deceive us, thereby producing Bad Science or False Science or Pseudoscience, which is the philosophies of men masquerading as "science".

If we can use statistics and a science experiment to demonstrate that there is a causal relationship between our independent variable and our dependent variable, then we can begin to conclude that there is a certain level of causation or "truth" in our science experiment, especially if the results can be replicated on demand within other science laboratories.

If we establish causation within our science experiment, then that means that there is a causal relationship between our independent variable and our dependent variable; but, that does NOT mean that our personal interpretation of the scientific data is automatically true. For every piece of verified or replicated scientific data that we generate, there are literally an infinite number of different conflicting personal interpretations that can be given to that one specific piece of verified scientific data. If you don't believe me, then study the hundreds of different interpretations of quantum mechanics that have been developed.

Quantum mechanics and quantum field theory are verified and proven Science. WE KNOW from science experiments, observations, experiences, replication, and verification that quantum mechanics and quantum field theory are REAL and TRUE; however, there is NO agreement in the science community among the scientists as to what quantum mechanics means or how quantum mechanics should be interpreted. How's that possible? It's possible due to a phenomenon that I have observed which I call the Denialistic Philosophies of Science.

The Denialistic Philosophies of Science are a *special case* or a *special pleading*. Each denialistic philosophy denies the existence of one thing or another. Some of the Denialistic Philosophies of Science deny the existence of lots of different things all at once. They all deny the existence of the evidence that falsifies them.

For example, the Materialists, Physicalists, Atomists, and Physical Reductionists deny the existence of the non-physical, the transdimensional, the spiritual, or the quantum. These philosophers and "scientists" start with the conclusion that the non-physical or the transdimensional does not exist. They deny the existence of the non-physical or the transdimensional. Materialism is therefore one of the denialistic philosophies of science.

The Naturalists, Darwinists, and Nihilists deny the existence of the supernatural, the transdimensional, the spiritual, or the quantum. These philosophers and "scientists" start with the conclusion that the supernatural or the transdimensional does not exist. Naturalism is therefore one of the denialistic philosophies of science. By definition, Materialism and Naturalism deny the existence of the quantum; yet, massless, entropyless, non-physical quantum waves or photons have been experienced, observed, replicated, and verified zillions of zillions of times thereby falsifying materialism, naturalism, and their derivatives which claim that the non-physical or the quantum does not exist.

The Denialistic Philosophies of Science are a *special case* or a *special pleading*. Their denials or their hidden premises are assumed to be axiomatically true or automatically true; and then, we are taught in our college classes that the Denialistic Philosophies of Science cannot be falsified. These people lie to us and try to convince us that their theories and their chosen beliefs cannot be proven false; therefore, they must be true. They cheat! These people use a wide variety of logic fallacies to convince us that they are right.

For example, the second law of thermodynamics denies the existence of syntropy or conservation, as well as the quantum fields. The second law of thermodynamics is assumed to be axiomatically true or automatically true; and then, we are taught in our college classes that it can't be falsified. They turn it into a LAW so that it can't be falsified, even though it is in fact demonstrably false. In other words, they grant it a *special pleading* or a *special exemption* so that it can't be falsified by scientific evidence. This is the prime example of Bad Science that is assumed to be axiomatically true. They take a falsehood and make it into a LAW or make it axiomatically true so that it can't be falsified with scientific evidence.

They go through this same process or methodology with all of the other Denialistic Philosophies of Science. They take their denial or their hidden premise, and then they declare it to be axiomatically true. In the complete absence of verification, they *jump straight to the conclusion* that their denials are true. They cheat! They break the rules of science and statistics; and then, they call it the Scientific Method.

Using *circular reasoning*, which is a logic fallacy, the Materialists, Naturalists, Darwinists, Nihilists, Behaviorists, and Atheists define "science" as Materialism, Naturalism, Darwinism, Nihilism, Behaviorism, and Atheism so that their "science" will always be true, and so that their "science" cannot be falsified. We are taught in our science classes in our

public schools that these things cannot be falsified because they are axiomatically true or automatically true. Alas, Materialism, Naturalism, Darwinism, and the other Denialistic Philosophies of Science are based upon a "denial of the evidence" that falsifies them. These people deny the existence of the evidence that falsifies their pre-chosen beliefs.

All of these different philosophies of science use CHANCE as their only creative mechanism. For these people, Creation by Chance and Creation by Random Disorder are their usual suspects, which means that chance is the only creative mechanism that these people have chosen to believe in. However, Creation by Chance violates the Primary Axiom of Science and Statistics which states that chance cannot do causation; therefore, Creation by Chance or Chance Causation is an *oxymoron* and automatically false.

Notice how I keep identifying and falsifying the usual suspects until the truth finally starts to rise to the top. In Real Science and True Statistics, we get at the truth through the process of identifying and eliminating all the falsehoods. If we successfully identify and eliminate all the falsehoods, then only the truth will remain; and, the truth that remains will consist of everything that has ever been experienced, observed, replicated, or verified.

The Denialistic Philosophies of Science are granted a *special pleading* or a *special exemption* and made untouchable. In other words, their denials or hidden premises are defined to be axiomatically true or LAW; and then, we are told that they cannot be falsified or are not allowed to be falsified. This methodology is the very definition of Bad Science because it can't be falsified, and we don't know how bad it is because we are taught in our public schools instead that their denials are the truth. Remember, these people deny the existence of the evidence that falsifies their beliefs. In other words, they cheat.

These people literally define their denials, hidden premises, deceptions, and lies as being axiomatically true or automatically true. We allow them to get away with it because we don't know what they are doing and because they claim to be "scientists". We assume that they are right, when in fact, we should be assuming that they are wrong. We should assume that their denials are false, pending verification or observational evidence proving that they are true. Instead, these people declare their denials or their hidden premises to be axiomatically true, and we let them get away with it because we don't understand the Philosophy of Science.

The Denialistic Philosophies of Science VIOLATE the "presumption of falsehood" rule that should reign supreme in Science, by turning their denials or their hidden premises into AXIOMS and LAWS. This is what they do, is it not? This is what has been experienced and observed.

In Real Science and True Statistics, our theories, hypotheses, philosophies, interpretations, and ideas are assumed to be false – pending observational verification or experimental confirmation. The Denialistic Philosophies of Science turn all of this on its head by assuming or concluding that their denials or hidden premises are true – before doing any actual science or experimentation. Then they call this process "The Scientific Method" in order to give it the official stamp of authenticity and truth. This is what they do, is it not?

The nihilists through the second law of thermodynamics turned their denials into an axiom or a LAW. The second law of thermodynamics denies the existence of the conservation of energy and psyche, conserved quanta (or psyches), the conservation of quantum information within conserved quanta, and the conserved perpetual motion machines that we call the quantum fields. The second law denies the existence of all the order, organization, structure, and syntropy that exists at the quantum level. The second law of thermodynamics denies the existence of conserved order and organization at the quantum level, because it denies the existence of the quantum level or the conserved level. The denialistic philosophies deny the existence of the scientific evidence that falsifies them.

Darwinism or the Theory of Evolution is ALL of the Denialistic Philosophies of Science rolled into one. Darwinism is the KING of the Denialistic Philosophies of Science. Darwinism is materialism, naturalism, nihilism, behaviorism, hard determinism, the second law of thermodynamics, creation by chance, creation by random disorder, and atheism. Darwinism or the Theory of Evolution is all of the denialistic philosophies combined into one. The Theory of Evolution has been declared to be axiomatically true or automatically true. It is treated and portrayed as if it were LAW; and, we are told in our public classrooms that it cannot be falsified. We have been lied to and deceived; and, we don't even know it because nobody understands any of this, and nobody is looking for it. Nobody understands any of this and nobody knows that they are being lied to, because nobody understands the Philosophy of Science and how it works.

The ultimate result of not knowing anything about the Philosophy of Science, including the denialistic philosophies, is that millions of people have believed in a lie for the past 150 years, when in fact spontaneous generation, abiogenesis, chemical evolution, creation by chance, and the theory of evolution were falsified in 1859 by Louis Pasteur; and, they have been false ever since.

In True Science or Real Science, however, our denials should be considered to be automatically false, until we find some observational evidence, experiential evidence, scientific evidence, replicated evidence, or verified evidence to support them. Alas, our denials can NEVER be observed NOR verified. It is philosophically and logically impossible to verify our denials. Our denials and the Denialistic Philosophies of Science, therefore, VIOLATE one of the primary tenets of science and statistics – the presumption of the falsehood of our philosophies or theories, until verified evidence or supporting evidence has been experienced and observed.

Atheism is the philosophy of science or the scientific theory that denies the existence of God as one of its hidden premises and as its scientific conclusion. The non-existence of God cannot be observed or verified; therefore, this scientific theory is automatically false according to the Presumption of Falsehood tenet within science and statistics, thereby leaving the existence of God open for verification instead. The atheists experience the non-existence of God in their own personal lives, and they claim that God's non-existence has been replicated in their own lives; but, their denial or their hidden premise is falsified every time that someone else sees God or experiences God in their own lives. Denials and the Denialistic Philosophies of Science are easily FALSIFIED with observational evidence, experiential evidence, scientific evidence, as well as verified evidence. ONLY one of these is needed to falsify a denial or a denialistic philosophy.

In contrast, a denial can NEVER be observed NOR verified. A denial has to be taken on blind faith alone as being real and true. Denials cannot be verified; therefore, they should be considered to be automatically false, according to the rules of science and statistics. The "presumption of falsehood" tenet within Science and Statistics applies to every denial that has ever been created. Our denials are automatically false, because there will never be any observational evidence to support them. Notice how I keep identifying and falsifying the usual suspects until the truth finally starts to rise to the top.

The ultimate result of not knowing anything about the Philosophy of Science, including the denialistic philosophies, is that billions of people have believed in a lie for thousands of years on this planet, when in fact the Conserved Quanta, the Controlling Psyches within Nature, Universal Consciousness, or the Gods have always existed and will always exist. Psyches are conserved quanta, and thanks to the First Law of Thermodynamics or the Conservation of Energy, those massless, entropyless, non-physical intelligences, psyches, or conserved quanta have always existed, and they will always exist. Conserved quanta cannot be made, and they cannot be destroyed, because the energy or the quantum information within them is always conserved. Atheism denies the existence of all of this. Atheism is the ultimate result of denying what has been experienced and observed and verified.

Denials are easily falsified with observational evidence or experiential evidence. The atheists literally take a monumental blind leap of faith into nothing. Whereas, the thousands of times that our resurrected Lord Jesus Christ has been experienced and observed after He rose from the dead automatically FALSIFIES their atheistic denials or their blind leap of faith into nothing.

https://www.youtube.com/results?search_query=NDE+Jesus

Evidence cannot be refuted or falsified! Evidence can only be clarified, replicated, or verified. You can't refute or falsify the thousands of observations and experiences that have taken place with our resurrected Lord Jesus Christ. You can deny their existence, refuse to accept them, reject the evidence, and pretend that it doesn't exist. But, you can't refute it or falsify it, because it has been entered into evidence.

Instead of refuting the evidence, which can't be done, we can instead start to compare the evidence with the rest of the evidence, and then keep doing so until we achieve a preponderance of the evidence.

My best friend has seen Jesus Christ. He KNOWS that our resurrected Lord Jesus Christ exists. He has compared his experiences with those on YouTube, and he has told me about the ones that match with his own experiences or VERIFY his own experiences. Truth verifies truth.

Every now and then, we would watch an NDE account on YouTube, and my friend would say, "Oh, that one's fiction. It's a story that they made up, because that's not the way NDEs and out-of-body experiences really work." My best friend used his own personal out-of-body experiences and his experiences in the presence of Christ to help him to identify the ones on YouTube that really happened, and to separate those from the fictional stories and the lies. Truth verifies truth through a preponderance of the evidence; and over time, the preponderance of the evidence will eventually identify and falsify the deceptions and the lies. This is the way that Science or Knowledge should work, rather than denying the existence of the evidence that we personally don't like. My best friend used others' experiences to verify or to confirm his own. This is the way that Science should be done!

By definition, one-time events cannot be replicated, and they can at times be difficult to verify; but, those difficulties do not make them automatically false as the materialists, naturalists, and atheists claim. Evidence cannot be refuted. Evidence has to be explained. Evidence has to be dealt with. By pursuing a preponderance of the evidence, evidence and the interpretation of that evidence can be clarified or enhanced over time.

Observational evidence as well as experiential evidence cannot be refuted or falsified. The only way to falsify evidence is through a preponderance of other evidence. Some evidence might be hard to replicate or verify; but, evidence cannot be refuted. Evidence has to be considered and compared until we achieve a preponderance of the evidence. Of course, when it comes to evidence, the easier it is to replicate the better, and the more that it has been verified the better; but, one-time events or the "perfect storm" cannot be falsified or refuted because they are evidence, and evidence cannot be refuted or falsified. Evidence has to be clarified or explained or replicated or verified, because it can't be refuted. We keep comparing evidence with other evidence, until we achieve a preponderance of the evidence. That's the only way to deal with evidence. When it comes to evidence, in Science and Statistics, we are supposed to deal with it rather than dismissing it, hiding it, or destroying it as the materialists and naturalists do.

WE KNOW that certain phenomena and events are true because they have been experienced and observed by everyone. WE KNOW that massless, entropyless, non-physical photons, magnetic waves, microwaves, radio waves, and x-rays exist because they have been experienced and observed by everyone; and, their proven and verified existence FALSIFIES materialism, naturalism, and their derivatives which claim that the non-physical does not exist. In contrast, one-time events are by definition impossible to replicate and difficult to verify. When it comes to the one-time events, their truthfulness is established through a preponderance of the evidence.

Just because the earthquake or the volcano happened only once in that area, it doesn't mean that that one-time event is automatically false because it can't be replicated on demand. However, if a phenomenon can be replicated on demand, then WE KNOW that the phenomenon is real and truly exists. ALL of the quantum phenomena or non-physical phenomena that we have discovered can be replicated on demand; therefore, WE KNOW that they are real and truly exist. WE KNOW that non-physical physics, transdimensional physics, or quantum mechanics is TRUE because everything within it has been replicated and verified. They say that quantum field theory is the best proven and most verified Science on this planet. The proven and verified existence of massless, entropyless, non-physical quantum waves and quantum fields FALSIFIES Materialism, Naturalism, Darwinism, Nihilism, and the Second Law of Thermodynamics which claim that the non-physical does not exist.

Do you see how that works?

One-time events can be difficult to verify; but, when thousands of people have seen, talked with, touched, embraced, experienced, and observed our resurrected Lord Jesus Christ, then we can safely state that that particular type of "one-time event" or that particular phenomenon has been repeatedly verified by lots of different people. Verification increases the likelihood that specific events are real and truly happened. Since millions of different people have had out-of-body experiences, then that repeated verification of that specific phenomenon greatly increases the likelihood that that type of phenomenon is real and truly exists.

https://www.youtube.com/results?search_query=NDE+Jesus

1 Corinthians 15: 6: Resurrected Lord Jesus Christ seen by over 500 people.

3 Nephi 11: 1: Resurrected Lord Jesus Christ seen by over 2500 people.

Doctrine and Covenants 76: 22-24:

And now, after the many testimonies which have been given of him, this is the testimony, last of all, which we give of him: "That he lives!" For we saw him, even on the bright hand of God; and we heard the voice bearing record that he is the Only Begotten of the Father — that by him, and through him, and of him, the worlds are and were created, and the inhabitants thereof are begotten sons and daughters unto God.

A Modern-Day Encounter with God the Father and Jesus Christ

When the light rested upon me I saw two Personages, whose brightness and glory defy all description, standing above me in the air. One of them spake unto me, calling me by name and said, pointing to the other — "This is My Beloved Son. Hear Him!"

Thousands of different people have encountered, observed, experienced, spoken with, and touched our resurrected Lord Jesus Christ; therefore, they know that He truly exists. The best and fastest way to find and know the truth is to live it, experience it, and observe it for yourself, or to choose to trust someone who has.

You can easily falsify denials; but, you can't refute evidence. You can falsify personal interpretations of the evidence; but, you can't falsify experiential evidence or observational evidence. Evidence has to be dealt with. Evidence has to be explained, clarified, replicated, and verified because it can't be falsified or dismissed from evidence. The worst that we can do with evidence is to reject it, or to refuse to believe it, or to deny its

existence, or to falsify it through a preponderance of other evidence; but, the evidence still exists whether we know it or not.

When it comes to evidence, you can use other evidence or the preponderance of the evidence to clarify it, mitigate it, or demonstrate that it is most likely true or false. In contrast, we can't use our denials to prove anything, as the materialists and naturalists try to do. We can't use our denials as scientific evidence, as Darwinists and Atheists try to do. When it comes to Real Science and True Statistics, you have got to have evidence to support your case; but, our *denials of evidence* or *rejection of evidence* by definition and in principle have NO evidence to support them. There is NO evidence to support our denials.

When it comes to evidence, you can *reject it* or *deny its existence* or *assume that it doesn't exist*, but you can't refute it. *Rejecting evidence, denials used as scientific evidence, denying the existence of things we personally don't like, using our denials or hidden premises as our scientific conclusion*, and *assuming that evidence does not exist* are logic fallacies. These are just some of the logic fallacies upon which materialism, naturalism, and the denialistic philosophies are based.

Remember, you can't find and know the truth by *rejecting it* or by *denying its existence* as the materialists and naturalists try to do. The truth can ONLY be based upon evidence. The truth is always based upon the things that have been experienced or observed. The truth can never be based upon our denials, according the "presumption of falsehood" rule in science and statistics. If we are doing Science and Statistics correctly, then our denials are presumed to be false, because there will never be any observational evidence or verified evidence to support our denials. Notice how I keep identifying and falsifying the usual suspects until the truth finally starts to rise to the top. Rinse and repeat!

Our denials cannot be verified; but, they can be falsified! Therefore, our denials and the Denialistic Philosophies of Science are automatically false and presumed false according to the rules for science and statistics, because there can never be any observational evidence to support our denials. Our denials can only be taken on blind faith alone as being real and true. Since our denials cannot be verified with observational evidence, our denials are worthless and should not be used as scientific evidence, as the materialists and naturalists do with their denials.

The Denialistic Philosophies of Science erroneously use their denials as their hidden premises, as scientific evidence, as their scientific conclusion, and as scientific proof that their scientific conclusions are true. *Begging the question* (which is another logic fallacy), the denialistic philosophies use their denials as their hidden premises, and then use those same denials or hidden premises as their scientific conclusion as well. Using your premise as your scientific conclusion is a logic fallacy known as *begging the question*, wherein, they use their hidden premises or their denials as scientific proof that their conclusion is true.

The hidden premise in materialism denies the existence of the non-physical; and, materialism concludes as its "axiomatic scientific conclusion" that the non-physical does not exist. They are using their hidden premise or their denial as their scientific conclusion. That's *begging the question*. That's cheating! It is a deception and a lie. Our denials are literally deceptions and lies. Our denials are self-deceptions that we are then turning into hidden premises and using as scientific proof that our pre-chosen conclusions are true. In yet another *special pleading*, and as the premier example of *circular reasoning*, the Denialistic Philosophies of Science use their hidden premises or their denials as their scientific evidence and their scientific conclusion. They cheat!

The hidden premise in naturalism denies the existence of the supernatural or the quantum; and, naturalism concludes as its "axiomatic scientific conclusion" that the quantum or the supernatural does not exist. They are using their hidden premise or their denial as their scientific conclusion. That's *begging the question*.

The ultimate result of not knowing anything about the Philosophy of Science, including the denialistic philosophies, is that billions of people have believed in materialism, atomism, or physicalism for thousands of years on this planet – when in fact, the massless, entropyless, chargeless, non-physical photons or quantum waves have always existed during the entire time and have falsified materialism and naturalism for the billions or trillions or zillions of years that the photons have been in existence in this part of our multiverse. Materialism and naturalism have always been false for all eternity – for the entire eternity that massless, entropyless, non-physical conserved quanta have been in existence. That's the ultimate result of not knowing anything about the Philosophy of Science, including the Denialistic Philosophies of Science – billions of people on this planet have believed that the lies or the denials or the hidden premises within Materialism, Naturalism, Darwinism, Nihilism, Behaviorism, Hard Determinism, and Atheism are true. Their ignorance feeds the lies that they have chosen to believe in.

The hidden premise in nihilism or the second law of thermodynamics denies the existence of syntropy or the quantum laws of conservation; and, nihilism concludes as its "axiomatic scientific conclusion" that the quantum fields, the law of quantum information conservation within conserved quanta, and the conservation of energy or information or intelligence do not exist. The second law of thermodynamics starts with the conclusion that quantum or conserved order and organization do not exist. They are using their hidden premise or their denial as their scientific conclusion. That's *begging the question*.

One of the hidden premises in atheism denies the existence of Intelligences, Psyches, Designers, Creators, and Gods; and, atheism concludes as one of its "axiomatic scientific conclusions" that Conserved Quanta, Psyches, Designers, Creators, and Gods do not exist. They are using their hidden premise or their denial as their scientific conclusion. That's *begging the question*.

Darwinism is all of these different denialistic philosophies combined into one. Furthermore, Darwinism or the Theory of Evolution is Creation by Chance. Creation by chance or chance causation VIOLATES the Primary Axiom of Science and Statistics which states that chance cannot do causation. Darwinism is doubly false because it is Creation by Chance, as well as being the KING of the Denialistic Philosophies of Science.

Notice how I keep identifying and falsifying the usual suspects until the truth finally starts to rise to the top. The Denialistic Philosophies of Science are falsified by conserved quanta, massless and entropyless photons or quantum waves, the quantum fields, quantum mechanics, action at a distance, quantum field theory, and everything else that they deny the existence of. The denialistic philosophies are falsified by the things that have been experienced, observed, replicated, or verified. In contrast, their denials can never be verified nor proven true.

The Philosophy of Science and the denialistic philosophies may in fact be the best or the coolest thing that I have ever discovered in science and philosophy due to its massive scientific explanatory power. It teaches us how to separate the lies from the truth. Once we realize that our denials are automatically false, then a whole host of lies within "science" go out the window into the garbage can where they belong. As you can probably tell, I'm no longer afraid of the truth like I used to be when I was a materialist, naturalist, nihilist, and atheist. I'm now able to tell it as I see it, because I know for a fact that the denialistic philosophies are deceptive lies.

According to the "presumption of falsehood" rule in science and statistics, our denials are automatically false, because there will never be and can never be any observational evidence to support them. Our denials cannot be verified; therefore, our denials are automatically false and are presumed to be false according to the rules for science and statistics.

This is what you need to know as way of introduction to the Denialistic Philosophies of Science such as Materialism, Naturalism, Darwinism, Nihilism, Behaviorism, Hard Determinism, Atomism, Physical Reductionism, Classical Realism, Atheism, the Theory of Evolution, Creation by Chance, and the Second Law of Thermodynamics. Their denials or their hidden premises are automatically false, as well as being demonstrably false; therefore, their denials – which they also use as their scientific conclusions – are also false. The Denialistic Philosophies of Science are automatically false because they violate the rules of science and statistics, and because there will never be any observational evidence to verify their hidden premises or their denials. The denialistic philosophies deny the existence of the evidence that falsifies them.

Now that I know about the Denialistic Philosophies of Science, and now that I know that their denials or their hidden premises are false, these people can no longer trick me and deceive me, because I know what they are doing and know how they are doing it. The Denialistic Philosophies of Science are automatically false, because there will NEVER be any observational evidence or verified evidence to support the truthfulness of their denials. It's philosophically and logically impossible to prove a negative true or to prove a denial to be true. Their denials cannot be verified; therefore, their denials are automatically false according to the "presumption of falsehood" tenet or rule within Science and Statistics. Their denials are automatically false – pending verification; and, their denials can never be verified.

Since I KNOW that their denials are false, I eliminate the denialistic philosophies first thing when doing Science, so that I find myself looking for the True Cause of the event instead. I have discovered that the true cause of each event or phenomenon that is caused ends up being some type of eternal and everlasting Conserved Quantum, Intelligence, or Psyche. Conserved Quanta or Conserved Psyches are the only thing that we know of that are capable of designing, creating, and doing causation at every level of existence because the intelligence or the information or the memories within them are conserved.

Energy contains information or memories; and, energy is conserved. This is how our memories survive the death of our physical brain and show up in our after-death life review. Our conserved memories or our personal conserved quantum information are being stored within our psyche or our conserved quantum. This is the only thing that makes logical sense and explains the quantum evidence that we have experienced and observed. For example, the physical laws and the physical constants are conserved quantum information, that survive the death of our physical brain. This is what we have experienced and observed, is it not? Conserved quantum information has to be stored within conserved quanta. There is no other way that it can be done. Psyches or intelligences are conserved quanta.

As you can tell, I'm no longer afraid to have this conversation, because I'm no longer a materialist, naturalist, nihilist, or atheist.

Notice how I keep identifying and falsifying the usual suspects until the truth finally starts to rise to the top. If we successfully identify and eliminate everything that is false, then ONLY the truth will remain; and, the truth that remains will consist of everything that has ever been experienced, observed, replicated, or verified by Someone Psyche or Someone Intelligent somewhere, somehow, sometime. This is what I have experienced and observed. This is the greatest Philosophy of Science that has ever been developed. It immediately separates the truths from the lies in science and philosophy.

Mark My Words

Conserved Means Entropyless

Have you ever wondered what "conserved" truly means when it comes to physics?

Conserved means entropyless or syntropic. Conserved means that all of the entropy in the system has gone to zero and ceased to exist.

So, what is entropy?

Entropy is a catch-all term that is used to describe a variety of different physical phenomena.

Entropy is the aging process, which means that entropy consists of all the different clocks or timing mechanisms that are built into physical matter or a physical atom. Entropy is exclusively a physical phenomenon. Entropy doesn't exist at the quantum level in the Ageless Realm, or the Conserved Realm, or the Entropyless Realm.

Entropy is often defined as a transfer of heat from a hot system to a cold system. NO heat or NO thermodynamics, then NO entropy. There is NO entropy at absolute zero, both at the physical level and at the quantum level. Entropy goes to zero and ceases to exist, when the temperature or heat goes to absolute zero and ceases to exist.

Entropy is also mass or resistance to acceleration. A transdimensional or non-physical system, that has NO mass or NO resistance to acceleration, has NO entropy. That means that a transdimensional system, or a quantum system, or a massless system, or a non-physical system is conserved because it is entropyless. When mass goes to zero, entropy goes to zero.

Entropy is purely a physical phenomenon. Entropy doesn't exist at the quantum level in the Quantum Realm or the Massless Realm. Whenever mass or resistance to acceleration goes to zero, entropy goes to zero and ceases to exist. This truth is called the Quantum Law of Thermodynamics, which states that there is no mass, no resistance to acceleration, no thermodynamics, and therefore NO entropy at the quantum level in the Transdimensional Non-Physical Realm. Entropy doesn't exist at the quantum level, which means that everything is conserved at the quantum level.

Powerful. Logical. Rational. Obvious. Simple. Real. True.

These ideas are demonstrably true. By studying the equations for heat and entropy, one can see with his own eyes that this is true.

The equation for heat (**Q**):

Q = mcΔT

The typical equation for entropy:

S = Q/ΔT

Put the equation for heat into the equation for entropy, and then solve for entropy, in order to find another powerful and true definition for entropy:

S = mcΔT/ΔT

Solving this equation, we get the following: **S = mc**.

Entropy (**S**) is the heat storage capacity of mass (**mc**). Simple. Parsimonious. Logical. Real. True.

Entropy is mass's heat storage capacity. No mass, no heat, or no heat storage capacity, then NO entropy. Entropy (**S**) goes to zero whenever mass (**m**), heat (**Q**), thermodynamics (**Q** or **T**), temperature (**T**), or mass's heat storage capacity (**mc**) goes to zero.

Scientists lie, but the equations don't lie.

S = mc.

Entropy is mass's heat storage capacity. Entropy goes to zero when mass goes to zero. This is one of the greatest scientific discoveries in the whole universe, due to its explanatory power.

Conserved means entropyless.

—

True Definitions for Entropy

Here we have presented five true and demonstrable definitions for entropy.

1. Conserved or Syntropic or Exergic is the antonym for Entropy. Conserved means entropyless. Entropy means non-conserved. Entropy, mass, thermodynamics, and heat are not conserved. This is the Ultimate Law of Thermodynamics, which tells us what is not being conserved. Conserved means entropyless or syntropic. Conserved means that all of the entropy in the system has gone to zero and ceased to exist. This scientific truth explains why we exist. Energy is conserved. The Quantum Realm or the Transdimensional Non-Physical Realm is conserved.

2. Entropy is the aging process or the different clocks that are built into a physical atom. Entropy doesn't exist at the quantum level in the Transdimensional Realm, or the Quantum Realm, or the Ageless Realm, or the Conserved Realm. Conserved or Syntropic means eternal and everlasting. Entropy is not conserved. Entropy is temporary.

3. Entropy is the transfer of heat from a hot system to a cold system. NO heat, then NO entropy. When heat or temperature or thermodynamics goes to zero, then entropy goes to zero. When the temperature is absolute zero, then the amount of entropy in the system is absolute zero, whether we are talking about a transdimensional system or a physical system.

4. Entropy is mass or resistance to acceleration. NO mass, then NO entropy. Whenever mass goes to zero and ceases to exist, entropy goes to zero and ceases to exist. This is obviously true because it has been experienced and observed. Photons or quantum waves are obviously massless, heatless, chargeless, entropyless, conserved or pure energy, and have NO resistance to acceleration. Photons, quantum waves, virtual particles, thoughts, and superconductors function best and most efficiently at absolute zero, where entropy has gone to zero and ceased to exist. This is what has been experienced and observed.

5. Entropy is mass's heat storage capacity. **S = mc**. No mass, no heat, or no heat storage capacity, then no entropy. This is obviously true. The equations don't lie. Scientists lie, but the equations don't lie. This is one of the greatest scientific discoveries in the universe due to its explanatory power. It explains everything that has ever been experienced and observed. It explains why we are

here and why we exist. Everything at the quantum level or the massless level or the entropyless level is being conserved, in one form or another. Entropy is purely a physical phenomenon. Entropy ceases to exist at the quantum level, or the massless level, or the conserved level. Entropy is not conserved.

—

The Mass-Based Definition for Entropy

Here are the details for the mass-based definition for entropy:

Q = mcΔT

Heat (**Q**) is equal to mass (**m**) times that mass's heat storage capacity (**c**) times the change in temperature (**ΔT**).

S = Q/ΔT

Entropy (**S**) is equal to heat (**Q**) divided by the change in temperature (**ΔT**). Combining these two equations into one, we get the following equation. Notice that the temperature (**T**) is irrelevant, because the temperature cancels out.

S = mcΔT/ΔT

Solving this equation, we get the following: **S = mc**.

Entropy (**S**) is the heat storage capacity of mass (**mc**).

Simple. Parsimonious. Demonstrable. Replicable. Verifiable. Logical. Conservative. True.

This is the mass-based definition for entropy. It's there in plain sight just waiting for someone to find it and present it to the universe.

Notice that the temperature (**T**) is irrelevant, because the temperature cancels out. This process works and this equation holds true no matter what the temperature might happen to be. It even works at absolute zero Kelvin (**T or K = 0**). Everything continues to function perfectly fine at the quantum level, even at absolute zero. In fact, by studying photons and superconductors, we observe that absolute zero maximizes the efficiency and functionality of quantum objects such as photons, quantum waves, superconductors, virtual particles (thoughts), and quantum fields. Absolute zero is maximum efficiency and maximum order at the quantum level thanks to the Quantum Fields and the Conservation of Energy.

S = mc

This simple equation for entropy is possibly the greatest scientific discovery of all time. It's demonstrably true. It falsifies the second law of thermodynamics, which claims that entropy is conserved. Entropy is NOT disorder as the second law of thermodynamics erroneously claims. Entropy is mass's heat storage capacity.

No mass, then no entropy. No heat, then no entropy. Mass, heat, entropy, resistance to acceleration, and physical matter are NOT conserved. Their amount is constantly changing. Only the underlying energy is conserved.

c = the specific heat capacity of the mass under consideration

m = mass

Entropy (**S**) is mass's heat storage capacity (**mc**). No heat storage capacity, then no heat. No mass and no heat, then NO entropy. NO thermodynamics, then NO entropy. When temperature goes to absolute zero, entropy goes to absolute zero. Simple. Logical. Parsimonious. True. Science doesn't get any better than this.

This is what has actually been experienced and observed.

Entropy (**S**) equals mass (**m**), particularly the heat storage capacity of mass (**mc**). Entropy is mass's heat storage capacity (**S = mc**). Entropy is also resistance to acceleration or mass. This is the mass-based definition for entropy. No mass, then no entropy. It's that simple.

Conserved means entropyless.

According to the Ultimate Law of Thermodynamics, mass, heat, entropy, and thermodynamics are NOT conserved, whereas, everything in the Ageless Realm, the Massless Realm, the Transdimensional Realm, or the Quantum Realm is conserved in one form or another. According to the Quantum Law of Thermodynamics, there is no entropy and no thermodynamics at the quantum level in the Transdimensional Realm or the Massless Realm. This is obviously true, or we wouldn't exist in the first place.

Mark My Words

Transforming the Physical into the Non-Physical

If you don't understand how to transform the physical into the non-physical, then you don't understand Science, because Nature's Psyche or Nature's Intelligence within our suns or stars is constantly transforming physical atoms INTO non-physical quantum waves or photons all the time. Throughout the whole process, the total amount of energy is constantly conserved. This phenomenon has actually been experienced and observed; therefore, WE KNOW that it is real and truly exists, which makes it Science or Knowledge or Truth.

How would you transform a physical atom into something non-physical? This is Real Science, and you need to know how to do it, or you will never be able to understand Science.

In order to transform a physical atom into a massless, entropyless, chargeless, heatless, ageless, and conserved quantum wave or photon, YOU MUST transform the mass or the resistance to acceleration INTO infinite acceleration instead. A photon or a quantum wave goes from zero to the speed-of-light instantly – that's infinite acceleration. The ONLY way you can successfully achieve infinite acceleration is to temporarily remove the mass or the resistance to acceleration.

Simple. Logical. Parsimonious. Obvious. Real. Observed. Experienced. True.

The psyche or the intelligence within the quantum wave or the photon doesn't have to limit itself to the speed-of-light if it doesn't want to. Not only is a photon or a quantum wave capable of infinite acceleration, it is also capable of an infinite velocity or omnipresence. In other words, a photon or a quantum wave can quantum tunnel or teleport to its destination instantly, if it chooses to do so. Photons or quantum waves that travel faster than the speed-of-light are called tachyons. Tachyons are spirit matter. They can quantum tunnel or teleport at will, or they can choose to create a reference frame within their own dimension, phase, or timeline and give the appearance that they are standing still.

In order to transform a physical atom into a massless, entropyless, chargeless, heatless, ageless, resistanceless, and conserved quantum wave or photon, YOU MUST also transform the entropy (mass's heat storage capacity), the internal clocks (the aging process), the stored heat, or the non-conserved parts of a physical atom INTO something that is entropyless, ageless, heatless, and conserved instead. You must know how to eliminate entropy and transform that entropy INTO syntropy or exergy instead. You must accept the FACT that photons or quantum waves are entropyless, ageless, massless, heatless, chargeless, resistanceless, syntropic, exergic, and conserved. If you are unable to accept this obvious truth, then you will NEVER be able to understand Science and how it works.

This process is an integral part of Quantum Field Theory. You have to understand Quantum Field Theory and you have to KNOW how to transform a physical atom into something non-physical, or you will never know how Science works in the real world.

How would you transform something non-physical into a physical atom? This is Real Science, and you need to know how to do it, or you will never be able to understand Science.

In order to transform a massless, entropyless, chargeless, heatless, ageless, resistanceless, and conserved quantum wave or photon or tachyon into a physical atom, YOU MUST transform the infinite acceleration and the infinite velocity INTO mass or resistance to acceleration instead. When a photon or a quantum wave chooses to STOP, it goes from the speed-of-light to zero instantly – that's infinite resistance to acceleration. The ONLY way you can successfully achieve infinite resistance to acceleration or infinite deceleration is to stop that photon deliberately or to localize it in space-time instantly.

Simple. Logical. Parsimonious. Obvious. Real. Observed. Experienced. True.

The psyche or the intelligence in the photon, or the quantum wave, or the tachyon (spirit matter) can choose to stop its motion or infinite acceleration or omnipresence instantly. This phenomenon has actually been experienced and observed, has it not? We are seeing the light all the time because the psyches or the intelligences within the photons are choosing to STOP rather than passing through us as if we weren't even there. Quantum waves or photons don't have to stop if they don't want to. They can pass through us unphased as if we weren't there, if they choose to do so.

Whenever a photon or a quantum wave CHOOSES to stop, it transforms its omnipresence and infinite acceleration INTO some type of mass, heat, entropy (mass's heat storage capacity), clocks (aging process), or resistance to acceleration instead. This has actually been experienced and observed, has it not? Typically, a photon of light CHOOSES to transform itself into heat when it stops, rather than transforming itself into mass or physical matter. We feel the heat on our skin, but we don't typically observe mass or physical matter forming onto our skin. But, there's nothing (except for God) to stop a photon or a quantum wave from transforming itself into a physical atom or into mass if it wants to do so.

We certainly can't choose what a photon or a quantum wave will become when it finally CHOOSES to stop and transform itself into something else instead. God is the only one that we know of who can do that type of thing. God can transform tachyons, or spirit matter, or photons, or quantum waves INTO physical atoms at will because He has been observed doing so. God can command the psyches or the intelligences within the photons, tachyons, or spirit matter to transform themselves INTO physical atoms anytime and anywhere, and they will do so because they have covenanted with God to obey His commands. That's why He is God, and we are not.

This whole process is called the Perpetual Motion Cycle ($E = mc^2$). It's the Perpetual Motion Cycle because it has always existed, and it will always exist, because it is constantly conserved. The total amount of energy in the system never changes. The energy is simply changed from one form to another and then back again; and all the while, the total amount of energy is constantly conserved.

The Perpetual Motion Cycle is a cycle.

God or Nature's Psyche is constantly transforming physical atoms INTO massless, resistanceless, heatless, chargeless, entropyless, infinite acceleration, and conserved photons or quantum waves instead, within our suns or our stars. Then whenever a photon or a quantum wave CHOOSES TO STOP, God or the intelligence or the psyche within that photon or quantum wave chooses what form that former quantum wave will now become. When a photon or a quantum wave chooses to stop, it typically transforms itself into localized heat, mass, entropy (mass's heat storage capacity), resistance to acceleration, or clocks (the aging process) instead. A photon or a quantum wave chooses to stop, and then it chooses what it wants to become now that it has stopped.

This is the Perpetual Motion Cycle ($E = mc^2$). It will never get old, and it will never wear out, because it is constantly conserved. This is what has actually been experienced and observed; therefore, WE KNOW that it is real and truly exists. The Perpetual Motion Cycle is essentially a conserved or a transdimensional or a non-physical or a quantum phenomenon, which is why the Materialists, Naturalists, Nihilists, and Atheists have never discovered it. They aren't looking for it, and they don't want it to be true.

The Perpetual Motion Cycle is perfectly conserved order and organization at the quantum level, which means that the Perpetual Motion Cycle is syntropic, exergic, eternal, everlasting, and conserved. When it comes to physics, the Perpetual Motion Cycle explains everything that has ever been experienced and observed. The Perpetual Motion Cycle explains why we exist in the first place.

Notice that the proven and verified existence of the Perpetual Motion Cycle FALSIFIES the Second Law of Thermodynamics, which erroneously claims that the Perpetual Motion Cycle does not exist and that the total amount of disorder in the universe is constantly increasing and can never go to zero. They both can't be true because they are mutually exclusive. They falsify each other. The second law of thermodynamics was designed to falsify the Perpetual Motion Cycle and to convince people that it does not exist. Since the Perpetual Motion Cycle is obviously true, that makes the second law of thermodynamics obviously false.

Mark My Words

The Theory of Evolution Proves that God Exists

The Materialists, Naturalists, Darwinists, Nihilists, and Atheists have *jumped to the conclusion* that CHANCE can design and create anything, if given enough time to do so. That's what they teach, preach, and believe, is it not?

They have NEVER tested that hypothesis, because they can't. Chance doesn't do anything without an intelligent assist from some type of intelligent being. By definition and by axiom, CHANCE cannot design and create anything because CHANCE cannot do choice and causation. Once chance starts doing causation, then it is no longer chance! This is the fundamental axiom of Science and Statistics. Chance and causation are mutually exclusive. This is the Truth in Science and Statistics, whether we realize it or not. Chance cannot do causation. Saying that it can doesn't change the fact that it can't. Scientists are lying to us whenever they say that CHANCE can do choice, design, or causation if given enough time to do so.

The inability of the Theory of Evolution to meet its burden of proof through a preponderance of the evidence ends up becoming Scientific Proof of God's Existence or Scientific Proof of God's Necessity, whether we realize it or not. This simple observation is based upon the "best explanation rule" or the "parsimony rule" in Science, sometimes known as Occam's Razor.

Creation by Chance, or Causation by Chance, or Creation by Random Disorder, or Creation by Entropy, or Creation Ex Nihilo, or Creation by Blind Luck –

in other words, the Theory of Evolution – is NOT the best, nor the simplest, nor the most logical, nor the most rational, nor the most parsimonious explanation that can be given to scientific theories, scientific evidence, science experiments, and scientific data because chance of any kind cannot do causation. A LIE, such as Creation by Chance or the Theory of Evolution, cannot serve as the best explanation or the most parsimonious explanation for the scientific evidence. That's just logical common sense.

I wrote a huge book on this topic.

The Theory of Evolution Proved to Me that God Exists

https://www.amazon.com/gp/product/B01HZYBZ7K/

Most people don't understand this concept because they are not trained in our public schools to think of parsimony in terms of "observational evidence" or "meeting one's burden of proof". Instead, we are trained to blindly accept CHANCE as the most parsimonious, most logical, most rational, and best explanation for the design and the creation of our amino acids, our proteins, our genes, our genomes, and the life forms on this planet. It's a lie, but we believe that it is true because our college professors constantly tell us that it is true. It's the modern-day form of brain washing. If are told the lie long enough, we eventually come to believe that it is true. The Behaviorists, Atheists, Darwinists, and Naturalists know this, so that's why they keep lying to us.

Mark My Words

The Second Law of Thermodynamics Is Obviously False

The second law of thermodynamics states that the total amount of entropy or disorder in our universe is constantly increasing and that it can never decrease and go to zero. This claim is obviously false.

The very fact that we exist is scientific proof that this claim is false. We haven't observed anything that the second law of thermodynamics predicts. We don't observe proton decay. We don't observe an ever-encroaching gray goo coming in at us from all sides. We don't observe ever-increasing disorder or chaos. We don't even observe heat death. Heat death doesn't exist. Heat death or absolute zero represents Maximum Efficiency when it comes to energy, photons, quantum waves, psyches, quantum consciousness, quantum information, quantum fields, and superconductors. Does it not? These quantum objects function perfectly and function best at absolute zero. We don't observe anything that the second law of thermodynamics predicts, which is why it is false.

Instead, we observe constantly conserved order and organization thanks to the massless and entropyless Quantum Fields, the Conservation of Energy and Psyche, the Conservation of Quantum Information within Psyche, and the Perpetual Motion Cycle $E = mc^2$. The observation of these things falsifies the second law of thermodynamics as well as creation by chance, or creation by entropy, or creation by random disorder.

The ONLY part of the second law of thermodynamics that is demonstrably true is the obvious fact that physical matter cannot be used to design and run perpetual motion machines. Physical matter wasn't designed to work that way. Physical matter has physical limitations which prevent it from functioning as a usable and dependable perpetual motion machine in the real world. In other words, the ONLY part of the second law that is demonstrably true is the "thermodynamics"; and, thermodynamics or heat has absolutely nothing to do with "disorder" or "chaos".

Isn't that ironic? Disorder and the "disorder definition for entropy" have absolutely nothing to do with thermodynamics or heat. Disorder doesn't produce heat, and disorder doesn't eliminate heat. There's NO correlation between the two, which means that "disorder" is the wrong definition for thermodynamics or entropy.

The sun is often portrayed as maximum chaos or maximum disorder, and there's tons of mass and heat and therefore "entropy" that's associated with our sun. Heat death doesn't maximize disorder as the second law of thermodynamics claims. There's NO correlation between heat and disorder!

For contrast, notice that there is NO mass, heat, or entropy associated with photons, quantum waves, and the quantum fields, which is why these massless and entropyless quantum objects represent perfectly conserved syntropic Order and Organization at the quantum level in the Spirit World or the Quantum Realm. The Quantum Fields are massless, entropyless, syntropic, exergic, and conserved perpetual motion machines. The Perpetual Motion Cycle $E = mc^2$ runs on energy, which is why it is constantly and forever conserved. The Conservation of Energy and Psyche, as well as the Conservation of Quantum Information within Psyche or Intelligence, is what makes these quantum perpetual motion machines possible at the quantum level in the Transdimensional Non-Physical Realm. Quantum objects don't need heat as a lubricant to make them run – in fact, these perpetual motion machines run best and most efficiently at "heat death" or absolute zero. This is what we have experienced and observed, is it not?

The second law of thermodynamics will never be true as long as the Quantum Fields exist. You would have to destroy the Quantum Fields in order to return our universe to the Chaos Realm from whence it was originally organized. You would have to destroy the Quantum Fields in order to make the second law of thermodynamics even remotely true. Even then, the second law of thermodynamics would still be false thanks to the Conservation of Energy and Psyche. Photons, Quantum Waves, Quantum Fields, Conservation of Energy and Psyche, and Syntropy ARE scientific proof that the second law of thermodynamics is false.

When our scientists finally figure out how things really work in the real world, they will vote to replace the falsified second law of thermodynamics with Quantum Field Theory, Conservation of Energy and Psyche, and the Conservation of Quantum

Information within Psyche. They will replace the second law of thermodynamics with the things that have actually been experienced and observed.

According to the Philosophy of Science, there are many different Scientific Methods or scientific methodologies that we can use in our pursuit of the truth. I used a few of them in this essay. I use many more of them in my books.

ALL of the different Scientific Methods can be used to FALSIFY the Theory of Evolution, the Second Law of Thermodynamics, Materialism, Naturalism, Darwinism, Nihilism, Atheism, Creation by Chance, Creation Ex Nihilo, or anything else that is by definition produced by chance alone because chance by definition cannot do causation. When we scientists FALSIFY a theory or an idea, we do indeed PROVE that it is false. Negating the Consequent or Falsifying a Theory is a philosophically valid and logically sound Scientific Method than can be used to PROVE that our ideas and theories are false.

Therefore, it is NO exaggeration and it is NO joke whenever I state the fact that the Scientific Methods PROVE that the Theory of Evolution and the Second Law of Thermodynamics are false, because that's precisely what they were designed to do.

OBSERVATION is one of the Scientific Methods; and, what do we observe?

We observe that we exist.

The very fact that we exist is scientific proof that the second law of thermodynamics is false, because the second law of thermodynamics predicts or says that we shouldn't exist. It's obvious that the second law is false.

If disorder or entropy were truly a conserved substance as the second law of thermodynamics claims, then we wouldn't exist. There has to be some type of Syntropy within the system in order for us to exist in the first place. That massless entropyless Syntropy or Conservation exists and is stored within the Energy, the Quantum Waves, the Quantum Information, the Psyches or Intelligences, the Photons, and the Quantum Fields. These things are different types of non-physical immaterial perpetual motion machines. The Quantum Realm or Psyche Realm is the Perpetual Motion Realm or the Energy Realm. It will never get old, never wear out, and never die because it is constantly conserved. It is eternal and everlasting. It has always existed, and it will always exist. Energy cannot be made, and it cannot be destroyed. It's that simple.

You would have to destroy the Energy, the Psyches or Quantum Consciousness, the Quantum Information within Psyches or Intelligences, the Quantum Realm, the Perpetual Motion Cycle $E = mc^2$, and the Quantum Fields in order to make the second law of thermodynamics true.

Mark My Words

Evolution Is Entropy or Creation by Random Disorder

The Materialists, Naturalists, Darwinists, Nihilists, Behaviorists, and Atheists axiomatically define entropy as "disorder".

They call this falsehood or lie the Second Law of Thermodynamics, which axiomatically and erroneously states that the total amount of entropy or disorder in our universe is constantly increasing and that it can never decrease and go to zero. This falsehood predicts that we shouldn't be here, yet here we are. Furthermore, according to the equations for heat and entropy, as mass goes to zero, entropy goes to zero along with it. Entropy can and does go to zero whenever mass or heat goes to zero. Therefore, the second law of thermodynamics is false.

There is NO CORRELATION between entropy and disorder. Disorder doesn't cause heat, and disorder doesn't eliminate heat either. Disorder is the wrong definition for entropy; yet, that definition persists and is dominant anyway because it is false. It is its falsehood that makes it preferable to everything else, because above everything else the Materialists and Naturalists are trying to trick us and deceive us just as they have been tricked and deceived. The corollary to the Second Law of Thermodynamics states that entropy, chance, or disorder can design and create anything, if given enough time to do so. The Theory of Evolution has the same exact corollary as the Second Law of Thermodynamics. They are the same thing in the end. They are the same deception and lie.

These people have chosen to believe that entropy or disorder can design and create things if given enough time to do so. For the Materialists, Naturalists, and Darwinists, both the theory of evolution and the second law of thermodynamics are Creation by Entropy, Creation by Disorder, or Creation by Chance. They are the same thing. They are magic! The Atheists are superstitious because they believe in Creation Ex Nihilo or Creation by Magic.

Once I realized and understood what the Materialists, Naturalists, and Darwinists are teaching and preaching, I immediately realized that in their worldview, evolution is entropy – the false definition for entropy. In other words, according to the Darwinists, evolution is Creation by Entropy, Creation by Random Disorder, or Creation by Chance. Evolution works like magic.

There are a couple of things that you need to know about evolution, or creation by chance. First of all, extinction, genetic deterioration, or genetic entropy is the dominant law in biology – NOT spontaneous generation, abiogenesis, or chemical evolution. Evolution doesn't work as advertised. Evolution cannot design and create. Evolution or genetic mutations can only maim, kill, and destroy. Evolution produces extinction – NOT creation and life. The theory of evolution, spontaneous generation, chemical evolution, or abiogenesis was falsified in 1859 by Louis Pasteur, and it has been false ever since. Second of all, the Primary Axiom of Science and Statistics axiomatically states that chance is not causation. In other words, chance cannot do choice, correlation, design, causation, or creation. Chance and causation are mutually exclusive, which means that they falsify each other. Once chance starts doing causation, then it is no longer chance but has become some type of deliberate causation or intelligent design and creation instead. The Primary Axiom of Science and Statistics falsifies the theory of evolution and the second law of thermodynamics.

The theory of evolution is just as false as the second law of thermodynamics because they are the same thing in the end. They both are design and creation by chance, disorder, chaos, death, or entropy. Evolution is entropy – the false definition for entropy. Therefore, evolution or creation by chance is just as false as creation by entropy or creation by disorder, because they are the same thing after all. They are nothing but superstition or magic, because chance cannot design and create things at will, no matter how much time it is given to do so. The "magic" or quantum mechanics exists at the quantum level, NOT the physical level. At the physical level, creation by chance, the theory of evolution, and the second law of thermodynamics are physically impossible. This is what we have actually experienced and observed.

The second law of thermodynamics states that the total amount of entropy or disorder in our universe is constantly increasing and that it can never decrease or go to zero. This has NEVER been experienced NOR observed. It's nothing but science fiction. ALL we ever observe in this universe is constantly conserved Order and Organization thanks to the massless and entropyless Quantum Fields, the Conservation of Energy and Psyche, the Conservation of Quantum Information within Psyche, and the Perpetual Motion Cycle $E = mc^2$. These things falsify the second law of thermodynamics.

Evolution is entropy, particularly the "creation by chance" or "creation by random disorder" definitions for entropy. Evolution is the second law of thermodynamics' definition for entropy. Evolution is Materialism, Naturalism, Darwinism, Nihilism, Atheism, Creation by Chance, and Creation by Random Disorder. So is the second law of thermodynamics. That fact automatically falsifies the Theory of Evolution because the second law of thermodynamics is demonstrably false. We don't observe any of the things that the second law of thermodynamics predicts! We have never observed nor experienced Creation by Chance or Creation Ex Nihilo or Creation by Random Disorder. The very fact that we exist is scientific proof that the second law of thermodynamics is false. The very fact that proteins, genes, and genomes exist is scientific proof that the Theory of Evolution is false.

There are literally dozens of different, contradictory, mutually exclusive, self-falsifying and self-defeating definitions for entropy within "science"; and, that's a serious problem. There is NO standardized definition for entropy within science that is backed up by hard observational evidence. Entropy can literally be anything that you want it to be. That's what makes entropy so confusing. There is NO agreed upon definition for entropy. Nobody knows what entropy is. Entropy as the second law of thermodynamics is a thought experiment only – it has no real-world evidence to back it up or sustain it, just like the Theory of Evolution.

The second law of thermodynamics and the theory of evolution are the same falsehood or lie masquerading as science. They are both Creation by Chance, Chance Causation, Chance Causality, the Null Hypothesis, and produced by chance alone. They are automatically FALSE because chance cannot do design, creation, choice, or causation. Once we have established choice or causation, then we are no longer looking at chance. Chance and causation are mutually exclusive. Choice and causation are mutually exclusive. ONLY psyche, intelligence, or quantum consciousness can do choice or causation at the quantum level in the quantum realm. It's the only thing that's been observed doing so.

You see, there is NO way to use the Scientific Method to run a True Science Experiment with either the Theory of Evolution or the second law of thermodynamics, because there is NO way to vary the amount of disorder or chance in order to observe its effect on the amount of disorder or chance. This is a *tautology* or *circular reasoning*. This is a logic fallacy. Entropy defined as "disorder" or "chance" is self-defeating and self-falsifying. There's literally NO way to vary the amount of disorder or chance in order to observe its effect on the amount of disorder or chance. Entropy as disorder or chance is a FAILED and FALSIFIED thought experiment. You can't use disorder or chance as the independent variable, and you can't use them as the dependent variable either. Once you start manipulating disorder or chance, then they are NO longer disorder or chance but have become some type of causation instead because YOU are manipulating them.

Do you see how that works?

It FALSIFIES the Theory of Evolution and the Second Law of Thermodynamics. The very fact that we can't use the Scientific Method and run a True Science Experiment on disorder or chance is scientific proof that the Theory of Evolution and the Second Law of Thermodynamics ARE FALSE.

There's just NO way on this planet to use the Scientific Method or a True Science Experiment to demonstrate and prove and verify Creation by Chance, Creation by Random Disorder, Creation by Entropy defined as disorder or chance, Creation Ex Nihilo, the Theory of Evolution, or the Second Law of Thermodynamics. ALL of these are produced by chance alone. All of these are defined as Creation by Chance or Creation by Random Disorder. There's no way to manipulate chance or disorder as an independent variable, and then measure its effect on chance or disorder as the dependent variable. Both the Theory of Evolution and the second law of thermodynamics are a dead-end where the Scientific Method and a True Science Experiment are concerned, because chance cannot do causation.

Chance and causation are mutually exclusive. They falsify each other. Once chance or random disorder starts to do something, then it is no longer chance, but has become some type of causation instead. Once YOU start manipulating chance or disorder, then they are NO longer chance or true disorder. Should a system somehow magically achieve maximum disorder, it has no place to go but upwards into some type of order or organization or correlation. Once correlation or causation has been established, then chance or maximum disorder has completely ceased to exist.

These realties and truths FALSIFY Creation by Chance, Creation by Random Disorder, Creation by Entropy, the Theory of Evolution, and the Second Law of Thermodynamics. It's that simple.

Mark My Words

Science and the Scientific Method Falsify Naturalism and Darwinism

For me personally, one of the most interesting and significant scientific discoveries that came my way was the result of studying the Philosophy of Science for a year or more.

I discovered that by *negating the consequent*, the Scientific Method falsifies Materialism, Naturalism, Darwinism, and Atheism. Think about it, when done right, Science and the Scientific Method prove that Materialism, Naturalism, Darwinism, Nihilism, and the Theory of Evolution are false. That's huge! It's a game-changer!

I wonder why nobody has ever thought of it before. I wish I would have known how to falsify Materialism, Naturalism, Darwinism, Atheism, and the Theory of Evolution when I was first in college forty years ago. It would have saved me a lot of trouble, time, and grief.

Materialism, Naturalism, Darwinism, and the Theory of Evolution are DEAD; and now, I can use the Scientific Method to prove that it is so. That's a monumental scientific discovery, in my not so humble opinion. I found that one very satisfying when it was first revealed to me.

This amazing development in the Philosophy of Science is documented in one of my most-ignored books, *Using the Scientific Method to Eliminate the Usual Suspects and to Prove the Truth*.

I eventually realized that if you successfully use the Scientific Method to eliminate everything that is false – such as Materialism, Naturalism, Darwinism, Nihilism, Atheism, and the Theory of Evolution – then ONLY the truth will remain. It's amazing to study and observe what remains, after these things have been eliminated from science. Their opposite remains. The truth remains. Quantum Mechanics or Supernatural Mechanics remains. Psyche or Quantum Non-Local Consciousness remains. Intelligence remains.

Logic or common sense also FALSIFIES Materialism, Naturalism, Darwinism, Nihilism, and Atheism.

Materialism is the claim that the quantum or the non-physical does not exist. I never found any evidence that verifies or confirms that claim. Nobody has. Instead, the verified and proven existence of Quantum Mechanics and Action at a Distance falsifies that claim. The verified and proven existence of Dark Energy, Dark Matter or Spirit Matter, Magnetism, Gravity, Thoughts, and Light falsifies Materialism or Physicalism.

Naturalism is the claim that the quantum or the supernatural does not exist. I never found any evidence that verifies or confirms that claim. Nobody has. Instead, the verified and proven existence of Quantum Mechanics and Action at a Distance falsifies that claim. Action at a Distance is Supernatural. Quantum Mechanics is Supernatural. The verified and proven existence of Quantum Mechanics falsifies Naturalism.

Nihilism is based upon the claim that an after-life or spirit world does not exist. I've never found any evidence that verifies or confirms that claim. Nobody has. Instead, the veridical experiences and observations from Near-Death Experiences (NDEs), Out-of-Body Experiences (OBEs), Shared-Death Experiences (SDEs), and our after-death Life Reviews falsify that claim. Millions of different people have had an NDE or some type of OBE. The repeatedly verified and proven existence of NDEs falsifies Nihilism. We are going somewhere after all, when we die. Our thoughts and memories and personality continue to exist after our physical brain is dead and gone. That's what all the evidence is telling us. Nihilism is also the philosophical belief that the Psyche, Spirit, or Soul does not exist and that life, therefore, has no meaning or purpose. I never found any evidence that verifies or confirms that claim. Nobody has.

Atheism is the claim that God does not exist. I never found any evidence that verifies or confirms that claim. Nobody has. Instead, the verified and proven existence of the Biblical God Jesus Christ falsifies that claim. Thousands of different people have seen Him in the flesh after He rose from the dead; and, thousands of different people have seen Him, embraced Him, and conversed with Him during their NDEs because He is the being of light and love whom most people encounter during their NDEs. In fact, during their NDEs, Jesus Christ has come and gotten some of the Atheists out of Hell, just for the asking.

Darwinism is a creative combination of Materialism, Naturalism, Nihilism, and Atheism. When those go down, Darwinism or the Theory of Evolution goes down with them as fruit from the poisoned tree. Darwinism is the belief that the rocks or raw physical matter designed and created and produced all of the proteins and their matching genes. Darwinism is a philosophical belief in spontaneous generation. Darwinism is Atheism – the creation of something by "nothing". It can't be done. It's physically impossible.

Creation ex nihilo is also a type of Atheism or Darwinism – the creation of something from "nothing" – magic. Creation ex nihilo is just as absurd as Darwinism or Atheism. Creation ex nihilo has NEVER been experienced nor observed, and it never will be because it is impossible. Not even God can do the impossible.

Back in 2012, I was a Materialist, Naturalist, Nihilist, and Atheist. I knew what an atheist was; but, I didn't know what those other words meant. Getting a proper definition for Materialism, Naturalism, Darwinism, Nihilism, Atheism, and Creation Ex Nihilo was one of my all-time greatest scientific discoveries.

Why?

Well, once you understand the major premises or hidden assumptions upon which Materialism, Naturalism, Darwinism, Nihilism, and Atheism are based, you just sort of automatically KNOW that they are false because you KNOW that their major premises or hidden assumptions have never been experienced nor observed. Science is observation and experience, not wishful thinking, blind-faith, and dogma. Materialism, Naturalism, Darwinism, Nihilism, Atheism, and Creation Ex Nihilo are self-defeating or self-falsifying because their major premises or hidden assumptions have never been experienced nor observed; and, they never will be.

Think about it!

Examine the hidden premises of Materialism, Naturalism, Darwinism, Nihilism, and Atheism as if you were a scientist and a logician. You will quickly observe that these falsified philosophies are self-defeating, once you know what their hidden premises are.

Let me demonstrate.

How are you going to observe that **God does not exist**? How do you go throughout the whole multi-verse, every dimension, every realm, every existence, every time, every atom, and every alternative reality verifying that God does not exist? If you can do that, then you are a God, and God does in fact exist.

How do you observe and verify that the **spirit world does not exist**? In order to do that, you would have to go to the spirit world and observe that it does not exist, and then return to tell about it. That's precisely what people do to verify that the spirit world does indeed exist; and, millions have done so.

How do you observe and verify that **Quantum Mechanics or Supernatural Mechanisms do not exist**? You can't. It's impossible. Instead, the repeated observation and repeated verification of Action at a Distance, the Quantum Zeno Effect, and Quantum Tunneling proves that Supernatural non-physical mechanisms do in fact exist. Even observing the effects of dark matter, dark energy, gravity, magnetism, wind, and light on physical matter proves that non-physical phenomena or supernatural phenomena do in fact exist. You can't use the repeated observation of non-physical phenomena as scientific proof that the **non-physical does not exist**.

How do you prove that the proteins and their matching genes spontaneously generated out of thin air? The phenomenon would have to be experienced and observed; and, it hasn't. In fact, **spontaneous generation**, chemical evolution, macro-evolution, creation ex nihilo, materialism, naturalism, and the theory of evolution were falsified in 1859 by Louis Pasteur – the very same year that Charles Darwin published "On the Origin of Species". We've known that the theory of evolution is false from the very beginning of the theory.

The false is falsified by the truth; and, the truth is repeatedly experienced, verified, and observed by Someone Psyche or Someone Intelligent in any realm of existence.

Your college professors and public schoolteachers are trying to hide this information from you. In fact, they will be fired and lose their jobs if they teach any of this information to you. Materialism, Naturalism, Darwinism, Nihilism, and Atheism are religions; and, religions protect themselves by censoring, banning, blocking, ridiculing, restricting, and firing their opposition. The tables have turned, and the Atheists now control our school-grounds; but, it's just another religion. All we did is replace one religion with a different religion. We've made no social progress where the truth is concerned. In fact, we've gone backwards further into the dark ages as a result. In order to get the truth into our lives, we have to eliminate everything that is false and everything that has been falsified, starting with Materialism, Naturalism, Darwinism, Nihilism, Atheism, Creation Ex Nihilo, and the Theory of Evolution.

Remember, if you successfully eliminate everything that is false, then only the truth will remain. It's fascinating to study and observe what remains after you have falsified and eliminated Materialism, Naturalism, Darwinism, Nihilism, Atheism, Creation Ex Nihilo, and the Theory of Evolution. The truth remains.

Mark My Words

—

Source

Using the Scientific Method to Eliminate the Usual Suspects and to Prove the Truth.

https://www.amazon.com/dp/B01J6STHP0

https://www.amazon.com/dp/1521133581

The Scientific Method Proves That the Theory of Evolution Is False

https://www.amazon.com/dp/B01IAAIRT2

https://www.amazon.com/dp/1521133611

The Scientific Methods

Due to the wide variety of different logic fallacies which are built directly into the Scientific Methods, it is technically impossible to find the truth and to KNOW the truth using the Scientific Methods.

The original meaning for the word "Science" was knowledge; however, it is technically impossible to prove truth or to find knowledge using Science and the Scientific Methods.

Since this is the case, doesn't that technically mean that Science and the Scientific Method are absolutely worthless? It would seem that way, wouldn't it? What good is it if it can't be used to prove the truth?

Well, all is not lost. Science and the Scientific Methods do have their uses, as the Lived Experiences of the human race testify.

We can use Scientific Methods to falsify theories or to prove theories false!

Science is very good at falsifying theories or proving theories false if we let it do so. Most scientists, though, don't let it do so, because of their prejudices or confirmation biases. But, I personally no longer suffer from those kinds of illusions and delusions, because I am no longer a Materialist or an Atheist. I'm now free to let science tell me whatever it wants to tell me.

We can use Science and the Scientific Methods to falsify theories because *negating the consequent* is philosophically and logically sound!

The argument structure for falsification, or *negating the consequent*, looks like this:

Scientific Hypothesis: If Theory X is true, then we will observe Y.

Scientific Observation: We don't observe Y.

Scientific Conclusion: Therefore, Theory X is false, and Theory X has been falsified.

That's the Scientific Method in action, doing what it was meant to do, falsifying a theory!

Let me provide an example to show how this works in practice.

Scientific Definition of the following Theories: Materialism, Naturalism, Darwinism, and the Theory of Evolution are defined as "Design and Creation by Physical Matter". The Materialists, Naturalists, and Darwinists contend that raw inert physical matter designed, created, and manufactured the first genomes and the first living cells, because according to these people physical matter is the only thing that exists and therefore the only thing that could have designed, programmed, engineered, created, and manufactured the first genomes and the first life forms on this planet. We choose to take these people at their word whenever they claim to believe that physical matter is the only thing that exists. They really believe that! These people truly believe that the rocks designed and created the first genomes and life forms. These people call it Science.

Premise or Axiom or Law: The Scientific Methods can be used to falsify theories and to prove them false.

Scientific Hypothesis: If Materialism, Naturalism, Darwinism, and the Theory of Evolution are true, then we will observe the rocks and physical matter designing, creating, and manufacturing genomes and life forms from scratch.

Scientific Observations: We NEVER observe the rocks designing and creating anything. They can't design and create anything at all.

Scientific Conclusion: Therefore, Materialism, Naturalism, Darwinism, and the Theory of Evolution are false, and have been successfully falsified.

That's the Scientific Method in action, doing what it was meant to do, falsifying a theory or two!

I just falsified Materialism, Naturalism, Darwinism, and the Theory of Evolution! I love doing that! I wish I would have known how to do that forty years ago!

Furthermore, Atheism is "Creation by Nothing"; and, we all know that nothing cannot design and create anything.

Scientific Hypothesis: If Atheism is true, then we will observe nothing designing and creating something.

Scientific Observations: We NEVER observe nothing designing and creating anything.

Scientific Conclusion: Therefore, Atheism is false, and has been successfully falsified.

I just falsified Atheism, using the Scientific Method.

Best of all, using the Scientific Methods to falsify theories or to prove theories false IS philosophically and logically sound! In other words, I HAVE IN FACT FALSIFIED Materialism, Naturalism, Darwinism, Atheism, and the Theory of Evolution. The debate has ended. I have just PROVEN Materialism, Naturalism, Darwinism, Atheism, and the Theory of Evolution FALSE, using the Scientific Method. We now KNOW that these theories are false, because we KNOW why they are false. They are FALSE because the rocks and blind luck have never been caught in the act of design and creation; and, they NEVER will be. Case closed!

Since we now KNOW that Materialism, Naturalism, Darwinism, Atheism, and the Theory of Evolution are FALSE, we must look someplace else if we want to find the truth. So, where else are we going to look for the truth?

Well, it's my contention that THE TRUTH cannot be falsified by the Scientific Methods. When we are dealing with THE TRUTH, the Scientific Methods will continuously verify the truth for us, as long as we have our definitions for everything correct so that we are getting the correct interpretation of the Scientific Evidence. Using the Scientific Methods, we can center in on THE TRUTH through a process of elimination by using the Scientific Methods to falsify the falsehoods such as Materialism, Naturalism, Atheism, and Darwinism.

So, what's the opposite of Materialism and Naturalism? Since these have been falsified by the Scientific Method, then their opposite is most likely true – through a process of elimination. If you knock down all the falsehoods, THE TRUTH will be the only thing left standing. It's elementary my dear friend!

Psyche or Non-Local Consciousness is the opposite of Materialism. Spirit, spirituality, non-locality, the supernatural, and God are the opposite of Naturalism and Atheism.

I recently discovered that Lived Experience or Psyche Experience is in fact the BEST and the most efficient way of finding and KNOWING the TRUTH. Lived Experience or Direct Observation is vastly superior to philosophical guesswork, scientific hypotheses, and philosophical interpretations of scientific data. In fact, the BEST definition for Science is "Observation". We observe things, so that we can come to understand things and know things.

For example, we KNOW from the Lived Experiences of the human race that Psyche or Non-Local Consciousness exists, because people have lived it and experienced it firsthand for themselves during their Out-of-Body Experiences and Near-Death Experiences. We don't have to run any science experiment on Psyche, because we already KNOW that it exists. KNOWING is vastly superior to hypothesizing, theorizing, experimenting, and philosophizing.

We KNOW from observation and lived experience that Psyche or Intelligence can design, program, engineer, manufacture, and create anything that it sets its mind to. Psyche can do science! Psyche has been caught in the act! We KNOW this is the TRUTH, because it has NEVER been falsified! It's impossible to use the Scientific Methods to falsify THE TRUTH!

Intelligent Design, Creation by Psyche, and Manufacturing by Human Beings or Human Psyches have been observed and verified trillions of trillions of times with an infinite more times to go. THE TRUTH is always verified by our Scientific Observations and NEVER once falsified, because THE TRUTH can never be falsified by our Scientific Methods or Observations! God is KNOWN by living Him and experiencing Him. God is KNOWN by observing Him! THE TRUTH IS KNOWN by living it and experiencing it for yourself, or by choosing to trust someone who has. This is the way Science should be and should be done!

We KNOW from Observation and Lived Experience that Psyche exists and that Psyche or Intelligent Beings can design and create anything that they set their minds to. This Reality has been verified trillions of trillions of times and has NEVER been falsified. THE TRUTH cannot be falsified by the Scientific Methods, because it will always be true, stay true, and hold true. In contrast, Materialism, Naturalism,

Atheism, and Darwinism have been falsified thousands of different ways and trillions of different times.

That's the way you do Science; and, that's the way Science should be done!

But, guess what's going to happen now. It happened before, and it's going to happen some more. The Materialists, Naturalists, Darwinists, and Atheists are going to chime in and sound off and assure you authoritatively as scientists and PhDs that I don't understand Science, that I don't understand the Scientific Method, and that I don't understand the power of Evolution. They are going to tell you that I'm an idiot, because that's what these people do. I KNOW because I have lived it and experienced it. I KNOW, because I used to be a Materialist, Nihilist, and Atheist and was in complete denial of Reality at the time. I KNOW how it goes, because I have lived it and experienced it. These people have to DENY it, or they will no longer be Materialists, Naturalists, Darwinists, and Atheists. But, I have observed that the Materialists, Naturalists, Darwinists, and Atheists are always wrong.

Consequently, I finally realized that if I wanted to KNOW the TRUTH, then I had to look someplace else besides Materialism, Naturalism, Darwinism, and Atheism because these philosophies have in fact been falsified by Science and the Scientific Methods thousands of different ways and trillions of different times. In fact, the major premises or the primary assumptions of Materialism, Naturalism, Darwinism, and Atheism are impossible to verify! They have NEVER been verified, and will NEVER be verified, because they can't be. They have to be taken on blind faith as being true, because they are in fact FALSE and have been falsified by the Scientific Method.

That's the way Science and the Scientific Methods work or should work!

Cool, huh?

I bet you'll never get anything like this from your Science teacher or college professor, because they don't have anything like this to give you.

Mark My Words

Truth in Science

While meeting together in a group, a friend asked me why I'm so passionate about Truth in Science. My response to her came rushing forth, and I couldn't hold it back.

Because I have experienced firsthand how bad it can get when the Scientists, Medical Doctors, and Psychiatrists LIE to us in order to make money at our expense. It can kill you and destroy your life. My children disowned me, and my wife and I were separated for seven years. I attempted suicide. The third leading cause of death is death by medical

doctor or death by psychiatrist. These people can kill you and will kill you if you let them.

They got me addicted to prescription drugs. The psychiatrists lied to me and told me that they weren't giving me enough Valium to become addicted to it. It was a lie. Due to my genetic history from my mother's side, I was immediately addicted. When I finally decided to go cold turkey and quit, it took six solid months of hallucinations and delusions and suicidal ideation — six solid months of withdrawal — before I started thinking rationally and logically once again. Six months is a long time to go through withdrawal. That's addiction! They kept lying to me and telling me that people can't become addicted to Valium. It's a lie. I later discovered that Valium is the third most addictive substance on the planet. The LIES in Science can kill you.

Anytime that I would complain about the side effects, they would give me another mind-altering drug to try to mask the symptoms of the other drugs. They had me nested six deep and cycling through sleeping pills. I went insane. The technical or medical term for it is Substance-Induced Psychosis.

The people on the television and the radio knew my name, and they were telling me that I am worthless, that I'm going to hell, and that I need to kill myself right now. I was psychotic. I was delusional and hallucinating. I had to kill myself. I overdosed on bottles of the sleeping pills, almost succeeded in killing myself, and ended up in the looney bin or psych ward once again, where they immediately got me addicted to Valium all over again. For each week I was on Valium, I was looking at six seeks of withdrawal symptoms which included psychosis, hallucinations, and delusions. Withdrawal was a bitch — every single time that I tried to get off the Valium, or the Xanax, or the Halcion.

Scientists lie. Scientists, medical doctors, psychiatrists, and pharmacologists will lie to you in order to make money off you. Their lies can kill you. That's why I'm so passionate about Psychology and Truth in Science. Their lies can kill you. Prescription drugs can kill you. I want to protect people from that kind of abuse if I can, because it can destroy your life.

That was my response to my friend and is a summation of the discussion that the seven of us had together when she asked me that question. Another person in the group had experienced the same thing from Phenobarbital for her seizures — hallucinations and suicidal thoughts to the extreme. Their junk can kill you.

The psychiatrists and medical doctors and pharmacologists only have ONE TOOL in their arsenal, and they overuse it and abuse it, because that's how they make their money. Their drugs can kill you. These people will kill you if you let them.

Most people could care less about Truth in Science. In fact, it seems as if most people prefer to be lied to when it comes to Science. The truth can be painful.

But, I'm passionate about Truth in Science no matter what anyone else thinks about the topic. I live for it!

I'm trying to become the honest scientist — the one that you can trust to tell you the truth no matter how brutal that truth might be. That's my competitive advantage.

Mark My Words

Promoting Truth in Science

My purpose or mission in life is to find and promote the Truth in Science. That starts by identifying and eliminating everything that is false. I'm passionate about identifying and stamping out the lies.

I'm actively developing and promoting Truth in Science and Freedom of Science. We currently don't have either our in our society.

There is a massive paradigm shift taking place within Science on the fringe, and you will hear nothing about it in our public schools, because our atheistic and naturalistic and materialistic college professors will be the last people on the planet to understand it, embrace it, adopt it, and start teaching it. If you want to know about it, you have to find it on your own. My websites are dedicated precisely to that purpose.

The change is huge, but it is slow in coming. However, we have progressed far enough in science that it's now actually possible to find and know the truth through science, if one is willing to look for it and embrace it when he or she finds it. It's not like it was fifty years ago when we didn't have a clue where to look to find the truth in science. The truth in science is now starting to achieve critical mass.

I'm the Science Detective. I'm the Science Author on Amazon. The New Paradigm is my specialty. I'm passionate about Truth in Science.

https://amazon.com/author/science

I think people like you need to know, ponder, and study the paradigm shift that is taking place within science. We scientists are currently switching over from philosophy and wishful thinking to the things that have actually been experienced and observed and verified. We are switching to a science that is based upon the observations and lived experiences of the human race as a whole.

The following is a list of the websites that I'm associated with. Click on the websites and study the articles. Search for the free books. I leave it up to you to find better evidence to support the different concepts that I present. Some of these websites are a work-in-progress, and I'm still finishing some of the associated books, but it's time for me to start sharing what I have learned.

There's a lot for you to study and think about on my websites, and you don't have to buy one of my books to get started.

Mark My Words

Truth in Science

Scientists Lie

https://markme.website/scientists-lie/

Origin Science

https://origin-science.org/

https://science-2-0.com/

Syntropy

https://syntropy.website/

https://syntropy.site/

Falsifying the Second Law of Thermodynamics

https://quantum-law-of-thermodynamics.com/

https://ultimate-law-of-thermodynamics.com/

https://evolution-is-entropy.com/

Falsifying the Theory of Evolution with Science

https://markme.website/

Quantum Mechanics

https://quantum-neuroscience.com/

The Ultimate Model of Reality

https://psyche-ontology.com/

https://ultimate-model-of-reality.com/

https://bio-psycho-social.com/

https://mypsyche.us/

Philosophy of Science

https://philosophy-of-science.com/

https://science-2-0.com/

Computer Science

https://allthings.computer/

https://allthings.computer/phpbb3/index.php

Scientific Proof of God's Existence

https://scientific-proof-of-god.com/

Finding a Religion That's Compatible with the Truths in Science

https://god-is-light.com/

https://ldssoul.com/

http://ldssoul.com/forum/index.php

Dangerous Science — the Stuff that Can Kill You

https://tripping.website/

Science Fiction

https://stealth-ascendancy.com/

My Twitter Account

https://twitter.com/Mark_Me_Words

My Facebook Account

https://www.facebook.com/MarkMyScience/

Author's Home Page

https://markme.us/

Enjoy!

Mark My Words

Truth in Science by Mark My Words

Learn something new that you have never thought about before. Learn something new every day.

Mark My Words

https://origin-science.org/truth/

https://markme.us/truth/

https://markme.website/truth/

The Next Essay in this Truth in Science Series

https://markme.website/scientists-lie/

Scientists Lie

Scientists lie.

Scientists lie whenever they try to convince us that chance can do causation.

The math doesn't lie.

That's why I find mathematical proofs so convincing. I'm a statistician, mathematician, and logician as well as being a philosopher and a scientist. In subsequent parts of this essay, I will use simple math and statistics to demonstrate that scientists lie whenever they make chance the cause of their scientific theories, because chance cannot do causation of any kind. For now, I will start by introducing the Philosophy of Science that supports this truth assertion.

Most of my books are about the Philosophy of Science.

I studied and wrote about the Philosophy of Science for years trying to find a true and realistic Philosophy of Science that is actually based upon the observations and lived experiences of the human race as a whole.

So, what did I learn?

Well first of all, I learned that there is a fatal flaw built directly into the Scientific Method that prevents us from using the Scientific Method to prove anything true. This is the *affirming the consequent* or the *jumping to conclusions* logic fallacy.

There are many different versions of the Scientific Method and many different Scientific Methods or scientific methodologies. The following version of the standard Scientific Method, that shows up in many of our college textbooks, makes the *affirming the consequent* logic fallacy somewhat obvious. You still have to use your brain in order to see it, though.

the Scientific Method

1) Form a **HYPOTHESIS**.

2) Select a SCIENTIFIC METHOD or Scientific Methodology to **TEST** the hypothesis.

3) Run the Science Experiment; and then, **Observe** and **Measure** the **RESULTS**.

4) Find the BEST **INTERPRETATION** or the BEST **EXPLANATION** for the Scientific Data, the Scientific Evidence, the Scientific Observations, and the **RESULTS** of the Science Experiment.

Notice that Science, the Scientific Method, and Science Experiments are OBSERVATION. Observation is step three of this process.

If something has NEVER been experienced NOR observed – not even by God – then it doesn't exist. This is Logic 101, and the very foundation of the Scientific Method. We NEED observations in order for it to be science. It has to be caught in the act, or it isn't science. It's nothing but pure speculation instead. Scientific evidence has to be eye-witness evidence! It has to be experienced and observed in order for it to be real and true.

There has to be Someone Intelligent or Someone Psyche who is there to make the observations, or Science and the Scientific Methods don't work. There has to be someone there to do the Science, otherwise science and the Scientific Methods don't do anything.

Creation by Chance or Chance Causation has NEVER been experienced NOR observed because chance cannot do causation. Chance and causation are mutually exclusive. They falsify each other. It has to be experienced and observed in order for it to be real and true. Causation by Chance, or Creation by Chance, or Creation Ex Nihilo (Atheism) has never been experienced nor observed, which means that it isn't science.

If you want the truth, it is important to allow all of the evidence into evidence – and that includes ALL the times when mass, heat, and entropy have gone to zero and ceased to exist. We can't have the truth if people are constantly deleting, changing, hiding, and destroying evidence, observations, or the lived experiences of the human race.

Science IS observation, experience, and knowledge. It has to be experienced and observed, or there's no way to know that its real and true. It has to be experienced and observed in order for it to be science or knowledge.

For example, our resurrected Lord Jesus Christ has been experienced and observed both in the flesh and during our near-death experiences. He is the being of light and love that people encounter during their near-death experiences and out-of-body experiences. He has been experienced and observed many times by thousands of different people both in the flesh and in the spirit; therefore, WE KNOW that He is real and that He truly exists; and, that's the ONLY way that we can KNOW that He is real and truly exists. He has to reveal Himself to some of us from time to time, or we would NEVER know that He exists. That's just the way it is. Science is observation. If there are no observations, then it isn't science. The ONLY way that we can KNOW that God exists is to see Him and touch Him for ourselves, or to choose to trust someone who has. This is logical common sense, and everyone overlooks it all the time, because they want their personal speculations and wishful thinking to be true instead.

If it's never been experienced nor observed, then it is completely worthless as scientific evidence and shouldn't be used as scientific evidence. A scientific theory has to be experienced and observed by someone, sometime, somewhere; or, there's no way that it could ever possibly be true. That's what's wrong with the Many Worlds Interpretation of Quantum Mechanics. Nobody has ever observed it or experienced it, which makes it worthless.

Creation by Chance, or Chance Causation, or Creation Ex Nihilo (Atheism) has NEVER been experienced NOR observed because chance cannot do causation; therefore, WE KNOW that it isn't real and that it doesn't exist. God can't even do Creation Ex Nihilo, so why believe that evolution or entropy can? The Gods work with what already exists, and then try to raise it to a higher level of being or existence. The LAW of Conservation of Energy tells us that the Gods can't do Creation Ex Nihilo. Therefore, the Conservation of Energy Law also tells us that evolution, chance, and entropy can't do Creation Ex Nihilo either. This is how we use the Scientific Methods to find and know the truth – by studying what has actually been experienced and observed. The Conservation of Energy and the Perpetual Motion Cycle $E = mc^2$ have been experienced and observed; therefore, we KNOW that they are real and true. The same thing can be said for Quantum Mechanics and the Quantum Fields.

OBSERVATION or First-Hand Experience is one of the Scientific Methods. There are many different Scientific Methods or scientific methodologies, and OBSERVATION is the best. A phenomenon has to be experienced and observed, or it can't possibly be real and true.

The QUICKEST and BEST way to find and know the truth is to live it, experience it, and observe it for yourself, or to choose to trust someone who has. This is infinitely faster, better, and more reliable than any science experiment. Furthermore, Science and every science experiment have to be based upon observations and eye-witness evidence, or it isn't science. Observation is the best form of Science, because we KNOW that it is real and true.

Observation is an integral and essential part of the Scientific Method. No observations, then no science. Then we are looking at nothing but philosophy, guesswork, and wishful thinking instead. Creation by Chance or Creation Ex Nihilo has never been experienced nor observed, therefore, it can't possibly be real and true. Both the theory of evolution and the second law of thermodynamics are Creation by Chance or Creation Ex Nihilo or Spontaneous Generation or Magic, which means that they can't possibly be real or true. It's that simple. They have never been experienced nor observed.

The second law of thermodynamics states that the total amount of entropy or disorder in our universe is constantly increasing and that it can never decrease and go to zero. This claim is obviously false. The very fact that we exist is scientific proof that this claim is false. We haven't observed anything that the second law of thermodynamics predicts. We don't observe proton decay. We don't observe an ever-encroaching gray goo coming in at us from all sides. We don't observe ever-increasing disorder or chaos. We don't even observe heat death. Heat death doesn't exist. Heat death or absolute zero represents Maximum Efficiency when it comes to photons, quantum waves, psyches, quantum consciousness, quantum information, quantum fields, and superconductors. Does it not? We don't observe anything that the second law of thermodynamics predicts, which is why it is false.

Instead, we observe constantly conserved order and organization thanks to the massless and entropyless Quantum Fields, the Conservation of Energy and Psyche, the Conservation of Quantum Information within Psyche, and the Perpetual Motion Cycle $E = mc^2$. The observation of these things falsifies the second law of thermodynamics as well as creation by chance, or creation by entropy, or creation by random disorder.

Based upon the Observation Version of the Scientific Method, the second law of thermodynamics is false. Furthermore, entropy defined as "disorder" has absolutely nothing to do with thermodynamics or heat. There's NO correlation between the two, which means that there's no way in this universe that the second law of thermodynamics could ever be true. Likewise, based upon the Creation by Chance or the Creation by Random Disorder definition for entropy and the second law of thermodynamics, we automatically KNOW that the second law is false because chance of any kind cannot do choice, design, creation, nor causation. Everything that exists falsifies the second law of thermodynamics.

This is what we want – Truth in Science and Truth through Science – not all the fiction, deceptions, and lies that we typically get from our scientists. Anytime a scientist or college professor tells us that chance can design and create anything if given enough time to do so, we automatically KNOW that he or she is lying to us because chance cannot do causation. Chance and causation are mutually exclusive.

The primary axiom of Statistics states that chance cannot do causation. Chance and causation are mutually exclusive. Chance can NEVER produce a Real Effect. Once it starts doing so, then it is no longer chance. Chance or the Null Hypothesis produces a correlation coefficient of ZERO. By definition, the size of the effect (coefficient of determination) from chance is also ZERO. Chance doesn't determine anything, because it produces no effect. Chance is the Null Hypothesis, which means that chance doesn't do anything. In other words, chance does null or nil. The Null Hypothesis means that there is NO causal relationship between our independent variable and our dependent variable. The Null Hypothesis means that our independent variable has NO EFFECT on our dependent variable. Chance or the Null Hypothesis cannot do causation. Once it starts producing an effect, then it is no longer chance. This is the heart of Science and Statistics whether we realize it or not.

Therefore, Materialism, Naturalism, Darwinism, Nihilism, Atheism, Chemical Evolution, Spontaneous Generation, Abiogenesis, Macro-Evolution, Creation by Random Disorder, Creation by Entropy, Creation Ex Nihilo, the Theory of Evolution, and the Second Law of Thermodynamics are AUTOMATICALLY FALSE because they are by definition Creation by Chance or Causation by Chance. They are

AUTOMATICALLY FALSE because they have never been experienced nor observed, because chance cannot do creation, choice, or causation. Once it starts doing so, then it is no longer chance. These falsified philosophies of science or interpretations of science are AUTOMATICALLY FALSE because these people teach, preach, and believe that Chance designed, and created, and produced everything that exists in this universe. That's what they teach and believe, is it not?

Lack of observational support is one of the ways in which the Scientific Method and a Scientific Theory FAILS to meet its burden of proof. Then people start making up anything they want in order to explain the evidence that they are observing, when they fail to observe the things that they expect to happen by chance alone. There have to be OBSERVATIONS in order for there to be True Scientific Evidence. Science is observation, experience, and knowledge after all – or it should be. OBSERVATIONS are absolutely essential when it comes to Science, Science Experiments, and the Scientific Method. Without them, we don't have science but are dealing with pure speculation or guesswork instead.

However, it is in the fourth and final step of the Scientific Method where the whole process typically FAILS thanks to all of the false and falsified personal INTERPRETATIONS that we fallible human beings choose to give to the scientific data, scientific evidence, or the results of our science experiments.

Materialism, Naturalism, Darwinism, Nihilism, and Atheism are faulty and falsified philosophies of science or interpretations of science. They are NOT science. They are interpretations of science or philosophies of science. They are AUTOMATICALLY FALSE because they use chance exclusively as their creative mechanism, and chance can never do choice or causation. Once it starts doing so, then it is no longer chance.

According to the Philosophy of Science, thanks to the *affirming the consequent* logic fallacy that's built into the Scientific Method, there are theoretically an infinite number of different, contradictory, mutually exclusive, self-defeating, and self-falsifying INTERPRETATIONS that can be given to each piece of scientific evidence or scientific data that we develop or produce.

This is also where the Materialists, Naturalists, Darwinists, Nihilists, and Atheists FAIL because they are constantly giving faulty and falsified INTEPRETATIONS to all of the different science experiments that they conduct.

They directly employ the *affirming the consequent* or the *jumping to conclusions* logic fallacy to their scientific data in order to prove that the theory of evolution and the second law of thermodynamics are true. These people *jump to the conclusion* that chance can design and create anything that it sets its mind to, if given enough time to do so. That's NOT science. That's cheating! But they don't even know it, because they don't understand the Philosophy of Science. These people don't study the Philosophy of Science in school because the Philosophy of Science PROVES that Materialism, Naturalism, Darwinism, Nihilism, Atheism, the Theory of Evolution, and the Second Law of Thermodynamics ARE FALSE. These people toss out and reject everything that falsifies their theories and their ideas. That's how they do "science". They cheat!

The Materialists, Naturalists, Darwinists, Nihilists, and Atheists have *jumped to the conclusion* that CHANCE can design and create anything, if given enough time to do so. That's what they teach, preach, and believe, is it not? They have NEVER tested that hypothesis, because they can't. Chance doesn't do anything without an intelligent assist from some type of intelligent being. By definition and by axiom, CHANCE cannot design and create anything because CHANCE cannot do choice and causation. Once chance starts doing causation, then it is no longer chance! This is the fundamental axiom of Science and Statistics. Chance and causation are mutually exclusive. This is the Truth in Science and Statistics, whether we realize it or not. Chance cannot do causation. Saying that it can doesn't change the fact that it can't. Scientists are lying to us whenever they say that CHANCE can do choice, design, or causation if given enough time to do so.

The inability of the Theory of Evolution to meet its burden of proof through a preponderance of the evidence ends up becoming Scientific Proof of God's Existence or Scientific Proof of God's Necessity, whether we realize it or not. This simple observation is based upon the "best explanation rule" or the "parsimony rule" in Science, sometimes known as Occam's Razor.

Creation by Chance, or Causation by Chance, or Creation by Random Disorder, or Creation by Entropy, or Creation Ex Nihilo, or Creation by Blind Luck – in other words, the Theory of Evolution – is NOT the best, nor the simplest, nor the most logical, nor the most rational, nor the most parsimonious explanation that can be given to scientific theories, scientific evidence, science experiments, and scientific data because chance of any kind cannot do causation. A LIE, such as Creation by Chance or the Theory of Evolution, cannot serve as the best explanation or the most parsimonious explanation for the scientific evidence. That's just logical common sense.

I wrote a huge book on this topic.

The Theory of Evolution Proved to Me that God Exists

https://www.amazon.com/gp/product/B01HZYBZ7K/

Most people don't understand this concept because they are NOT trained in our public schools to think of parsimony in terms of "observational evidence" or "meeting one's burden of proof". Instead, we are trained to blindly accept CHANCE as the most parsimonious, most logical, most rational, and best explanation for the design and the creation of our amino acids, our proteins, our genes, our genomes, and the life forms on this planet. It's a lie, but we believe that it is true because our college professors constantly tell us that it is true. It's the modern-day form of brain washing. If are told the lie long enough, we eventually come to believe that it is true. The Behaviorists, Atheists, Darwinists, and Naturalists know this, so that's why they keep lying to us.

Pagano, R.R. (2010). *Understanding Statistics in the Behavioral Sciences* (9th ed.). Belmont, CA: Wadsworth.

Every experiment exists to give the null hypothesis a chance to be rejected. (*Understanding Statistics*, p. 285.)

By definition, the Null Hypothesis in a science experiment is produced by chance alone. The Null Hypothesis represents CHANCE, and the Alternative Hypothesis represents CAUSATION. Chance and causation are mutually exclusive.

Statistics was designed to help scientists to produce a more realistic and demonstrably true INTERPRETATION of their scientific data and scientific evidence. This is done by using Statistics to falsify and reject the Null Hypothesis, which is produced or "caused" by chance alone.

The whole purpose of Statistics, Science, the Scientific Method, and Science Experiments is to falsify and reject the Null Hypothesis, Creation by Chance, Chance Causality, Causation by Chance, or anything that is produced by chance alone. The whole purpose of Statistics, Science, the Scientific Method, and Science Experiments is to falsify and reject Materialism, Naturalism, Darwinism, Nihilism, Atheism, Chemical Evolution, Spontaneous Generation, Abiogenesis, Creation by Death or Natural Selection, Creation by Random Chaos or Random Chance, the Theory of Evolution, the Second Law of Thermodynamics, or anything else that is produced by chance alone.

Therefore, WE KNOW that scientists and college professors are lying to us whenever they make the claim that Chance or the Null Hypothesis can do causation. They don't know what they are talking about whenever they make this claim, and they are making this claim all the time in everything that they do within our public schools. These people are lying to us all the time because they have convinced themselves that chance can do causation at will, and they are wrong.

Mark My Words

Correlation Is Not Causation

In Statistics and in all of our Science Classes in college, they repeatedly keep saying that **Correlation Is NOT Causation**.

This is one of the fundamental axioms of Statistics and Science – correlation is not causation. They typically introduce correlation early in a statistics class, right after they get done with the definitions, because correlation is such a powerful scientific tool. Correlation also shows up in every Science class whenever they start talking about the difference between observational studies and science experiments. Correlation is the scientific study of the relationship between two variables.

Pagano, R.R. (2010). *Understanding Statistics in the Behavioral Sciences* (9th ed.). Belmont, CA: Wadsworth.

> **Aside from the practical utility of using a relationship for prediction, why would anyone be interested in determining whether two variables are related? One important reason is that if the variables are related, it is possible that one of them is the cause of**

the other. As we shall see later in this chapter, the fact that two variables are related is not sufficient basis for proving causality.

Nevertheless, because correlational studies are among the easiest to carry out, showing that a correlation exists between the variables is often the first step toward proving that they are causally related. Conversely, if a correlation does not exist between the two variables, a causal relationship can be ruled out. (*Understanding Statistics*, p. 114.)

Did you catch that?

In a science experiment, if there is NO correlation between your independent variable and your dependent variable, then a causal relationship can be ruled out. That's the end of it.

If there is no correlation then there is no causation.

That's powerful Science!

That's powerful Philosophy of Science.

Yet, what do we observed from the false and falsified philosophies of science or interpretations of science?

We OBSERVE the Materialists, Naturalists, Darwinists, Nihilists, and Atheists taking correlation of any kind and turning it into causation. They notice a correlation between the fossil record and life forms, and then they *jump straight to the conclusion* that Evolution or Random Chance produced the fossil record and the life forms. This is what they do, do they not? They *jump to the conclusion* that CHANCE caused or produced everything that exists in this universe. They literally turn correlation into causation, and they do this every day of their lives. Do they not? Is that not what we have experienced and observed from these people? They take the slightest correlation, and then they turn it into causation.

They call it a Scientific Inference. They infer that Evolution or Random Chance produced the fossil record, the proteins, the genomes, the brains, and the life forms because they notice a correlation; and then, they employ the *begging the question* logic fallacy and USE their Scientific Inference as evidence and proof that their conclusions are true.

That's how they trick themselves and deceive themselves. The Darwinists and Naturalists do this every day of their lives, and they don't even know it.

Pinel, J. (2014). Biopsychology (9th ed.). New York: Pearson.

Scientific inference is the fundamental method of biopsychology and of most other sciences [like the theory of evolution] **– it is what makes being a scientist fun.**

The empirical method that biopsychologists and other scientists use to study the unobservable is called <u>scientific inference</u>.

The fact that the neural mechanisms of behavior [and evolution] **cannot be directly observed and must be studied through scientific inference is what makes biopsychological research such a challenge – and as I said before, so much fun.** (*Biopsychology*, p. 13.)

Unbelievable!

This person literally turned *Scientific Inferences* or *Begging the Question* INTO an empirical method or a scientific method. It violates everything that science and statistics stands for! This is Bad Science! This is the very definition of Bad Science. This is what the Materialists, Naturalists, Darwinists, Nihilists, and Atheists do every day of their lives, whether they know it or not. They turn their Conclusions or their Scientific Inferences into evidence and PROOF.

Scientific Inference is a logic fallacy – the logic fallacy that we call *begging the question*.

The whole of Materialism, Naturalism, Darwinism, Nihilism, Atheism, the Theory of Evolution, and the Second Law of Thermodynamics are nothing more than a Scientific Inference. They are philosophies of science or interpretations of science. They are NOT science, because they *start with the conclusion* that CHANCE ALONE produced everything that exists in this universe. These people don't do science. They *jump straight to the conclusion* that CHANCE caused everything that exists. That's not science. That's philosophy, religion, and dogma.

Kalat, J. W. (2008). *Introduction to Psychology* (9th ed.). Belmont, CA: Wadsworth, Cengage Learning. **"The brain is the product of Evolution."** (p. 12).

How do you know? Because I say so!

Pinel, J. (2014). *Biopsychology* (9th ed.). New York: Pearson. **"Genetic endowment is a product of its evolution."** (p. 24).

Siegler, R., DeLoache, J., & Eisenberg, N. (2010). *How Children Develop*, (3rd ed.). New York: Worth.

The obvious question is: Why does the human brain – the product of millions of years of evolution – take such a devious developmental path, producing a huge excess of synapses, only to get rid of a substantial portion of them? The answer appears to be evolutionary economy. (p. 110).

Appears to be!

Notice how they turned correlation into causation! They turned appearances or inferences into causation. That's how they do their science. What's obvious is that they are *begging the question*. They used appearances and inferences as scientific evidence and as proof that their pre-chosen conclusions are true.

Notice how they ALWAYS *jump to conclusions* and *start with the conclusion* that Evolution or Random Chance produced the human brain. That's *begging the question*, and that's a logic fallacy.

Well, let's *jump to some conclusions* ourselves, and see what we get. Anyone can do it. It takes no expertise to make Scientific Inferences. Scientists do so all the time. That's how they lie to us and get away with it. They call it a Scientific Inference.

Notice that for millions of years while evolution was making our brains, we didn't have a brain. So, how did we survive and become the fittest animals on the planet when we didn't have a brain for millions of years? It also took millions of years for evolution to make our eyes, so for millions of years we were both dumb and blind. So, how did we survive for millions of years while we were blind and didn't have a brain? How did we breathe? How did we walk? How did we eat? How did we reproduce? We didn't have eyes or a brain for millions of years, so how did we survive and become top of the food chain? The Theory of Evolution is self-defeating and self-falsifying, and they don't even know it. It consists of nothing but lies or logic fallacies. It has no substance.

By using logic fallacies, we can produce any conclusion or any interpretation of the scientific data and scientific observations that we want. It's called Scientific Inference, and it's *begging the question*. It's turning correlation into causation!

Begging the question MEANS that you take your Conclusion or your Scientific Inference, and then you USE your Conclusion or Scientific Inference as PROOF or as EVIDENCE that your Conclusion or Scientific Inference is true. It's *circular reasoning*, which is yet another logic fallacy. That's how the Darwinists and Naturalists do their science. That's how they trick themselves and deceive themselves. They are cheating and lying, and most of them don't even know it. Yes, there are those who KNOW that they are lying to us, but most of them have been tricked and deceived just as we have been tricked and deceived by them.

Correlation is not causation. But whenever the Materialists, Naturalists, Darwinists, Nihilists, and Atheists see even the slightest amount of correlation, they automatically *jump to the conclusion* that CHANCE caused that correlation. Technically, they are right – chance most likely produced the correlation that they are observing – but they are completely WRONG to assume that CHANCE can do causation. That's where they FAIL – by turning correlation into causation; and, they do it every day of their lives until they finally figure out that it's wrong. I KNOW because I used to be a materialist, naturalist, nihilist, and atheist. I KNOW how they think and act, because I used to be one of them.

It's when these people *jump to conclusions* or *beg the question* and conclude that CHANCE caused the fossil record, the proteins, the genomes, the brains, and the life forms – that's where they go WRONG and FAIL. That's how they produce their faulty interpretations of science or their falsified philosophies of science in the first place. They *beg the question* and USE their Scientific Inferences or their Conclusions as PROOF or as EVIDENCE that their Conclusion or Scientific Inference is true. They equate correlation with causation, and they don't even know that they are doing it.

The axiom stating that correlation cannot do causation is the foundation for every science experiment.

The Null Hypothesis in every science experiment has a correlation coefficient equal to ZERO or close to ZERO. The Null Hypothesis is a correlation coefficient of ZERO or NO causation. A correlation coefficient of ZERO between our independent variable and our dependent variable means that a causal relationship can be ruled out. If there is NO correlation between our variables, then there is NO causation or NO causal relationship between our variables. The Null Hypothesis means that there is ZERO correlation between the independent and dependent variables. The Null Hypothesis means that the independent variable produces NO EFFECT on the dependent variable. The Null Hypothesis means that there is NO causal relationship between the independent variable and the dependent variable. It's that simple. That's how we do Science and Science Experiments! Or, at least that's how we should do them.

But, that's NOT how the Materialists, Naturalists, Atheists, and Darwinists do and interpret their science experiments. Instead, these people *jump straight to the conclusion* that CHANCE caused or produced everything that exists in our universe, including the proteins, genes, genomes, eyes, brains, and life forms. They cheat. They turn correlation into causation every time. They literally use the Null Hypothesis as PROOF that their Alternative Hypothesis (Causation by Evolution or Causation by Chance) is true. That's what they do, is it not?

If they notice the slightest correlation, they turn it into causation. Then after they have run their science experiments and have discovered some type of causation within their data, they *jump to the conclusion* that their Alternative Hypothesis is true, and that CHANCE is the cause of their Alternative Hypothesis. This too is how they FAIL and end up with the faulty conclusion that CHANCE caused the results of their science experiment. They literally use the Null Hypothesis as their Alternative Hypothesis.

The Null Hypothesis in a science experiment represents CHANCE, and the Alternative Hypothesis in a science experiment represents CAUSATION. Chance and causation are mutually exclusive, which means that the Null Hypothesis and the Alternative Hypothesis are mutually exclusive.

The Null Hypothesis means that there is ZERO correlation between the independent variable and the dependent variable! The Null Hypothesis therefore means that there is NO causal relationship between the independent variable and the dependent variable. The Null Hypothesis means that the independent variable has NO EFFECT on the dependent variable.

The Materialists, Naturalists, Darwinists, Nihilists, and Atheists toss all of this out the window and *start with the conclusion* that CHANCE can design and create anything if given enough time to do so. That's what they teach, preach, and believe. Is it not? These people literally use the Null Hypothesis to PROVE the truthfulness of their Alternative Hypothesis. They literally use the Null Hypothesis as their Alternative Hypothesis. That's *begging the question*. They *jump straight to the conclusion* that CHANCE produced the results of their science experiment, which means that they *start with the conclusion* that CHANCE caused the results of their science experiment. When it matters most, these people equate chance with causation every time. That's how they choose to interpret their science. This is what they do, is it not?

By definition or by axiom, the Null Hypothesis is produced by chance alone; and therefore, a Null Hypothesis produces ZERO correlation between the independent variable and the dependent variable, thereby ruling out any causal relationship between the independent variable and the dependent variable. The Null Hypothesis states that our independent variable has NO EFFECT on our dependent variable. In truth, the Null Hypothesis states that chance cannot do causation. By definition, the Null Hypothesis or Chance produces ZERO correlation. The Null Hypothesis or Chance can't even do proper correlation, let alone causation. Powerful, is it not? That's how we separate the truths from the lies in our science experiments while interpreting the results of our science experiments.

But what do we OBSERVE from the Materialists, Naturalists, Darwinists, Nihilists, and Atheists? Whenever they see some type of causation within the data from their science experiment, they automatically *jump to the conclusion* that Evolution or Random Chance CAUSED their Alternative Hypothesis to be true. These people literally USE the Null Hypothesis (or Chance) as their Alternative Hypothesis every time, and they always *jump to the conclusion* that CHANCE produced everything that we experience and observe in this universe.

These people are lying to us and deceiving us all the time, and I finally figured out how they are doing it and getting away with it. Here I put the truth and the lies side-by-side in the hope that some of us will learn how to discern the difference between the two.

Scientists lie, and we need to learn how to detect when they are lying to us and when they are not. Here in this short essay, we produced a powerful and incontrovertible TEST of truth assertions that quickly tells us when our college professors and the scientists are lying to us. Anytime a scientist or college professor is using Chance or the Null Hypothesis to do causation or to prove that chance is doing causation, then you automatically KNOW that he or she is lying to you, because Chance or the Null Hypothesis by definition cannot do causation. Chance and causation are mutually exclusive and exhaustive. They falsify each other.

That's what I got from my Statistics class. Did you get that from your Statistics class too? If not, why not? What were they trying to hide from you?

This simple TEST for truth assertions is also a Scientific Method. There are many different types of Scientific Methods or scientific methodologies that we can use to help us to get at the truth. This is one of them. I automatically KNOW that these people are lying to us whenever they start using chance to do causation. Now you KNOW it too.

They can't deceive you anymore, once you know how they are doing it and how they have been getting away with it.

Knowing about this won't help you in any of your Science classes, because these people want you to accept the lie and believe the lie. They will expect you to keep lying right along with them. They will demand nothing less. They will fail you if you refuse to go along. That's how they operate. But, knowing that chance can never do causation will help you to KNOW immediately whenever these people are lying to you, or trying their best to teach you the truth.

Mark My Words

The Null Hypothesis and the Alternative Hypothesis

We've already mentioned in this essay series the Null Hypothesis and the Alternative Hypothesis, which are an integral part of every science experiment. We saw how the Materialists, Naturalists, Darwinists, Nihilists, and Atheists implicitly and often unknowingly use the Null Hypothesis as their Alternative Hypothesis by making correlation or chance DO causation.

We developed a powerful TEST for truth assertions that lets us KNOW immediately when the scientists and our college professors are lying to us. Whenever these people start using Chance or the Null Hypothesis to do causation, then we automatically KNOW that they are lying to us and trying to trick us and deceive us.

Simple. Powerful. Parsimonious. Logical. True.

Now let's study Statistics, the Scientific Method, Science Experiments, the Null Hypothesis, and the Alternative Hypothesis for a few minutes to see what else we can learn.

> **In any experiment, there are two hypotheses that compete for explaining the results: the *alternative hypothesis* and the *null hypothesis.* The alternative hypothesis is the one that claims the difference in results between conditions is due to the independent variable.**
>
> **The null hypothesis is set up to be the logical counterpart of the alternative hypothesis such that if the null hypothesis is false, the alternative hypothesis must be true. Therefore, these two hypotheses must be mutually exclusive and exhaustive.**
>
> **The null hypothesis asserts that the independent variable has *no effect* on the dependent variable. You can see that these two hypotheses are mutually exclusive and exhaustive. If the null hypothesis is false, then the alternative hypothesis must be true.**
>
> **As you will see, we always first evaluate the null hypothesis and try to show that it is false. If we can show it to be false, then the alternative hypothesis must be true.**

We always evaluate the results of an experiment by assessing the null hypothesis. The reason we directly assess the null hypothesis instead of the alternative hypothesis is that we can calculate the probability of chance events, but there is no way to calculate the probability of the alternative hypothesis. We evaluate the null hypothesis by assuming it is true and testing the reasonableness of this assumption by calculating the probability of getting the results *if chance alone is operating*. The null hypothesis is evaluated by assuming chance alone is responsible for the differences in scores between conditions.

Every experiment exists to give the null hypothesis a chance to be rejected.

The null hypothesis is having a correlation equal to ZERO.
(*Understanding Statistics*, pp. 242, 261, 285, 336.)

The Null Hypothesis represents chance, and the Alternative Hypothesis represents causation. Chance and causation are mutually exclusive, which means that the Null Hypothesis and the Alternative Hypothesis are mutually exclusive.

This is the KEY to separating the truths from the falsehoods in every Science.

Every science experiment is designed to reject Chance or the Null Hypothesis so that we can establish Causation or the Alternative Hypothesis instead. The purpose of an experiment is to find the True Cause of an event. The Null Hypothesis represents chance, and the Alternative Hypothesis represents causation. Chance and causation are mutually exclusive and falsify each other. These must be mutually exclusive and exhaustive, which means that chance can NEVER be used as causation in a True Science Experiment. To do so invalidates the results of our science experiment. Chance can NEVER be used as a causal explanation in Science and Statistics. To do so means that we are no longer doing science.

Yet, we constantly see the Materialists, Naturalists, Darwinists, Nihilists, and Atheists using chance and making chance the ONLY CAUSE of all the different things that exist in this universe. These people lie, cheat, and deceive because they don't understand Science and the Philosophy of Science and how they are supposed to work and meant to work. These people deliberately use CHANCE as the cause of everything that exists in this universe, and these people are constantly teaching and preaching that CHANCE can design and create anything if given enough time to do so. That's what they teach and believe, is it not? They are NOT doing science – they are doing the philosophy of science or a faulty and falsified interpretation of science.

Science exists to eliminate "chance" as the causal mechanism for all the different things that we experience and observe. The purpose of a science experiment is to find the True Cause of an event, and chance or the null hypothesis can NEVER be it, because by definition chance cannot produce a Real Effect and because by definition chance cannot be the True Cause of anything. Chance and causation are and must be mutually exclusive and exhaustive, otherwise, our science experiment is going to be falsely interpreted. Within a True Science

Experiment and a True Science, we can NEVER make chance the cause or the causal explanation of what we are observing and experiencing. To do so falsifies our personal interpretation of the scientific evidence. Once you make chance the cause of what you are observing or proposing, then your scientific inference or causal explanation is automatically false and falsified. It's that simple.

The whole purpose of a science experiment is to falsify, reject, and eliminate anything and everything that is produced by chance alone. The whole purpose of a science experiment is to falsify, reject, and eliminate the Null Hypothesis, Chance Causality, Creation by Chance, Creation by Random Disorder, or Chance Causation. The whole purpose of a science experiment is to falsify, reject, and eliminate Materialism, Naturalism, Darwinism, Nihilism, Atheism, Behaviorism, Determinism, Atomism, Physicalism, Creation by Chance, Creation Ex Nihilo, Creation by Random Disorder, Creation by Entropy, Chemical Evolution, Abiogenesis, Spontaneous Generation, Macro-Evolution, Creation by Death or Natural Selection, the Theory of Evolution, the Second Law of Thermodynamics, and everything else that is produced by chance alone. These things are automatically FALSE because by definition they were produced by chance alone. They ARE the Null Hypothesis because they are "caused" by chance alone. They are precisely the thing that we trying to identify and eliminate from Science because they were produced by chance alone.

Everything that is produced by chance alone goes into the garbage can where it belongs. That's how Science and Science Experiments work, or at least that's how they are supposed to work. We reject the Null Hypothesis because it was "caused by chance alone" or produced by chance alone! WE KNOW that chance can never do causation. We are searching for the True Cause of our scientific evidence, and by decree, chance can never be it. Chance cannot be the True Cause of anything. By definition, chance cannot produce a Real Effect. If it's producing a Real Effect, then it is no longer chance. When running a science experiment, we are looking for the True Cause of an event, and by axiom or by decree chance can never by it. Remember, chance cannot be the True Cause of anything. If it's causing something, then it is no longer chance! Chance and causation are mutually exclusive. Chance cannot do choice, creation, or causation.

This is the KEY to separating the truths from the falsehoods in Science.

The primary axiom of Science, Statistics, the Scientific Method, and Science Experiments is that chance cannot do causation. Chance and causation are mutually exclusive. If chance is causing something to happen, then it is no longer chance but has become some type of intelligent and deliberate causation or choice instead.

Remember, chance cannot do design, creation, causation, nor choice. Once it starts doing so, then it is no longer chance!

In his book, Pagano repeatedly emphasizes that we USE Statistics to eliminate the faulty interpretations that we want to give to the scientific evidence. The FATAL FLAW in the Scientific Method consists of all the faulty and falsified personal interpretations of the scientific data that we choose to give to the results of our science experiments. Scientists lie, especially to themselves. Scientists trick themselves and deceive themselves. Self-deception works, and it works every time. Nobody is immune. Statistics was designed to help scientists to get closer to the truth by removing and eliminating "chance" as a causal mechanism. The fundamental axiom of Statistics states that chance cannot do causation. Chance cannot produce a Real Effect! Chance or the Null Hypothesis exists to be rejected, falsified, and eliminated from Science, so that we can establish Causation or the Alternative Hypothesis instead.

If we verify or retain the Null Hypothesis or Causation by Chance in our science experiment, then WE KNOW that our independent variable has NO EFFECT on our dependent variable; and therefore, WE KNOW that there is NO causal relationship between our independent variable and our dependent variable, because WE KNOW that chance cannot do causation.

This is how we identify and eliminate the things that are false in science. We identify and eliminate everything that was produced by chance alone. We identify and eliminate Materialism, Naturalism, Darwinism, Nihilism, the Theory of Evolution, the Second Law of Thermodynamics, and their derivatives. That's how we get to the truth in Science, by eliminating everything that is false! It works because *negating the consequent* is philosophically valid and logically sound. This is the pinnacle of the Philosophy of Science whether we realize it or not.

>**Statistics uses probability, logic, and mathematics as ways of determining whether or not observations made in the real world or laboratory are due to random happenstance or perhaps due to an orderly effect one variable has on another. Separating happenstance, or chance, from cause and effect is the task of science, and statistics is a tool to accomplish that end.**
>
>**Occasionally, data will be so clear that the use of statistical analysis isn't necessary. Occasionally, data will be so garbled that no statistics can meaningfully be applied to it to answer any reasonable question. But I will demonstrate that most often statistics is useful in determining whether it is legitimate to conclude that an orderly effect has occurred. If so, statistical analysis can also provide an estimate of the size of the effect.** (*Understanding Statistics*, p. xxvii.)

Separating chance from causation IS the task of Science, the Scientific Method, Science Experiments, and Statistics!

According to the Philosophy of Science, the whole purpose for statistics and the Scientific Method is to ELIMINATE chance or happenstance as a causal source or a causal agent. Chance is an invalid cause or a false cause according to statistics and the Scientific Method.

In contrast, the Materialists, Naturalists, Darwinists, Nihilists, and Atheists have *jumped to the conclusion* that CHANCE designed and created everything in this physical universe. That is a faulty and falsified interpretation of scientific evidence, yet these people continue to insist that CHANCE made everything that exists. Therefore, every piece of scientific data and scientific evidence that these people encounter gets their personal interpretation that CHANCE ONLY designed it and created it. They are wrong, but that doesn't stop them from *jumping to that conclusion*. *Jumping to conclusions* is a logic fallacy, but they don't know that, because they do it all the time and have been taught to do it in their college science classes.

Materialism, Naturalism, Darwinism, Nihilism, Atheism, Behaviorism, Determinism, Physical Reductionism, Chemical Evolution, Macro-Evolution, Creation by Death (natural selection), Spontaneous Generation, Abiogenesis, the Theory of Evolution, and the Second Law of Thermodynamics ARE Creation by Chance, or Causation by Chance, or the Null Hypothesis. Therefore, if we falsify Creation by Chance or Chance Causality, then we have successfully falsified these FALSE religions, or FALSE philosophies, or FALSE personal interpretations of science.

The whole purpose of Statistics is to FALSIFY random happenstance, chance causality, causation by chance, creation by random disorder, or Creation by Chance. The whole purpose of Science and Statistics is to find the True Cause of an event, instead. By definition, in principle or practice, chance can NEVER be the True Cause of anything. Chance cannot produce a Real Effect. Separating happenstance or chance from "cause and effect" is the task of Science, the Scientific Method, Science Experiments, and Statistics.

In other words, the whole purpose of Science and Statistics is to identify Chance Causality or Creation by Chance and eliminate them from Science. The whole purpose of Science and Statistics is to identify Materialism, Naturalism, Darwinism, Nihilism, Atheism, the Theory of Evolution, and the Second Law of Thermodynamics; and then eliminate them from Science because they were produced by chance alone.

> **By far, most of the applications of inferential statistics are in the area of hypothesis testing. Scientific methodology depends on this application of inferential statistics. Without objective verification, science would cease to exist, and objective verification is often impossible without inferential statistics. You will recall that at the heart of scientific methodology is an experiment. Usually, the experiment has been designed to test a hypothesis, and the resulting data must be analyzed.**
>
> **Occasionally, the results are so clear-cut that statistical inference is not necessary. However, such experiments are rare. Because of the variability that is inherent from subject to subject in the variable being measured, it is often difficult to detect the effect of the independent variable without the help of inferential statistics. In this chapter, we shall begin the fascinating journey into how experimental design, in conjunction with mathematical analysis, can be used to verify truth assertions or hypotheses, as we have been calling them.** (*Understanding Statistics*, p. 239.)

Without objective verification, our scientific conclusions are completely worthless. If we are using Chance or the Null Hypothesis as our conclusion, then our Science and our Science Experiment are completely worthless. If we are using Chance or the Null Hypothesis as our independent variable or our dependent variable, then our Science, our Conclusions, and our Science Experiment are completely worthless. Without objective verification, science ceases to exist! The Materialists, Naturalists, Darwinists, and Atheists are no longer doing science because they cannot meet their burden of proof and because they cannot provide objective verification of their personal beliefs. These people are no longer doing science because they have chosen to believe that CHANCE designed and created everything that exists in this universe. That's NOT science. That's philosophy of science or a personal interpretation of science. It's NOT science. These people don't do science in the true sense of the word.

Chance can never be the cause of what we are observing because chance cannot do causation and because chance cannot produce a Real Effect. If chance is producing a Real Effect, then it is no longer chance. It has become some type of causation instead. Chance and causation are mutually exclusive. They falsify each other. They preclude each other from happening. They are also exhaustive. It's either chance or causation. It's never something in between. Chance cannot do causation.

Pagano makes Statistics an integral part of the Scientific Method, particularly the fourth and final "interpretation phase" of the Scientific Method. The purpose of Statistics is to prevent us from getting faulty interpretations of our scientific results. The purpose of Statistics is to prevent us from using CHANCE as our causal mechanism or our causal explanation!

Powerful, is it not?

A lot of lights went on when this finally hit home.

Scientists lie, but if used properly, the math doesn't lie. We can calculate the probability of chance events, and if calculated properly, it immediately becomes obvious that CHANCE cannot do creation or choice or causation. Chance has no mechanism in place for doing so.

https://ultimate-model-of-reality.com/the-probability-of-protein-synthesis-by-chance-alone/

https://ultimate-model-of-reality.com/a-binomial-distribution-table-falsifies-creation-by-chance/

Scientists are lying to us whenever they try to convince us that chance can do causation.

This is a powerful scientific test for truth assertions!

Creation by Chance, Chance Causality, or Causation by Chance cries out for empirical verification. These are truth assertions from the Materialists, Naturalists, Darwinists, and Atheists; and, these truth assertions need to be experimentally verified like anything else. The problem is that they can't be verified by a science experiment because they are in fact Creation by Chance or the Null Hypothesis to begin with. BOTH the Null Hypothesis and Creation by Chance are by definition caused by chance alone. They ARE the same thing. Creation by Chance IS the Null Hypothesis. Materialism, Naturalism, Darwinism, Nihilism, Atheism, the Theory of Evolution, and the Second Law of Thermodynamics ARE the Null Hypothesis or Creation by Chance. They are precisely what we are trying to identify and eliminate from science in the first place.

The way we get at the truth through Science and Statistics and the Scientific Method is to use them to falsify and eliminate the obvious lies – starting with Materialism (creation by rocks), Naturalism (creation by death or natural processes), Darwinism (creation by chance), Atheism (creation ex nihilo), and the Second Law of Thermodynamics (creation by random disorder or entropy). Science and Statistics and the Scientific Method were in fact designed to identify these falsehoods and then eliminate these kinds of things from Science. The whole purpose of Science and Statistics is to identify Creation by Chance of any type, and then eliminate it from Science.

Instead, we see the Materialists, Naturalists, Darwinists, Nihilists, and Atheists making CHANCE the cause of everything that exists in complete violation of the rules of Science and Statistics.

We see from Statistics that CHANCE and CAUSATION are mutually exclusive. Creation by Chance precludes or falsifies Causation. Two events are mutually exclusive if both cannot occur together at the same time. Two events are mutually exclusive if the occurrence of one event precludes or prevents the occurrence of the other event. Chance and Causation are mutually exclusive. Once it becomes causation, then it is no longer chance. Once we establish that our results were produced by chance alone, then we KNOW that we are NOT looking at any type of causation. We automatically KNOW that Materialism, Naturalism, Darwinism, Nihilism, Atheism, Behaviorism, Determinism, Physical Reductionism, the Theory of Evolution, and the Second Law of Thermodynamics CANNOT DO CAUSATION because they are by definition Creation by Chance or the Null Hypothesis, which means that they were produced by chance alone.

Materialism, Naturalism, Darwinism, Nihilism, Atheism, the Theory of Evolution, and the Second Law of Thermodynamics ARE Creation by Chance and by definition ARE caused by chance alone. They ARE the Null Hypothesis. They ARE Creation by Chance or Chance Causation. They ARE what we are trying to falsify and eliminate from Science. They cannot produce results or Real Effects. Chance Causation or Causation by Chance IS an oxymoron – a contradiction in terms. It's self-falsifying because CHANCE cannot do creation or causation! Chance cannot do choice. Materialism, Naturalism, Darwinism, Atheism, and their derivatives ARE Chance Causation, and they are automatically FALSE because chance cannot do creation nor causation.

Here we just used the Rules of Statistics to falsify these lies. Anything that is caused by chance alone is automatically FALSE because chance can't cause anything to happen. Chance and causation are mutually exclusive. If we are looking at causation or have found causation, then we KNOW that chance is no longer involved in the situation. CHOICE is coming into play instead. PSYCHE and CHOICE are the same thing – they do causation, both at the physical level and the quantum non-physical level!

In contrast, if we have established that our interpretations of the scientific evidence were "caused by chance alone" or "produced by chance alone", then we automatically KNOW that there is NO causal relationship between the independent variable and the dependent variable because chance cannot do causation. Once we establish that we are looking at the Null Hypothesis or Chance Causality, then we KNOW that we are NOT looking at any type of causation. Chance cannot do causation. Chance and causation are mutually exclusive. The one precludes the other.

That's one of the reasons why WE KNOW that Materialism, Naturalism, Darwinism, Nihilism, Atheism, the Theory of Evolution, and the Second Law of Thermodynamics ARE FALSE, because they were produced by chance alone and are therefore the Null Hypothesis. Since they ARE the Null Hypothesis or Chance, they cannot be the Alternative Hypothesis or Causation, which means that they cannot be the True Cause of what we are observing and experiencing. These things can NEVER be the True Cause of anything because they were produced by chance alone, and chance cannot do causation. Since they are Creation by Chance or produced by chance alone, they cannot be the True Cause of the events that we have chosen to study. It's that simple! In every science experiment, we are looking for the True Cause of the event, and the Null Hypothesis or Chance can NEVER be it.

Simple. Logical. Rational. Parsimonious. True.

See all the different problems, controversies, debates, and lies that we can solve and eliminate simply by choosing to use Statistics, Science, and the Scientific Methods the way that they were designed to be used – to identify and eliminate from science any type of Chance Causality, Creation by Chance, or the Null Hypothesis!

Mark My Words

Truth in Science – Proof of Concept

Scientists lie.

That's part of the reason why I find the Statistics and the Math so convincing, because I KNOW that scientists lie. Scientists have been lying to me all of my life, and now I KNOW how they did it and got away with it. I wanted to believe their lies. Self-deception works, and it works every time. Nobody is immune.

Let me provide a couple of examples so that you can see it too.

We can calculate the probability of chance events, so let's do so, and see what we get. Let's produce some Truth in Science. Let's produce some Truth through Science.

I got the following off the internet from Uncle Google.

Thyroid releasing hormone or TRH should be the smallest protein in the human body, with 234 amino acids. (> 100 amino acids is a protein.) The smallest polypeptide in the human body should be insulin, with 54 amino acids. (10-100 amino acids is a polypeptide.)

Let's run with this for a minute and see what we get.

Proteins are made from amino acids. In the human body, there are 21 different amino acids that are used to make proteins.

Now, let's run our science experiment.

First, we have to start with a God-given ocean full of the correct 21 amino acids. If there are no coins, and if there's nobody there to toss the coins and count the results, CHANCE doesn't do anything. Games of chance, or chance of any kind, requires an intelligent assist in order to get them going in the first place. If there are NO amino acids or if we have an ocean full of the incorrect amino acids, nothing is ever going to happen, and we will NEVER have any results. We will NEVER be able to produce a protein by chance alone if there are NO God-given amino acids to start with. That's just the way it is. This is Logic 101, and everyone overlooks it.

Remember, you have to give CHANCE something to work with, to begin with, or it will never do anything. Chance requires a set-up or an intelligent assist. Someone Psyche or Someone Intelligent has to be there to make the coins, toss the coins, count the heads and tails, and recognize when there are all heads or all tails. If that person isn't there to make the coins and to toss the coins and to count the coins, then chance doesn't do anything.

Now, let's assume for sake of argument that we have convinced God to stir this ocean full of the correct 21 amino acids, so that CHANCE actually has a chance to connect 54 different amino acids together every minute on average. God has to stir the pot or the ocean, or nothing is ever going to happen, and CHANCE can do nothing to produce anything. This is logical common sense. There has to be some action – someone has to toss the coins – or chance isn't going to do anything. Someone Psyche or Someone Intelligent has to CHOOSE to toss the coins or to stir the amino acids or to line up the amino acids, or CHANCE isn't going to do anything. Someone also has to make the coins or make the amino acids in the first place, or chance isn't going to do anything.

Remember, Someone Psyche or Someone Intelligent has to toss the coins or stir that ocean full of amino acids, or chance isn't going to do anything. Games of chance require an intelligent assist or an intelligent set-up, or nothing will ever happen.

Remember, we have to give CHANCE something to work with, or CHANCE will never do anything. That's just the way it is. Games of chance have to be set up by an Intelligent Person in the first place, or nothing will ever happen.

Furthermore, let's say for sake of argument that CHANCE is able to miraculously line up 54 amino acids into some kind of sequence every minute on average by chance alone, while God continues to stir this ocean full of the correct 21 amino acids for all eternity. Let's say that we get an "event" or a toss of those 54 "coins" or 54 amino acids lined-up every minute on average, and let's say that God is there to recognize the single event when the toss of those amino acids produced insulin by chance alone. We've got to make sure that that ocean full of the correct 21 amino acids is big enough that it will actually happen by chance alone, and we have to make sure that God is there to see it happen when it happens, so that He can recognized that CHANCE just produced insulin.

We also need some sort of timing mechanism or "rate of coin toss" so that we can calculate how long on average it will take for CHANCE to produce an insulin polypeptide by chance alone. We have set one minute on average for the rate of our coin toss or the rate of the amino acid sequencing and testing. That's a pretty good rate considering the fact that God has to check that complete ocean-full of amino acids every minute to see if chance has produced insulin, or not. This is a game of chance after all, so we need some type of Referee to determine if the game has been won.

We have to start with some type of given for the sake of argument, so that we can see with our own eyes what CHANCE can really do.

How long is it going to take on average for CHANCE to produce insulin by chance alone, if this stirred ocean of the correct amino acids is randomly assembling 54 amino acids into some kind of sequence every minute on average?

We can calculate the probability of chance events.

Assuming that 54 amino acids are placed into a random sequence and "tested" or "observed" once every minute on average, the probability of making insulin by chance alone from an ocean full of the correct amino acids is:

$(1/21)^{54}$

This is equal to: 3.98252109 e -72

Our sample size is 54 amino acids long. Our target is an insulin polypeptide. The probability of getting the correct amino acid in each spot is 1/21. This probability ONLY holds true if there are ONLY the correct 21 amino acids in our ocean full of amino acids. If there is anything else in that ocean such as water molecules, then the probability drops to 1/22, because water is not an amino acid, and we are trying to produce insulin by chance alone.

Now, it's time for that leap of faith. There is always a leap of faith when it comes to any game of chance. I have to find some way to bring the passage of time into the equation. Therefore, I convert our probability to minutes, on the assumption that it takes a minute on average to toss and then count 54 amino acids at a time, within this ocean full of the correct 21 amino acids. I assume that we get a successful toss of 54 amino acids every minute for each increment in our probability equation. It's just a guestimate, but an extremely generous one, giving chance the full benefit of the doubt. We want chance to succeed after all.

This is "3.98252109 e 72" minutes on average – give or take a few trillion years – for CHANCE to produce an insulin polypeptide by chance alone, starting with a God-given ocean full of the correct 21 amino acids and after convincing God to keep stirring that ocean so that the amino acids can randomly assemble into "testable" sequences every minute on average by chance alone. God also has to preserve those amino acids in existence while He stirs them so that they don't disintegrate and become something else, and thereby decrease our chances of getting insulin by chance alone.

Furthermore, we need God there to identify the insulin molecule or insulin polypeptide when CHANCE finally produces it. If God isn't there to recognize and secure the insulin polypeptide when it is finally made by chance alone, then the stirring process will eventually destroy that polypeptide and the searching process will have to start all over again.

Chance doesn't do anything without Someone Psyche or Someone Intelligent there to design and run the game of chance in the first place. Someone has to toss the coins or the amino acids, or nothing will ever happen. This is Logic 101, and everyone completely overlooks it, because they have erroneously convinced themselves that CHANCE can design and create anything that it sets is mind to, if given enough time to do so. They are WRONG. Chance doesn't do anything without an Intelligent God there to design and run the game of chance in the first place.

Now we take our "3.98252109 e 72" minutes guestimate, divide that by 60 minutes, divide that by 24 hours, and divide that by 365 days to get the number of YEARS that it would take on average under a best-case scenario to produce an insulin polypeptide by chance alone.

We end up with "7.57709492 e 66" YEARS on average for CHANCE to produce an insulin polypeptide by chance alone, starting with a God-given ocean full of the correct amino acids and assuming a successful event or a successful toss of the amino acids every minute on average. That's NEVER going to happen! Our physical universe hasn't been around long enough for that to have happened by chance alone.

This can't be done by chance alone, even with a God-given ocean full of the correct 21 amino acids. Even with God constantly stirring the pot or tub or ocean, it would take trillions of trillions of trillions of trillions of years for such a thing to happen by chance alone. It's just not going to happen. Our observable universe hasn't been around long enough for this to happen by chance alone. Insulin was designed and made by God. Without God to design it, make it, fine-tune it, field-test it, and recognize it for what it is, insulin wouldn't exist. Only God could recognize that He had actually created insulin and KNOW what it is good for and how to use it.

God is not dumb enough to stir an ocean full of the correct amino acids for trillions of trillions of trillions of trillions of YEARS on the off chance that CHANCE will produce an insulin polypeptide by chance alone. God will just grab the correct amino acids that He needs, assemble them into insulin, and be done with it in a matter of minutes.

CHANCE cannot make an insulin polypeptide by chance alone. God had to design and make the insulin, or it wouldn't exist.

Technically, insulin isn't a protein. It's a polypeptide that is 54 amino acids long. From now on, whenever you think of insulin, remember that God designed and made the stuff in the first place. We KNOW for a fact that chance didn't do the job, so that leaves God as the only other plausible candidate. Someone made the stuff, or it wouldn't exist. Anything that is obviously made obviously has a Maker who made it. Insulin was obviously made. Therefore, insulin obviously has a Maker. God made insulin; and, there's no chance that I got that wrong.

Whenever we eliminate Chance or the Null Hypothesis, then that leaves some type of Causal Agent as our Alternative Hypothesis or Causal Explanation, because chance cannot do causation. This is how Science Experiments work!

Here we just falsified and rejected the Null Hypothesis (which is produced by chance alone) and accepted the Alternative Hypothesis (which is some type of causation by a causal agent). In this case, the Alternative Hypothesis is that God designed and created insulin, because chance could never have done the job.

We just rejected the Null Hypothesis which states that insulin was produced by chance alone, which means that we now have to accept our Alternative Hypothesis which states that insulin was produced by some type of Intelligent Agent, or Causal Agent, or Person. This is the way that Science, the Scientific Method, Science Experiments, and Statistics work, or the way they should work!

This is Science in action. This is a True Science Experiment. This is Truth in Science. We just found the truth by rejecting the lie. We just discovered that insulin cannot be produced by chance alone. This is a significant and monumental scientific discovery, is it not? This changes everything, does it not?

When we finally have the Truth in Science, everything changes right before our very eyes.

This is the pinnacle of the Philosophy of Science.

We just developed a powerful and incontrovertible TEST for truth assertions that works every time. Whenever a college professor or scientist tells you that CHANCE can design and create anything if given enough time to do so, then you automatically KNOW that he or she is lying to you.

Creation by Chance or Chance Causation is the Null Hypothesis in any science experiment, and it's precisely what we are trying to falsify and eliminate from science.

If someone is telling you that CHANCE made your eyes, your brain, your genes, your proteins, your genome, or your body, then you automatically KNOW that they are lying to you and trying to deceive you. It doesn't matter how many degrees they have behind their name or how much authority they have behind their name, they are lying to you and trying to trick you and deceive you.

Powerful! Is it not?

This is Truth in Science. This is Truth through Science. This is the pinnacle of the Philosophy of Science – a true and valid Philosophy of Science.

—

Let's try another one.

THR is the smallest protein in the human body.

The probability of making THR (thyroid releasing hormone) by chance alone is:

$(1/21)^{234}$

When I put this into my graphing calculator, the answer was ZERO (0). Literally. The answer is ZERO. This is NEVER going to happen ever! This is beyond what a graphing calculator can handle. This is beyond what God could do by chance alone!

The probability of making THR by chance alone is ZERO. You can wait for all eternity, and it will NEVER happen, even with a God-given tub full of the correct amino acids or with an ocean full of the correct amino acids. It's NEVER going to happen. Making THR by chance alone is impossible. It can't happen by chance alone. It requires some kind of intelligent and deliberate intervention from God Himself.

Here we just falsified and rejected the Null Hypothesis (which is produced by chance alone) and accepted the Alternative Hypothesis (which is some type of causation by a causal agent). In this case, the Alternative Hypothesis is that God designed and created THR, because chance could never have done the job. This is Science in action. This is a True Science Experiment. This is Truth in Science.

THR is the smallest protein in the human body. THR can never be made by chance alone. Every time you think of any protein, remember that God designed it, created it, made it, fine-tuned it, and field-tested it. Anything that is obviously made obviously has a Maker who made it. Proteins were obviously made. Therefore, proteins obviously have a Maker who made them.

God made your proteins. Each protein is one of God's Signatures. People are looking for the Signature of God in nature and science, and there it is in each one of your proteins.

Your genome is also God's Signature. God had to make a matching gene for each of the proteins that He designed and made. God made your proteins and your genes. This is obviously true. According to the statistics and the probabilities, there is NO chance that I am wrong. God made your proteins and your genes.

This is one of the most convincing Scientific and Statistical Proofs of God's Existence that I have encountered so far. I just used Statistics, Probability, a Scientific Method, and Science to prove that God exists; and, there's NO chance that I'm wrong.

Remember, whenever a scientist tries to convince you that CHANCE can design, create, choose, and do causation, then you AUTOMATICALLY KNOW that he or she is lying to you, because chance of any kind cannot do causation. Chance cannot produce a Real Effect. Chance cannot make choices or do causation. Chance and causation are mutually exclusive. They falsify each other.

There's NO way in this universe that CHANCE could ever have designed and made your proteins and your genes. That's just the way it is. The math doesn't lie! The scientists will lie and tell you that CHANCE can produce anything if given enough time to do so. They are wrong. They are lying to you.

Chance cannot design and produce anything of significance; and, chance of any kind including games of chance REQUIRE an intelligent assist from an Intelligent Being to get them going in the first place. Someone has to make the coins, toss the coins, count the coins, and then recognize the times when the coin tosses produce all heads or all tails. Without Someone Psyche or Someone Intelligent there to run the game of chance in the first place, CHANCE is never going to do anything. Chance just isn't that reliable. Chance doesn't do causation. Chance doesn't have a mind of its own.

Without God there to recognized that CHANCE has produced insulin or THR and to secure and use that new protein when CHANCE produces it, nothing will ever come of those newly made polypeptides and proteins. Without God there to grab them and use them, they will eventually be destroyed by chance alone.

We just PROVED that a functional protein cannot be made by chance alone, because chance cannot do creation or causation.

Cool, huh?

This is a very powerful test for truth assertions in Science!

Anytime a scientist or college professor is telling you that chance can design and create anything if given enough time to do so, then you automatically KNOW that that person is lying to you.

There I just set you free.

Can you feel the truth of it?

They can no longer trick you and deceive you. You are free now to go and find the True Cause of the different events that we humans experience and observe.

Scientists and college professors lie all the time.

The fact that they truly believe that chance can design and create things doesn't make their lies any more true than they were before. Their sincerity, their belief, or their complete ignorance doesn't make their lies true, just because they believe that they are true.

The Materialists, Naturalists, Darwinists, Nihilists, and Atheists are making truth assertions all the time. Some of them KNOW that they are lying to you, but most of them don't. Most of them have been tricked and deceived just as we have been tricked and deceived. But the fact that they don't know that they are lying doesn't automatically make their lies true. They are still lying to you and deceiving themselves whenever they make the claim that chance can produce anything if given enough time to do so.

In this particular essay, we've used Statistics, a Scientific Method, Science, and a Science Experiment to PROVE that chance cannot even design and create the smallest polypeptide or the smallest protein in the human body, let alone a genome, a brain, or a life form.

Here in this essay, we found the Truth in Science. We discovered that CHANCE cannot design and create anything, and we discovered that scientists and college professors are lying to us whenever they say that it can.

Mark My Words

Using the Scientific Method to Find the Truth

Now, we are ready to use the *negating the consequent* version of the Scientific Method to falsify and eliminate the theory of evolution and the second law of thermodynamics from Science.

Let's use the Scientific Method for what it was designed for. Let's falsify some theories. Unlike *affirming the consequent*, *negating the consequent* or *falsifying a theory* is philosophically and logically sound. Here's how *falsifying a theory*, or *negating the consequent*, works in principle using the Scientific Method:

Scientific Hypothesis: If Theory X is true, then we will observe Y.

Scientific Observations: We don't observe Y.

Scientific Conclusion: Therefore, Theory X is false and has been falsified by the Scientific Method.

This IS the Scientific Method!

This IS a syllogistic form of the Scientific Method. This IS the *negating the consequent* version of the Scientific Method. This IS the reliable and the true version of the Scientific Method – the one that actually works and produces real results. I stated before that there are many different versions of the Scientific Method and many different methodologies or Scientific Methods. People would know this if they had ever studied the Philosophy of Science.

This IS the philosophically valid and logically sound version of the Scientific Method called *negating the consequent*. By *negating the consequent*, we can use the Scientific Method TO PROVE that our theories and ideas are false, and we have in fact PROVEN that they are false. This is the most powerful form of the Scientific Method.

If the second law of thermodynamics predicts that we should be observing proton decay and ever-increasing disorder and chaos throughout our universe, yet we don't observe anything of the sort, then the second law of thermodynamics is false and has been falsified by the Scientific Method. It's that simple! There we just used the *negating the consequent* version of the Scientific Method to falsify the second law of thermodynamics! It's that simple! We have in fact PROVEN that the second law of thermodynamics is FALSE. We don't observe ANY of the things that the second law predicts, therefore, WE KNOW that it is false and KNOW that it has been falsified by the Scientific Method. It's that simple.

Powerful! Is it not?

This is the most powerful form of the Scientific Method!

The things that have NEVER been experienced NOR observed by anyone, not even God, are automatically false and have automatically been falsified by the Scientific Method. Science is observation, experience, and knowledge. If something has NEVER been experienced NOR observed by anyone, then it is false and doesn't exist. That's *negating the consequent* in action! That's the Scientific Method in action! We can use the Scientific Methods TO PROVE that our theories and ideas are false, and we have in fact PROVEN that they are false.

Brutal!

The Many World or Multi-Me interpretation of Quantum Mechanics predicts that thousands or millions of us should be encountering other versions of ourselves from other universes. The fact that nobody has ever encountered another version of himself from another universe is scientific proof that the Many Worlds interpretation of Quantum Mechanics is false. Not even out-of-body explorers, near-death experiencers, or the Biblical God Jesus Christ have ever mentioned having encountered other versions of themselves from other universes or other dimensions.

If it has NEVER been experienced NOR observed by anyone, then there is no way that it could ever possibly be true. That's just the way it is. And, that's the Scientific Method in action, whether we realize it or not. Science is observation and experience and knowledge, or it should be. If it has never been experienced nor observed, then it doesn't exist, and it isn't true.

Here's how *negating the consequent* or *falsifying a theory* works in practice:

Scientific Hypothesis: If the Theory of Evolution and the Second Law of Thermodynamics are true, then we will observe CHANCE doing chemical evolution, abiogenesis, and spontaneous generation all around us all the time in real time. If the Theory of Evolution and the Second Law of Thermodynamics are true, then we will observe CHANCE producing brand

new genes, proteins, genomes, and life forms from molecules and atoms out of thin air and out of the swamps all around us all the time in real time.

Scientific Observations: We have NEVER observed CHANCE designing and creating proteins, genes, genomes, and life forms; and, we NEVER will because CHANCE cannot do causation or choice. The chemical evolution, abiogenesis, or spontaneous generation of proteins and genes from atoms by chance alone is physically impossible and has NEVER been observed. Chance cannot do causation or creation. It's never been observed doing so.

The Scientific Conclusion: Therefore, the Theory of Evolution, Materialism, Naturalism, Darwinism, Atheism, the Second Law of Thermodynamics, Chance Causation, and Creation by Chance are false and have been falsified by the Scientific Method or by the lack of Scientific Observations to support them.

This IS the Scientific Method in action.

I just used the Scientific Method TO PROVE that the Theory of Evolution, the Second Law of Thermodynamics, and Creation by Chance or Causation by Chance are FALSE. I just used the Scientific Method to falsify these things, which means that I used the Scientific Method TO PROVE that they are false.

This IS the valid and sound version of the Scientific Method that's called *negating the consequent* or *falsifying a theory*. We have in fact used the Scientific Method to PROVE beyond a shadow of a doubt that the Second Law of Thermodynamics and the Theory of Evolution are FALSE. We have *negated the consequent* or FALSIFIED these theories. We have PROVEN that they are false.

Materialism, Naturalism, Darwinism, Nihilism, Atheism, the different versions of the Theory of Evolution, and the Second Law of Thermodynamics ARE AUTOMATICALLY FALSE because by definition they were produced by chance alone, and we all should know that chance cannot do causation or creation. Creation by Chance or Causation by Chance has NEVER been experienced NOR observed, which means that it is FALSE. It's that simple. This is Truth in Science. This is Truth through Science. This is the pinnacle of the Philosophy of Science.

Here in these essays, I have used Statistics and a couple of different versions of the Scientific Method to PROVE that the second law of thermodynamics and the theory evolution are false. I have also PROVEN that Materialism, Naturalism, Darwinism, Nihilism, Atheism, Determinism, Behaviorism, and every other form of Creation by Chance or Creation Ex Nihilo is FALSE. Chance cannot do causation. Chance and causation are mutually exclusive. They falsify each other.

Materialism, Naturalism, Mechanism, Darwinism, Nihilism, Atheism, Creation by Death or Natural Selection, Boltzmann Brains, Chemical Evolution, Spontaneous Generation, Macro-Evolution, the Theory of Evolution, Creation by Random Disorder or Creation by Entropy, and the Second Law of Thermodynamics ARE AUTOMATICALLY FALSE because they are Creation by Chance and because they have NEVER been experienced NOR observed.

We can calculate the probability of chance events. Statistics is the calculation of the probability of chance events.

After making these kinds of calculations, we have observed that CHANCE ALONE will never assemble amino acids into a functional protein. Protein synthesis can't happen by chance alone. Our proteins and their matching genes were designed, and made, and used by God. It's that simple. Here in these science experiments we just used Statistics and the Scientific Methods to find the truth, whether we realize it or not.

—

I ran a similar antagonistic thought experiment in one of my books. Let's end with this one just for the fun of it.

For sake of argument, let's say that we take world-famous atheist Richard Dawkins, grind him up in a blender into individual atoms, and dump the contents of that blender into a tub. According to Richard Dawkins, he is all there in that tub. There is nothing of Richard Dawkins to be found anywhere else in any reality or universe. So, how long is it going to take for Chance or Evolution or Entropy to reassemble Richard Dawkins from the contents of that tub? He's all there, is he not? So, how long is it going to take for chance or evolution or entropy to reassemble him? According to Richard Dawkins, Stephen Hawking, Ludwig Boltzmann, and Charles Darwin, chance, or evolution, or entropy can do the job if given enough time to do so. So, how long is it going to take? How long will it take Chance or Evolution or Entropy to reassemble Richard Dawkins from the contents of that tub?

We can calculate the probability of chance events, so let's run the calculation and see what we get.

If you have been following along, then you already know the answer to this question. If God is willing to stir that tub full of Dawkins for all eternity, then it should on average take $(10)^{234}$ minutes – give or take a few trillion years – for chance or evolution or entropy to assemble that tub full of Dawkins into the simplest protein in the human body by chance alone.

Do you see the problem?

We ground him down into individual atoms, NOT individual amino acids. It's going to take even longer to assemble the 21 different amino acids from scratch so that we can start our science experiment in the first place.

So, do you see the true answer to this question?

You guessed it!

IT'S NEVER GOING TO HAPPEN.

It's impossible to assemble functional proteins from an ocean full of the correct amino acids by chance alone, which means that it is IMPOSSIBLE for evolution, entropy, or chance to produce a functional protein from physical atoms by chance alone. Chance cannot do choice or causation.

THEREFORE, IT'S NEVER GOING TO HAPPEN.

It can't be done by chance alone, which means that it was never done that way in the first place. Using Statistics, Probabilities, and the Scientific Method, we have just PROVEN that a functional protein cannot be produced by chance alone, because chance cannot do choice, creation, or causation.

God designed, made, field-tested, and assembled your proteins along with the matching genes to go along with those proteins. Proteins, genes, genomes, and life forms are God's Signature. God's Signature is written on every protein, gene, and cell in your physical body. This is Scientific Proof of God's Existence whether we realize it or not. This is Truth in Science whether we realize it or not.

Here we just used the Scientific Method to find the truth.

Mark My Words

Using the Scientific Method to Eliminate the Usual Suspects and to Prove the Truth

https://markme.us/using-the-scientific-method/

I'm using Science and the Scientific Methods to PROVE the truth.

How is that done, since it's impossible to use the Scientific Method to prove anything true, thanks to the *affirming the consequent* logic fallacy that's built directly into the Scientific Method? Well, where there's a will, there's a way. Some of us have found a legitimate work-around that actually works.

You see, you can USE the Scientific Method to prove things false. It's called *negating the consequent*. It works, and it's logically sound. If you successfully use the Scientific Method to find and eliminate everything that is false, then ONLY the truth will remain. That's precisely what I have done in my books. I keep eliminating everything that is false so that ONLY the truth remains; and, the truth that remains has ended up being the things that we constantly and repeatedly experience and observe, so that WE KNOW that they are true.

That's how you use Science and the Scientific Methods to PROVE the truth. We use Statistics and the Scientific Method to prove things false, so that we can eliminate them from science. After we have eliminated everything that is false from science, then ONLY the truth remains. And, the truth consists of the things that we constantly experience and observe collectively as a race, so it's impossible to eliminate them from science. Only the Materialists, Naturalists, Darwinists, and Atheists deliberately identify and eliminate observed evidence from science. In contrast, the smart ones among us retain and embrace the things that have actually been experienced and observed, because they are obviously true.

I learned this little trick from Sherlock Holmes.

How often have I said to you that when you have eliminated the impossible [and the false]**, whatever remains, however improbable, must be the truth? — Sherlock Holmes.**

I'm a scientist, and I turned myself into a detective. I identify and eliminate everything that is false – everything that has NEVER been experienced NOR observed – so that ONLY the truth remains. That's what I do over and over again in ALL of my books. I'm the Scientist Detective. I'm also the science author on Amazon.

https://amazon.com/author/science

https://ultimate-model-of-reality.com/contact/

https://markme.us/truth/

I'm trying to set people free. I'm trying to find and fix everything that's wrong with Science. I'm trying to identify and eliminate the deceptions and the lies. I want to find and know the truth. I'm trying to find a true and realistic Philosophy of Science or Interpretation of Science. I'm trying to help people. I'm trying to help other people to avoid the mistakes that I have made. I want better for them than what I was able to find for myself when I first went looking for the truth. That's what the following book and my other books are all about.

Using the Scientific Method to Find the Truth

https://www.amazon.com/gp/product/B01J6STHP0/

The original title of this book was, "Using the Scientific Method: To Eliminate the Usual Suspects and to Prove the Truth". Nobody got it.

The original idea was to get at the truth and to "prove the truth" by falsifying and eliminating everything that is demonstrably false or wrong. If you eliminate the usual suspects, then you end up with the true culprit or the true causal agent instead. That's the idea that I was trying to get across. This idea comes from crime scene investigation. This idea comes from Sherlock Holmes. People didn't get it because they had erroneously been taught that the Scientific Method can't be used to prove anything. Well, they are wrong. The Scientific Method can be used to prove things false or to falsify our incorrect theories, ideas, and beliefs.

We can use the Scientific Methods TO PROVE that certain ideas are wrong or false. We can use the Observations and the Experiences of the human race TO PROVE that certain ideas are false or wrong. Whenever we falsify something, we have indeed PROVEN that it is false.

If we successfully eliminate everything that is false from Science, then ONLY the truth will remain. This is logical common sense. This is deductive reasoning through a process of elimination! It's a real thing; and, the Materialists, Naturalists, Darwinists, Nihilists, and Atheists have convinced themselves that it doesn't exist. The reason they have problems with this is because Science and the Scientific Methods FALSIFY Materialism, Naturalism, Darwinism (Creation by

Chance), Nihilism, and Atheism (Creation Ex Nihilo) right and left, coming and going.

The Materialists, Naturalists, Darwinists, Nihilists, and Atheists have serious problems wrapping their minds around the idea that Science and the Scientific Methods can be used to Prove the Truth through a process of eliminating the false, because these people have been erroneously taught in school that Science and the Scientific Methods can't be used to prove anything. These people prefer and believe the lies that they are told in our public schools, rather than finding and knowing the truth for themselves.

Although it is technically true – due to the *affirming the consequent* logic fallacy – that it's impossible to use the Scientific Methods to prove anything true, there is a legitimate work-around that actually produces the same result. This work-around is a philosophically *valid* and an observationally *sound* way to find the truth through a process of elimination. This is deductive reasoning through a process of elimination. It works!

The way this is done in practice is that we use the Scientific Methods and *negating the consequent* to falsify and eliminate everything that is false such as Materialism, Naturalism, Darwinism, Atheism, and their derivatives so that ONLY the Truth remains.

Think about it logically.

If you successfully eliminate everything that is false – everything that has NEVER been experienced NOR observed by anyone, not even God – then ONLY the Truth will remain, because ONLY the observed and the verified and the proven will remain. The Truth is whatever has been experienced, lived, and observed by Someone Psyche or Someone Intelligent somewhere sometime. Science is observation and experience, or it should be.

If you use Science and the Scientific Methods to eliminate everything that is false, then ONLY the truth will remain. This is logical common sense. It's fascinating to study what remains after we have eliminated everything that is false. The experienced and the observed remain. The false is falsified by the truth, and the truth is constantly and repeatedly experienced and observed.

Even though the Scientific Methods cannot be used to prove anything true, they can indeed be used to prove things false. When you falsify something, you do in fact proven that it is false. The Scientific Methods along with Direct Observations are extremely powerful tools for falsifying False Philosophies and False Theories such as Materialism, Naturalism, Darwinism, Nihilism, and Atheism.

It works.

How often have I said to you that when you have eliminated the impossible [and the false]**, whatever remains, *however improbable*, must be the truth? — Sherlock Holmes.**

This is deductive reasoning through a process of elimination! It's a legitimate scientific method that helps us to get at the truth. We eliminate everything that has NEVER been experienced NOR observed, and soon we are left ONLY with the things that have been experienced and observed – and soon we are left with the truth. The truth consists of whatever has been experienced and observed by someone sometime in the past.

It's taken me five years to develop this message and to decide what I'm about or what I believe in. It's here now. It's coming online.

Let's recap.

We find the truth by eliminating everything that is false. The truth is what remains after we have used Science and the Scientific Methods to falsify and eliminate the deceptions and the lies.

Due to the *affirming the consequent* logic fallacy, which is built into the Scientific Methods, you CANNOT USE the Scientific Methods to prove anything true directly, which means that the Scientific Methods cannot be used for a direct proof of God's existence. This reality has got most people believing that we can't use the Scientific Methods to prove anything at all. I have had many Scientists, Materialists, and Atheists tell me this. They have told me that Science and the Scientific Methods can't be used to prove anything. THEY ARE WRONG!

It's annoying, because many of the Atheists, Naturalists, and Materialists use this false belief as a crutch or a shield to hide behind, as a reason for believing in the impossible and the patently absurd such as Creation by Chance or Creation Ex Nihilo, and as a reason for doing sloppy science. These people don't understand the Scientific Method nor the Philosophy of Science, if they truly believe that the Scientific Methods can't be used to prove anything. Science and the Scientific Methods can be used TO PROVE that our theories and ideas are false. It's called "falsifying a theory". It's an integral part of the Philosophy of Science!

Nevertheless, there are millions of Scientists and Darwinists out there in the world right now who are using their ignorance of the Scientific Method and their belief that the Scientific Methods can't prove anything as an excuse or a reason for continuing to hope that someday – some way – someday – science will finally find a way to prove the theory of evolution true. It's never going to happen, because the Scientific Methods can't be used directly to prove anything to be true – certainly not something that is demonstrably false. We've NEVER observed chemical evolution, macro-evolution, creation by death or natural selection, abiogenesis, spontaneous generation, NOR creation by chance in action because these things are physically impossible. We can use Science and the Scientific Methods to falsify them, but we can't use science nor the Scientific Methods to prove that they are true, as the Naturalists and Darwinists try to do.

Science is observation and experience. We can use our collective observations and our personal experiences TO PROVE that the second law of thermodynamics, the theory of evolution, creation ex nihilo, abiogenesis, spontaneous generation, chemical evolution, macro-evolution, and creation by chance ARE FALSE. We can never use the Scientific Methods to prove that they are true. They have NEVER been experienced NOR observed, so there's no way to verify them or to demonstrate that they are true, but we can falsify them!

So, how do you use the Scientific Methods as a proof of God's existence, if the Scientific Methods can't be used to prove anything to be true? That is the million-dollar question, isn't it; and, I get the impression that I'm the ONLY person on the planet who has solved this one!

Science is observation and experience. We can use the observations and the experiences of the human race as a whole to demonstrate and prove that Jesus Christ rose from the dead and that Jesus Christ is the being of light and love who is waiting for us on the other side after we die. We USE the collective observations and experiences of the human race as a whole to demonstrate and indirectly prove the things that are real and truly exist. THE TRUTH has been experienced and observed by someone, somewhere, sometime. If something has NEVER been experienced NOR observed by anyone, not even God, then there's no way in the universe that it could possibly be true.

The quickest and best way to find and know the truth is to live it, experience it, and observe it for yourself, or to choose to trust someone who has. This is infinitely faster and better than Science and the Scientific Method. Through observation and experience, we go directly to KNOWING the truth. That's a lot faster and whole lot more productive than running a science experiment, analyzing the data, and providing dozens of false interpretations for that data.

https://www.youtube.com/watch?v=jWj1hazNE08

https://www.youtube.com/results?search_query=NDE+Jesus

https://www.youtube.com/results?search_query=NDE+Howard+Storm

Combine the observations and the experiences of the human race with the FACT that the Scientific Methods can be used to falsify Materialism, Naturalism, Darwinism, Nihilism, and Atheism, then before we know it, we find ourselves staring at the truth. The truth is what remains after everything that is false has been eliminated.

You have to realize and accept the fact that the Scientific Methods can indeed be used to PROVE theories false! It's called falsifying a theory or *negating the consequent*; and, it's philosophically and logically sound. When you successfully *negate the consequent*, you have in fact PROVEN a theory false, or falsified a theory. This IS Philosophy of Science 101, and the core essential understanding of the Scientific Method one must have if one is going to use the Scientific Methods to PROVE anything. You have to KNOW that the Scientific Methods can be used to PROVE things false. You have to KNOW that the Scientific Methods can be used to falsify theories, hypotheses, and philosophical concepts that are in fact false.

Another essential aspect of the Philosophy of Science that everyone completely overlooks is that chance can NEVER do causation. Chance and causation are mutually exclusive. They falsify each other. If you are observing causation or can demonstrate causation, then YOU KNOW that there is not one bit of chance involved in your science experiment. If you are seeing the Null Hypothesis, Creation by Chance, or something that is produced by chance alone, then YOU KNOW that your independent variable has NO correlation with and NO causal relationship with your dependent variable. Chance and causation are mutually exclusive. They falsify each other. Chance cannot do causation!

Therefore, every type of Creation by Chance or Chance Causation is automatically false because it is the Null Hypothesis. It's precisely what we are trying to eliminate from science. Materialism, Naturalism, Darwinism, Nihilism, Atheism, the Theory of Evolution, and the Second Law of Thermodynamics ARE Creation by Chance and ARE by definition caused by chance alone. They ARE the Null Hypothesis. They are precisely what we are trying to identify and eliminate from science.

The fundamental law of Statistics and Science is that chance cannot produce a real effect.

If we are detecting any kind of an effect or any kind of causation, then we are no longer dealing with chance. We are dealing with psyche, intelligence, causation, or choice instead. Chance and causation are mutually exclusive. Chance cannot do causation. ONLY psyche, or intelligence, or life force, or quantum consciousness can do choice or causation at every level of existence and reality. This is what we have experienced and observed; therefore, WE KNOW that it is true.

This is the Ultimate Truth when it comes to the Philosophy of Science. Chance cannot do causation. Chance cannot produce a real effect. Chance and causation are mutually exclusive.

Finally, you have to be willing to use the Scientific Methods, statistics, and scientific observations to **falsify** Materialism, Naturalism, Darwinism, Nihilism, Scientism, and Atheism. If you are unwilling to do this, then you will NEVER be able to develop a Scientific Proof of God's Existence, and you will NEVER be able to find the truth. This is crucial. You have to be open-minded enough and smart enough to use the Scientific Methods for what they are good for and for what they were designed for, if you are going to get Science and the Scientific Methods to PROVE anything to you.

We get at the truth through a process of elimination! And, the falsehoods or lies are the things that we are using the Scientific Methods and Science to eliminate. Science and Science Experiments were designed to identify, falsify, and eliminate the Null Hypothesis or Creation by Chance or Chance Causality. The whole purpose of a science experiment is to eliminate the Null Hypothesis or Creation by Chance so that we can find the True Cause of the dependent variable instead.

I eventually realized that Materialism, Darwinism, Naturalism, and Atheism CANNOT BE used to prove anything, because they are FALSE. We cannot use anything that is false to prove something to be true; so, in the end, the Materialists and Atheists were right in that THEIR THEORIES cannot be used to prove anything, because their theories are false. However, once we know that ALL of their theories are false, we can indeed use their atheistic and materialistic falsehoods, deceptions, and lies to point us towards THE TRUTH. I do it all the time now.

Nowadays, I actively USE the Materialists, Darwinists, Naturalists, and Atheists to point me to the things that they don't want me seeing, reading, and understanding because I KNOW that the Science and the Theories which the Materialists and Atheists have formally rejected are in fact THE TRUTH and will always end up being THE TRUTH. I learned to use their atheistic theories and materialistic theories to point me to THE TRUTH. Genius, huh? Consequently, I'm always using their falsified theories to point me to the truth.

Remember, if we successfully eliminate everything that is false, then ONLY the truth will remain. Furthermore, the truths that remain end up being the things that have actually been experienced and observed.

Science, the Scientific Methods, and Science Experiments are completely worthless unless we have the true interpretation or the true meaning of the scientific data or scientific evidence. Even something as simple and powerful as the bidirectional Perpetual Motion Cycle ($E = mc^2$) has been completely hidden from the world by the Materialists, Naturalists, Darwinists, Nihilists, and Atheists because they don't know what it is and don't know that it exists, and because they don't believe in the non-physical stuff like energy, photons, quantum waves, quantum fields, quantum consciousness, action at a distance, and psyche.

I'm NO Einstein! I used to be an Einstein, but I'm not any longer. I no longer believe in Classical Realism, Materialism, Naturalism, Creation by Chance, and the Second Law of Thermodynamics. Einstein believed in these things. I do not. I've used Science to falsify them. I'm NO Einstein. I had to work hard at it, but I'm now seventy years ahead of Einstein.

I believe in Quantum Mechanics, Quantum Field Theory, the Perpetual Motion Cycle, the massless and entropyless Quantum Fields, the Conservation of Energy and Psyche, the Conservation of Quantum Information within Psyche, Quantum Non-Locality, Quantum Non-Physicality, Quantum Entanglement, Quantum Complementarity, the Transdimensional Realm, the Non-Physical Realm or the Quantum Realm, massless and entropyless Energy, the Quantum Zeno Effect (psyche-to-psyche telepathy), Quantum Waves, the transformation of the Wave Function into Mass, the transformation of Mass into Quantum Waves or Photons, Virtual Particles or Non-Physical Thoughts and Memories, After-Death Life Reviews, Quantum Tunneling (teleportation of physical matter), Quantum Superposition, Quantum Phase-Shifting, Quantum Omnipresence, Quantum Omniscience, Quantum Consciousness or Psyche, Holography or Quantum Information Storage within Psyche or Light, Syntropy, Exergy, Action at a Distance, and everything else that has been experienced and observed.

I believe in all the things that have been experienced and observed. I believe in all the things that Einstein didn't believe in. I'm NO Einstein. Not anymore. I'm seventy to a hundred years ahead of Einstein because that's about as long as it has taken us to find the truth since Einstein.

The science community as a whole still doesn't have a True Interpretation of Quantum Mechanics and Quantum Field Theory. Until they do, they will have no clue what's really going on at the quantum level in the Energy Realm, Spiritual

Realm, Non-Physical Realm, Transdimensional Realm, or Quantum Realm. Without a True Interpretation of Quantum Mechanics, our science community has gone nowhere for the past hundred years. They've made no progress whatsoever!

I have interacted with the Naturalists and Atheists on social media long enough to know that they will twist my words and say that I'm denying the existence of physical matter. I'm not a solipsist. I'm a Realist. I believe in everything that has been experienced and observed – nothing more and nothing less. Physical matter has obviously been experienced and observed.

I just no longer believe in the unscientific, fictional, unrealistic, unobserved, and unproven tripe that the Materialists, Naturalists, Darwinists, and Atheists are trying to feed us. These people are trying to convince us that the non-physical does not exist. They are obviously wrong. Photons are massless or non-physical. I no longer believe in Creation by Chance, Creation Ex Nihilo, Creation by Random Disorder, Creation by Death or Natural Selection, or Creation by Magic either. I no longer believe in their deceptions and their lies. I have found the truth instead. I only believe in the things that have been experienced and observed.

I believe in the stuff that has been experienced or observed first-hand as well as the stuff that has been demonstrated and proven to be real in nature or in a science lab. I believe in everything that has been experienced and observed by someone, sometime, somewhere because it really truly exists.

I no longer believe their atheistic claims that the non-physical quantum waves do not exist because the non-existence of the non-physical has never been experienced nor observed. Photons in their massless, entropyless, chargeless, non-physical, quantum, wave-like format are obviously immaterial, omnipresent in scope or range, and non-physical. I don't believe in the non-existence of these non-physical massless photons because their non-existence has never been experienced nor observed by anyone. The proven and verified existence of massless, entropyless, non-physical photons, quantum waves, virtual particles (thoughts or quantum information), and quantum fields FALSIFY Materialism, Naturalism, Darwinism, Nihilism, Atheism, the Theory of Evolution, and the Second Law of Thermodynamics which claim that these non-physical things do not exist.

Obviously, I am not the first scientist on this planet to discover and use $E = mc^2$. However, I may be the first person on the planet to figure out what it truly is and how it truly works at the quantum level. $E = mc^2$ is the Perpetual Motion Cycle. Some type of Perpetual Motion Cycle, Exergy, or Syntropy has to exist, or the second law of thermodynamics would actually be true, and nothing that we see around us would exist. We wouldn't exist if the second law of thermodynamics were actually true. The very fact that we exist, the sun exists, lifeforms exist, and this physical universe exists is scientific proof that the second law of thermodynamics is false.

Some type of eternal, everlasting, and conserved Perpetual Motion Cycle or Syntropy has to exist, or nothing on this planet would exist, and the second law of thermodynamics would actually be true. The Perpetual Motion Cycle, Exergy, Syntropy, and the massless entropyless non-physical Quantum Fields and Photons

PROVE that the second law of thermodynamics is false. The false is falsified by the truth, and the things that are true are the things that are constantly and repeatedly experienced and observed.

In Science, we want the truth, not all the deceptions and lies that currently exist in science. Science and the Philosophy of Science are in serious need of a massive upgrade so that they include Quantum Mechanics, Non-Physical Mechanics, Energy Mechanics, or Spiritual Mechanics as a given. We are not there yet in the science community at large. Most of our scientists are still stuck in the dark ages with Classical Realism and Classical Mechanics and Naturalism. Until they come in out of the dark, they will have no clue as to what's really going on at the quantum level in the non-physical realm. A massive re-education program is underway, and I was NOT immune. I, too, in 2012-2013 used to be a materialist, naturalist, nihilist, and atheist until the scientific evidence convinced me that I was wrong. It's been a slow process of re-education. Although collectively as a whole, my re-education during the past six years has been massive – as it should be. I'm a scientist. We are supposed to be falsifying our theories and replacing them with the truth. That's what I have been doing during the past six years of my life. My books are a testament to that reality. I'm trying to find and fix everything that's wrong with science so that we have the truth instead.

https://amazon.com/author/science

https://ultimate-model-of-reality.com/contact/

https://markme.us/truth/

Do you see how I use science, deduction, and logic to eliminate the falsehoods and to demonstrate or prove the truth? That's what scientists are supposed to do. They are supposed to identify, falsify, and eliminate the deceptions and the lies so that ONLY the truth will remain. We can indeed use the Scientific Methods, Science, Statistics, and Common-Sense to prove things false – starting with Materialism, Naturalism, Darwinism, Nihilism, Atheism, the Theory of Evolution, and the Second Law of Thermodynamics. That's what we are supposed to be doing. We are supposed to be falsifying and eliminating these things from Science so that ONLY the truth will remain. It's not what we are doing, but it is what we are supposed to be doing as scientists.

The truth that remains actually ends up being the things that have been experienced and observed somewhere, sometime, by someone.

https://www.youtube.com/results?search_query=NDE+Jesus

https://www.youtube.com/watch?v=YbEojX4iaGQ&list=PLV33Nfr_WaaDcscy-1AKwMV3dZTNLOdH0

What good are Science and the Scientific Methods if they can't be used to find and know the truth? What good are Science and the Scientific Methods if they can't be used to prove anything? What good are Science and the Scientific Methods if we are always giving them a FALSE interpretation or a materialistic and naturalistic interpretation? What good are Science and the Scientific Methods if we are constantly censoring evidence, deleting evidence, hiding evidence, and destroying evidence? What good are Science and the Scientific Methods if we constantly refuse to eliminate the things that have been falsified?

Materialism, Naturalism, Darwinism (Creation by Chance), Nihilism, Atheism (Creation Ex Nihilo), the Theory of Evolution, and the Second Law of Thermodynamics HAVE BEEN FALSIFIED by Science, the Scientific Methods, Statistics, Observations, Knowledge, and the collective Experiences of the human race; yet, we continue to retain these falsified philosophies, continue to pretend that they are "science", and continue to insist that they are true. It's the greatest scam in human history, and we continue to fall for it because we desperately want it to be true.

Remember, Materialism, Naturalism, Darwinism, and their derivatives are NOT science – they are philosophies of science or interpretations of science. Science as a whole is in desperate NEED of a NEW and IMPROVED Philosophy of Science that has actually been experienced and observed. That's what I have been trying to provide in all of my different books about Science.

Whenever the Materialists, Naturalists, Darwinists, Nihilists, and Atheists tell us that Science and the Scientific Methods can't be used to prove anything, they are lying to us and deceiving themselves. They are correct in that Materialism, Naturalism, Darwinism, Nihilism, and Atheism CAN'T BE USED to find, know, or prove the truth because these things are false and because these falsified ideas are NOT science. Materialism, Naturalism, Darwinism, Nihilism, and Atheism are philosophies of science. They are not science. They are religions, belief systems, dogma, worldviews, or philosophies of science. They are faulty and falsified interpretations of science.

The problem that these people are having is that by *jumping to conclusions* they have erroneously defined "science" as Materialism, Naturalism, Darwinism, Nihilism, and Atheism; and then, they naturally and automatically observe that Materialism, Naturalism, and their derivatives can't be used to prove anything true. They never once realize, though, that the reason that Materialism, Darwinism, and Naturalism can't be used to prove anything true is because they are demonstrably false. We can't use falsehoods to prove the truth, unless we are eliminating those falsehoods in order to get at the truth. That's just the way it works.

I have had Materialists and Atheists tell me that theories cannot be used to prove anything, because they are theories. That was kind of confusing and frustrating. There are some really strange ideas and information floating around concerning Science and the Scientific Method. I think what these people were trying to tell me is that Science and the Scientific Methods cannot be used to prove anything. These people ARE WRONG! What good is Science and the Scientific Methods if they can't be used to find and know the truth? What good is Science and the Scientific Methods if they can't be used to PROVE things? The way we PROVE things in Science is to experience them, live them, and observe them first-hand, or to choose to trust someone who has. The fastest and most efficient way to find and know the truth is to live it, experience it, and observe it for yourself.

I have had multiple Scientists, Atheists, Naturalists, and Materialists tell me that I don't understand Science and the Scientific Method. I have observed that the Materialists, Naturalists, Nihilists, and Atheists are always wrong. It's these people who really don't understand Science and the Scientific Methods. These people don't have a clue. I KNOW, because I used to be one of them. There used to be a time when I believed them, because I used to be a materialist, naturalist, nihilist, and atheist. These people had successfully convinced me that I would never be able to use Science nor Theories nor Methods to prove anything. I believed them, because I used to be a practitioner and adherent of Scientism. I saw no other way and knew of no other way.

These people had convinced me that Science and the Scientific Methods could never be used to prove Materialism, Naturalism, Darwinism, Nihilism, and Atheism false. THEY WERE WRONG! Falsifying Theories is precisely what the Scientific Method is good for and precisely what the Scientific Method was designed for! Once I stopped believing these people and their lies, I started using the Scientific Methods, Observational Methods, Statistics, and other methods to falsify Materialism, Naturalism, Darwinism, Nihilism, Atheism, and Creation by Chance all the time.

After I repeatedly caught these people trying to trick us and deceive us, I stopped trusting them. Since then, Science and Logic and the Scientific Methods have proven to me that the Materialists and the Atheists are ALWAYS wrong. Once I got rid of My Materialism and My Atheism, Science and Scientific Methods have been proving things to me right and left. Materialism and Naturalism were holding me back and stunting my intellectual growth; and, I certainly am not the first victim of these false and unproductive and unscientific ideologies.

Materialism, Naturalism, Darwinism, Nihilism, Atheism, the Theory of Evolution, Creation by Disorder, Creation by Chaos, and the Second Law of Thermodynamics ARE the Null Hypothesis, the Control Group, or Creation by Chance Alone. They are precisely what we are trying to falsify and eliminate from science by doing science experiments in the first place. These atheistic and naturalistic philosophies are automatically FALSE because they are the Null Hypothesis or Creation by Chance.

A lot of this is semantics, but it does have real world applications. For example, in statistics we say that the Null Hypothesis or Creation by Chance is never proven to be true. It's been said that you can't prove anything in science or statistics. That claim is NOT true. You can use science and the Scientific Methods to prove that your theories and ideas are false. In other words, we can USE science, statistics, the Scientific Method, and science experiments TO PROVE that Materialism, Naturalism, Darwinism, Nihilism, Atheism, the Theory of Evolution, the Second Law of Thermodynamics, the Null Hypothesis, and Creation by Chance ARE FALSE. Whenever we falsify these theories, philosophies, and ideas we have in fact PROVEN that they are false. That's the way Science and Statistics work, whether we know it or not.

Science is not as gimped nor crippled as the Darwinists, Naturalists, and Atheists claim that it is. Throughout all of my different books, I have used Science,

Observations, Experiences, Science Experiments, and the Scientific Method to FALSIFY Materialism, Naturalism, Darwinism, Nihilism, Atheism, the Theory of Evolution, the Second Law of Thermodynamics, Creation Ex Nihilo, and Creation by Chance. That's what Science was designed to do. I have used Science TO PROVE that these things are false. That's what scientists are supposed to do; and, I am a scientist.

Therefore, another legitimate and useful title for the book mentioned at the beginning of this essay is, "Using the Scientific Method: To Falsify the Lies". The KEY point here is that we learn to USE the Scientific Methods to help us get at the truth.

I love science. I've been a scientist all of my life. I want to figure out how everything works. I want to find and know the truth. I want to find, fix, and repair everything that's wrong with science by identifying it and replacing it with the truth.

That's why I find it discouraging, disappointing, confusing, and frustrating to observe that most people on this planet couldn't care less. They have absolutely no interest in science. They could care less how things work. They find it boring. They have no interest whatsoever in finding and knowing the truth. They will pay money to be lied to; but, they have absolutely no interest whatsoever when it comes to investing in the truth or finding the truth.

It's ironic, but it's true. The truth doesn't sell. You can't even give it away! Nobody wants it. This is what I have experienced and observed.

Maybe someday that will happen – maybe someday we will collectively start to pursue the truth and be willing to eliminate the false, when we become better people. But right now, it's not happening in the real world that we call science. It's not happening within our society either. Instead, everyone is actively embracing and promoting the deceptions and the lies. They will pay huge amounts of money to be lied to, because the deceptions and the lies are what they want most. That's just the way it is, and I had to get used to it.

Mark My Words

The Probability of Protein Synthesis by Chance Alone

Throughout known life, there are 22 genetically encoded (proteinogenic) amino acids, 20 in the standard genetic code and an additional 2 that can be incorporated by special translation mechanisms. In eukaryotes, there are only 21 proteinogenic amino acids, the 20 of the standard genetic code, plus selenocysteine.

The Materialists, Naturalists, Darwinists, Nihilists, Atheists, Determinists, Physical Reductionists, Atomists, Behaviorists, and Classical Realists claim that proteins and their matching genes formed by chance alone.

Let's examine the probability of protein synthesis happening by chance alone.

First, we have to start with a God-given tub full of the correct 21 different amino acids that are used to form proteins, so that CHANCE actually has something to work with. CHANCE can't produce a tub full of the correct 21 different amino acids to start with. God has to do that for CHANCE so that CHANCE has a chance of making a protein by chance alone to begin with. Without that God-given tub full of the correct 21 different amino acids, there's no chance that CHANCE will ever be able to produce a protein by chance alone. CHANCE needs the building blocks for proteins already in place to begin with, or nothing is ever going to happen, either purposefully or by chance. This is logical common sense. Only the irrational would disagree with it. We have to start with a pot full or a tub full of the correct amino acids as a God-given fact; otherwise, there can be NO protein synthesis. No amino acids, then no protein synthesis. It's that simple.

Starting with that God-given tub full of the correct 21 amino acids, the probability of getting two correct amino acids in a row by chance alone is $(1/21) * (1/21) =$ "0.0022675737". That's two-tenths of a percent, which some people would consider to be within the realm of possibility. It's highly improbable, but it's possible for chance to get two amino acids in the correct sequence.

Let's back up a second. The odds of getting the correct amino acid in the first spot the first time is $(1/21)$, which is "0.0476190476". The odds are against that happening, but clearly it's possible to get the correct amino acid in the first spot the first time, assuming that God has already made that amino acid and placed it into the pot, or the tub, or the tube.

The probability of getting three correct amino acids in a row is $(1/21) * (1/21) * (1/21) =$ "0.0001079796998". That's 1/10000, which is still within the realm of detectable possibility according to the typical binomial distribution table, which takes things out to .0001 before the table drops to zero probability. It's extremely improbable for chance to get three correct amino acids in a row, but the hopeful and those operating on blind faith alone will claim that it's theoretically possible. "There's still a chance", they will say.

The probability of getting four correct amino acids in a row is $(1/21) * (1/21) * (1/21) * (1/21) =$ "0.000005141890467". That's five chances in a million; and, here we enter into the realm of impossibility. Technically, that's never going to happen, and it gets worse from there. At some point we have to call it a loss, or call it an impossibility; and for me, this is where it starts to happen.

Here, time really begins to be a factor. Five chances in a million are not good odds. If four amino acids in the wild find each other one time per year by sheer luck alone, it's going to take nearly a million years on average for the correct four to end up in the right sequence, and then the wind or the sun or the water will immediately destroy them anyway. Even with a God-given tub full of the correct 21 amino acids, God is going to have to stir the pot, or nothing is going to happen. The amino acids will just sit there and do nothing for all eternity, unless someone is stirring the pot. That same person also has to keep that tub full of amino acids in existence for millions of years so that chance has a chance of combining some of them into a useful sequence that is four amino acids long.

Remember, there are NO God-given tRNA molecules, mRNA molecules, and ribosomes in this tub to carry the amino acids to the appropriate ribosome and to stitch the amino acids together into a protein at the ribosome. We are relying upon chance alone to put these amino acids together into proteins. It's going to take a while to do so.

https://en.wikipedia.org/wiki/Protein_biosynthesis

The probability of CHANCE getting five correct amino acids in a row is "2.4485 e -7". That's never going to happen. There's just no chance that CHANCE can do the job that needs to be done. So, the best that CHANCE can do by chance alone is a protein that is four or five correct amino acids in a row; and, we are being extremely generous here, giving chance the full benefit of the doubt. The whole process is improbable to begin with, and it soon becomes impossible, even with a God-given tub full of the correct 21 amino acids to start with.

Remember, this process starts with a God-given tub full of the correct 21 amino acids, and with NOTHING else in the tub besides those correct amino acids. If you add anything else into the tub, then the probability drops to zero even quicker than it would with a tub full of the correct 21 amino acids. Furthermore, God has to keep stirring this pot or this tub continuously so that the different amino acids are shifting position and coming into contact with each other. If God isn't there to stir the pot, then nothing is ever going to happen.

God would have to make all of the amino acids in the first place and put them into close proximity to each other so that chance can then combine them into something unusable that is four or five amino acids long. One website stated that the shortest protein in the human genome is 44 amino acids long, and that's 21^{40} more than what chance can do by itself. Think about it. It's true. Chance just isn't all that reliable. The house always wins, which means that the Materialists, Naturalists, Darwinists, and Atheists always lose because they are relying exclusively upon chance alone to get the job done.

It's NEVER going to happen!

It can't happen. It's impossible. It's impossible to do protein synthesis by chance alone. That's just the way it is.

The claims and the definitions vary from one website to the next, but I also found this on Uncle Google:

Thyroid releasing hormone or TRH should be the smallest protein in the human body, with 234 amino acids. (> 100 amino acids is a protein.) The smallest polypeptide in the human body should be insulin, with 54 amino acids. (10-100 amino acids is a polypeptide.)

Let's run with this for a minute and see what we get.

The probability of making insulin by chance alone is:

$(1/21)^{54}$

This is equal to: 3.98252109 e -72

This can't be done by chance alone, even with a God-given tub full of the correct amino acids. Even with God constantly stirring the pot, it would take trillions of trillions of years for such a thing to happen by chance alone. It's just not going to happen. Our observable universe hasn't been around long enough for this to happen by chance alone. Insulin was designed and made by God. Without God to design it, make it, fine tune it, and field test it, insulin wouldn't exist.

The probability of making THR by chance alone is:

$(1/21)^{234}$

When I put this into my graphing calculator, the answer was ZERO (0). Literally. The answer is ZERO. This is NEVER going to happen ever! This is beyond what a graphing calculator can handle. This is beyond what God could do by chance alone!

The probability of making THR by chance alone is ZERO. You can wait for all eternity, and it will NEVER happen, even with a God-given tub full of the correct amino acids or with an ocean full of the correct amino acids. It's NEVER going to happen. Making THR by chance alone is impossible. It can't happen by chance alone. It requires some kind of intelligent and deliberate intervention.

If it were possible to make genes and proteins by chance alone, then we should also be finding computer chips and computers and calculators that were made by chance alone, given all the sand and minerals that we have on this planet, and the constant stirring that is taking place. It is obvious that genes, proteins, genomes, life forms, computer chips, and computers can ONLY be made through intelligent design and intelligent intervention. It is obvious that these things cannot arise through chance alone.

It is obvious that God made your polypeptides and your proteins! Each protein or polypeptide that you encounter is God's Signature. Proteins are impossible to make by chance alone. It can't be done, which means that it wasn't done that way. God designed and made your proteins and polypeptides. God also made the matching genes to go along with those polypeptides and proteins. The probability of anything like that arising by chance alone is ZERO. It's impossible. It's NEVER going to happen.

Your genome is God's Signature, and that Signature is written on every cell in your body. Polypeptides and proteins are also God's Signature, and that Signature is also written on you and throughout you. You were made by God whether you realize it or not. According to the probabilities and the statistics, there is no chance that I am wrong.

I just used Science and Statistics to PROVE that God exists. God has to exist, or genomes and proteins wouldn't exist. God has to exist, or you wouldn't exist. It's that simple. I've been doing this for a few years now, and I'm getting better at it. I'm not afraid of the truth anymore, like I used to be when I was a materialist, naturalist, nihilist, and atheist. I simply tell it as it is. Your proteins and genes were deliberately and intelligently MADE. Anything that is obviously made obviously has a Maker who made it. Our genes and proteins were obviously MADE; therefore, our genes and proteins obviously have an Intelligent Maker who designed them and made them. It's obvious that this is true.

Because I'm a mathematician, statistician, logician, and quantum theorist, I understand the math, and I understand probability and chance. For me personally, this is one of the most convincing Scientific and Statistical Proofs of God's Existence that I have ever encountered; and, I have encountered dozens of them, which I found convincing.

Mark My Words

A Binomial Distribution Table Falsifies Creation by Chance

https://ultimate-model-of-reality.com/a-binomial-distribution-table-falsifies-creation-by-chance/

A binomial distribution table FALSIFIES chance causation, creation by chance, or chance causality of any kind, except for the very simplest of sequences.

We can calculate the probability of chance events.

https://origin-science.org/Binomial-Distribution-Table

The probability of tossing two heads in a row is 0.25, according to a binomial distribution table. That's doable, and realistic.

The probability of tossing three heads in a row is 0.125. That too is realistically possible.

The probability of tossing four heads in a row is 0.0625. The probability of five heads in a row is 0.0312. The probability of tossing six heads in a row is 0.0156. The probability of tossing seven heads in a row is 0.0078. That's getting down there. That's less than one percent, and it gets much worse from there. You can toss seven fair coins a hundred times, and it's possible that you will NEVER get seven heads to come up. Chance just isn't that reliable! If you need seven heads to win the jackpot, chances are good that you will run out of time or money before you achieve your goal.

According to a binomial distribution table, it is impossible to get fifteen heads in a row or fifteen tails in a row from a fair coin by chance alone. The probability of doing so has effectively dropped to zero.

And, that's with a binomial distribution – heads or tails – two possibilities for each trial or event. When it comes to proteins and amino acids, there are 21 different amino acids that are used to make proteins, which means that the sequence that can be produced by chance alone drops significantly lower than a sequence fifteen units long, as I will demonstrate later on in this essay.

Remember, according to a binomial distribution table, it is statistically impossible to get fifteen heads in a row with a fair coin, which also means that it is statistically and realistically impossible to get fifteen correct amino acids in a row, when trying to build a protein by chance alone.

When dealing with probability, when do you declare it a lost cause? When do you cash out and cut your losses? When do you declare a "low probability" to be synonymous with ZERO, IMPOSSIBILITY, or NO CHANCE?

On the typical binomial distribution table, they cut it off at 1 chance in 10,000. Beyond that, they call it impossible. Beyond a probability of 0.0001, they declare the chance or probability to be 0.0000, or they declare the event to be impossible, because it is.

Nobody has time enough to waste to make something happen by chance alone!

I'm a statistician and mathematician among other things. I have the formula for computing the probabilities – which are found on a binomial distribution table – to great precision on my graphing calculator.

The probability of getting 15 heads in a row with a fair coin is:

0.00003051757813

That's three chances in 100,000.

That means that if you toss 15 fair coins one hundred thousand times in a row, the chance or the probability that you will get ALL heads is three times, or three trials, or three tries in 100,000.

Time starts to be a factor here.

Assuming that you have 15 fair coins to start with, and assuming you can toss those fair coins and count the result once every minute, it will take you 100,000 minutes or 70 days of tossing those coins continuously in order to end up with fifteen heads, three separate times on average. Are you really going to sit there and toss 15 coins for 70 days around the clock on the hope of getting 15 heads in a row, two or three times during that 70-day period of time? That's why tossing 15 heads in a row with a fair coin is both statistically and realistically impossible. Lack of time prevents us from doing so. It's still theoretically possible on paper, but it completely impossible in reality.

Now swap those coins for amino acids instead.

According to a binomial distribution table and a Sign Test based upon the Correct or Incorrect amino acid coming up by chance alone, the probability of getting 15 correct amino acids in a row is also:

0.00003051757813

There's NO chance that it will ever happen.

The Materialists, Naturalists, Darwinists, Nihilists, and Atheists will complain and say, "But there's still a chance!"

NO, THERE ISN'T.

Are you really going to sit there and keep stirring the pot or keep tossing those amino acids for 70 days just so that you can get fifteen of them to accidentally line up in the correct sequence? You see, when it comes to CHANCE, you actually have to have a real person there who is willing to keep tossing the coins, or to keep tossing the amino acids, or keep stirring the amino acids until the correct sequence comes up by chance alone. Without a person there to toss the coins or to stir the amino acids, NOTHING is ever going to happen. This is Logic 101, and everyone completely overlooks it.

There really is nothing like chance causation or creation by chance, because you still need a person, psyche, or intelligence there to keep tossing the coins or to keep stirring the amino acids until "chance" produces the correct results. You also NEED a person there to recognize that fifteen heads in a row has been achieved or to recognize that the correct amino acid sequence has been achieved. And, with the amino acid "protein" that is produced by "chance", you also NEED a person, psyche, intelligence, or quantum consciousness there to make sure that that new "protein" is protected and used, or that "protein" which was produced by chance alone will be of no value to anyone.

There is no such thing as pure chance or chance causation!

"Chance" of any kind needs an intelligent set-up or an intelligent assist. In the case of coins, "chance" needs an intelligent person, psyche, or quantum consciousness to make the coins, toss the coins, count the coins, and recognize when the fifteen coins have produced fifteen heads. "Chance" doesn't do anything and can't do anything without an intelligent person or psyche there to toss the coins or to line up the amino acids into sequences.

You have to have a person who is willing to find and/or make the coins and the amino acids in the first place. Without any coins or amino acids, there's nothing to toss; and therefore, there is NOTHING for chance to do! Then that same person has to be dumb enough and have time enough to keep tossing the fifteen coins or the fifteen amino acids for 70 days hoping that they will fall out just right. Finally, that same Person, Psyche, or Intelligence has to be there to recognize that the correct amino acid sequence has been achieved by chance alone. Without that intelligent person there to recognize that the correct amino acid sequence has been achieved, they will just keep tossing the coins or stirring the amino acids for all eternity with NO RESULTS. You NEED Someone Psyche or Someone Intelligent there to recognize and use that correct amino acid sequence when CHANCE finally produces it.

You see, there really is NO such thing as pure chance or chance causation, which is why they say that the Null Hypothesis in a science experiment was by definition produced by chance alone. A Null Hypothesis means that the independent variable had NO effect on the dependent variable, which means that there is NO causal relationship between the two.

Chance cannot produce causal relationships!

Chance cannot do causation!

Chance has NO mechanism in place for recognizing when it has succeeded.

Once we have established a causal relationship between the independent variable and the dependent variable, we are no longer looking at chance but are looking at causation instead.

Chance and causation are mutually exclusive!

Even games of chance require an intelligent assist from some type of person, psyche, quantum consciousness, or life force.

When it comes to protein synthesis and the origin of life, we are dealing with 21 amino acids. God would have to make all of the amino acids in the first place and then put them into close proximity to each other, for there to be a chance that CHANCE can combine them into a protein that is 15 amino acids long. No God-given amino acids, then NO proteins. It's that simple.

But, let's say that we have a God-given tub full of the correct 21 amino acids, God is still going to have to keep stirring that pot or that tub constantly, until the correct 15 amino acids miraculously line up in the correct sequence. If nobody stirs the pot, then nothing is ever going to happen! Amino acids are 3D and not 2D like coins. It's going to take a lot of work, stirring, and luck to get fifteen 3D amino acids to line up in the correct sequence by chance alone. There has to be Someone to stir the pot so that chance can do its job! There has to be someone to toss the coins, so that chance can do its job. If there is NO psyche, or intelligence, or quantum consciousness there to toss the coins or to stir the pot, nothing is ever going to happen. It's that simple!

Even then, there is still NO CHANCE that it will ever happen. There's NO chance that we will ever get 15 amino acids in the correct sequence by chance alone.

Why?

Well you see, by definition, chance or chemical evolution is blind. It has NO way of knowing that it has finally achieved the correct sequence of 15 amino acids. Chance has NO use for a correct sequence of fifteen amino acids. Chance wouldn't know what to do with it. Chance wouldn't even know that it has it. "Chance" or "chemical evolution" will just keep stirring the pot or the tub until that correct sequence of 15 amino acids is destroyed. There's nothing there in that tub to recognize that it has a correct sequence of fifteen amino acids. Recognition requires psyche, or intelligence, or quantum consciousness. A sequence of fifteen correct amino acids is absolutely worthless without some Intelligent Psyche there who KNOWS how and where to use that protein or sequence that was allegedly produced by chance alone.

Let's say that chance miraculously produces a gene. That gene is absolutely worthless without some Intelligent Psyche who KNOWS how, where, and why to use that gene. The same reality applies to proteins.

Creation by chance alone is impossible. Chance of any kind always requires an intelligent assist of some kind to take advantage of it.

Remember, there's nothing there in the tub to protect that correct sequence of amino acids, once it is assembled by chance alone. Soon, the sun, wind, water, or chance will destroy that correct sequence of 15 amino acids, and the whole process will have to start all over again. Nothing will ever be accomplished by chance alone. There is no such thing as Chance Causation. There is no such thing as Creation by Chance. Someone Intelligent has to be there to recognize that the correct amino acid sequence has been achieved; and then, that same Someone has to protect and use that correct amino acid sequence, or nothing will ever come of it. Chance cannot do design and creation by chance alone. It's that simple.

A binomial distribution with a sign test is a very rough and crude estimate in comparison to the true probability of getting fifteen correct amino acids into the right sequence by chance alone. The true probability of getting fifteen amino acids in the correct order by chance alone is much lower than what we got from the binomial distribution table and sign test, as I will demonstrate in the next section.

For now, all you really need to know is that it is statistically and realistically IMPOSSIBLE to get fifteen heads in a row, with a fair coin. It's theoretically possible on paper, but in the real world, nobody is going to toss and count fifteen fair coins for a solid month around the clock on the off chance that he or she will get fifteen heads somewhere along the way. In real world practice, it's never going to happen.

The same can be said for fifteen amino acids. Nobody is going to keep sequencing fifteen amino acids randomly and checking to see if the sequence is right, on the off chance that he or she will end up with the correct sequence somewhere along the way. And, without that intelligent person, psyche, life force, or quantum consciousness there to toss and count the coins or the amino acids in the first place, NOTHING will ever happen for all eternity, because someone intelligent has to be there to toss the coins and count the number of heads for "chance" to do its job to begin with. Someone Psyche or Someone Intelligent has to be there to recognize the correct amino acid sequence for what it is and take advantage of it, once chance has finally produced it; otherwise, nothing will ever come of it.

Chance Cannot Do Causation!

Chance and causation are mutually exclusive.

This is the fundamental axiom of Science, Statistics, the Scientific Methods, and Science Experimentation. Chance cannot do causation!

Furthermore, we are dealing with 21 amino acids here as well as proteins, and NOT coins. Amino acids are more complex than coins. Amino acids and proteins are more like 21-sided dice than coins. Amino acids and proteins are 3D. The chances of success go down drastically when we start dealing with 3D objects instead of 2D coins.

Let's say that we can get God to give us a tub full of the correct amino acids, and nothing else but the correct amino acids in that tub; and, let's say that you can convince God to keep that tub in existence and keep stirring the contents of that tub for all eternity; and, let's say that God has time to keep stirring that tub for all eternity so that CHANCE can do its work in a timely fashion – how long is it going to take on average for CHANCE to produce a correct amino acid sequence 15 amino acids long, if CHANCE is putting 15 amino acids together every minute?

This is a legitimate statistics problem, and it does have a real answer that can be found. We can calculate the probability of chance events.

There are 21 different amino acids that are used to make proteins, and we are looking for a correct sequence that is in fact 15 amino acids long – 15 trials or 15 events. The Multiplication Rule in Probability applies here. We are looking at 15 separate events combined into one result. The answer is (1/21) multiplied by itself fifteen times – one time for each trial, or event, or amino acid in the sequence.

The true probability of getting 15 correct amino acids in a row by chance alone is:

$(1/21)^{15}$

This is equal to: 1.46794769 e -20

YIKES!

That's one solid chance in 100,000,000,000,000,000,000 attempts or trials, with fifteen amino acids being lined up and checked for correctness in each attempt, trial, or event.

That's NEVER going to happen!

Look what happens when we switch from fifteen coins to fifteen 21-sided dice instead!

We definitely entered into the realm of impossibility!

We start with the estimated 100,000,000,000,000,000,000 minutes that it will take on average for "chance" to find or produce the correct amino acid sequence that is fifteen amino acids long, by chance alone, if CHANCE is successfully lining up 15 amino acids every minute. We have to start with some sort of given, or there is nothing for chance to do.

Let's start with a God-given ocean full of the correct 21 amino acids. And, let's convince God to keep stirring this ocean for all eternity so that chance can do its job. Producing the correct amino acid sequence by chance alone will NEVER happen unless God keeps stirring the pot. Let's say that chance can miraculously sequence 15 amino acids randomly every minute given this ocean full of the correct amino acids. We start with the estimated 100,000,000,000,000,000,000 minutes that it will take for chance to do its job. Then we divide that huge number by 60 minutes, 24 hours, and 365 days in order to find the number of years that God will have to stir the contents of that pot or tub so that CHANCE can produce a "protein" that is 15 amino acids long by chance alone.

Assuming that God can miraculously keep getting 15 amino acids to line up into a sequence every minute to produce a single trial or event every minute, it will take God 190,285,875,190,000 YEARS on average, of stirring that same ocean full of the correct amino acids, in order to get the correct amino acid sequence fifteen amino acids long that we are looking for – one time. Furthermore, God is going to have to be on hand the whole time to recognize and use the correct amino acid sequence when it is finally achieved by chance alone. If God isn't there to recognize and use that correct amino acid sequence when CHANCE finally produces it, then nothing will come of it.

That's 190 trillion years on average of stirring the pot, for CHANCE to produce a correct or functional protein that is fifteen amino acids long by chance alone. The scientists estimate that our physical universe is only 13.8 billion years old. Is God really going to be dumb enough to stir a pot or a tub full of the correct amino acids for 190 trillion years on the off chance that He will get a correct sequence of fifteen amino acids by chance alone? Would you be willing to do that? Would you be able to do that? Of course not! So, why believe that CHANCE can do it at will? It's superstition to believe that CHANCE can design and create things at will.

And, what did we get from all of this hypothetical stirring?

We got a "protein" that was 15 amino acids long. We got nothing! There is NO protein that is 15 amino acids long! Proteins are much longer than that. 190 trillion years of stirring the pot, and we still have nothing to show for it.

Are you starting to get a sense for why Creation by Chance or Causation by Chance is NOTHING but superstition, and fiction, and wishful thinking?

Creation by Chance, *Causation by Chance*, *Chance Causality*, and *Wishful Thinking* are logic fallacies. These are just a few of the dozens of different logic fallacies upon which the Theory of Evolution and the Second Law of Thermodynamics are based. They are both Creation by Chance, which means that they are both automatically FALSE. Chance cannot do causation! Chance is precisely the thing that we are trying to identify and eliminate from Science by running science experiments in the first place! Creation by Chance is nothing but science fiction. Creation by Chance is magic at best and a deceptive lie at worst.

I got the following off the internet from Uncle Google.

Thyroid releasing hormone or TRH should be the smallest protein in the human body, with 234 amino acids. (> 100 amino acids is a protein.) The smallest polypeptide in the human body should be insulin, with 54 amino acids. (10-100 amino acids is a polypeptide.)

Let's run with this for a minute and see what we get.

The probability of making insulin by chance alone is:

$(1/21)^{54}$

This is equal to: 3.98252109 e -72

This can't be done by chance alone, even with a God-given ocean full of the correct amino acids. Even with God constantly stirring the pot or tub or ocean, it would take trillions of trillions of trillions of years for such a thing to happen by chance alone. It's just not going to happen. Our observable universe hasn't been around long enough for this to happen by chance alone. Insulin was designed and made by God. Without God to design it, make it, fine-tune it, and field test it, insulin wouldn't exist. Only God could recognize that He had actually created insulin and KNOW what it is good for and how to use it.

Technically, insulin isn't a protein. It's a polypeptide that is 54 amino acids long. From now on, whenever you think of insulin, remember that God designed and made the stuff in the first place. We KNOW for a fact that chance didn't do the job, so that leaves God as the only other plausible candidate. Someone made the stuff, or it wouldn't exist. Anything that is obviously made obviously has a Maker who made it. Insulin was obviously made. God made insulin; and, there's no chance that I got that wrong.

Here we just falsified and rejected the Null Hypothesis (which is produced by chance alone) and accepted the Alternative Hypothesis (which is some type of causation by a causal agent). In this case, the Alternative Hypothesis is that God designed and created insulin, because chance could never have done the job. This is Science in action. This is a True Science Experiment.

The probability of making THR by chance alone is:

$(1/21)^{234}$

When I put this into my graphing calculator, the answer was ZERO (0). Literally. The answer is ZERO. This is NEVER going to happen ever! This is beyond what a graphing calculator can handle. This is beyond what God could do by chance alone!

The probability of making THR by chance alone is ZERO. You can wait for all eternity, and it will NEVER happen, even with a God-given tub full of the correct amino acids or with an ocean full of the correct amino acids. It's NEVER going to happen. Making THR by chance alone is impossible. It can't happen by chance alone. It requires some kind of intelligent and deliberate intervention from God Himself.

Here we just falsified and rejected the Null Hypothesis (which is produced by chance alone) and accepted the Alternative Hypothesis (which is some type of causation by a causal agent). In this case, the Alternative Hypothesis is that God designed and created THR, because chance could never have done the job. This is Science in action. This is a True Science Experiment.

THR is the smallest protein in the human body. THR can never be made by chance alone. Every time you think of any protein, remember that God designed it, created it, made it, fine-tuned it, and field tested it. Anything that is obviously made obviously has a Maker who made it. Proteins were obviously made.

God made your proteins. Each protein is one of God's Signatures. People are looking for the Signature of God in nature and science, and there it is in each one of your proteins.

Your genome is also God's Signature. God had to make a matching gene for each of the proteins that He designed and made. God made your proteins and your genes. This is obviously true. According to the statistics and the probabilities, there is NO chance that I am wrong. God made your proteins and your genes.

This is one of the most convincing Scientific and Statistical Proofs of God's Existence that I have encountered so far. I just used Statistics, Probability, and Science to prove that God exists; and, there's NO chance that I'm wrong.

One day in 2015, I finally decided to go looking for Scientific Proof of God's Existence, and eventually I found what I was looking for. Did I not? It's obvious that I found Scientific Proof of God's Existence; and, there's NO chance that I got it wrong.

This is what I got from my statistics class.

It's hiding in plain sight where nobody can see it, because nobody is looking for it and nobody wants it.

I bet you never got anything like this from your statistics class, and neither did your statistics professor, because you weren't looking for it and didn't want to find it in the first place. No seeking, then no finding.

Yet, it's there and it's obvious for anyone who is willing to look and see.

Creation by Chance, or Chance Causality, or Causation by Chance is IMPOSSIBLE.

That means that Materialism, Naturalism, Darwinism, Nihilism, Atheism, Creation Ex Nihilo, Creation by Random Disorder or Entropy, Creation by Death or Natural Selection, Abiogenesis, Spontaneous Generation, Chemical Evolution, Macro-Evolution, the Theory of Evolution, and the Second Law of Thermodynamics ARE IMPOSSIBLE because they are Creation by Chance, Chance Causality, Chance Causation, or caused by chance alone.

There's the truth, whether you want it or not. Scientists lie, but the math doesn't lie. I'm a scientist, mathematician, statistician, and logician. I believe the math, and I KNOW that it is true. We can calculate the probability of chance events; and, WE KNOW for a fact that chance cannot do causation. Chance and causation are mutually exclusive.

We can calculate the probability of chance events; and, CHANCE cannot produce fifteen heads in a row in a timely fashion with any kind of reliability, with a fair coin. If chance can't do fifteen heads in a row in a timely fashion, then chance can NEVER produce a protein or a gene. There is the truth that everyone is searching for and that nobody is willing to accept. God made your proteins and your genes. Anything that is obviously made obviously has a Maker who made it. Genes, proteins, quantum fields, quantum waves, and physical atoms are obviously made. Therefore, it is obvious that they have a Maker who made them. The whole purpose of Science, Statistics, Science Experiments, and the Scientific Method is to find the TRUE CAUSE of each event that we experience or observe.

Chance is NEVER the True Cause of anything! Chance cannot do causation.

Now, the only thing left to do is to figure out which God is the God who made your proteins and your genes. I will leave that up to you to figure out for yourself, because it will be meaningless to you unless you discover it for yourself and want to become a part of it. I believe that answer can be found as well, if you are willing to look for it with an open mind and heart. It took a while, but I eventually found the answer to that question also, after I finally decided to go looking for it. The answer is there to be found, if one is willing to look for it and is able to accept it when he finds it. God has revealed Himself to us millions of times in thousands of different ways. It's subtle most of the time, but it's there to be found for those who are willing to look and see.

Mark My Words

The Perpetual Motion Cycle

Once we have the True Meaning or the True Definition for a scientific theory or a scientific concept, then even the simplest of mathematical equations can reveal the profoundest truths, that have been hidden from the world for the duration of human history.

$E = mc^2$ is one such equation.

Obviously, I am not the first person on the planet to discover and use $E = mc^2$. However, I may be the first person to discover what it truly means and what it truly is.

$E = mc^2$ is the Perpetual Motion Cycle.

This Perpetual Motion Cycle is the heart of modern-day physics, and we don't even know it because we are still caught-up in or stuck in the dark ages with Materialism, Naturalism, Darwinism, Nihilism, Atheism, Creation Ex Nihilo, Classical Realism, Creation by Chance, the Theory of Evolution, and the Second Law of Thermodynamics.

Most people on this planet don't even know it, but the Perpetual Motion Cycle $E = mc^2$ FALSIFIES Materialism, Naturalism, Darwinism, Nihilism, Determinism, Behaviorism, Determinism, Classical Realism, Atheism or Creation Ex Nihilo, the Theory of Evolution, and the Second Law of Thermodynamics, which ARE Creation by Chance, Creation by Disorder, Creation by Chaos, Creation by Death, or Creation by Entropy.

The false is falsified by the truth, and the truth is repeatedly and constantly experienced and observed.

$E = mc^2$ is Quantum Mechanics, particularly Quantum Field Theory. $E = mc^2$ is Conserved! The Quantum Fields ARE perpetual motion machines. The Quantum Fields operate on the Perpetual Motion Cycle and are an integral part of the Perpetual Motion Cycle. Once the Quantum Fields were made by Nature or Nature's Psyche, they have always been conserved thanks to $E = mc^2$ and the Conservation of Energy, the Conservation of Quantum Information, and the Conservation of Psyche or Quantum Consciousness.

It's Nature's Psyche or Nature's Intelligence who makes the quantum waves or the photons in the first place. Later, it is Nature's Psyche or Nature's Intelligence who collapses the wave function and transforms infinite acceleration or omnipresent quantum waves or photons INTO mass, heat, resistance to acceleration, and mass's heat storage capacity (entropy) instead.

According to the Perpetual Motion Cycle $E = mc^2$ and Quantum Field Theory, the Gods or the Controlling Psyches had to design and make the massless, heatless, and entropyless Quantum Fields BEFORE they could make and sustain mass, resistance to acceleration, heat, and mass's heat storage capacity which is entropy. The proven and verified existence of the Quantum Fields FALSIFIES the second law of thermodynamics, which says that they don't exist and can't exist. The second law of thermodynamics will NEVER be true as long as the Quantum Fields exist. The Quantum Fields are pure Syntropy or pure Exergy. The Quantum Fields are perfect Order and Organization. The Quantum Fields are massless, invisible, non-physical, immaterial, entropyless, syntropic, exergic, and conserved Perpetual Motion Machines. Their very existence falsifies the second law of thermodynamics.

According to the Law of Psyche, each psyche or intelligence or life force or quantum consciousness has a certain amount of energy that's under its control, and that controlling psyche can form or transform the energy under its control into anything that it wants that energy to be, anytime and anywhere that it chooses to do so. This is the Perpetual Motion Cycle $E = mc^2$ in action. The controlling psyche CHOOSES what form the energy under its control will be. Psyches can also coordinate their actions transpersonally or telepathically at the quantum level through thoughts or quantum waves. This is the way things really work at the quantum level in the quantum realm, or the non-physical realm, or the spiritual realm. This is what has actually been experienced and observed.

According to Quantum Field Theory, particles are born, and particles die. In other words, particles or quanta of any kind are made, and they can be unmade or reabsorbed back into the quantum fields from whence they came, anytime and anywhere that their controlling psyche CHOOSES to make them or transform them or dissolve them. All the while, their energy is conserved. This is the Perpetual Motion Cycle $E = mc^2$ in action.

The Perpetual Motion Cycle has been experienced, and observed, and verified. That means that it is real and truly exists. The Perpetual Motion Cycle $E = mc^2$ works perfectly and eternally, both coming and going, thanks to the Conservation of Energy and Psyche, as well as the Conservation of Quantum Information.

Currently in our sun, Nature or Nature's Psyche is transforming mass, heat, and entropy into massless, heatless, chargeless, and entropyless photons or quantum waves. Nature's Psyche is transforming resistance to acceleration, mass's heat storage capacity (entropy), mass, and heat into infinite acceleration instead. Photons and quantum waves go from zero to the speed-of-light instantly. That is infinite acceleration. Photons and quantum waves also CHOOSE their ultimate velocity, which can be less than the speed-of-light, the speed-of-light, or an infinite velocity which we call omnipresence or quantum tunneling. This is the part of the Perpetual Motion Cycle where Nature or Nature's Psyche transforms mass, heat, and entropy into massless, heatless, chargeless, and entropyless photons, quantum waves, and infinite acceleration instead. This is happening all the time through our sun or within our sun.

Likewise, any time that a photon, virtual particle, or quantum wave CHOOSES TO STOP, that photon or quantum wave transforms its omnipresence and infinite acceleration INTO mass, heat, resistance to acceleration, or mass's heat storage capacity (entropy) instead. The controlling psyche within a photon or a quantum wave can transform that photon or quantum wave into ANYTHING that it wants that photon or quantum wave to be, anytime and anywhere that it CHOOSES to do so. In Feynman Diagrams, photons or quantum waves are constantly transforming themselves into electrons and positrons, or quarks and gluons. Massless, heatless, chargeless, and entropyless photons and quantum waves are constantly transforming themselves INTO mass, heat, and entropy (mass's heat storage capacity) all the time. All the while, the energy is conserved!

This, too, is a part of the Perpetual Motion Cycle. It's happening all the time. This, too, has been experienced and observed. A miniature Big Bang happens every time a photon CHOOSES to stop and land on our skin. That photon transforms itself into mass or heat. We typically feel the heat, when a photon lands on our skin. However, a photon or quantum does NOT have to transform itself into heat if it doesn't want to. It can transform itself into mass, resistance to acceleration, and mass's heat storage capacity (entropy) instead, if it chooses to do so. A photon or quantum wave doesn't have to stop if it doesn't want to. A photon or quantum wave can pass through water, glass, our earth, our sun, our physical body, or a black hole as if they weren't even there, if a quantum wave or a photon chooses to do so. This too has been experienced and observed.

The whole Perpetual Motion Cycle $E = mc^2$ has been experienced and observed; and, it FALSIFIES the second law of thermodynamics which claims that the amount of disorder or entropy is constantly increasing and that it can never decrease and go to zero.

ALL of the conservation laws falsify the second law of thermodynamics! The second law is a violation of the Conservation of Energy or the First Law of

Thermodynamics. According to the second law of thermodynamics, nothing should exist. Everything should be random disorder or random chaos, or there should be nothing at all. The second law of thermodynamics FAILS TO PREDICT what we are actually experiencing and observing. Everything that exists falsifies the second law of thermodynamics! The very fact that you exist is scientific proof that the second law of thermodynamics is false.

Do we really observe proton decay?

Do we really observe ever-increasing disorder and chaos? Do we really observe an ever-encroaching gray goo coming in at us from all sides as the second law of thermodynamics predicts? Or do we observe constantly conserved order and organization as Quantum Field Theory and the Perpetual Motion Cycle predict?

The one that we actually experience and observe is the one that's actually real and true. This is the pinnacle of Science 2.0. Science 2.0 allows all of the evidence into evidence, and then it pursues a preponderance of that evidence. Science 2.0 is observation and experience. The Perpetual Motion Cycle has been experienced and observed; therefore, we know that it is real and truly exists.

We have NEVER observed an ever-increasing amount of substance coming into existence out of thin air from nothing as the second law predicts. We have NEVER observed creation ex nihilo, creation by entropy, creation by disorder, or creation by chance as the second law predicts. We have never observed the proton decay that the second law predicts. We have NEVER observed anything that the second law of thermodynamics predicts. Everything that exists FALSIFIES the second law of thermodynamics.

But, we have observed the Perpetual Motion Cycle $E = mc^2$ in action, both coming and going. The Perpetual Motion Cycle and constantly conserved Order and Organization at the quantum level FALSIFY the second law of thermodynamics which predicts and claims that they do not exist. We have indeed observed the Perpetual Motion Cycle and the massless, heatless, and entropyless photons, quantum waves, and quantum fields in action. We do indeed observe constantly conserved Order and Organization thanks to Quantum Field Theory, the Quantum Fields, the Perpetual Motion Cycle, the Conservation of Quantum Information, and the Conservation of Energy and Psyche.

The one that has been experienced and observed is the one that's actually real and true. The second law of thermodynamics is a con and a scam that is falsified by everything that exists and by everything that has been experienced and observed. The second law of thermodynamics is FALSIFIED by the Perpetual Motion Cycle $E = mc^2$, as well as Quantum Mechanics and Quantum Field Theory. The second law of thermodynamics will NEVER be true as long as the Quantum Fields or the Perpetual Motion Cycle exists.

My ultimate goal is to identify and fix everything that is wrong with Science; and, that process starts with and includes replacing the theory of evolution, the second law of thermodynamics, creation ex nihilo, and creation by chance WITH the Conservation of Energy and Psyche, the Conservation of Quantum Information, the Conservation of Order and Organization, the Law of Psyche, Quantum Mechanics,

Quantum Field Theory, and the Perpetual Motion Cycle instead. We replace the things that have NEVER been experienced NOR observed with the things that have actually been experienced, observed, verified, and proven true instead. That's what Science 2.0 is all about, and that's part of the reason why I chose to upgrade my science to Science 2.0. One day I simply realized that science as a whole is in massive need of a serious upgrade. Since then, I have tried to do just that.

    Mark My Words

Where's the Science?

During my writing career, my focus has been on Science, the Philosophy of Science, Logic, Psychology, and the Scientific Methods. My recent degree work has been in Psychology, with an emphasis on the Philosophy of Science and the Scientific Methods. I personally have come to believe that Psychology, the study of the Human Psyche, is the final frontier in Science. Psyche or Non-Local Consciousness is the Science that most of the scientists have seemed to avoid or ignore, thus making Psyche the Ultimate Science.

In some of the reviews for my books, I have had people ask, "Where's the Science?"

I find it fascinating that some people can't find any science in any of my books. What's their problem? I assume it's because they have a very narrow definition for Science and Methodology, in comparison to mine.

I consider Psychology, the Study of the Human Psyche, to be a Science. They don't. I consider Genetic Entropy and the Mathematical Modeling of Natural Selection to be Science, and they don't. I KNOW that Quantum Mechanics falsifies Materialism and Naturalism, and they don't. (Of course, it is also obvious that some of these people haven't even read the book that they are critiquing. They are judging the book based upon its cover or title. Materialists and Atheists tend to do that a lot. I KNOW, because I used to be one of them.)

The original definition for Science was "Knowledge". BEFORE the Scientific Method was developed, Science was the pursuit of Truth and Knowledge; and, Science or Knowledge was based upon Direct Observation or Lived Experience. Science consisted of the things that we KNEW to be TRUE, because we had lived them and experienced them for ourselves.

Through Lived Experienced or Direct Observation, we human beings or human psyches can go directly to KNOWING the TRUTH in any realm of existence or any reality. Therefore, I treat Lived Experiences or Direct Observations as Scientific Evidence, because Lived Experience is the BEST source of Truth and Knowledge. Lived Experience is an infinitely superior source of Scientific Evidence or Knowledge than the philosophical musings and philosophical interpretations which we manufacture through the various different Scientific Methods.

Due to the *affirming the consequent*, *begging the question*, and *jumping to conclusions* logic fallacies which are built into the Scientific Method, the Scientific Methods are fundamentally flawed and cannot be used to prove the truth directly.

So, where's the science?

Well, according to the Philosophy of Science, there is NO scientific proof of truth that can be had from the Scientific Methods, so there really is NO science that can be developed directly from the Scientific Methods. Consequently, Lived Experience or Direct Observation IS SCIENCE, and is infinitely superior to the Scientific Methods, when it comes time to find the truth, prove the truth, and know the truth. The BEST way to KNOW the TRUTH and to prove the truth is to live it and experience it for yourself, or to choose to trust someone who has. Proof of the Truth can be had through Lived Experience or Direct Observations; but, proof of the truth is impossible through the Scientific Methods thanks to all the different logic fallacies that are built directly into the Scientific Method.

Science doesn't have to be complicated in order to be Science, but it does have to be TRUE to count as Science or Knowledge, in my humble opinion.

Since the Scientific Methods can NEVER be used to prove the truth, does that mean that Science and the Scientific Methods are worthless?

Not necessarily!

The Scientific Methods can be used to PROVE theories false. It's called "Falsifying a Theory" or "Negating the Consequent"; and, it's logically and philosophically sound. This is the advantage that accrues from understanding the Philosophy of Science and how the Scientific Methods really work.

Here, let's do some REAL SCIENCE for once. This may in fact be the very first time that you will have seen REAL SCIENCE in action, in a way that's philosophically and logically sound.

Here's the Science!

Scientific Axiom or Scientific LAW: We can use Science and the Scientific Methods to falsify theories or to PROVE theories false, because *negating the consequent* is philosophically and logically sound!

The argument structure for falsification, or *negating the consequent*, looks like this:

Scientific Hypothesis: If Theory X is true, then we will observe Y.

Scientific Observation: We don't observe Y.

Scientific Conclusion: Therefore, Theory X is false, and Theory X has been falsified by the Scientific Method.

That's the Scientific Method in action, doing what it was meant to do, falsifying a theory!

Let me provide an example to show how this works in practice.

Scientific Definition of the following Theories: Materialism, Naturalism, Darwinism, and the Theory of Evolution are defined as "Design and Creation by Physical Matter". The Materialists, Naturalists, and Darwinists contend that raw inert physical matter designed, created, and manufactured the first genomes and the first living cells, because according to these people physical matter is the only thing that exists and therefore the only thing that could have designed, programmed, engineered, created, and manufactured the first genomes and the first life forms on this planet. We choose to take these people at their word whenever they claim to believe that physical matter is the only thing that exists. They really believe that! These people truly believe that the rocks designed and created the first genomes and the first life forms on this earth. These people call it Science.

Scientific Premise or Axiom or LAW: The Scientific Methods can be used to falsify theories and to prove them false.

Scientific Hypothesis: If Materialism, Naturalism, Darwinism, and the Theory of Evolution are true, then we will observe the rocks and physical matter designing, creating, and manufacturing genomes and life forms from scratch.

Scientific Observations: We NEVER observe the rocks designing and creating anything. Raw physical matter cannot design and create anything at all.

Scientific Conclusion: Therefore, Materialism, Naturalism, Darwinism, and the Theory of Evolution are FALSE, and have been successfully falsified by the Scientific Method and by the Scientific Observations of the human race.

There's the Science!

I just falsified Materialism, Naturalism, Darwinism, and the Theory of Evolution for REAL! Science doesn't have to be complicated in order to be REAL and TRUE.

People ask, "Where's the Science?"

Well, there's the Science; and, it's based upon the Lived Experiences or the Direct Observations of the human race as a whole. I have in fact FALSIFIED Materialism, Naturalism, Darwinism, Creation by Rocks, and the Theory of Evolution; and, I used the Scientific Method, Science, and the collective Scientific Observations of the human race to do so! I love doing that! I wish I would have known how to do that forty years ago! But, the Materialists and Atheists don't teach you this kind of stuff in school, for some strange reason.

Let's run Atheism through the same logically sound Scientific Method, which we call *Negating the Consequent*.

Atheism is "Creation by Nothing"; and, we all know that nothing cannot design and create anything.

Scientific Hypothesis: If Atheism is true, then we will observe nothing designing and creating something.

Scientific Observations: We NEVER observe nothing designing and creating anything. Creation Ex Nihilo, or Atheism, or Creation by Nothing is patently absurd.

Scientific Conclusion: Therefore, Atheism is false, and has been successfully falsified by the Scientific Method and by the Scientific Observations of the human race.

There's the Science! I just falsified Atheism, using the Scientific Method.

Best of all, using the Scientific Methods to falsify theories or to prove theories false IS philosophically and logically sound! In other words, I HAVE IN FACT FALSIFIED Materialism, Naturalism, Darwinism, Atheism, and the Theory of Evolution. The debate has ended. I have just PROVEN Materialism, Naturalism, Darwinism, Atheism, and the Theory of Evolution FALSE, using the Scientific Method.

There's the Science! People keep asking me, "Where's the Science?" Well, there's the Science!

We now KNOW that these theories and philosophies are false, because we KNOW why they are false. They are FALSE because the rocks and blind luck have never been caught in the act of design and creation; and, they NEVER will be. Case closed!

Science IS Observation, or at least it should be. What do we OBSERVE happening around us right now? Does it fit with Materialism, Naturalism, and Darwinism? Do these philosophies of science match with Reality? Or, do the Scientific Methods point us to a better and more realistic Philosophy of Science and Paradigm?

Since we now KNOW by the Scientific Method that Materialism, Naturalism, Darwinism, Atheism, Creation by Rocks, and the Theory of Evolution are FALSE, we must look someplace else if we want to find the truth. So, where else are we going to look for the truth?

Well, we can always turn to the Lived Experiences or the Direct Observations of the human race, including our Spiritual Experiences, Psyche Experiences, Non-Local Experiences, Out-of-Body Experiences (OBEs), Near-Death Experiences (NDEs), Psychic Experiences, Shared-Death Experiences (SDEs), and Theophanies or Revelations of the Biblical God Jesus Christ, which the Materialists and Naturalists choose to reject and ignore.

Our Lived Experiences also include our physical experiences.

Remember, Lived Experience or Direct Observation IS Scientific Evidence; and, through Lived Experiences, we human beings or human psyches can go directly to KNOWING the TRUTH in any realm of existence, without having to use the Scientific Methods in order to do so. Knowledge of the Truth IS Science; and,

Lived Experience or Direct Observation IS the BEST source of Knowledge, or Scientific Evidence, or Truth.

A Psyche Ontology based upon the Lived Experiences or Scientific Observations of the human race is the Ultimate Paradigm or the Ultimate Model of Reality. The Scientific Methods actually point us to a Psyche Epistemology, or Psyche Experiences, or Lived Experience, or Direct Scientific Observation as the BEST and most reliable source of Scientific Evidence in any realm of existence, including the physical realm and the spirit realm. A Psyche Epistemology based upon the collective Lived Experiences or Scientific Observations of the human race IS the Ultimate Philosophy of Science. Through Lived Experiences, or Psyche Experiences, or Direct Scientific Observations of different realms of existence, human beings or human psyches can go directly to KNOWING the TRUTH by living the truth and experiencing the truth for itself.

Science, in order to be Knowledge, should be TRUE; and, in order to be TRUE, Science must match with Reality or the Lived Experiences of the human race. Materialism, Naturalism, Atheism, Nihilism, and Darwinism are FALSE, because they don't match with Reality, meaning that they don't match with the Lived Experiences and the Direct Observations of the human race.

We NEVER observe the rocks designing and creating anything! That's what the collective Observations of the human race are telling us. Something is said to be Scientifically Accurate if it matches with Reality or the Lived Experiences of the human race. Materialism and Naturalism are unscientific, because they don't match with Reality nor the Lived Experiences of the human race as a whole. It's that simple!

So, where's the Science?

The REAL SCIENCE is recorded in the Lived Experiences and the Direct Observations of the human race! Lived Experience IS Scientific Evidence, the best type of Scientific Evidence! We have NEVER observed the rocks nor raw physical matter designing and creating anything; therefore, we simply KNOW that Materialism, Naturalism, and Darwinism are FALSE. Through Lived Experience or Direct Observation, human beings or human psyches can go directly to KNOWING the TRUTH in any realm of existence or any reality, simply by living the truth and experiencing the truth for itself. Cool, huh?

Since Materialism and Naturalism have been FALSIFIED trillions of different times in thousands of different ways by the Scientific Methods, Scientific Observations, and the Lived Experiences of the human race, WE NEED a New and Better Paradigm or Philosophy of Science that is actually based upon the Lived Experiences or the Scientific Observations of human beings or human psyches, if we want to find and KNOW the TRUTH in all realms of existence.

I'm talking about a Paradigm Shift here, wherein a Psyche Epistemology, Psyche Experiences, or Lived Experiences become our primary source of Scientific Evidence and Knowledge. I'm talking about the End of Materialism and the End of Naturalism, because these philosophies of science have been falsified. As

Scientists, we should switch from a Materialistic Ontology to a Psyche Ontology, because Materialism and Naturalism have been falsified.

In a Physical Ontology or a Materialistic Ontology, physical matter is the fundamental unit of reality; but, that type of ontology does not match with REALITY and has been falsified. In a Psyche Ontology, Non-Local Consciousness, Intelligence, or Psyche becomes the fundamental unit of reality. A Psyche Ontology actually matches with REALITY and the Lived Experiences of the human race, because ONLY Psyche can do ontology, lived experience, existence, and reality in any realm of existence that we can possibly imagine! ONLY Psyche can do Scientific Observations in a Non-Local Realm, Non-Physical Realm, Transdimensional Realm, or Spiritual Realm.

SCIENCE is all about the observations or lived experiences! Or, at least it should be.

We human beings or human psyches can generate a ton of faulty hypotheses and faulty conclusions, including false interpretations of scientific evidence; but, the collective experiences or collective observations of the human race form our Reality and make up our Existence as a species.

What do we observe?

Collectively as a race, we have OBSERVED that the rocks cannot design, program, engineer, field-test, fine-tune, manufacture, deploy, and create genomes and life forms.

Where's the Science?

Our Science is stored in the Lived Experiences or Direct Observations of the human race as a whole. That's where our Science or our Knowledge resides. It has been estimated that there are now 13 million different Near-Death Experiences on record. A Near-Death Experience is also an Out-of-Body Experience. That's where our Next Generation Science is to be found – in our Lived Experiences or Psyche Experiences.

The Materialists, Naturalists, Nihilists, Darwinists, and Atheists make the scientific claim or the scientific hypothesis that ONLY physical matter exists. Materialism, Naturalism, and their derivatives ARE "Creation by Rocks" or "Spontaneous Generation". These people have chosen to believe that somehow, some way, the rocks or raw physical matter is able to come alive spontaneously all on its own. Do their beliefs match with Reality or the Lived Experiences of the human race? Do we observe the rocks and raw physical matter spontaneously coming alive?

What do we observe?

That's the question that we should be constantly asking ourselves, if we want to do Science for REAL. Our Science or Knowledge resides in our collective OBSERVATIONS as a race, or at least it should.

Darwinism, the Theory of Evolution, Materialism, Naturalism, and even Atheism can all be defined as "Design and Creation by Rocks".

What does the Science or the Observations tell us about this?

Can the rocks design and create?

NO, they can't.

So, what's the opposite of Darwinism?

Intelligent Design Theory, or Design and Creation by Intelligent Beings, is the opposite of Darwinism and the Theory of Evolution. Darwinism IS Creation by Rocks. The opposite of that is Creation by Intelligent Psyches. The rocks or raw physical matter have NEVER been caught in the act of design and creation. In fact, in 1859, Louis Pasteur falsified Spontaneous Generation or Creation by Rocks. Louis Pasteur used a scientific method to falsify Materialism, Darwinism, Naturalism, Creation by Rocks, and the Theory of Evolution the very same year that Charles Darwin published "On the Origin of Species". How ironic is that?

Where's the Science?

Louis Pasteur did the Science in 1859 and falsified Materialism, Naturalism, Darwinism, Creation by Rocks, and Spontaneous Generation in 1859. There's the Science! Do you believe the Science, the REAL SCIENCE, the collective observations of the human race? Most people don't; but, a few of us do.

Our Lived Experiences or Scientific Observations tell us clearly and conclusively that the rocks cannot design and create, that the rocks cannot spontaneously generate and come alive, and that the rocks can't do Programming and Science. A genome is a radically advanced computer program. A genome is software. A living cell is radically advanced hardware. The rocks have NEVER been caught in the act of designing, engineering, and manufacturing hardware. The rocks have NEVER been caught in the act of doing computer programming or writing software such as genomes. ONLY Intelligent Psyches or Intelligent Beings can do such a thing. That's what the Sciences or the Observations of the human race are telling us.

So, where's the Science?

The Science resides in our observations as a race; and, ALL of the Science is telling us that the rocks cannot design and create, and that Materialism and Naturalism are false.

Furthermore, the Materialists and Naturalists find themselves caught in a variety of chicken-egg situations or a Catch-22 paradoxes, since these people have chosen to believe that only physical matter exists. Their chosen philosophical assumption is unsustainable. There is NO evidence and will NEVER be any evidence to support their claim that the Non-Physical, or the Non-Local, or the Spiritual does not exist. There is NO evidence and will NEVER be any evidence to support their materialistic claim that Psyche or Intelligence does not exist.

Instead, ALL of the Scientific Evidence that we have on hand as a race tells us clearly and conclusively that Materialism and Naturalism are FALSE. Reality itself tells us that Materialism and Naturalism are false. The existence of Light and Gravity falsifies Materialism and Naturalism. The existence of Forces, Fields, Magnetism, Space, Time, Dark Energy, and Dark Matter falsifies Materialism and Naturalism. Dark Matter is Spirit Matter. These things are Supernatural; and, therefore, these things falsify Naturalism. The existence of Thought, Dreams, Consciousness, Life, Psyche, and Quantum Non-Locality falsifies Materialism and Naturalism. Thoughts and Dreams are spiritual experiences, because the contents of our Thoughts and Dreams cannot be recorded by our physical instruments. We KNOW from the Lived Experiences or Scientific Observations of the human race that Psyche, or Intelligence, or Non-Local Consciousness does in fact exist. Psyche or Intelligence IS the opposite of Materialism and Naturalism.

There are materialistic Neuroscientists who worship their brain scans and truly believe that the neurons in the physical brain are Consciousness or God. I'm a psychologist and I find the brain scans fascinating too; but, the human psyche has to tell you truthfully what he or she was thinking about during the brain scan BEFORE you can KNOW what the brain scans really mean. KNOWLEDGE always comes through Psyche or Thought.

The brain scans can't tell us one single thing about what's happening in the Non-Local Realm, or the Psyche Realm, or the Spirit World! In contrast, the human psyche can! The human psyche can tell you what the brain scans mean and what the person was thinking about during each part of the brain scan. After all, the human psyche has to interface through the physical brain in order to successfully interact with the physical world; but, the human psyche can also function alone all by itself in the Non-Local Realm or the Spirit World according to the Scientific Evidence from Near-Death Experiences (NDEs) and Out-of-Body Experiences (OBEs).

While being subjected to brain scans, many people have lied to the researcher and thought about something other than what the researcher told them to think about. The brain scans couldn't detect that these people were lying or thinking about something else besides what the researcher wanted them thinking about. The brain scans can detect that the Psyche is thinking and that the Psyche is interfacing with the physical brain, but the brain scans can't detect what these People or Psyches are thinking about.

Where's the Science?

The SCIENCE is in the human psyche, because ONLY the human psyche can do Science in any realm of reality that we can imagine. The Science is in the Lived Experiences or the Scientific Observations of the human race.

Remember: Science, the REAL SCIENCE, is found in the human psyche. ONLY the human psyche can do Science. The human psyche is functional and can do science in any realm of existence, or any Reality. The brain scans can't. The brain scans only work in the physical realm and actually need the human psyche in order to get the correct interpretation of the brain scans. ONLY Psyche can do interpretation of Scientific Data in any Reality or any realm of existence.

There's the Science!

Every time I pick up a book or article about Science, Light, Quantum Mechanics, Quantum Objects, Quantum Entanglement, Quantum Telepathy, Quantum Nonlocality, Trans-Dimensionality, Action at a Distance, Forces, Fields, Magnetism, the Zero-Point Field, Conscious Observers, Non-Local Consciousness, Particle Physics, Space, Gravity, Spirit Matter, Dark Matter, Dark Energy, Living Light, Faster than Light Travel, Cosmological Constants, Intelligence, Thoughts, Dreams, Psychology, Philosophy, Psyche, Life, Consciousness, Mind-Over-Matter, the Word of Command, the Placebo Effect, Time, Space-Time, Universal Constants, Physical Laws, Supernatural Healings, the Occult, Causality, Cosmic Fine-Tuning, Genomes, Origins of Physical Life Forms, Spirituality, Heaven, Hell, Revelations, Visions, NDEs, OBEs, SDEs, Spiritual Experiences, the Supernatural, or God, I quickly realize that NONE of these things are possible if Materialism is true. Materialism or Naturalism precludes and excludes the existence of these kinds of things.

Materialism and Naturalism are exclusionary philosophies. If Materialism were really true, it would prevent these kinds of Non-Physical things from happening and make them impossible. You and I would not exist if Materialism or Naturalism were 100% true. In fact, physical matter wouldn't exist if Materialism were 100% true, because ONLY Psyche or Non-Local Consciousness can convert spirit matter into physical matter according to Quantum Mechanics or Spiritual Mechanics. Since these Non-Physical things have been experienced and thereby proven to exist, we KNOW that Materialism and Naturalism are false. The existence of Light falsifies Materialism and Naturalism; and, spirit matter or non-local matter is comprised of light.

There are dozens of different Non-Physical Phenomena and Spiritually Based Non-Local Science Disciplines, which demonstrate clearly and conclusively that Materialism and Naturalism are FALSE and that seem to imply that the Materialists and Naturalists as a collective whole are trying to deceive us, as they have been deceived.

Truth can be KNOWN directly by living it and experiencing it for ourselves. Truth or Knowledge falsifies lies such as Materialism, Naturalism, and Darwinism. Materialism, Naturalism, and Evolution would not exist without some Intelligent Psyche to have created these philosophical concepts in the first place; and, Psyche IS the opposite of Materialism and Naturalism.

There's the Science; and, the WHOLE of it tells us that Materialism and Naturalism are FALSE!

Once again from a different slightly perspective, Materialism, Naturalism, the Theory of Evolution, Creation by Rocks, and their derivatives are based upon an illogical grand-father paradox or Catch-22.

Examine the Science or the Logic that comes from our Lived Experiences or the Scientific Observations of the human race!

Where's the Science?

Our Science, the REAL SCIENCE, is found in our collective OBSERVATIONS as a race. The REAL SCIENCE is found in the human psyche.

WE HAVE OBSERVED that, according to Quantum Mechanics and the Quantum Law of Complementarity, some kind of Psyche, Non-Local Consciousness, Conscious Observer, or Word of Command is needed to convert spirit matter into physical matter. Without God's Psyche or some kind of Intelligent Psyche, there would be NO physical matter – it would still all be dark matter or spirit matter. Without physical matter, there would be NO DNA, NO Planets, NO Genomes, and NO Physical Life Forms. It's elementary, my dear friend.

As Scientists, WE HAVE OBSERVED a thousand different examples of Cosmic Fine-Tuning, or what we typically call "Physical Constants". Each one of these observations is a miniature scientific proof of God's existence.

Why?

It's because ONLY Psyche or Intelligence can do field-testing, organization, engineering, manufacturing, science, and fine-tuning in any realm of existence or any Reality. God's Psyche must exist, or this universe would be nothing but chaos. Psyche or Intelligence is our primary source of order and organization – the whole becomes greater than the sum of the parts under the direction of Psyche or Intelligence. Furthermore, the Biblical God Christ Jehovah has told dozens of different people in person that HE created the heavens, this earth, and all of the life forms on this earth. HIS is the most parsimonious and logical explanation that has been given to us so far, because ONLY Intelligent Psyches can do design, creation, organization, engineering, science, and fine-tuning. Raw Physical matter has NEVER been caught in the act of doing Science, Engineering, nor Fine-Tuning. Follow the evidence. That's what I finally chose to do.

WE OBSERVE that Evolution (genetic change from one generation to the next), Natural Selection, and Random Mutations did not exist and could not exist until AFTER God or some Intelligent Psyche designed, programmed, created, manufactured, and deployed the first genomes and the first life forms to begin with. In other words, some person or some Psyche had to DO the Science and create and manufacture the genomes and the life forms BEFORE evolution, natural selection, and random mutations could be brought into existence and come into play. This means that evolution, natural selection, and random mutations could NOT have designed and created the genomes and life forms on this planet, because these physical reactions or physical processes did NOT exist until after some Intelligent Psyche designed and created the genomes and the life forms in the first place. That's what our Common Sense and Scientific Observations are telling us.

Genomes ARE the pinnacle of Computer Science! Our physical bodies and the living cell ARE God's hardware. ONLY Intelligence or Psyche can design and manufacture hardware. Our genome is God's program, or God's software, or God's

Signature. ONLY Psyche or Intelligence can do programming and write software such as genomes. ONLY Psyche can enliven or animate physical matter. That's what our Lived Experiences or Scientific Observations are telling us has to be true. REAL SCIENCE matches with Reality or the Lived Experiences of the human race.

There's the Science!

Do you believe it?

I do.

But, it wasn't always that way for me. There used to be a time when I was a Materialist, Nihilist, and Atheist, and refused to think about these kinds of things. I KNOW how it goes, because I have been there and done that too. It's part of my Lived Experience.

It can be fascinating to study the psychology of the Materialists, Naturalists, Nihilists, and Atheists. We Materialists and Atheists are in denial. I KNOW, because that's the way it was for me when I was a Materialist and Atheist. Materialism, Naturalism, and Atheism of any kind are based upon a refusal to look at evidence – a refusal to look at, study, and accept the Lived Experiences or the Scientific Observations of the human race as a whole. Materialism, Darwinism, Atheism, and Naturalism are based upon wishful thinking and blind faith; and, that's NOT science! That's philosophy and religion.

The TRUTH is KNOWN by living it and experiencing it for yourself, or by choosing to trust someone who has.

Once you start accepting NDEs, OBEs, SDEs, and other Spiritual Experiences into evidence and start treating these Psyche Experiences or Lived Experiences as Scientific Evidence, then that's the end of Materialism and Naturalism. Atheism, Materialism, Naturalism, Darwinism, and Nihilism cannot survive in the Light of Truth and Knowledge; and, KNOWLEDGE or SCIENCE comes directly from the Lived Experiences or the Psyche Experiences of the human race. Lived Experience IS Scientific Evidence; and, there's where you will find the Science when you finally decide to go looking for it.

Summarizing the Different Types of Syntropy

There's physical matter, and then there's everything else.

Entropy seems to be restricted exclusively to physical matter. Entropy is a function of physical matter. Everything else seems to be based upon Syntropy.

What a fascinating observation to make!

ALL of those invisible and intangible forces, fields, and quantum waves are based upon Syntropy. They NEVER burn out, and they NEVER wear out, because they are eternal and everlasting, without a beginning of days or an end of years.

There's physical matter or entropy; and then, there's everything else which is based upon Syntropy. There's classical physics, and then there's Quantum Mechanics or Transdimensional Physics. The one is based upon entropy, and the other is based upon Syntropy. This just might be the most interesting and useful scientific observation of all-time. Its explanatory power is through the roof, is it not? Syntropy has to exist, or entropy and physical matter would not exist.

Remember, Syntropy has to exist, or we would not exist. Our very existence here in this physical universe in these physical bodies tells us that someone somewhere KNOWS how to do Syntropy or Organization where physical universes, genomes, proteins, eyes, brains, and physical bodies are concerned.

Organization is Syntropy. Psyche is Syntropy. Quantum Mechanics is Syntropy. The Priesthood Power of God is Syntropy. Spirit Matter is Syntropy. God is Syntropy. Quantum Mechanics or Syntropy is the Priesthood Power of God. Syntropy is a new and better way of doing science, explaining science, and understanding science. Syntropy is eternal and everlasting.

Syntropy means "without beginning of days or an end of years". Syntropy is Eternal Life. Syntropy or Eternal Existence is the opposite of creation ex nihilo. Creation ex nihilo or the catholic (universal) version of creation is Atheism – creation by nothing from nothing. I'm not a creationist. I'm definitely not that type of creationist. Creation ex nihilo is impossible. Not even God could do creation ex nihilo, because it's impossible. Creation ex nihilo is Atheism and magic. Creation ex nihilo is also unscientific, because it's impossible. Creation ex nihilo is the exact opposite of Syntropy. The one is magic, and the other is organization and science.

God is a scientist, and not a magician. God organized this physical universe from pre-existing Dark Matter or pre-existing Spirit Matter. Remember, it's the physical matter that is the anomaly – everything else is based upon Syntropy. The physical matter is visible to us with our physical eyes; and, everything else is not. We detect all the dozens of different invisible forces, fields, and laws by the effect that they have on physical matter. Physical matter or entropy is the anomaly. Everything else, or Syntropy, or Quantum Mechanics is the norm.

Syntropy concerns the things that are without a beginning of days or an end of years. Syntropy is the opposite of Materialism, Naturalism, Atheism, Entropy, and Classical Physics. Syntropy is the original prime construct from which this physical universe and entropy were made.

Our physical universe had a beginning. It was ordered and organized by **someone** who has NO beginning and will have no end – Syntropy, Psyche, or God. Our physical universe or physical matter was built, or made, or organized from **something** that had NO beginning and will have no end – Spirit Matter, Dark Matter, or Quantum Objects. Creation ex nihilo is falsified by Syntropy and all of that pre-existing pre-physical Spirit Matter or Dark Matter.

Observation or Premise: Anything that was obviously made obviously had a Maker or Creator who made it.

Observation or Premise: A genome was obviously made.

Logical Conclusion: Therefore, a genome obviously has a Maker or a Creator who designed it, programmed it, engineered it, field-tested it, fine-tuned it, made it, manufactured it, and deployed it.

To make or create something is to infuse it with Order or Syntropy. Organization, Engineering, Manufacturing, and Creation are different types of Syntropy. Syntropy is the opposite of spontaneous generation or magic. There is NO such thing as spontaneous generation, chemical evolution, creation ex nihilo, or macro-evolution. These things are prevented from happening by entropy or the second law of thermodynamics. These things are prevented from happening by physical matter. It requires some kind of Syntropy or Psyche in order to design and create and make things from scratch.

You know what this means, don't you?

It means that evolution can't happen. Evolution is impossible. The different types of evolution are prevented from happening by entropy and physical matter. Entropy and physical matter actually falsify the Theory of Evolution. Random mutations are entropy. Random chance or random diffusion is based upon entropy. Natural selection leads us towards entropy, death, and extinction. The chemical evolution of genes and proteins from atoms is prevented from happening by entropy. Spontaneous generation, abiogenesis, chemical evolution, and macro-evolution are prevented from happening by entropy or physical matter. Physical matter imposes a structure or an order that prevents evolution of any type from happening. Physical matter is the basis of the ultimate consensus reality.

It would be unpredictable chaos or evolution if it weren't for the order, structure, and limitations being imposed upon us by physical matter or by consensus. Natural evolution would not be a good thing. It would be chaos. That's why God, physical matter, genes, and entropy prevent any type of macro-evolution from happening. It's bad enough as it is that our genes are allowed to mutate. Think of all the problems which that causes! Now multiply that by a billion to imagine what it would be like if all the other types of evolution were allowed to happen.

Remember, classical physics or entropy prevents evolution. It was designed to do so. This just might be the most significant scientific discovery of all time.

It actually requires some type of Syntropy, Psyche, or Intelligence in order to be able to design, create, program, organize, engineer, field-test, fine-tune, manufacture, and bring physical things into existence. Some type of Syntropy is needed in order to bring life into existence, especially physical life. Psyche, Intelligence, or Quantum Non-Local Consciousness is LIFE, which means that it is Syntropy.

Remember, Syntropy has to exist, or we wouldn't exist. It's time that we, as scientists, finally start to use Syntropy, Psyche, and Quantum Mechanics to explain how things really work in every realm of existence, including both this physical realm and the non-local quantum realm. It's time for us to take Quantum Mechanics or Syntropy seriously, instead of ridiculing it, mocking it, and dismissing it as pseudo-science. It's time for us to upgrade our science to Science 2.0 and start allowing ALL of the evidence into evidence.

God's Psyche or God's Intelligence is Syntropy. It had no beginning and it will have no end. Your Psyche, or Intelligence, or Quantum Non-Local Consciousness is Syntropy. It had no beginning and it will have no end.

In contrast, entropy is death – physical death. It has an end. Physical matter or entropy is the anomaly. WE KNOW it is so, because everything else that we have encountered is based upon Syntropy – it NEVER ages, and it never wears out. Remember, Syntropy never ages and it never wears out. It is without a beginning of days or an end of years. Syntropy is eternal and everlasting. Compared with that, physical matter or entropy is indeed the anomaly or the odd ball.

Physical bodies and physical universes begin and end thanks to both syntropy (the beginning of a physical universe and a physical body) and entropy (the ending of a physical universe and physical body); but, Syntropy or Quantum Mechanics or Psyche has no beginning and will have no end, because it is eternal and everlasting. Syntropy is light, life, love, psyche, intelligence, power, order, organization, and God. The Atonement of Christ is Syntropy. It has to be, because its purpose is to help us overcome entropy, corruption, sin, and death.

The obvious fact – that some type of Syntropy MUST exist or we would not exist – makes logical sense to me. The NEED for Syntropy is so glaringly obvious that Syntropy MUST exist; and therefore, its existence MUST be true. WE KNOW that some type of Syntropy exists, because entropy exists. We know that Syntropy exists because physical matter exists. There's physical matter or entropy; and then, there's Syntropy. We have observed that everything else that we have encountered besides physical matter is based upon Syntropy – it never ages nor gets old. It's eternal and everlasting.

Like I said, this just might be the most significant and useful scientific discovery of all time. Its explanatory power is off the charts.

If physical matter or entropy were the ONLY thing that exists, as the Materialists and Naturalists claim, then it would be physically impossible for our genes, proteins, eyes, brains, and physical bodies to exist. Entropy prevents physical matter from organizing spontaneously into functional genes, proteins, eyes, brains, and physical bodies. Evolution is entropy, and that means that the different types of evolution or the different types of entropy prevent the Theory of Evolution from becoming true.

If entropy prevailed and there were no Order or no Syntropy, then the stars, planets, and galaxies would not form and would not exist. The Syntropy, or the available energy, or the zero-point field, the Higgs field, or the love has to exist, or there would be no stars and no suns. There would be no physical life anywhere without some type of Syntropy or Order in this physical universe. This is rationally obvious to anyone like me who spends even the smallest amount of time thinking about it.

Thanks to Syntropy or Quantum Mechanics, we KNOW that Materialism, Naturalism, Darwinism, Nihilism, and Atheism are false and have been falsified. Thanks to Syntropy, entropy is not the end. You will go on long after your physical brain is dead and gone. Thanks to Syntropy, your consciousness or psyche is eternal and everlasting. You have always existed, and you will always exist. Psyche is Syntropy. Relationships are Syntropy. Love is Syntropy. God is Syntropy. The best part of you is Syntropy, and it's capable of counteracting the effects of entropy and surviving the death of your physical brain.

Mark My Words

The Ultimate LAW of Thermodynamics

Read this carefully, because it's the KEY to everything else in Science.

The first law of thermodynamics is a version of the law of conservation of energy, adapted for thermodynamic systems. The law of conservation of energy states that the total energy of an isolated system is constant; energy can be transformed from one FORM to another, but it can be neither created nor destroyed.

https://en.wikipedia.org/wiki/First_law_of_thermodynamics

Once I fully understood and accepted this simple axiom or Law, instantly everything became obvious and clear.

Energy or Syntropy can be neither created nor destroyed. However, Energy or Syntropy can be transformed from one FORM to another FORM. The FORM is never conserved. Physical matter and entropy are different FORMS of Energy, and the FORM is never conserved. Only the underlying Energy or Syntropy is being conserved. This is the Ultimate Law of Thermodynamics. Physical matter and entropy are never conserved. The FORM is never conserved.

Energy can be transformed.

Transformed by whom?

Who triggers this transformation? Who decides what FORM Energy will take? Someone Psyche has to make this decision, or nothing will happen, and physical matter will never be made.

Physical matter and entropy were made.

Made by whom?

Made from what?

Scientific Observation: Anything that was obviously made obviously had a Maker who made it.

Scientific Observation: Physical matter, physical laws, physical constants, physical restrictions, space-time, locality, time, and entropy were obviously made.

Scientific Conclusion: Therefore, it is logical and rational to conclude that physical matter, physical laws, physical constants, physical restrictions, space-time, locality, time, and entropy obviously have a Maker who made them.

The fatal flaw of the Materialists, Naturalists, Darwinists, Nihilists, and Atheists is that they teach and truly believe that only physical matter and entropy exist, which means that these people erroneously teach that physical matter and entropy are conserved. They are wrong. These people deliberately misinterpret the First Law of Thermodynamics, which states that Energy or Syntropy is conserved; and, these people get the rest of us to misunderstand and misinterpret the First Law of Thermodynamics by successfully convincing us that only physical matter and entropy exist and that consequently physical matter and entropy are being conserved.

One of my greatest scientific discoveries came when I first realized that physical matter and entropy are NOT conserved. The amount of physical matter, entropy, and space-time changes over time as God sees fit. Space, time, locality, entropy, spirit matter, and physical matter are different FORMS of energy. The FORM is never conserved. The FORM is constantly changing. It's the underlying Energy or Syntropy that's being conserved according to the First Law of Thermodynamics and $E = mc^2$. This scientific discovery is the answer to life, the universe, and everything. It explains everything.

Remember, physical matter, entropy, locality, and space-time are NOT conserved. Their amount is constantly changing. This is the Ultimate Law of Thermodynamics.

Physical matter, entropy, locality, and space-time are MADE, which means that they can be unmade or disassembled. With God's help, we humans make brand new physical matter in our particle accelerators. We humans also destroy physical matter or unmake physical matter in our atomic bombs. The physical matter and the associated entropy are NOT conserved. Their quantity is constantly changing. Remember, Energy is constantly changing FORM while at the same time being conserved. That's the way that science really works contrary to what the Materialists, Naturalists, Darwinists, and Atheists claim to be true. The false is falsified by the truth; and, the truth is repeatedly experienced and observed.

Physical matter, physical laws, physical constants, physical restrictions, space-time, space, locality, time, and entropy were MADE which means that they have some kind of Maker. It also means that they can be disassembled or destroyed. They were made from Energy by God, which means that God can cause them to end or unmake them. They are different FORMS of Energy, and the FORM is never conserved. Only the underlying Energy or Syntropy is conserved. Physical matter and entropy are different FORMS of Energy, which means that physical matter and entropy are never conserved. The FORM is never conserved. This is the Ultimate Law of Thermodynamics.

My purpose in life is to identify everything that is false, eliminate it from science, and replace it with the truth. The materialistic and naturalistic idea that physical matter and entropy are conserved is a false idea and a falsified idea. I have eliminated it from my science and replaced it with the truth – namely, that it's the underlying Energy or Syntropy that's actually being conserved and not the physical matter or the entropy. All I want is the truth because everything else will mess you up in the end. It's the underlying Energy or Syntropy that ends up being eternal and everlasting, not the physical matter or the entropy. Physical matter and entropy are different FORMS of Energy, and the FORM is never conserved. Energy is Syntropy, and Energy is conserved.

Observation and experience have proven to us that the Ultimate Law of Thermodynamics is true. Physical matter, entropy, locality, time, space, and space-time are MADE, which means that they can also cease to exist anytime that God decides to put a stop to them.

According to the Theory of Relativity, when a photon or tachyon travels at the speed-of-light or faster, TIME STOPS. Entropy is a function of time. Entropy measures the passage of time. Entropy is an aging process. Locality is associated with space. Entropy is associated with time. Entropy is the passage of time. According to the Theory of Relativity, when a photon or tachyon travels at the speed-of-light or faster, ENTROPY STOPS, or ENTROPY CEASES TO EXIST. From our entropic perspective, it took that photon 13.4 billion years to reach us. From the photon's syntropic perspective at the speed-of-light, it experienced NO passage of time; and, it arrived the very moment it launched. In other words, the photon and the tachyon didn't age. They didn't experience the passage of time. Entropy ceased to exist while they were traveling at the speed-of-light or faster. They literally quantum tunneled to their destination from their perspective. Remember, entropy is variable; and, entropy or the passage of time ceases to exist at the speed-of-light or faster. In other words, entropy or the passage of time is not conserved at the speed-of-light or faster. Instead, entropy ceases to exist; and, everything becomes syntropic instead at velocities and frequencies faster than the speed-of-light.

Once again, the Materialists, Naturalists, Darwinists, Nihilists, and Atheists erroneously teach that nothing can travel faster than the speed-of-light because these people erroneously teach that only physical matter and entropy exist. These people are right in that physical matter or entropic matter cannot travel faster than the speed-of-light; however, these people are wrong in that spirit matter, quantum matter, syntropic matter, spiritual matter, tachyons, and psyche can and do travel faster than the speed-of-light; and, they can also quantum tunnel at will. At velocities equal to or greater than the speed-of-light, TIME STOPS; and, entropy or the passage of time ceases to exist. There is no entropy in the Syntropic Realm. Objects existing at frequencies faster than the speed-of-light experience no entropy because they experience NO passage of time. Entropy ceases to exist at velocities and frequencies faster than or equal to the speed-of-light, according to the Theory of Relativity.

God deliberately created physical matter and entropy (the passage of time) for us in order to slow things down for us so that we can live them, observe them, experience them, learn from them, experiment with them, and remember having done so. At velocities equal to or faster than the speed-of-light, there is NO entropy because there is NO passage of time. Time stops. Time or entropy ceases to exist. Time, entropy, and the passage of time were made by God, which means that they can also cease to exist or come to a stop. I have experienced and observed that TIME STOPS and entropy (the passage of time) temporarily ceases to exist, just before you are quantum tunneled to a different location on this earth. God has complete control over Quantum Mechanics including quantum tunneling. From the photon's perspective, it quantum tunnels or teleports to its destination because it experiences no passage of time or no entropy during its voyage. From my perspective, time stopped, and I experienced no passage of time when God quantum tunneled me and the car I was driving to safety. God can quantum tunnel physical matter anywhere in this universe instantaneously simply by stopping the passage of time or temporarily removing the entropy from that physical matter.

Remember, there's nothing sacred about physical matter and entropy. It's the Energy or the Syntropy that's being conserved. Energy can be organized into many different FORMS, including physical matter, spirit matter, space-time, physical laws, physical constants, space, time, locality, and entropy. Remember, it's the Energy or the Syntropy that's actually being conserved, not the physical matter or the entropy. This is the Ultimate Law of Thermodynamics. It explains everything that we humans have experienced and observed. It corrects one of the fatal flaws in traditional science – namely, the erroneous belief that physical matter and entropy are being conserved. The Ultimate Law of Thermodynamics states that physical matter, entropy, locality, and space-time are NOT conserved. It's the underlying Energy or Syntropy that's being conserved, and NOT the physical matter or the entropy. Physical matter and entropy are just different FORMS of energy. Energy can be made to take on many different FORMS; and, the FORM is never conserved. It's the Energy or the Syntropy that's actually being conserved and NOT the FORM. The Ultimate Law of Thermodynamics is the answer to life, the universe, and everything. It explains everything.

This Ultimate Law of Thermodynamics is one of the greatest scientific discoveries of all time; and, it's only made possible by eliminating Materialism, Naturalism, Darwinism, Nihilism, Behaviorism, Determinism, Physical Reductionism, and Atheism from science. If you successfully eliminate everything that is false and everything that has been falsified, then only the truth will remain. This is Logic 101. You start by eliminating Materialism, Naturalism, Darwinism, Nihilism, and Atheism from science; and then you see where that gets you. It's interesting to study and observe what remains after you have successfully removed Materialism, Naturalism, Darwinism, Atheism, and their derivatives from science. The experienced and the observed remain. The truth remains.

Remember, the Biblical God Jesus Christ has been experienced and observed both in the flesh and during our near-death experiences AFTER He rose from the dead. Science is observation and experience.

The Materialists, Naturalists, Darwinists, Nihilists, and Atheists define "science" as Materialism, Naturalism, Darwinism, Nihilism, and Atheism. Based upon all these different scientific discoveries that I have made during the past couple of years, I have upgraded my science to Science 2.0; and, I have redefined Science as "observation and experience". Observation and experience is a much better way to do science because the "non-existence of things" will never be experienced nor observed. It's impossible to experience and observe the "non-existence of something", including the non-existence of God. As scientists, we should go with what has been experienced and observed, not the wishful thinking of the Naturalists, Darwinists, and Atheists.

Physical matter and entropy were obviously made from Energy. Physical matter had physical limitations, entropy, locality, sub-light velocities, mass, time, space, and physical restrictions programmed into it. Physical matter is Organized Energy, and it was organized by God. Since physical matter and entropy are never conserved, according to the Ultimate Law of Thermodynamics, God must of necessity exist in order to have organized, created, or made the first particles of physical matter from Raw Energy.

Scientific Observation: Anything that was obviously organized obviously had an Organizer who organized it.

Scientific Observation: Physical matter, physical laws, physical constants, space-time, physical restrictions, space, time, locality, and entropy were obviously organized and made from Energy. These things are different forms of Energy. Physical matter is Organized Energy.

Scientific Conclusion: Therefore, it is logical and rational to conclude that physical matter, physical laws, physical constants, space-time, physical restrictions, space, time, locality, and entropy obviously have an Organizer who organized them and made them from Energy.

This truth has been experienced and observed. The false is falsified by the truth; and, the truth is repeatedly experienced and observed.

The Ultimate Law of Thermodynamics states that the FORM is never conserved, which means that the different forms of Energy such as physical matter and entropy are never conserved. Their quantity or amount is constantly changing. It's the underlying Energy or Syntropy that's always being conserved. It's the Energy, Syntropy, or Life Force that ends up being eternal and everlasting, not entropy or death. That is what has actually been experienced and observed.

Materialism, Naturalism, Darwinism, Nihilism, Behaviorism, Determinism, Physical Reductionism, and Atheism were designed to prevent us from discovering the Ultimate Law of Thermodynamics. Entropy is death; and, the Scientific Naturalists and Atheists erroneously teach that entropy or death is conserved. These people are wrong. Life after death has been experienced and observed.

The Psyche, Syntropy, Energy, or Life Force is always conserved; but, the entropic matter or physical matter is never conserved. Physical matter and entropy come and go as God sees fit. Physical matter and entropy are made, which means that they can be unmade or disassembled. In contrast, Psyche, Syntropy, Energy, or Life Force can be neither created nor destroyed. Psyche, Syntropy, Energy, and Life Force are eternal and everlasting, without a beginning of days or an end of years, because they are constantly being conserved. God's Psyche or God's Intelligence is being conserved. Your psyche or your intelligence is being conserved. In contrast, physical matter and entropy are never conserved. This is the Ultimate Law of Thermodynamics.

The Ultimate Law of Thermodynamics is the answer to life, the universe, and everything. It has been experienced and observed.

Mark My Words

The Quantum LAW of Thermodynamics

Psyche is the innate intelligence within all the different forms of energy which gives that energy the inherent ability to understand, follow, and obey God's Laws and God's Commands. Energy is intelligent, psychic, conscious, sentient, perceptive, alive, aware, and conserved. This is what has been experienced and observed.

The Quantum Law of Thermodynamics states that Heat Death is impossible at the psyche level or the quantum level because Psyche, Life Force, or Energy is always conserved in the Psyche Realm or the Quantum Realm. The Quantum Law of Thermodynamics states that there is NO Thermodynamics and NO Heat Flow at the quantum level or the psyche level. Thermal equilibrium, entropy, and heat death DO NOT EXIST and do not apply at the quantum level or the psyche level. This means that entropy or death is NOT conserved. This is what has been experienced and observed by Out-of-Body Explorers and those who have had a Near-Death Experiences.

The Quantum Law of Thermodynamics FALSIFIES Materialism, Naturalism, Darwinism, Nihilism, Behaviorism, Determinism, Physical Reductionism, Classical Physics, the Second Law of Thermodynamics, and Atheism which claim that entropic physical matter is the ONLY thing that exists, and that entropy or death is conserved. The Quantum Law of Thermodynamics ultimately falsifies or trumps the Second Law of Thermodynamics. At the quantum level, Maxwell's Demon wins the battle after all and entropy is defeated. The Quantum Law of Thermodynamics is based upon the First Law of Thermodynamics or the idea that Psyche or Energy is always conserved. The Conservation of Psyche or the Conservation of Energy holds true on both sides of the veil – both in the Quantum Realm and in the Physical Realm.

When it comes to the Quantum Law of Thermodynamics, there is nothing to prove because it has already been experienced and observed. Nevertheless, ALL of the zeroes and infinities that we encounter in math is proof that the Quantum Law of Thermodynamics or Syntropy is true. Think about it. Syntropy is ZERO entropy. Syntropy is also Eternal Life or Infinite Life. Psyche or the Life Force is syntropic – it is eternal and everlasting. It cannot be made, and it cannot be destroyed. Psyche or Energy is always conserved on both sides of the veil – both on the spiritual side and the physical side. Syntropy means the Conservation of Psyche or the Conservation of Energy. This is what has been experienced and observed.

Remember, the Quantum Law of Thermodynamics states that Heat Death is impossible in the Quantum Realm or the Psyche Realm. Within the Quantum Realm or the Psyche Realm, entropy or the second law of thermodynamics does not exist; and, Syntropy, Psyche, or Eternal Life reigns supreme. This is what has been experienced and observed. Syntropy or the Quantum Law of Thermodynamics is the answer to life, the universe, and everything. It explains everything that has ever been experienced and observed.

Mark My Words

What Do Genes Code For?

What is science trying to tell us? Science is observation and experience. What are observation and experience trying to tell us?

Let's start this adventure by asking, "What do the genes code for?" A lot can be learned by asking and answering this simple question, especially when we apply knowledge from computer science to gene expression. Let's look at the science.

> **Genes code for proteins**. Genes contain the coded formula needed by the cell to produce proteins. Proteins are the most common of the complex molecules. Proteins and genes are molecules.
>
> Gene expression is protein synthesis – the production of proteins from genes.
>
> Research in the 1970s showed that numerous **non-coding** sequences — **introns** — are also found within genes, interrupting the protein-coding regions, or exons. It is estimated that only about five percent of human **DNA** encodes protein.
>
> The genome of an organism is inscribed in DNA, or in some viruses within RNA. The portion of the genome that **codes for a protein** or an RNA is referred to as a **gene**. Those **genes** that **code for proteins** are composed of tri-nucleotide units called codons, each coding for a single amino acid.
>
> In this article, we'll take a closer look at the **genetic code**, which allows DNA and RNA sequences to be "decoded" into the amino acids of a protein.
>
> There are an estimated **19,000-20,000** human protein-coding genes. The estimate of the number of human genes has been repeatedly revised down

from initial predictions of **100,000** or more as genome sequence quality and gene finding methods have improved and could continue to drop further.

The **2.9 billion** base pairs of the haploid human genome correspond to a maximum of about **725 megabytes** of data since every base pair can be coded by 2 bits.

The haploid human genome (**23 chromosomes**) is estimated to be about 3.2 billion bases long and to contain 20,000–25,000 distinct protein-coding genes. That means there is roughly **800 megabytes** of data storage capacity within the human genome.

Transcription of messenger RNA and translation of messenger RNA into proteins both occur on the time scale of **1 minute** for a protein of typical length. However, longer transcripts and bigger proteins take proportionally longer to make. The largest protein in the human body is titin. It would take approximately an hour to translate its ~30,000 amino acids.

Transfer RNA **activation** refers to the attachment of an amino acid to its Transfer RNA codon.

In the genetic code, codons made of three bases specify an amino acid. With three bases, there are **64** possible permutations. With three codons corresponding to STOP codons, this leaves **61** combinations that can code for an amino acid.

In a mammalian cell, there can be as many as **10 million** ribosomes.

There are 37.2 trillion cells in your body.

In summary, for a typical human of **70** kg, there are almost **7*10²⁷** atoms (that's a **7** followed by **27** zeros!) Another way of saying this is "**seven billion billion billion**." Of this, almost **2/3** is hydrogen, **1/4** is oxygen, and about **1/10** is carbon. These **three** atoms add up to **99**% of the total!

1 Petabyte = 1000000000000000B = 10^{15} bytes = 1000 terabytes.

1 Yottabyte = 1000^8 bytes = 10^{24} bytes = 1000000000000000000000000 bytes = 1000 zettabytes = 1 trillion terabytes.

A yottabyte is one septillion bytes. To save all those bytes you need a data center as big as the states of Delaware and Rhode Island combined. It doesn't seem like much, until they tell you the price tag: $100 trillion.

It would take thousands of yottabytes of usable information storage to handle the command and control for all the different atoms in your physical body, telling them where to go and what to do when they get there.

— Uncle Google

For command and control and the transfer of usable information, we require Bytes at a minimum because a Bit doesn't mean anything except ON or OFF. That's why information storage estimates are done in bytes and not bits. A byte is where physical information storage and program codes begin to be useful.

I have been a computer scientist all of my adult life for decades of my life; and, memory storage capacity has been of supreme concern for me in my work throughout the years. I understand how physical information storage works. Data storage at the physical level is highly inefficient because it is based upon atoms; and, atoms take up a lot of physical space. There's an infinitely better and more efficient way to store data and information – at the quantum level or the psyche level.

Scientific Observations: Any type of programming code requires a Programmer to code it or make it. Any type of hardware requires an Engineer to design it and make it.

Scientific Observations: Genes and a genome are obviously programming code. Genes code for proteins; and, a genome is both software and hardware. A genome is hardware made from DNA molecules.

Scientific Conclusion: Our genes and genomes obviously have a Programmer who programmed them; and, our genes and genomes obviously have an Engineer who made them.

Look at the numbers above from Uncle Google. What are they trying to tell us?

In order to control all of the atoms within a physical body individually at the physical level would require at least thousands of yottabytes of physical data storage and some way to transmit that information between the atoms and molecules. Such a thing is physically impossible. It can't be done, which means that it isn't being done – not at the physical level.

However, the psyche or intelligence who is controlling the atoms within your physical body is indeed controlling, and transmitting, and storing yottabytes of information. How's that being done? Where is that information being stored? How is this information being organized, stored, accessed, and transmitted? Who is doing all of this information storage and organization? Who knows what it all means, or who understands it all? Who knows how to access it all and use it all? It can't be our genes, because there are NO genes within our atoms!

The Materialists, Naturalists, Darwinists, Nihilists, Behaviorists, Determinists, Physical Reductionists, Evolutionists, Classical Physicists, and Atheists assure us that it's all being handled by our genes at the physical level.

Are they right, or is there something important that they are missing?

Scientific Observation: Anything that was obviously made obviously had a Maker who made it.

Scientific Observation: Proteins and their matching genes were obviously made. A genome was obviously made.

Scientific Conclusion: Therefore, it is logical and rational to conclude that proteins, their matching genes, and the genomes had a Maker who made them.

Who coded the genomes?

I'm a computer programmer and computer scientist; and, I know for a fact that elegant programming code doesn't write itself. This reality and truth has been experienced and observed.

Scientific Observation: Programming code obviously requires an Intelligent Programmer to write it, program it, and make it.

Scientific Observation: A genome is the ultimate software or programming code that has ever been made or written. In fact, a genome is both hardware and software simultaneously.

Scientific Conclusion: A genome obviously has an Intelligent Programmer who wrote it, programmed it, and made it.

Now, with this information securely in place, let's study protein synthesis or gene expression. We are particularly interested in trying to figure out how all of that information and data being used in protein synthesis is being transmitted and received by the individual atoms, molecules, enzymes, and proteins within your physical cells. The atoms and molecules are communicating with each other wirelessly. How do they do that? Well, they aren't doing it at the physical level because action at a distance is physically impossible.

Protein Synthesis

These online articles provide some of the details for protein synthesis or gene expression:

> https://gizmodo.com/5557676/how-much-money-would-a-yottabyte-hard-drive-cost
>
> https://en.wikipedia.org/wiki/Protein_biosynthesis
>
> https://en.wikipedia.org/wiki/Gene_expression
>
> https://en.wikipedia.org/wiki/Amino_acid_activation
>
> https://en.wikipedia.org/wiki/Transcription_(biology)
>
> https://en.wikipedia.org/wiki/Messenger_RNA
>
> https://en.wikipedia.org/wiki/Transfer_RNA

Discussing Protein Synthesis or Gene Expression

Our genes code for proteins.

How do our genes get translated into proteins?

It's a complex process that's been called Gene Expression or Protein Synthesis.

Transcription enzymes are molecules.

Transcription factors or transcription enzymes read genes and produce messenger RNA molecules from those genes. How do the transcription enzymes know that the cell needs a new protein? How do the transcription enzymes know which type of protein is needed and where it is needed? How is that information transmitted and understood? How do the transcription enzymes know which one of the 20,000 genes needs to be read in order to produce that newly requisitioned protein, and where that specific gene is located within the genome? How do the transcription enzymes go directly to the specific gene that needs to be read? Where does that wireless command-and-control information come from?   How did the transcription enzymes learn how to read, translate, and transcribe the DNA and the genes? What makes the transcription enzymes move? Targeting and docking require some kind of command and control, do they not? What are the transcription molecules using to see, read, and perceive the genome and their environment? They go to the specific gene that they need to read, know where it is, know how to get there, and recognize it when they see it. How are these transcription molecules doing that? It's as if they have eyes and a brain; yet, they are only molecules. These individual molecules are smarter and more intelligent than we are. How's that possible?

If I told you to make a specific protein, how long would it take for you to find the correct gene and then transcribe that gene into an mRNA molecule? The mapping of our human genome took decades. The transcription enzymes typically do transcription in a matter of minutes. They KNOW precisely which protein has been requisitioned by the cell, and they KNOW precisely where its associated gene is located within the genome. How do they KNOW? These are molecules. How would you transmit information and knowledge to atoms and molecules? It's NOT going to be done at the physical level. It's not going to be done by our genes! It has to be done at the quantum level or the psyche level. When it comes to Science, this is where the tires really hit the pavement. It does indeed generate a lot of friction among the science community at-large. The Materialists and Naturalists assure us that the quantum level or the psyche level does not exist; therefore, these people have no idea how all this information is being transmitted between atoms and molecules, or by whom.

Why are the introns there? Who knows what they mean and how to use them? Do they serve any purpose? God knows! I don't. Random chance or entropy clearly couldn't have designed and produced a genome, so who did? It couldn't be evolution because evolution is entropy or death. There's no intelligence or life in the various different types of evolution.

The introns are skipped during the encoding or transcription of messenger RNA, which is used by ribosomes to produce proteins. How does this invisible psyche or force know the difference between the introns and the exons? Who or what is telling the transcription enzymes to skip the introns? How does each transcription enzyme KNOW which gene to read and where that gene is located in the genome? It goes straight to what it needs and reads only what it needs. How does it do that? How does it know to skip the introns and only read the exons? How is that information being transmitted, received and understood? Who is handling the logistics for all of this within a living cell?

How is all this 3D addressing for all the different genes being stored? It's not being stored within the genes – not at the physical level at least. By definition, in principle, the genes cannot communicate with each other and other molecules at the physical level. This communication has to be taking place at the quantum level or the psyche level because it isn't taking place at the physical level. There are NO wires in our brains! There are NO wires connecting the individual atoms together into a communication network. There's NO Wi-Fi for genes and molecules and atoms at the physical level because action at a distance is physically impossible according to classical physics; therefore, if wireless communication between atoms exists, it exists at the quantum level or the sub-atomic level.

Telepathy is Wi-Fi___33 at the quantum level or the psyche level. Thoughts and memories are quantum waves – they survive the death of our physical brain.

How does a cell decide that it needs a new protein? Who issues that instruction or command to produce a new protein? How is that information or command set transmitted from molecule to molecule? How do the transcription factors or transcription enzymes know in advance, at a distance, telepathically which gene to read and where that gene is located in the genome? How does Command and Control KNOW where each gene is located and what each gene does? How would you send messages and commands to atoms and molecules knowing that you can't do so at the physical level because it's physically impossible at the physical level?

How do these molecules communicate with each other and coordinate with each other? How do the transcription enzymes do targeting and docking and manipulation? From whom are these molecules getting their instructions and location addressing? The transcription enzymes KNOW precisely what part of the chromosome to unzip. How do they know? How did they get this information? How is that information transmitted to the chromosome making it unzip? Where is all this information being stored, since our 750-megabyte genome clearly doesn't have enough memory storage capacity to store and transmit petabytes of data and information?

The Materialists, Naturalists, and Darwinists assure us that it's all being controlled by our genes; but, command and control can't be taking place within our genes because there are no genes within atoms! How are genes able to control atoms at a distance at the physical level? They can't! It's physically impossible! There's NO physical mechanism for doing so!

In a eukaryotic cell, messenger RNA is transcribed in the nucleus; and then, it is transported or moved through the nuclear membrane to the cytoplasm and then moved to a specific ribosome where it is translated into a protein. Who tells the mRNA molecule how to get out of the nucleus and into the cytoplasm where the ribosomes reside? Who is telling each mRNA molecule which one of the 10 million ribosomes to move towards? What is making the mRNA molecule move? Most of the ribosomes are conveniently located near where the constructed protein is needed and will be used. How is all of this invisible non-physical information being transmitted and received by the different atoms and molecules?

Furthermore, someone psyche or someone intelligent has to requisition and make the necessary amino acids and the tRNA codons BEFORE they can be combined or activated and turned into transfer RNA. The transfer RNA is made in the nucleus of the cell from DNA. The protein itself is constructed from dozens to thousands of tRNA molecules outside of the nucleus at some distant ribosome. How do the activation enzymes KNOW in advance which amino acid and codon combination will be needed in the future at some distant ribosome? Who tells the activation enzymes which type of tRNA molecule to make? There are theoretically 64 different types of tRNA molecules that can be made. Who tells the newly made tRNA molecules where to go? Each one has to go to a specific ribosome outside the nucleus in the cytoplasm. There are over 10 million different ribosomes in the cell. How are the tRNA molecules told which ribosome to move towards? How is this information transmitted to the tRNA molecule? Who or what makes the tRNA molecules move? Who is handling the logistics or the Command and Control for the activation and transportation of tRNA molecules?

The transfer RNA and messenger RNA are made within the nucleus; and, the ribosomes reside outside the nucleus. Who is doing all of this telepathic communication and action at a distance making sure that the tRNA molecules, mRNA molecules, and ribosomes coordinate with each other and work together like a finely tuned machine? How do each of the manufactured transfer RNA molecules KNOW in advance which ribosome to move towards? A tRNA molecule moves towards ONE of 10 million different ribosomes. How does it know which ribosome to move towards? Who is telling each tRNA molecule where to go after it is made? Who or what makes them move?

In eukaryotic cells, tRNAs are transcribed by RNA polymerase III as pre-tRNAs in the nucleus. The ribosomes reside outside of the nucleus. Who is handling the logistics for all of this? It can't be entropy or random chance! By definition, in principle, entropy or random chance can't do logistics or Command and Control. It can't be our genes because the genes have NO physical means for commanding and controlling tRNA and mRNA molecules – not at the physical level at least. This information exchange has to be taking place wirelessly or telepathically at the quantum level or the psyche level because it is physically impossible at the physical level.

Who is requisitioning each and every one of the billions of different tRNA molecules that are being made within a cell in real time as they are needed? Who is getting the correct amino acids there on time for the activation process? Transfer RNA is made in the nucleus of a cell and has to make its way outside the nucleus to ONE of 10 million different ribosomes for protein synthesis. Who is handling the logistics for all of this and getting each tRNA molecule to the correct place at the correct time? Who is requisitioning each tRNA molecule and amino acid in the first place and getting all the atoms and amino acids and codons into place for that physical tRNA activation process? Who designed and created this whole process? There's NO such thing as spontaneous generation or creation ex nihilo.

Remember, each transfer RNA molecule that is made is sent to only ONE of the 10 million different ribosomes within the cell. Who chooses its destination? Who tells each tRNA molecule where its target ribosome is located? How does a tRNA molecule do this targeting and docking procedure? How does the tRNA know how to do so? Who or what makes the messenger molecules move? Just like in this essay, this protein synthesis and tRNA manufacturing process is done over and over and over again trillions of times within each living cell every day of your life.

The tRNA and mRNA molecules are manufactured within the nucleus of the cell. They can literally go anywhere within the cell after they are manufactured and released from the nucleus; so, why do they pick that one specific ribosome to migrate towards after they have been manufactured or made? Who or what makes these tRNA and mRNA molecules move? How do they know where to go? Someone intelligent or someone psyche has to be coordinating the long-distance interaction between the mRNA molecule, the ribosome, and all the dozens or hundreds of tRNA molecules that are incoming towards the same ribosome. Someone has to handle the logistics and keep all of this straight! How are all of these molecules and ribosomes communicating with each other and coordinating with each other at a distance? Action at a distance and telepathy are quantum mechanical processes, NOT physical processes.

Scientific Observation: Anything that was obviously made obviously had a Maker or Someone Intelligent who made it. This scientific truth is obviously true because it has been experienced and observed trillions of different times by billions of different people. You could say that this particular premise is axiomatic or LAW.

Scientific Observation: Genes, proteins, transcription enzymes, activation enzymes, transfer RNA, messenger RNA, and ribosomes were obviously made. They didn't just spontaneously generate out of thin air ex nihilo. Spontaneous generation or creation ex nihilo is physically impossible. Spontaneous generation, abiogenesis, chemical evolution, or macro-evolution is prevented from happening by entropy or the second law of thermodynamics. This is what has been experienced and observed. Entropy or death cannot design and create.

Scientific Conclusion: Therefore, it is logical and rational to conclude that genes, proteins, transcription enzymes, activation enzymes, transfer RNA, messenger RNA, and ribosomes have Someone Intelligent or Someone Psyche who made them in the first place. Anything that was obviously made obviously had a Maker who made it. This is Observational Science 101.

Who is storing the 3D addressing information for each of the 10 million different ribosomes within a cell? Who is making these assignments or handling the logistics for all of this? Who is making sure that each of the trillions of different tRNA molecules are making it to the correct ribosome on time? The tRNA molecules and mRNA molecules could go anywhere after leaving the nucleus of the cell; so, why do they go to that one specific ribosome? What's making these messenger molecules move? They are "transporting". They are moving. Why? How?

Molecules normally just sit there and vibrate under the effects of Brownian motion; so, what's making the atoms and molecules within a living cell move towards a specific targeted destination? It's as if these things are alive while they are within a living cell. How's that possible? It's not possible according to the Materialists and Naturalists! It's definitely NOT possible at the physical level!

How is all that information being transmitted in advance from the mRNA molecule to the individual tRNA molecules that will be docking with the different ribosomes in the future? How is this communication at a distance between atoms and molecules taking place? The tRNA molecules are being told where to go. How do they receive, translate, and understand this information? Who taught them this language? The activation enzymes are also being told in advance which amino acid to bind with which codon during the production of transfer RNA molecules. The amino acids and codons are being told where to go too. Who is coordinating all of this so that it works flawlessly rather than gumming up and screeching to a halt? Who enlivens and activates these formerly inert and inactive atoms and molecules once they are drawn into a living cell?

Where did the first ribosomes come from? Who tells the different proteins to self-assemble into ribosomes? Ribosomes are made from proteins; and, proteins are made by ribosomes? So, where did the ribosomes come from in the first place? Who made the proteins that were used to construct the first ribosomes? Who made the first ribosomes? Who taught the ribosomes how to manufacture proteins from mRNA molecules and tRNA molecules? Who taught the mRNA and tRNA to go to the ribosomes rather than someplace else? Who designed and produced this gene expression or protein synthesis process in the first place? Or did it all just spontaneously generate out of thin air as the Materialists and Naturalists claim?

Who is making all of these assignments and timing all of these appointments? Who is overseeing all of this? Remember, this is atoms and molecules that we are talking about here, NOT genes! There aren't any genes within an atom! Who is doing all of this communication between atoms and molecules? How is this information transmitted? How do the atoms and molecules translate and understand this information? How do the atoms and molecules know where to go? They can go anywhere; yet within a living cell, they go precisely where they are told to go! How is all this Command and Control being done? Who is doing it?

Command and Control obviously needs a Commander of some sort; or, nothing would happen!

Amino acids and codons are made from atoms. Who or what is telling each atom that it has an appointment at a specific location at a specific time for amino acid synthesis and tRNA codon synthesis? Who designed and made all of this fine tuning? Atoms can be assembled into anything, theoretically; so, who or what is assembling these particular atoms into amino acids and codons; and, how does it know how to do so? Outside of a living cell, atoms NEVER self-assemble into functional proteins and their matching genes. They can't because it's physically impossible. That's why the protein synthesis process was designed and created in the first place.

Fine-tuning requires a Fine Tuner; or, there would be NO fine-tuning! Random chance can't do fine-tuning. Random chance produces disorder, entropy, chaos, and death. Fine-tuning is Syntropy.

Design and Creation require a Designer and a Creator. The protein synthesis process was obviously designed and created. Therefore, it is obvious that the whole gene expression process has a Designer and Creator who made it.

How do the activation enzymes KNOW in advance at a distance which amino acid to combine with which codon during the activation process and the production of transfer RNA? How are those activation enzymes reading that future mRNA molecule from a distance so that they KNOW in advance which type of transfer RNA molecule to make? Who or what is doing all of the Command and Control for the production of transfer RNA; and, how are all of these petabytes of information being transmitted to the different individual atoms? Who is keeping track of it all? How are the atoms able to communicate with each other and coordinate with each other? What makes these atoms and molecules move towards each other and bind with each other? In a beaker, the atoms and molecules do nothing but Brownian motion; but, within a living cell the very same atoms and molecules literally come alive with purpose and activity! They combine together on command.

How? Why?

Who is transmitting all of these different commands and instructions to the individual atoms and molecules? Who is keeping it straight and making it work? Why do the atoms and molecules choose to obey while they are within a living cell, while the very same atoms don't do anything but Brownian motion and random diffusion while they are outside a living cell? How do the atoms and molecules KNOW that they are within a living cell? How are the atoms able to understand these instructions and how do the atoms see and know where to go? Where is all of this information being stored in the first place? There aren't any genes within an atom!

The Materialists, Naturalists, Darwinists, and Atheists assure us that only physical matter or physical atoms exist; but, clearly this is NOT the case. Clearly these people are wrong! Materialism, Naturalism, and their derivatives are based exclusively on entropy. Entropy is death. Entropy or death cannot design and create.

How would you communicate between atoms and make atoms do things and construct things if given the assignment to do so?

How do the activation enzymes know at a distance in advance which amino to combine with which codon so that the transfer RNA molecules arrive in the correct order and the correct number at each ribosome for RNA translation or protein synthesis? Protein synthesis happens rapidly, which means that the transfer RNA molecules arrive pre-sorted in the correct sequence BEFORE they get to the ribosome for RNA translation into a functional protein. There's NO hunting and pecking and randomness going on throughout protein synthesis – there's ONLY purposeful activity. How's that possible? It's NOT possible at the physical level.

How is this information transmitted from molecule to molecule and by whom? Who or what makes these individual molecules move or swim through the cytoplasm to the specific ribosome where they are needed in a nice and orderly manner? How does each tRNA molecule know which of the 10 million ribosomes it is supposed to move towards? Who is storing and organizing all of this information? Who set up this information network and logistics highway in the first place? Who is controlling all of this, and how is all of this information being transmitted to the molecules and being received and understood by the molecules? Who taught the atoms and the molecules how to communicate with each other?

Who is handling all of the telepathic communication and coordination taking place between the targeted ribosome, the incoming mRNA molecule, and the hundreds of different incoming tRNA molecules? The Materialists and Naturalists and Darwinists will tell you that it is evolution and/or the genes who are handling all of this. Are they right? Do they know what they are talking about? Or are they making things up out of thin air?

Some people have read the first couple of pages of some of my books and then asked, "Where's the Science?" Do you see any science in any of this? The Materialists and Naturalists will tell you that none of what I have written here is real science. These people will tell you that I don't understand science. They have told me that I don't understand the power and the beauty and the ability of evolution to be able to design and create. These people have told me that what I have written here is NOT science. Instead, they will assure you that Materialism, Naturalism, Darwinism, and Atheism ARE science. Are they right, or are they employing *circular reasoning* in order to make their case? *Circular reasoning* or *begging the question* is a logic fallacy the last time I checked.

Entropy is death. Materialism, Naturalism, Darwinism, Nihilism, Atheism, Classical Physics, and the Theory of Evolution are based exclusively on entropy. These people teach that ONLY physical matter and entropy exist. Evolution is entropy. Evolution is death. The theory of evolution is Creation by Entropy or Creation by Death. These people will assure you that evolution, entropy, or death can design and create anything it wants to at will. These people are wrong! Entropy or death cannot design and create. It's physically impossible. Furthermore, entropy or death cannot do protein synthesis or gene expression. Design and creation require some type of Intelligence, Psyche, or Syntropy, and so does protein synthesis.

Could You Do the Logistics for Protein Synthesis?

You are intelligent. Could you handle and do the logistics for protein synthesis within a single living cell in your body?

If I were to give you the assignment to handle the logistics for 10 million ribosomes, millions of messenger RNA molecules, billions of transfer RNA molecules, billions of amino acids and codons, and trillions of atoms within a single living cell, would you be able to handle the logistics for all of that and keep all of it straight in your head? Would the individual atoms and molecules and proteins make it to their assignments on time and arrive at the correct location in time?

Of course not!

A single physical brain doesn't have anywhere near enough memory storage capacity and processing power for such a task; yet, someone psyche or someone intelligent is doing all of this with ease at the quantum level or the psyche level for the trillions of different cells within your physical body; otherwise, a cell would be nothing but entropy, disorder, and random chaos. A cell would be dead or entropic without all of this Psychic Command and Control. Whoever is handling the logistics for all of this is infinitely more intelligent and infinitely more powerful than we are. Yet, the Materialists, Naturalists, Darwinists, and Atheists want you to believe that all of this happens by random chance! Do you believe them? I don't. Not anymore. I used to be a Materialist, Naturalist, Nihilist, and Atheist until I actually started looking at the science and the evidence.

It's physically impossible for a single human being to handle all of the logistics for even one single living cell; yet, these little critters (the atoms and molecules) know precisely where they are supposed to go and show up on time like clockwork as if they are infinitely more intelligent and skillful than our smartest human being or most powerful computer. Outside of a living cell, the atoms and molecules are dumb as rocks; but, inside a living cell, these very same atoms and molecules are infinitely more intelligent and competent than even our best and brightest human beings. How do the atoms and molecules get so smart once they find themselves within a living cell? What's going on here?

The Materialists, Naturalists, Darwinists, and Atheists assure us that the genes are handling all of this. Are they right? Or, are they trying to trick you and deceive you?

Who is handling the timing and the logistics for protein synthesis or gene expression and keeping it purring like a finely tuned machine?

Remember, anything that is obviously fine-tuned obviously had a Fine Tuner who fine-tuned it! The logistics for protein synthesis has obviously been fine-tuned. Therefore, protein synthesis or gene expression obviously has a Fine Tuner.

Remember, any type of Command and Control obviously needs a Commander. The logistics for protein synthesis is obviously being commanded and controlled, or nothing would happen in the first place except Brownian motion and random diffusion. The logistics for protein synthesis obviously has a Commander or a Psyche who is in control of it all.

The Materialists, Naturalists, Darwinists, Nihilists, Behaviorists, Determinists, Physical Reductionists, Evolutionists, Classical Physicists, and Atheists assure us that protein synthesis is happening by random chance and was produced by random chance. Are they right? Or did they get their science wrong?

According to these people, only physical matter and entropy exist. Entropy is death. Materialism, Naturalism, and the Theory of Evolution are Creation by Entropy or Creation by Death. According to these people, entropy or death can design and create at will. Are they right? Or did they get their science wrong?

Do you see any signs of intelligence or psyche behind any of this, or do you still believe that it all happens by random chance? Entropy is death. Do you really believe that entropy or death can do protein synthesis? Do you really believe that entropy or death is running the show? The Theory of Evolution is Creation by Entropy or Creation by Death. The Physicalists, Darwinists, and Naturalists assure us that only physical atoms and entropy exist. Their science is based exclusively on entropy or death. The Materialists and Naturalists tell us that there is no such thing as Psyche or Syntropy. Do you think they are right? Do you really think that Creation by Death or the Theory of Evolution works as advertised? Or do you think that someone intelligent is running the show?

It's as if there is some type of Overlord Psyche or Command and Control Psyche; and, there also seems to be some type of intelligence or psyche within each and every atom too! Someone intelligent or someone psyche is overseeing all of this and controlling all of this or it would be nothing but Brownian motion, random diffusion, entropy, chaos, and death. Death can't design and create. These very same atoms and molecules in a beaker don't do anything but Brownian motion and random diffusion; but, within a living cell they literally come alive with purpose and activity. How's that possible? The answer isn't found within our genes. The atoms don't have genes – at least not at the physical level! Who knows what the atoms have at the quantum level or the psyche level, though? God knows! I certainly don't! For all we know, there is a quantum computer within each and every physical atom handling all of the data processing and memory storage at the quantum level which isn't possible at the physical level.

When the Human Psyche and the Human Spirit leave the physical body, then all of these different physical processes shut down and stop working; and, entropy, random diffusion, chaos, and death take over instead. Entropy is death. Syntropy is the opposite of entropy. Syntropy is life. Syntropy is eternal life. Psyche is Syntropy – Psyche is eternal and everlasting, without a beginning of days or an end of years. It requires some type of Syntropy in order to produce life. Some type of Syntropy MUST exist, or all of that subsequent entropy and life wouldn't have been possible in the first place. While doing science, that is what I have experienced and observed.

Syntropy really is the answer to life, the universe, and everything – but only if you choose to believe that it is real and true. You can also go with Materialism, Naturalism, Darwinism, Nihilism, and Atheism; and remain in ignorance if that's what you want to do. The choice is yours to make. Nobody else can do that for you. Science is choice. Science is observation and experience.

Do you believe the science – the observations and the experiences; or, do you believe the philosophical speculation and wishful thinking of the Materialists, Naturalists, Darwinists, Nihilists, and Atheists? Beliefs are chosen into existence by the Human Psyche. Your beliefs and actions are chosen into existence by your Psyche.

Choice is purely a function of Psyche or Intelligence. The rocks and entropy haven't been caught in the act of doing science and choice. Raw physical matter has no choice. According to the Materialists, Naturalists, and Behaviorists, the rocks and entropy cannot do choice. They are right. There is no choice at the physical level. Physical matter simply does what it is told to do. Told by whom? It is physically impossible for the rocks and entropy to do science and make choices. This is what has been experienced and observed. Science IS observation and experience. It requires some type of Syntropy or Psyche in order to be able to do science and make choices. It takes some type of Syntropy, or Psyche, or Intelligence in order to make life.

Materialism and Naturalism cannot explain how protein synthesis is made to work within a living cell; but, Quantum Mechanics, Syntropy, and Psyche certainly can.

Remember, gene expression and protein synthesis NEVER happen outside of a living cell. They can't. It's physically impossible. Outside of a living cell, protein synthesis is prevented from happening by random diffusion, entropy, or the second law of thermodynamics. That is what has been experienced and observed.

Protein synthesis requires some kind of Intelligent, Purposeful, and Syntropic Command and Control at the quantum level in order to work right at the physical level; otherwise, it would be nothing but random chaos, random diffusion, and Brownian motion within a cell; and, nothing would ever get done.

Mark My Words

—

Source

The Scientific Method Proves That the Theory of Evolution Is False

> https://www.amazon.com/dp/B01IAAIRT2

I Am Not a Creationist: So What Am I?

> https://www.amazon.com/dp/B071XTM8XY

The Second Comforter: Supping with Our Resurrected Lord Jesus Christ

> https://www.amazon.com/dp/B01IAKHTY6

Entropy Falsifies Evolution

One of the most fascinating and useful of my scientific discoveries is that Evolution of any kind is based upon entropy. Evolution is Entropy. Random mutations are entropy; and, natural selection leads to death and extinction which is entropy.

Of course, it's obvious that Materialism, Naturalism, Darwinism, Nihilism, Atheism, Behaviorism, Scientism, Determinism, Physical Reductionism, and Classical Physics are based exclusively on entropy. What isn't immediately obvious is the true ramifications of such a scientific observation.

Entropy cannot do the Quantum Mechanical or the Supernatural. In other words, entropy cannot do design and creation. Entropy cannot do order and organization. Entropy cannot do Psyche or Intelligence. Entropy cannot do Syntropy. Entropy can only do death, which means that entropy cannot do life.

Chemical Evolution is the spontaneous generation of genes and proteins from atoms. In order for something like Chemical Evolution to have happened for real, it would have required some kind of Quantum Mechanical or Supernatural intervention because entropy prevents Chemical Evolution from taking place naturally in the wild. Chemical Evolution would have required some type of Syntropy, or Creation, or Engineering, or Intelligent Intervention in order to make it happen for real. The fact that Chemical Evolution or Spontaneous Generation is prevented from happening by entropy or random diffusion is scientific proof that the Theory of Evolution is false. The fact that Spontaneous Generation has been falsified by Science itself is scientific proof that the Theory of Evolution is false.

Macro-Evolution is one species giving birth to a completely different genetically incompatible species – two cats giving birth to a dog, or two chimp-like ancestors giving birth to chimpanzees and human beings. Macro-Evolution is prevented from happening both by genetics and by random mutations or entropy. In other words, Macro-Evolution is physically impossible. There's no way in the universe that Macro-Evolution, or Random Mutations, or Entropy would be able to produce a genetically compatible Mr. and Mrs. Mutant at the same time in the same place over and over and over again year after year for millions of years. It's physically impossible, which means that it didn't happen because it couldn't happen. Genetics or standardization prevents Macro-Evolution; and, Random Mutations also prevent Macro-Evolution. Macro-Evolution never happened, because it's physically impossible and prevented from happening by both genetics and entropy. The fact that Macro-Evolution of any kind is prevented from happening by entropy or random mutations is scientific proof that the Theory of Evolution is false.

This is one of my most significant and life-changing scientific discoveries, especially since I used to be a Materialist, Naturalist, Nihilist, and Atheist. These philosophies are based exclusively on entropy; and, entropy cannot design and create anything at all. Materialism, Naturalism, and their derivatives such as Darwinism or the Theory of Evolution are self-defeating and self-falsifying because they are based upon entropy and entropy prevents them from happening in the first place. Proteins and the matching genes to go along with those proteins don't just spontaneously generate out of thin air from scratch as the Naturalists and Darwinists claim, because entropy prevents them from doing so. Entropy prevents spontaneous generation or macro-evolution from happening in the first place. When it comes to Evolution, entropy provides the seeds for its destruction or the seeds for its falsification.

Fascinating, is it not?

Well, I think it is because I used to be a Materialist, Naturalist, Nihilist, and Atheist; and at times, I was even on the fence when it came to Darwinism and the Theory of Evolution.

Evolution is entropy. Entropy or evolution cannot do Syntropy or Quantum Mechanics. Entropy or evolution cannot do Syntropy or Psyche and Intelligence. Evolution or entropy cannot do Syntropy or Organization and Life. Entropy or evolution cannot do Science, Manufacturing, Engineering, Fine-Tuning, Construction, Creation, and Life. Entropy produces death and extinction, NOT life. Random mutations introduce entropy into our genomes; and, natural selection facilitates death or entropy rather than order, organization, life, and creation. Furthermore, natural selection doesn't touch our genes. Under the stress and pressures of random mutations and natural selection, entropy leads every physical life form towards death and extinction.

Entropy cannot produce Syntropy; but, Quantum Mechanics and Psyche certainly can. Entropy cannot design and create; but, Psyche or Intelligence certainly can. Psyche or Intelligence has been caught in the act of design and creation; and, entropy has not. This means that entropy or evolution could NEVER have created or produced our genome; but, some type of Intelligent Being or God-Like Psyche certainly could have.

We KNOW that some type of Syntropy must exist because entropy exists; and, we also KNOW that entropy falsifies Evolution or prevents the different types of Evolution from becoming actual or real. Therefore, some type of Syntropy or Organizing Force had to be used in order to produce our proteins, genes, and genomes in the first place.

Here we have a prime example of how Science itself FALSIFIES Materialism, Naturalism, Darwinism, Nihilism, Atheism, and the Theory of Evolution. This is one of my most significant scientific discoveries. It was hiding there in plain sight where nobody seemed to be able to see it or understand it. Evolution is entropy, and entropy prevents the Theory of Evolution from becoming true.

Mark My Words

Genetic Entropy

Genetic Entropy is one of my most favorite scientific discoveries. This scientific discovery doesn't belong to me, though. It originates with John Sanford.

After reading his book, *Genetic Entropy*, I simply KNEW that the theory of evolution is false because I now KNOW why it is false.

The theory of evolution is typically defined as Creation by Mutation/Selection – particularly, 'the origin of species by means of natural selection'. The first part of the title of Darwin's book is, "On the Origin of Species by Means of Natural Selection".

Creation by Natural Selection IS science fiction. Natural selection doesn't touch our genes! It can't. It's physically impossible for natural selection to get at our genes and change them. Natural selection cannot design and create anything, let alone a genome.

In truth, natural selection is NOT the mechanism of change behind the theory of evolution. It's the random mutations that produce genetic change, NOT natural selection! Natural selection or survival of the fittest doesn't do anything. It just waits for you to die. Natural selection is entropy or death. They built a whole "science" on a fictional, immaterial, invisible process that doesn't even touch our genes – natural selection! And, they literally give natural selection ALL the credit for designing, programming, creating, and producing our genomes and our physical bodies. For these people, natural selection is their god. They worship it with a passion. Natural selection is a man-made god, an idol.

Natural Selection: The evolutionary process by which heritable traits that best enable organisms to survive and reproduce in particular environments are passed to ensuing generations.

Everyone who has taken introductory psychology has learned that nature and nurture together form who we are. As the area of a rectangle is determined by both its length and its width, so do biology and experience together create us.

As *evolutionary psychologists* remind us, our inherited human nature predisposes us to behave in ways that helped our ancestors survive and reproduce. We carry the genes of those whose traits enabled them and their children to survive and reproduce. Thus, evolutionary psychologists ask how natural selection might predispose our actions and reactions when dating and mating, hating and hurting, caring and sharing. Nature also endows us with an enormous capacity to learn and to adapt to varied environments. We are sensitive and responsive to our social context.

To explain the traits of our species, and all species, the British naturalist Charles Darwin (1859) proposed an evolutionary process. Follow the genes, he advised. Darwin's idea, to which philosopher Daniel Dennett (2005) would give "the gold medal for the best idea anybody ever had," was that natural selection enables evolution.

Natural selection implies that certain genes — those that predisposed traits that increased the odds of surviving long enough to reproduce and nurture descendants — became more abundant.

Natural selection, long an organizing principle of biology, has recently become an important principle for psychology as well. *Evolutionary psychology* studies how natural selection predisposes not just physical traits suited to particular contexts — polar bears' coats, bats' sonar, humans' color vision — but also psychological traits and social behaviors that enhance the preservation and spread of one's genes. We humans are the way we are, say evolutionary psychologists, because nature selected those who had our traits — those who, for example, preferred the sweet taste of nutritious, energy-providing foods and who disliked the bitter or sour flavors of foods that are toxic. Those lacking such preferences were less likely to survive to contribute their genes to posterity.

As mobile gene machines, we carry not only the physical legacy but also the psychological legacy of our ancestors' adaptive preferences. We long for whatever helped them survive, reproduce, and nurture their offspring to survive and reproduce.

"The purpose of the heart is to pump blood," notes evolutionary psychologist David Barash. "The brain's purpose," he adds, is to direct our organs and our behavior "in a way that maximizes our evolutionary success. That's it." (*Social Psychology*, p. 8, 159.)

Everything they wrote here is false or incomplete.

It's NOT a rectangle, it's a triangle! It's not just nature and nurture that form us. There's an essential third component!

Do you know what it is?

NATURE vs. NURTURE vs. NIRVANA: An Introduction to Reality

    https://www.amazon.com/dp/B01JWRCSVA

    https://www.amazon.com/dp/1521132615

The third component is deliberately eliminated from science by the Materialists, Naturalists, Darwinists, Nihilists, and Atheists. These people state that it does not exist. The BioPsychoSocial Model tells us that it does exist. Somebody is right, and somebody is wrong. They both can't be right.

These people have been teaching for over 150 years that Natural Selection made you, that evolution made you; but, that's physically impossible. Natural selection can't make anything. Natural selection and the theory of evolution are based exclusively on entropy. Natural selection results in entropy or death. Evolution is entropy. Entropy is death. Entropy cannot make anything at all. Entropy or death can only destroy. The different types of evolution can only destroy. The different types of evolution or entropy can only produce death and extinction.

Natural selection doesn't predispose anything! Natural selection doesn't organize anything! It can't. Natural selection doesn't touch our genes! Natural selection doesn't endow us with anything! Natural selection has NO ability to learn anything. Natural selection doesn't enable anything. Natural selection doesn't do anything. Natural selection is supposed to be dumb and blind without a soul or a mind, according to the Darwinists. Creation by Natural Selection wins the rotten tomato for the most stupid, illogical, irrational, and ineffective idea ever created.

The theory of evolution is correlational, NOT observational. NO type of evolution has ever been caught in the act of design and creation. It's physically impossible for natural selection and random mutations to design and create something. Entropy prevents them from doing so. Chemical evolution or macro-evolution is prevented from happening by random diffusion or entropy. Macro-evolution is also prevented from happening by genetics. The genes are there to prevent macro-evolution from happening. Evolution of any type is entropy and death. Death cannot create life!

In fact, evolution (genetic change), random mutations, and natural selection didn't even exist until AFTER God designed, programmed, engineered, field-tested, fine-tuned, manufactured, created, and produced the proteins and their matching genes in the first place.

The theory of evolution is a fictional story that they made up out of thin air after-the-fact to fit the facts. NO part of it can actually design and create. It's a fictional story, not science.

We are sensitive and responsive to our social context, NOT our genes!

Who is this **WE** that they keep talking about in our Social Psychology textbooks? **WE** can't be our society, environment, or social context that **WE** are sensitive to and responsive to! And, it's definitely NOT our genes. Our genes aren't sensitive and responsive to anything according to the Evolutionists! The genes, natural selection, and random mutations are supposed to be dumb and blind without a soul or a mind. Our genes can't be sensitive nor responsive to anything.

Personal pronouns imply a person or a psyche – NOT our genes (nature) and NOT our environment (nurture).

Natural selection and your genes DO NOT and CANNOT pass your Psyche Legacy (psychological legacy) from one generation to the next! Natural selection doesn't touch your genes, and it definitely doesn't touch nor change your Psyche either. It's science fiction to imply that it does. The theory of evolution is science fiction. "Design and creation by natural selection" is science fiction. The idea that your genes carry your "longings" or "desires" from one generation to the next is science fiction. There's no such thing as genetic memory, at least not at the physical level. All of our thoughts and memories are carried as quantum waves from one generation to the next through our Psyche or Quantum Non-Local Consciousness.

Your genes don't care whether you live or die. Only YOU care whether you live or die. Your Psyche cares whether you live or die; but, your genes do not. In order for your genes to care, they would have to have some sort of Psyche, or Intelligence, or Consciousness, or Awareness. But, if your genes have a Psyche, then the very existence of that Psyche falsifies Materialism, Naturalism, and Darwinism which claim that Psyche does not exist. Physicalism, Naturalism, and the Theory of Evolution are self-defeating. They don't work as advertised because they can't work as advertised.

Natural selection results in entropy and death, NOT X-Men and new unique life forms. Natural selection cannot design and create and program genomes. Natural selection doesn't touch our genes.

Random mutations are also entropy; but, at least random mutations by definition in principle actually change our genes. However, random mutations cannot design and create anything either.

Entropy is death. Death cannot design and create new unique genomes and life forms. That's physically impossible! Mutation and Selection can only produce entropy or death. They are based exclusively on entropy or death. Death cannot design and create life. Death can only end life.

Remember, the theory of evolution is Creation by Entropy or Creation by Death. That's NEVER going to work because it's physically impossible!

Isn't it refreshing to finally have access to the truth, rather than all the science fiction that the Evolutionists have been feeding us throughout our lives?

Well, I think it is.

I used to be a Materialist, Naturalist, Nihilist, and Atheist until I finally started to study the evidence. The evidence and the truth set me free! The Science and the Scientific Evidence convinced me that God must exist in order to have done all the Science and Fine-Tuning which natural selection and evolution could NEVER have done.

Mark my Words

—

Can Natural Selection Create?

The Physicalists, Naturalists, Darwinists, Nihilists, and Atheists teach that natural selection can create anything that it sets its mind to. These people teach that natural selection made you.

Are they right?

They are not!

They are deceiving themselves and trying to trick us and deceive us as well.

Creation by Natural Selection is demonstrably false, which means that it has been falsified by Scientific Evidence. Natural selection doesn't do anything. Natural selection doesn't touch our genes, nor does it pre-determine our future. Natural selection doesn't have a mind. There is NO intelligence or psyche within natural selection. There may (or may not) be some type of intelligence or psyche within our genes; but, there is NOTHING there when it comes to Natural Selection or Evolution. Natural selection is a fictional concept that they made up out of thin air. It doesn't really exist as a person or an entity. The same can be said of evolution. I'm not the only scientist to have figured this out by now. Natural selection is worthless as a creative agent and can't function as a creative agent.

[Editorial Note: I have written permission from John C. Sanford to use all of the quotes from John C. Sanford which I use in my books, so long as I cite the sources which I have done.]

START OF THE QUOTE FROM "GENETIC ENTROPY" BY JOHN SANFORD — USED BY PERMISSION FROM THE AUTHOR JOHN SANFORD.

Chapter 9: Can Natural Selection Create?

Newsflash — Mutation/Selection cannot even create a single gene.

We have been examining the problem of genomic degeneration and have found that deleterious mutations occur at a very high rate. Natural selection can only eliminate the worst of these, while all the rest accumulate — like rust on a car. Might beneficial mutations at other sites in the genome compensate for this continuous and systematic erosion of genetic information? The answer is that beneficial mutations are much too rare, and they are much too subtle to keep up with such relentless and systematic erosion of information. This is carefully documented by Sanford et al. (2013), and Montañez et al. (2013). It is very easy to systematically destroy information, but apart from the operation of intelligence it is very hard (arguably impossible) to create information.

This problem overrides all hope for the forward evolution of the whole genome. However, some limited traits might still be improved via Mutation/Selection. Just how limited is such progressive ("creative") Mutation/Selection? By now it should be clear that random spelling errors in an instruction manual could never give rise to an airplane component (say a molded aluminum part), which then resulted in a significantly improved overall performance of a jet plane. Not even with an unlimited number of flight trials/crashes and an unlimited budget. So, it is certainly reasonable to ask the parallel biological question, "Could Mutation/Selection create a single functional gene from scratch?"

A gene is like a book, book chapter, or an executable program — and minimally consists of a text string with 1,000 characters. Mutation/Selection could not create a single gene because of the enormous preponderance of deleterious mutations, even within the context of a single gene. The net information must always still be declining, even within a single gene or linkage block. Even if a gene was 50% established, deleterious mutations would degrade the completed half of the gene much faster than beneficials could create the missing half of the gene. However, to better understand the limits of forward selection, let us for the moment discount all deleterious mutations and only consider beneficial mutations. Could Mutation/Selection then create a new and functional gene?

    1. Defining our first desirable mutation. The first problem we encounter in trying to create a new gene via Mutation/Selection is defining our first beneficial mutation. By itself, no particular nucleotide (A, T, C or G) has more value than any other, just as no letter in the alphabet has any particular meaning outside of the context of other letters. So, selection for any single nucleotide can never occur except in the context of the surrounding nucleotides (and in fact, within the context of the whole genome). A change of a single letter within a word or chapter can only be evaluated in the context of the surrounding block of text. This brings us to an excellent example of the principle of "irreducible complexity" within the genetic realm. In fact, it is irreducible complexity at its most fundamental level. We immediately find we have a paradox. To create a new function, we will need to select for our first beneficial mutation, but we can only define that new nucleotide's value in relation to its neighbors — and we are going to have to be changing most of those neighbors also. We create a circular path for ourselves. We will keep destroying the "context" we are trying to build upon. This problem of the fundamental inter-relationship of nucleotides is called epistasis. True epistasis is almost infinitely complex, and virtually impossible to analyze, which is why geneticists have always conveniently ignored it. Such bewildering complexity is exactly why language and information (including genetic language and genetic information) can never be the product of chance, but always requires intelligent design. The genome is literally a book, written literally in a language, and short sequences are literally sentences. Having random letters fall into place to make a single meaningful sentence, by accident, would require more tries (more time), than earth history can provide (i.e., "methinks it is like a weasel" would take $27 \wedge 28$ tries — that is 10 followed by 40 zeros). The same is true for any functional string of nucleotides. If there are more than a dozen nucleotides in a functional string, we know that realistically they will never just "fall into place". This has been mathematically demonstrated repeatedly. But as we will soon see, neither can such a sequence arise by selecting one nucleotide at a time. A pre-existing "concept" is required as a framework upon which a sentence or a functional sequence must be built. Such a concept can only pre-exist within the mind of the author. Starting from the very first mutation, we have a fundamental problem even in trying to define what our first desired beneficial mutation should be.

    2. Waiting for the first mutation. Let's assume we can know the first desired mutation. How long do we have to wait for it to happen? Human evolution is generally assumed to have occurred in a small population of about 10,000 individuals. The mutation rate for any given nucleotide, per person per generation is exceedingly small (very roughly about one mutation per 30 million individuals, for a given nucleotide site). Within a population of

10,000, one would have to wait 3,000 generations (at least 60,000 years) to expect a specific nucleotide to mutate. But two out of three times, it will mutate into the "wrong" nucleotide. So, to get a specific desired mutation at a specific site just in one individual will take three times as long, or at least 180,000 years. Once the mutation arises in one individual, it has to become "fixed" (such that each individual in the population will eventually have a double dose of that mutation). Because a newly arisen mutation arrives in a population as just a single copy, it arrives on the brink of extinction. The vast majority of new mutations soon drift back out of the population, even the ones that are beneficial. So, any specific desired mutation must arise many times before it "catches hold" in the population. Only if the mutation is dominant and has a very distinct benefit does selection have any reasonable chance to rescue it from random elimination via drift. According to population geneticists, apart from effective selection, in a population of 10,000, our given new mutant has only one chance in 20,000 (the total number of non-mutant nucleotides present in the population) of NOT being lost via drift. Even with some modest level of selection operating, there is a very high probability of random loss, especially if the mutant is recessive or is weakly expressed (we actually know that most mutations will be both recessive and nearly neutral). Therefore, even a beneficial mutation will be randomly lost due to genetic drift most of the time. Our numerical simulations suggest a weakly beneficial mutant will be lost about 99 out of 100 times. So, a typical mildly-beneficial mutation must happen about 100 times before it is likely to "catch hold" within the population. So, on average, in a population of 10,000 we would have to wait $180,000 \times 100 = 18$ million years to stabilize our first desired beneficial mutation, to begin building our hypothetical new gene. So, in the time since we supposedly evolved from chimp-like creatures (6 million years), there would not be enough time to realistically expect our first desired mutation to go to fixation in the genomic location where our required gene is hopefully going to arise. A vast amount of mutations would arise during 18 million years, but only once would that specific nucleotide mutate to that specific new nucleotide — such that it's not lost due to genetic drift and is fixed.

   3. Waiting for the other mutations. After our first desired mutation has been found and fixed, we need to repeat this process for all the other nucleotides encoding our hoped-for gene. A gene is minimally 1,000 nucleotides long. More realistically, a human gene is on average about 50,000 nucleotides long, when regulatory elements and introns are included. To be extremely generous we will only consider a gene of 1,000 nucleotides (and we assume each nucleotide is by itself selectable). If this process was a straight, linear, and sequential process, it would require about 18 million years $\times$ 1,000 = 18 billion years to create the smallest possible gene. This is more than the time since the reputed Big Bang! So, it is a gross understatement to say that the rarity of desired mutations limits the rate of evolution. Furthermore, single nucleotides do not carry any information by themselves, and cannot be selectively favored. Specified information requires many characters (minimally, a sentence or similar text string is needed). Like

any message, a genetic message which specifies some life function requires many nucleotides to reach its "functional threshold". Functional threshold is the minimal number of characters (or nucleotides) needed to convey a meaningful message. Below the functional threshold, individual letters or nucleotides have no benefit and cannot be favored by selection. This means that realistically, waiting time will be much, much longer — because no selection can happen until the minimum string of nucleotides falls into place by chance. If the functional threshold for selection is 12 (no selection until all 12 letters are in place), the waiting time in our hypothetical human population becomes trillions of years.

Sanford, John (2015-02-23). Genetic Entropy (Kindle Locations 1684-1755). FMS Publications. Kindle Edition. USED BY PERMISSION.

END QUOTE.

—

Trillions of years!

Well, that's the END of the Theory of Evolution, isn't it?

The Darwinists NEVER use their God-given brains to stop and think about these kinds of things. At the best possible average pace, with God making sure that there are NO deleterious mutations and NO devolution taking place, it would take on average 18 million years to fixate and stabilize a SINGLE beneficial mutation through "Natural Means" into a population of 10,000 apes which God has already designed and created in the first place and kept alive and functional during those 18 million years, just so that population of 10,000 God-created apes can achieve their first beneficial mutation through "Natural" Hands-off Mutation and Selection. 18 million years on average per beneficial mutation! Think about it!

If those apes need 1,000 such beneficial mutations in order to become men, then you are looking at 18 billion years on average to produce those targeted 1,000 beneficial mutations; and, that's with a population of 10,000 apes that God has already designed and created in the first place and that God is making sure receive ONLY beneficial mutations and NO devolution or deleterious mutations. And, that's also with God keeping that population of 10,000 apes alive during those 18 billion years so that they can indeed "evolve" their necessary 1,000 beneficial mutations and become men all on their own through "Natural Means".

Furthermore, it has been estimated that it would in fact take at least 20 million such beneficial mutations to convert chimpanzees into humans through "Natural Means". With that targeted goal in mind and assuming NO deleterious mutations or extinctions along the way, how long would it take on average to convert 10,000 chimpanzees into 10,000 humans using Natural Selection and Random Mutations to do the job? So, what do you get if you multiply 20 million beneficial mutations with 18 million years per beneficial mutation? At the BEST possible pace, with God keeping those 10,000 chimpanzees alive all along the way, and with God making sure that there is NO devolution, NO extinction, and NO deleterious mutations taking place, the quickest on average that Mutation and Selection could convert a chimpanzee into a human through "Natural Means" is 360 trillion years.

John Sanford isn't exaggerating whenever he says that it could take trillions of years for Mutation/ Selection to design and create something useful "naturally". And, it really isn't Natural Evolution if God has to design and create the 10,000 chimpanzees in the first place, and then keep them alive for 360 trillion years by blocking ALL deleterious mutations and preventing ALL extinctions that might take place during that period of time, just so He can convert 10,000 chimpanzees into 10,000 humans "naturally" or through "Natural Means".

Think about it! At the BEST possible average pace, it would take at least 360 trillion years for Mutation and Selection to convert a population of 10,000 chimpanzees into 10,000 humans through "Natural Means", with God keeping those 10,000 mutants alive and preventing deleterious mutations during the whole time. And, that's with 10,000 chimpanzees that God has already designed and created in the first place! How old did they say our universe is? How long would it take to convert a bacterium into a human through "Natural Means" when billions of beneficial mutations are needed? Wouldn't it be easier and faster to just let God design and create those 10,000 humans in the first place?

YES, it would be!

The Darwinists and Materialists NEVER stop and use their God-given brains to think about and calculate these kinds of things. You will NEVER get these kinds of calculations and truths from the Darwinists because they don't DO this kind of science. It's too difficult and painful for them. I'm willing to wrap my mind around these kinds of things. Your typical Darwinist isn't. Your typical Darwinist is afraid of it because they don't want to be proven wrong. For the Materialists and Darwinists, ignorance is bliss! But, ignorance is the reason why the Darwinists and Materialists truly believe that Mutation/Selection can design and create anything that it sets its mind to. The rest of us KNOW BETTER!

If you think about it, this is radically advanced science — the best that humans are able to come up with! Can you see and understand now why the 9th Chapter of "Genetic Entropy" put an END to the Theory of Evolution for me? It's because I understood what John Sanford was talking about and chose to believe that it is true. Now the onus is on you.

What do the Darwinists typically do when presented with these kinds of Statistical Models of the Mutation/Selection Process?

Assuming that they don't go head-in-the-sand and actually study them instead, the Darwinists try to shave the figures in half or by one-tenth, which is exactly what they do when designing Mutation/Selection Models of their own. They cheat. They choose parameters that are scientifically inaccurate and don't match with Reality in order to shave those estimates down to something that they might be willing to accept. They keep shaving and cheating until they get the numbers that they want, and then they call the results "Science".

If they really take John Sanford's Models seriously, the Darwinists will demand that God artificially accelerate the Mutation/Selection Process so as to make it possible for the Theory of Evolution to be true. But, even if God were to speed up the process a thousand-fold, it's still going to take 360 billion years for chimpanzees to evolve into humans through "Natural Means"; and, the Darwinists are still going to complain, even though we are starting with Chimpanzees that God has designed and created in the first place.

It can be fascinating and entertaining to watch a Darwinist try to shave trillions of years off a Statistical Estimate that he doesn't like, all in an attempt to increase the possibility that the Theory of Evolution might be true.

And, that's just the beginning of the problems for the Theory of Evolution. It only gets worse from there on forward, because John Sanford actually has TEN points in chapter 9 of "Genetic Entropy", each of which decreases the likelihood of Mutation/Selection creating anything at all, even if it has an infinite number of years to do so. Evolution by random mutations and evolution by natural selection CANNOT design, create, and deploy anything! It has been conclusively and finally demonstrated that it is so. It has been empirically and logically observed to be so. The Theory of Evolution doesn't work and can't create new unique genomes from scratch, so we have no choice but to declare the whole thing to be FALSE.

Many different scientists taught me that Random Mutations and Natural Selection (Evolution) cannot design and create genomes and life forms. They met their burden of proof and demonstrated to me that it must be so. (See the partial list of Reference Materials below for a selection of some of the best scientific evidence that falsifies the Theory of Evolution.)

I chose to believe the scientific evidence rather than the claims of the Materialists, Naturalists, and Darwinists who teach that Psyche, Intelligence, and Syntropy do not exist.

Evolution Is Creation by Death

I did make a scientific discovery in recent days (May 2018) that I think has merit and value to the Scientific Community.

I do seem to be the first person on the planet to realize that NO type of evolution can do selection. Selection of any kind involves choice; and, choice is exclusively the product of a Psyche or a Mind. Evolution of any type by definition, in principle, is dumb and blind without a soul or a mind. The very definition of Evolution eliminates its ability to do selection, a priori. Evolution can't do selection because evolution can't do Psyche or Choice. Evolution doesn't exist as a Psyche, Person, Intelligence, Mind, or Soul capable of doing choice or selection.

Furthermore, any attempt to imbue evolution with a soul, psyche, or mind automatically FALSIFIES Materialism, Naturalism, Darwinism, Nihilism, and the Theory of Evolution which claim that Psyche or Syntropy does not exist. The theory of evolution is a non-starter because evolution of any kind cannot do selection or choice.

Always remember, creation by natural selection is science fiction and wishful thinking because evolution of any kind cannot do selection or choice.

Materialism, Naturalism, Darwinism, Nihilism, Behaviorism, Determinism, Physical Reductionism, Atheism, Classical Physics, and the Theory of Evolution are based exclusively on entropy. Entropy is death. Death cannot design, create, and produce life. Such a thing has never been experienced nor observed. Darwinism or the theory of evolution is Creation by Death, or Creation by Entropy. Creation by Death or the theory of evolution is impossible. It can't happen, which means that it didn't happen.

Therefore, the theory of evolution cannot be used to explain the origin of life. The most that the theory of evolution can explain is death, extinction, devolution, and genetic entropy. Evolution of any kind is a function of death or entropy. Evolution is entropy. Evolution is death. Evolution or death, of any kind, can't be used to explain the origin of life. The truthfulness of this reality becomes obvious, once a person realizes and accepts the fact that the Theory of Evolution, Physicalism, Naturalism, Atheism, Nihilism, Classical Physics, and Darwinism are based exclusively on entropy. Entropy or death cannot produce life. The different types of evolution or entropy cannot produce life. It's impossible for entropy or death to produce life.

At times I've wondered why nobody else has been able to see and understand these obvious truths; but then, I used to be a Materialist, Naturalist, Nihilist, and Atheism, and there I find my answer. At the time, I wasn't able to see nor understand these obvious scientific truths because I didn't want to see them, understand them, nor accept them. No seeking, then no finding. I wasn't looking for any of this, so I never found it. I only found it after I started looking for it. I had convinced myself that this type of information doesn't exist. Self-deception works, and it works every time, especially when it comes to scientists like me.

The axiom stating that evolution is dumb and blind without a soul or a mind, if taken as being true, prevents evolution of any type from being able to do selection. In other words, the very definition of evolution as being dumb and blind FALSIFIES the Theory of Evolution by preventing evolution of any kind from being able to do selection or choice. Evolution is entropy; and, entropy is death. Death cannot do selection or choice. Death cannot do life. Entropy or death can only destroy. That is what has been experienced and observed.

Meanwhile, the Materialists, Naturalists, Darwinists, Nihilists, and Atheists DEMAND that you accept on blind faith that their claims – that death, entropy, or evolution can produce life – are true.

We scientists have FALSIFIED the Theory of Evolution trillions of times in thousands of different ways, but we choose to ignore the evidence because it isn't telling us what we want to hear. That's the way we do science in this world – by ignoring the evidence, discounting the evidence, banning the evidence, and destroying the evidence. We do our science this way so that we can prove to ourselves and to others that the Theory of Evolution is true.

There is another way to do science, though – a better way of doing science. I call it Science 2.0; and, it involves allowing ALL of the evidence into evidence. Once we choose to do so, then ALL of the evidence that we have on hand as a race FALSIFIES the claims of Materialism, Naturalism, Darwinism, Nihilism, Behaviorism, Determinism, Physical Reductionism, and Atheism which claim that this evidence does not exist.

Once we have eliminated all of the falsehoods such as Materialism, Naturalism, Atheism, Nihilism, and Darwinism, then we are left staring at THE TRUTH, which is that ONLY Psyche can design, program, engineer, field-test, fine-tune, manufacture, create, and do science.

By eliminating all of the falsehoods or pseudo-sciences, it becomes obvious that God's Psyche must of necessity exist in order to have DONE all of the Science that needed to be done, which evolution and the rocks could NEVER have done. Remember, evolution and the rocks can't do science because they can't do selection or choice.

The observation and realization, that evolution of any type cannot do selection, just might be one of my greatest scientific discoveries even if I don't end up being the first person on the planet to have made this discovery.

Obviously, everyone across the world is now starting to make these kinds of scientific discoveries right and left because we have finally started to take our blinders off. More and more of us are willing to see, which makes us able to see. Nowadays, it's obvious that the Theory of Evolution is false; whereas, we couldn't see it before because we didn't want to see it, and we didn't know where to look.

All you want is the truth. Everything else is worthless in the end.

Mark My Words

—

Source

Science 2.0: I Upgraded My Science
https://www.amazon.com/dp/B0771K6WTX

The Scientific Method Proves That the Theory of Evolution Is False
https://www.amazon.com/dp/B01IAAIRT2

https://www.amazon.com/dp/1521133611

NATURE vs. NURTURE vs. NIRVANA: An Introduction to Reality
https://www.amazon.com/dp/B01JWRCSVA

https://www.amazon.com/dp/1521132615

Myers, D. G. (2010). *Social Psychology* (10th ed.). New York: McGraw-Hill.

Reference Materials

Wells, J. (2000). *Icons of Evolution: Science or Myth? Why Much of What We Teach About Evolution Is Wrong*. Washington, DC. Regnary.

Sanford, J. (2014). *Genetic Entropy* (4th ed.). Cornell University: FMS Foundation.

Sanford, J. C., Marks, R. J., Behe, M. J., Dembski, W. A., & Gordon, B. L. (Eds.). (2013). *Biological Information: New Perspectives*. Hackensack, NJ: World Scientific.

Meyer, S. C. (2010). *Signature in the Cell: DNA and the Evidence for Intelligent Design*. New York: HarperCollins.

Meyer, S. C. (2013). *Darwin's Doubt: The Explosive Origin of Animal Life and the Case for Intelligent Design*. New York: HarperCollins.

Mark My Words. (2016). *The Scientific Method: Proves That the Theory of Evolution Is False*. Kindle. Retrieve from: https://www.amazon.com/dp/B01IAAIRT2

Mark My Words. (2016). *The Theory of Evolution Proved to Me that God Exists: Why I Am No Longer an Atheist and Why I No Longer Believe in the Theory of Evolution*. Kindle. Retrieve from: https://www.amazon.com/dp/B01HZYBZ7K

Kin Selection

Natural Selection is science fiction. By definition, in principle, Natural Selection can't do CHOICE or selection. Natural selection can't do what they say it does. By definition, in principle, natural selection or survival of the fittest is supposed to be dumb and blind without a soul, psyche, or mind. Therefore, natural selection can't do choice or selection. Instead, natural selection results in entropy or death. Natural selection, or entropy and death, cannot design and create and produce anything. "Design and creation by Natural Selection" is prevented from happening by entropy, random diffusion, or the second law of thermodynamics. Random mutations are also entropy. Entropy is death. The theory of evolution is Creation by Entropy, or Creation by Death. That's not going to work because it's physically impossible. Death or entropy cannot design and create life. This reality is obviously true.

Do you want the truth, or do you want the science fiction?

Can you handle the truth? Most of our scientists can't.

Kin Selection is another smoking gun where Psyche is concerned.

Kin Selection: The idea that evolution has selected altruism toward one's close relatives to enhance the survival of mutually shared genes.

Our genes dispose us to care for relatives. Thus, one form of self-sacrifice that *would* increase gene survival is devotion to one's children. Compared with neglectful parents, parents who put their children's welfare ahead of their own are more likely to pass their genes on. As evolutionary psychologist David Barash wrote, "Genes help themselves by being nice to themselves, even if they are enclosed in different bodies." Genetic egoism (at the biological level) fosters parental altruism (at the psychological level). Although evolution favors self-sacrifice for one's children, children have less at stake in the survival of their parents' genes. Thus, according to the theory, parents will generally be more devoted to their children than their children are to them. (*Social Psychology*, p. 452.)

By definition, in principle, evolution is dumb and blind without a soul or a mind. This means that evolution of any type cannot DO choice or selection. It's a deceptive lie to say that evolution can do choice or selection. It's science fiction.

Without realizing it, these people have painted themselves into a corner with this one. The idea of genes helping each other even if they are enclosed in different bodies is physically impossible. There's NO physical mechanism in place whereby the genes can communicate with each other, especially the genes in different bodies! Such an idea is ludicrous. If the genes are communicating with each other and recognizing each other, then they are doing so at the quantum level or the psyche level because they can't do so at the physical level.

The ONLY way to make Kin Selection true is if we all axiomatically agree in advance that the genes (and evolution) are psychic and are therefore capable of determining telepathically which genes are related to them and which genes are not. In order for the genes to be nice to each other, they have to know each other, perceive each other, recognize each other, and show favoritism to each other. They have to be psychic and have some kind of psyche. The genes can't be dumb and blind if we want Kin Selection to work as advertised. The genes have to communicate with each other, know each other, and recognize each other in order for them to be able to help themselves and be nice to themselves. The genes also have to be subliminally communicating their desires to the Human Psyche in order to make the Human Psyche be nice to the genes too. The only way to make Kin Selection work as advertised is if we agree in advance that the genes have some sort of Psyche or Mind and are telepathically connected with each other.

However, if we agree in advance axiomatically that genes have a psyche, that genes perceive each other telepathically, that genes know each other and recognized each other, and that genes are therefore psychic, then we have in fact

FALSIFIED Materialism, Naturalism, Darwinism, Nihilism, Atheism, even the Theory of Evolution in the process.

These people personify the genes – imbue them with Psyche and Intelligence – and in the process falsify the major premises of Materialism, Naturalism, Darwinism, Nihilism, and Atheism which state that Psyche or Syntropy does not exist. Thereby, these falsified philosophies or falsified religions end up being self-defeating. The truth cannot be built upon falsehoods

When it comes to Science, we observe that Psyche, Quantum Mechanics, or Syntropy always ends up being the best possible explanation that can be given to ALL of the evidence that we are observing and experiencing, including the physical evidence. Remember, entropy and physical matter would not exist without a massive initial infusion of Syntropy, Intelligence, Power, Quantum Mechanics, or Psychic Intervention somewhere sometime along the way.

Remember, perception is a function and a product of Psyche. At the physical level, the genes have no way of knowing or perceiving which genes are related to them and which genes are not; and at the physical level, the genes have NO way to pass that information on to the physical body or physical brain even if the genes were to know who is related to them and who is not. At the physical level, the genes have no way to perceive each other, thereby falsifying the claims of Kin Selection. By restricting and limiting everything to the physical level, the Materialists and Naturalists automatically falsify anything and everything that needs Psyche or Intelligence or Perception in order to become true – things such as Kin Selection and Creation by Mutation/Selection.

Remember, evolution can't do selection! By definition, in principle, evolution cannot do CHOICE! Selection requires some type of choice, and choice requires some type of Psyche or Mind! By definition, in principle, evolution is dumb and blind without a soul or a mind. The theory of evolution is self-defeating because evolution of any kind can't do selection or choice. Evolution doesn't exist as some type of Psyche or Person who is capable of making choices or doing selection. The false is falsified by the truth; and, the truth is repeatedly experienced and observed. The theory of evolution is obviously false because evolution of any type by definition in principle cannot do selection, psyche, or choice.

Evolution of any type cannot do science, but the Human Psyche and God's Psyche certainly can. They've been caught in the act of doing so.

With just a bit of scientific observation, it's easy to see that Kin Selection is a *fictional ad hoc just-so story* that they made up out of thin air after-the-fact to match with what we have experienced and observed from Intelligent Beings or the Human Psyche. They took Kin Selection, a philosophical idea or religious idea, and they personified it, humanized it, anthropomorphized it, and deified it. Kin Selection and the Theory of Evolution are man-made idols or man-made gods. The Theory of Evolution is our modern-day form of idolatry, which is a belief in false gods that are incapable of delivering the goods.

The theory of evolution is self-defeating because it can't do what they say it does. The genes can't be nice to each other without a psyche or a soul; and, if the genes have a psyche or a soul, then the very existence of Psyche or Syntropy or Soul falsifies Materialism, Naturalism, Darwinism, Nihilism, Behaviorism, Determinism, Physical Reductionism, Atheism, and the Theory of Evolution. The false is falsified by the truth; and, the truth is repeatedly experienced and observed.

Mark My Words

—

Source

God Is in the Light: God is light, and in Him is no darkness at all.

https://www.amazon.com/dp/B07168S37N

Reference

Myers, D. G. (2010). *Social Psychology* (10th ed.). New York: McGraw-Hill.

Proof of God's Existence

Observation or Premise: Anything that was obviously made obviously had a Maker or Creator who made it.

Observation or Premise: A genome was obviously made.

Logical Conclusion: Therefore, a genome obviously has a Maker or a Creator who designed it, programmed it, engineered it, field-tested it, fine-tuned it, made it, manufactured it, and deployed it.

—

This syllogism is both a Philosophical Proof of God's Existence and a Scientific Proof of God's Existence. It works. It's true. Every aspect of it is true, because both the premises and the conclusion have been experienced and observed in real life by real people. This Scientific Proof of God's Existence is both philosophically *valid* and observationally *sound*. It, and similar ones like it, convinced me that God does indeed exist.

Variations on this syllogism were my first Scientific Proof of God's Existence; and, I first built one based upon the non-existence of abiogenesis, or the non-existence of macro-evolution. After science itself proved to me that God must of necessity exist, then I simply KNEW that God exists, even though I still have yet to get to know God.

In 2016, when I first realized that NO type of evolution – chemical evolution, micro-evolution, macro-evolution, spontaneous generation, abiogenesis, random mutations, or natural selection – is capable of designing, creating, and implementing genomes from atoms or from scratch, I just KNEW that God exists because I KNEW why He must of necessity exist. God must exist in order to have done ALL of the science, organization, proteins, and genomes which the different types of evolution could NEVER have done.

In a very real sense, the Theory of Evolution proved to me that God exists. In other words, the inability of the different types of evolution to design, create, manufacture, and deploy proteins and the matching genes to go along with those proteins BECAME convincing scientific proof of God's necessity, which effectively became Scientific Proof of God's Existence.

Based upon this same exact theme, I developed many other Scientific Proofs of God's Existence and eventually used the Scientific Method itself to falsify Materialism, Naturalism, Darwinism, Atheism, and the Theory of Evolution.

I kept a log of my ongoing discoveries in the following books:

1. **Summary Of: The Theory of Evolution Proves that God Exists**

    https://www.amazon.com/dp/B01GQCWED6

    https://www.amazon.com/dp/1521130485

2. **The Theory of Evolution Proved to Me that God Exists: Why I Am No Longer an Atheist and Why I No Longer Believe in the Theory of Evolution**

    https://www.amazon.com/dp/B01HZYBZ7K

    https://www.amazon.com/dp/1521131228

3. **The Scientific Method Proves That the Theory of Evolution Is False**

    https://www.amazon.com/dp/B01IAAIRT2

    https://www.amazon.com/dp/1521133611

4. **Using the Scientific Method to Eliminate the Usual Suspects and to Prove the Truth**

    https://www.amazon.com/dp/B01J6STHP0

    https://www.amazon.com/dp/1521133581

This was an era of transition for me. I used to be a Materialist, Naturalist, Nihilist, and Atheist. To develop a Scientific Proof of God's Existence that I actually KNEW to be true, as well as using the Scientific Methods to falsify Materialism, Naturalism, Darwinism, Atheism, and the Theory of Evolution represented a MAJOR paradigm shift in my way of thinking and doing science.

Once I finally KNEW the Truth, there was no going back to the deceptions and the lies; and, I was able to progress forward from there and make additional adjustments and improvements to my science, my worldview, and my philosophical underpinnings.

Although it is technically true – due to the *affirming the consequent* logic fallacy – that it's impossible to use the Scientific Methods to prove anything true, there is a legitimate work-around that actually produces the same result. This work-around is a philosophically *valid* and an observationally *sound* way to find the truth through a process of elimination.

The way this is done in practice is that we use the Scientific Methods and *negating the consequent* to falsify and eliminate everything that is false such as Materialism, Naturalism, Darwinism, Atheism, and their derivatives so that ONLY the Truth remains.

Think about it logically.

If you successfully eliminate everything that is false – everything that has NEVER been experienced nor observed by anyone, not even God – then ONLY the Truth will remain, because ONLY the observed and the verified and the proven will remain. The Truth is whatever has been experienced, lived, and observed by Someone Psyche sometime somewhere. Science is observation, or it should be.

Sherlock Holmes taught me how this process of elimination works in principle or practice.

How often have I said to you that when you have eliminated the impossible [and the false]**, whatever remains, however improbable, must be the truth? — Sherlock Holmes.**

Sherlock Holmes and the Philosophy of Science taught me how to use the Scientific Method to falsify the Theory of Evolution, Materialism, Naturalism, and Atheism; and, observations from real life people like you and me falsified Nihilism for me.

If you successfully eliminate everything that is false, then ONLY the Truth will remain; and, that's precisely how we use the Scientific Methods to "prove" the truth, by using the Scientific Methods to eliminate everything that is false or everything that has NEVER been experienced nor observed. It works!

Mark My Words

—

Source

Science 2.0: I Upgraded My Science.

https://www.amazon.com/dp/B0771K6WTX

Convincing Proof of God's Existence

I used to be a Materialist, Naturalist, Nihilist, and Atheist back in 2012. I was a practitioner of Scientism – the philosophical belief that science and the Scientific Method is the ONLY way to find and know the truth. I also believed that Natural Selection could do what they say it does. I simply knew that Natural Selection exists and that its effects are real. I had bought into the party line.

At the time, I interacted with Materialists, Naturalists, Darwinists, Nihilists, and Atheists online. I considered them kindred spirits. They were like me. I felt as if I were a part of them. They were my people.

One of my atheistic friends told me that there is NO scientific proof of God's existence and that there will NEVER be any scientific proof of God's existence. I believed him. I'd never found any convincing proof of God's existence. I figured that God's existence had to be taken on blind faith as being real and true. That is what the Atheists teach.

But one has to consider the messenger in order to truly understand the message.

We defined "science" as Materialism, Physicalism, Naturalism, Darwinism, Nihilism, and Atheism. Obviously, Atheism, Naturalism, or "science" will never prove to anyone that God exists. Science defined as Materialism, Naturalism, and Atheism defines God out of existence, a priori, by definition and in principle.

We were RIGHT! Science as we defined "science" would NEVER produce any scientific proof of God's existence.

How you choose to define things makes all the difference in the world! It's a choice! And, our choices have consequences whether we realize it or not. Our beliefs are chosen into existence.

Shifting One's Perspective

Now, consider what happens when we choose to define "science" as Fine-Tuning.

Science is Fine-Tuning.

Wow!

Instantly, we are looking at thousands if not millions of different Scientific Proofs of God's Existence. Each ACT of Fine-Tuning or Science is a miniature scientific proof of God's existence.

Examine the following syllogism. It's a Philosophical Proof of God's Existence. Because it's a syllogism, if the premises are true, then the conclusion has to be true as well.

Scientific Observation or Premise: Anything that was obviously fine-tuned obviously had a Fine-Tuner who fine-tuned it.

Scientific Observations: A genome was obviously fine-tuned. A protein and a gene were obviously fine-tuned with a specific purpose in mind. Our physical universe and the physical constants were obviously fine-tuned. Your computer and your car were obviously fine-tuned. This computer program that I'm using was obviously fine-tuned. Your physical body was clearly fine-tuned for specific purposes. The neurotransmitters within your brain were clearly and obviously fine-tuned. It's undeniable.

Scientific Conclusion: ALL of these things were obviously fine-tuned which means that they obviously have a Fine-Tuner who fine-tuned them.

If you choose to define Science as Fine-Tuning, then instantly this syllogism and philosophical proof of God becomes a convincing, valid, solid, and sound Scientific Proof of God's Existence.

Instantly, we have Scientific Proof of God's Existence; and, it happens automatically by choosing to define Science as Fine-Tuning rather than choosing to define science as Materialism, Naturalism, Darwinism, Nihilism, and Atheism.

Do you see how that works?

It works wonderfully, but only if you choose to allow it to work. It's a choice! Our beliefs are chosen into existence.

However, if you continue to define science as Materialism, Naturalism, Darwinism, Nihilism, and Atheism, then you will NEVER have any scientific proof of God's existence. That's just the way things work.

There's no compelling reason to define science as Materialism, Naturalism, Darwinism, Nihilism, and Atheism. These things are not the best explanation, anyway. I mean think about it! Who did ALL of that Fine-Tuning or Science BEFORE we human beings arrived on the scene?

When we are talking about the physical constants, this physical universe, and physical matter, we are talking about a Fine-Tuner who was doing Science or Fine-Tuning BEFORE this physical universe began. That's the very definition of a God, is it not?

By choosing to define Science as Fine-Tuning, instantly we have compelling and believable and convincing Scientific Proof of God's Existence. That's one of the reasons why I go overboard trying to define everything and compare everything. I learn best and learn most through comparison and contrast.

Mark My Words

—

Source

Scientific Proof of God's Existence: Finding God Where the Atheists Refuse to Look for Him

https://www.amazon.com/dp/B07B26CRHX

Defining Syntropy

So, what do we observe?

Science is all about observation and experience, or it should be! Philosophical speculation is worthless if it isn't backed up by observations and experience.

So, what do we observe?

We observe entropy.

We observe things running down, wearing out, and ceasing to exist. We observe disease, death, cancer, corruption, and extinction – all the result of entropy or random mutations. We see disorder, chaos, and death ever on the increase.

We observe the stars burning out and blowing up. We observe our physical universe moving slowly towards heat death.

We observe that Materialism, Naturalism, Darwinism, Nihilism, Atheism, Classical Physics, and the Theory of Evolution are based exclusively on entropy. Materialism and Naturalism are entropy. Nihilism is entropy. Physical matter is subject to entropy. Evolution is entropy. The whole thing is moving towards entropy and heat death.

So, where did all the Order, Organization, and Syntropy come from in the first place? There had to be an initial infusion of Syntropy, or all of this subsequent entropy wouldn't have been possible. This is Logic 101.

Where's the Syntropy? It should be observable too. Syntropy has to exist, or there wouldn't be all this entropy. So, what is Syntropy?

Syntropy has a number of different definitions.

> **The negentropy has different meanings in information theory and theoretical biology. In a biological context, the negentropy (also negative entropy, syntropy, extropy, ectropy, or entaxy) of a living system is the entropy that it exports to keep its own entropy low; it lies at the intersection of entropy and life. In other words, negentropy is reverse entropy. It means things becoming more orderly. By 'order' is meant organization, structure, and function: the opposite of randomness or chaos.**

https://en.wikipedia.org/wiki/Negentropy

Syntropy is typically defined as being the opposite of entropy. So, what's entropy, and what's the opposite of entropy?

Entropy is corruption, random mutation, evolution, disease, disorder, death, and extinction. The different types of evolution are entropy, and the different types of evolution are prevented from happening by entropy. The Theory of Evolution is based upon entropy or death – that's never going to work.

This means that Syntropy must be some type of incorruption, healing, order, organization, life, light, intelligence, atonement, resurrection, immortality, eternal life, mercy, grace, and love. It sounds like heaven to me. It sounds like God.

Syntropy is order, organization, structure, life, mercy, grace, and love – the opposite of randomness, chaos, entropy, death, extinction, classical physics, materialism, naturalism, evolution, and creation ex nihilo or atheism. Syntropy is NOT magic. Syntropy has always existed and will always exist. Syntropy is eternal and everlasting, without a beginning of days or an end of years. Psyche is Syntropy. Quantum Mechanics is a type of Syntropy. The Priesthood Power of God is Syntropy. Relationships or friendships are a type of Syntropy. Love is Syntropy – the more you give away, the more you have. According to the Near-Death Experiencers, all of the different universes are based upon Love or Syntropy. The whole shebang is based upon Love, God's Love.

In contrast, creation ex nihilo is atheism – the creation of something from nothing by nothing. Creation ex nihilo is impossible. Creation ex nihilo is magic; and, there's no such thing as magic. Creation ex nihilo is spontaneous generation; and, spontaneous generation or evolution was falsified in 1859 by Louis Pasteur, the very same year that Charles Darwin published "On the Origin of Species".

Syntropy is what chemical evolution and macro-evolution pretend to be but are not. Instead, spontaneous generation, or chemical evolution, or macro-evolution is prevented from happening by entropy. Creation ex nihilo is based upon nothing; whereas, Syntropy is based upon Quantum Mechanics, Psyche, Relationships, Life, and Love.

Remember, Syntropy MUST of necessity exist, or all of that subsequent entropy would not have been possible. In the beginning, our physical universe had to receive a huge infusion of Syntropy, or none of that subsequent entropy could have happened or been possible. Some type of Syntropy or God MUST exist, or we would not exist as physical beings. Entropy or physical matter cannot design and create. Entropy can only corrupt and destroy. Entropy moves everything towards disorder and death. In contrast, Syntropy moves everything towards order, life, and love. Syntropy is LIFE and LOVE.

John 10: 10:

The thief cometh not, but for to steal, and to kill, and to destroy: I am come that they might have life, and that they might have it more abundantly.

To have life more abundantly!

How?

SYNTROPY!

How?

Atonement, Resurrection, Priesthood Power, and Eternal Life.

Entropy cometh to steal, kill, and destroy.

Jesus Christ or Syntropy came that we might have life and have it more abundantly. Syntropy is like mercy, grace, and love – the more you give away, the more you have. Syntropy is expansive and eternal. Entropy deflates and destroys.

Materialism, Naturalism, Darwinism, Nihilism, Behaviorism, Determinism, Physical Reductionism, Behaviorism, Atheism, Classical Physics, Psychiatry, and Pharmacology are based exclusively on entropy. Entropy is death and extinction. In contrast, everything good is based upon Syntropy, Relationships, Quantum Mechanics, Psyche, or Love. The drugs can't do relationships or psyche therapy; but, the various different types of Syntropy certainly can. Syntropy is psyche therapy. Syntropy is friendship and love.

Remember, everything we observe is based upon quantum laws and physical laws. The quantum laws and physical laws were ordered or organized by God, or Syntropy, or Psyche, or Intelligence, or Life. They work because God or Syntropy makes them work.

Syntropy is Divine Light, Life Energy, Order, Organization, Atonement, Life, the Spark of Life, Life Force, Psyche, Intelligence, Quantum Mechanics, the Priesthood Power of God, and Reverse Entropy. The Atonement of Christ is Syntropy. It reverses the effects of death and entropy.

Syntropy means "without beginning of days or an end of years". Syntropy means eternal and everlasting. Syntropy means immortal. Psyche is Syntropy. Quantum Mechanics is Syntropy. Spirit Matter or Dark Matter is Syntropy. God's Priesthood Power is Syntropy. The Light of Christ, or Dark Energy, or the Zero-Point Field is Syntropy. God is Syntropy. They are without a beginning of days or an end of years.

Death, extinction, or natural selection is entropy.

Random mutations are entropy.

Classical physics is entropy. Materialism and Naturalism are based upon entropy. Darwinism is based exclusively on entropy. Entropy cannot design and create. It can only destroy.

The chemical evolution of genes and proteins from atoms is prevented from happening by entropy. Macro-evolution, chemical evolution, abiogenesis, and spontaneous generation are physically impossible because they are prevented from happening by random diffusion, chaos, and entropy.

Evolution of any kind is entropy, randomness, chance, or chaos. The opposite of evolution, random mutations, and natural selection are design, planning, teleology, purpose, programming, creation, order, organization, life, law, quantum mechanics, psyche, intelligence, and Syntropy. The Theory of Evolution is based upon entropy, and entropy prevents evolution of any kind from happening in the first place. The opposite of the Theory of Evolution is Syntropy.

Remember, Syntropy or Quantum Mechanics is incomplete, ineffective, illogical, inefficient, meaningless, purposeless, and worthless if it is defined by, restrained by, limited by, restricted by, and polluted by Entropy, Chaos, Randomness, Materialism, Naturalism, Darwinism, Nihilism, and Atheism. Syntropy is the opposite of entropy. Syntropy is the opposite of Materialism and Naturalism. Syntropy is effectively the opposite of Classical Physics and Determinism.

To those of us who are restricted by, limited by, and restrained by entropy, physical bodies, and this physical universe, Quantum Mechanics or Transdimensional Physics seems to be magic; but, Quantum Mechanisms or Syntropic Mechanisms are the LAW and therefore are the NORM where the spirit world, or transdimensional world, or quantum world, or the prime construct is concerned. Syntropy, or Power, or Light brings order out of randomness and chaos. In other words, Syntropy reverses entropy or Syntropy overcomes entropy.

Syntropy is the Prime Construct. Syntropy is the organizing force of the multiverse. Syntropy is God's Love. Syntropy is Quantum Mechanics or Transdimensional Physics. Action at a Distance is based upon Syntropy. Syntropy is eternal and everlasting. The Priesthood Power of God is described as being "without beginning of days or an end of years". The Priesthood Power of God is Syntropy or Quantum Mechanics. Syntropy renews itself and enhances itself as it renews us and enhances us. The more God gives away, the more He has when He is done.

Syntropy or Quantum Mechanics ends up being Scientific Proof of God's Existence. Entropy or physical matter ends up being Scientific Proof of God's Existence.

Does the Second Law of Thermodynamics Prove the Existence of God?

- By John M. Cimbala

Professor of Mechanical Engineering

The Pennsylvania State University

In this short article, I summarize my ideas about the second law of thermodynamics, and why I believe it points to a creator God.

This article also appears in the book ***In Six Days - Why Fifty Scientists Choose to Believe in Creation***, edited by John F. Ashton, and published by Master Books, Green Forest, AR. Copyright 2000 by John F. Ashton. The book is available on-line at Creation Ministries International and at Answers in Genesis.

A formal definition of the second law of thermodynamics is "In any closed system, a process proceeds in a direction such that the unavailable energy (the entropy) increases." In other words, in any closed system, the amount of disorder always increases with time.

Things progress naturally from order to disorder, or from an available energy state to one where energy is more unavailable. A good example: a hot cup of coffee cools off in an insulated room. The total amount energy in the room remains the same (which satisfies the first law of thermodynamics). Energy is not lost, it is simply transferred (in the form of heat) from the hot coffee to the cool air, warming up the air slightly. When the coffee is hot, there is available energy because of the temperature difference between the coffee and the air. As the coffee cools down, the available energy is slowly turned to unavailable energy. At last, when the coffee is room temperature, there is no temperature difference between the coffee and the air, i.e. the energy is all in an unavailable state. The closed system (consisting of the room and the coffee) has suffered what is technically called a "heat death." The system is "dead" because no further work can be done since there is no more available energy. The second law says that the reverse cannot happen! Room temperature coffee will not get hot all by itself, because this would require turning unavailable energy into available energy.

Now consider the entire universe as one giant closed system. Stars are hot, just like the cup of coffee, and are cooling down, losing energy into space. The hot stars in cooler space represent a state of available energy, just like the hot coffee in the room. However, the second law of thermodynamics requires that this available energy is constantly changing to unavailable energy. In another analogy, the entire universe is winding down like a giant wind-up clock, ticking down and losing available energy. Since energy is continually changing from available to unavailable energy, someone had to give it available energy in the beginning! (I.e. someone had to wind up the clock of the universe at the beginning.)

Who or what could have produced energy in an available state in the first place? Only someone or something not bound by the second law of thermodynamics. Only the creator of the second law of thermodynamics could violate the second law of thermodynamics and create energy in a state of availability in the first place.

As time goes forward (assuming things continue as they are), the available energy in the universe will eventually turn into unavailable energy. At this point, the universe will be said to have suffered a heat death, just like the coffee in the room. The present universe, as we know it, cannot last forever. Furthermore, imagine going backwards in time. Since the energy of the universe is constantly changing from a state of availability to one of less availability, the further back in time one goes, the more available the energy of the universe. Using the clock analogy again, the further back in time, the more wound up the clock. Far enough back in time, the clock was completely wound up.

The universe therefore cannot be infinitely old. One can only conclude that the universe had a beginning, and that beginning had to have been caused by someone or something operating outside of the known laws of thermodynamics.

Is this scientific proof for the existence of a Creator God? I think so. Evolutionary theories of the universe cannot counteract the above arguments for the existence of God. Evidence such as this helped to convince me to believe in God, and to accept His plan of salvation through His son Jesus Christ. For further details about my conversion to Christianity, I have written a short testimony.

http://www.personal.psu.edu/faculty/j/m/jmc6/second_law.html

I believe 100% that this is the way things truly are. Some type of Syntropy, Psyche, Intelligence, or God MUST EXIST in order to have provided that initial infusion of Order, Organization, Available Energy, or Syntropy into this physical universe so that all of the subsequent entropy would have been possible. Entropy or physical matter couldn't have provided that initial infusion of Syntropy. WE KNOW for a fact that that Available Energy or Syntropy had to come from someplace else besides this physical universe, because this physical universe is based upon entropy, and restricted to entropy, and can only produce entropy.

Somebody somewhere knows how to wind up a universe, or we wouldn't be here. If the original construct were nothing but physical matter or entropy, then we wouldn't be here. Heat death would prevail. Someone Psyche has to know how to reverse entropy, or we wouldn't be here. There has to be Syntropy somewhere in some kind of Syntropy Realm or Quantum Realm, or we wouldn't exist. It's time that we, as scientists, finally choose to face reality and accept the fact that the ONLY reason we exist as physical beings is due to some type of Syntropy, which was infused into this physical universe at the beginning of this physical universe.

Entropy of Any Kind Cannot Design and Create

Entropy cannot design, create, and originate anything. It's physically impossible. There's no such thing as the spontaneous generation of Order and Syntropy from entropy and physical matter. That's physically impossible. Entropy is a dead-end. Entropy is randomness, disorder, heat death, extinction, evolution, death, and chaos.

Evolution is entropy. The Theory of Evolution, Materialism, Naturalism, Atheism, Nihilism, and Classical Physics are based exclusively upon entropy and restricted to entropy. Evolution or entropy couldn't have made us. It's physically impossible. The different types of evolution are prevented from happening by entropy. There has to be some counterforce opposing entropy, or we wouldn't exist. Someone Psyche had to load this physical universe with Syntropy, or the subsequent entropy would not have been possible.

Syntropy is available energy. The available energy had to come from someplace, or we wouldn't exist. It had to come from someplace besides this physical universe, because this physical universe is based on entropy. The Syntropy came first, and the Syntropy has always existed. If it didn't, then we wouldn't exist. This is Logic 101.

Syntropy or Quantum Mechanics is the BEST, most-realistic, most-explanatory, and most all-inclusive way to support, validate, and understand science. Syntropy is the BEST way to explain the order and organization within this physical universe. According to the Big Bang Theory, this physical universe should be nothing more than one big homogenous ball of expanding gas with every atom moving away from every other atom.

It is not!

Why?

The answer is Syntropy. Someone Psyche has infused Syntropy or Order into certain parts of this physical universe so as to bring order, organization, life, stars, planets, and galaxies into existence. This Syntropy could only have come from the quantum realm or the psyche realm, because Syntropy does NOT exist natively or naturally here in this physical realm. The physical is ruled by entropy. The very existence – the observed and proven existence – of Quantum Mechanics, or Transdimensional Physics, or Syntropy FALSIFIES Materialism, Naturalism, Darwinism, Nihilism, Behaviorism, Determinism, Scientism, Atheism, and even Classical Physics. The very existence of Syntropy negates or counteracts entropy. Thanks to Syntropy, our physical universe is NOT a closed system.

Syntropy is like love – the more of it you give away, the more of it that you have when you are done. Syntropy, order, or love is completely different than entropy, corruption, disorder, and death.

It's interesting to observe that the Materialists, Naturalists, Darwinists, Nihilists, and Atheists have succeeded in eliminating Syntropy from our vocabulary. Our spell-checkers have never seen the word "Syntropy". According to the Materialists and Naturalists, Syntropy or Psyche does not exist. These people don't believe in Syntropy or the Atonement of Christ. These people don't believe in Syntropy, or the Non-Physical, or Action at a Distance, or Quantum Mechanics. It's impossible to believe in, understand, and accept something that you have never heard of and refuse to learn about. You will never find Syntropy if you never go looking for it. I KNOW, because I used to be a Materialist, Naturalist, Nihilist, and Atheist; and, I had never heard of Syntropy, and didn't believe that such things exist. Syntropy is a new discovery for me. I'd never really thought of it, until the past year or so; and, I'm 57 years old.

When it comes to Syntropy, I practically had to invent the term, just so that I would have a word to use to describe the phenomenon. The whole thing started when I asked myself what the opposite of entropy is. Everything has its opposite. I realized that the different types of evolution are entropy and are prevented from happening by entropy. There's no such thing as spontaneous generation or macro-evolution thanks to entropy. The Theory of Evolution is impossible thanks to entropy. The Materialists, Naturalists, Darwinists, and Atheists are prevented from understanding and accepting Syntropy, thanks to entropy. Entropy is so dominant in our society that practically no one has ever heard of Syntropy. Your typical scientist won't have a clue when it comes to Syntropy; and, they're supposed to be the best and brightest among us.

Summarizing Syntropy

There's physical matter, and then there's everything else.

Entropy seems to be restricted exclusively to physical matter. Entropy is a function of physical matter. Everything else seems to be based upon Syntropy.

What a fascinating observation to make!

ALL of those invisible forces, fields, and quantum waves are based upon Syntropy. They NEVER burn out, and they NEVER wear out, because they are eternal and everlasting, without a beginning of days or an end of years.

There's physical matter or entropy; and then, there's everything else which is based upon Syntropy. There's classical physics, and then there's Quantum Mechanics or Transdimensional Physics. The one is based upon entropy, and the other is based upon Syntropy. This just might be the most interesting and useful scientific observation of all-time. Its explanatory power is through the roof, is it not? Syntropy has to exist, or entropy and physical matter would not exist.

Remember, Syntropy has to exist, or we would not exist. Our very existence here in this physical universe in these physical bodies tells us that someone somewhere KNOWS how to do Syntropy or Organization where physical universes, genomes, proteins, eyes, brains, and physical bodies are concerned.

Organization is Syntropy. Psyche is Syntropy. Quantum Mechanics is Syntropy. The Priesthood Power of God is Syntropy. Spirit Matter is Syntropy. God is Syntropy. Quantum Mechanics or Syntropy is the Priesthood Power of God. Syntropy is a new and better way of doing science, explaining science, and understanding science. Syntropy is eternal and everlasting.

Syntropy means "without beginning of days or an end of years". Syntropy or Eternal Existence is the opposite of creation ex nihilo. Creation ex nihilo or the catholic (universal) version of creation is Atheism – creation by nothing from nothing. I'm not a creationist. I'm definitely not that type of creationist. Creation ex nihilo is impossible. Not even God could do creation ex nihilo, because it's impossible. Creation ex nihilo is Atheism and magic. Creation ex nihilo is also unscientific, because it's impossible. Creation ex nihilo is the exact opposite of Syntropy. The one is magic, and the other is organization and science.

God is a scientist, and not a magician. God organized this physical universe from pre-existing Dark Matter or pre-existing Spirit Matter. Remember, it's the physical matter that is the anomaly – everything else is based upon Syntropy. The physical matter is visible to us with our physical eyes; and, everything else is not. We detect all the dozens of different invisible forces, fields, and laws by the effect that they have on physical matter. Physical matter or entropy is the anomaly. Everything else, or Syntropy, or Quantum Mechanics is the norm.

Syntropy concerns the things that are without a beginning of days or an end of years. Syntropy is the opposite of Materialism, Naturalism, Atheism, Entropy, and Classical Physics. Syntropy is the original prime construct from which this physical universe and entropy were made.

Our physical universe had a beginning. It was ordered and organized by **someone** who has NO beginning and will have no end – Syntropy, Psyche, or God. Our physical universe or physical matter was built, or made, or organized from **something** that had NO beginning and will have no end – Spirit Matter, Dark Matter, or Quantum Objects. Creation ex nihilo is falsified by Syntropy and all of that pre-existing pre-physical Spirit Matter or Dark Matter.

Observation or Premise: Anything that was obviously made obviously had a Maker or Creator who made it.

Observation or Premise: A genome was obviously made.

Logical Conclusion: Therefore, a genome obviously has a Maker or a Creator who designed it, programmed it, engineered it, field-tested it, fine-tuned it, made it, manufactured it, and deployed it.

To make or create something is to infuse it with Order or Syntropy. Organization, Engineering, Manufacturing, and Creation are different types of Syntropy. Syntropy is the opposite of spontaneous generation or magic. There is NO such thing as spontaneous generation, chemical evolution, creation ex nihilo, or macro-evolution. These things are prevented from happening by entropy or the second law of thermodynamics. These things are prevented from happening by physical matter. It requires some kind of Syntropy or Psyche in order to design and create and make things from scratch.

You know what this means, don't you?

It means that evolution can't happen. Evolution is impossible. The different types of evolution are prevented from happening by entropy and physical matter. Entropy and physical matter actually falsify the Theory of Evolution. Random mutations are entropy. Random chance or random diffusion is based upon entropy. Natural selection leads us towards entropy, death, and extinction. The chemical evolution of genes and proteins from atoms is prevented from happening by entropy. Spontaneous generation, abiogenesis, chemical evolution, and macro-evolution are prevented from happening by entropy or physical matter. Physical matter imposes a structure or an order that prevents evolution of any type from happening. Physical matter is the basis of the ultimate consensus reality.

It would be unpredictable chaos or evolution if it weren't for the order, structure, and limitations being imposed upon us by physical matter or by consensus. Natural evolution would not be a good thing. It would be chaos. That's why God, physical matter, and entropy prevent any type of macro-evolution from happening. It's bad enough as it is that our genes are allowed to mutate. Think of all the problems which that causes! Now multiply that by a billion to imagine what it would be like if all the other types of evolution were allowed to happen.

Remember, classical physics or entropy prevents evolution. It was designed to do so. This just might be the most significant scientific discovery of all time.

It actually requires some type of Syntropy, Psyche, or Intelligence in order to be able to design, create, program, organize, engineer, field-test, fine-tune, manufacture, and bring physical things into existence. Some type of Syntropy is needed in order to bring life into existence, especially physical life. Psyche, Intelligence, or Quantum Non-Local Consciousness is LIFE, which means that it is Syntropy.

Remember, Syntropy has to exist, or we wouldn't exist. It's time that we, as scientists, finally start to use Syntropy, Psyche, and Quantum Mechanics to explain how things really work in every realm of existence, including both this physical realm and the non-local quantum realm. It's time for us to take Quantum Mechanics or Syntropy seriously, instead of ridiculing it, mocking it, and dismissing it as pseudo-science. It's time for us to upgrade our science to Science 2.0 and start allowing ALL of the evidence into evidence.

God's Psyche or God's Intelligence is Syntropy. It had no beginning and it will have no end. Your Psyche, or Intelligence, or Quantum Non-Local Consciousness is Syntropy. It had no beginning and it will have no end.

In contrast, entropy is death – physical death. It has an end. Physical matter or entropy is the anomaly. WE KNOW it is so, because everything else that we have encountered is based upon Syntropy – it NEVER ages, and it never wears out. Remember, Syntropy never ages and it never wears out. It is without a beginning of days or an end of years. Syntropy is eternal and everlasting. Compared with that, physical matter or entropy is indeed the anomaly or odd ball.

Physical bodies and physical universes begin and end thanks to both syntropy (the beginning of a physical universe and a physical body) and entropy (the ending of a physical universe and physical body); but, Syntropy or Quantum Mechanics or Psyche has no beginning and will have no end, because it is eternal and everlasting. Syntropy is light, life, love, psyche, intelligence, power, order, organization, and God. The Atonement of Christ is Syntropy. It has to be, because its purpose is to help us overcome entropy, corruption, sin, and death.

The obvious fact – that some type of Syntropy MUST exist or we would not exist – makes logical sense to me. The NEED for Syntropy is so glaringly obvious that Syntropy MUST exist; and therefore, its existence MUST be true. WE KNOW that some type of Syntropy exists, because entropy exists. We know that Syntropy exists because physical matter exists. There's physical matter or entropy; and then, there's Syntropy. We have observed that everything else that we have encountered besides physical matter is based upon Syntropy – it never ages nor gets old. It's eternal and everlasting.

Like I said, this just might be the most significant and useful scientific discovery of all time. Its explanatory power is off the charts.

If physical matter or entropy were the ONLY thing that exists, as the Materialists and Naturalists claim, then it would be physically impossible for our genes, proteins, eyes, brains, and physical bodies to exist. Entropy prevents physical matter from organizing spontaneously into functional genes, proteins, eyes, brains, and physical bodies. Evolution is entropy, and that means that the different types of evolution or the different types of entropy prevent the Theory of Evolution from becoming true.

If entropy prevailed and there were no Order or no Syntropy, then the stars, planets, and galaxies would not form and would not exist. The Syntropy, or the available energy, or the zero-point field, or the love has to exist, or there would be no stars and no suns. There would be no physical life anywhere without some type of Syntropy or Order in this physical universe. This is rationally obvious to anyone like me who spends even the smallest amount of time thinking about it.

Thanks to Syntropy or Quantum Mechanics, we KNOW that Materialism, Naturalism, Darwinism, Nihilism, and Atheism are false and have been falsified. Thanks to Syntropy, entropy is not the end. You will go on long after your physical brain is dead and gone. Thanks to Syntropy, your consciousness or psyche is eternal and everlasting. You have always existed, and you will always exist. Psyche is Syntropy. Relationships are Syntropy. Love is Syntropy. God is Syntropy. The best part of you is Syntropy, and it's capable of counteracting the effects of entropy and surviving the death of your physical brain.

Mark My Words

—

Source

I Am Not a Creationist: So What Am I?

https://www.amazon.com/dp/B071XTM8XY

Syntropy in Defense of Quantum Mechanics: The Answer to Life, the Universe, and Everything

https://www.amazon.com/dp/B07BPT3W8R/

References with a Similar Theme

God Is in the Light: God is light, and in Him is no darkness at all.

https://www.amazon.com/dp/B07168S37N

Quantum Mechanics from a Non-Physical Spiritual Perspective.

https://www.amazon.com/dp/B01J023TGU

https://www.amazon.com/dp/1521132380

Quantum Neuroscience: The Answer to Life, the Universe, and Everything.

https://www.amazon.com/dp/B079Z6QQQB

Science 2.0: I Upgraded My Science.

https://www.amazon.com/dp/B0771K6WTX

Scientific Proof of God's Existence: Finding God Where the Atheists Refuse to Look for Him

https://www.amazon.com/dp/B07B26CRHX

Scientific Proof of God's Existence: A Primer

https://www.amazon.com/dp/B071713NNL

https://www.amazon.com/dp/1521325170

Syntropy Is Order, Power, Organization, Creation, and Life

One of my most useful scientific discoveries is the fact that Quantum Mechanics or Syntropy is the NORM. Psyche and Spirit Matter are based upon Quantum Mechanics or Syntropy. The primal construct or primal reality is Quantum Mechanics or Syntropy – it's without a beginning of days or an end of years. Quantum Mechanics and Syntropy have NO physical limitations. This reality has been experienced and observed. Observation and experience are Science. Quantum Mechanics is one of our best-proven and most-used Scientific Disciplines.

So, why physical matter, or whence physical matter?

Physical matter was not a part of the Primal Construct. Physical matter is not a part of the NORM. God infused a part of his power, energy, psyche, and glory into each particle of physical matter thereby converting it from spirit matter to physical matter.

Why would God do such a thing?

The goal was to create the Ultimate Consensus Reality. A physical reality is the Ultimate Consensus Reality. It is highly reliable, dependable, and predictable. It's ordered and lawful. It keeps the Commandments of God – the physical laws. In a physical reality, I can depend on this paper being here on my computer and cloud drive tomorrow when I go looking for it. I can depend upon my physical body being here tomorrow as well in basically the same shape it was yesterday.

God deliberately infused a part of Himself into each particle of physical matter in order to minimize the uncertainties natively and naturally associated with spirit matter. When it comes to physical matter – all of the physical laws, the sub-light speed limit associated with the Theory of Relativity, and the deliberately short De Broglie Wavelength associated with physical matter greatly limit and reduce the Quantum Uncertainties which are a native, inherent, and original part of spirit matter.

A ton of interesting science, knowledge, and truth can be acquired by making a detailed comparison between Syntropy and Entropy. What we are in fact doing is comparing Quantum Physics with Classical Physics. We are comparing the Primal Construct with the Ultimate Consensus Reality. The one has always existed, and the other one was made by God to reduce the Quantum Uncertainties to a minimum.

Do you see how that works and why it is necessary?

The Ultimate Consensus Reality, a physical reality, wouldn't be possible without the physical laws and classical physics which were deliberately designed and put into place by God to greatly reduce and minimize the Quantum Uncertainties that are an integral and native part of spirit matter to begin with.

God deliberately and purposefully keeps the physical atoms within your physical body from quantum tunneling away from you at will. That's what the physical laws or the Commandments of God do. They greatly reduce and limit the Quantum Uncertainties that are a native and original part of spirit matter.

There's tons of scientific evidence proving that the Physical Constants were precisely fine-tuned by some kind of Fine-Tuner, which means that the Physical Constants were deliberately made by Someone Psyche or Someone Intelligent in order to reduce and minimize the effects of Quantum Uncertainty and thereby greatly increase the reliability, dependability, and predictability of our Physical Consensus Reality. It was done with a purpose in mind. It works because God makes it work.

> **Scientific Observation or Premise: Anything that was obviously fine-tuned obviously had a Fine-Tuner who fine-tuned it.**

> **Scientific Observations: A genome was obviously fine-tuned. A protein and a gene were obviously fine-tuned with a specific purpose in mind. Our physical universe and the physical constants were obviously fine-tuned. Your computer and your car were obviously fine-tuned. This computer program that I'm using was obviously fine-tuned. Your physical body was clearly fine-tuned for specific purposes. The neurotransmitters within your brain were clearly and obviously fine-tuned. It's undeniable.**

Scientific Conclusion: ALL of these things were obviously fine-tuned which means that they obviously have a Fine-Tuner who fine-tuned them.

This is one of the most convincing Scientific Proofs of God's Existence that I have encountered so far. It works because the Premises and the Conclusion have been experienced and observed. It works because it is Science.

Syntropy versus Entropy

Let's compare Syntropy with Entropy and see what we can learn.

Syntropy is Order, Organization, Creation, and Life. Syntropy is one of my greatest scientific discoveries. Quantum Mechanics is Syntropy. Spirit Matter or Dark Matter is Syntropy. Magnetism, Dark Energy, and Gravity are some sort of Syntropy. The Strong and Weak Nuclear Forces are a type of Syntropy. The Zero-Point Field is Syntropy. Quantum Waves are Syntropy. Thoughts and Memories are Syntropy. Psyche or Quantum Non-Local Consciousness is Syntropy. Syntropy is eternal and everlasting, without a beginning of days or an end of years.

Entropy is disorder, randomness, chaos, disease, and death. Materialism, Naturalism, Atheism, Darwinism, Nihilism, Classical Physics, and the Theory of Evolution are based exclusively upon entropy. Entropy cannot design and create. Entropy is death and extinction.

Random mutations are entropy. Natural selection or survival of the fittest results in death or entropy. Entropy cannot design and create. Entropy is death. Death cannot design and create. Chemical Evolution, Macro-Evolution, Spontaneous Generation, Abiogenesis, and Creation Ex Nihilo are prevented from happening by entropy and physical laws. Chemical Evolution or Macro-Evolution is physically impossible thanks to entropy. Evolution of any kind is based upon entropy, and entropy prevents the Theory of Evolution from becoming true.

The very existence of entropy FALSIFIES Materialism, Naturalism, Darwinism, Chemical Evolution, Macro-Evolution, and the Theory of Evolution because entropy proves that these things cannot design and create anything at all. Furthermore, the existence of entropy proves that some sort of Syntropy must exist. All of that subsequent entropy would not have been possible without a massive infusion of Syntropy to begin with. This is logical common sense.

The Fruits of Entropy

One of my greatest scientific observations and scientific discoveries came to me when I finally realized that Materialism, Naturalism, and Darwinism have been falsified trillions of different times in thousands of different ways. That was a major conceptual breakthrough for me because I used to be a Materialist, Naturalist,

Nihilist, and Atheist. Materialism, Naturalism, Darwinism, Nihilism, Behaviorism, Determinism, Physical Reductionism, and Atheism are based exclusively upon entropy; and, entropy prevents the Theory of Evolution from becoming true.

Materialism, Naturalism, and Darwinism are BEST defined as "Design and Creation by Physical Matter" or "Creation by Entropy". These people literally believe and teach that the rocks – physical reactions – or entropy designed, programmed, created, engineered, field-tested, manufactured, and deployed ALL of the different genomes and life forms on this planet. That's what these people really believe, and that's what these people teach in all of our public classrooms.

Ironically, Louis Pasteur falsified Materialism, Naturalism, Darwinism, and the Theory of Evolution in 1859 by demonstrating that the rocks – raw physical matter – cannot design and create. Entropy cannot design and create. It's physically impossible! That was the same year that Charles Darwin published, "On the Origin of Species". Louis Pasteur proved the Theory of Evolution or Creation by Rocks false the same year that Charles Darwin introduced his theory to the world. We've known from the very beginning of the theory that the Theory of Evolution is false because entropy cannot design and create. Remember, Science and the Scientific Methods can be used to prove theories false. It's called falsifying a theory; and, that's what the Scientific Methods do, or are supposed to do.

Technically, the Scientific Methods can't be used to prove anything true. Due to a wide variety of different logic fallacies which are built directly into the Scientific Methods, the Scientific Methods cannot be used to prove a theory true, as the Materialists, Naturalists, and Darwinists try to do.

Instead, we can use the Scientific Methods to falsify theories. Using the Scientific Methods to prove our theories false is philosophically and logically sound. In other words, if you falsify a theory by *negating the consequent*, you have in fact proven that theory false.

Here's how it works in principle:

Scientific Hypothesis: If Theory X is true, then we will observe Y.

Scientific Observations: We don't observe Y.

Scientific Conclusion: Therefore, Theory X is not true.

Here's how it works in practice:

Scientific Hypothesis: If the Theory of Evolution, Materialism, Naturalism, and Darwinism are true, then we will observe the rocks and physical reactions designing, creating, and manufacturing genomes and life forms from scratch.

Scientific Observations: We have NEVER observed the rocks and physical reactions designing and creating genomes and life forms; and, we NEVER will. They can't. It's physically impossible and prevented from happening by entropy.

The Scientific Conclusion: Therefore, the Theory of Evolution, Materialism, Naturalism, and Darwinism are not true. They have been successfully falsified by the Scientific Method.

See: Slife, B. D. & Williams, R. N. (1995). Science and Human Behavior. In *What's Behind the Research? Discovering Hidden Assumptions in the Behavioral Sciences*, (pp. 167–204). Thousand Oaks, CA: SAGE Publications.

http://mypsyche.us/wp-content/uploads/2017/04/Science.pdf

I just successfully falsified Materialism, Naturalism, Darwinism, and the Theory of Evolution; and, I used the Scientific Method to do so! Best of all, my argument is philosophically and logically sound. I have in fact falsified Materialism, Darwinism, Naturalism, and the Theory of Evolution for REAL! I just used the Scientific Method to PROVE these things false. I successfully used the Scientific Methods for what they are good for – falsifying theories that are false. I wish I would have known how to do that forty years ago. It would have saved me a lot of confusion and grief.

In contrast, there is NO way to falsify theories that are true, such as "Design and Creation by Intelligent Beings or by Intelligent Psyches". Instead, true theories are continuously verified over and over again. "Design and Creation by Psyche or by Intelligence" has been observed and verified and experienced trillions of trillions of different times and ways, with an infinite number of more to go. See how that works? That's Science and the Scientific Methods in action for REAL, doing what they are best at!

Entropy and physical matter cannot design and create; but, Syntropy, Psyche, Intelligence, and Quantum Mechanics certainly can. They've been caught in the act of doing so.

This has been and is one of my greatest scientific discoveries and scientific observations of all time, during the whole of my science career.

Quantum Mechanics and Syntropy trump classical physics and entropy. In fact, classical physics and entropy are a very small sub-set of Quantum Mechanics and Syntropy. Quantum Mechanics is supernatural in nature and origin. 95% of our universe is supernatural and non-physical – Dark Energy and Dark Matter. Only 5% of our universe is physical matter. The physical matter and entropy are a very small sub-set of the Quantum Mechanics and Syntropy. There are NO physical limitations and NO entropy within spirit matter, psyche, and quantum mechanisms.

I have observed that there are literally thousands of different ways to falsify Creation by Rocks – or Materialism, Naturalism, Darwinism, and the Theory of Evolution; and, they are all philosophically and logically sound. Thanks to the Scientific Methods and scientific observations, we KNOW that Creation by Rocks, Materialism, Naturalism, and Darwinism are false because we KNOW why they are false – rocks and physical reactions cannot design and create. Entropy cannot design and create. It's elementary my dear friend.

There's NO way that Evolution (Mutation and Selection) could have designed, programmed, engineered, created, and manufactured the first genomes and life forms because Evolution didn't even exist until AFTER God had designed, created, and deployed the first genomes and life forms in the first place.

I have observed that the Scientific Methods or Observation have been used thousands of different ways to falsify Materialism, Naturalism, Darwinism, and even Atheism. Atheism is Creation by Nothing, or Creation by Chance. The Atheists really truly believe that Nobody and Nothing designed and created everything. Nobody, Nothing, and Chance are the holy trinity of Atheism. These people believe in those false gods or idols with a passion. Materialism, Naturalism, and Atheism are in fact our modern-day form of idolatry; and, these people are idolaters. These people worship the rocks and physical reactions and entropy as if these things were God.

Ironically, God must of necessity exist in order to have done ALL of the science that needed to be done, which the rocks or raw physical matter could never have done. Only Psyche or some type of Syntropy can design and create; and, God's Psyche is the only one we know of who was there at the time and could have done the job.

Finally, as a capstone, the Biblical God Jesus Christ has told us repeatedly in the Bible, Book of Mormon: Another Testament of Jesus Christ, the Doctrine and Covenants, and the Pearl of Great Price that HE designed and organized the heavens, this earth, and all of the life forms on this earth. The Biblical God confessed to doing the job.

https://www.lds.org/scriptures?lang=eng

I'm not going to apologize for finding, learning, and knowing the truth. That's what we scientists are supposed to do, is it not? I wouldn't be doing my job if I kept feeding you the falsehoods, the falsified, and the lies; now would I?

For me personally, falsifying Materialism and Darwinism became my first really convincing Scientific Proof of God's Existence. After falsifying Materialism, Naturalism, and Darwinism, I simply KNEW that God exists. I was finally willing to follow the evidence, wherever it might lead me.

I discovered that entropy or physical matter cannot design and create and manufacture anything at all. That reality is science and truth; and, it is set in stone where our physical universe is concerned.

Mark My Words

—

Source Material

Mark My Words. (2016). *The Theory of Evolution Proved to Me that God Exists: Why I Am No Longer an Atheist and Why I No Longer Believe in the Theory of Evolution*. Kindle.

https://www.amazon.com/dp/B01HZYBZ7K

Mark My Words. (2016). *The Scientific Method: Proves That the Theory of Evolution Is False*. Kindle.

https://www.amazon.com/dp/B01IAAIRT2

Physical Matter and Entropy Were Made

You have to know and understand what the Materialists, Naturalists, Darwinists, Nihilists, Behaviorists, Determinists, and Atheists TEACH and BELIEVE before you can figure out what's wrong with it. These people teach and believe that physical matter or entropy is the ONLY thing that exists. Do they not? Consequently, these people literally teach and believe that physical matter and entropy are conserved. Many of these people have no clue that they have chosen to believe in the Conservation of Entropy or the Conservation of Physical Matter; or, they don't completely understand the full ramifications of their chosen beliefs. I KNOW because I didn't understand the full ramifications of my beliefs when I was a Materialist, Naturalist, Nihilist, and Atheist.

The Physicalists and Naturalists teach and believe that physical matter and entropy are eternal and everlasting and that the whole universe is going to end in Heat Death. Entropy is death; and, these people teach and believe that death or entropy is conserved. They teach and believe that death is eternal and everlasting. This is what they teach and believe, is it not? I would guess that most of these people don't fully realize that they believe in the Conservation of Entropy, the Conservation of Death, and the Conservation of Physical Matter. Nevertheless, they do. That's the net result of choosing to believe that ONLY entropic physical matter exists.

Based upon $E = mc^2$, these people teach and believe that energy is synonymous with physical matter and that physical matter is conserved. Consequently, these people define the First Law of Thermodynamics as the "Conservation of Physical Matter". Based upon their Supremacy of Entropy Doctrine, these people teach and believe that entropy or death is conserved. Consequently, these people define the Second Law of Thermodynamics as the "Conservation of Entropy".

THEY ARE WRONG! These people got their science wrong.

Read this carefully, because it's the KEY to everything else in Science.

The first law of thermodynamics is a version of the law of conservation of energy, adapted for thermodynamic systems. The law of conservation of energy states that the total energy of an

isolated system is constant; energy can be transformed from one FORM to another but can be neither created nor destroyed.

https://en.wikipedia.org/wiki/First_law_of_thermodynamics

Once I fully understood and accepted this simple axiom or LAW, instantly everything became obvious and clear. I'm a scientist. My goal is to figure out what everything is and how it works.

Energy, Psyche, or Syntropy can be neither created nor destroyed. However, Energy or Syntropy can be transformed from one FORM into another FORM. The FORM is never conserved. Physical matter and entropy are different FORMS of Energy, and the FORM is never conserved. Only the underlying Energy, Psyche, or Syntropy is being conserved. This is the Ultimate Law of Thermodynamics. Remember, physical matter and entropy are never conserved. The FORM is never conserved.

Energy can be transformed.

Transformed by whom?

Who triggers this transformation? Who decides what FORM Energy will take? Someone Psyche has to make this decision, or nothing will happen, and physical matter will never be made.

Physical matter and entropy were made.

Made by whom? Made from what?

Scientific Observation: Anything that was obviously made obviously had a Maker who made it.

Scientific Observation: Physical matter, physical laws, physical constants, physical restrictions, space-time, locality, time, and entropy were obviously made.

Scientific Conclusion: Therefore, it is logical and rational to conclude that physical matter, physical laws, physical constants, physical restrictions, space-time, locality, time, and entropy obviously have a Maker who made them.

https://en.wikipedia.org/wiki/Atomic_orbital

https://syntropy.site/wp-content/uploads/2018/08/Atomic-Orbital.pdf

https://en.wikipedia.org/wiki/Electron

https://syntropy.site/wp-content/uploads/2018/08/Electron.pdf

https://en.wikipedia.org/wiki/Energy

https://syntropy.site/wp-content/uploads/2018/08/Energy.pdf

Can you sense any order, organization, structure, or law within the Atomic Orbitals or the Electron Orbitals? Why do the electron orbitals take on their characteristic, lawful, and predictable patterns and shapes rather than being totally random and chaotic like the Materialists, Darwinists, Naturalists, and Atheists say that everything should be? What's making the electrons do that? Where's the map, rules, and laws for electron orbitals being stored? How is this information or knowledge being accessed and used by the electrons? It's as if the electrons are conscious, sentient, aware, purposeful, deliberate, and alive. They act differently depending upon where they are in the world or what assignment they have been given to perform. The Atomic Orbitals give structure, size, order, organization, law, and shape to atoms. Do they not? Physical atoms are comprised of different forms of energy.

Physical matter is made from different FORMS of energy. The underlying energy is eternal and everlasting; but, the FORM of that energy is not conserved. Remember, physical matter, physical laws, and entropy are made from different forms of energy. The FORM is never conserved. The FORM can be changed. The different "particles" are made from different waves of energy, or different forms of energy.

The following is the Science that I pulled from the internet. I eliminated the falsehoods and the falsified as I went along. It gives a much better picture of truth and reality than what we can get from the Materialists, Naturalists, and Atheists.

Most of an atom's mass is comprised of the kinetic energy and binding energy of the quarks, NOT the quarks themselves; but, even quarks are made from energy. It's ALL made from different FORMS of energy. Mass is just bound or confined energy. Mass is bound energy, frozen energy, confined energy, or condensed energy. Energy, and the psyche or intelligence within that energy, is the fundamental unit of reality. ALL energy is psychic or intelligent, conscious, alive, sentient, perceptive, and aware.

Quantum Field Theory defines "elementary particles" as excitations in the quantum fields that fill our entire universe. An electron is an excitation or a quantum wave in the electron field. They are NOT particles after all. They are quantum waves, excitations in fields, or quantum wave packets of energy called quanta. It's ALL energy. It's all comprised of different FORMS of energy.

Even in a vacuum, the electron field is there. Even in a planet or a star, the electron field is there. Fields are omnipresent. Add some energy to that field at a particular spot; and, it vibrates producing a quantum or a "particle". The vibration is your "particle", or the electron in the case of an electron field. "Particles" are energy vibrations or quantum waves of energy. Every "elementary particle" is a vibration in its own unique quantum field; and, these vibrations and fields interact with each other transferring energy, momentum, and charge between the different "particles" and fields. This explains how the quantum, the spiritual, or the sub-atomic really works. It's all comprised of intangible, invisible, non-physical waves, forces, fields, and energy.

In physics, a force is any interaction that, when unopposed, will change the motion of an object. A force is comprised of energy. A force is just one form of energy. There are many others.

A field is something that is present everywhere in space and time, and it can have waves in it. "Particles" are made from waves of energy rippling in a quantum field. "Particles" are quantum waves of energy within a quantum field of energy. All the different fields permeate the whole of space and time. They are highly organized energy. The very existence of these different invisible, intangible, non-physical forces and fields falsifies the claims of Materialism, Naturalism, and Atheism which claim that these invisible, intangible, non-physical forces and fields do not exist. Psyche is a force and a field within energy itself. For all we know, Psyche IS Energy, and Energy IS Psyche. ALL the different FORMS of energy are psychic, intelligent, conscious, sentient, perceptive, alive, and aware.

There are many different types of forces and fields; and, each one of them is made from energy. Each type of field was made with a different purpose, goal, or functionality in mind. In quantum field theory, what we perceive as "particles" are excitations of the quantum field itself. Each quantum field was made to function in a specific manner and to perform a specific function. The quantum fields co-exist with each other and interact with each other.

The electric field is a part of nature that is found everywhere. At any given point in space, and at any particular time, you can measure it. If it's non-zero on average in some region, it can have physical effects, such as making your hair stand on end or causing a spark. The electric field can also have waves, in which the size of the field repeatedly becomes larger and smaller — visible light is such a wave, as are X-rays and radio waves, and all the other things we collectively call "electromagnetic waves". Magnets are invisible, intangible, non-physical waves of energy.

Waves are made from energy, and so are quantum fields. They are all different FORMS of energy.

What is a particle?

A "particle" is a quantum wave rippling in a quantum field. It's ALL energy. It's highly organized energy. A quantum field is comprised of organized energy; and, a quantum wave is also comprised of organized energy. They are all different forms of energy. A quantum or a "particle" is a quantum wave of energy or a packet of energy.

Photons are light waves. Einstein called these energy packets photons, and these are now recognized as a fundamental particle. Quanta or "particles" are energy packets. Quanta are waves of energy in a quantum field. They are caused or made. Quanta have a beginning which means that they have someone or something that causes them to begin.

The least-intense possible wave that a quantum field can have is called a "quantum" or a "particle". A "particle" or "wave of energy" behaves in accordance with your typical idea of a "physical particle", functioning as a unit, moving in a straight line and bouncing indivisibly off of things, which is why we call it a "particle". A quantum or "particle" has a minimum energy intensity and therefore functions as a unit even though it is in truth comprised of quantum waves of energy propagating through a quantum field of energy.

A quantum is a packet of energy – it functions as a unit. A quantum is also a quantum wave rippling in a quantum field. A quantum is all-or-nothing. It functions as a unit or a whole even though it is comprised of energy waves within a quantum field. The wave is either there, or it isn't. A quantum wave functions as if it were a packet of energy or a "particle" of energy. Atoms, with their neutrons, protons, and electrons, are not particles at all but pure waves of energy. A "particle" is a quantum wave of energy propagating through a quantum field of energy.

In the case of an electric field, its particles are called "photons"; they represent the dimmest possible flash. Your eye can absorb light one photon at a time. Quanta or quantum waves function as a unit or a whole even though they are comprised of waves of energy rippling within a quantum field of energy. It's ALL comprised of energy.

Higgs bosons are ripples or waves of the Higgs field. By finding the Higgs boson, we have evidence that the field itself exists – like seeing water waves and concluding there must be an ocean beneath. The Higgs field is extremely important in particle physics. It's no exaggeration to say that it makes your existence possible. Without it, atoms would not exist; electrons would zip away from nuclei at the speed of light. The value of the Higgs field determines what kinds of nuclei are stable and some of the differences between matter and antimatter. Almost all "particles" or quanta that we know of are affected in some way by the Higgs field. The Higgs field provides structure, stability, order, and organization to all the other fields. The Higgs field shepherds the other fields. God is in the Higgs field.

Organized fields of energy are made. Organized waves of energy or organized packets of energy are made or caused to begin. Quantum fields and quantum waves are made. Quanta or "particles" are made. Anything that was obviously made obviously has a Maker who designed it, made it, and caused it to begin.

A field is something that is present everywhere in space and time. The different fields are omnipresent. Fields are organized energy. They each have their own specific purpose and function. Quanta or "particles" are packets of energy or organized waves of quantum energy that are rippling through quantum fields of energy.

The Higgs field (unlike most of the elementary fields of nature) has a non-zero average value throughout the entire universe. Because it does, many particles have mass, including the electron, quarks, and the W and Z particles of the weak interactions. If the Higgs field's average value were zero, those particles would be massless or very light. That would be a disaster; atoms and atomic nuclei would disintegrate. Nothing like human beings, or the earth we live on, could exist without the Higgs field having a non-zero average value. Our lives truly depend upon it.

Can you sense any order, organization, planning, intention, structure, purpose, or meaning in any of this? Or are you like the Materialists, Naturalists, Darwinists, Nihilists, and Atheists and simply believe that it all came about by random chance and blind luck? Anything that was obviously organized obviously has an Organizer who organized it. This is Logic 101.

The Materialists, Naturalists, Darwinists, Nihilists, Behaviorists, Determinists, Physical Reductionists, and Atheists are OBVIOUSLY WRONG whenever they claim that entropic physical matter is the only thing that exists, and that entropic physical matter is the fundamental unit of reality. The whole of Science and Quantum Field Theory proves that they are wrong.

Energy is a conserved quantity; the law of conservation of energy states that energy can be converted in form, but not created or destroyed. That would make energy, or the psyche within that energy, the fundamental unit of reality.

Adapted from Uncle Google

See in particular:

https://en.wikipedia.org/wiki/Energy

https://en.wikipedia.org/wiki/Quantum_field_theory

https://profmattstrassler.com/articles-and-posts/the-higgs-particle/the-higgs-faq-2-0/

https://wikis.utexas.edu/display/utatlas/Higgs+boson+FAQ

https://www.youtube.com/watch?v=kixAljyfdqU

https://www.youtube.com/playlist?list=PLsPUh22kYmNBpDZPejCHGzxyfgitj26w9

Particles or quanta are constructs. They are instantiated or caused to begin. They are made. Technically, physical matter does not exist. It's an illusion. Physical matter is simply quantum waves rippling through quantum fields. It's all made from energy. It's the underlying energy that truly exists and is timeless and eternal. Physical matter is made from different FORMS of energy, which means that entropic physical matter can be disassembled and formed into something else instead.

Psyche is the innate intelligence within all the different FORMS of energy which gives that energy the inherent ability to understand, follow, and obey God's Laws and God's Commands. Energy, and the psyche or intelligence within it, is eternal and everlasting. It cannot be made, and it cannot be destroyed. Energy of any kind is syntropic, sentient, conscious, perceptive, intelligent, psychic, alive, and aware. Energy, and the psyche within it, is always conserved.

We have the things that were made – quanta, particles, forces, and fields; and, we have the things that cannot be made and cannot be destroyed – energy, psyche, intelligence, or life force. The energy isn't made, it's TRANSFORMED; whereas, everything else is MADE from energy. Energy or Psyche is the fundamental unit of reality – not entropic physical matter.

Do you see how that works?

It's the answer to life, the universe, and everything. It explains everything that has ever been experienced or observed.

It's all there hiding in plain sight where the Atheists and Naturalists cannot see it nor find it because they refuse to go looking for it.

Time is a construct. Entropy is a function of time, which means that entropy is a construct. Entropy is constructed from energy which means that it can be deconstructed and turned back into raw energy. Entropy is death. The way you convert entropic physical matter into syntropic physical matter is to remove the entropy or the death from the entropic physical matter. Resurrection from death has been experienced and observed. The way that resurrection works is that the entropy or the death is removed from the entropic physical matter; and thereby, that physical matter is converted into syntropic physical matter or immortal physical matter instead.

The Materialists, Naturalists, Darwinists, Nihilists, and Atheists will assure you that everything I have written here is wrong. They will tell you that I don't understand science. They define science as Materialism, Naturalism, Darwinism, Nihilism, Scientism, and Atheism. These people have told me many times that I don't understand science. They will tell you that there is nothing fundamental about energy, and that psyche does not exist. It's all over the internet. This is what these people teach and believe. These people will tell you that energy and consciousness is an emergent property of something much more fundamental. What's more fundamental than energy and consciousness? According to these people, physical matter is the fundamental unit of reality; and, energy or consciousness is just an emergent property of entropic physical matter. I think I understand their science quite well, don't you think? I KNOW what they teach and believe. I used to be a Materialist, Naturalist, Nihilist, and Atheist after all.

I eventually discovered that the Materialists, Naturalists, Darwinists, Nihilists, and Atheists are almost always wrong, especially when it comes to the fundamentals. The observations and experiences of the human race as a whole convinced me that I was wrong. Nowadays, I define Science as observation and experience; and, the observations and experiences of the human race FALSIFY

Materialism, Naturalism, Darwinism, Nihilism, Atheism, Classical Physics, and their derivatives.

The fact that Materialism, Naturalism, Darwinism, Nihilism, Atheism, and their derivatives FALSIFY my current theories and ideas tells me that I'm finally on the right track and have finally found the truth. These people have a knack for finding the truth and then formally rejecting it. One of my greatest scientific discoveries came when I first realized that the truth is invariably the opposite of what the Materialists, Naturalists, and Atheists have chosen to believe. These people reject everything that has been experienced and observed by Someone Psyche, and then they preach the opposite instead as if it were true. The fact that Materialism and Naturalism FALSIFY my current theories and ideas is positive proof that I have finally found the truth.

The fatal flaw of the Materialists, Naturalists, Darwinists, Nihilists, and Atheists is that they teach and truly believe that only physical matter and entropy exist, which means that these people erroneously teach and believe that physical matter and entropy are conserved. They are wrong. These people deliberately misinterpret the First Law of Thermodynamics, which states that Energy or Syntropy is conserved; and, these people get the rest of us to misunderstand and misinterpret the First Law of Thermodynamics by successfully convincing us that only physical matter and entropy exist and that consequently physical matter and entropy are being conserved. Since these people believe that ONLY physical matter or entropy exists, these people literally and erroneously define the First Law of Thermodynamics as the "Conservation of Physical Matter" and the Second Law of Thermodynamics as the "Conservation of Entropy". Entropy is death; and, these people literally but erroneously teach that entropy or death is conserved. That's what they teach, is it not?

One of my greatest scientific discoveries came when I first realized that physical matter and entropy are NOT conserved. The amount of physical matter, entropy, and space-time changes over time as God sees fit. Space, time, locality, entropy, spirit matter, and physical matter are different FORMS of energy. The FORM is never conserved. The FORM is constantly changing. Its amount or quantity is constantly changing. **It is the underlying Energy, Psyche, or Syntropy that's being conserved according to the First Law of Thermodynamics and $E = mc^2$** — NOT the physical matter and NOT the entropy. Physical matter and entropy are NOT conserved. This is the Ultimate Law of Thermodynamics.

The Ultimate Law of Thermodynamics differentiates between what is being conserved and what is not conserved. This scientific discovery is the answer to life, the universe, and everything. It explains everything that has ever been experienced and observed. Consequently, the greatest scientific discovery in applied science that fallen, mortal, physical beings like us could ever make would be to find a way to remove entropy from entropic physical matter. That would indeed result in the answer to life, the universe, and everything. Would it not?

Remember, physical matter, entropy, locality, and space-time are NOT conserved. Their amount is constantly changing. Only the underlying Energy or

Psyche is conserved. Psyche is the innate intelligence within all the different forms of energy which gives that energy the inherent ability to understand, follow, and obey God's Laws and God's Commands. In contrast, physical matter and entropy are made from different FORMS of energy. The FORM is never conserved. Energy's FORM can always be transformed into something else. Energy, and the psyche or the intelligence within that energy, is much smaller and more fundamental than physical matter. Physical matter is comprised of different forms of energy. This is the Ultimate Law of Thermodynamics. This is what has been experienced and observed. Physical matter and entropy are NOT conserved.

Physical matter, entropy, locality, and space-time are MADE, which means that they can be unmade or disassembled. With God's help or permission, we humans make brand new physical matter in our particle accelerators. We humans also destroy physical matter or unmake physical matter in our atomic bombs. The physical matter and the associated entropy are NOT conserved. Their quantity is constantly changing. Particles of physical matter are popping in-and-out of "existence" all the time. They don't literally cease to exist as the Materialists and Naturalists claim. Instead, they change FORM and switch dimensions; but, their underlying energy is always conserved. Electrons phase-shift and quantum tunnel both of which are quantum mechanical processes that have been experienced and observed. They don't cease to exist while they are phase-shifted into a different dimension. Their energy is always conserved.

There's NO such thing as Creation Ex Nihilo. The First Law of Thermodynamics and the Ultimate Law of Thermodynamics FALSIFIES Creation Ex Nihilo.

Remember, Energy is constantly changing FORM while at the same time being conserved. Both syntropic physical matter and entropic physical matter are made from different FORMS of energy. That's the way that science really works contrary to what the Materialists, Naturalists, Darwinists, and Atheists claim to be true. The false is falsified by the truth; and, the truth is repeatedly experienced and observed. Energy or psyche is conserved.

Physical matter, physical laws, physical constants, physical restrictions, space-time, space, locality, time, and entropy were MADE from energy which means that they have some kind of Maker. It also means that they can be disassembled or destroyed. They were made from Energy by God, which means that God can cause them to end or unmake them. They are different FORMS of Energy, and the FORM is never conserved. Only the underlying Energy, Psyche, or Syntropy is conserved. Physical matter and entropy are different FORMS of Energy, which means that physical matter and entropy are never conserved. The FORM is never conserved. This is the Ultimate Law of Thermodynamics.

My purpose in life is to identify everything that is false, eliminate it from science, and replace it with the truth. The materialistic and naturalistic idea that physical matter and entropy are conserved is a false idea and a falsified idea. I have eliminated it from my science and replaced it with the truth – namely, that it's the underlying Energy, Psyche, or Syntropy that's actually being conserved and NOT the physical matter NOR the entropy. All I want is the truth because everything else will mess you up in the end. It's the underlying Energy or Syntropy that ends up being eternal and everlasting, not the physical matter nor the entropy. Physical matter and entropy are different FORMS of Energy, and the FORM is never conserved. Energy is Syntropy, and Energy is conserved. Syntropy is the First Law of Thermodynamics; and, it tells us that physical matter and entropy are NOT conserved.

Observation and experience have proven to us that the Ultimate Law of Thermodynamics is true. Physical matter, entropy, locality, time, space, and space-time are MADE, which means that they can also cease to exist anytime that God decides to put a stop to them. Only the underlying Psyche or Energy is conserved. It cannot be made, and it cannot be destroyed, which means that Energy or Psyche is conserved.

According to the Theory of Relativity, when a photon or tachyon travels at the speed-of-light or faster, TIME STOPS. Entropy is a function of time. Entropy measures the passage of time. Entropy is an aging process. Locality is associated with space. Entropy is associated with time. Entropy is the passage of time. According to the Theory of Relativity, when a photon or tachyon travels at the speed-of-light or faster, ENTROPY STOPS, or ENTROPY CEASES TO EXIST. From our entropic perspective, it took that photon 13.4 billion years to reach us. From the photon's syntropic perspective at the speed-of-light, it experienced NO passage of time; and, it arrived the very moment it launched. In other words, the photon and the tachyon didn't age. They didn't experience the passage of time. Entropy ceased to exist while they were traveling at the speed-of-light or faster. Entropy was NOT conserved. These objects literally quantum tunneled to their destination from their perspective. Remember, entropy is variable; and, entropy or the passage of time ceases to exist at the speed-of-light or faster. In other words, entropy or the passage of time is not conserved at the speed-of-light or faster. Instead, entropy ceases to exist; and, everything becomes syntropic instead at velocities and frequencies faster than the speed-of-light.

Once again, the Materialists, Naturalists, Darwinists, Nihilists, and Atheists erroneously teach that nothing can travel faster than the speed-of-light because these people erroneously teach that only physical matter and entropy exist. These people are right in that physical matter or entropic matter cannot travel faster than the speed-of-light; however, these people are wrong in that spirit matter, quantum matter, syntropic matter, spiritual matter, tachyons, and psyche can and do travel faster than the speed-of-light; and, they can also quantum tunnel at will. At velocities equal to or greater than the speed-of-light, TIME STOPS; and, entropy or the passage of time ceases to exist. There is no entropy in the Syntropic Realm. Objects existing at frequencies faster than the speed-of-light experience no entropy because they experience NO passage of time. Entropy ceases to exist at velocities and frequencies faster than or equal to the speed-of-light, according to the Theory of Relativity.

God deliberately created physical matter and entropy (the passage of time) for us in order to slow things down for us so that we can live them, observe them, experience them, learn from them, experiment with them, and remember having done so. At velocities equal to or faster than the speed-of-light, there is NO entropy because there is NO passage of time. Time stops. Time or entropy ceases to exist. Time, entropy, and the passage of time were made by God, which means that they can also cease to exist or come to a stop. I have experienced and observed that TIME STOPS and entropy (the passage of time) temporarily ceases to exist, just before you are quantum tunneled to a different location on this earth. God has complete control over Quantum Mechanics including quantum tunneling. From the photon's perspective, it quantum tunnels or teleports to its destination because it experiences no passage of time or no entropy during its voyage. From my perspective, time stopped, and I experienced no passage of time when God quantum tunneled me and the car I was driving to safety. God can quantum tunnel physical matter anywhere in this universe instantaneously simply by stopping the passage of time or temporarily removing the entropy from that physical matter.

Remember, there's nothing sacred about physical matter and entropy. It's the Energy, Psyche, or Syntropy that's being conserved. Energy can be organized into many different FORMS, including physical matter, spirit matter, space-time, physical laws, physical constants, space, time, locality, and entropy. Remember, it's the Energy or the Syntropy that's actually being conserved, not the physical matter nor the entropy. This is the Ultimate Law of Thermodynamics. It explains everything that we humans have experienced and observed. It corrects one of the fatal flaws in traditional science – namely, the erroneous belief that physical matter and entropy are being conserved. The Ultimate Law of Thermodynamics states that physical matter, entropy, locality, and space-time are NOT conserved. It's the underlying Energy or Syntropy that's being conserved, and NOT the physical matter nor the entropy. Physical matter and entropy are just different FORMS of energy. Energy can be made to take on many different FORMS; and, the FORM is never conserved. It's the Energy, Psyche, or Syntropy that's actually being conserved and NOT the FORM. The Ultimate Law of Thermodynamics is the answer to life, the universe, and everything. It explains everything.

This Ultimate Law of Thermodynamics is one of the greatest scientific discoveries of all time; and, it's only made possible by eliminating Materialism, Naturalism, Darwinism, Nihilism, Behaviorism, Determinism, Physical Reductionism, and Atheism from science. If you successfully eliminate everything that is false and everything that has been falsified, then only the truth will remain. This is Logic 101. You start by eliminating Materialism, Naturalism, Darwinism, Nihilism, and Atheism from science; and then you see where that gets you. It's interesting to study and observe what remains after you have successfully removed Materialism, Naturalism, Darwinism, Atheism, and their derivatives from science. The experienced and the observed remain. The truth remains.

Remember, the Biblical God Jesus Christ has been experienced and observed both in the flesh and during our near-death experiences AFTER He rose from the dead. Science is observation and experience.

The Materialists, Naturalists, Darwinists, Nihilists, and Atheists define "science" as Materialism, Naturalism, Darwinism, Nihilism, and Atheism. Based upon all these different scientific discoveries that I have made during the past couple of years, I have upgraded my science to Science 2.0; and, I have redefined Science as "observation and experience". Observation and experience are a much better way to do science because the "non-existence of things" will never be experienced nor observed. It's impossible to experience and observe the "non-existence of something", including the non-existence of God. As scientists, we should go with what has been experienced and observed, not the wishful thinking of the Naturalists, Darwinists, and Atheists.

Physical matter and entropy were obviously made from Energy. Physical matter had physical limitations, entropy, locality, sub-light velocities, mass, time, space, and physical restrictions programmed into it. Physical matter is Organized Energy, and it was designed and organized by God. Since physical matter and entropy are never conserved, according to the Ultimate Law of Thermodynamics, God must of necessity exist in order to have organized, created, or made the first particles of physical matter from Raw Energy.

Scientific Observation: Anything that was obviously organized obviously had an Organizer who organized it.

Scientific Observation: Physical matter, physical laws, physical constants, space-time, physical restrictions, space, time, locality, and entropy were obviously organized and made from Energy. These things are different forms of Organized Energy. Physical matter is Organized Energy.

Scientific Conclusion: Therefore, it is logical and rational to conclude that physical matter, physical laws, physical constants, space-time, physical restrictions, space, time, locality, and entropy obviously have an Organizer who organized them and made them from Energy.

This truth has been experienced and observed. The false is falsified by the truth; and, the truth is repeatedly experienced and observed.

The Ultimate Law of Thermodynamics states that the FORM is never conserved, which means that the different forms of Energy such as physical matter and entropy are never conserved. Their quantity or amount is constantly changing. "Particles" or quanta are being made and annihilated all the time. There's nothing sacred or inviolate about particles of matter. It's the underlying Energy or Syntropy that's always being conserved. It's the Energy, Syntropy, or Life Force that ends up being eternal and everlasting, not entropy or death. That is what has actually been experienced and observed.

Materialism, Naturalism, Darwinism, Nihilism, Behaviorism, Determinism, Physical Reductionism, and Atheism were designed to prevent us from discovering the Ultimate Law of Thermodynamics. Entropy is death; and, the Scientific Naturalists and Atheists erroneously teach that entropy or death is conserved. These people are wrong. Life after death has been experienced and observed.

The Psyche, Syntropy, Energy, or Life Force is always conserved; but, the entropic matter or physical matter is never conserved. Physical matter and entropy come and go as God sees fit. Physical matter and entropy are made, which means that they can be unmade or disassembled. In contrast, Psyche, Syntropy, Energy, or Life Force can be neither created nor destroyed. Psyche, Syntropy, Energy, and

Life Force are eternal and everlasting, without a beginning of days or an end of years, because they are constantly being conserved. God's Psyche or God's Intelligence is being conserved. Your psyche or your intelligence is being conserved. In contrast, physical matter and entropy are never conserved. This is the Ultimate Law of Thermodynamics.

    The Ultimate Law of Thermodynamics is the answer to life, the universe, and everything. It has been experienced and observed.

    Mark My Words

PART I — GENETIC ENTROPY PUT AN END TO THE THEORY OF EVOLUTION FOR ME

In my Pursuit of the True Reality of All Things, I had to be willing to abandon any residual belief in the Theory of Evolution, BEFORE I was able to find THE TRUTHS that I sought.

Belief in a LIE cannot help us to find THE TRUTH. That's just the True Reality of the situation.

There was a time in my life when I was a Materialist and an Atheist. I realized one day that I didn't like where that road was taking me, so I decided to turn around and go the other way. My life has been getting better and better ever since.

The purpose of the Scientific Method is to help us to find THE TRUTH, through a preponderance of the evidence.

the Scientific Method has no value to us if we use it to convince ourselves that a LIE is TRUE, as the Materialists and Darwinists always seem to do.

That's what I discovered during my Pursuit of the True Reality of All Things, and during my usage of the Scientific Method.

Mutation/Selection Cannot Design and Create Anything!

In order for the Theory of Evolution and Darwinian goo-to-you evolution to be true, it would have to be able to design and create anything and everything, without any help from any of us including God. But, Evolution can't do so, because it's not alive and it can't create. Evolution has no hands and no mind. It can't design and create anything! Evolution or Change ONLY works on the things which God has already designed and created in the first place.

Chance Mutations and Selection working together cannot design, program, test, engineer, refine, build, and then implement a gene into a living organism from scratch. Mutation and Selection cannot even design and build a functional protein or a functional information-rich RNA strand, without direct hands-on intelligent intervention from an intelligent human being. Only a blind and loyal Darwinist will actually believe that Darwinian Chance is real and true, and that chance can truly create everything from scratch.

Human beings are just on the verge of finally being able to do these things, so what makes the Darwinists and Materialists believe that evolution and chance had mastered the design and creation and construction of genomes and life forms billions of years ago? The Darwinists have way too much blind faith in the creative powers of chance and evolution!

EVERY science experiment which the Darwinists intelligently design and implement proves the need and THE TRUTH of the Intelligent Design Theory. The Darwinists can't win, because EVERY science experiment which they intelligently design and run proves Evolution by Chance to be false and proves evolution by Intelligent Design to be true.

The message which I get from the Intelligent Design people is that the only type of evolution that actually works is Intelligently Designed Evolution. Think about it, because it is true. The only kind of evolution that actually works is some kind of Intelligently Designed Evolution or Genetic Engineering or Creation by Intelligent Beings.

What's missing from the Theory of Evolution is the hands-off, undirected, no man-handling Abiogenesis or self-assembly of genomes and biological information, which Evolution NEEDS in order to do what the Darwinists say that it did.

But, it's common knowledge and common-sense logic that Random Chance or Random Mutations cannot design, program, test, engineer, refine, redesign, build, retest, rebuild, and then implement ANYTHING NEW into the wild. Chance cannot create or build anything! If you try to get Chance to build you something like a functional genome from scratch, you will be waiting for all eternity, and it still won't happen.

Now think about this: Chance is in fact THE CREATIVE ARM of the Theory of Evolution! What are the chances that Evolution designed, programmed, engineered, and implemented all of the different genomes and all of the different life forms on this planet?

THERE IS NO CHANCE AT ALL!

For me, Chapter 9 of the book "Genetic Entropy" was officially the END of the Theory of Evolution — the final nail in its coffin. After that, there was no going back! The scientist in me won't let me go back. After that I just KNEW that the Theory of Evolution or Creation by Evolution is false and cannot do and will never be able to do any of the things that the Darwinists say that it does.

I have posted the following quote from John C. Sanford all over the place in my Amazon Reviews. Take note of the books that I choose to quote and PROMOTE, because these are the books and the authors who taught me how and why the Theory of Evolution is FALSE. Thanks to these people I am no longer an Atheist and no longer believe in the Theory of Evolution.

The book, "Genetic Entropy 4th Edition", is the Top One on my list when it comes to evolution. If you can only afford one, then buy this one, read it, read it again, study it and ponder it until you understand it, and then memorize it. If you do, if you really understand this book, then you will know precisely why Creation by Darwinian Chance and Creation by Mutation/Selection could NEVER be true. I found this book to be essential reading, if you want to know THE TRUTH about the Theory of Evolution!

[Editorial Note: I have written permission from John C. Sanford to use all of the quotes from John C. Sanford which I use in this book, so long as I cite the sources which I have done.]

START OF THE QUOTE FROM "GENETIC ENTROPY" BY JOHN SANFORD — USED BY PERMISSION FROM THE AUTHOR JOHN SANFORD.

Chapter 9: Can Natural Selection Create?

Newsflash — Mutation/Selection cannot even create a single gene.

We have been examining the problem of genomic degeneration and have found that deleterious mutations occur at a very high rate. Natural selection can only eliminate the worst of these, while all the rest accumulate — like rust on a car. Might beneficial mutations at other sites in the genome compensate for this continuous and systematic erosion of genetic information? The answer is that beneficial mutations are much too rare, and they are much too subtle to keep up with such relentless and systematic erosion of information. This is carefully documented by Sanford et al. (2013), and Montañez et al. (2013). It is very easy to systematically destroy information, but apart from the operation of intelligence it is very hard (arguably impossible) to create information.

This problem overrides all hope for the forward evolution of the whole genome. However, some limited traits might still be improved via Mutation/Selection. Just how limited is such progressive ("creative") Mutation/Selection? By now it should be clear that random spelling errors in an instruction manual could never give rise to an airplane component (say a molded aluminum part), which then resulted in a significantly improved overall performance of a jet plane. Not even with an unlimited number of flight trials/crashes and an unlimited budget. So, it is certainly reasonable to ask the parallel biological question, "Could Mutation/Selection create a single functional gene from scratch?"

A gene is like a book, book chapter, or an executable program — and minimally consists of a text string with 1,000 characters. Mutation/Selection could not create a single gene because of the enormous preponderance of deleterious mutations, even within the context of a single gene. The net information must always still be declining, even within a single gene or linkage block. Even if a gene was 50% established, deleterious mutations would degrade the completed half of the gene much faster than beneficials could create the missing half of the gene. However, to better understand the limits of forward selection, let us for the moment discount all deleterious mutations and only consider beneficial mutations. Could Mutation/Selection then create a new and functional gene?

1. Defining our first desirable mutation. The first problem we encounter in trying to create a new gene via Mutation/Selection is defining our first beneficial mutation. By itself, no particular nucleotide (A, T, C or G) has more value than any other, just as no letter in the alphabet has any particular meaning outside of the context of other letters. So, selection for any single nucleotide can never occur except in the context of the surrounding

nucleotides (and in fact, within the context of the whole genome). A change of a single letter within a word or chapter can only be evaluated in the context of the surrounding block of text. This brings us to an excellent example of the principle of "irreducible complexity" within the genetic realm. In fact, it is irreducible complexity at its most fundamental level. We immediately find we have a paradox. To create a new function, we will need to select for our first beneficial mutation, but we can only define that new nucleotide's value in relation to its neighbors — and we are going to have to be changing most of those neighbors also. We create a circular path for ourselves. We will keep destroying the "context" we are trying to build upon. This problem of the fundamental inter-relationship of nucleotides is called epistasis. True epistasis is almost infinitely complex, and virtually impossible to analyze, which is why geneticists have always conveniently ignored it. Such bewildering complexity is exactly why language and information (including genetic language and genetic information) can never be the product of chance, but always requires intelligent design. The genome is literally a book, written literally in a language, and short sequences are literally sentences. Having random letters fall into place to make a single meaningful sentence, by accident, would require more tries (more time), than earth history can provide (i.e., "methinks it is like a weasel" would take $27 \wedge 28$ tries — that is 10 followed by 40 zeros). The same is true for any functional string of nucleotides. If there are more than a dozen nucleotides in a functional string, we know that realistically they will never just "fall into place". This has been mathematically demonstrated repeatedly. But as we will soon see, neither can such a sequence arise by selecting one nucleotide at a time. A pre-existing "concept" is required as a framework upon which a sentence or a functional sequence must be built. Such a concept can only pre-exist within the mind of the author. Starting from the very first mutation, we have a fundamental problem even in trying to define what our first desired beneficial mutation should be.

2. Waiting for the first mutation. Let's assume we can know the first desired mutation. How long do we have to wait for it to happen? Human evolution is generally assumed to have occurred in a small population of about 10,000 individuals. The mutation rate for any given nucleotide, per person per generation is exceedingly small (very roughly about one mutation per 30 million individuals, for a given nucleotide site). Within a population of 10,000, one would have to wait 3,000 generations (at least 60,000 years) to expect a specific nucleotide to mutate. But two out of three times, it will mutate into the "wrong" nucleotide. So, to get a specific desired mutation at a specific site just in one individual will take three times as long, or at least 180,000 years. Once the mutation arises in one individual, it has to become "fixed" (such that each individual in the population will eventually have a double dose of that mutation). Because a newly arisen mutation arrives in a population as just a single copy, it arrives on the brink of extinction. The vast majority of new mutations soon drift back out of the population, even the ones that are beneficial. So, any specific desired mutation must arise many times before it "catches hold" in the population. Only if the mutation is

dominant and has a very distinct benefit does selection have any reasonable chance to rescue it from random elimination via drift. According to population geneticists, apart from effective selection, in a population of 10,000, our given new mutant has only one chance in 20,000 (the total number of non-mutant nucleotides present in the population) of NOT being lost via drift. Even with some modest level of selection operating, there is a very high probability of random loss, especially if the mutant is recessive or is weakly expressed (we actually know that most mutations will be both recessive and nearly neutral). Therefore, even a beneficial mutation will be randomly lost due to genetic drift most of the time. Our numerical simulations suggest a weakly beneficial mutant will be lost about 99 out of 100 times. So, a typical mildly-beneficial mutation must happen about 100 times before it is likely to "catch hold" within the population. So, on average, in a population of 10,000 we would have to wait $180,000 \times 100 = 18$ million years to stabilize our first desired beneficial mutation, to begin building our hypothetical new gene. So, in the time since we supposedly evolved from chimp-like creatures (6 million years), there would not be enough time to realistically expect our first desired mutation to go to fixation in the genomic location where our required gene is hopefully going to arise. A vast amount of mutations would arise during 18 million years, but only once would that specific nucleotide mutate to that specific new nucleotide — such that it's not lost due to genetic drift and is fixed.

3. Waiting for the other mutations. After our first desired mutation has been found and fixed, we need to repeat this process for all the other nucleotides encoding our hoped-for gene. A gene is minimally 1,000 nucleotides long. More realistically, a human gene is on average about 50,000 nucleotides long, when regulatory elements and introns are included. To be extremely generous we will only consider a gene of 1,000 nucleotides (and we assume each nucleotide is by itself selectable). If this process was a straight, linear, and sequential process, it would require about 18 million years $\times 1,000 = 18$ billion years to create the smallest possible gene. This is more than the time since the reputed Big Bang! So, it is a gross understatement to say that the rarity of desired mutations limits the rate of evolution. Furthermore, single nucleotides do not carry any information by themselves, and cannot be selectively favored. Specified information requires many characters (minimally, a sentence or similar text string is needed). Like any message, a genetic message which specifies some life function requires many nucleotides to reach its "functional threshold". Functional threshold is the minimal number of characters (or nucleotides) needed to convey a meaningful message. Below the functional threshold, individual letters or nucleotides have no benefit and cannot be favored by selection. This means that realistically, waiting time will be much, much longer — because no selection can happen until the minimum string of nucleotides falls into place by chance. If the functional threshold for selection is 12 (no selection until all 12 letters are in place), the waiting time in our hypothetical human population becomes trillions of years.

Sanford, John (2015-02-23). Genetic Entropy (Kindle Locations 1684-1755). FMS Publications. Kindle Edition. USED BY PERMISSION.

END QUOTE.

—

Trillions of years!

Well, that's the END of the Theory of Evolution, isn't it?

The Darwinists NEVER use their God-given brains to stop and think about these kinds of things. At the best possible average pace, with God making sure that there are NO deleterious mutations and NO devolution taking place, it would take on average 18 million years to fixate and stabilize a SINGLE beneficial mutation through "Natural Means" into a population of 10,000 apes which God has already designed and created in the first place and kept alive and functional during those 18 million years, just so that population of 10,000 God-created apes can achieve their first beneficial mutation through "Natural" Hands-off Mutation and Selection. 18 million years on average per beneficial mutation! Think about it!

If those apes need 1,000 such beneficial mutations in order to become men, then you are looking at 18 billion years on average to produce those targeted 1,000 beneficial mutations; and, that's with a population of 10,000 apes that God has already designed and created in the first place and that God is making sure receive ONLY beneficial mutations and NO devolution or deleterious mutations. And, that's also with God keeping that population of 10,000 apes alive during those 18 billion years so that they can indeed "evolve" their necessary 1,000 beneficial mutations and become men all on their own through "Natural Means".

Furthermore, it has been estimated that it would in fact take at least 20 million such beneficial mutations to convert chimpanzees into humans through "Natural Means". With that targeted goal in mind and assuming NO deleterious mutations or extinctions along the way, how long would it take on average to convert 10,000 chimpanzees into 10,000 humans using Natural Selection and Random Mutations to do the job? So, what do you get if you multiply 20 million beneficial mutations with 18 million years per beneficial mutation? At the BEST possible pace, with God keeping those 10,000 chimpanzees alive all along the way, and with God making sure that there is NO devolution, NO extinction, and NO deleterious mutations taking place, the quickest on average that Mutation and Selection could convert a chimpanzee into a human through "Natural Means" is 360 trillion years.

John Sanford isn't exaggerating whenever he says that it could take trillions of years for Mutation/ Selection to design and create something useful "naturally". And, it really isn't Natural Evolution if God has to design and create the 10,000 chimpanzees in the first place, and then keep them alive for 360 trillion years by blocking ALL deleterious mutations and preventing ALL extinctions that might take place during that period of time, just so He can convert 10,000 chimpanzees into 10,000 humans "naturally" or through "Natural Means".

Think about it! At the BEST possible average pace, it would take at least 360 trillion years for Mutation and Selection to convert a population of 10,000 chimpanzees into 10,000 humans through "Natural Means", with God keeping those 10,000 mutants alive and preventing deleterious mutations during the whole time. And, that's with 10,000 chimpanzees that God has already designed and created in the first place! How old did they say our universe is? How long would it take to convert a bacterium into a human through "Natural Means" when billions of beneficial mutations are needed? Wouldn't it be easier and faster to just let God design and create those 10,000 humans in the first place?

The Darwinists and Materialists NEVER stop and use their God-given brains to think about and calculate these kinds of things. You will NEVER get these kinds of calculations and truths from the Darwinists, because they don't DO this kind of science. It's too difficult and painful for them. I'm willing to wrap my mind around these kinds of things. Your typical Darwinist isn't. Your typical Darwinist is afraid of it, because they don't want to be proven wrong. For the Materialists and Darwinists, ignorance is bliss! But, ignorance is the reason why the Darwinists and Materialists truly believe that Mutation/Selection can design and create anything that it sets its mind to. The rest of us KNOW BETTER!

If you think about it, this is radically advanced science — the best that humans are able to come up with! Can you see and understand now why the 9th Chapter of "Genetic Entropy" put an END to the Theory of Evolution for me? It's because I understood what John Sanford was talking about and chose to believe that it is true. Now the onus is on you.

—

What do the Darwinists typically do when presented with these kinds of Statistical Models of the Mutation/Selection Process?

Assuming that they don't go head-in-the-sand and actually study them instead, the Darwinists try to shave the figures in half or by one-tenth, which is exactly what they do when designing Mutation/Selection Models of their own. They cheat. They choose parameters that are scientifically inaccurate and don't match with Reality in order to shave those estimates down to something that they might be willing to accept. They keep shaving and cheating until they get the numbers that they want, and then they call the results "Science".

If they really take John Sanford's Models seriously, the Darwinists will demand that God artificially accelerate the Mutation/Selection Process so as to make it possible for the Theory of Evolution to be true. But, even if God were to speed up the process a thousand fold, it's still going to take 360 billion years for chimpanzees to evolve into humans through "Natural Means"; and, the Darwinists are still going to complain, even though we are starting with Chimpanzees that God has designed and created in the first place.

It can be fascinating and entertaining to watch a Darwinist try to shave trillions of years off a Statistical Estimate that he doesn't like, all in an attempt to increase the possibility that the Theory of Evolution might be true.

And, that's just the beginning of the problems for the Theory of Evolution. It only gets worse from there on forward, because John Sanford actually has TEN points in chapter 9 of "Genetic Entropy", each of which decreases the likelihood of Mutation/Selection creating anything at all, even if it has an infinite number of years to do so. Evolution by random mutations and evolution by natural selection CANNOT design, create, and deploy anything! It has been conclusively and finally demonstrated that it is so. It has been empirically and logically observed to be so. The Theory of Evolution doesn't work and can't create new unique genomes from scratch, so we have no choice but to declare the whole thing to be FALSE.

This was ONLY three pages from "Genetic Entropy", a 245-page book. But, these three pages were enough to convince me that the Theory of Evolution is FALSE and that there is NO SUCH THING as Darwinian goo-to-you evolution.

Molecules-to-man evolution NEVER happened and can NEVER happen. Macro-evolution of any kind is scientifically IMPOSSIBLE. One of my goals in this book is to eliminate the impossible, so that we are left staring at THE TRUTH. Creation by Evolution, of any kind, is IMPOSSIBLE; therefore, evolution of any kind must be eliminated as a possible cause or source for the origin of life on this planet.

If you want to know more, then go out and buy a copy of John Sanford's book for yourself. In my book here, the most I can do is to tell you where to go and look; and, I do make it a point to PROMOTE the books and the authors that I found most helpful and useful during my research. Clearly, I have given this whole thing some study and thought.

I was impressed with John Sanford's research and input. You will NEVER get this kind of information from a Darwinist because the Materialists don't want you seeing and learning this kind of scientific information.

John Sanford has done extensive modeling of Mutation/Selection using Mendel's Accountant, the most honest and accurate scientific model of the Mutation/Selection process. Therefore, John Sanford KNOWS precisely what he is talking about, and the Darwinist's don't.

Add to that the fact that Darwinists and Evolutionists have been searching for over 150 years for one single sign of Mutation/Selection creating ANYTHING substantial and new, or adding any useful information to ANYTHING at all; and, they have observed absolutely NOTHING in terms of Mutation/Selection originating and adding new biological information, deploying new unique genomes, or creating new life forms from scratch. Instead, the Scientists have observed that the process of speciation from the original genome that God gave to an organism typically results in a LOSS of information from that organism's genome. In other words, evolution by natural selection moves an organism towards its eventual extinction.

Evolution by natural selection cannot create or originate anything, but it does cause our eventual extinction due to usable information being selected away from the original genome which God gave us. Yet, the Darwinists try to attribute EVERYTHING to Mutation and Selection!

Nevertheless, EVERY BENEFICIAL MUTATION that the Darwinists point us to took place in a genome and an organism which God designed and created in the first place!

The Scientists have also observed that evolution by random mutations causes cancer, because the deleterious mutations outnumber the hypothetical beneficial mutations thousands to one (possibly millions to one) in every single organism. In other words, evolution by random mutations causes cancer and also moves an organism towards its eventual extinction. That's what evolution does for us!

—

ANOTHER QUOTE FROM "GENETIC ENTROPY":

> Therefore, if we assume man evolved from a chimp-like creature, during that process there must have been about 20 million nucleotide fixations within both the human and chimp lineages, but natural selection could only have selected for about 1,000 of these. All the rest would have had to have been fixed by random drift, resulting in millions of nearly-neutral deleterious substitutions. The result? A maximum of 1,000 beneficial fixations and millions of deleterious fixations. This would not just make us inferior to our chimp-like ancestors, in 6 million years it would obviously have killed us!
>
> Sanford, John (2015-02-23). Genetic Entropy (Kindle Locations 1918-1922). FMS Publications. Kindle Edition. USED BY PERMISSION.

END QUOTE.

Ouch! Not only would deleterious mutations make us inferior to our chimp-like ancestors but also we would have gone extinct during the process!

"Genetic Entropy" — buy his book, read it, and keep reading it until you actually understand it. His book is foundational fundamental core material when it comes to understanding why Mutation/Selection cannot design and create anything at all. His book is as important as anything that I will write in this book of mine. I want to go out of my way to quote and PROMOTE his book; and, the author John C. Sanford has given me written permission to do so.

So, what did we learn?

If we had actually evolved from chimpanzees or a common ancestor during the past six million years, due to the number of deleterious mutations that would have accumulated along the way, the human species (and chimpanzee species) would have gone extinct during that period of time. Mutation/Selection CANNOT do what the Darwinists say that it does. It can't design and create anything at all!

Let's say that again! Mutation/Selection cannot design and create anything at all! It doesn't work that way. Mutation/Selection might in fact be the creative arm of the Theory of Evolution and the Darwinists can lie through their teeth all day long telling us that Mutation/Selection does design and create anything and everything; but, the fact that Mutation/Selection can't design and create anything at all is scientific proof that the Theory of Evolution is FALSE.

Design implies a mind, and manufacture implies hands; and, Mutation/Selection has NONE.

—

Another quote from "Genetic Entropy":

> I have seen estimates of the ratio of deleterious-to-beneficial mutations ranging from one thousand to one up to one million to one. I believe the best estimates are closer to one million to one (Gerrish and Lenski, 1998). The actual rate of beneficial mutations is so extremely low as to thwart any actual measurement (Bataillon, 2000; Elena et al., 1998). Therefore, I cannot draw a small enough curve to the right of zero to accurately represent how rare such beneficial mutations really are.
>
> Sanford, John. Genetic Entropy (Kindle Locations 448-451). FMS Publications. Kindle Edition. USED BY PERMISSION OF THE AUTHOR.

Beneficial mutations are extremely rare, and ONLY God knows which ones are actually going to be beneficial!

A mutating species could go extinct due to deleterious mutations a MILLION times over, before it finally gets a beneficial mutation that actually makes it a better species. I can already hear some Darwinist saying, "YOU LIAR! A million times? That's an obvious over-exaggeration!" But unknowingly, it will once again be the Darwinist who has gotten his facts wrong, because it only takes ONE extinction to put an end to an evolving species. We don't need a MILLION extinctions to put an end to Darwinian Evolution or "Darwinian Change by Mutation/Selection". We only need ONE.

A blindly loyal Darwinist will never be able to see this or understand it, but I'm sure that the rest of us will.

There is absolutely NO WAY for any of us to have evolved from apes through the Mutation/Selection process. Every species on this planet would have gone extinct by now if we were trying to get those species to evolve into another species through random mutations and natural selection. Mutation/Selection eventually proves FATAL for evolving species, and extinction is always the result. There is absolutely NO WAY possible for Mutation and Selection working together to have designed and created anything at all, because they don't have that kind of power and ability in them.

—

That's Powerful Science, Isn't It?

For me this was the END of the Theory of Evolution. It would take trillions of years to create a single functional gene through mutation and natural selection; and, it would require the prior existence of a population of 10,000 living organisms of the same species that God has already created in the first place, and God maintaining that population of 10,000 organisms in existence for trillions of years, in order to create a single gene exclusively through the Mutation/Selection process within that pre-existing population. You see, it takes infinitely more blind faith to believe that Mutation/Selection created all the life on this planet, than it does to just simply believe that God did it all in the first place.

Mutation/Selection should be eliminated as The Source of all the genomes and life forms on this planet, because "Creation by Mutation and Selection" is logically, mathematically, and scientifically IMPOSSIBLE. Mutation and Selection cannot design and create genes and genomes! Therefore, Mutation/Selection should be eliminated as the Designer and the Creator of all the life on this planet, because Creation by Evolution of any kind is IMPOSSIBLE. Enough said!

—

Now take note, the Darwinists whom I have encountered trash-talk and dismiss John Sanford calling him a YEC (Young Earth Creationist). I can't figure out if he is truly a YEC or not, because he talks about millions of years in his hypothetical models as the quote above clearly indicates, and I caught him mentioning 13.8 billion of years for the age of the universe in his books. But, it doesn't matter if he is a YEC or isn't a YEC, because John Sanford taught me THE TRUTH about Evolution and Mutation/Selection and the fact that neither of these things can design and create anything at all!

The Darwinists can't dismiss the Science, because it speaks for itself once you understand it; so, they waste all of our time trying to dismiss and destroy the man. But, the Darwinists really have no other choice because the Whole of Science stands firmly against them, once you get a proper and correct interpretation of that scientific evidence. So, all of their Darwinian efforts will indeed be focused upon trying to destroy and dismiss the man. They have produced a well-orchestrated smear campaign against him; but, I have learned that whenever I encounter anything that the Darwinists don't want me seeing or reading, then that's exactly the thing that I should be seeing and reading!

John Sanford's books provide POWERFUL and CONVINCING scientific, empirical, mathematically, statistical, and logical evidence proving that Creation by Darwinian Chance, Creation by Mutation/Selection, and goo-to-you evolution are FALSE. When it comes to the Theory of Evolution, you really don't need to know more than that, unless you are trying to get a college degree so that you can preach Evolutionary Dogma and Lies in our public schools.

My recommendation is to buy all of John Sanford's books and read them, because the Darwinists clearly don't want you seeing them and reading them.

To Summarize

Whether Mutation and Selection were only given 6,000 years to design and create everything or an infinite amount of time to design and create everything, it makes no difference whatsoever either way, because even if Darwinian Chance and "Mutation and Selection" were given an infinite amount of time to design and create anything at all, they would still be unable to do so because they don't have hands and minds and because they don't exist or live in the first place.

The Darwinists can bad-mouth John Sanford all they want, but that only points us to the man and his books, and it gives me more incentive to support and PROMOTE the man and his books! For once in my life, I have finally found someone who is telling me THE TRUTH about Mutation/Selection and the Theory of Evolution! I could never have gotten any of that from a Darwinist!

Furthermore, for the killing blow, random mutations and natural selection by definition in principle cannot start to work and do anything at all, until after God has created the first living organism. So, there is absolutely NO way possible for Mutation/Selection to have created the first living organism, because Mutation/Selection doesn't come into play and can't come into play until after God has designed and created the first living organisms, in the first place.

Since Mutation and Selection working together cannot design and create anything at all, this becomes yet another convincing Scientific Proof of God's Existence, because God MUST EXIST in order to have done all the Science and Creation that Evolution or Chance could never have done.

—

I Enter into Evidence

I have given Macro-evolution another name which I use quite often in my essays, namely "Creation by Darwinian Chance" or simply "Darwinian Chance". It's Magic! The Darwinists invoke Magic as the creative arm or the creative entity, where the Theory of Evolution is concerned. Macro-evolution IS Creation by Evolution.

John Sanford's research and books kill Darwinian Chance dead!

Now I enter into evidence three books from John C. Sanford that are essential reading if you want to know THE TRUTH about Evolution:

"Genetic Entropy 4th Edition".

"Biological Information - New Perspectives A Synopsis and Limited Commentary" by Dr. John Sanford.

John Sanford, along with 28 other PhD scientists, also had a hand in producing "Biological Information: New Perspectives". It's a big 584-page science book that costs $178.00 new on Amazon. It, too, puts an END to the Theory of Evolution.

I own multiple copies of all three.

See for links: http://www.markme.us/id/

Both singly and combined together, these books put an end to the Theory of Evolution. And, isn't that all you really want, THE TRUTH about Evolution?

I eventually observed and learned that we will NEVER hear THE TRUTH about the Theory of Evolution from a Darwinist, because they have NONE to give. How can they give us THE TRUTH about Darwinian Chance when the thing doesn't even exist in the first place? How can the Darwinists give us THE TRUTH about Macro-evolution when it never happened in the first place and never will?

Although I have been given written permission from John C. Sanford to quote all I want from these three books, I have done so sparingly only quoting the bits that pertained directly to the questions and ridiculing which I was getting from the Darwinists online while I was interacting with them.

I USED the Darwinists to point me to what I should be studying and learning about the Theory of Evolution; and, the Darwinists tended to point me directly to John C. Sanford, Stephen C. Meyer, Jonathan Wells, Michael Denton, Harun Yahya, Hugh Ross, William Dembski, and Werner Gitt as the people and the books that they didn't want me seeing and reading. These were the people and the books that the Darwinists ridiculed the most and spent most of their time trying to debunk; and, the Darwinists pointed me directly to them by attacking these men and their books in public. Hopefully, the Darwinists will someday see fit to add this book and "Mark My Words" to their list of banned reading material.

Interesting, is it not, how I USED the Darwinists to point me directly to the people and the books that the Materialists didn't want me seeing and reading? The Darwinists give me lemons, and then I turn around and make lemonade!

I can be just as devious as they are, but in a good way. ☺

—

A Free Tip Which Will Make Your Purchase of This Book Worth Every Cent You Paid

The $178.00 "Biological Information: New Perspectives" science book can be found and read for free online at this link!

http://www.worldscientific.com/worldscibooks/10.1142/8818

"Biological Information - New Perspectives a Synopsis and Limited Commentary" by Dr. John Sanford" can be read for free at this link!

http://www.biologicalinformationnewperspectives.org/#!synopsis/c1294

My number ONE book about Evolution which I recommend that you buy and read until you understand it is "Genetic Entropy" by John C. Sanford. Alas, that one is not available for free, and will set you back ten dollars on Kindle; but, it's worth every penny! It's best to get it on Kindle, because it's in its 4th edition and Sanford is constantly updating it as additional Science becomes available.

You will notice that I promote these books heavily in almost all of my responses to the Darwinists online. The Rational Wiki (Atheists' Wiki) has dedicated space to these books, so these books must be doing their job revealing THE TRUTH about Evolution to the world, which the Atheists are quick to reject and try to debunk.

The typical response that I have seen from the Materialists and Darwinists regarding Sanford's books usually start with their attempt to explain to me why he isn't a Real Scientist, how he is a Young-Earth Creationist, how he hasn't done Real Science for decades, how he's nuts because he believes in God, and that Sanford and the other 28 PhD scientists which he is currently working with should be ignored and dismissed because they haven't been sanctioned or permitted by the Darwinists to speak on the subject of evolution.

Whenever the Materialists actually talk about the books or the quotes which I have provided them from Sanford's books, the Darwinists proceed to try to explain to me why Sanford isn't a Real Scientist and why he must be wrong as a result; and then, they always end up succeeding in explaining to me why they are wrong and why John Sanford is right after all. Their Darwinian interpretations of the Scientific Evidence are ALWAYS WRONG thanks to the fact that Darwinian Chance doesn't exist and couldn't have designed and created anything in the first place!

If you can actually get the Darwinists to talk about the "science", it quickly becomes apparent why their interpretation of "science" is wrong, because it's based exclusively on logic fallacies. Once you get into the science that is typically associated with the Theory of Evolution, it quickly becomes clear that the Theory of Evolution or the Primary Axiom IS illogical, unintuitive, and impossible — unless, of course, you are religiously devoted to the Primary Axiom and taking it all on blind faith as the Darwinists are, then it will all make perfect logical sense to you. Self-deception works, and it works every time!

Their Darwinian argument always goes like this:

> "All of the life that you can see around you is evidentiary scientific proof that the Theory of Evolution really works and is really true, because Evolution designed and created all of the life which you see around you. Since Evolution is true, you did indeed descend from apes just like we said you did; and, the fact that you descended from apes is scientific proof that the Theory of Evolution is true. The fact that Mutation and Selection have been observed in the wild is scientific proof that Darwinian Chance can design and create anything; and, the existence of Darwinian Chance is scientific proof that the Theory of Evolution is true. The fact that you exist is scientific proof that Macro-evolution can do everything that it sets its mind to; and, Macro-evolution is true because it designed and created you. The fact that each of

the fossils on Darwin's Tree of Life were found for real in the fossil record is scientific proof that the Theory of Evolution is true, because Evolution produced all of the different fossilized creatures that are presented to us on Darwin's Tree of Life. Evolution proves to us that God does not exist; and, the fact that we all know that God does not exist is scientific proof that the Theory of Evolution is true."

That's the sum total of their presentation and case, both from the Darwinists and the Atheists at Rational Wiki and elsewhere. I think that covers it all. I have never seen anything different.

If their case or presentation makes perfect logical sense to you and you can't see anything wrong with it, then you are a Darwinist, or an intellectually fulfilled Atheist, or you write for the Rational Wiki.

If you can see the logic fallacies or KNOW why these claims are false, then you are a Real Philosopher and/or a Real Scientist.

—

Abductive Reasoning and a Process of Elimination

This book of mine about evolution is a detective story — a search for THE TRUTH.

How often have I said to you that when you have eliminated the impossible, whatever remains, *however improbable*, must be THE TRUTH? — Sherlock Holmes

My primary goal in this book is to eliminate the impossible so that we are left with THE TRUTH.

You will also notice that I play around with deductive reasoning wherein I use premises, and then I draw a single conclusion from those premises.

It is said that Sherlock Holmes uses abductive reasoning. What is abductive reasoning? Abductive reasoning can be defined as "inference to the best explanation". If you notice carefully, in my essays, thoughts, and ideas, I am always searching for the best explanation or the best interpretation of the scientific evidence and the other evidence available to us. I go with the best and get rid of all the rest.

As with inductive reasoning, with abductive reasoning, the conclusion isn't 100% sure. The conclusion is the "best guest", or the "best explanation", or the "most likely candidate", or the "most logical suspect", or the "most reasonable conclusion", or the "most logical conclusion" that can be drawn from the available evidence.

For example, if I were to abduce or adduce through a process of observation, experience, scientific data, logic, and the elimination of falsehoods that ONLY Intelligent Beings can design and create new and functional genomes and therefore new and functional life forms, then our best guess or most logical conclusion is that some kind of intelligent being designed and created ALL of the genomes and life forms on this planet. It's the most reasonable and the most logical conclusion that can be drawn from the evidence that we have at hand.

The more that I successfully eliminate anything that could NEVER have done the job of design and creation; and, the more that I successfully provide observational, scientific, experiential, and logical evidence of intelligent beings designing and creating new and functional things in this physical realm, the greater the probability is that my conclusion (that ONLY intelligent beings can design and create) is in fact correct.

In this book, I successfully eliminate the impossible; and, in my own life I have successfully observed and experienced first-hand the design and creational powers of intelligent beings even while writing this book; therefore, I am now 100% certain that intelligent beings designed, programmed, created, manufactured, and deployed all of the genomes and life forms on this planet, because intelligent beings are the ONLY thing that could have done so. Nothing else qualifies!

So, where does the uncertainty of abductive and inductive reasoning come into all of this?

Well, even though I am 100% certain through "abductive reasoning" and a "process of eliminating the impossible" that intelligent beings designed and created all of the life on this planet, the evidence that I have observed and experienced first-hand cannot tell me precisely who this person or who these people were. They could in fact be highly advanced benevolent aliens, or they could be malevolent aliens putting together a future food supply or future hunting grounds. They could also be powerful telekinetic spirits who have the ability to use their minds to organize physical matter into any pattern that they desire. We just don't know, because the evidence that we have at hand can't tell us precisely who these intelligent beings were who designed and created all of the life on this planet. We know they must exist, but the evidence can't tell us who they are.

If you notice carefully, the people who have started to buy into my arguments will eventually ask me the question, "Which of all the thousands of gods that human beings have imagined or created out of thin air is indeed the one that designed and created it all?" Sometimes they are serious, and sometimes they are mocking me, but that's the question that I'm eventually asked. If they conclude like I do that ONLY intelligent beings can design and create, then the next logical question to ask is, "Who was that Masked Man who designed and created all of the life on this planet?"

I can't answer their question directly from the scientific evidence and observational evidence that I have on hand; and, they seem to instinctively know that it might be so.

I KNOW through "abductive reasoning" and the "process of eliminating the impossible" that these intelligent beings MUST EXIST, because ONLY intelligent beings can design and create new genomes and new life forms. It's a logical deduction. However, I can't use the evidence which I personally have experienced to pin the deed onto a particular person, at least not in any manner that would actually be convincing to someone else. When it comes to the spiritual, we each have to experience it first-hand for ourselves in order to know that it is real and true.

So, how do we identify or unmask THE PERSON who designed and created it all?

For that, we have to look for a confession. We have to look for a person who has actually confessed to doing the job of designing and creating all of the life on this planet. So far, I have found only two people who confessed to doing the job.

We have Allah, or the person whom the Christians would call God the Father, as one candidate who confessed to doing the job of design and creation on this planet, in the Koran.

And, the other person I found is Jesus Christ, the Son of God the Father; and, He too confessed to organizing this universe and this earth, and confessed to designing and creating all of the life on this planet, as found and documented in the Bible, Book of Mormon: Another Testament of Jesus Christ, Doctrine and Covenants, and Pearl of Great Price.

If I were forced to pick only one intelligent being for the job, I would go with Jesus Christ as the most likely candidate, because I have at hand the most evidence to support that particular conclusion. I have multiple different copies of His written confession from multiple different sources. Jesus Christ is the best and most likely candidate due to vast number of different times that He has actually confessed to doing the job. Confessions are admissible in a court of law; and, I always go with the best evidence that I have at hand.

However, if I am permitted to select multiple intelligent beings, then I would in fact go with Jesus Christ the Son of God, God the Father or Allah, and the Council of the Gods (the Elohim) as the designers and creators of all the genomes and life forms on this planet, because there is written evidence that they too had a hand in the design and creation process which went into our universe.

—

Now Let's Look at What the Opposition Has to Offer

I am interested in comparative religion studies wherein I make some kind of attempt to compare the different Religions in the world — what they believe and what they do. Obviously, since I spent time as an Atheist and a Materialist, I have a great deal of interest in the beliefs and actions found in the Atheistic, Materialistic, Darwinistic, and Naturalistic Religions. And, don't let anyone fool you! They ARE Religions, or worldviews.

So, how do these compare with Christianity, Islam, and Judaism?

Some of the most illogical and irrational things that I have ever read in my life have come from the Atheists and their Rational Wiki.

The Atheistic and Materialistic claim that NOTHING designed and created everything in this universe is THE MOST illogical, irrational, unscientific, and nonsensical claim that mankind has ever devised. It can't be tested, replicated, observed, demonstrated, questioned, or examined in any way, shape, or form.

When was the last time that you got NOTHING to do something for you? Are you still waiting for NOTHING to do your homework, make you a computer, write you a program, get you a job, repair your genome, and build you a house? If you are waiting for NOTHING to do these things for you, then you are going to be waiting forever for NOTHING.

In my arguments, I take a rational and logical reasoned approach towards eliminating everything that is IMPOSSIBLE, so that the only thing that remains must be THE TRUTH.

It is IMPOSSIBLE for NOTHING to have designed and created EVERYTHING.

It is IMPOSSIBLE for DARWINIAN CHANCE or CHANCE of any kind to have designed and created EVERYTHING. The "creative element" in Random Mutations is synonymous with Chance; Chance is synonymous with NOTHING; and, Creation by Chance is synonymous with Magic.

Creation by Macro-evolution or Abiogenesis is IMPOSSIBLE!

It is IMPOSSIBLE for MUTATION/SELECTION (Micro-evolution) to design and create new functional proteins, new functional genes, new functional genomes, and new living functional organisms from scratch.

It is IMPOSSIBLE for MUTATION/SELECTION to encode new functional and useful genetic information or genetic programming into a DNA strand and into the correct place within a genome. It is IMPOSSIBLE for MUTATION/SELECTION to know where to place the beneficial mutations, if there are any.

MUTATION/SELECTION cannot make ANY adjustments to genomes until AFTER God has designed, built, and then deployed those genomes in the first place. Design and Creation by MUTATION/SELECTION is IMPOSSIBLE!

It's Logical Common Sense. It's Science!

What are you left with after you have eliminated the IMPOSSIBLE?

How often have I said to you that when you have eliminated the impossible, whatever remains, *however improbable*, must be THE TRUTH? — Sherlock Holmes

After you have eliminated the IMPOSSIBLE, then you are left with THE TRUTH. You find THE TRUTH staring you in the face.

From the Wikipedia: "Holmes' primary intellectual detection method is abductive reasoning (seeking the best explanation for the available evidence). Holmesian deduction consists primarily of observation-based inferences."

My deduction method consists primarily of logical common sense combined with a process of elimination. I eliminate the ILLOGICAL and the IMPOSSIBLE so that we are left with THE TRUTH. I pursue parsimony — the simplest and most likely explanation or interpretation of the evidence. I infer to the best explanation or the best interpretation of the evidence, whether that evidence is scientific or any other kind of evidence. The hope is to zero in on THE TRUTH through a process of elimination.

THE TRUTH is all we really want after all, unless you are an Atheist or Materialist of course.

AGAIN, the Atheistic and Materialistic claim that NOTHING designed and created everything in this universe is THE MOST illogical, irrational, unscientific, and nonsensical claim that mankind has ever devised. When was the last time that you got NOTHING to do something for you?

What are the odds that NOTHING designed and created EVERYTHING?

NOT A CHANCE!

—

Waiting for Nothing to Do Something

The Darwinists, Materialists, and Atheists spend all of their time waiting for NOTHING to do something. It's a 150 years later, and the Darwinists are still waiting.

According to the Darwinists and Materialists, the Theory of Evolution and Creation by Darwinian Chance is obviously true or self-evidently true, and the Theory of Evolution is unquestionably axiomatic.

According to the Darwinists, the Theory of Evolution is true, because the Theory of Evolution is true. Enough said!

That's the extent of their reasoning on the subject, because they can't take it any further, because the Scientific Evidence, Observational Evidence, and Common-Sense Logic as a whole stand against them and tells us that they are WRONG.

If you are a Real Rational Logical Philosopher and/or a Real Scientist then you just KNOW that the Darwinists have NO Real Science to support their claims, ONLY their faulty philosophical interpretation of the Scientific Evidence.

Their FALSE philosophical interpretation of the scientific evidence and their logic fallacies are the ONLY things that the Darwinists have, which tell them that the Theory of Evolution must be true.

I can see what's wrong with the Materialists' Presentation every time, because I am a philosopher and a scientist. I have yet to FAIL to figure out for myself what's wrong with the Darwinists' claims, thanks to the Real Scientists and the Real Philosophers whom I have chosen to study, learn from, and read.

Whenever the Darwinists have asked me questions which I couldn't answer scientifically, I found most of the answers in one of John C. Sanford's three books, which he had a hand in writing and producing.

Whenever the Atheists and the Darwinists try to take me on philosophically, books from William Lane Craig, John C. Lennox, and Frank Turek have helped me to put the Darwinists into their graves.

Good enough!

And sometimes, good enough really is good enough.

We have NOTHING to fear from the Materialists, Atheists, and Darwinists, because they have NOTHING to offer us.

Creation by Chance, Creation by Evolution, Creation by NOTHING, and Creation by Natural Reactions IS AN ABUSE of the Scientific Method; and, the Theory of Evolution IS Creation by Chance, or Creation by Evolution, or Creation by Natural Reactions such as Random Mutations or Natural Selection.

That's what I discovered during my Pursuit of the True Reality of All Things, and during my usage of the Scientific Method.

PART II — MATERIALISM IS USELESS AND FALSE

In my Pursuit of the True Reality of All Things, I had to be willing to abandon any residual belief in Materialism, Darwinism, Naturalism, Behaviorism, and Atheism, BEFORE I was able to find THE TRUTHS that I sought.

Belief in a LIE cannot help us to find THE TRUTH. That's just the True Reality of the situation.

There was a time in my life when I was a Materialist and an Atheist. I realized one day that I didn't like where that road was taking me, so I decided to turn around and go the other way. My life has been getting better and better ever since.

The purpose of the Scientific Method is to help us to find THE TRUTH, through a preponderance of the evidence.

the Scientific Method has no value to us if we use it to convince ourselves that a LIE is TRUE, as the Materialists and Darwinists always seem to do.

I had to get rid of My Materialism, My Atheism, Darwinism, and The Theory of Evolution BEFORE I could successfully start to study the Spiritual Sciences and achieve noticeable results.

That's what I discovered during my Pursuit of the True Reality of All Things, and during my usage of the Scientific Method.

Materialism Is a Denial of Reality

Materialists have a hard time seeing the Light.

You see, Materialistic Atheists in principle have conceptual problems understanding and accepting the reality of the Quantum Sea of Light or the Zero-Point Field of Light, because such a thing is sub-quantum, immaterial, non-physical, holographic, conscious, and spiritual. Such a concept is hypothetical and experimental and often mathematical; and, the Materialists deliberately cripple themselves a priori by excluding anything non-physical or spiritual from their worldview, their personal religion, and their SCIENTIFIC EXPLORATION.

Instead, the Materialists will be found online mocking the Quantum Sea of Light or the Light of Christ, because they don't understand it and can't understand it. It's beyond their current level of comprehension and acceptance.

If you don't ever look for it, then you will never find it. That's the reality of Science! That's the reality of Life! The Materialists never look for anything sub-quantum or spiritual, so they never find anything spiritual or immaterial because they prevent themselves from doing so.

Ironically, ALL light is immaterial, non-physical, and spiritual. Light of any kind has NO mass and NO matter. Light is immaterial, non-physical, and spiritual. That's why some scientists and theoreticians have chosen to call the "Zero-Point Field" the "Quantum Sea of Light", because they truly believe that the Zero-Point Field IS some type of light. In fact, some of the scientists have taken to calling it the "Zero-Point Field of Light". Spirit is light. Light is spirit. Light is immaterial. The existence of Light PROVES that Materialism is FALSE. The existence of Time PROVES that Materialism is FALSE, but that's another story.

Obviously, such concepts make Materialists extremely nervous, because the existence of the Zero-Point Field and the existence of Light proves empirically that Materialism is wrong. Light is spiritual, and spirit is light. And, the existence of Light proves that the Materialists are wrong at a basic fundamental level. The Materialists literally refuse to see the light.

Can you see how Materialists and Atheists completely cripple themselves and their chances for discovering something new simply by the philosophical point of view that they have chosen to believe in? The Materialists literally make the "existence of Light" completely out of the question, when it comes to their personal worldview!! The Materialists DEFINE light or the spiritual completely out of existence, hobbling themselves in the process. Very unscientific!

It has been hypothesized by some scientists and mathematicians that this Zero-point Field of Light exists throughout ALL of that "empty space" between the nucleus of an atom and that atom's electrons. In fact, many scientists believe that it is this Quantum Sea of Light which actually holds the electrons in their orbits and prevents the electrons from crashing into the nucleus of their atom. The Materialists say that 99.999% of an atom is empty space; and, others say that that space really isn't empty because it is filled with a Quantum Sea of Light or a Zero-Point Field (Forces and Fields of Spirit, Consciousness, and Light) which actually gives the atom structure and substance.

So, who is right — the Materialists or the Spiritualists?

My bet is now on the Spiritualists, because I have slowly discovered that there are millions of non-physical or immaterial things that still have substance or influence — things like forces, fields, waves, strings, thoughts, dreams, dark energy, and LIGHT. The existence of radio waves and television waves PROVES Materialism FALSE!

Searching for and finding evidence for God's existence is the most interesting thing that we can study and learn about. It can be fascinating to identify all of the different kinds of "breadcrumb trails" that God deliberately left behind for us to find and follow, which will lead us directly to Him who is waiting for us at the end of each trail.

It reminds me of the board game "Clue". At the beginning of your search, it's not the least bit clear "Who Dunnit". But, through a process of elimination and through a process of asking questions and getting answers, you finally settle upon convincing proof of "Who Dunnit". At the end of your search, you just KNOW "Who Dunnit"; and, you can even identify who was lying to you along the way if someone did in fact lie to you.

At the end of my Search for Reality, I did indeed discover that it was the Materialists and the Darwinists who have been lying to me all my life. Materialism is demonstrably FALSE, which means that it is a LIE.

—

Spirit Is Light, and Light Is Spirit

The Materialists DENY the existence of Light. That's how "in the dark" the Materialists really are!

Don't let them fool you into believing that they make a special exception for Spirit or Light, because I have caught the Materialists online mocking and ridiculing the Quantum Sea of Light, the Zero-Point Field of Light, and the Light of Christ. The Materialists also mock and deny the existence of the Living Lights, or what we typically call Spirits.

Materialism IS by definition the philosophical belief or religion, which teaches and believes that Matter (and the associated energy which matter can be converted into) is the ONLY thing that exists in the universe.

The ONLY light that the Materialists might be willing to make an exception for is the visible light that they can see with their eyes, possibly trying to associate it with energy; but they can't have it BOTH WAYS! That's cheating! The Materialists can't deny the existence of Spirit yet at the same time accept the existence of Light, because Spirit is Light! So technically, a hard-core Materialist will DENY the existence of all light if he is true to his creed, because the very existence of light PROVES that Materialism is FALSE.

Are you starting to see how STUPID and ILLOGICAL Materialism really is?

Materialism only makes sense to an Atheist, because he needs Materialism so that he can deny the existence of Spirit, the Spiritual, the Living Lights, the Holy Spirit, and God. Often, you will actually catch an Atheist looking at the Light and literally denying that it exists. The Materialists are online right now mocking and ridiculing and laughing at the different types of light that they can't see with their physical eyes.

Spirit IS Light. Light IS Spirit. It's just that — what we typically classify as Spirits — exist or live at a frequency of light which isn't visible to the naked human eye. Typically, the entities that we call Spirits are in fact Living Lights. The Living God has a Spirit and a Spirit Body. It's the Spirit or the Light that gives us Life. That's why the Biblical God is often called The Living God. He IS Life, and He gives Life! He is the Father of Lights.

God the Father and Jesus Christ have gained the types of Spirits and Spiritual Power that has the ability to levitate and manipulate physical matter at the atomic level. Their thoughts penetrate into all of that empty space between quarks and between the nucleus of an atom and its electrons. Their thoughts reside in the

Quantum Sea of Light or the Zero-Point Field of Light, what they call the Light of Christ.

There are different types of "spirit" or light.

There seems to be some type of spirit matter, which is not alive but is simply acted upon. The Spirits that are alive are Living Lights. We humans are the spirit children of God the Father and Heavenly Mother. They sired and birthed our Spirit or Light. At our very core, we are Living Lights.

—

It has also been hypothesized by some Scientists that the Zero-Point Field or the Quantum Sea of Light exists in all that "empty space" between us and the Andromeda Galaxy, and between us and the end of this Universe. Those who are of this opinion truly believe that the Zero-Point Field has become something like a Cosmic Internet or a Cosmic Consciousness which pervades the whole of this Universe and ALL of this Physical Reality.

In the Doctrine and Covenants, the Biblical God Jesus Christ calls it the "Light of Christ" that fills all things and fills the immensity of space, confirming that this Zero-Point Field or Quantum Sea of Light or Dark Energy does indeed exist universally within every particle of matter and within all of that "empty space" in this universe. Christians as a whole tend to call this Quantum Sea of Light or the Zero-Point Field the Holy Ghost or the Holy Spirit. Why? It's because light is immaterial, non-physical, and spiritual; and, even we humans have found different ways to communicate through light. The Holy Ghost is a communicator and a revelator and a recorder, among other things.

The very existence of light of any kind proves Materialism wrong. Spirit is Light; and, Light is Spirit.

Some scientists have hypothesized that this Quantum Sea of Light or Zero-Point Field is in fact what makes up gravity, because as we all know or should know, each atom within our physical body is connected gravitationally with every other atom in this universe, including the atoms on the opposite side of this universe. It's all interconnected gravitationally, instantaneously, and simultaneously from one end of the universe to the other through the Zero-Point Field or Quantum Sea of Light or Gravity.

To the religious believers, particularly the Christians, this Quantum Sea of Light is the Spirit Realm — the Realm of the Living Lights. Out-of-Body Travelers KNOW from direct personal experience that when they are IN SPIRIT, they can travel to anywhere on this earth or to anywhere in this universe instantaneously at the speed of thought. They think of a place, and their spirit is just there instantaneously. There is no speed limit when it comes to Pure Consciousness, Pure Spirit, or Pure Light.

David Bohm has called matter "frozen light" or "condensed light".

If you take Pure Light or Pure Consciousness or Pure Thought and slow it down a tiny fractional bit, then it becomes spirit matter — something a bit different than Spiritual Consciousness or Pure Thought which has no speed limit. The associated or linked spirit matter lags just a tiny little bit behind the Thought. It's barely perceptible. If you slow light down a lot more, then it becomes packets of photons and can only travel at the "speed of light". There are visible photons and invisible photons, and God has slowed them down to what we call the "speed of light". If you slow light down to zero speed, then it becomes frozen light or physical matter. Everything is made up of light.

Are you starting to see the light?

This is COOL STUFF to study and learn about! The Theory of Evolution pales in comparison! PURE LIGHT or TRUE LIGHT is the stuff that God and Spirits are made of!

For me, this is the most interesting chapter in this whole book, because here we are talking about things that truly matter — the things of Eternal importance!

—

Physical Matter Is Enlivened by Light

A huge portion of the "Pistis Sophia" is all about Light. It can be found for free online by looking for it!

Physical matter is enlivened or animated by Spirits or Living Lights.

In his book, "Temple and Cosmos" page 152, Hugh Nibley emphasizes this REALITY.

> "Matter without light is inert and helpless," says the Pistis Sophia.
>
> The rays from the worlds of light stream down to the earthly world, for awakening mortals.
>
> Sometimes the column of light joins heaven to earth, as in our Facsimile No. 2 (a very important principle), even as the divine plan is communicated to distant worlds by a spark. According to Carl Schmidt, it is the dynamics of light from one world that animates another.
>
> The spark is also called "the drop"; the Egyptians call it the prt ("drop"). It is the divine drop of light that man brought forth with him from above, the spark that reactivates bodies that have become inert by the loss of former light. It's like a tiny bit of God himself.
>
> Christ calls upon the Father to send light to the apostles.

End of quotes from "Temple and Cosmos".

The Living Light or Living Spirit is also at times called The Spark.

The Transformer movies talk about the AllSpark. It's this AllSpark which gives the Transformers a Spirit, or Life, or Living Light.

From the Wikipedia, "The AllSpark is an ancient artifact or object capable of creating new Transformer life by bestowing machinery with sparks."

God the Father (as well as the Biblical God Jesus Christ) is often called THE LIVING GOD, because He and His Spirit is the AllSpark from which all of our Spirits or Living Lights derive.

These are the things which the Materialists actively ridicule, mock, and laugh at online whenever they start trying to preach their religion and ram it down our throats. Yet, in comparison to this, Materialism is extremely boring!

—

Dark Energy

Not all of what makes up this universe is matter!

Astrophysicists and Astronomers have taken to calling the Quantum Sea of Light or the Zero-Point Field by the name "Dark Energy", and they have demonstrated empirically through observational measurements that this Dark Energy comprises 72.1 percent of this universe! This Dark Energy or Spirit Realm or Spiritual Construct comprises 72.1 percent of our Universe's Cosmic Density! All of that empty space isn't empty after all! This Dark Energy or Quantum Sea of Light is stretched throughout the whole of this universe, or it is stretching and expanding the whole of it!

On page 36 of "Why the Universe Is the Way It Is" by Hugh Ross, he put up a table entitled, "Inventory of All the Stuff that Makes Up the Universe". That table tells us what God put into this universe, giving us the chance to speculate why it is there and what it all means. It's a listing of what comprises this Universe's Cosmic Density.

First of all, we have Dark Energy (the self-stretching property of the cosmic space surface), which many believe to be the Zero-Point Field or the Quantum Sea of Light that stretches an atom thus keeping an atom from imploding. This Dark Energy makes up 72.1 percent of his universe's "cosmic density"; and, Dark Energy completely fills the immensity of space. Dark Energy or PURE LIGHT is the Universal Construct. This is the Light of Christ or the Spirit Realm or the Realm of the Living Lights, where the Living Lights (your spirit and mine) live or reside naturally. Dark Energy or the Zero-Point Field or the Quantum Sea of Light IS the Primal Construct of this physical universe. Dark Energy is the forces and fields that keep an atom and a proton from imploding into a singularity and taking all the rest of us with it.

There is no such thing as empty space, because the whole of space is based-upon and upheld or sustained by Dark Energy and is completely filled with Dark Energy or this Zero-Point Field or Quantum Sea of Light. This Dark Energy or Light of Christ is the Universal Construct that prevents this universe from imploding and keeps this universe stretching and expanding. That's why this Dark Energy comprises 72.1 percent of this universe's cosmic density, because it is literally stretched throughout the whole of this universe's "empty space" keeping this universe expanded like a balloon. If God were to pull out the Dark Energy or Light of Christ from this universe, then this universe would deflate or implode back into the singularity from whence it came.

Imagine what you could do if you were to set up a Network or an Internet onto this thing, as God has already done. God is in the Light, particularly the Quantum Sea of Light. In the Doctrine and Covenants Sections 88 and 93, the Biblical God calls this thing the Light of Christ, and God says that it fills the immensity of space. This Light of Christ or Quantum Sea of Light is what God uses to communicate with us, and this Dark Energy is how God hears all of our prayers and thoughts instantly and simultaneously. The dude has some serious bandwidth!

This Dark Energy or Zero-Point Field is indeed the Cosmic Consciousness or the Cosmic Internet. God can literally beam thoughts into our minds, into all of that "empty space" between the nucleus of an atom and its orbiting electrons. Cool, huh? Of course, that might also explain why God is bothered by and complains about all of our unholy and impure thoughts. He wants us to learn how to control ourselves so that He doesn't have to listen to all of our rot whenever we get our thoughts into a rut.

This Dark Energy goes by many different names! It MUST BE extremely important if it comprises 72.1 percent of this universe; and, it is important and essential because Dark Energy or PURE LIGHT is the thing that keeps this universe expanding and expanded. If you were to pull out this PURE LIGHT, or Zero-Point Field of Light, or Quantum Sea of Light, then this whole universe would implode into a singularity.

This PURE LIGHT or Dark Energy doesn't seem to have any speed limit, being simultaneously everywhere at all times — in all things, through all things, and filling the immensity of space. This PURE LIGHT or PURE THOUGHT or PURE CONSCIOUSNESS is the Primal Construct. This whole universe is ONE GREAT THOUGHT.

—

Laughing at What They Know Not

In their complete and total ignorance, the Materialists and Atheists mock and laugh at the Quantum Sea of Light or Light of Christ, not knowing what it is or what it does. They mock it and ridicule this Dark Energy or Light of Christ because it is immaterial or spiritual, and they can't get their hands on it. Yet, these Materialists and Atheists are mocking and laughing at something that comprises 72.1 percent of this universe's cosmic density, because their deliberate ignorance makes them dense enough to do so.

Ironically, these very same Materialists and Atheists are NEVER caught laughing at and mocking Evolution or Chance, even though Evolution and Chance are infinitely more insubstantial, immaterial, non-existent, and laughable than Dark Energy or the Zero-Point Field can even begin to be!

Why do they do it? Why mock something like the Zero-Point Field or the Light of Christ for which there is abundant necessity and evidence, yet admire and worship something like Creation by Evolution or Creation by Chance for which there is absolutely NO evidence whatsoever? Why the selective amnesia or hypocrisy?

These Materialists, Atheists, and Darwinists admire and worship Creation by Chance or Creation by Evolution, because these non-existent things are telling the Materialists exactly what they want to hear, that God does not exist. That's what this whole controversy is all about. The Materialists are more than willing to believe in immaterial and non-existent things like Creation by Evolution or Creation by Chance, so long as those non-existent immaterial things are telling them that God does not exist and that they don't have to repent of their sins.

Interesting is it not, how human psychology works? We are in denial about the things that we don't want to know about or hear about. The Materialists are in denial about 95% of this universe's cosmic density because it is spiritual, non-physical, and immaterial! How dense can they get? Well, they can get pretty dense, because they want to be that way and they want it that way.

They don't want to know about any of this because it suggests that God might exist. It suggests that they might be facing some kind of Final Judgment someday and have to answer to God for all their sins and misdeeds. Materialism is a denial of Reality, because the Materialists don't want to have to face this Reality.

This Light of Christ or Dark Energy IS what keeps this universe stretching and expanding! If God were to pull this Dark Energy or Light of Christ out of this universe, then this universe would implode into the singularity from whence it came.

The astrophysicists call it Dark Energy because it can't be seen with the naked eye; but, it really is comprised of invisible Light of some kind. I find myself wanting to call this stuff PURE LIGHT, the kind of light that isn't weighed down or slowed down by extra baggage. This PURE LIGHT might come in two flavors or types — a sentient living active type and a non-sentient inert reactive type. I'll mention more of this possibility in a later part of this essay.

This Light of Christ or Dark Energy IS also the thing that keeps protons, neutrons, and electrons from decaying and imploding into singularities. Something is preventing those protons from decaying! This Light of Christ or Dark Energy IS the thing that provides forces and fields to keep electrons from merging with the nucleus of their atom. This Light of Christ or Dark Energy must of necessity exist, or we would not exist as physical mortal beings, because it is this Dark Energy or Light of Christ that prevents EVERYTHING from imploding into a singularity. Something or Someone HAS TO BE doing that, otherwise we all would still be in that singularity from whence we came!

If we were to take a Materialist or Atheist and pull ALL of this Light of Christ out of his physical body, his physical body would implode into a singularity and he would cease to exist — only his spirit or consciousness would remain.

Yet, day-in and day-out online, you will see these very same Materialists and Atheists laughing at and mocking the Quantum Sea of Light, the Zero-Point Field, Dark Energy, or the Light of Christ because in their deliberate ignorance the Materialists do not know what it is, or what it does, or why it is necessary for it to exist. They laugh at what they know not! But, in comparison to all of this information about Light, the Theory of Evolution is extremely boring!

Materialism is a very shallow, limited, and even laughable worldview or religion, because it is so immaterial and unsubstantial.

—

Exotic Dark Matter

Next on the list of items that make up our Universe's Cosmic Density, we have what they call Exotic Dark Matter (particles and/or energy that weakly interacts with ordinary particles and light).

For anyone who has studied physics, metaphysics, and the nature of a spirit body, they quickly realize that "spirit" comes in at least two types, spirit matter and spirit energy, because physicists have demonstrated that matter of any kind can be converted into energy, and energy can be converted back into matter.

Many of us believe this Exotic Dark Matter to be spirit matter — the material and energy that makes up our spirit bodies. Exotic Dark Matter or Spirit Matter comprises 23.3 percent of this universe's cosmic density. Matter of any kind is the kind of stuff that God designed to be acted upon. Matter simply reacts to whatever a living spirit or a living light tells it to do. Matter is reactionary. Thought or Consciousness or Intelligence or Life is what makes that matter react. Consciousness or Life is what does the choosing, desiring, and willing. Matter of any kind was designed to obey the commands of Living Spirit or Living Light. That's why your physical body does what your spirit tells it to do, as long as your physical body is functioning properly.

There are different types of Spirit or Light, many different types.

Apparently, this Spirit Matter is NOT alive, because it is a more refined type of matter, and matter was designed by God to be acted upon, not to act. It's the LIGHT that is active, alive, sentient, aware, and conscious; and, not the matter.

This Exotic Dark Matter or Spirit Matter is said to be concentrated into halos around galaxies preventing the Ordinary Matter in galaxies from flying apart and dispersing evenly or homogenously throughout the whole universe.

After the Big Bang 13.8 billion years ago, all physical matter should have distributed itself evenly throughout the universe because every particle was moving away from every other particle; and, this universe should be nothing more than one big homogeneous ball of gas as a result. But, someone forced huge chunks of that Ordinary Matter to gather into galaxies and then later to gather into stars and planets. Someone was driving the bus. The Biblical God says that He is the one who did this for us. Who else was living back then who could have done the job?

It's the SPIRIT or the LIGHT which controls and manipulates matter — both physical matter and spirit matter. SPIRIT or LIGHT moves this matter and puts this matter where it wants this matter to be. Your spirit or mind might have had a hand in gathering ordinary physical matter into galaxies, stars, and planets; and, your spirit or light or consciousness might have had a hand in organizing and moving all of that Exotic Dark Matter or Spirit Matter into halos around the different galaxies to keep the galaxies from flying apart.

Exotic Dark Matter or Spirit Matter or some kind of "Immaterial Matter" comprises 23.3 percent of this universe's cosmic density, so it must have some purpose, or some use, or some reason for being. There are probably other uses for it that we haven't even thought of yet. But for me, the first thing that comes to mind is that it's massive enough and influential enough to keep Ordinary Matter in check or in line. Something or Someone has to be keeping galaxies from flying apart and becoming homogenous.

When it comes to speed of travel, this Exotic Dark Matter or Spirit Matter seems to lag ever so slightly behind the SPEED OF THOUGHT.

Of course, there is another possibility to consider. I like to think that there are always possibilities. It's possible that this Spirit Matter actually doesn't travel at all. In other words, it's possible that it is PURE THOUGHT or PURE LIGHT or PURE CONSCIOUSNESS that does the traveling, and then that PURE THOUGHT builds its reality out of the Spirit Matter that exists at its chosen destination.

Out-of-Body Travelers have stated that there are consensus realities that are designed, built, and maintained by spirit beings through a collective consciousness or a common consent; and then, there are non-consensus realities in the spiritual realms where the reality actually molds itself to your thoughts and your desires.

So, is it the Exotic Matter or Spirit Matter that travels thus creating that slight lag, or is it the PURE THOUGHT or PURE LIGHT that does the actual instantaneous traveling with the lag being caused by the fact that a SPIRIT or PURE LIGHT has to build or construct its reality out of spirit matter when that SPIRIT reaches its chosen destination? Obviously, questions still remain; and, only God knows the answers for sure.

Exotic matter, dark matter, spirit matter, consensus matter, and non-consensus matter — that's way too much matter for the Materialists! Imagine it! Matter that the Materialists cannot get their hands on; therefore, it must not exist. Yet, it shows up in the Cosmic Density of this Universe. So, something's the matter with Materialism.

Any way you choose to look at it, this is RADICALLY ADVANCED SCIENCE! Materialism pales in comparison! Materialism is a DENIAL OF REALITY.

Ordinary Matter

Next on the chart in Hugh Ross' book is Ordinary Dark Matter (quantum particles that interact strongly with light). These little buggers seem to be controlled by sentient living light! It's the Sentient Living Light that's driving the bus; and of course, you make busses out of Ordinary Matter.

Ordinary Matter is the stuff that the Materialists and Atheists worship and admire, because they can get their hands on it. They make idols out of it!

This Ordinary Dark Matter is what our consciousness and our spirits act upon and control here in the physical realm. Ordinary Dark Matter comprises 4.35 percent of our universe's cosmic density. This Ordinary Dark Matter is what our physical bodies are made of. However, the astrophysicists like to break this Ordinary Dark Matter into three types. The type here called Ordinary Dark Matter, which seems to be distributed unorganized throughout space, the type that becomes Ordinary Bright Matter (stars), and the type that goes into making earths and physical bodies like ours. I just call the whole thing "Ordinary Matter".

Then on the chart in Hugh Ross's book comes the Ordinary Bright Matter (stars and star remnants). These make up 0.27 percent of the universe's cosmic density. This "star stuff" that Carl Sagan worshipped only makes up 0.27 percent of this universe. Carl Sagan said that this Ordinary Matter is all that there ever was and all that there will ever be. Impressive ignorance was it not?

And finally, on Hugh Ross's chart, we have planets, humans, and other living organisms — the part that the Materialists really choose to believe in, which is a subset of Ordinary Dark Matter; and, we make up only 0.0001 percent of this universe's cosmic density. In a sense you could clearly say that the Materialists only believe in 0.0001 percent of this universe — the stuff that they can really get their hands on. That's how limited the Materialistic and Atheistic worldview can become! Even if you want to be generous to them, the Materialists still only believe in 4.62 percent of this universe — the Ordinary Matter. The Materialists DENY the existence of over 95 percent of this Universe and its Cosmic Density. The Materialists are so dense, that they deliberately eliminate 95 percent of this universe from their worldview or religion. Impressive ignorance, is it not? That's a really HUGE denial of Reality! No wonders the Atheists and Materialists are always in a State of Denial.

The Materialists deny the reality of all the rest of this universe's cosmic density, because they can't see it with their physical eyes nor get their hands on it with their physical instruments. When it comes to cosmic density, the Materialists are out of their depth and really dense! Materialism is a very shallow and BORING philosophy of life or world religion.

A Multitude of Different Kinds of Light or Luminosity

There is now another possibility to consider. I like to think that there are always possibilities. There is the possibility that there might in fact be something in existence that DOES NOT register at all in this Universe's Cosmic Density. Think about it! This would be the kind of stuff that dreams are made of.

If Dark Energy, comprising 72.1 percent of this universe, is made up of PURE LIGHT or PURE SPIRIT, then anything that exists in this universe that doesn't register at all in this Universe's Cosmic Density would be the PUREST OF LIGHT or the PUREST OF CONSCIOUSNESS AND THOUGHT. Think about it! PUREST THOUGHT, or PUREST LIGHT, or PUREST CONSCIOUSNESS, or PUREST INTELLIGENCE — the hypothetical stuff that exists but doesn't weigh in at all in our Universe's Cosmic Density.

So now we come to the stuff that is truly immaterial, non-physical, and completely spiritual or sub-spirit-matter because it doesn't figure at all into the universe's cosmic density! This stuff is truly the stuff that dreams are made of.

This would be the PUREST and most refined of all existence. This stuff IS thought, consciousness, the construct of light, awareness, or Intelligence. This stuff IS TRUTH, what we truly are at our very core. Obviously, this PUREST STUFF can travel faster than the "speed of light". In fact, many people have experienced the out-of-body situation, where their spirit or consciousness can travel instantly to any place in the universe or anywhere on this world simply by thinking about it or willing it to be so. Others while IN SPIRIT have actually experienced being two or more places at once, simultaneously. Think about it! Fascinating, is it not?

Consciousness or Thought is like Gravity and exists simultaneously everywhere in this universe — all interconnected in some way and instantly accessible. Some of the scientists call this Cosmic Consciousness or Universal Consciousness. The Doctrine and Covenants and Book of Abraham call this PUREST of stuff "INTELLIGENCE" or "LIGHT AND TRUTH". Others call it The Mind of God.

Light itself is immaterial and has no mass. Light PROVES that Materialism is false. Thanks to light, all you have to do is open your eyes and see that Materialism is wrong. But, what is light made of? The Doctrine and Covenants and Book of Abraham at times seem to indicate that LIGHT or SPIRIT is made-up-of and/or occupied-by Consciousness, or Intelligence, or Thought.

LIGHT or SPIRIT and SPIRIT MATTER seem to be made-up-of and/or driven by Intelligence, Consciousness, or Thought. Each photon of light is conscious and aware of its environment, and it knows its destination before it launches into space, because the quantum fields through which it travels are omnipresent and have no physical limitations. The quantum fields are like "wires" at the subquantum level, which means that energy, information, messages, thoughts, quantum waves, packets of energy, and quanta can travel at an infinite velocity through the quantum fields, because there are no "spacetime gaps" or "distance gaps" at the quantum level like there is at the physical level between physical atoms.

Which came first, Intelligence or Light, Consciousness or Spirit? We don't know. They are probably co-eternal and intricately bound to each other. Is there really any difference between them? We don't know, but there might be, especially if one registers in our Universe's Cosmic Density and the other one does not. If Thought, or Consciousness, or INTELLIGENCE does not weigh into our Cosmic Density, then how would we ever know it to be so unless God were to actually reveal that knowledge to us? God hasn't told me anything about it. How about you?

Both spirit matter and consciousness seem to be made up of light — possibly different types of light. Thoughts or Intelligence or Awareness seems to be made up of some kind of Light also. Forces, and fields, and gravity are theoretically made up of different types of light; and, most light is not visible to the naked eye. Spirit matter and spirit bodies are made up of some type of light. The vibrating strings in String Theory are also made up of light. There are different types of light and different frequencies of light.

David Bohm said that matter is condensed or frozen light. Photons are concentrated light, and during the concentration process that particular type of light slows down to what we call the "speed of light". Light slows down the more "material" or "Materialistic" that it becomes — the closer it comes to manifesting in our physical realm. Slow light down to ZERO SPEED, and it becomes physical matter. But, there are other types of light which have no speed limit — PURE LIGHT or PURE THOUGHT, the kind of light that makes up Dark Energy and 72.1 percent of this Universe's Cosmic Density, and possibly the kind of light that doesn't weigh-into our Universe's Cosmic Density at all.

Spiritualism is the belief that LIGHT, or SPIRIT, exists distinct from matter and that SPIRIT or LIGHT or ENERGY is the fundamental component or primal construct of reality.

Which types of LIGHT show-up within our universe's cosmic density, and which types do not? We don't know. The best we can do is to speculate that Thought, Consciousness, Sentience, Self-awareness, or INTELLIGENCE might be the type of light that does NOT weigh into our universe's Cosmic Density. When we climb down that rabbit hole, we speculate and assume that we will find THOUGHT or INTELLGENCE or THE PUREST OF LIGHT at the very bottom of it, but it too will indeed be some type of LIGHT — SENTIENT LIVING LIGHT.

—

Speculating about the Purest Light

I'm going to speculate here that there IS something even more refined than Dark Energy or PURE SPIRIT or PURE LIGHT. This PUREST LIGHT would be what Dark Energy, or the Zero-Point Field, is made-of or made-by. This PUREST LIGHT would in fact be PURE SENTIENCE or PURE THOUGHT or PURE INTELLIGENCE or PURE LIFE.

I went down the rabbit hole, and now I'm trying to figure out how far down it goes!

There appears to be a type of light or consciousness or thought that is even more primal and elemental than Dark Energy or the Quantum Sea of Light — a type of light or intelligence that doesn't even show up in our Universe's Cosmic Density. It's possible that there is NO sentience, or awareness, or life anywhere in our Universe's Cosmic Density. It's possible that PURE LIFE or the PUREST OF LIGHT might not register at all or weigh-in at all in our Universe's Cosmic Density. I like to consider this possibility, because it has certain advantages.

You see, the Biblical God has revealed to us in the Doctrine and Covenants Section 93 that Intelligence or Consciousness is what He calls "LIGHT AND TRUTH". Your thoughts or consciousness at a very fundamental level is nothing more than light and what you TRULY ARE; and as we have already established, light or consciousness fills the immensity of space, and light is immaterial and non-physical; and, this sentient or conscious or intelligent type of Light theoretically might not figure into the cosmic density of this universe at all. This THOUGHT LIGHT or INTELLIGENT LIGHT or PUREST LIGHT would be trans-dimensional or extra-dimensional, and not a part of this physical universe at all. There would be no way to detect it from within this universe — the best we could do is to observe its effects or its influence. It would truly be the Driver of the bus.

According to this theory, Dark Energy or the Zero-Point Field would not be sentient and self-aware. Instead, this Dark Energy or PURE LIGHT or Quantum Sea of Light would be part of the primal construct of this universe comprising 72.1 percent of this universe — kind of like a network or telephone exchange.

This Dark Energy or Light of Christ could in fact have been designed and created by THOUGHT LIGHT or INTELLIGENT LIGHT to be used both as a stabilizing or structural feature of physical matter and as a universal communication system. This Dark Energy or Light of Christ could in fact be the Utility System that was built into this universe — the wiring and plumbing so to speak.

The REAL SENTIENCE or the PUREST LIGHT, Consciousness or Intelligence, wouldn't even register at all in this universe's cosmic density but would in fact permeate or occupy the whole of this universe simultaneously. It would be trans-dimensional or extra-dimensional. It would be THE LIGHT and THE LIFE of this universe.

Heady stuff, huh? It leaves Materialism behind in the dust! Doesn't it?

—

At your very essence, at your very eternal immaterial core, you ARE conscious sentient light — YOU ARE PURE LIFE or the PUREST OF LIGHT and INTELLIGENCE. You have always existed, and you will always exist. The Gods might be able to disassemble your spirit body, but they can't touch or destroy your inner consciousness, or light, or intelligence. The Gods can't force your inner light to be good or to be evil, either. You choose that for yourself.

The Biblical God calls this stuff, this Innermost Light, "INTELLIGENCE" in His scriptures, particularly the Doctrine and Covenants and the Book of Abraham. God defines Intelligence as "Light and Truth". Your Immortal Immaterial Intelligence is what you TRULY are at your very core. Your Intelligence IS your light, and your life, and your truth.

As mentioned, Consciousness or Thought is made up of a special kind of light, and that kind of light has no speed limit. That's why God can hear your thoughts and prayers instantaneously no matter where He might happen to be in this universe at any given point in time. Let's see Materialism do that!

Thought or Consciousness is trans-dimensional or extra-dimensional and thus has NO physical limitations or spiritual limitations either. Thought or Consciousness existed before the spiritual, which existed before the physical.

1 John 1: 5: "God is light, and in Him is no darkness at all." If you notice carefully, God has the type of Spirit or Light that can actually levitate, move, and manipulate physical matter telekinetically.

God would be the GREATEST LIGHT, the MOST GLORIOUS LIGHT, or THE PUREST OF PURE LIGHTS. God would be THE LIGHT and THE LIFE of this Universe. This universe would be a GREAT THOUGHT in the Mind of God.

Have you heard of any of this before? It all fits together perfectly, doesn't it?

One of my gifts or talents is an ability to notice patterns or trends; and, I love a good mystery. What greater mystery is there than trying to figure out how the non-physical realms or the spiritual realms might be organized and setup by God? This is something that a Materialist or an Atheist won't touch with a ten-foot pole, which makes it even more appealing once you have finally seen the Light.

I know something they don't. I learned something that they refuse to look at and study. How cool is that? It makes a person feel good, especially after taking a ton of ridicule and abuse from the Atheists, Darwinists, and Materialists online. The Materialists are the "blind and the ignorant" calling all of the rest of us blinder and stupider and "ignoranter" than they are.

I will leave it up to you to decide for yourself which one of us has our head in the sand. But, I got tired of interacting with the Materialists, because they are so shallow and dense most of the time; and, their constant ridicule, name-calling, and contempt grates on you after a while. You find yourself wishing at times that you could pull the Quantum Sea of Light out of them just to see what happens to them. Would it still be considered murder if you were to kill them by removing something from them that they say does not exist?

When it comes to the PUREST OF PURE LIGHTS, I'm definitely not it. I still have a bit of the devil in me; and, that dude seems to be dark light or black light. Still, the boy in me (who liked to fry ants under a magnifying glass) would like to see what would happen to a couple of these Materialists if God were to pull the "unseen", the "untouchable", or the "immaterial" out of them. It would be fun to watch them implode, wouldn't it? For them, the immaterial really wouldn't exist anymore, now would it? And neither would their physical matter. Their physical matter would implode — no longer being sustained or upheld by spirit or light. It would be interesting to see what part (if anything) of them remained, if God were to pull the spiritual or the light completely out of them.

Is there something more primal than SPIRIT or LIGHT that might in fact remain if the SPIRIT or the LIGHT were to be pulled out of them? I like to think that there are possibilities.

—

Materialism Is the Study of Reactions

Think about it logically and dispassionately.

Materialism of any kind is the Study of Reactions.

Whether we are talking about chemical reactions or reactions from natural selection and random mutations and other natural processes, we are NOT talking about design and creation!

Creation by Reactions IS scientifically and logically IMPOSSIBLE!

Reactions do NOT design and create. They instead REACT to designers, engineers, manufacturers, and creators.

Darwinism or the Theory of Evolution is the Study of Reactions. The Atheistic Darwinists try to remove the Designers and the Creators from the equation, and they try to get us to focus exclusively on the REACTIONS, trying to trick us in the process into believing that Reactions can somehow magically design and create.

But the Darwinists and the Materialists are WRONG.

ONLY ACTORS — designers, creators, manipulators, manufacturers, builders, engineers, planners, programmers with hands and spirits and minds — can design and create. Manipulation and manufacture imply HANDS or some kind of Telekinesis. Designing, programming, planning, and organizing imply SPIRITS or MINDS. It's Actors with Hands and Minds who do the actual designing and creating — NOT Materialistic reactions! Chemical reactions and evolutionary reactions CANNOT design and create anything at all because they are purely physical reactions, and there are NO hands or minds there. Reactions cannot design and create! And, the Materialists LIMIT themselves and try to limit us purely to PHYSICAL REACTIONS, telling us that nothing else exists.

Materialism of any kind is a very limiting and UNSCIENTIFIC philosophical worldview or Religion. The Materialists try to tell us that man-made idols and physical reactions can design and create. Materialism is just a modern-day form of idolatry — a worship of false gods. It's a fairy tale! Everyday online, we see Materialists, Darwinists, and Atheists worshipping chemical reactions, evolutionary reactions, and their own knee-jerk reactions.

God designed and organized two types of things in this Universe — the living things that were designed and organized and meant to ACT, and the inert physical or Materialistic processes and physical matter which was designed and created and meant to REACT. It is the living spirit that ACTS or CREATES; and, it is the physical matter and the physical processes that REACT or are ACTED UPON. Physical Reactions cannot design and create, although the Materialists and Darwinists will assure us that they do.

The Materialists and Darwinists limit themselves exclusively to the things that were designed and meant to REACT, while completely ignoring the spirit or life or living light which God organized and intended to do all of the ACTION or CREATION in this physical universe.

The Materialists are right about one thing. Physical Processes, Chemical Reactions, and all other Things That React CANNOT give revelations and visions, CANNOT function as Lawgivers, CANNOT function as Holy Spirits, CANNOT perform atonements or reconciliations with God or provide salvation for our immortal spirit, and CANNOT predict or prophecy of future events that they will have a hand in bringing about. But the Materialists don't seem to realize or understand that (for the very same reasons) Physical Reactions and Chemical Reactions CANNOT function in the role of Designers, Creators, Revelators, Spirits, Causal Agents, Planners, Actors, and GOD.

Materialism is a reaction, and thus doesn't require a lot of active thought, which is why God tells us that we must worship Him with all of our heart, might, mind, and strength. God calls us to action, because God organized us to be Actors and Creators!

Lehi was a prophet and a scientist, and the Biblical God Christ Jehovah revealed the following things to him 600 B.C. Lehi wrote these words:

> 2 Nephi 2: 13-16:
>
> 13 And if ye shall say there is no law, ye shall also say there is no sin. If ye shall say there is no sin, ye shall also say there is no righteousness. And if there be no righteousness there be no happiness. And if there be no righteousness nor happiness there be no punishment nor misery. And if these things are not there is no God. And if there is no God we are not, neither the earth; for there could have been no creation of things, neither to act nor to be acted upon; wherefore, all things must have vanished away.
> 14 And now, my sons, I speak unto you these things for your profit and learning; for there is a God, and he hath created all things, both the heavens and the earth, and all things that in them are, both things to act and things to be acted upon.
> 15 And to bring about his eternal purposes in the end of man, after he had created our first parents, and the beasts of the field and the fowls of the air, and in fine, all things which are created, it must needs be that there was an opposition; even the forbidden fruit in opposition to the tree of life; the one being sweet and the other bitter.
> 16 Wherefore, the Lord God gave unto man that he should act for himself. Wherefore, man could not act for himself save it should be that he was enticed by the one or the other.

Source:

https://www.lds.org/scriptures/bofm/2-ne/2?lang=eng

The whole chapter is a fascinating read, especially if you are scientifically inclined and want to know how things really work.

There are two different types of THINGS in this universe, the People or Living Spirits or Living Lights who were organized and meant to be ACTORS and CREATORS, and then the physical things that were designed and created to be ACTED UPON or TO REACT. As their Creator and Organizer, the Biblical God knows about these THINGS, which is why He periodically revealed these things to His chosen prophets, turning them into scientists in the process.

ACTION and REACTION! ACTORS and THINGS TO BE ACTED UPON! That's what this physical universe is all about.

Materialism is the Study of Reactions — chemical reactions, physical reactions, and evolutionary reactions.

Theology is the Study of Designers, Actors, and Creators. Theology and/or Spiritualism is also the Study of Spirits, Living Lights, Angels, and Gods.

Design and Creation by Reactions IS scientifically and logically IMPOSSIBLE. Thus, Creation by Chance Reactions and Creation by Evolutionary Reactions is scientifically and logically IMPOSSIBLE. Chance and Evolution cannot design and create, because they are purely REACTIONS! There's NO Actor there!

How often have I said to you that when you have eliminated the impossible, whatever remains, *however improbable*, **must be THE TRUTH? — Sherlock Holmes**

My goal in this book is to eliminate the IMPOSSIBLE so that we are left staring at THE TRUTH. Reactions CANNOT design, create, and act; therefore, they should be ELIMINATED from Science as possible Designers and Creators and Causal Agents. It's only logical. THIS IS SCIENCE! This is SCIENTIFICALLY ACCURATE because it matches with REALITY! Evolution and Chance have NEVER been caught in the act of designing and creating anything! Intelligent Beings have! Therefore, Creation by Chance and Creation by Evolution should be eliminated from Science as possible Designers and Creators, because Evolution and Chance CANNOT design and create anything at all. It's elementary my Dear Reader!

Once you have eliminated the IMPOSSIBLE, the Materialism, then it all points to God as the Designer and Creator of this universe and all the genomes and life in this universe, because God is a Living Designer, Actor, and Creator. This makes logical sense to me; whereas, Creation by Chance or Creation by Evolution or Creation by Reactions does not. God makes sense to me as the Origin and/or the Originator of all the life in this universe. Evolution does not. Materialism is FALSE; therefore, Materialism is UNSCIENTIFIC! Materialism is NOT SCIENCE. It is a philosophy or religion. The scientist in me wants to go with what's actually POSSIBLE! Materialism is NOT it!

You will have to decide for yourself if I met my Burden of Proof, or not. Obviously, my goal here is to go with the preponderance of the evidence. I employ a type of deductive reasoning in which I attempt to eliminate the IMPOSSIBLE PREMISES so that we are left staring at THE TRUTH or the CORRECT CONCLUSION.

My goal here was to put an end to My Materialism and My Atheism making it impossible for me to go back. I have accomplished my goal. I have seen too much, experienced too much, and learned too much to go back to Darwinism and My Atheism. I have decided which side of the fence I want to stand on and defend. I have made my choice; and now, I'm determined to make the best of it.

—

Providing a Capstone to Spiritualism and a Gravestone for Materialism

The Materialists, Darwinists, and Atheists can be extremely stupid, ignorant, uneducated, and dense when it comes to the parts of SCIENCE that they have chosen not to believe in. I know, because I used to be one of them.

If you don't ever go searching for the spiritual, then you will never find it. The spiritual hides behind a Materialist's unbelief.

Spiritualism is the view that Spirit or Living Light is the prime element of reality.

Materialism is a denial of that reality! The Materialists try to convince us that there is no such thing as consciousness or living light. The Materialists deny Spirit, Light, and Life.

Materialism is a stupid and idiotic idea — the dumbest, the most limiting, and the shallowest idea that human beings have ever created out of thin air. There's no substance to it!

But, there is an important distinction that should be made here.

Just because Materialism is the dumbest idea that has ever been devised, it doesn't automatically follow that Materialists and Darwinists and Atheists are the most stupid people that you will ever encounter. You see, I used to be an Atheist and a Materialist; and, although I did indeed harbor some stupid ideas at the time, I'm not exactly stupid or dense as you might be able to sense. Materialists and Atheists can repent or change their minds, especially once they realize how stupid and limiting and boring Materialism and Atheism really are. Even Materialists and Darwinists can learn to be more well-rounded people if they truly want to do so. Once they get over their fear of God, their fear of Spirituality, their fear of Repentance, their fear of God's Commandments, their fear of God's Scriptures, and their fear of Religion, then anything is possible.

Atheism and Materialism are made up of a great deal of suppressed or subconscious fear and prejudice; and, sometimes it isn't all that well-hidden. I have watched the Atheists and Darwinists in action during various debates, and I can sense a great deal of fear coming off them or from them — it typically manifests itself as anger, hatred, vitriol, pride, ridicule, contempt, and rage. Richard Dawkins has often been described as "everyone's favorite madman". He can't seem to control himself at times. Sometimes he acts as if he is a trapped animal.

In contrast, LOVE or CHARITY casteth out all fear. These kinds of Christian or Charitable people are actually kind and fun to be around.

—

I sense that there is still a lot that we don't know about the unseen realms, but it's amazing what we can piece together from the Revelations of God — particularly those found in the Doctrine and Covenants and the Pearl of Great Price.

God has deliberately limited the scope and range of our consciousness, spirit bodies, and physical bodies; but, He has placed no limits on Himself. As mortal beings, we are in a probationary state or a testing state of existence, to prove to ourselves who and what we really are at our innermost core. God will have a proven people before sharing His powers and abilities with them.

The book "The Holographic Universe" by Michael Talbot speculates a lot about how our Consciousness or Thoughts or Intelligence is holographic in nature, thus allowing our Consciousness or Intelligence to store an infinite amount of information in basically no space at all. Cool, huh?

—

Materialists can mock the Zero-Point Field or the Quantum Sea of Light all they want, but it only serves to demonstrate and reveal to all the rest of us the depth of their ignorance, which they have imposed upon themselves. Materialists become very shallow people indeed; and, they do it deliberately to themselves! There's no depth to Materialism, and Darwinism is purely a Materialistic philosophy.

The Materialists only permit themselves to think about half of science at most and only 5% of reality at most. Materialism is extremely shallow, boring, and limited. Materialism is like putting on blinders before going sight-seeing. The Materialists and Atheists do this to themselves, because they are trying to hide from God. Materialists are like the infant toddler who covers his eyes truly believing that if he can't see it, then the scary monster doesn't exist anymore.

I sometimes have a hard time believing that I used to be an Atheist and a Materialist. Materialism is UNSCIENTIFIC or ANTI-SCIENCE, irrational, illogical, unrealistic, stupid, limited, shallow, and boring; but, you only realize that after you have gotten rid of it.

—

Ironically, we Materialists and Atheists are more than willing to believe in things that are IMPOSSIBLE or things that DON'T EXIST, as long as those things are telling us that God does not exist and that we will never have to repent of our sins or suffer for our sins.

We Materialists and Atheists desperately want something to tell us that God does not exist and that we will never have to answer to God for our sins, vices, and misdeeds; so, we create out of thin air something that does not exist — something like Creation by Chance or Creation by Evolution or Creation by Reactions — to tell us that God does not exist, so that we don't have to think about God and our Final Judgment anymore. Materialism is a state of denial! Materialism is one of the ways that we tell ourselves that we can sin as much as we like, and we will never have to answer for it. KNOW THYSELF!

—

The things I mention in this chapter are really cool stuff — COOL SCIENCE — but you are not going to be able to see it or understand it or accept it until long after you have chosen to open your mind and your eyes enough to actually be able to see it. In my case, I had to get rid of My Materialism and My Atheism, before I was able to see and understand and accept any of these things. I'm a completely different person now than I was five years ago.

Here's one of my favorite quotes from Master Yoda:

> Master Yoda: "Size matters not. Look at me. Judge me by my size, do you? Hmm? Hmm? And well you should not. For my ally is the Force, and a powerful ally it is. Life creates it, makes it grow. Its energy surrounds us and binds us. Luminous beings are we, not this crude matter. You must feel the Force around you; here, between you, me, the tree, the rock, everywhere, yes. Even between the land and the ship."

Science Fiction isn't ALWAYS fiction! In fact, some of the coolest Science Fiction ends up becoming Science Fact. Life or Thought creates this FORCE or FIELD.

There's a force or a field within us connecting every particle in our physical bodies with every other particle in this universe. Some of the scientists call this the Zero-Point Field or the Quantum Sea of Light or Gravity. Astrophysicists call it Dark Energy because it can't be seen with the naked eye but can only be detected indirectly within the Cosmic Density of this Universe and fills the immensity of space. Christians call it the Holy Spirit or the Light of Christ. I have taken to calling it PURE LIGHT. Yoda called it The Force.

Granted, you might not be able to use This Force to lift X-Wing fighters out of a swamp; but, we definitely can use This Force to pray to God and get an instantaneous answer in return. This Force is like a Cosmic Internet or a Cosmic Telephone System. The Holy Ghost can speak thoughts and answers to our prayers directly into our minds using This Force or This Cosmic Phone System.

Luminous beings are we — spiritual beings or beings of light — not this crude matter which the Materialists limit themselves to and worship. The Materialists and Atheists actually believe that we cease to exist when our physical matter dies. As Master Yoda implies, that's a crude understatement of the True Reality of things.

It doesn't matter if he is a fictional character, or not. I'm with Yoda on this one.

As you can tell, I got a little bit excited about all of this, and I studied it and kept studying it until I finally understood it. I wanted to figure out how it all fits together. The more I studied it, the more I concluded that it must be true (God's Truth), because I have never found a better explanation for life, the universe, and everything. It's infinitely better and more informative than "42".

People will MOCK you for finding any of this information useful, informative, interesting, or believable. It's what they do, because they don't have anything better going on in their lives. Just ignore them. The goal here is to acquire information that will make your life better, happier, more interesting, more productive, and more successful.

Information and speculation about the non-physical, unseen, spiritual realms are INFINITELY more interesting than Materialism can even begin to be. So, let them MOCK. Meanwhile, go and learn something new that you have never thought about before. It will make your life a lot more fun and interesting.

—

Light, Truth, and Intelligence

For those who are still interested in knowing more, I tried to save you some time by providing you with some of the revelations from the Biblical God explaining to us what Light or Intelligence really is. Don't let the Materialists or Atheists make you ashamed to learn about and study these kinds of things. They just want to keep you in the dark where they currently are. I know, because I used to be an Atheist and a Materialist. Just ignore them; and go out and learn something new. Open your eyes and see the LIGHT! In fact, one of my mottos is to "Learn Something New Every Day"; and, I just as well start with the spiritual and non-physical things because they are the most interesting.

I'm a completely different person now than I was when I was an Atheist and a Materialist. Some people think I'm a better person. Some people even post and say that they admire my courage or knowledge.

Notice how consistent all of the following is, both with itself and with what I have presented before this:

1 John 1: 5: "God is light, and in Him is no darkness at all." If you notice carefully, God has the type of Spirit or Light that can actually levitate, move, and manipulate physical matter telekinetically.

D&C 88: 7-13:
7 Which truth shineth. This is the light of Christ. As also he is in the sun, and the light of the sun, and the power thereof by which it was made.
8 As also he is in the moon, and is the light of the moon, and the power thereof by

which it was made;
9 As also the light of the stars, and the power thereof by which they were made;
10 And the earth also, and the power thereof, even the earth upon which you stand.
11 And the light which shineth, which giveth you light, is through him who enlighteneth your eyes, which is the same light that quickeneth your understandings;
12 Which light proceedeth forth from the presence of God to fill the immensity of space —
13 The light which is in all things, which giveth life to all things, which is the law by which all things are governed, even the power of God who sitteth upon his throne, who is in the bosom of eternity, who is in the midst of all things.

Cool, huh?

This Light of Christ is in EVERYTHING giving it substance, structure, presence, and girth! It's the power by which things are made! It's the power by which things are sustained. It's the power that keeps protons from decaying and imploding. This LIGHT is in everything sustaining and upholding everything, and it proceeds forth from the presence of God. If this Light of Christ weren't there, then this whole universe would implode back into the singularity from whence it came. If God were to remove His influence and His presence and His Light from this universe, then it would implode back into the singularity from whence it came.

This Dark Energy or Light of Christ is what gives this universe presence, substance, girth, structure, expansion, LAW, order, contact with God, and life. That's why this Zero-Point Field comprises 72.1 percent of this Universe's Cosmic Density; and, the Biblical God Jesus Christ knew about its existence long before we did, which is why Jesus Christ was able to tell Joseph Smith about it over a 150 years ago!

God is the ULTIMATE SCIENTIST!

This Light of Christ is like the Dark Energy that the Astrophysicists talk about, or the Quantum Sea of Light that the New Agers talk about, or the Zero-point Field or Gravity Field or Strong Forces and Weak Forces or Holographic Universe that physicists talk about. This Light of Christ is also the Life Force. There's also talk of Quantum Consciousness, Morphic Fields, and a Universal Consciousness. ALL of this stuff is immaterial or non-physical! It all exists as immaterial SPIRIT or LIGHT! God knows what He is talking about, because He's the one who set it all up.

The Materialists don't have the prerequisites for this kind of SCIENCE course! It's RADICALLY MORE ADVANCED than the Materialists can even begin to understand. So, all you will see them doing is mocking it because they don't understand it — what it is, what it does, and why it is necessary for it to exist. Just ignore them, because they don't have a clue what's really going on in this universe. How can the Materialists know what's going on around them and within them, because they refuse to look at it and learn anything new from it! All they can do is mock it and ridicule it. They definitely are NOT going to learn something from it. The Materialists are the worst kind of scientists, if they can be called scientists at all.

Abraham 3: 11-12:

11 Thus I, Abraham, talked with the Lord, face to face, as one man talketh with another; and he told me of the works which his hands had made;
12 And he said unto me: My son, my son (and his hand was stretched out), behold I will show you all these. And he put his hand upon mine eyes, and I saw those things which his hands had made, which were many; and they multiplied before mine eyes, and I could not see the end thereof.

Abraham 3: 17-19:

17 Now, if there be two things, one above the other, and the moon be above the earth, then it may be that a planet or a star may exist above it; and there is nothing that the Lord thy God shall take in his heart to do but what he will do it.
18 Howbeit that he made the greater star; as, also, if there be two spirits, and one shall be more intelligent than the other, yet these two spirits, notwithstanding one is more intelligent than the other, have no beginning; they existed before, they shall have no end, they shall exist after, for they are gnolaum, or eternal.
19 And the Lord said unto me: These two facts do exist, that there are two spirits, one being more intelligent than the other; there shall be another more intelligent than they; I am the Lord thy God, I am more intelligent than they all.

Abraham 3: 21-28:

21 I dwell in the midst of them all; I now, therefore, have come down unto thee to declare unto thee the works which my hands have made, wherein my wisdom excelleth them all, for I rule in the heavens above, and in the earth beneath, in all wisdom and prudence, over all the intelligences thine eyes have seen from the beginning; I came down in the beginning in the midst of all the intelligences thou hast seen.
22 Now the Lord had shown unto me, Abraham, the intelligences that were organized before the world was; and among all these there were many of the noble and great ones;
23 And God saw these souls that they were good, and he stood in the midst of them, and he said: These I will make my rulers; for he stood among those that were spirits, and he saw that they were good; and he said unto me: Abraham, thou art one of them; thou wast chosen before thou wast born.
24 And there stood one among them that was like unto God, and he said unto those who were with him: We will go down, for there is space there, and we will take of these materials, and we will make an earth whereon these may dwell;
25 And we will prove them herewith, to see if they will do all things whatsoever the Lord their God shall command them;
26 And they who keep their first estate shall be added upon; and they who keep not their first estate shall not have glory in the same kingdom with those who keep

their first estate; and they who keep their second estate shall have glory added upon their heads for ever and ever.

27 And the Lord said: Whom shall I send? And one answered like unto the Son of Man: Here am I, send me. And another answered and said: Here am I, send me. And the Lord said: I will send the first.

28 And the second was angry, and kept not his first estate; and, at that day, many followed after him.

Online Source for the Book of Abraham:

https://www.lds.org/scriptures/pgp/abr?lang=eng

God ORGANIZED these Intelligences in the beginning. God did NOT create them! These Intelligences ARE eternal and indestructible. They are trans-dimensional or extra-dimensional and preceded spirit matter and physical matter. It's these Intelligences who designed and created spirit matter and then later physical matter. Spirit Matter and Physical Matter are hosts for Intelligences — the things that Intelligences act upon. The Apostle Paul called our Physical Bodies the Temples of the Gods, another way of saying that our physical bodies are tabernacles or temples for our Intelligences or Living Lights.

God is God because He is more intelligent than all of the rest of us combined.

—

Doctrine and Covenants 93:29: Man was also in the beginning with God. Intelligence, or the light of truth, was not created or made, neither indeed can be.

These Intelligences cannot be created or made. They can only be organized.

Doctrine and Covenants 93:30: All truth is independent in that sphere in which God has placed it, to act for itself, as all intelligence also; otherwise there is no existence.

There is NO existence or life without these Intelligences. It's these Intelligences that bring spirit matter and physical matter into existence and enlivens these different types of matter.

Doctrine and Covenants 93:36: The glory of God is intelligence, or, in other words, light and truth.

Doctrine and Covenants 130:18: Whatever principle of intelligence we attain unto in this life, it will rise with us in the resurrection.

Our Intelligence doesn't die when our physical body dies. Our Intelligence doesn't die should our spirit body die.

Doctrine and Covenants 84:45: For the word of the Lord is truth, and whatsoever is truth is light, and whatsoever is **light is Spirit**, even the Spirit of Jesus Christ.

Whatever is Light is Spirit. Spirit is Light, and Light is Spirit.

Doctrine and Covenants 93:28: He that keepeth God's commandments receiveth truth and light, until he is glorified in truth and knoweth all things.

Doctrine and Covenants 88:7: Which truth shineth. This is the light of Christ. As also he is in the sun, and the light of the sun, and the power thereof by which it was made.

This Light of Christ is what keeps our sun and our earth from imploding into a singularity.

Doctrine and Covenants 93:37: Light and truth forsake that evil one.

Doctrine and Covenants 93:40: But I have commanded you to bring up your children in light and truth.

Psalms 43:3 O send out thy light and thy truth: let them lead me; let them bring me unto thy holy hill, and to thy tabernacles.

Light and Truth, or Intelligence or Thought, is the primal construct.

John 3:21: But he that doeth truth cometh to the light, that his deeds may be made manifest, that they are wrought in God.

Doctrine and Covenants 93:9: The light and the Redeemer of the world; the Spirit of truth, who came into the world, because the world was made by him, and in him was the life of men and the light of men.

This Spirit of Truth is the light and life of men, and also the Redeemer of the world.

Doctrine and Covenants 93:39: And that wicked one cometh and taketh away light and truth, through disobedience, from the children of men, and because of the tradition of their fathers.

Doctrine and Covenants 93:42: You have not taught your children light and truth, according to the commandments; and that wicked one hath power, as yet, over you, and this is the cause of your affliction.

Doctrine and Covenants 88:6: He that ascended up on high, as also he descended below all things, in that he comprehended all things, that he might be in all and through all things, the light of truth.

This Light of Truth or Light of Christ is in all things and through all things. The Light of Christ IS the Dark Energy that expanded our universe and keeps our universe expanding.

Doctrine and Covenants 124:9: And again, I will visit and soften their hearts, many of them for your good, that ye may find grace in their eyes, that they may come to the light of truth, and the Gentiles to the exaltation or lifting up of Zion.

Joseph Smith History 1:25: So it was with me. I had actually seen a light, and in the midst of that light I saw two Personages, and they did in reality speak to me; and though I was hated and persecuted for saying that I had seen a vision, yet it was true; and while they were persecuting me, reviling me, and speaking all manner of evil against me falsely for so saying, I was led to say in my heart: Why persecute me for telling THE TRUTH? I have actually seen a vision; and who am I that I can withstand God, or why does the world think to make me deny what I have actually seen? For I had seen a vision; I knew it, and I knew that God knew it, and I could not deny it, neither dared I do it; at least I knew that by so doing I would offend God, and come under condemnation.

Ether 4:12: And whatsoever thing persuadeth men to do good is of me; for good cometh of none save it be of me. I am the same that leadeth men to all good; he that will not believe my words will not believe me — that I am; and he that will not believe me will not believe the Father who sent me. For behold, I am the Father [of Truth], I am the light, and the life, and THE TRUTH of the world.

Satan is known as the father of lies. Jesus Christ is the Father of Truth.

Alma 38:9: And now, my son, I have told you this that ye may learn wisdom, that ye may learn of me that there is no other way or means whereby man can be saved, only in and through Christ. Behold, he is the life and the light of the world. Behold, he is the word of truth and righteousness.

Jesus Christ, our Anointed Savior, is the light, life, and truth of this world. If you notice carefully, God has the type of Spirit or Light that can actually levitate, move, and manipulate physical matter telekinetically.

Doctrine and Covenants 85:7: And it shall come to pass that I, the Lord God, will send one mighty and strong, holding the scepter of power in his hand, clothed with light for a covering, whose mouth shall utter words, eternal words; while his bowels shall be a fountain of truth, to set in order the house of God, and to arrange by lot the inheritances of the saints whose names are found, and the names of their fathers, and of their children, enrolled in the book of the law of God.

Online Source for the Doctrine and Covenants:

https://www.lds.org/scriptures/dc-testament?lang=eng

Online Source for the Book of Mormon:

https://www.lds.org/scriptures/bofm?lang=eng

Online Source for the Pearl of Great Price:

https://www.lds.org/scriptures/pgp?lang=eng

—

Intelligence

Intelligence has several meanings, three of which are:

(1) It is the light of truth which gives life and light to all things in the universe. It has always existed.
(2) The word intelligences may also refer to spirit children of God.
(3) The scriptures also may speak of intelligence as referring to the spirit element that existed before we were begotten as spirit children.

Intelligence cleaveth unto intelligence: D&C 88:40.
Intelligence was not created or made: D&C 93:29.
All intelligence is independent in that sphere in which God has placed it: D&C 93:30.
The glory of God is intelligence: D&C 93:36-37.
Intelligence acquired in this life rises with us in the resurrection: D&C 130:18-19.
The Lord rules over all the intelligences: Abr. 3:21.
The Lord showed Abraham the intelligences that were organized before the world was: Abr. 3:22.

—

The Light of Christ or Dark Energy

Dark energy within astrophysics is the Light of Christ or the Matrix of Quantum Fields. It has also been called the Quantum Sea of Light or the Zero-Point Field. It's all part of the same system yet broken down into different aspects. Gravity or spacetime seems to be part of the same universal system as well. Gravity is like sin – it has weight. Even though the quantum fields are massless, entropyless, non-physical, and intangible, they have presence and substance that is detectable as dark energy. The quantum fields are the luminiferous ether or the universal fabric from which everything is made. The quantum fields are the medium through which the conserved quanta (psyches or intelligences) communicate with each other using quantum waves to do so. There are no physical limitations, resistance to acceleration, mass, thermodynamics, heat, or entropy at the quantum level or energy level where everything is conserved to one extent or another. Quantum waves are WiFi at the quantum level. Energy contains information, and the quantum information within that energy can be conserved or remembered within conserved quanta or psyches. The conserved quanta are the transceivers; and, the quantum fields are the universal medium through which this quantum information is transmitted. Energy is infinitely malleable, and the information within it can be transformed into anything anytime by the psyches or intelligences or conserved quanta who are in control of that energy. This is what has been experienced and observed. Quantum mechanics is God's priesthood power.

Dark matter is spirit matter or phase-shifted matter. The Gods make everything spiritually before they phase-shift it into our physical reality. This is what has been experienced and observed. The galaxies are born all at once, and then they start to wind up and become subject to physical laws and physical limitations from that point onward. A lot more galaxies can be made and phase-shifted into our physical reality from all of that dark matter that currently exists within our universe. When it comes to our physical universe, according to the LAW of the Conservation of

Energy, the Gods can dissolve or disassemble the whole thing instantaneously, absorb the physical atoms and suns and black holes back into the quantum fields from whence they came, and then start over again brand new from a blank slate anytime that they choose to do so. That's what you can do when you have complete control over space, time, physical matter, and quantum mechanics. You can do anything you want whenever you want. You can destroy just as easily as you can create. That's why the Gods have to prove themselves trustworthy before they are made into Gods. You don't want a devil with the power of the Gods, or he could destroy the universe at will and throw the whole thing back into Chaos.

https://ldssoul.com/the-birth-of-our-earth/

As you can tell, I have given this whole thing some research and thought over the past few years. I now know what everything is and how it works (February 2020), because I'm no longer afraid to have the conversation or to do the research or to think about it because I'm no longer a materialist, naturalist, nihilist, or atheist. These concepts are no longer out of reach for me; and, by allowing quantum mechanics, quantum fields, and conserved quanta in to play, we can literally explain everything that comes our way. I now know what everything is and how it works. I'm running out of questions to ask.

The Light of Christ or the Quantum Fields

Divine energy, power, or influence that proceeds from God through Christ and gives life and light to all things. It is the law by which all things are governed in heaven and on earth (D&C 88:6-13). It also helps people understand gospel truths and helps to put them on that gospel path which leads to salvation (John 3:19-21; 12:46; Alma 26:15; 32:35; D&C 93:28-29, 31-32, 40, 42).

The light of Christ should not be confused with the Holy Ghost. The light of Christ is not a person. It is an influence that comes from God and prepares a person to receive the Holy Ghost. It is an influence for good in the lives of all people (John 1:9; D&C 84:46-47).

One manifestation of the light of Christ is conscience, which helps a person choose between right and wrong (Moro. 7:16). As people learn more about the gospel, their consciences become more sensitive (Moro. 7:12-19). People who hearken to the light of Christ are led to the gospel of Jesus Christ (D&C 84:46-48).

The Lord is my light: Ps. 27:1.
Let us walk in the light of the Lord: Isa. 2:5; (2 Ne. 12:5).
The Lord shall be an everlasting light: Isa. 60:19.
The true Light lighteth every man that cometh into the world: John 1:4-9; (John 3:19; D&C 6:21; D&C 34:1-3).
I am the light of the world: John 8:12; (John 9:5; D&C 11:28).
Whatsoever is light, is good: Alma 32:35.
Christ is the life and the light of the world: Alma 38:9; (3 Ne. 9:18; 3 Ne. 11:11; Ether 4:12).

The Spirit of Christ is given to every man that he may know good from evil: Moroni 7:15-19.
That which is of God is light, and groweth brighter and brighter until the perfect day: D&C 50:24.
The Spirit giveth light to every man: D&C 84:45-48; (D&C 93:1-2).
He that keepeth God's commandments receiveth light and truth: D&C 93:27-28.
Light and truth forsake that evil one: D&C 93:37.

—

Source for these Scriptures and Ideas:

https://www.lds.org/?lang=eng

If you notice carefully, God has the type of Spirit or Light that can actually levitate, move, and manipulate physical matter telekinetically.

—

Abductive Reasoning seeks for the best and most parsimonious interpretation or explanation for the available evidence. I also employ deductive reasoning. By eliminating ALL of the FALSE PREMISES, we can deduce the CORRECT CONCLUSION. It ALL points to God.

—

Our Spirits or Our Living Lights Seek Nirvana, Bliss, and Peace

In my books, I tended to define NIRVANA as the bliss, happiness, will, and blessedness of our own internal Light or Spirit. Spirit IS Light. Light IS Spirit.

It is the Spirit or Living Light or NIRVANA within each one of us that has the inherent ability to design and create — NOT our physical matter! Our physical matter can't do anything except to react to whatever our Spirits tell it to do.

A rock has physical matter, but a rock does NOT have the kind of spirit within it that has the ability to design and create new things. Even though a rock has a lot of physical matter, you will NEVER catch a rock in the act of designing and creating new things from scratch. It isn't the physical matter that designs and creates! Physical matter was designed to be molded and acted upon by Spirits or Living Lights. Physical matter was designed to be occupied and enlivened by Spirits or Living Lights. Inert physical matter cannot design and create!

There are different types of Spirit or Light. While in Spirit, people have seen the consciousness or awareness of rocks and nails. But, there isn't a designing and creating type of Spirit within a rock. There isn't a loving and charitable Spirit within a rock either. A rock is a spiritual recording device, nothing more.

It's our Human Spirit or Living Light which gives us our desires, intelligence, will, and ability to choose — NOT our physical matter. Living Intelligent Spirits like ours are Actors, Designers, and Creators. In contrast, physical matter was designed to be acted upon. Spirits or Living Lights act. Physical Matter reacts and is acted upon.

The reason we humans can design and create is that Our Spirits inherited that particular ability from the Parents of Our Spirits — God the Father and Heavenly Mother. The Spirits or Living Lights within human beings is unique among God's creations, because we are the Children of God.

Our Living Light or Spirit or NIRVANA enlivens our physical matter, giving that formerly inert physical matter purpose, intelligence, desires, goals, and in the case of humans the ability to design, plan, repent, and create.

—

Matching with Reality Makes It Scientifically Accurate

Something is said to be Scientifically Accurate if it matches with Reality.

What's really cool is that the Scriptures from the Biblical God match perfectly with our current understanding of SCIENCE! It's as if the Scriptures from the Biblical God match perfectly with Reality and thus are completely Scientifically Accurate!

What are the odds?

God thought of these things and experienced these things BEFORE we did, which is why the Biblical God was able to tell Joseph Smith about them over 150 years ago before our scientists even began to think about them.

God is the ULTIMATE SCIENTIST! And God is NOT a Materialist, or an Atheist, or a Darwinist!

God KNOWS that He didn't have to rely upon or wait upon evolution, chance, random errors, and random reactions to design and create things for Him. God is like us. If we want something done, we don't wait for evolution, chance, random blind luck, or chemical reactions to do it for us. We just go out and do it — do the Science or do the Manufacturing or do the Construction or do the Work that we want to have done.

We are Actors and Creators and Builders, not blind random processes or Materialistic reactions. We have an inner light or an inner spirit which gives us the ability to design and create. Evolution does not!

If you notice carefully, God has the type of Spirit or Light that can actually levitate, move, and manipulate physical matter telekinetically. Intelligent Glorified God-like Spirits can design and create and manipulate physical matter with their minds. Evolution cannot!

Creation by Evolution, Creation by Chance, and Creation by Random Reactions IS IMPOSSIBLE! The impossible should be eliminated from Science as being unscientific. That's what the True Science and the ULTIMATE SCIENTIST is telling us.

Materialism Is THE LIE That Everyone Wants to Believe In

The Materialists and Atheists won't be able to teach you anything about the Spiritual Sciences because they are unaware that there is such a thing. How can the Materialists and Atheists teach you something that they don't know anything about?

Once you get rid of any residual Materialism, the True Reality of EVERYTHING suddenly becomes obvious and clear. It's like taking the blinders off of your eyes!

Darwinism is a form of Materialism.

Materialism is THE LIE that millions of people want to believe in.

Materialism is THE LIE that Satan whispers into our ears, because Satan wants us to believe that he doesn't exist, that God doesn't exist, and that the Holy Spirit doesn't exist. THE LIE is the opposite of THE TRUTH.

Jesus Christ says that He is The Way, THE TRUTH, and The Life; and that, none of us will get back into Heaven and the Presence of the Father except through Him.

Materialism is the MOST SUCCESSFUL LIE in human history, because it is THE LIE that most people have chosen to believe in and desperately want to be true.

Materialism is THE BANE and THE CURSE against Scientific Exploration and Scientific Discovery, especially when it comes to the Spiritual Sciences.

Isn't it ironic and interesting how the Materialists and Darwinists employ a non-existing, immaterial, spiritual, non-physical, philosophical, non-entity as their designer, creator, and manufacturer within Materialism and the Theory of Evolution? That's a contradiction in terms, and it VIOLATES the Law of Non-Contradiction. It's also a CATEGORY ERROR to turn philosophical and scientific concepts such as evolution and chance and natural reactions into PEOPLE who can design and create and do science at will!

Materialism was designed to keep us in the Dark Ages, especially where spirituality or righteousness of any kind is concerned. Materialism was designed to drive God out of our lives! Get rid of your Materialism and suddenly the whole world becomes a completely different place, a much better place! Materialism will NEVER lead us to God's Mercy and God's Love, because Materialism is a deliberate and knowing rejection of God and a rejection of ALL the good that God has to offer us.

I slowly got rid of My Atheism and My Materialism, and my life has gotten better and more interesting ever since! Materialism and Atheism are BORING!

I once was blind, but now I see.

Keep the best and get rid of all the rest! That's what I try to do.

In a sense, that's what the Scientific Method is all about — keep THE TRUTH, and get rid of all the false premises, false hypotheses, and deliberate lies such as Materialism, Creation by Evolution, and Creation by Natural Reactions.

PART III — THE SCIENTIFIC METHOD

In my Pursuit of the True Reality of All Things, I had to be willing to follow the evidence, ANY EVIDENCE, wherever it might lead me, BEFORE I was able to find THE TRUTHS that I sought.

Belief in a LIE cannot help us to find THE TRUTH. That's just the True Reality of the situation.

There was a time in my life when I was a Materialist and an Atheist. I realized one day that I didn't like where that road was taking me, so I decided to turn around and go the other way. My life has been getting better and better ever since.

While trying to employ the Scientific Method in my Pursuit of the True Reality of All things, the SCIENCE and the EVIDENCE pointed me to additional Spiritual Sciences and Scriptural Evidence, which I had refused to look at while I was a Materialist and an Atheist

I had to be willing to look at ALL the Evidence, BEFORE I was able to find THE TRUTHS which I went searching for. We will never find THE TRUTH, if the only thing we are willing to look at, accept, and believe-in is the LIES. That's just the True Reality of the situation.

SCIENCE IS THE PURSUIT OF THE TRUTH; and, we use the Scientific Method in our Pursuit of THE TRUTH, whether those TRUTHS are spiritual in nature or physical in nature. In fact, the very best and most interesting Scientists develop Scientific Methodologies to study the Spiritual Aspects of our PHYSICAL REALITY.

The purpose of the Scientific Method is to help us to find THE TRUTH, through a preponderance of the evidence.

the Scientific Method has no value to us if we use it to convince ourselves that a LIE is TRUE, as the Materialists and Darwinists always seem to do.

That's what I discovered during my Pursuit of the True Reality of All Things, and during my usage of the Scientific Method.

The Scientific Method Proves That the Theory of Evolution Is False

I am brutal and unmerciful when it comes to Darwinism and the Theory of Evolution.

Whenever I'm dealing with the Theory of Evolution, which IS Creation by Evolution, I'm totally offensive and go straight at the thing and slit its throat, letting it bleed out and die — by demonstrating clearly and conclusively that evolution of any kind cannot design, create, manufacture, do science, or DO any Creative Acts at all.

Those of you – who are smart enough to catch onto the fact that there is NO way to slit evolution's throat – have just PROVEN to yourself why evolution or chance of any kind cannot design anything, program anything, create anything, manufacture anything, or do any science whatsoever. Evolution isn't alive. Evolution and Chance have NO soul.

ONLY intelligent soulish creatures can design and create and manufacture! Evolution has NO soul, and NO physical presence either, which means that evolution and random chance couldn't have designed and created and manufactured ALL of the genomes and ALL of the life forms on this planet as the Darwinists claim they did, because evolution cannot DO SCIENCE and evolution CANNOT CREATE!

Evolution and Chance cannot employ or use the Scientific Method to get things done!

—

Macro-Evolution IS Creation by Evolution

Macro-evolution IS spontaneous generation, or Creation by Chance and Creation by Evolution. The Theory of Evolution IS Creation by Evolution.

Spontaneous generation or anomalous generation is the formation of living organisms without descent from similar organisms. Typically, the idea was that certain life forms such as fleas could arise from inanimate matter such as dust, or that maggots could arise from dead flesh. Macro-evolution or Spontaneous Generation IS life arising magically or spontaneously from inanimate matter.

Macro-evolution IS your spaghetti dinner magically coming alive on you, sprouting wings, and flying away quoting Shakespeare as it goes. Macro-evolution is watching the Flying Spaghetti Monster come alive on you right before your very eyes in real time!

Louis Pasteur proved Spontaneous Generation and Macro-evolution FALSE in 1859, the very same year that Charles Darwin published, "On the Origin of Species" and started the evolution revolution. The Theory of Evolution was FALSE to begin with!

When it came to Macro-evolution or Spontaneous Generation, THE TRUTH of the situation and the falsehood were presented to us at the exact same time so that we each could choose which idea that we wanted to accept and believe in. Life is all about making a decision or a choice as to what it is that we really want to believe and believe in.

In 1859, Louis Pasteur used the Scientific Method to present to us THE TRUTH that Macro-evolution or Spontaneous Generation IS FALSE.

Charles Darwin, with NO scientific evidence to support his speculation or claim and with the Fossil Record directly opposing him, presented to us the Falsehood — Creation by Chance and Creation by Evolution, which was eventually called The Theory of Evolution.

People have been deciding for themselves which one they want to believe in, ever since then.

I have chosen to go with the SCIENCE, THE SCIENTIFIC EVIDENCE, and the Scientific Method on this one. You can and will go with whatever it is that you personally want to believe in.

—

Creation by Evolution and Creation by Chance Have Been FALSIFIED

Remember, games of chance require an intelligent assist in order to exist. Games of chance do not and cannot design and run themselves. Someone psyche or someone intelligent has to make the coins, decide to toss the coins, count the coins, and determine whether they are all heads or all tails. Every game of chance requires an intelligent assist in order to exist, including the genetic lottery that we call evolution or random mutations. God had to design, make, field-test, and fine-tune the genes, proteins, and genomes BEFORE evolution, natural selection, and random mutations could exist in the first place. First things first. Without an intelligent assist, games of chance cannot exist.

The Scientific Method is the process of eliminating false premises, false hypotheses, and false ideas. the Scientific Method is deductive reasoning through a process of eliminating False Premises or Impossible Hypotheses. the Scientific Method is abductive reasoning through a process of finding the Best Explanation, the Best Interpretation, and the Most Parsimonious Conclusion for Scientific Evidence and Scientific Data. While using the Scientific Method to DO SCIENCE, we strive to eliminate the FALSE PREMISES in the hope of eventually arriving at the CORRECT CONCLUSION. That process is called deductive reasoning. the Scientific Method is logical and scientific.

I go straight at Creation by Evolution or Creation by Chance and strangle the thing dead by demonstrating clearly, logically, and conclusively that evolution and chance cannot design, create, manufacture, or do science. Evolution and Chance to not qualify as rational and logical designers, creators, scientists, and manufacturers.

Applying the Scientific Method rationally and logically to evolution and chance will make it obvious and clear why these philosophical concepts cannot be designers, creators, manufacturers, and scientists. Evolution and Chance cannot manufacture, design, create, or do SCIENCE.

According to the Scientific Method, Creation by Evolution and Creation by Chance are FALSE PREMISES or FALSE HYPOTHESES and should be eliminated from Science because they are logically and scientifically IMPOSSIBLE.

You can RUN THE SCIENCE EXPERIMENT and PROVE to yourself right now that evolution and chance cannot design and manufacture anything at all, by trying to get evolution or chance to design and create something for you right here, right now. Call out to evolution and chance right now and tell them what you want them to design and create for you.

How's that going for you? Where do you even start? How do you get in contact with them in order to hire them to do the job?

At least with the Biblical God, you can call out to Him in prayer and He will at times respond to you.

Design implies a spirit or a mind; and, evolution has no soul and no mind. Creation implies hands, manufacturing, manipulation, or at the very least telekinesis; and, evolution or chance has no hands and can't DO SCIENCE!

The Theory of Evolution is DEAD. In fact, it was never alive to begin with!

Evolution and Chance have no hands, no soul, no intelligence, and no mind; therefore, they cannot DO SCIENCE! Only living Intelligent Beings can DO SCIENCE — people like the Biblical God and human beings. Evolution and Chance do not qualify! They are not people. Evolution and Chance are NON-ENTITIES; consequently, they cannot use the Scientific Method to get things done.

ONLY people can design, manufacture, create, and DO SCIENCE!

Since Evolution and Chance cannot design, manufacture, and create, the Theory of Evolution is FALSE; and, evolution is not and cannot be the source of life on this planet, as the Darwinists claim that it is. Both logically and scientifically, evolution CANNOT be the source or the origin of all of the genomes and life forms on this earth. ONLY some type of Intelligent Being could actually fill that role.

Creation by Evolution and Creation by Chance ARE IMPOSSIBLE; and, The Theory of Evolution IS Creation by Chance or Creation by Evolution!

Creation by People or Creation by Intelligent Beings is REAL and TRULY HAPPENS for REAL.

The BEST scientific evidence is simple OBSERVATION; but, evolution and chance have NEVER BEEN OBSERVED doing design, manufacturing, and creation!

the Scientific Method tells us clearly and conclusively that Creation by Evolution and Creation by Chance are FALSE PREMISES or FALSE HYPOTHESES.

Follow the Logic and the Evidence Wherever They Might Lead You

Once I realized that the Theory of Evolution IS Creation by Chance or Creation by Evolution, then I suddenly KNEW why the Theory of Evolution is FALSE.

By Thinking Critically and Evaluating the Evidence, which the Scientific Method demands that we do, I have PROVEN to myself that the Theory of Evolution is FALSE, because I have PROVEN in every way that I can possibly imagine that evolution and chance cannot design, create, manufacture, or DO CREATIVE ACTS. Evolution and chance cannot use the Scientific Method to get things done. People or Intelligent Beings CAN!

Evolution and chance cannot PRODUCE CREATIVE OBSERVABLE RESULTS!

There IS much stronger evidence supporting "Creation by Intelligent Beings" than there is supporting Creation by Evolution or Creation by Chance!

Creation by Intelligent Beings can be DEMONSTRATED through a preponderance of the evidence to be REAL and TRUE. Creation by Intelligent Beings is demonstrable, observable, replicable and repeatable on demand, predictive, parsimonious, falsifiable, meets its burden of proof, and produces RESULTS. Creation by Intelligent Beings is REAL SCIENCE, because it has been OBSERVED IN ACTION and AGREES WITH OBSERVATIONS! Creation by Intelligent Beings has been PROVEN to be TRUE! It's as SURE and as TRUE as anything possibly can be!

In contrast, Creation by Evolution and Creation by Chance can be demonstrated through a preponderance of the evidence to be IMPOSSIBLE and FALSE. There are NO OBSERVATIONS supporting Creation by Chance or Creation by Evolution. Macro-evolution or Spontaneous Generation has NEVER been replicable or "consistently repeatable" in any way, shape, or form — thus making Macro-evolution or Creation by Evolution BAD SCIENCE, and IMPOSSIBLE to observe and verify scientifically using the Scientific Method.

Macro-evolution IS Creation by Evolution, and it has been demonstrated to be FALSE through a preponderance of the evidence! The Theory of Evolution or Creation by Evolution has NO scientific evidence supporting it, because it can't have any scientific evidence supporting it, because evolution of any kind cannot create anything at all! There is NO way to catch evolution or chance IN THE ACT of designing and creating something, because they cannot design and create.

Creation by Intelligent Beings is clearly SUPERIOR to Creation by Evolution, Creation by Chance, and the Theory of Evolution. It's SIMPLER and more PARSIMONIOUS to let God do all of the design, creation, manufacture, and science that needs to be done, rather than sitting around waiting for evolution and chance to do it all. Waiting for evolution and chance to do something is like waiting for a train that will never come.

GOOD SCIENCE demands that Chance be eliminated as an explanation for any scientific measurements or Scientific Evidence, because Chance is NOT predictive. By definition, in principle, Creation by Chance IS BAD SCIENCE, because Chance cannot be used to reliably PREDICT anything at all! The Theory of Evolution IS Creation by Chance!

REAL SCIENCE or GOOD SCIENCE allows us to make predictions. We cannot use Creation by Evolution or Creation by Chance to make PREDICTIONS! Evolution and Chance are unpredictable, and they are not replicable on demand.

In contrast, I PREDICT that YOU will witness and observe Intelligent Beings designing, creating, driving, manufacturing, and doing some kind of science or work every single day of your life that you are awake, sober, and aware of your surroundings. REAL SCIENCE, GOOD SCIENCE, MAKES PREDICTIONS which can be verified or falsified. I PREDICT that YOU will in fact see and observe Intelligent Beings in action! NONE of our observations have ever disconfirmed this PREDICTION — namely, that Intelligent Beings can do, and do in fact do, and WILL DO design, creation, manufacturing, construction, and SCIENCE! This PREDICTION has been tested and verified an infinite number of times; therefore, we have high confidence that Creation by Intelligent Beings is REAL and TRUE.

Creation by Intelligent Beings wins by a process of elimination and by a preponderance of the evidence. The PREPONDERANCE OF THE EVIDENCE stands firmly in favor of Creation by Intelligent Beings.

Some CONCLUSIONS and some PREDICTIONS are backed up by much stronger evidence than others. We actually HAVE EVIDENCE against the Theory of Evolution or Creation by Evolution! We actually HAVE EVIDENCE, lots of evidence, for Creation by Intelligent Beings!

If Evolution and Chance cannot design, create, manufacture, or do science, then the Theory of Evolution or Creation by Evolution is BY DEFINITION FALSE. According to the Scientific Method, Creation by Chance and Creation by Evolution should be eliminated, because they can't design and create. We have tons of convincing evidence against Creation by Evolution; therefore, the Theory of Evolution or Creation by Evolution has been successfully FALSIFIED!

There's NO WAY for the Darwinists and Materialists to get around this REALITY, because there is NO WAY to get blood from a stone!

—

The Scientific Method

the Scientific Method typically goes through the following sequence in an attempt to arrive at THE TRUTH.

1) Form a **HYPOTHESIS**.
2) Select a SCIENTIFIC METHOD or Scientific Methodology to **TEST** the hypothesis.
3) Run the Science Experiment; and then, **Observe** and **Measure** the **RESULTS**.

4) Find the BEST **INTERPRETATION** or the BEST **EXPLANATION** for the Scientific Data, the Scientific Evidence, the Scientific Observations, and the RESULTS of the Science Experiment.

We can indeed **HYPOTHESIZE** that Evolution or Chance might be able to design and create something NEW and USEFUL from scratch.

However, the Theory of Evolution or Creation by Evolution FAILS at STEP 2, because there is NO way to design and run a SCIENCE EXPERIMENT, where in fact we get to **TEST** and OBSERVE Evolution or Chance IN ACTION designing and creating NEW genetic information, NEW genomes, and NEW life forms from scratch. According to the Scientific Method, Creation by Evolution or the Theory of Evolution doesn't even get off the launching pad, because evolution cannot design and create! There is NO WAY to **TEST** Creation by Evolution and observe it in action!

Once again, there is NO way to **TEST** and OBSERVE "Design by Evolution" or "Creation by Evolution"! Can you think of a way to catch evolution or chance IN THE ACT of DESIGN and CREATION, which doesn't involve some kind of magic, sorcery, trickery, lying, or wishful thinking?

I can't!

We can't get our hands on evolution and chance and make them do things for us! Evolution and chance cannot use or employ the Scientific Method to get things done! Evolution and chance CANNOT DO SCIENCE, and a lot of SCIENCE needed to be done in order to produce the first genome and the first life form!

Creation by Evolution cannot be **TESTED** and VERIFIED, making Creation by Evolution or the Theory of Evolution BAD SCIENCE!

There is NO Scientific Methodology that can be created and used to **TEST** the actions of something that DOES NOT EXIST — something like Creation by Evolution or Creation by Chance. Consequently, there is NO way to **MEASURE THE RESULTS** of Creation by Evolution and Creation by Chance, because Chance and Evolution cannot design, or create, or manufacture, or produce creative **RESULTS** for us to examine and measure.

When it comes to the Scientific Method, **MEASURABLE RESULTS** are what we are after!

There is NO WAY possible to provide **MEASURABLE RESULTS** for Creation by Chance or Creation by Evolution. Nobody can think of a way to do so, because Evolution and Chance cannot design, manufacture, create, or **PRODUCE MEASURABLE RESULTS** demonstrably, predictably, and reliably.

Natural Selection and Random Mutations ONLY work on the genomes and life forms that God has already designed and created in the first place. Natural Selection and Random Mutations cannot WORK on dead, inanimate, inert molecules and matter. Mutation and Selection cannot design, program, manufacture, create, or DO SCIENCE!

According to the Scientific Method, Creation by Mutation/Selection IS A FALSE PREMISE or a FALSE HYPOTHESIS, because Mutation/Selection cannot design and create.

According to the Scientific Method, Evolution and Chance cannot DO MANUFACTURING and SCIENCE, because Evolution and Chance cannot design and create!

Finally, there is NO way to **INTERPRET RESULTS** which cannot be PRODUCED in the first place. Chance and Evolution cannot be made or forced to PRODUCE **RESULTS** for us to **INTERPRET, EXPLAIN,** and EXAMINE! That kind of SCIENCE is ONLY DONE by Intelligent Creative Living Beings.

When it comes to Creation by Evolution or the Theory of Evolution, we can't **TEST** it; we can't examine any **RESULTS** from it; and, we can't find anything from it to **INTERPRET** either, because evolution of any kind cannot design and create anything at all.

the Scientific Method demonstrates clearly and conclusively that Creation by Evolution or the Theory of Evolution IS FALSE!

The purpose of the Scientific Method is to help us to find THE TRUTH, through a preponderance of the evidence.

the Scientific Method has no value to us if we use it to convince ourselves that a LIE is TRUE, as the Materialists and Darwinists always seem to do.

That's what I discovered during my Pursuit of the True Reality of All Things, and during my usage of the Scientific Method.

—

APPLYING the Scientific Method TO CREATION BY EVOLUTION

There are standardized RULES OF SCIENCE which Scientists are expected to follow and obey, while using the Scientific Method to falsify or verify a chosen HYPOTHESIS or PREMISE. Creation by Evolution or the Theory of Evolution VIOLATES all of these standardized RULES OF SCIENCE. In contrast, Creation by Intelligent Beings is in COMPLETE COMPLIANCE with all of these standardized RULES OF SCIENCE!

When it comes to design, creation, manufacture, and doing science, Chance and Evolution CANNOT PRODUCE RESULTS! Therefore, Creation by Chance and Creation by Evolution ARE NOT REPLICABLE! Something must be able to PRODUCE RESULTS and must be able to MAKE CORRECT PREDICTIONS in order to be Replicable and Repeatable. GOOD SCIENCE is replicable or reproducible on demand, which means that Creation by Evolution or the Theory of Evolution is BAD SCIENCE.

Something that CANNOT PRODUCE RESULTS cannot be the least bit LOGICAL, SCIENTIFIC, PREDICTIVE, OBSERVABLE, TESTABLE, or PARSIMONIOUS. Creation by Evolution or Creation by Chance IS ALWAYS the least parsimonious EXPLANATION or INTERPRETATION that can be given to evidence, especially Scientific Evidence, because these philosophical concepts CANNOT PRODUCE RESULTS or DO SCIENCE.

The lack of REPLICABILITY is a Major Reason to be skeptical of the Theory of Evolution or Creation by Evolution. Creation by Evolution and Creation by Chance should NOT be taken seriously because they are NOT REPLICABLE on demand, because they CANNOT PRODUCE RESULTS, because they CANNOT DO SCIENCE, and because they CANNOT be used to MAKE CORRECT PREDICTIONS.

The Theory of Evolution or Creation by Evolution IS FALSE, because it DOESN'T WORK, HAS NEVER BEEN OBSERVED IN ACTION, DOESN'T PRODUCE CREATIVE INTELLIGENT DECIPHERABLE RESULTS, DOESN'T MAKE CORRECT PREDICTIONS, VIOLATES the Scientific Method, DOESN'T COMPLY WITH THE STANDARD RULES OF SCIENCE, IS NOT PARSIMONIOUS, IS BASED UPON FALSE ASSUMPTIONS, FAILS TO MEET ITS BURDEN OF PROOF, and IS NOT REPLICABLE ON DEMAND. Creation by Evolution or the Theory of Evolution IS BAD SCIENCE!

Creation by Evolution and Creation by Chance are FALSE because they VIOLATE these standardized RULES OF SCIENCE and thus VIOLATE the Scientific Method. This means that Creation by Evolution or The Theory of Evolution is UNSCIENTIFIC!

Just because the Atheists, Darwinists, and Materialists want the Theory of Evolution to be true doesn't make it true! Just because the Atheists, Darwinists, and Materialists don't want God to exist doesn't make it so!

According to the Scientific Method, Creation by Chance and Creation by Evolution ARE UNSCIENTIFIC because they ARE UNTESTABLE; and, the Theory of Evolution IS Creation by Evolution. A hypothesis or theory MUST BE TESTABLE if we expect to be able to apply the Scientific Method to it!

We cannot develop TESTING METHODS in order to TEST Evolution and Chance in the PROCESS of designing, creating, and manufacturing NEW genomes and NEW life forms from scratch. In contrast, we can in fact, design and create TESTING METHODS to OBSERVE and TEST Intelligent Beings in the PROCESS of designing, creating, manufacturing, and doing SCIENCE.

Our CONFIDENCE in the CONCLUSION that Intelligent Beings can design and create IS extremely high.

—

the Scientific Method DEMANDS THAT WE ELIMINATE THE FALSE AND THE IMPOSSIBLE

According to the Scientific Method, INCORRECT THEORIES and FALSIFIED THEORIES should be eliminated from SCIENCE.

The Darwinists and Materialists have REJECTED "Creation by Intelligent Beings" and have replaced it with Creation by Evolution or Creation by Chance. Such an action is ILLOGICAL and UNSCIENTIFIC — it's purely philosophical or religious in nature and VIOLATES the Scientific Method!

The philosophical claims and religious beliefs of the Darwinists, Materialists, and Atheists ARE NOT Scientific Evidence; consequently, there is NO Scientific Evidence whatsoever supporting Creation by Chance or Creation by Evolution! There can't be because Chance and Evolution cannot design and create.

the Scientific Method and the Rules of Science state that FALSIFIED THEORIES, like Creation by Evolution and Creation by Chance, should be eliminated from SCIENCE; but, we see the Darwinists and Materialists doing the exact opposite of what the Scientific Method and the Rules of Science demand that they do.

In SCIENCE, the burden of proof is on anyone who makes a claim, such as the claim that Evolution and Chance can design and create functional genomes and new unique life forms. That claim should be demonstrable if it is true! But, there is no way to DEMONSTRATE that Creation by Evolution or Creation by Chance is real and true, because we can't force Evolution and Chance into a science lab and make them design and create something for us.

What do we observe instead? We observe the Darwinists and Materialists trying to shift their burden of proof onto their opposition!

Creation by Chance and Creation by Evolution should be DEMONSTRABLE if they are really true; but in fact, they FAIL on the merits of their own case. We can DEMONSTRATE clearly and conclusively that evolution and chance cannot design and create anything at all. Only the Darwinists and the Materialists cannot see the obviousness of it all.

The PREPONDERANCE OF THE EVIDENCE stands firmly against Creation by Evolution and Creation by Chance! By definition in principle, Creation by Evolution and Creation by Chance ARE logically and scientifically IMPOSSIBLE, and therefore FALSE.

The simplest kind of evidence is careful OBSERVATION; but, Creation by Evolution and Creation by Chance have NEVER BEEN OBSERVED in action!

Evolution and Chance modify and degrade things; but, they NEVER design and create and manufacture NEW things from scratch, because they can't! Eventually, as scientists, we want to move beyond OBSERVATIONS to the point where we do in fact have EXPLANATIONS; however, this IS impossible when it comes to the Theory of Evolution because we have YET TO OBSERVE Creation by Chance or Creation by Evolution in ACTION!

There's NO WAY for Creation by Evolution or Creation by Chance to meet its burden of proof, because Evolution and Chance cannot do creation and cannot do science! Creation by Evolution and Creation by Chance are oxymorons and OBVIOUSLY FALSE, because evolution and chance cannot Do Creation. Creation by Chance and Creation by Evolution VIOLATE the Law of NonContradiction, because they ARE incongruous contradictions! Creation by Evolution and Creation by Chance also VIOLATE the Rules of Science and the Scientific Method! And to the horror of many of us, the Theory of Evolution IS Creation by Evolution or Creation by Chance. The Theory of Evolution is BAD SCIENCE!

Anything that is an oxymoron, such as Creation by Evolution or Creation by Chance, IS by definition in principle NON-EXISTENT NONSENSE, meaning that this "pointedly foolish" incongruous "contradiction in terms" is automatically and axiomatically FALSIFIED and FALSE. This Truth and Reality IS both logical and scientific. SCIENCE and OBSERVATION and EXPERIENCE and LOGIC, individually and collectively, HAVE PROVEN to us that Creation by Evolution and Creation by Chance have NEVER happened, can NEVER happen, and will NEVER happen.

The Theory of Evolution IS Creation by Evolution or Creation by Chance; therefore, the Theory of Evolution or Creation by Evolution is IMPOSSIBLE NON-EXISTENT NONSENSE, scientifically and logically and axiomatically FALSE and FALSIFIED, and has NEVER happened and will NEVER happen.

We must eliminate the IMPOSSIBLE, if we want to get at THE TRUTH. the Scientific Method demands no less from us!

I have used the Scientific Method to PROVE to myself that the Theory of Evolution or Creation by Evolution is FALSE. You can do the same for yourself as well, if you want to do so.

—

SUMMARIZING WHAT I HAVE LEARNED FROM the Scientific Method

The purpose of the Scientific Method is to help us to find THE TRUTH, through a preponderance of the evidence.

the Scientific Method has no value to us if we use it to convince ourselves that a LIE is TRUE, as the Materialists and Darwinists always seem to do.

That's what I discovered during my Pursuit of the True Reality of All Things, and during my usage of the Scientific Method.

Design, Manufacture, and Creation by "Evolution or Chance" IS axiomatically IMPOSSIBLE and scientifically FALSIFIED. Chance and Evolution CANNOT DO SCIENCE, CANNOT DO DESIGN, and CANNOT DO CREATION!

The Theory of Evolution or Creation by Evolution has to go, because it is UNSCIENTIFIC and demonstrably FALSE. By definition in principle, Creation by Evolution or Creation by Chance IS IMPOSSIBLE and UNSCIENTIFIC and VIOLATES the Scientific Method.

One of the goals of the Scientific Method is to eliminate the False Premises, the Falsified Premises, and the Impossible Premises. According to the Scientific Method, Creation by Evolution and Creation by Chance have to be ELIMINATED from SCIENCE because they are false, unscientific, and impossible — if we are to be Scientists.

Creation by Evolution and Creation by Chance ARE FALSE and FALSIFIED PREMISES; therefore, they are also FALSE and FALSIFIED CONCLUSIONS.

Science IS knowledge, and I KNOW that the Theory of Evolution is FALSE because I know why it is FALSE. the Scientific Method has PROVEN to me that The Theory of Evolution or Creation by Evolution IS FALSE. Creation by Evolution or Creation by Chance ALWAYS provides a FALSE interpretation for ANY Scientific Evidence! It is IMPOSSIBLE for something that is false to provide the Best Explanation or the Best Interpretation for Scientific Evidence; and, the Scientific Method is all about finding the Best Interpretation or the CORRECT EXPLANATION for Scientific Evidence!

In contrast, Creation by Intelligent Beings ALWAYS matches with REALITY and thus ALWAYS provides the BEST EXPLANATION or the BEST INTERPRETATION for any kind of evidence, including Scientific Evidence. Something is said to be Scientifically Accurate if it matches with REALITY! Creation by Intelligent Beings matches with REALITY; therefore, it is Scientifically Accurate, Real, and TRUE.

OBSERVE IF IT IS NOT SO!

Creation by Evolution or the Theory of Evolution is BAD SCIENCE, and BAD RELIGION too. Creation by Evolution or Creation by Chance IS FALSE because evolution and chance cannot design, manufacture, create, or DO anything at all!

The Theory of Evolution IS Creation by Chance or Creation by Evolution.

When I applied the Scientific Method to Creation by Evolution, it became CLEAR and OBVIOUS how and why The Theory of Evolution MUST BE FALSE.

Evolution and Chance CANNOT DO SCIENCE or CREATION! Intelligent Beings CAN! What more do we need to say?

—

Eliminate the Impossible Premises

How often have I said to you that when you have eliminated the impossible, whatever remains, *however improbable*, must be THE TRUTH? — Sherlock Holmes

Evolution and Chance MUST BE ELIMINATED from Science, because Creation by Evolution and Creation by Chance ARE IMPOSSIBLE!

Once we have eliminated Evolution and Chance as our designers and creators, then whatever remains, however improbable or unlikable, MUST BE THE TRUTH!

THIS IS SCIENCE!

This is Abductive Reasoning through a process of eliminating the impossible, and eliminating the least parsimonious explanations, and eliminating the worst ideas produced by mankind. Abductive Reasoning seeks for the best and most parsimonious interpretation or explanation for the available evidence. Creation by Evolution or Creation by Chance is NOT it! Creation by Intelligent Beings IS!

This is also Deductive Reasoning. By eliminating ALL of the FALSE PREMISES, we can deduce the CORRECT CONCLUSION. Creation by Evolution and Creation by Chance should be eliminated as PREMISES because they are IMPOSSIBLE and demonstrably FALSE, which means that Creation by Evolution and Creation by Chance cannot be used as the CONCLUSION either! the Scientific Method demands no less, because the Scientific Method IS DEDUCTIVE REASONING.

The CORRECT CONCLUSION is that evolution of any kind cannot design, manufacture, and create; but, Intelligent Beings certainly can. Intelligent Beings can DO SCIENCE. Evolution and Chance cannot.

So, the question is, "Do YOU want THE TRUTH? Or, do you want the Darwinian Lies, because the Darwinian Lies are telling you exactly what you want to hear — namely, that God does not exist and that you don't have to repent of your sins because there will be no Final Judgment?"

I'm ONLY interested in THE TRUTH, and I KNOW that I have found THE TRUTH where the Theory of Evolution is concerned. I have NO doubts whatsoever!

Creation by Evolution is IMPOSSIBLE and FALSE.

Creation by Intelligent Beings is REAL and TRUE.

IT'S SCIENCE! IT'S LOGICAL!

What you do with this information is up to you.

Mark My Words!

—

The Existence of God IS Demonstrable

The claim that "God does NOT exist" cannot be demonstrated. It should be DEMONSTRABLE, if it is true!

People have an obligation to present evidence to support the claim that "God does NOT exist", if they are in fact making this claim. Yet, what do we observe? We observe people making this claim without providing one single shred of evidence to support their claim. That's cheating; and, that's UNSCIENTIFIC!

The ONUS or burden of proof is always on the prosecutors; therefore, the burden of proof should in fact be on the people making the claim that "God does NOT exist". Yet, what do we observe? We observe the Atheists, Materialists, and Darwinists trying to shift their burden of proof to their opposition. That's cheating, and that's UNSCIENTIFIC!

In contrast, the Existence of the Biblical God is indeed demonstrable!

If one were to input the Bible, the Book of Mormon: Another Testament of Jesus Christ, the Doctrine and Covenants, and the Pearl of Great Price as THE PREMISES, then it IS logical to CONCLUDE that the Biblical God does indeed exist, that the

Biblical God has indeed revealed Himself to mankind, and that the Biblical God continues to reveal Himself to mankind in our day and time.

The Existence of the Biblical God is demonstrable through a preponderance of the evidence, because the Biblical God went out of His way to meet His burden of proof.

There's lots of convincing evidence for the Existence of the Biblical God. There's lots of convincing evidence which demonstrates that the Biblical God has indeed revealed Himself to mankind. You just have to be willing to go looking for that evidence, and then be willing to study that evidence when you have finally found it.

Science can indeed explain how it might be possible for God to speak to a person, or how it might be possible for God to appear to a person and shake hands with that person. As demonstrated within the Scriptures which were produced by the Biblical God, thousands of different people have seen, talked with, and touched the Biblical God, our resurrected Lord Jesus Christ. There is some replication taking place there. Some of us have had our prayers answered, and we know it. Once again, there is some replication taking place there.

In contrast, there IS NO evidence and can be NO evidence to support the claim that "God does NOT exist". That's the REALITY of the situation. Science cannot explain the existence or the actions of NOTHING, or a NON-ENTITY that does not exist. Philosophically and logically it is IMPOSSIBLE to prove that something does not exist.

ALL of the evidence is on the side of the Biblical God demonstrating to us that HE does exist, that HE has indeed revealed Himself to mankind, and that HE continues to reveal Himself to mankind.

The Existence of the Biblical God MAKES Creation by Evolution or the Theory of Evolution superfluous and unnecessary.

—

Removing Accountability and Responsibility Is Sinful and Wrong

As found online at Google from the first thing that came up:

> There's a famous passage from "The Grand Inquisitor" section of Dostoevsky's "The Brothers Karamazov" in which Ivan Karamazov claims that, "If God does not exist, then everything is permitted. If there is no God, then there are no rules to live by, no moral law we must follow; we can do whatever we want."

As is my way, I prefer to adjust this quote and say, "If God does not exist, then nobody will be held accountable for their choices, actions, and sins. If God does not exist, then we can get away with murder and genocide, and we will never be held accountable for it."

For many of us sinners, this is a very pleasing and desirable point of view — the idea that we will never be held accountable for any of our choices, decisions, sins, addictions, vices, adultery, murder, or immoral actions of any kind. And, this is one of the main reasons why many Atheists and Darwinists so desperately want the Theory of Evolution to be true. The Darwinists will even lie at times to themselves and to us in an attempt to prove that the Theory of Evolution is true. It has been demonstrated hundreds of times that this reality is so.

Most Darwinists lie to themselves in order to prove to themselves that the Theory of Evolution is true, so that they can confidently declare that God does not exist and really believe it. This is what the Theory of Evolution is all about. In contrast, if a person truly believes in God and knows that God exists, then that person has absolutely no need whatsoever for the Theory of Evolution.

The most horrific evils, the greatest and worst and most abundant evils of the 20th century, stem directly from Darwinism and Atheism. The Fascistic Darwinists and Militant Atheists exterminated over 200 million people during the 20th century.

Creation by Evolution and Creation by Chance are demonstrably FALSE. The Theory of Evolution IS Creation by Evolution or Creation by Chance. The Theory of Evolution is possibly the greatest LIE ever perpetrated against the human race. It's certainly the LIE that the most people seem to have bought into, because they desperately want it to be true.

—

Seek the Most Parsimonious Interpretation or Explanation

Some people have dedicated their whole lives to defending and promoting the Theory of Evolution. Others have spent their whole lives trying to debunk it.

Wouldn't it have been easier and a whole lot simpler to just let God design and create all of the genomes and life forms in the first place?

Creation by Intelligent Beings is the most parsimonious, most logical, most scientific, and BEST interpretation or explanation that can be given to evidence, any type of evidence, including Scientific Evidence.

Creation by Evolution and Creation by Chance are ALWAYS the least parsimonious explanation, because it will ALWAYS be easier and simpler to just let God, or the Intelligent Beings, do all of the science, design, creation, and manufacturing which needs to be done. You can wait ALL eternity for Evolution or Chance to design and create and manufacture something from scratch for you, and it will NEVER happen.

In contrast, Design and Creation by Intelligent Beings is an infinitely more believable and parsimonious explanation for getting Science done than Creation by Evolution or Creation by Chance will ever be. If you need something designed and manufactured right now, to whom are you going to turn, the Intelligent Beings OR Evolution and Chance? Which one do you think is the most likely to get the job done?

The Theory of Evolution IS Creation by Evolution or Creation by Chance! The Darwinian idea that evolution or chance can design, create, and manufacture anything that it sets its mind to IS possibly the dumbest, least parsimonious, most illogical, and most unscientific idea that human beings have ever come up with. I can't think of anything worse! Creation by Evolution is UNSCIENTIFIC, ILLOGICAL, and IMPOSSIBLE.

Since evolution and chance could NEVER have designed and created any of the genomes or life forms on this planet, it is obviously clear that God must of necessity exist in order to have done all of the DESIGN, SCIENCE, CREATION, and MANUFACTURING which Chance and Evolution could NEVER have done.

This is SCIENCE!

This is abductive reasoning through a process of seeking the best possible explanation for the evidence. This is deductive reasoning through a process of eliminating the False Premises or the Impossible Premises. Evolution and Chance have to be eliminated as Designers, Creators, Manufacturers, and Scientists, because these philosophical concepts CANNOT DO SCIENCE and CANNOT CREATE!

The purpose of the Scientific Method is to help us to find THE TRUTH, through a preponderance of the evidence.

the Scientific Method has no value to us if we use it to convince ourselves that a LIE is TRUE, as the Materialists and Darwinists always seem to do.

That's what I discovered during my Pursuit of the True Reality of All Things, and during my usage of the Scientific Method.

—

The Theory of Evolution Proves that God Exists

I wrote an 875-page tome, which goes into great detail regarding this particular subject. It's entitled, "The Theory of Evolution Proves that God Exists: Why I Am No Longer an Atheist and Why I No Longer Believe in the Theory of Evolution".

Source for the book:

https://www.amazon.com/dp/B01HZYBZ7K

If you follow along in that book and understand the various concepts as they are presented to you, then you will KNOW why I am no longer an atheist and why I no longer believe in the Theory of Evolution.

If God exists or can be demonstrated to exist, then there is no further need for the Theory of Evolution.

Mark My Words!

—

Show NO Mercy towards Creation by Evolution

The Atheists and Darwinists will show no mercy towards you; but, continue to show mercy and compassion towards them, because they desperately need it. I know because I used to be one of them.

Meanwhile, show no mercy whatsoever towards Darwinism, the Theory of Evolution, Creation by Chance, and Creation by Evolution because those things are demonstrably FALSE and clearly EVIL.

We are doing a person a great service any time that we can successfully explain to him, or her, why the Theory of Evolution MUST BE FALSE and how the Theory of Evolution has been FALSIFIED by the Scientific Method. Once they have successfully eliminated the FALSEHOODS and LIES from their lives, then they are free to pursue THE TRUTH.

THE TRUTH is the only thing that has value to us in the end, whether the Darwinists and Materialists know that to be true or not.

The Theory of Evolution IS Creation by Chance or Creation by Evolution.

When I applied the Scientific Method to Creation by Evolution, it became CLEAR and OBVIOUS how and why The Theory of Evolution MUST BE FALSE.

Evolution and Chance CANNOT DO SCIENCE, DESIGN, MANUFACTURING, or CREATION. Intelligent Beings reliably CAN. Parsimony and checkmate!

Mark My Words!

The Scientific Method

It is said that Scientists, at least the good ones, seldom speak of proof — they always leave room for doubt or for further upcoming discoveries.

In contrast, the Darwinists and Materialists often say that they have proven the Theory of Evolution to be true, but all that really means is that they have convinced themselves that the theory is true. However, their alleged "proof" doesn't apply to the rest of us who KNOW that the Theory of Evolution is FALSE and why it is FALSE.

Still, the claim that nothing can be proven to be true is also false. For example, it can be proven to be true that if you keep sticking paper clips into electrical sockets, you are going to get electrocuted unless you take steps to protect yourself. It can also be proven to be true that if you stick your hand on a hot stove, you are going to get burned. There are some scientific claims that can indeed be proven to be true beyond a shadow of a doubt.

However, whenever the scientists start saying that nothing can be proven to be true, what most of them (being Materialists) are trying to say is that God or Spirit cannot be proven to exist. They are right in one respect — there is NO way for one person to prove to another person that God exists. However, they are wrong in that God can appear to you or speak to you directly and thereby prove to you that He exists. In other words, the existence of God can be proven to you through an appearance of God to you.

However, God doesn't seem to be in the habit of appearing to anyone and everyone; so, the question becomes, "Are there other ways to prove that God exists besides direct personal appearances to you?"

And now we come to an extremely important part of the Scientific Method and Scientific Methodologies — namely, demonstrable proof through a preponderance of the evidence.

There are many things that cannot be proven directly through empirical observation or scientific experimentation, yet they are still true and still exist. These are things like crime scene investigation, thoughts, dreams, altered states of consciousness, revelations from God, spirituality, out-of-body experiences, near-death experiences, psychic experiences, contact with the dead, and God. These things cannot be proven through replicable direct observation nor through scientific experimentation; but, they can indeed be proven through direct personal experience and a preponderance of the evidence, as long as we are willing to accept that evidence as being true.

For example, even though it is impossible to replay the murder and thereby prove who did it, it is indeed possible through a preponderance of the evidence to prove beyond all reasonable doubt who committed the murder.

Likewise, it is impossible to rerun the tape and observe who designed and created this universe and all the life on this planet; but, it is possible to demonstrate or prove through a preponderance of the evidence that it was the Biblical God who did the job, especially when He starts appearing to lots of different people confessing to having done the job. Confessions are admissible in a court of law.

—

I am a scientist. I study and try to employ Psychology and Philosophical Logic. Psychologists consider themselves to be scientists, and they use the Scientific Method in order to pursue and document their science. However, Psychologists and professors of The Science of Philosophy run into a difficult and insurmountable problem whenever they try to train their scientific instruments onto something like "Psyche" or "Spirit" or "Thought". These things by definition in principle cannot be detected, recorded, or measured with physical instruments.

When dealing with the 'non-physical", the scientists are forced to abandon the Standard Materialistic Scientific Testing Methodology and forced to develop and to use different Testing Methodologies within the Scientific Method, in order to explore and do science whenever they are dealing with the non-physical. In other words, because of the existence of "Psyche" or "Thought" and other things non-physical like Light, the scientists were forced to develop additional Scientific Testing

Methodologies before they could then examine these things scientifically using the Scientific Method. Intrigued? I know that I was when I was first introduced to this scientific reality.

The best professional discussion regarding this scientific reality, that I have encountered so far, is in the book "Introduction to Psychology 9th Edition" by James W. Kalat. He is an evolutionist and tries his best to cater to the Naturalists, but he can't avoid this scientific reality and was forced to deal with it in his book. Chapter 2, "Scientific Methods in Psychology", is particularly revelatory.

I quote from page 29:

> "The philosopher Karl Popper argued that because no observation proves a theory to be correct, the purpose of research is to find theories which are incorrect. That is, the point of research is to falsify the incorrect theories, and a good theory is one that withstands all attempts to falsify it. In other words, it wins by a process of elimination. A well-formed theory, therefore, is falsifiable — that is, stated in such clear, precise terms that we can see what evidence could count against it — if of course such evidence existed. Falsifiable means that we can imagine something that would count as evidence against the theory. However, when Popper wrote that research is always an attempt to falsify a theory, he went too far."

END QUOTE.

Why did Karl Popper go too far?

It's because there are some things in this world that are not falsifiable. The existence of Mind, Psyche, Consciousness, or Spirit is not falsifiable nor is it detectable in a science lab. Therefore, some adjustments must be made to the Scientific Method or Scientific Methodologies so that we can actually consider Psychology (the Study of the Human Spirit or Human Mind) to be a Science, and then do real Scientific Research into Psychology.

Karl Popper went too far by limiting science to Naturalism or Materialism, and ONLY to the things which can be falsified or proven false using our physical scientific instruments! Too much Materialism and not enough logic and science!

There are things in this world that cannot be falsified nor detected with scientific instruments; yet, we want to employ science in order to discover and identify and explain these things. That means that scientists had to develop new and different and more sophisticated Scientific Methodologies that could then be employed within the Scientific Method in order to explore the non-physical realities of our existence.

Psychology considers itself to be a Science; and, psychologists and philosophers like me consider themselves to be scientists. Do you see the problem here that we have to deal with? Probably not if you are a Naturalist or a Darwinist or an Atheist and have deliberately blinded yourself to the non-physical. Psychologists run into difficulties and effectively kill their Science dead if they deliberately limit themselves to Naturalism and limit themselves exclusively to materialistically falsifiable evidence.

Psychologists were forced to develop other Scientific Methodologies that they could then use to study "Psyche" or "Thought" scientifically using the Scientific Method.

There's no way to determine what a person is thinking, with our scientific instruments. If you want to know what a person is thinking, you have to ask them. Our physical instruments can indeed detect brain activity, but they cannot detect what thoughts or experiences are taking place within that brain activity. We can't read or detect a person's mind or thoughts with our scientific instruments. Mind, or consciousness, or thoughts, or spirit are not detectable with physical instruments; and thus, they are not falsifiable, because they can't be detected in the first place. How can you falsify something if you can't detect it?

The original definition for Psychology was "the study of the spirit" or "the study of the mind". Literally, the study of the psyche! The human psyche is not detectable with our physical scientific instruments, so how can you study such a thing scientifically? How can you study the non-physical or the spiritual scientifically?

Well, first of all, you have to drop any pretense to Naturalism. Naturalism or Materialism is a bane to science — it stunts and kills scientific exploration and scientific discovery across a wide range of different sciences and scientific endeavors.

Second, you have to shift over to a different mode of scientific evidence and scientific discovery. You have to shift over to the "Burden of Proof" method of doing the Scientific Method, or the "preponderance of the evidence" method of doing the Scientific Method, whenever you want to discuss or pursue scientific discoveries regarding the non-physical or the spiritual realms, because we can no longer rely upon the ability to falsify the evidence using the various physical instruments at our disposal.

My psychology texts tell me that in order to study such a thing as spirit or mind scientifically, the scientists have to get very creative with their scientific experiments. They have to step up their game! In contrast, the Naturalists and Materialists simply run away and hide, by stating that Intelligent Design, or The Study of the Spirit, or the Study of Thoughts and the Mind, or the study of God are NOT SCIENCE and NOT SCIENTIFIC.

I can't count the number of times that I have heard the "NOT SCIENCE" claim from a Naturalist or a Darwinist or an Atheist. It's a dodge! But, they really have nothing better to offer in defense of their beliefs. They have to "in principle" delete and deny anything that they can't detect with their physical instruments, because they truly believe that the physical is all that there is and all that exists. It's a very crippling and limiting worldview that the Naturalists and Atheists adopt, so that they don't have to deal with things like spirit, spirituality, intelligent designers, Judgment, and God.

In contrast, the message which I get from the Intelligent Design people is that the only type of evolution that actually works is Intelligently Designed Evolution. Think about it, because it is true. The only kind of evolution that actually works is Intelligently Designed Evolution or Genetic Engineering. It is the Genetic Engineering, which is the Real Science, and not Creation by Darwinian Chance.

The Naturalists and the Atheists try to take all of the fun out of life, and they try to force us not to look at the spiritual side of life. After all, think about it, happiness, joy, pleasure, love, friendship, and feelings are to one degree or another triggered by non-physical thoughts. The effects or the results of those feelings can indeed be detected with physical instruments, but the precise thought that triggered them cannot. A person could in fact be exhibiting a great deal of happiness and joy while he is thinking about killing his worst enemy. Physical instruments cannot detect or determine the thought that is causing the happiness or joy. It could be anything! You never know what the person is really thinking about, unless he or she tells you honestly what he or she is thinking about.

So, how do psychologists study such a thing scientifically since it is impossible to get our scientific instruments to detect thoughts, or spirit, or consciousness, or mind, or psyche? How can you study these things if they can't be falsified? Those are questions that psychologists keep asking themselves all the time, because Psychology is a science and Psychologists consider themselves to be scientists.

On page 29 of "Introduction to Psychology 9th Edition" by James Kalat, we have this quote:

> "Instead of insisting that all research is an effort to falsify a theory, another approach is to discuss Burden of Proof, the obligation to present evidence to support one's claim. In science, the Burden of Proof is on anyone who makes a claim that should be demonstrable if it is true."

These claims about Psyche or Thought or Light or the Existence of God should be demonstrable if true. Burden of Proof is the obligation to present evidence to support one's claim! Burden of Proof or meeting one's Burden of Proof is a different type of Scientific Methodology or Scientific Testing within the Scientific Method. It is a legitimate Scientific Methodology that is used all the time in criminal cases and courts of law. Since we have no ability to read a prisoner's thoughts with our scientific instruments, we have to develop other ways of getting at THE TRUTH — we have to develop other Scientific Methodologies in addition to the "Falsifiable Materialistic Methodology". Burden of Proof is a very powerful and essential Scientific Methodology. The goal is to go with the Preponderance of the Evidence. Juries do so all the time. Whether a person is guilty or innocent is demonstrable, at least in theory.

I eventually learned that we can, and we should, be applying the Scientific Method to Religion and Spirituality just as much and just as often as we apply the Scientific Method to the physical realities of our lives. the Scientific Method loses much of its value unless we learn to apply it to our chosen Religion and our Spiritual Pursuits. We should also be applying Rationality and Logic to our chosen Religion and Spiritual Pursuits.

Intrigued?

Well, I certainly was the first time that all of this started to dawn on me.

I decided for myself that if I'm going to do that religion thing, then I'm going to go with the best and jettison all the rest. I eventually chose the religion that has the most depth, is the most interesting, makes the most demands, makes the most sense, and promises the most rewards. I haven't found anything else that even comes close to it. I have learned to go with the best and get rid of all the rest. I wanted the best for me and mine.

A question that I love to ask myself from time to time is, "Has God met His Burden of Proof?" I believe that the Biblical God has. Why? Well, let's run this little thought experiment.

If you input the Bible, the Book of Mormon: Another Testament of Jesus Christ, the Doctrine and Covenants, and the Pearl of Great Price as the PREMISES, then it is logical to CONCLUDE that the Biblical God, our Resurrected Lord Jesus Christ, exists and really did rise from the dead and continues to appear to people in the flesh here and now in the physical world. The conclusion follows logically from the evidence or from the premises. That's deductive reasoning! If the premises are true, then the CONCLUSION must be true also. The Scriptures that the Biblical God had a hand in writing and producing prove to us that He is real and that He truly exists, just as this book here proves to you that I am real and that I truly exist. The conclusion follows logically from the premises. That's a form of SCIENCE! It's a form of meeting one's Burden of Proof! That's deductive reasoning.

Cool, huh?

The Biblical God isn't exactly stupid, you know! He wants us to find Him and get to know Him; so, He deliberately left behind traces of His existence for us to find. Obviously, He did so on His own terms and not ours, but that doesn't change the fact that He has spoken to us and given us commandments which He wants us to find and obey, because He knows that keeping His commandments will eventually lead us back to Him.

I believe that the Biblical God has met His Burden of Proof within the Bible, the Book of Mormon, the Doctrine and Covenants, and the Pearl of Great Price. You might choose to believe differently which is fine if it works for you, but it won't change what I have experienced first-hand and now believe and know to be true. Within those books, the preponderance of the evidence states, demonstrates, and proves that God exists. The existence of the Biblical God is demonstrable through the Scriptures and through the modern-day revelations of God and revelations from God!

Belief in God and a belief in God's existence are not falsifiable; but, they are indeed demonstrable through a preponderance of the evidence! The scientific prediction is that if you take God's Word to His Chosen Prophets seriously and treat His Word as truth, God's Truth, and keep His Commandments, then you will find God in the process.

Jesus Christ through the Book of Mormon even asks you to run that science experiment for yourself, a number of different places within that book. Jesus Christ makes the same request of us in the Bible, asking us to make an experiment on His words to see if obedience to His words produces desirable fruits as a result.

Jesus Christ is the ULTIMATE SCIENTIST, and He actually asks us to experiment with His teachings and His words. Experimentation is an essential part of the Scientific Method! God doesn't want us taking Him on blind faith. God wants us digging in and doing the scientific research, running experiments on His Words, and praying for ourselves so that we find Him for ourselves when we are done!

I believe that the Biblical God deliberately and knowingly meets His Burden of Proof through a preponderance of the evidence in the Bible, Book of Mormon: Another Testament of Jesus Christ, Doctrine and Covenants, and Pearl of Great Price. God employed a legitimate Scientific Methodology to prove to us through a preponderance of the evidence that He exists. He had to do so, or NONE of us would have ever heard of Him and NONE of us would ever know that He exists.

There's actually some demonstrable proof for God's existence. In contrast, there is NONE for the Theory of Evolution or Creation by Evolution.

God has to find some way to reveal Himself to us, or He would remain forever concealed and unknown. However, God's existence is not falsifiable. There's no way for us to trap God in a cage and then examine Him and falsify Him with our scientific instruments. God is not replicable in a science lab either. However, the existence of God is indeed demonstrable through a preponderance of the evidence.

This concept of Burden of Proof and the Preponderance of the Evidence changed the way that I look at the Scientific Method and Scientific Methodologies for the better. Whenever scientists encounter something that isn't falsifiable or directly observable on demand, then the goal becomes to make it demonstrable through a preponderance of the evidence.

A lot of things in Psychology, Forensics, Religion, Theology, and Crime Scene Analysis are not falsifiable or reliably replicable. However, they are demonstrable by a preponderance of the evidence. Police and Forensic Scientists examining crimes scenes have to meet their Burden of Proof through a preponderance of the evidence in order to convict a criminal of a crime. Scientists dealing with spirit or mind or "thoughts" have to meet their Burden of Proof also through a preponderance of the evidence. The Biblical God deliberately went out of His way to meet His Burden of Proof by appearing to His chosen prophets and speaking to them, as recorded in His Scriptures — the Bible, Book of Mormon, Doctrine and Covenants, and Pearl of Great Price. Even God had to meet His Burden of Proof, if He were to expect any of us to actually believe that He exists.

However, when it comes to some of the other areas of Psychology and Philosophy and Theology, it can be difficult (or impossible) at times for the scientists to meet their Burden of Proof, but that doesn't stop them from trying.

—

The purpose of the Scientific Method is to help us to find THE TRUTH, through a preponderance of the evidence.

the Scientific Method has no value to us if we use it to convince ourselves that a LIE is TRUE, as the Materialists and Darwinists always seem to do.

That's what I discovered during my Pursuit of the True Reality of All Things, and during my usage of the Scientific Method.

Hugh Ross PhD and his "Reasons to Believe" team have developed Scientific Proofs of God's existence, through a wide variety of different sciences and scientific evidence. I believe that they successfully meet their Burden of Proof from time to time. Their different books using science to prove that God exists can be a fascinating read. I have learned to trust Hugh Ross to give me a scientifically accurate interpretation of Biblical verses.

Gerald L. Schroeder PhD does the same thing with some of his books, including one of his books entitled "The Science of God".

The Science of God is not falsifiable. However, it is demonstrable. It's based upon the pursuit of the preponderance of the evidence. Drop in sometime and read some of the books from Hugh Ross or Gerald L. Schroeder, and then decide for yourself whether they meet their Burden of Proof, or not.

When it comes to God, each person has to take a leap of faith — either to conclude that He does not exist or to conclude that He does exist. However, a study of the scientific evidence for God's existence, the philosophical evidence for God's existence, and the revelatory evidence for God's existence does indeed make it easier to be sure that your leap of faith is based upon True Evidence, rather than being based upon false evidence or no evidence.

Since everyone has to take a leap of faith when it comes to God, you want to make sure that you are taking the right one. A demonstrable preponderance of the evidence and the Scientific Method makes that possible. The Biblical God deliberately meets His Burden of Proof through the philosophical evidence, the scientific evidence, and the revelatory theological evidence.

There is NO evidence that proves that God does not exist. There can be NO evidence that proves that God does not exist — it's philosophically impossible. ALL of the evidence is on God's side, and it proves through a demonstrable preponderance of the evidence that God exists. And, to some of us, the Biblical God will choose to make an appearance in the flesh; and, then we will have an even more direct and sure knowledge that God is real and truly exists.

Because the Biblical God deliberately met His Burden of Proof in His Scriptures and because God has deliberately left behind traces of Himself in Science and in Nature, we can in fact use the Scientific Method to demonstrate through a preponderance of the evidence that God does indeed exist.

Let the research and experimentation begin!

The True Nature of Proof

What is the Nature of PROOF?

Something becomes PROVEN for you only after YOU have chosen to believe that the associated supporting evidence is true. Proof or conversion takes place on an individual basis!

Therefore, from my own personal perspective, whenever I have chosen to believe the Scientific Evidence which the Cutting-Edge Research Scientists have presented to me, then that particular concept has become PROVEN to me through the simple process of choosing to believe that the associated evidence and the associated interpretation of the evidence ARE TRUE.

Consequently, when I finally chose to believe the Scientific Evidence showing me clearly and conclusively that Evolution by Spontaneous Generation and Evolution by Abiogenesis have NEVER been observed anywhere on this planet and cannot be replicated or reproduced in any Science Lab, then I have chosen to believe that Scientific Evidence to be true. Suddenly by that personal choice of mine, it has been PROVEN to ME that Macro-evolution or goo-to-you evolution CANNOT happen and is therefore FALSE, because I chose to believe the Scientific Evidence which tells me that molecules-to-man evolution cannot happen and is false.

Nothing becomes PROVEN for YOU, until after you choose to believe that the associated evidence is true. For you, NONE of these things which I mention will have been proven true, if you refuse to look at and believe and trust the associated Scientific Evidence which I have seen and know to be true.

True faith or true belief is based upon true evidence. False faith or false belief is based upon false evidence. I discovered that the Theory of Evolution is based upon a mass array of FALSE EVIDENCE. Therefore, those who choose to believe in the Theory of Evolution have a false belief or a false faith where the Theory of Evolution is concerned.

I choose to believe the Scientific Evidence that declares that Macro-evolution IS impossible and does not exist. I choose to believe in the Scientific Evidence which declares to us that Micro-evolution (Natural Selection and Random Mutations) ARE NOT SUFFICIENT enough to create new biological information, new unique genomes, and new unique life forms. By making a choice to believe in the TRUE EVIDENCE or the most recent Scientific Evidence, for me it has been PROVEN beyond a shadow of a doubt that the Theory of Evolution or Creation by Darwinian Chance is false and does not hold water.

TRUTH is knowledge of things as they REALLY ARE. A thing is said to be scientifically accurate if it matches with Reality. It has been suggested to me many times that God knows which concepts are TRUE and which ones are FALSE; therefore, another possible definition for TRUTH is things as God KNOWS them to be. If you can get God to reveal to you what He knows, then you will know THE TRUTH.

People believe in the Biblical God because of the abundant amounts of evidence that He left behind which proves that He exists. If there were no evidence for His existence, then nobody would have ever heard of Him and nobody would even know that He exists. The Biblical God deliberately left eye-witness evidence behind that proves that He exists; but, for you personally, the existence of God will remain unproven until AFTER you have chosen to believe that the evidence which the Biblical God has provided to us is real and true. This is something that only YOU can do for yourself – nobody else can do it for you. YOU are the only one who can PROVE to yourself based upon the evidence that God really does exist.

It doesn't become proven for YOU until long after you have chosen to believe that the evidence is TRUE. If you are not the trusting type, then don't be too surprised if others get there before you do.

Talking About What Chance Can Do for You

Here, I will let the professionals tell you about what Chance can do for you.

BEGINNING OF QUOTE:

> Our next question is crucial. How much influence or effect does chance have on the coin's turning up heads? My answer is categorically, "None whatsoever." I say that emphatically because there is no possibility, real or imagined, that chance can have any influence on the outcome of the coin toss.
>
> Why not? Because chance has no power to do anything. It is cosmically, totally, consummately impotent. Again, I must justify my dogmatism on this point. I say that chance has no power to do anything because it simply is not anything. It has no power because it has no being. I've just ventured into the realm of ontology, into metaphysics, if you please. Chance is not an entity. It is not a thing that has power to affect other things. It is no thing. To be more precise, it is nothing. Nothing cannot do something. Nothing is not. It has no "isness." Chance has no isness. I was technically incorrect even to say that chance is nothing. Better to say that chance is not. What are the chances that chance can do anything? Not a chance. It has no more chance to do something than nothing has to do something. It is precisely at this point that equivocation creeps (or rushes) into the use of the word chance. The shift from a formal probability concept to a real force is usually slipped in by the addition of another seemingly harmless word, by. When we say things happen "by chance," the term by can be heard as a dative of means. Suddenly chance is given instrumental power. It is the means by which things come to pass. This "means" now assumes a certain power to effect change. Something that in reality is nothing now has the ability or power to do something.
>
> Sproul, R. C.; Mathison, Keith (2014-08-12). Not a Chance: God, Science, and the Revolt against Reason (Kindle Locations 195-208). Baker Publishing Group.

END OF QUOTE.

The Darwinists take a non-entity, Chance, and magically transmute it into a being that can act as a causal agent. Something that in reality is nothing suddenly and magically in the hands of the Atheistic Darwinists gains the power and the ability to do everything and create everything. It's MAGIC!

> "When scientists attribute instrumental power to chance, they have left the domain of physics and resorted to magic. Chance is their magic wand to make not only rabbits but entire universes appear out of nothing." (Sproul, R. C.; Mathison, Keith. "Not a Chance". Kindle Locations 229-231.)

In the hands of a Darwinist and Evolutionist, Chance is their magic wand to make entire genomes and entire life forms appear out of nothing. The Theory of Evolution is nothing more than a different version of Creation Ex Nihilo. Creation out of nothing is FALSE no matter where you might encounter it. It's illogical, and not the least bit scientific, because it is magic!

There is NO such thing as creation out of nothing — creation ex nihilo.

The singularity that preceded our big bang and this universe contained within it everything in this universe — ALL time, ALL space, ALL consciousness or Intelligence, ALL spirit matter, ALL light, ALL matter, ALL energy, ALL judgement, ALL mercy, ALL law, ALL potential order, ALL futurity, ALL of YOU, and everything else that would ever exist in this universe. That singularity was SOME THING!

That singularity did NOT originate by Chance. Nothing originates by Chance! Someone put that singularity together, removed all of the entropy from it, and filled it full of potential BEFORE triggering it and creating this universe.

—

The Theory of Evolution Violates the Law of NonContradiction

BEGINNING OF QUOTES:

> To argue that something comes from nothing requires the denial of the law of noncontradiction. The law states simply that A cannot be A and non-A ($\neg$ A) at the same time and in the same relationship. Something can be A and B at the same time but not in the same relationship. I can be a father (A) and a son (B) at the same time, but not in the same relationship. For something to come from nothing it must, in effect, create itself. Self-creation is a logical and rational impossibility. For something to create itself it must be able to transcend Hamlet's dilemma, "To be, or not to be." Hamlet's question assumed sound science. He understood that something (himself) could not both be and not be at the same time and in the same relationship. For something to create itself, it must have the ability to be and not be at the same time and in the same relationship. For something to create itself it must be before it is. This is impossible. It is impossible for solids, liquids, and gasses. It is impossible for atoms and subatomic particles. It is impossible for light and heat. It is impossible for God. Nothing anywhere, anytime, can create itself.

Sproul, R. C.; Mathison, Keith (2014-08-12). Not a Chance: God, Science, and the Revolt against Reason (Kindle Locations 256-264). Baker Publishing Group. Kindle Edition.

—

Chance is not an entity.
Nonentities have no power because they have no being.
To say that something happens or is caused by chance is to suggest attributing instrumental power to nothing.
Something caused by nothing is in effect self-created.
The concept of self-creation is irrational and violates the law of noncontradiction.
To persist in theories of self-creation one must reject logic and rationality.

Sproul, R. C.; Mathison, Keith (2014-08-12). Not a Chance: God, Science, and the Revolt against Reason (Kindle Locations 266-271). Baker Publishing Group. Kindle Edition.

END OF QUOTES.

Over and over again, the Materialists will tell you that God does not exist and that this universe and all the life on this planet has no need for a Creator; and then, they will turn around and tell you that Chance or Evolution designed and created it all.

The Materialistic Darwinists designed the Theory of Evolution to eliminate the need for a Creator; and, the Darwinists openly and boldly say that the Theory of Evolution eliminates any need for a Creator, mocking and ridiculing Creationists in the process. Then these very same Darwinists turn around and invoke Chance and submit Random Mutations as the creative element in the Theory of Evolution. The Atheistic Darwinists tell us that Evolution and Chance created it all, thus making Evolution or Chance the Creator of All things, while at the same time assuring us that there is NO Creator and no such thing as a Creator. This completely violates the Law of Noncontradiction! It IS a blatant contradiction!

The whole Theory of Evolution is based exclusively upon logic fallacies such as these. If the Theory of Evolution is based exclusively upon logic fallacies and falsehoods, then it is by definition in principle FALSE. The Theory of Evolution goes nowhere in the realm of logic and rationality and science, because the whole Theory of Evolution or Creation by Evolution is FALSE to begin with!

Abiogenesis is creation out of nothing, or Creation by Chance. Macro-evolution is Evolution by Abiogenesis or Creation by Chance. Evolution by Abiogenesis and Macro-evolution are Spontaneous Generation. Spontaneous Generation out of nothing, life spontaneously springing from non-life or life creating itself, is Macro-evolution; and, Macro-evolution violates the Law of Noncontradiction.

Something cannot be created out of nothing, because that also violates the Law of Noncontradiction! Nothing cannot make itself into something! Chance cannot design and create anything! It's logical. It's science. Science is the search for suitable, logical, and sufficient causes. Chance is not a suitable, logical, or sufficient cause for anything! Chance intrinsically violates the Law of Noncontradiction by introducing an unlimited number of contradictions into any subject, equation, or concept. Don't believe me? Then try getting Chance to design and create something for you and see how long you have to wait before Chance gets the job done.

Invoking Chance as a Creator and a Causal Agent produces an infinite number of logic fallacies and contradictions into your life and existence, and completely destroys science, rationality, and logic in the process. Yet, the Darwinists invoke Chance as the creative "entity" in the Theory of Evolution. The Darwinists deify Chance and imbue Chance with creative God-like powers in order to eliminate the need for a Creator and a God. That's the very definition of contradiction!

Employing Chance as the creator of this universe and as the creator of all the life on this planet violates the Law of Noncontradiction, and it rejects logic and scientific rationality in the process. To persist in Evolution by Abiogenesis, Evolution by Spontaneous Generation, or Macro-evolution one MUST reject logic and rationality!

Without any backing from Logic and Rationality and Science, the Theory of Evolution devolves into nothing more than a philosophical assumption.

—

Evolution relies upon CHANCE to do its MAGIC.

Chance is synonymous with Magic.

Since the Theory of Evolution relies exclusively upon Chance to do ALL of its creation, the Theory of Evolution is logically FALSE, because Chance cannot design and create anything! The Theory of Evolution has no foundation upon which to build. Chance is not a person or an entity. Chance is NOT a causal agent of any kind. Chance cannot do any of the things that the Atheistic Darwinists say that it does, because Chance does not in reality exist. Try getting Chance to build you a house, a computer, a car, a bridge, a genome, or a road. How long will you have to wait until Chance builds one of these things for you? You can wait for all eternity, and it will NEVER happen!

It's only logical. It's science. Science is the search for sufficient and adequate causes. Chance is NOT a sufficient reliable cause for anything! You can test it in real time and see that it is so. Try getting Chance to create something for you, anything, and see how well that goes for you. Try getting Chance to do something for you or to answer your prayers.

Since Chance is NOT a suitable or sufficient replacement for the Biblical God, God must of necessity exist in order to do ALL of the different things that Chance could never have done.

But, can you get an Atheistic Darwinist to understand and believe any of this?

Not a chance!

Each Scientific Discipline IS Proof of God's Existence

EACH Scientific Discipline is an example of God hiding in plain sight where nobody can see Him and nobody can find Him, unless we go looking for Him.

Wrap your mind around that reality, and suddenly an avalanche of truth will come rushing into your life!

I discovered that EACH Scientific Discipline is a witness or a testament of God's existence; but, with EACH Scientific Discipline, you have to become open-minded enough, smart enough, and educated enough to be able to see how God fits into that particular picture or discipline. It can take a while to come up to speed; and, you will NEVER see it or understand it if you don't want to see it or understand it. When I was an Atheist, I couldn't see any of this or understand any of this, because I didn't want to. A desire to see THE TRUTH and to find THE TRUTH, a desire to see God and find God, is the most powerful and useful gift that you can give to yourself, because it will open your eyes to vistas and realities that you have never considered before.

In my case, I was unlucky because I was a skeptic and an Atheist, but I was lucky in the fact that I am multi-disciplinary. I have education and experience in almost every Scientific Discipline. Once I was willing to open my mind and see, I slowly began to realize that EACH Scientific Discipline IS a proof of God's existence. God gave us Science so that we could prove to ourselves that He exists. That realization has changed my life for the better! I once was blind, but now I see!

The question, "Who Created God", is a smoke screen that the Atheists and Darwinists put up in order to hide their ignorance and hide the REAL Scientific Questions!

God does NOT need a creator, because God has always existed.

The Gods ARE the self-existing Ones, or the self-sustaining Ones.

In principle, by definition, the Gods have always existed; and thus, the Gods need NO creator.

God must exist, or you would not exist!

A God created the first physical body, and then inhabited it. There is NO other way for a genome and a physical body to come into existence! That first physical body didn't just pop into existence out of nothing! It had to have a Creator and it had to be designed and created. Physical objects need to be created or organized by intelligent spiritual beings, in order to take form.

EVERY living organism either has a Creator or Parents! That's what Science teaches us!

You have to have a Creator, or new physical things cannot come into existence. That's logical common-sense Science. Darwinists do a little magic trick by eliminating God as the Creator and putting Random Chance or Darwinian Chance in

His place as the creator of all things. The Darwinists lie and cheat in order to make their case for the Theory of Evolution.

Are you truly willing to believe what Science is telling you?

Without God, nothing would exist. That's what Science is telling us!

Resurrection from the Dead, Immortality, and Eternal Life are ONLY possible if God actually exists! Revelations from God and Revelations of God are ONLY possible if God exists. Your physical existence is ONLY possible if God exists. God must reveal Himself or remain forever concealed. God has revealed Himself to us throughout ALL the various Physical Sciences and Philosophical Sciences!

Those who have seen God know that He exists. Science is ALL about making personal observations. Observation — seeing and experiencing and understanding THE EVIDENCE — is what Science is all about!

I find the scientific proofs of God's existence a lot more convincing than the philosophical proofs of God's existence; and, despite what the Atheists and Skeptics will try to tell you, there are some very convincing proofs of God that come to us from scientific research and scientific evidence. It was a patent-holding scientist and a few other PhD scientists who convinced me that God really does exist — mostly by using Science itself to do so. Think about this for a while!

The very fact that we exist, you and I, is proof of God's existence.

There has been an infinite eternity that has gone on before us, before we ever arrived here on this earth.

According to the Law of Entropy, EVERYTHING should have burned out an eternity ago, and we should be sitting at heat death right now. Yet, here we are.

This reality is logical and empirical PROOF that there is something or someone out there who actually knows how to reverse entropy and has actually reversed entropy in order to produce our universe here, in the first place. We exist here and now, because something or someone out there KNOWS how to reverse entropy!

GOD MUST EXIST, OR YOU WOULD NOT EXIST!

Even the Theory of Evolution IS a proof of God's existence!

The fact that there is NO such thing as Macro-evolution, Evolution by Spontaneous Generation, or Evolution by Abiogenesis IS proof of God's existence! EVERY living organism in this Physical Realm needed a Creator or Parents! God must exist in order to have done all of the Science that evolution could never have done.

The fact that random mutations and natural selection CANNOT produce new genetic programming, new biological information, new biological designs, new genomes, and new life forms IS a proof of God's existence!

Why?

It's because ONLY a God could do these things for real!

The insufficiency and inability of evolution, natural selection, and random mutations to create new life forms or any life form IS a proof of God's existence!

Without God, nothing would exist.

Science properly understood IS a proof of God's existence!

This is a new and recent discovery for me. My Atheism, my unbelief, and my lack of knowledge prevented me from seeing and understanding that EACH Scientific Discipline is a proof of God's existence. It took me decades to start to see it and understand it; and even then, it still took additional years of study and prayer before I actually started to believe it to be true.

For me, this has been a hard-won victory. It didn't come easily; and, there were years when I didn't want it to come.

The fact that we can't get our hands on Darwinian Chance or Darwinian Evolution and slit its throat IS scientific proof enough that the thing isn't real or true. Think about it! The fact that we can't get a hold of Darwinian Evolution and crucify it to death IS scientific proof that the thing doesn't exist in the first place.

In contrast, they did indeed get their hands on the Designer and Creator of this universe, the Biblical God Jesus Christ; and, they did indeed crucify Him on a cross. It doesn't get more real or true than that!

The Darwinian claim that some kind of Chance designed and created it all BROKE the Theory of Evolution for me. That claim is so illogical and so unscientific that I can't stomach it anymore. I can't go back to my ignorance of this subject. I can't put that genie back into the bottle, nor do I want to.

If you don't want to understand these things, then you NEVER will. That's just the reality of the situation. God is easily and successfully hidden from those of us who don't want to find Him or know Him. God can hide in plain sight, and those of us who don't want to see Him will NEVER see Him.

Self-deception works, and it works every time! God set this whole thing up so that you actually have to work, and work very hard, in order to find Him and know that He exists. Knowledge comes while searching for it; and, there is a lot of hidden knowledge out there waiting for us to find it!

For some of us, it can take a lifetime to see, understand, discover, and accept what Science is really trying to tell us — that God exists.

The purpose of the Scientific Method is to help us to find THE TRUTH, through a preponderance of the evidence.

the Scientific Method has no value to us if we use it to convince ourselves that a LIE is TRUE, as the Materialists and Darwinists always seem to do.

That's what I discovered during my Pursuit of the True Reality of All Things, and during my usage of the Scientific Method.

My Scientific Theory

"A little science estranges a man from God. A lot of science brings him back." — ATTRIBUTED to Francis Bacon, but not written by Francis Bacon.

Science itself became the Main Cure for My Atheism!

I'm a scientist, and the Scientific Evidence eventually started to speak for itself.

My Scientific Theory: Once we arrive at THE TRUTHS hidden within ANY Scientific Discipline, then that Scientific Discipline suddenly becomes a proof of God's existence.

You have to get down into THE TRUTH of a Scientific Discipline, before it will start to testify of God's existence; but, once you do, suddenly you find yourself staring into the face of God.

That has happened to me a few times already with multiple different Scientific Disciplines. My Scientific Theory is that it's possible to achieve this with ANY Scientific Discipline. Now I have some work to do in order to determine if My Theory is true.

The Corollary:

God gave us Science so that we could prove to ourselves that He exists.

My Hypothesis:

Each Scientific Discipline IS a Proof of God's Existence!

EACH Scientific Discipline is an example of God hiding in plain sight where nobody can see Him and nobody can find Him, unless we actually go looking for Him.

My great discovery is that EACH Scientific Discipline is a witness or a testament of God's existence; but, with EACH Scientific Discipline, you have to become open-minded enough, smart enough, and educated enough to be able to see how God fits into that particular picture or discipline. It can take a while to come up to speed; and, you will NEVER see it or understand it if you don't want to see it or understand it.

When I was an Atheist, I couldn't see any of this or understand any of this, because I didn't want to. A desire to see THE TRUTH and to find THE TRUTH, a desire to see God and find God, is the most powerful and useful gift that you can give to yourself, because it will open your eyes to vistas and realities that you have never considered before.

Science, when properly understood, becomes a proof of God's existence!

Conclusion:

Every Scientific Discipline, when you finally get down to THE TRUTHS hidden within it, becomes a proof of God's existence. God gave us Science so that we could prove to ourselves that He exists. When you finally have THE TRUTH in hand, it will ALL fit together perfectly, and it will ALL point to God. This is My Scientific Theory. Now, I get to spend the rest of my life taking each Scientific Discipline and trying to see if I can prove my theory correct. I have already done so to my satisfaction with over a dozen different Scientific Disciplines, but I still have a lot more to go.

Once we understand what it's really all about, we suddenly realize that it all points to God.

God reveals Himself to us in two separate ways — through Revelation or Scripture, and through Nature or Science.

That's the way it is!

God has to find some way to reveal Himself to us, or He will remain forever concealed and forever unknown.

PART IV — USING THE SCIENTIFIC METHOD TO "PROVE" THAT GOD EXISTS

In my Pursuit of the True Reality of All Things, I had to be willing to follow the evidence, ANY EVIDENCE, wherever it might lead me, BEFORE I was able to find THE TRUTHS that I sought.

Belief in a LIE cannot help us to find THE TRUTH. That's just the True Reality of the situation.

There was a time in my life when I was a Materialist and an Atheist. I realized one day that I didn't like where that road was taking me, so I decided to turn around and go the other way. My life has been getting better and better ever since.

While trying to employ the Scientific Method in my Pursuit of the True Reality of All things, the SCIENCE and the EVIDENCE pointed me to additional Spiritual Sciences and Scriptural Evidence, which I had refused to look at while I was a Materialist and an Atheist.

I had to be willing to look at ALL the Evidence, BEFORE I was able to find THE TRUTHS which I went searching for. We will never find THE TRUTH, if the only thing we are willing to look at, accept, and believe-in is the LIES. That's just the True Reality of the situation.

SCIENCE IS THE PURSUIT OF THE TRUTH; and, we use the Scientific Method in our Pursuit of THE TRUTH, whether those TRUTHS are spiritual in nature or physical in nature. In fact, the very best and most interesting Scientists develop Scientific Methodologies to study the Spiritual Aspects of our PHYSICAL REALITY.

I slowly realized over the past year that the Scientific Method, combined with ELIMINATING the IMPOSSIBLE PREMISES, can in fact be used to demonstrate and even PROVE that God Must Exist.

GOD MUST EXIST in order to have DONE all of the SCIENCE, Manufacturing, Designing, and Creation which needed to be done 3.8 Billion Years ago, that the various Natural Processes could NEVER HAVE DONE at the time. In fact, BEFORE God designed and created the first life form, there were NO "Natural Processes" such as Mutation and Selection that could have done any design and creation.

I discovered that God is the most parsimonious and logical SOLUTION to the Origin of Life question, because SCIENCE, the Scientific Method, and Deductive Reasoning taught me that Natural Processes of any kind cannot Design and Create anything at all. ONLY Intelligent Beings of some kind have the inherent innate ability to Design, Create, and Manufacture physical objects from scratch.

The rocks that were around 3.8 billion years ago on this earth couldn't have designed and created the first functional information-rich genome and the first life form. That's logical common sense. I eliminated the rocks, water, elements, and sunshine as our Designer and Creator, because such things cannot reliably design and create. They have no heart, hands, mind, intelligence, spirit, or soul.

Creation by Chance or Creation by Evolution IS the same exact thing as Creation by Rocks. The Theory of Evolution IS Creation by Evolution, which is the exact same thing as Creation by Rocks or "Creation out of Thin Air".

SCIENCE and the Scientific Method taught me that Natural Processes such as evolution, natural selection, random mutations, and chance occurrences CANNOT deliberately and purposefully and reliably design, program, manufacture, test, inspect, improve, create, implement, and deploy new functional genomes and new unique life forms from scratch.

ONLY Intelligent Living Beings CAN deliberately and knowingly design, manufacture, and create! Natural Processes can't.

ONLY Intelligent People CAN DO SCIENCE! Natural Processes can't.

That's what I discovered during my Pursuit of the True Reality of All Things; and, I used the Scientific Method, Abductive Reasoning, and Deductive Reasoning in order to do so. I eliminated the IMPOSSIBLE, so that I was left with THE TRUTH.

The purpose of the Scientific Method is to help us to find THE TRUTH, through a preponderance of the evidence.

the Scientific Method has no value to us if we use it to convince ourselves that a LIE is TRUE, as the Materialists and Darwinists always seem to do.

That's what I discovered during my Pursuit of the True Reality of All Things, and during my usage of the Scientific Method.

My Method for Detecting THE TRUTH

He might only be a fictional character, but I followed Sherlock Holmes' Model for detecting THE TRUTH.

How often have I said to you that when you have eliminated the impossible, whatever remains, *however improbable*, must be THE TRUTH? — Sherlock Holmes

I eliminated the FALSE PREMISES and the IMPOSSIBLE PREMISES such as Creation by Chance, Creation by Rocks, Creation by Natural Processes, Creation by Mutation and Selection, Creation by Abiogenesis, Creation by Macro-evolution, Creation by Spontaneous Generation, Creation by Nothing, Creation by a mythical RNA World, and Creation by Evolution, because these things cannot Design, Create, or Manufacture anything at all.

Once I had eliminated the IMPOSSIBLE and the clearly FALSE, then there was only ONE CONCLUSION or ONE TRUTH that remained, namely Creation by Intelligent Beings.

It's elementary my Dear Reader!

Creation by Intelligent Beings is the ONLY thing that has withstood the TEST of TIME!

I did the best I could to employ the Scientific Method and to document my thoughts and reasoning within the essays of this book, and in my much larger book "The Theory of Evolution Proves that God Exists: Why I Am No Longer an Atheist and Why I No Longer Believe in the Theory of Evolution".

While reading this book, you will have to JUDGE for yourself whether I met my Burden of Proof, or not.

One thing I can say for sure, though, is that I used a Preponderance of the Evidence while choosing the CONCLUSIONS that I wanted to believe in. When it comes to Evidence of Any Kind, whether that Evidence be spiritual, psychological, revelatory, or physical in nature, I no longer have my head in the sand; and, I no longer refuse to "look and learn" like is used to do when I was a Materialist and an Atheist.

We will NEVER find THE TRUTH, if we refuse to look at THE TRUTHS which are currently available to us. Give me Evidence, lots and lots of Evidence! I'm now willing to follow that Evidence wherever it might lead me, even if it leads me to God.

Proof of God by the Impossibility of Creation by Evolution

This is a Scientific Proof of God's existence, assuming of course that you believe that evolution and chance mutations might in fact have something to do with Science.

—

Introducing the Scientific Method

the Scientific Method typically runs through the following sequence in an attempt to arrive at THE TRUTH.
1) Form a **HYPOTHESIS**.
2) Select a SCIENTIFIC METHOD or Scientific Methodology to **TEST** the hypothesis.
3) Run the Science Experiment; and then, **Observe** and **Measure** the **RESULTS**.
4) Find the BEST **INTERPRETATION** or the BEST **EXPLANATION** for the Scientific Data, the Scientific Evidence, the Scientific Observations, and the RESULTS of the Science Experiment.

—

the Scientific Method employs abductive reasoning, wherein the Researcher or Scientist or Detective searches for the BEST explanation for the Scientific Evidence and Other Evidence at hand, seeks for the MOST parsimonious and logical CONCLUSION for the available evidence, and strives for the BEST interpretation, or the BEST inference, or the MOST CORRECT CONCLUSION when it comes to any type of evidence.

the Scientific Method also employs deductive reasoning, wherein the Researcher or Scientist or Detective strives to ELIMINATE ALL of the False and Impossible Premises or Hypotheses in the hope of arriving at the CORRECT CONCLUSION.

Notice that Sherlock Holmes employs both abductive and deductive reasoning in an attempt to arrive at the CORRECT CONCLUSION. You could say that Sherlock Holmes strives to use the Scientific Method in order to deduce and adduce the CORRECT CONCLUSION.

**How often have I said to you that when you have eliminated the impossible, whatever remains, *however improbable*, must be THE TRUTH?
— Sherlock Holmes**

Sherlock Holmes actively eliminates the False and the Impossible PREMISES or HYPOTHESES, so that only the TRUE CONCLUSION will remain. ELIMINATE THE IMPOSSIBLE! That's what Deductive Reasoning and the Scientific Method is supposed to be about — arriving at the CORRECT CONCLUSION by eliminating ALL of the FALSE PREMISES and FALSE HYPOTHESES!

THAT PROCESS IS the Scientific Method!

When in doubt, Sherlock Holmes satisfies himself with the BEST POSSIBLE EXPLANATION, however improbable or unlikable or inconvenient that FINAL CONCLUSION might be. His goal once again is to arrive at the TRUE CONCLUSION by any METHOD that he can get to work for him. That's what Abductive Reasoning and the Scientific Method is supposed to be about — inferring the BEST POSSIBLE EXPLANATION from the available evidence at hand.

Whether we are dealing with abductive reasoning, deductive reasoning, or the Scientific Method, the goal is to arrive at the CORRECT CONCLUSION by any means or METHOD that will produce the results and give us THE TRUTH which we seek.

—

A Definition of Terms

Defining Creation by Evolution: Creation by Evolution is the idea or hypothesis that Evolution can design, program, manufacture, create, and deploy new unique functional genomes and new unique life forms from scratch. Creation by Evolution is the idea or hypothesis that Evolution can design, create, and manufacture anything that it sets its mind to.

Defining Creation by Chance: Creation by Chance is the idea or hypothesis that Chance designed and created this Universe, and that Chance designed, created, and manufactured ALL of the genomes and ALL of the life forms on this planet. Creation by Chance is the idea or hypothesis that Chance can design and create anything that it sets its mind to.

Defining Creation by Darwinian Chance: Creation by Darwinian Chance is the idea or hypothesis that Random Chance Mutations and Natural Selection working together can design, program, manufacture, create, and deploy new unique functional genomes and new unique life forms from scratch. Creation by Darwinian Chance is the idea or hypothesis that Mutation/Selection can design and create anything that it sets its mind to.

Defining Abiogenesis: Abiogenesis is the hypothetical process by which life arises naturally from non-living matter. Scientists speculate that life may have arisen as a result of random chemical processes happening to produce self-replicating molecules. Abiogenesis is life from lifelessness. Creation by Abiogenesis is the ideas or hypothesis that random atoms and random molecules can spontaneously put themselves together and come alive on us.

Defining Spontaneous Generation: Spontaneous generation or anomalous generation is the formation of living organisms without descent from similar organisms. Typically, the idea was that certain life forms such as fleas could arise from inanimate matter such as dust, or that maggots could arise from dead flesh. Macro-evolution or Spontaneous Generation IS life arising magically or spontaneously from inanimate matter. Creation by Spontaneous Generation is the idea or hypothesis that inanimate, inert, dead matter has the innate ability to assemble itself and come alive. Spontaneous Generation IS Self-Assembly.

Defining Macro-evolution: Macro-evolution IS Spontaneous Generation, Abiogenesis, or Creation by Chance and Creation by Evolution. The Theory of Evolution IS Creation by Evolution.

The Theory of Evolution: The Theory of Evolution IS Creation by Chance or Creation by Evolution. The Theory of Evolution in its rawest and most pristine form is the claim or the hypothesis that Evolution can design, program, manufacture, create, and deploy new unique functional genomes and new unique life forms from scratch. The Theory of Evolution is the idea or hypothesis that Evolution can design and create anything that it sets its mind to. The Materialists and many of the Darwinists believe that the Theory of Evolution or Creation by Evolution makes the Biblical God superfluous and unnecessary. In fact, they believe that since the Theory of Evolution is axiomatically true, the Theory of Evolution does in fact prove that God does not exist. The creators of the Theory of Evolution intended for it to be a complete replacement for the Biblical God and intended for Creation by Evolution to be proof that God does not exist.

—

Examining These Scientific Hypotheses

I have had a couple of skeptics ask me, "What does Spontaneous Generation have to do with evolution?" They were completely clueless. I have had Darwinists give Macro-evolution a FALSE definition in order to distance it from Spontaneous Generation and Abiogenesis.

Macro-evolution IS Evolution by Abiogenesis. Spontaneous Generation IS Evolution by Abiogenesis. Evolution by Spontaneous Generation IS Macro-evolution! Macro-evolution is Darwinian "goo-to-you evolution". Evolution by Abiogenesis is Darwinian molecules-to-man evolution. Creation by Darwinian Chance is Macro-evolution. Macro-evolution is also one species magically giving birth to a completely different species. Macro-evolution IS Creation by Evolution.

I have given Macro-evolution another name which I use quite often in my essays, namely "Creation by Darwinian Chance" or simply "Darwinian Chance". Calling Macro-evolution "Chance" or "Blind Luck" places it into its proper category and is quite useful, because we all KNOW from observation and common-sense logic that Chance of any kind cannot design and create anything at all.

The Darwinists and Materialists employ a CATEGORY ERROR, a severe logic fallacy, each time that they make the claim that Evolution or Chance can design and create anything that it sets its mind to.

The Darwinists and Materialists deliberately, but erroneously, place Evolution and Chance into the CATEGORY OF PEOPLE, who in fact have the innate inherent ability to design and create anything and everything that they set their minds to. This action on the part of the Materialists and Darwinists is a FATAL Category Mistake. It BROKE the Theory of Evolution for me, once I fully understood it.

When it comes to the Scientific Method, Creation by Evolution and Creation by Chance FAIL at STEP 2, the TESTING PHASE of the Scientific Method, because there is no way to TEST and OBSERVE evolution and chance IN THE ACT of designing, creating, and manufacturing new things from scratch!

SCIENCE and the Scientific Method cannot be used to TEST and OBSERVE Macro-evolution of any kind in ACTION designing and creating new functional genomes and new unique life forms. Macro-Evolution FAILS the Scientific Method, because Creation by Evolution is UNTESTABLE, UNMEASURABLE, and NON-SENSICAL. If something FAILS the Scientific Method, then it is by definition in principle FALSE. Creation by Macro-evolution IS FALSE, because Macro-evolution of any kind cannot design and create anything at all.

Step 2 of the Scientific Method is to RUN THE TEST or DO THE SCIENCE EXPERIMENT. There's no way to TEST Creation by Evolution and Creation by Chance. There's NO WAY to TEST Creation by Natural Processes either, because the very act of trying to make "Natural Processes" design and create make it so that we are no longer OBSERVING Natural Processes in action but are in fact OBSERVING man-made intelligently designed Science Experiments instead; and, we already KNOW that Intelligent Beings can design and create.

Evolution and Chance have NEVER BEEN OBSERVED in ACTION, designing, manufacturing, and creating new living life forms or new physical objects from scratch. Evolution and Chance and Natural Mindless Processes cannot be OBSERVED in ACTION deliberately designing and creating and manufacturing, because these things are NOT people and are NOT Intelligent Beings.

Step 3 of the Scientific Method is to OBSERVE and MEASURE the RESULTS of the Science Experiment. There is no way to do so when it comes to Creation by Evolution and Creation by Chance because they DO NOT PRODUCE RESULTS for us to measure!

Step 4 of the Scientific Method is to infer or adduce the BEST EXPLANATION and/or to deduce the CORRECT CONCLUSION for the Scientific Evidence or the SCIENTIFIC RESULTS that we have on hand.

Since there is no way possible to TEST Creation by Evolution and Creation by Chance and make them produce RESULTS, Creation by Evolution or Creation by Chance is the WORST possible EXPLANATION or INTERPRETATION for any kind of evidence, including Scientific Evidence. This means that Creation by Evolution or Creation by Chance is ALWAYS the WORST possible CONCLUSION that can be given to Scientific Evidence or evidence of any kind. Creation by Evolution and Creation by Chance ALWAYS FAILS to provide an interpretation or explanation that actually makes SENSE logically and scientifically! Creation by Evolution or Creation by Chance is ALWAYS the WORST POSSIBLE INFERENCE for the available evidence at hand.

Yet, the Materialists and Darwinists assure us every single day of our lives that Evolution and Chance can design and create anything that they set their minds to.

Running a Few Science Experiments

Evolution by Spontaneous Generation, Macro-evolution, and/or Evolution by Abiogenesis would be proven to exist for real if your spaghetti dinner were to suddenly come alive on you, sprout wings, and fly away quoting Shakespeare as it goes. That's what we would be witnessing every day of our lives, if Evolution by Abiogenesis were true and happened in the real world. Run the Science Experiment right now and try to get your lunch or your dinner to come alive on you. How's that going for you? You would become instantly famous around the world, if you could replicate such a feat on demand!

For another example, take a single living cell, poke a hole in it draining out its contents and letting it die; and then, wait for that cell to reassemble itself back into a living cell. You can wait for all eternity and that cell will NEVER reassemble itself, even though the contents of that cell are still there exactly where you killed it dead. Why? It's because Creation by Abiogenesis is FALSE – never happened, and it never will. Run the Science Experiment right now! Go step on that snail or step on that spider, and then demand that it reassemble itself and come back alive on you. How's that going for you? You would become instantly famous throughout the world if you could replicate such a feat on demand!

Take Richard Dawkins, run him through a blender, and then then empty all the contents into a tub; then wait for those contents to reassemble into Richard Dawkins. How long will you have to wait until that tub full of Dawkins reassembles itself into a living breathing Richard Dawkins? It will NEVER happen! That's the true nature of Evolution by Abiogenesis or Macro-evolution — it will NEVER happen, which also means that it NEVER happened. Run the Science Experiment right now! Put a peach or a strawberry or a banana into a blender and grind it all up to a pulp. Now, command the thing to reassemble itself into a peach, or strawberry, or banana. How is that going for you? Trust me, you would become instantly famous worldwide if you could replicate that particular feat on demand anytime that you wanted to do so.

—

Premises, Observations, Comparisons, and Scientific Evidence

1) There are standardized Rules of Science which Scientists are expected to follow and obey, while they are using the Scientific Method to do science.

Creation by Evolution or Creation by Chance VIOLATES all of these Rules of Science, making Creation by Evolution or the Theory of Evolution UNSCIENTIFIC. In contrast, Creation by Intelligent Beings COMPLIES WITH ALL of the Rules of Science and FULFILLS ALL of the requirements of the Scientific Method!

One of them is REAL SCIENCE, and one of them is fake pseudo-science.

The Darwinists and Materialists tell us that Creation by Evolution is real and true, and that Creation by Intelligent Beings is pseudo-science or a false science.

the Scientific Method, the Scientific Observations, and the Scientific Evidence tell us that the Darwinists are wrong and that the Darwinists are in fact lying to us and lying to themselves.

Creation by Evolution and Creation by Chance FAIL to meet their Burden of Proof EVERY TIME they are called upon to do so. When it comes to the design, manufacture, and creation of fully functional genomes, evolution and chance DO NOT WORK!

In contrast, Design and Creation by Intelligent Beings IS Demonstrable and PROVABLE on Demand ANYTIME we want to run that particular Science Experiment.

Creation by Evolution or Creation by Chance ALWAYS PROVES TO BE the FALSE PREMISE or the FALSE HYPOTHESIS, because evolution and chance cannot create.

In contrast, Creation by Intelligent Beings ALWAYS COMES THROUGH for us, because Creation by Intelligent Beings is REAL and TRUE and because Intelligent Beings can create. Creation by Intelligent Beings ALWAYS PROVES TO BE the CORRECT PREMISE and ALWAYS ends up being THE CORRECT CONCLUSION as well. Creation by Intelligent Beings is Demonstrable, Provable, and Replicable on Demand. Creation by Intelligent Beings is REAL and TRUE.

2) Evolution by Abiogenesis, Evolution by Spontaneous Generation, Self-assembly, and Macro-evolution have NEVER been OBSERVED in action anywhere on this planet. They have NEVER happened, and they will NEVER happen. There's no way for any of us to make them happen, and no way for us to observe them in action.

Louis Pasteur proved Spontaneous Generation and Macro-evolution FALSE in 1859, the very same year that Charles Darwin published, "On the Origin of Species" and started the evolution revolution. The Theory of Evolution was FALSE to begin with!

The first arm of the Theory of Evolution, Macro-evolution or Spontaneous Generation or Abiogenesis or Creation by Chance, is demonstrably and obviously FALSE. the Scientific Method has been employed in many different ways to FALSIFY Creation by Evolution and Creation by Chance. We have NO evidence and can have NO evidence that will ever support Creation by Evolution, because Evolution cannot design and create anything at all.

In contrast, we EACH PROVE to ourselves EVERY DAY through OBSERVATION, by the Scientific Method, and through common sense LOGIC that Creation by Intelligent Beings is REAL and TRUE. We have seen Creation by Intelligent Beings in ACTION!

Creation by Evolution or the Theory of Evolution IS ALWAYS the False Premise or the False Hypothesis.

In contrast, Creation by Intelligent Beings IS the TRUE PREMISE and the CORRECT CONCLUSION as well, because Creation by Intelligent Beings is REAL and TRULLY HAPPENS FOR REAL!

3) Creation by Chance is scientifically and logically IMPOSSIBLE! Yet, the Darwinists and Materialists assure us that Chance or Random Mutations designed and created ALL of the genomes and ALL of the life forms on this planet.

The Materialists and Darwinists are WRONG.

Chance has NEVER been caught in the ACT of doing anything! Chance isn't a person, and Chance is a non-existent non-entity. Chance cannot be relied up to do anything for us.

In contrast, Intelligent Beings have been caught in the ACT of designing, creating, and manufacturing new, unique, and interesting physical objects.

The Materialists tell us the Chance created this universe and all of the life forms on this planet; but, the Materialists are wrong, because Chance cannot design, manufacture, or create anything at all. Creation by Chance is an oxymoron.

There is NO way to employ the Scientific Method in an attempt to TEST Chance in the ACT of manufacture, design, and creation in order to OBSERVE and MEASURE how well Chance does when it comes time for Chance to design and create something new and unique.

Creation by Chance, or Creation by Darwinian Chance, or Creation by Random Mutations or the Theory of Evolution IS ALWAYS the False Premise or the False Hypothesis, because Chance cannot design and create.

In contrast, Creation by Intelligent Beings IS the TRUE PREMISE and the CORRECT CONCLUSION as well, because Creation by Intelligent Beings is REAL and TRULLY HAPPENS FOR REAL!

4) The Abiogenesis or Self-Assembly of new proteins and new genes cannot be replicated or reproduced in any science lab, because a collection of random molecules NEVER self-assemble into anything living and complex, because they can't. Molecules like amino acids and nucleic acids need coaxing from Intelligent Engineers in order to come together into functional Proteins and functional information-rich DNA.

Once again, Creation by Evolution or Creation by Chance FAILS to be the correct hypothesis and FAILS to be a True Premise.

In contrast, Intelligent Beings can ALWAYS be depended upon to design, create, and manufacture ANYTHING that they set their minds to.

5) Michael Denton in his science book, "Evolution: A Theory in Crisis", demonstrated convincingly and conclusively that there is NO such thing as Evolution by Abiogenesis, Macro-evolution, or Evolution by Spontaneous

Generation. I enter this Science book into evidence as proof that Macro-Evolution is IMPOSSIBLE, UNSCIENTIFIC, and FALSE.

Macro-evolution of ANY kind IS Creation by Evolution or Creation by Chance!

Creation by Evolution IS demonstrably, logically, rationally, scientifically, empirically, and observationally FALSE. There's NOTHING that the Materialists and Darwinists can do to make evolution and chance design and create!

the Scientific Method has been used in thousands of different ways to PROVE that Creation by Evolution and Creation by Chance ARE FALSE. There is NO way to use the Scientific Method to demonstrate that Creation by Chance or Creation by Evolution is real and truly happens for real.

In contrast, the Scientific Method can be and has been used an infinite number of times to demonstrate and PROVE that Creation by Intelligent Beings is REAL and TRUE.

6) Mutation/Selection cannot design, creation, manufacture, or DO SCIENCE!

Here I now enter into evidence the following books as SCIENTIFIC PROOF that Random Mutations and Natural Selection cannot design, create, or manufacture anything from scratch.

I now enter into evidence the books, "Genetic Entropy 4th Edition" by John Sanford, "Evolution: A Theory in Crisis" by Michael Denton, "Biological Information: New Perspectives" by Robert J Marks II (Author, Editor), John C Sanford (Author, Editor), Michael J Behe (Editor), William A Dembski (Editor), Bruce L Gordon (Editor), "Biological Information - New Perspectives A Synopsis and Limited Commentary" by Dr. John Sanford, "In the Beginning Was Information" by Werner Gitt, "Without Excuse" by Werner Gitt, and "Being as Communion: A Metaphysics of Information" by Professor William A. Dembski.

Each one of these books proves clearly and convincingly that Mutation and Selection working together CANNOT design, program, engineer, build, and create ANYTHING. Biological Information needs some kind of intelligence to design it, program it, engineer it, and create it. Biological Information and genomes require and demand creativity, and Evolution of any kind CANNOT CREATE!

Evolution is blind, without hands, and without a mind. Anything that has no hands and no mind and no intelligence — like a rock — will be incapable of designing and creating anything. Creation by Evolution is the same thing as Creation by Rocks. This is common sense logic. This is deduction and the process of elimination! This is SCIENCE and the Scientific Method in action!

the Scientific Method was designed and created to ELIMINATE the False Premises or the False Hypotheses; and, Creation by Chance or Creation by Evolution should be the first hypothesis send down the crapper, but the fact that we can't do so explains in clear detail why evolution and chance cannot design and create.

The goal of SCIENCE is to remove ALL of the FALSE PREMISES so that we are left with THE TRUTH. Creation by Evolution is both a False Premise and a False Conclusion; and, the Theory of Evolution IS Creation by Evolution.

The ONLY reason to believe in the Theory of Evolution is because you want to believe in it.

God must design, program, create, manufacture, and deploy ALL of the genomes and ALL of the life forms BEFORE Random Mutations and Natural Selection can start to do their thing.

Therefore, the second arm of the Theory of Evolution, Creation by Mutation and Selection also known as Micro-evolution, IS also FALSE and INSUFFICIENT to the task of designing, creating, and manufacturing new functional genomes and new unique life forms from scratch. ONLY Intelligent Living Beings, such as God, would in fact be capable of doing such a thing!

Once again, Creation by Evolution or Creation by Mutation/Selection FAILS to meet its Burden of Proof.

In contrast, we have ALREADY PROVEN to ourselves that Intelligent Beings or Intelligent People CAN design, program, manufacture, create, and deploy anything that they set their minds to. As a race, we human beings have within us the inherent POTENTIAL to design and create and manufacture even genomes and life forms, if we were willing to dedicate enough time and resources towards doing so.

7) Mathematicians and Scientists have run the numbers and some of them have concluded that it could take trillions of trillions of years for a single useful functional protein to self-assemble by natural random means. Those are not good odds. Why is this so hard? It's because ALL of the amino acids which go into that protein have to self-assemble first, before they can all magically get together and self-assemble into that single protein. The odds of this happening by random chance alone are like "infinity to one" against it ever happening, meaning that it can NEVER happen. The same exact odds apply to a bunch of DNA molecules putting themselves together and then magically assembling themselves into a usable functional gene. It can't happen because it's statistically impossible. It requires an Intelligent Living Being to put functional and useful programming code into the DNA. Useful functional programs, such as genomes, NEVER write or program themselves.

Once again, Creation by Evolution or Creation by Chance is PROVEN by the Scientific Method to be FALSE.

According to the SCIENCE and the Scientific Method only Intelligent Beings can design, create, and manufacture anything that they set their minds to.

8) SCIENCE, the Scientific Method, personal observation, common sense, and mathematics have proven conclusively and convincingly that there is NO such thing as Evolution by Abiogenesis, Evolution by Spontaneous Generation, and Macro-evolution. Abiogenesis has NEVER been observed in the wild. Macro-evolution has

NEVER happened, and it can NEVER happen. Macro-evolution or Abiogenesis or Darwinian Chance cannot design and create life.

Creation by Chance or Creation by Evolution, also known as The Theory of Evolution, ALWAYS PROVES TO BE the False Premise or the False Hypothesis or the False Explanation that is given to Scientific Evidence by the Darwinists.

In contrast, Creation by Intelligent Beings ALWAYS COMES THROUGH for us, because Intelligent Beings or Intelligent People can in fact design, create, and manufacture ANYTHING that they can successfully visualize designing, creating, and manufacturing.

One of these ALWAYS VIOLATES the Scientific Method! And, one of these IS ALWAYS IN COMPLETE COMPLIANCE with the Scientific Method and SCIENTIFIC OBSERVATIONS.

The Darwinists assure us that Creation by Evolution is real and true. The Materialists assure us that Creation by Chance is real and true. The Darwinists and Materialists also tell us that Creation by Intelligent Beings is pseudo-science and is FALSE.

You will have to decide for yourself if the Darwinists and Materialists are telling you THE TRUTH. You know where I stand on the subject, but my stand on the subject has nothing to do with you.

9) The Darwinists and Materialists will complain about these points and tell you that I am lying, and that I am distorting the Scientific Method and THE TRUTH.

The Darwinists and Materialists will tell you that I know nothing about SCIENCE and the Scientific Method, but that they do. The Darwinists will tell you that they are the experts on evolution and that I am not, and that I cannot be.

The Darwinists and Materialists ALWAYS TELL US that Creation by Intelligent Beings is pseudo-science and is FALSE. The Darwinists and Materialists ALWAYS ASSURE US that Creation by Chance and Creation by Evolution are real and true.

The Darwinists and Materialists ARE ALWAYS WRONG.

The Materialists, Naturalists, Atheists, and most of the Darwinists TELL US that God does NOT exist; but, can the Materialists and the Darwinists be trusted to tell us THE TRUTH? What can they know about God, when they refuse to look at the evidence that supports the existence of the Biblical God? What makes the Materialists and the Atheists the experts on God?

I think most of us KNOW where the Fault Lies and Who Lies.

10) We have successfully used the Scientific Method, Abductive Reasoning, and Deductive Reasoning to DEMONSTRATE that Creation by Evolution and Creation by Chance, also known as the Theory of Evolution, IS FALSE and why it IS FALSE.

We have also used the Scientific Method, Abductive Reasoning, and Deductive Reasoning TO PROVE beyond a shadow of a doubt through a preponderance of the evidence that design, creation, and manufacture by Intelligent Beings or Living Persons IS REAL and TRUE.

—

THE CONCLUSION

Creation by Evolution or Creation by Chance, also known as The Theory of Evolution, IS ALWAYS a False Hypothesis or a False Premise, making this thing a FALSE CONCLUSION or a FALSE INTERPRETATION as well. This IS what the Scientific Method is telling us! the Scientific Method was designed for the express purpose of eliminating ALL of the False Premises or False Hypotheses.

In contrast, Creation by Intelligent Beings or Creation by Intelligent People WILL ALWAYS make for a Good Premise, a Good Hypothesis, GOOD SCIENCE, and a GOOD CONCLUSION also. This is what SCIENTIFIC OBSERVATION and the Scientific Method are telling us!

Since Creation by Evolution and Creation by Chance ARE demonstrably, scientifically, logically, arguably, and obviously IMPOSSIBLE and FALSE, God must of necessity exist in order to have done ALL of the Science, Design, Manufacturing, and Creation which needed to be done 3.8 billion years ago when the bacteria came on the scene, 542 million years ago during the Cambrian Explosion, and in our recent era when humanoids, apes, and men came onto the scene or were planted onto this earth by the Gods.

We have successfully used SCIENCE and the SCIENTIFID METHOD to eliminate the IMPOSSIBLE PREMISES so that we are left staring at the ONLY POSSIBLE CORRECT CONCLUSION — namely, that God must exist in order to have DONE ALL of the SCIENCE and CREATION that has needed to be done during this earth's history BEFORE man arrived on the scene.

—

Reviewing Our CONSLUSION

Creation by Evolution or Creation by Chance IS IMPOSSIBLE and FALSE.

Creation by Intelligent Beings is REAL and TRUE.

God is a person. God is an Intelligent Being.

Since there is no such thing as Evolution by Abiogenesis, Evolution by Spontaneous Generation, Self-assembly, and Macro-evolution, it is scientifically obvious that Abiogenesis could not and did not assemble the first gene, the first protein, the first living cell, and the first life form.

Since Evolution by Abiogenesis and Macro-evolution and Darwinian Chance do not exist for real and could not have designed and created the first life forms in this universe, the only other thing that makes LOGICAL SENSE is that God designed and created the first DNA strand, the first proteins, the first genes, the first genome, the first living cell, and the first complex life form.

There is NO other logical explanation for the origin of life, because SCIENCE has demonstrated clearly and conclusively that Macro-evolution, Evolution by Abiogenesis, and Evolution by Spontaneous Generation could NEVER have done the job, because those things don't exist in the first place.

Once you have eliminated Darwinian Chance or Macro-evolution from the equation, whatever remains, however improbable, must be THE TRUTH.

The ONLY PERSON we know of who could have created the first life forms is the Biblical God; therefore, God must of necessity exist in order to do all of the different things that Macro-evolution could NEVER have done. God is a sufficient cause; and, Abiogenesis or Chance is not. Evolution and Chance cannot design, program, engineer, create, build, or deploy anything; so, God must have done it. This is logical common sense. This is deductive reasoning through a process of eliminating the Impossible Premises. This is abductive reasoning through a process of Inferring the Best Explanation for the evidence at hand. This is SCIENCE, because this is the Scientific Method in action.

Abductive Reasoning seeks for the best, most logical, and most parsimonious interpretation or explanation for the available evidence. I also employed deductive reasoning. By eliminating ALL of the FALSE PREMISES, we can deduce the CORRECT CONCLUSION.

The CORRECT CONCLUSION is that evolution of any kind cannot design and create; but, Intelligent Beings or Intelligent People certainly can.

Since evolution and chance and rocks could NEVER have designed, created, and manufactured the genomes and the life forms that needed to be designed and created 3.8 billion years ago, 542 million years ago during the Cambrian Explosion, and 6,000 years ago, God must of necessity exist in order to have done ALL of the design and creation and SCIENCE that needed to be done.

According to the Scientific Method and Deductive Reasoning and Abductive Reasoning, ONLY Intelligent Beings or Living Persons can design, create, and manufacture new physical items like genomes and life forms. The Biblical God IS the ONLY PERSON that we know of who has in fact repeatedly revealed Himself to us and repeatedly CONFESSED to designing, creating, and manufacturing ALL of the genomes and ALL of the life forms on this planet.

Hereby, we have successfully used the Scientific Method and SCIENCE to PROVE to ourselves beyond a shadow of a doubt and through a preponderance of the evidence that God must of necessity exist in order to have done all of the SCIENCE and CREATION that needed to be done BEFORE human beings arrived on the scene and started doing design and creation and SCIENCE of their own.

Unless we are Darwinists, Materialists, and Atheists, THE TRUTH is all we want.

In this particular PROOF OF GOD, we have successfully used the Scientific Method to prove beyond a shadow of a doubt and through a preponderance of the evidence that the Biblical God IS THE ONLY PERSON that we know of who could have designed, manufactured, and created ALL of the genomes and ALL of the life forms on this planet when that particular SCIENCE needed to be done.

In the Bible, the Book of Mormon: Another Testament of Jesus Christ, the Doctrine and Covenants, and the Pearl of Great Price, the Biblical God Jesus Christ has repeatedly REVEALED Himself to us and repeatedly CONFESSED to designing and creating This Universe, This Earth, and ALL of the Genomes and Life Forms on this planet. That's good enough for me!

I found this PROOF OF GOD by the Scientific Method to be sufficient for my needs.

—

You will have to judge for yourself if I know anything about SCIENCE, the Theory of Evolution, and the Scientific Method. The Darwinists and Materialists and Skeptics have accused me of being stupid and ignorant when it comes to these things. You will have to decide for yourself if they are right. That's something I can't do for you. All I can do is to present my case to the best of my ability and then let YOU make up your own mind as to what you want to believe to be true.

I found this PROOF OF GOD convincing and believable. You might not. Such is life.

—

The purpose of the Scientific Method is to help us to find THE TRUTH, through a preponderance of the evidence.

the Scientific Method has no value to us if we use it to convince ourselves that a LIE is TRUE, as the Materialists and Darwinists always seem to do.

That's what I discovered during my Pursuit of the True Reality of All Things, and during my usage of the Scientific Method.

"Adventures Beyond the Body: How to Experience Out-of-Body Travel" by William Buhlman.

This is one of my many Reviews PROMOTING the book "Adventures Beyond the Body" by William Buhlman. This book single-handedly put an end to Materialism for me!

This book is one of my top-ten most favorite books because of all the different things that it taught me about spirit, spirituality, and the non-physical realms. This is one cool book! I have purchased multiple paperback copies for me and my friends, and I also own a Kindle copy. My intent here is to point EVERYONE to this book and tell them to read it, especially if they have any lingering feeling that Materialism or Naturalism might be true.

Spirit or the spiritual has to be experienced first-hand in order to know that it is real and true. There's really no other way, unless you are willing to trust other people's experiences of the spiritual, which I am willing to do. I have learned to trust this person's spiritual experiences, because they fit in extremely well with everything else that I have learned about spirit and spirituality so far.

The Copyright for this book permits "quotations embodied in critical articles and reviews". So, that's what I am going to do here. I'm embodying quotations in this CRITICAL ARTICLE and this REVIEW of his book, "Adventures Beyond the Body" — except it should be noted that I'm NOT being critical of Buhlman and his book, but instead trying to promote it. I USE his books to CRITICIZE the Materialists and the Atheists who mock and ridicule spiritual things in their "critical reviews" online.

Because Buhlman doesn't seem to be promoting any particular Organized Religion, he comes across as an unbiased authority on spirit and out-of-body travel, which some Atheists and Materialists might actually appreciate, since most Atheistic Materialists are deathly afraid of Organized Religion. Buhlman seems to take a secular approach to spirituality, which is really cool if you happen to be someone who is on the fence about all of this.

For PROMOTIONAL purposes and preemptive defense, here are a couple of my most favorite quotes from Buhlman's book which are germane to the topic at hand. Whenever I'm taking ridicule and heat from the Materialists and Atheists online, I love to drop these quotes and this book on them:

> Twenty years ago, I firmly believed that the physical world we see and experience was the only reality. I believed what my eyes told me — life possessed no hidden mysteries, only countless forms of matter living and dying. The facts were clear; there was no evidence or proof of nonphysical worlds or our continued existence after death. I questioned the intelligence of anyone softheaded enough to accept the illogical concepts of heaven, God, and immortality. In my mind these were fairy tales created to comfort the weak and manipulate the masses. For me, life was simple to understand: the world consisted of solid matter and form, and the concepts of life after death and heaven were feeble human attempts to create hope where none existed.

I possessed the arrogant knowledge of a man who judges the world with his physical senses alone. I supported my conclusions with the overwhelming observations provided by science and technology. After all, if something mysterious was there, science would certainly be aware of it.

My firm convictions of reality and life continued until June of 1972. During a conversation with a neighbor, our discussion turned to the possibilities of life after death and the existence of heaven. I proceeded to present my agnostic viewpoints with vigor. To my surprise my neighbor didn't contest my conclusions; instead, he related an experience that he had had several weeks before. One evening just after drifting to sleep, he was shocked to discover himself floating above his body. Completely awake and aware, he became frightened and instantly fell back into his physical body. Excited, he told me it wasn't a dream or his imagination, but a fully conscious experience.

Intrigued by his experience, I decided to investigate this strange phenomenon for myself. After several days of research, I discovered numerous references to out-of-body experiences throughout history. With some searching I found a book on the topic that actually described how out-of-body experiences are induced. The entire subject seemed extremely weird, and I considered the book the result of an overly active imagination.

Out of curiosity, I decided to try one of the out-of-body techniques before sleep. After repeated daily attempts, I began to feel a little ridiculous. In three weeks, the only thing I experienced out of the norm was an increase in my dream recall. I became more and more convinced that this entire subject was nothing more than an intense or vivid dream stimulated by the so-called out-of-body techniques.

Then, one night about eleven o'clock I drifted to sleep during my out-of-body technique and began dreaming that I was sitting at a round table with several people. They all seemed to be asking me questions related to my self-development and state of consciousness. At that moment in the dream I began to feel extremely dizzy, and a strange numbness, like from Novocain, began to spread throughout my body. Unable to keep my head up, I passed out, hitting my head on the table. Instantly I was awake, fully conscious, lying in bed facing the wall. I could hear an unusual buzzing sound and felt somehow different. Extending my arm, I reached for the wall in front of me. I stared in amazement as my hand actually entered the wall; I could feel the vibrational energy of it as if I was touching its very molecular structure. Only then did the overwhelming reality hit me, *My God, I'm not in my body.*

Excited, my only thought was, *It's real. My God, it's real!* Lying in bed, I stared at my hand in disbelief. When I tried clenching my fist, I could feel the pressure of my grip; my hand felt completely solid, but the physical wall in front of me looked and felt like a dense, vaporous material with form.

Determined to stand, I began to move effortlessly to the foot of my bed, my mind racing with the reality of it all. Standing, I quickly touched my arms and legs, checking to see if I was solid, and to my surprise I was completely solid, completely real. But around me, the familiar physical objects in my room no longer appeared completely real or solid; instead, they now looked like three-dimensional mirages. Glancing down, I noticed a large lump in my bed. Amazed, I could see that it was the sleeping form of my physical body silently facing the wall.

As I focused my vision on the opposite side of the room, the wall seemed to fade slowly from view. In front of me I could see a wide, green field extending far beyond my room. Looking around, I noticed a figure silently watching me from about ten yards away. It was a tall man with dark hair, a beard, and a purple robe. Startled by his presence, I became frightened and instantly "snapped back" into my physical body. With a jolt I was in my body, and a strange feeling of numbness and tingling faded as I opened my eyes. Excited, I sat up, my mind exploding with the realization of what had just occurred. I knew it was absolutely real, not a dream or my imagination. My entire ego awareness had been present.

Suddenly, everything I had ever learned about my existence and the world around me had to be reappraised. I had always seriously doubted that anything beyond the physical world existed. Now my entire viewpoint changed. Now I absolutely knew that other worlds do exist and that people like myself must live there. Most important, I now knew that my physical body was just a temporary vehicle for the real me inside, and that with practice I could separate from it at will.

Excited about my discovery, I grabbed a pen and paper and wrote down exactly what had occurred. A flood of questions filled my mind. Why is the vast majority of the human race unaware of this phenomenon? Why aren't the various sciences and religions investigating it? Is it possible that this unseen world is the "heaven" referred to in religious texts? Why isn't our government exploring this apparent parallel energy world? Is it possible that our overwhelming dependence on physical perceptions has led us to overlook an incredible avenue of exploration and discovery?

As the initial shock of my first experience sank in, I realized that my life could never be the same again. The more I pondered the significance of my experience, the more profound I realized it to be. All my agnostic beliefs had been swept away in a single night. I knew that I had to reappraise everything that I had learned since childhood, everything that I had assumed to be true. My comfortable conclusions about science, psychology, religion, and my existence had obviously been based on incomplete information. I felt excited, but also uneasy — my familiar concepts of reality no longer seemed relevant. Increasingly, I felt in a void. On several occasions when I talked to friends about my experience, they found it too bizarre to take seriously. In 1972 the term *out-of-body experience* had not even been coined; back then, the most common description was astral projection. No one that I knew at the time had even heard of astral projection, and if you told people you had left your body, they immediately thought that you were on drugs or losing your mind. I quickly discovered that I had to keep my experiences to myself or face some degree of disbelief and even ridicule.

After my first out-of-body experience, my mind was overflowing with endless possibilities and questions. Desperate for information and guidance, I spent several weeks in libraries and bookstores searching for whatever knowledge was available on the topic. I quickly found that little was available; only a handful of books had been written on the subject, and some of these were decades old and out of print. By the end of July 1972, I realized that I was on my own.

I decided to focus on the one technique that had worked for me before. This technique involved visualizing a physical location that I knew well as I drifted off to sleep. As before, I pictured my mother's living room with as much detail as possible. At first it seemed difficult, but after a few weeks I could picture the room's details with increasing clarity; the furniture, patterns in fabrics, textures, even small imperfections in wood and paint began to be clear in my mind. I realized that the more I pictured myself within the room interacting with the physical objects, the more detailed my visualizations would become. With practice I learned to physically walk around the room and memorize specific items that it contained. I also learned the importance of "feeling" the environment with my mind: the feel of carpet on my feet; the sensation of sitting in a chair, walking, turning on a lamp, or even opening the door. The more detailed and involved I was within my visualization, the more effective were my results. Although it was challenging at first, after a while it became fun to make my visualizations come alive in my mind. At this point I decided to keep a journal to record my out-of-body experiences.

Buhlman, William L., "Adventures Beyond the Body: How to Experience Out-of-Body Travel" (pp. 3-8). HarperCollins. Kindle Edition.

The Materialists, Darwinists, and Atheists can be extremely stupid, ignorant, inexperienced, uneducated, and dense when it comes to the parts of SCIENCE that they have chosen not to believe in. I know, because I used to be one of them.

—

Journal Entry, October 2, 1982

> I hear the buzzing, engine-like sounds and will myself out-of-body. I step to the bedroom door and automatically request "Clarity now!" My vision improves and I step through the door, into the living room. Still feeling a little out of sync, I verbally repeat my request with more emphasis, "Clarity now!" I feel my awareness and vision snap into place.
>
> My thoughts are clear, and I make a verbal demand, "I need to see the form I'm in now!" Instantly I feel an intense sensation of being drawn within myself. I'm suddenly different, weightless as though I'm floating in space. As I look forward I see a sparkling, bluish white form. For some reason, I seem to know that I'm looking at my nonphysical body from a different perspective. I stare in amazement at this form before me that shines and flows with energy and light. It looks like an energy mold created from a million tiny points of light; it radiates a bluish glow but appears to have a defined outer structure. The body of light before me is naked and is identical

to my physical form. Even though my body looks firm, there is a noticeable energy motion and radiation present. I can see what appears to be an ocean of blue stars throughout my body. It's difficult to describe because the stars are stable, yet moving at the same time; the light and energy of my body appear to change and flow almost like the waves of an ocean.

As I stare at the body of light, it hits me that I must be in another body. Yet I can't perceive any form or substance; I'm like a viewpoint in space without shape or form of any kind. As I reflect upon my new state of being, I feel a sensation of rapid motion and I'm instantly back within my physical body.

Lying still and reviewing my experience, I'm struck by an inescapable conclusion: I must possess multiple energy-bodies. The form I just experienced was noticeably lighter (less dense) than even my second nonphysical body. I realize that the traditional view of our possessing two bodies — a physical body and a spiritual body — is far too simplistic; we are much more complex than this. Just as there are multiple nonphysical energy dimensions within the universe, each of us must consist of multiple energy-bodies or vehicles of expression.

Now I seriously wonder just how many nonphysical bodies or forms this involves. I suspect that there must be one within each dimension of the universe and that all of these are interrelated and connected, just as the physical body is connected to its first nonphysical (spiritual) body.

Buhlman, William L., "Adventures Beyond the Body: How to Experience Out-of-Body Travel" (pp. 34-35). HarperCollins. Kindle Edition.

END OF QUOTES.

—

If you want to know more, then go out and buy a copy of William Buhlman's book for yourself.

In my book here, the most I can do is to tell you where to go and look; and, I do make it a point to PROMOTE the books and the people that I found most helpful and useful during my research. Clearly, I have given this whole thing some study and thought. This book from William Buhlman gave me solid irrefutable evidence that Materialism is FALSE. Whenever an Atheist or Materialist is being caustic, mean, and rude online, I love to drop these two quotes on them. They don't know what to do with them except to mock them and dismiss them out-of-hand; or even better, they don't respond at all.

These two quotes from William Buhlman have become an essential part of my defense package whenever I find myself engaging the Borg (Materialists) online.

The Darwinists and Materialists have accused me of carpet bombing the threads that I started online. These two quotes are a couple of bombs that I love to drop on them from time to time!

Take note that William Buhlman has other books for sale, but this one was my favorite one so far.

William Buhlman discovered first-hand that Materialism and Atheism are false.

The only way to KNOW that the spiritual is real is to experience it first-hand for yourself, or to trust someone who has experienced it first-hand. I have had trust issues for most of my life, so it would have been nice to have had some first-hand spiritual experiences of my own like these that Buhlman has had. Unfortunately, most Atheistic Materialists like me block ourselves from pursuing and having spiritual experiences like Buhlman's because we don't believe that such a thing is possible, so we never even try it. Materialism is a curse that ends up being self-confirming and self-reinforcing — believing Materialism to be true makes it true in one's own life.

For most Materialists and Atheists, they need some kind of jump-start or extreme experience to get them into spirituality and pursuing God. But, once you have had some convincing spiritual experiences of your own, then typically you don't want to go back to the limitations and darkness and blindness of Materialism.

It can be a very interesting (and sometimes scary) to experience that internal paradigm shift from Materialism to Spirituality or Spiritualism. It's like a light going on, and then suddenly you just KNOW that Materialism is false and why it is false. When you see the Light, you just know that your whole life before as a Materialist and an Atheist was completely false and a self-deceptive lie. It can be a bit of a shock to the system at first; but then, you find yourself getting very excited about all of the new and interesting things that have suddenly opened up for you.

Once you know that Materialism is false, then you KNOW that Darwinism is false, because Darwinism is a Materialistic philosophy. Once you know that Darwinism is false, then you KNOW that goo-to-you Darwinian evolution is FALSE also because the Theory of Evolution is fruit from the poisoned tree.

When it comes to Science of any kind, Materialism IS the ultimate bane of that Science. Materialism IS the poisoned tree from which all falsehoods are produced. Materialism ALWAYS produces the WORST explanation or interpretation of scientific evidence.

I know that some of this makes the Materialists and Atheists and Darwinists uncomfortable, but who cares? What have the Materialists and Atheists done for us? They take without giving anything of value in return. What good have their lies and self-deceptions done for any of us? What thing of value have they actually contributed to our intellectual and spiritual growth? What thing of value have they done for society? During the 20th century, they exterminated at least 200 million of their opponents in the name of Darwinism (Fascism) and/or some kind of Militant Enforced Atheism (Communism). Materialism has not been good for society — it's too selfish to be of any value to the rest of us. And, Materialism IS BAD SCIENCE!

Creation by Purely Materialistic Processes is scientifically impossible, because Physical Processes require some kind of intelligent being to get them going, coax them or force them in the right direction, and then deliberately weed out the failures while keeping the successes. Scientists are required to DO SCIENCE or to DO MANUFACTURING in order to make Materialistic Processes organize into new and useful things.

Spirit beings or living beings design, program, plan, engineer, create, manufacture, and deploy new things — NOT MATERIALISTIC PROCESSES! There's no brain, no mind, no intelligence, and no hands for manipulation and manufacturing when it comes to purely physical processes or random luck. Spirit or sentience or life designs and creates new things, not random material processes. This IS a Scientific Reality, thus making Materialism scientifically FALSE.

If Materialism is FALSE, then God must of necessity exist in order to do all of the different things and all of the science that needed to be done which "Materialism" or random material processes could never have done. This is logical common sense. This is reasoning, deduction by a process of eliminating False Premises, and even science!

This Scientific Reality can be hard to put into words at times, but random material processes and random chemical reactions cannot design and create anything from scratch. It requires some kind of intelligent mind in order to design and create something new from scratch, and actually make it work. Anytime living breathing Scientists DO SCIENCE in a science lab, then it is NO longer Creation by Evolution or Creation by Natural Processes or Creation by Chance, but it is instead Creation by Intelligent Beings. Intelligent Beings can design and create and manufacture new and unique things — Evolution, Chance, Chemical Reactions, and Mindless Directionless Physical Processes cannot.

Darwinian molecules-to-man evolution IS scientifically impossible. Darwinian ape-to-man evolution IS scientifically impossible. Creation by Chance IS scientifically impossible. Abiogenesis IS scientifically impossible. Creation of new functional genomes through step-by-step Mutation/Selection IS scientifically impossible. Macro-evolution of any kind IS scientifically impossible. Creation of new functional genomes by Micro-evolution or Macro-evolution IS scientifically impossible. Creation by Evolution IS scientifically impossible. Technically, Design and Creation by any kind of Mindless Materialism or any kind of Mindless and Directionless Materialistic Process IS scientifically impossible. My goal here is to eliminate the impossible.

How often have I said to you that when you have eliminated the impossible, whatever remains, *however improbable,* **must be THE TRUTH? — Sherlock Holmes**

When you have eliminated Evolution of every kind from the design and creation process, then what remains? It's elementary my dear Watson.

Follow the evidence!

Keep the best and get rid of all the rest!

My Favorite Near-Death and Out-of-Body Testimonies

P. B. says:

Mark hallucinates: "Any time that you can demonstrate to yourself that spirit exists or that God exists, it reveals Materialism for the lie that it is."

Therefore, Materialism must be true, because nobody has ever objectively proved that spirit exists or that God exists.

—

Says who? You?

P. B., just because YOU haven't had any spiritual experiences doesn't mean that nobody else has. That's YOUR problem, not ours. When it comes to spiritual experiences, YOU have to go and get your own in order for YOU to know that they are real and true. YOU have to objectively experience them for yourself in order to prove to yourself that spirit exists, and that God exists.

When it comes to spiritual experiences, I have had sufficient for my needs; and apparently, you haven't. I have had enough spiritual experiences to KNOW that Materialism, Naturalism, and Darwinism are FALSE.

Remember, it's no longer blind-faith for the people who have seen and touched our resurrected Lord Jesus Christ. In contrast, you will NEVER get Darwinian Chance to make a physical appearance in the flesh.

P. B., anytime you use extreme modifiers such as NOBODY, it automatically makes your statement FALSE, because there are indeed people out there in this world who have experienced the spiritual first-hand, and there are also people who have seen and TOUCHED our resurrected Lord Jesus Christ. Just because you haven't doesn't mean that they haven't.

Instead, what you have done here is reveal to the WORLD how narrow and limited your Materialistic Worldview has made you, which is the exact point that I was trying to make in the first place.

My primary purpose here is to demonstrate that Materialism of any kind is a LIE, because there are people in this world who have experienced the spiritual and experienced God first-hand and KNOW that they are REAL and TRUE.

—

For your benefit, here are a few of my favorite near-death experiences and out-of-body spiritual experiences from YouTube:

Ian McCormack:
https://www.youtube.com/watch?v=HbTAmN4m2lQ

Dr. Mary Neal:
https://www.youtube.com/watch?v=DX473dF7ChY

John Ramirez:
https://www.youtube.com/watch?v=_M_8lI0-7b0

Dr. Richard Eby:
https://www.youtube.com/watch?v=IInRiIi-zQw

Howard Storm:
https://www.youtube.com/watch?v=UPj4wci_bcI

These people have told their story more than once, so after you have found and experienced my favorite version of their story, then you can explore the other versions of their story from the links that will pop up on the side.

—

I will place other favorite spiritual experiences at the following link, as time allows and when I come across them:

http://www.markme.us/forums/forum/favorite-spiritual-experiences/

You can lead a horse to water, but you can't make him drink. You can lead an Atheist to God, but you can't make him read and think.

—

The purpose of the Scientific Method is to help us to find THE TRUTH, through a preponderance of the evidence.

the Scientific Method has no value to us if we use it to convince ourselves that a LIE is TRUE, as the Materialists and Darwinists always seem to do.

That's what I discovered during my Pursuit of the True Reality of All Things, and during my usage of the Scientific Method.

The Inadequacy of Evolution Proves that God Must Exist

The Theory of Evolution proves that God Exists!

Who would've thought it possible?

Certainly not me back when I was an Atheist and in an Atheistic frame of mind.

However, the more I study the details and the "evidence" associated with the Theory of Evolution, the more Evolution itself has proven to me that God must exist in order to do all of the different things that Evolution could NEVER have done. The Darwinists demand that I study "their science" telling me that I need to study some Real Science. The more that I do as they say and study "their science", the more that "their science" proves to me that Evolution couldn't have done ANY of the creative acts which the Darwinists assure us that Evolution did.

In order for the Theory of Evolution and Darwinian goo-to-you evolution to be true, it would have to be able to design and create anything and everything, without any help from any of us including God. But, Evolution can't do so, because it's not alive and it can't create. Evolution has no hands and no mind. It can't design and create anything! Evolution or Change ONLY works on the things which God has already designed and created in the first place.

The Theory of Evolution is its own worst enemy. It's a counter-intuitive and illogical theory that violates the Scientific Method in a wide variety of different ways. The more that I study the Theory of Evolution the clearer it becomes to me that Evolution could never have done any of the miraculous and magical creative acts which the Darwinists say that Evolution did. The more that I study the Darwinian "sciences" or the Darwinian interpretations of the Scientific Evidence, the clearer it becomes to me that Darwinism and the Theory of Evolution are nothing more than pseudo-science masquerading as some kind of real science. Evolution CANNOT do the molecules-to-man evolution which the Darwinists say that it did! That's what studying the Darwinian "sciences" has taught me!

The Darwinists and Evolutionists tell me that Evolution designed and created my eyes and my brain. I don't see how Evolution could have ever done so, because Evolution is NOT a person, personality, living entity, or even a real thing. Evolution has no hands, mind, spirit, or intelligence.

Only a blind and loyal Darwinist will actually believe that Darwinian Chance is real and true, and that chance can truly create everything from scratch.

Who would have thought it possible that the Theory of Evolution could have proven to someone that God must exist? But, that's exactly what happened to me while studying the amazing, miraculous, and impossible claims of Darwinian Evolution.

—

The Battlefield

In this book, I document some of the reasons why I am no longer an Atheist, and why I no longer believe in the Theory of Evolution. THE TRUTH is all that you really want to know, because knowing THE TRUTH sets you free to make an informed decision or choice!

Darwinists and Evolutionists typically use the Theory of Evolution as a proof of God's non-existence.

Their Darwinian Thesis: If the Theory of Evolution is true, then God does NOT exist.

My Counter-Thesis: If the Theory of Evolution is FALSE in any way, then God must exist in order to have done all of the different things that Evolution could NEVER have done.

This book uses Common Sense and their "Darwinian Science" against them and ends up being a Scientific Proof of God's Existence as a result.

Ironically, you don't need to know much science to convince yourself through common sense how and why the Theory of Evolution must be FALSE. Why is this so? It's because one soon discovers that there is actually NO scientific evidence whatsoever supporting the Theory of Evolution, which means that Darwinism is purely a philosophical worldview or a religion. Consequently, Darwinism can be proven false and dispatched philosophically, because ALL of the Scientific Evidence proves that the Theory of Evolution is FALSE.

Once you know WHY evolution of any kind COULD NEVER HAVE DONE any of the miraculous, magical, and creative things that the Darwinists say that evolution did, then you KNOW ALL that you really need to know about the Theory of Evolution. You KNOW that Creation by Evolution is IMPOSSIBLE.

The failures, falsehoods, and inadequacies of Darwinian Evolution have proven to me that God must exist, in order to do all of the SCIENCE and other things which evolution could NEVER have done.

My Observation: Each time Evolution fails to do what the Darwinists say that it does or say that it did, that FAILURE becomes yet another miniature Scientific Proof of God's Existence, because God must of necessity exist in order to do all of the different things that Evolution did not do or could not do. This is logical common sense.

This REALITY or TRUTH is yet another example of God hiding in plain sight where the Atheists and the Darwinists cannot see Him or find Him. God is standing there staring them in the face, but they can't see Him because they don't want to see Him.

Materialism of any kind ends up becoming a self-fulfilling prophecy.

—

Applying the Scientific Method

The purpose of the Scientific Method is to help us to find THE TRUTH, through a preponderance of the evidence.

the Scientific Method has no value to us if we use it to convince ourselves that a LIE is TRUE, as the Materialists and Darwinists always seem to do.

That's what I discovered during my Pursuit of the True Reality of All Things, and during my usage of the Scientific Method.

the Scientific Method can be used and has been used TO PROVE beyond a shadow of a doubt through a preponderance of the evidence that Creation by Evolution, Creation by Chance, Materialism, and the Theory of Evolution are IMPOSSIBLE and FALSE.

The Theory of Evolution IS Creation by Evolution.

Meanwhile, all aspects of the Scientific Method and the Rules of Science have PROVEN definitively and conclusively that Design and Creation BY Intelligent Beings is REAL and TRUE.

What you do with this information is up to you.

ALL of the Evidence IS on the side of "Creation by Intelligent Beings". SCIENCE, the Scientific Method, ALL of our OBSERVATIONS, Abductive Reasoning, and Deductive Reasoning make it clear and obvious that it is so.

In contrast, there is NO Evidence and can be NO Evidence to support Creation by Evolution, Creation by Chance, Design and Creation by Natural Processes, and Creation by Materialism, because these things cannot Design, Manufacture, and Create! It's elementary my Dear Reader!

I'm going with the AVAILABLE EVIDENCE, the PREPONDERANCE OF THE EVIDENCE, the SCIENCE, and the Scientific Method on this one. By doing so, it has become abundantly obvious to me that GOD MUST EXIST in order to have done ALL of the SCIENCE, design, manufacturing, creation, and organizing which had to be DONE that Evolution and Chance could NEVER have done 3.8 billion years ago, when all of that WORK needed to be DONE. When it comes to Design and Creation, Evolution and Chance DO NOT WORK! I want something to WORK if I'm actually going to believe in it.

I have chosen to FOLLOW THE EVIDENCE wherever it might lead me, even if that Evidence should lead me to God. I'm no longer afraid of The Evidence like I used to be.

In the Bible, the Book of Mormon: Another Testament of Jesus Christ, the Doctrine and Covenants, and the Pearl of Great Price, the Biblical God Jesus Christ has repeatedly REVEALED Himself to us and repeatedly CONFESSED to designing and creating This Universe, This Earth, and ALL of the Genomes and Life Forms on this planet. That's good enough for me!

I have chosen to FOLLOW THE EVIDENCE wherever it might lead me.

You can do whatever you want to do. That's what this Life is all about — making a decision and a choice!

Have a nice day and a wonderful life.

Mark My Words!

July 2016

PART V — THE ULTIMATE MODEL OF REALITY

Introductory Note: This part entitled, "THE ULTIMATE MODEL OF REALITY", is PART I of my books, "Putting Psyche Back into Psychology: Restoring Science to Consciousness" and "The Ultimate Model of Reality: Psyche Is the Ultimate Cause". This part of those books serves as an introduction to those books. These books comprise my magnum opus – my primary contribution to philosophy, quantum realities, science, psychology, and the reality of our lived experiences as a race of human psyches.

My goal is to keep this part tight, as an introduction to psyche and ultimate causality; but, there is still a lot to cover, and there will be times when I will try to go in-depth and other times when I will duplicate some of what was said before and discuss it from a different perspective. In all things, it should point back to Psyche and a Psyche Ontology, when we are done. ALL of the evidence keeps pointing me to Psyche. It got to the point where I couldn't deny it anymore.

My thesis is that Psyche or Non-local Consciousness is the Ultimate Cause of everything that has ever been brought into existence or organized from scratch, including physical matter. In these books, I develop and present the Ultimate Cause Model of Reality for your consideration.

My Ultimate Personality Theory is that Psyche, Personality, Identity, and our Memories survive separation from our physical body, bodily death, and brain death according to the empirical evidence from Near-Death Experiences (NDEs), Out-of-Body Experiences (OBEs), Shared-Death Experiences (SDEs), and other types of Spiritual Experiences (Spiritual Empiricism and Lived Experiences).

In these books, "Putting Psyche Back into Psychology: Restoring Science to Consciousness" and "The Ultimate Model of Reality: Psyche Is the Ultimate Cause", I employ procedural evidence, validated scientific evidence, empirical evidence, experiential evidence, common sense, lived experiences, and knowledge as proof that Psyche is the Ultimate Cause of everything that is real and true, and as proof that Psyche is the Ultimate Cause of everything that has ever been brought into existence from scratch.

I do have a presence online:

The Associated Facebook Page:

https://www.facebook.com/MarkMyScience/

The Associated Twitter Page:

https://twitter.com/Mark_Me_Words

However, I don't participate all that much online nowadays because the Materialists and Atheists have labeled me as spiteful and vindictive and banned me from their websites. You don't last very long online with the message that I have to share with the world, before they find some way to shut you down and lock you out. It goes with the territory, because I'm not telling them what they want to hear. I'm telling them the truth instead; and, they don't like that.

Abstract – Psyche Is the Ultimate Cause

The purpose of this essay is to introduce a fifth cause or an Ultimate Cause into philosophy, metaphysics, science, application, lived experience, and psychology.

Aristotle proposed four main causes responsible for a physical object's existence – namely, **material cause** (the substance from which it is made or built), **efficient cause** (the motions needed to get it manufactured or built), the **formal cause** (the blueprint, plan, or design from which it was built), and the **final cause** (the purpose, goal, or reason for which it was built).

I introduce a fifth cause or an **ultimate cause**, which is the Person, Individual, Builder, Manufacturer, Mover, Agent, Contractor, Consciousness, Designer, Architect, Awareness, Meaning-Maker, Planner, Desirer, Observer, Intelligence, Life, Light, Soul, Spark, and Psyche – the personal, living, individual, immaterial, non-local consciousness **who** makes the ultimate decision or the final choice between competing alternatives. This fifth cause or Ultimate Cause is Psyche – or Living Supernatural Non-Local Immaterial Trans-Dimensional Quantum Consciousness. This Ultimate Cause is the Prime Mover, the first cause behind ALL other causes.

Ultimate Cause or **Psyche** is the originating cause behind every other cause and everything that has ever been ordered and organized, including the ordering or organizing of spirit matter, the calling into existence and organization of physical matter, the design and creation of physical universes, and the design, programming, engineering, and manufacturing of all genomes and life forms in a timely and efficient manner.

My goal in this essay is to present and develop The Ultimate Model of Reality. I want to subsume everything that is useful, efficacious, and true while at the same time eliminating the philosophies and pseudo-sciences that are not demonstrable nor evidential and have to be taken on blind faith as being true. In order to make a case for Psyche as the Ultimate Cause, one has to understand how and why Materialism and Naturalism are false. The false models of reality have to be eliminated in order to make room for the True One or the Real One. As a scientist and a philosopher, I have been looking for such a construct all my life. I had to make my own model of reality, because as far as I can tell such a thing doesn't exist yet. I introduce this Ultimate Cause Model of Reality for your consideration.

Keywords: ultimate cause, psyche, psychology, psyche-therapy, philosophy, science, fifth cause, formal cause, final cause, NDEs, OBEs, SDEs, lived experiences, spiritual experiences, spiritual empiricism, radical empiricism, phenomenology, hermeneutics.

Aristotle's Four Causes and a Fifth

Aristotle proposed four main causes responsible for a physical object's existence – namely, **material cause** (the substance from which it is made or built), **efficient cause** (the motions needed to get it manufactured or built), the **formal cause** (the blueprint, plan, or design from which it was built), and the **final cause** (the purpose, goal, or reason for which it was built).

Aristotle's four causes are typically defined in terms of what it took to get a physical object made. In other words, they are defined in a materialistic and mechanistic fashion. So, imagine my surprise when someone on Google defined **efficient cause** as the manufacturer or the construction crew who puts the physical matter together into useful items – for example, the carpenter is the **efficient cause** of the table, cabinets, and chair. Well, that definitely messed things up, because suddenly we were talking about **who** made the chair and not just the physical processes involved in building the chair, while we were talking about the **efficient cause** of the chair. But, it certainly caught my attention.

I'm not the first person to realize that some kind of Psyche or Ultimate Cause must exist behind all science, manufacturing, engineering, production, and creation; otherwise, there would be no science experiments, no manufacturing, and no creation of new objects taking place. Philosopher Edward Feser employed this unique definition for **efficient causality** in his books:

> The *material cause* or underlying stuff the ball is made out of is rubber; its *formal cause*, or the form, pattern, or structure it exhibits, comprises such features as its sphericity, solidity, and bounciness. In other words, the material and formal causes of a thing are just its matter and form, considered as two aspects of a complete explanation of it. Next we have *the efficient cause*, that which actualizes a potency and thereby brings something into being. In this case that would be the actions of the workers and/or machines in the factory in which the ball was made, as they molded the rubber into the ball. Lastly we have the *final cause* or the end, goal, or purpose of a thing, which in the case of the ball might be to provide amusement to a child. (Feser, *Aquinas*, p. 16).

After such repeated input from many different sources, I found myself adjusting or tweaking the definitions for Aristotle's four causes as follows:

Aristotle proposed four main causes – namely, **material cause** (physical matter), **efficient cause** (the manufacturer or construction crew who puts the physical matter together into useful items and forms), the **formal cause** (the blueprint, plan, or design), and the **final cause** (the purpose, goal, or reason for choosing to act, or for choosing to manufacture an item and choosing to use an item).

Then on my own, as my research continued, I just kind of automatically found myself extending the "**person**" or the "**who**" or the "**personality**" to formal cause and final cause, because it was so easy to do. Thus, **formal cause** became the Designer or the Architect behind the blueprints, plan, or design. And, **final**

cause became The Person or The Agent who wanted an item manufactured, for whatever reason or purpose that this person had in mind. In this manner, my concept of **ultimate cause** was born, because ultimate cause is the person or the psyche **who** is behind the efficient cause, the formal cause, and the final cause of any item, idea, or thought that has ever been manufactured or brought into existence.

It dawned on me one day like an epiphany that there is an **ultimate cause** (or psyche) behind each one of Aristotle's four causes, and suddenly it was clear to me that we needed a fifth cause. Ultimate cause explains **who** is responsible for Aristotle's four causes. Psyche or ultimate cause has always been there behind everything that has ever been designed, manufactured, and brought into existence; but, **ultimate cause** had never been defined nor employed like it should have been. I figured that it was time to rectify the situation, because the empirical evidence for the existence of psyche is literally exploding across this world right now. (Buhlman, 1996; Gibson, 2006; & Rivas, 2016).

The only thing that was lacking was that I didn't know how **material cause** fit in with **ultimate cause**, at first. This required a bit more thought. But, I knew that it must be so.

Eventually, based upon the need for a Conscious Observer in Quantum Mechanics in order to convert spiritual quantum waves into solid physical matter, it all became clear to me; and, I found myself applying **ultimate cause** or psyche to physical matter and **material cause** because of Quantum Mechanics or Spiritual Mechanics; and thus in my Ultimate Model of Reality, **material cause** became **The Person** or the Conscious Observer or the Psyche **who** collapses the quantum wave function thus changing a quantum object from some kind of non-local spiritual wave (or spirit matter) into a physical particle located here in our 3D physical space-time. According to Quantum Mechanics, Psyche or Non-Local Consciousness must of necessity exist in order to bring physical matter into existence in the first place. Without Psyche, there would be NO physical matter. This is what Science is trying to teach us! But, few people are actually listening.

Ultimate cause or a conscious observer is the only thing that can convert spirit matter into physical matter, according to the Science of Quantum Mechanics. Thus, I had finally identified the **ultimate cause** behind every material cause and behind physical matter itself – psyche or the conscious observer. Ultimate Cause, the material cause and efficient cause of physical matter, is the Conscious Observer or Psyche which Quantum Mechanics tells us must exist. (Van Lommel, 2010). According to Quantum Mechanics, if Psyche didn't exist, then there would be no physical matter!

After reading Pim van Lommel's book, *Consciousness Beyond Life: The Science of the Near-Death Experience*, for the first time in my life I finally understood Quantum Mechanics in a way that actually made logical sense to me – from a non-local, non-physical, spiritual perspective. I finally had something useful to build upon. If you are going to read one book about Quantum Mechanics, that's the book to read. Quantum Mechanics never made any sense from a physical or materialistic perspective; but, Quantum Mechanics makes a whole lot of sense from a non-local or a spiritual perspective. (Mark My Words, 2016, *Quantum Mechanics from a Non-Physical Spiritual Perspective*).

In consequence of all of this, within this Ultimate Cause Model of Reality, the **material cause** sort of merged with **efficient cause** into a combined **mechanistic cause** and became The Person or The Psyche **who** observed or called all the physical matter into existence in the first place and **who** organized or manufactured physical matter into computer components, plastic, woodworks, car parts, bricks, cement, lumber, steel, candy, genomes, and the other material things that went into the construction and manufacture of the physical objects which are currently around us. Thus, **material cause** became not only the substance or essence from which something is made but also **The Person** or the psyche behind the organization and the origination of that substance or essence, including planets, stars, galaxies, and physical matter. Ultimate cause is the person or the psyche behind everything that has ever been designed, organized, built, manufactured, and created including physical matter and physical universes. Ultimately, there is an **ultimate cause**, or psyche, or person behind each and every material cause, efficient cause, formal cause, and final cause.

Ultimate cause has always been there behind each and every one of Aristotle's four causes waiting to be found, identified, and used; but, ultimate cause is typically taken for granted and completely ignored as if it doesn't even exist. In fact, the Materialists and Atheists don't want psyche or ultimate cause to exist. Ultimate Cause or Psyche is the antithesis of Materialism and Naturalism. The existence of Psyche or ultimate cause actually proves Materialism and Naturalism false. Psyche by definition in principle is supernatural. Ultimate Cause and Naturalism are mutually exclusive. If one can be demonstrated to be true, then the other one has been demonstrated to be false. Now I finally had something upon which to build.

Eventually, Aristotle's four main causes came to mean – **material cause** (the person who brought the substances, building materials, or physical matter into existence or into some kind of order if they already existed), **efficient cause** (the person who organized the spiritual substances, building materials, and/or physical matter into useful and interesting forms and objects and life forms), the **formal cause** (the person who designed and created the blueprints, plans, software, and genomes), and the **final cause** (the person who chose to act on the plan in order to get things done for whatever purpose he or she desired).

I eventually started to interpret Aristotle's four cause in terms of **ultimate cause** – namely, the person or the psyche who did all of these different things. Our philosophical theories are supposed to be useful and compelling, after all; and, I find this Ultimate Model of Reality useful and compelling. For me, this became the Ultimate Personality Theory and the Ultimate Ontology, a Psyche Ontology wherein Psyche is the fundamental unit of reality.

So, there you have an explanation for the origin of this Ultimate Cause Model of Reality and this Ultimate Cause Personality Theory. I eventually observed that there is an **ultimate cause** or psyche behind each one of Aristotle's four causes; and thereby, I realized that there are excellent reasons for adding a fifth cause onto

Aristotle's four causes. Ultimate cause or this fifth cause answers the question of "**who**" did the other four causes. Psyche or ultimate cause can do science, design things, do programming, manufacturing, construction, engineering, thinking, writing, language, philosophy, and creation; whereas, the rocks or raw physical matter cannot. That's what Science is trying to teach us.

For thousands of years, the Philosophers and the Materialists have focused all of their attention on the material thing or the material object which was the cause or the reason for something's existence. But, Materialism is dying and coming to an end; so, it is time for a new focus, or a new paradigm, or a new model of reality. Without even realizing it, this world is slowly switching over to psyche or ultimate cause, within Science itself. It was doing so without me; and now, it can do so with me. There is a paradigm shift taking place right now, but it goes mostly unnoticed because it has been slow in coming. (For an example see, Tart, 2009, *The End of Materialism: How Evidence of the Paranormal Is Bringing Science and Spirit Together*. Tart's book was over 50 years in the making). Many people are ten or twenty years ahead of me in making this paradigm shift within their theorizing and scientific research; but, I finally got there in the end which is all that really matters.

Thanks to these recent insights, I have started to define Aristotle's four causes in terms of **ultimate cause**, as The Person or The Psyche **who** is behind each and every one of Aristotle's four causes. So, I find myself asking, "**Who** was the substance maker or substance organizer, and thus the true **material cause** behind this object or this particle of physical matter?" Or, "**Who** was the manipulator or manufacturer or true **efficient cause** behind this finished product?" Or, "**Who** was the Architect or Designer and thus the real **formal cause** behind this blueprint, plan, computer program, life form, or genome?" Or, "**Who** was the person or the psyche or the **final cause** who desired and requisitioned the final product, or the person **who** chose to bring this concept or idea into existence in the first place?"

In other words, I have redefined Aristotle's four causes in terms of psyche or **ultimate cause**, thereby creating a new, more expansive, more robust, and all-inclusive personality theory and model of reality in the process – the Ultimate Personality Theory and the Ultimate Cause Model of Reality. Ultimate cause attempts to answer the question of **who** performed or **who** is behind the material cause, the efficient cause, the formal cause, and the final cause – a question which the Materialists and Naturalists refuse to ask.

Psyche Is the Ultimate Cause

Psyche is the ultimate cause.

Personality is psyche; and, psyche or personality is the part of us that survives bodily death and brain death according to the Empirical Sciences of Near-Death Experiences (NDEs), Shared-Death Experiences (SDEs), Out-of-Body Experiences (OBEs), and other spiritual experiences (Spiritual Empiricism or Lived Experiences). As I see it, this is the Ultimate Personality Theory, an Ultimate Cause

Personality Theory. This is a personality theory that will actually survive one's bodily death and continue to be true in the afterlife.

I had cause for developing this Ultimate Model of Reality; and, I ended up doing it for a good cause. I was trying to get at the ultimate truth of all things, as any good scientist should. I was looking for the meaning of Life, the Universe, and Everything; and, it ended up being a whole lot more than just 42. It ended up being Psyche or Ultimate Cause.

Ultimate Cause or Psyche is the reason for our existence; and, our personality or psyche will continue to exist and continue to thrive long after our physical body and physical brain are dead and gone, according to the empirical evidence from NDEs, SDEs, OBEs, and other spiritual experiences.

This Ultimate Cause Model of Reality or Psyche Ontology is a holistic model of reality based upon the Lived Experiences of the human race. In order to make it possible, though, one has to be willing to get rid of ALL the exclusionary philosophies and exclusionary models of reality such as Materialism, Naturalism, Scientism, Nihilism, Darwinism, and Atheism – the models of reality which exclude things like psyche, spirit matter, and non-locality from consideration. Exclude the exclusionary, and suddenly you find yourself staring at the truth!

I observed and discovered that one has to exclude the Exclusionary Philosophies if he or she really wants to get at the truth of our existence. I also discovered during my research that Truth and Knowledge are comprised of our Lived Experiences, including our out-of-body experiences, theophanies, and spiritual experiences.

Ultimately, Truth and Knowledge are based upon the Lived Experiences of the human race and the human psyche, wherein people experience Reality directly for themselves. Experiencing Reality should take precedence over philosophical and scientific interpretations of reality.

Why?

In Philosophy and Metaphysics and Scientific Interpretation, they guess at the Truth. That's not a good foundation upon which to know the truth. It's much better to live the truth and experience the truth directly for yourself, or to choose to trust someone who has done so.

Furthermore, it has been shown that it is impossible to use the Scientific Methods to prove the Truth, because scientific methods begin and end with philosophical assumptions and philosophical interpretations of the data, and because scientific methods are based upon the *affirming the consequent* logic fallacy whereby it becomes possible to make almost any interpretation of the scientific evidence fit with any piece of scientific data that we produce through experimentation. Scientific methods cannot be used to prove the Truth, which means that scientific methods are no better at getting at the Truth than philosophical speculation is, because scientific methods are based exclusively upon the philosophical interpretation of scientifically derived evidence.

We can't get at the Truth through Philosophy, Science, and the Scientific Methods.

The ONLY way to KNOW the Truth is to experience the truth and live the truth directly for ourselves, or to trust the people who have done so. The ONLY way to KNOW the Truth is through Lived Experience or Psyche Experience, because Psyche is the only thing that can do living, and experiencing, and the knowing of Truth. A Psyche Ontology is the Ultimate Ontology because only the human psyche can do ontology, metaphysics, philosophy, and science. Furthermore, only Psyche can do Lived Experiences, because our psyche is the only part of us that's truly ALIVE, in every sense of the word.

Psyche is truly the Ultimate Cause!

Introduction to Ultimate Cause

In this essay, I develop my own Theory of Personality, which becomes an Ultimate Cause Personality Theory; and, I also develop the Ultimate Model of Reality which I call the Ultimate Cause Model of Reality or a Psyche Ontology. This Ultimate Model of Reality is the antithesis of Scientific Naturalism and Materialism, and therefore a fully complete synthesis with Science, Philosophy, Psychology, Observation, and Lived Experience as a whole. This Ultimate Cause Model of Reality becomes the full embodiment of Science – no limitations and nothing excluded from consideration, as Science should be.

It is my thesis that Psyche is Personality, and that Personality is Psyche; and, Psyche or Consciousness is the Ultimate Cause behind everything that has ever been created, organized, or brought into existence from scratch. Psyche is Intelligence, in other words Light, Knowledge, and Truth. Psyche is life. Psyche is the fundamental unit of our existence and our reality; and therefore, Psyche is the fundamental unit of all Knowledge and Truth, because ONLY Psyche can do knowledge, truth, and lived experiences – and then remember these things in order to be able to use them and share them with others.

Personality or Psyche is the part of us that survives separation from our physical body, separation from our spirit body, death of our physical body, and brain death according to the Empirical Sciences of Near-Death Experiences (NDEs), Shared-Death Experiences (SDEs), Out-of-Body Experiences (OBEs), and Spiritual Experiences (Spiritual Empiricism). (See Alexander, 2012: Buhlman 1996; Durham 1998; Gibson, 2006; Hinze, 1997; Long, 2016; Moody, 1975; Moody, 2010; Neal, 2011; Ring, 1999; Rivas, 2016; Sharkey, 2008; Storm, 2000; and van Lommel, 2010).

Quantum Mechanics tells us that the human psyche or the conscious observer must exist in order for physical matter to be brought into existence. (See van Lommel, 2010). NDEs, SDEs, OBEs, and Spiritual Empiricism tell us that the human psyche or non-local consciousness does exist. (See Buhlman, 1996; and Rivas, 2016). A strong Personality Theory requires a strong empirical base. Follow the evidence. That's what I finally chose to do.

My Ultimate Personality Theory is that psyche or personality is the part of us that survives brain death and bodily death, according to the empirical evidence from NDEs, SDEs, OBEs, and other spiritual experiences. Psyche is the ultimate memory storage device, because the physical body doesn't have lived experiences and doesn't make memories while the Psyche is away from or separated from the physical body, according to the Lived Experiences of the Out-of-Body Travelers.

The major premises of Materialism and Scientific Naturalism have no evidence supporting them and will never have any evidence supporting them; whereas, their antitheses "Ultimate Cause" and "Psyche as the Ultimate Cause" have tons of empirical evidence supporting their truthfulness and usefulness.

Theologians often talk about a Prime Mover, an Uncaused Cause, or a First Cause. These are subsumed by Ultimate Cause. **Psyche is the ultimate cause** behind everything that has ever been brought into existence from scratch, including physical matter and physical universes.

This Ultimate Personality Theory involves psyche or ultimate cause. After studying Personality Theories quite extensively, it is my conclusion that this fifth cause or Ultimate Cause is necessary to explain what's truly going on, in all the realities of our existence. This Ultimate Cause is in fact The Cause which the majority of the naturalistic theoreticians deliberately choose to avoid and reject, a decision and limitation that makes their theories and philosophies incomplete and unsatisfying in the end.

Psyche or Ultimate Cause is the choice-maker, the meaning-maker, the final-cause generator, the deal-breaker, and the tiebreaker affirming or rejecting each and every alternative which comes its way. Psyche, or Ultimate Cause, or Non-Local Consciousness is what brought ALL of the other causes into existence in the first place. Psyche by definition in principle can negate a decision, an opportunity, or a choice by choosing to say "No". That's what makes psyche so unique and powerful. Psyche can choose not to respond to a stimulus. Raw physical matter cannot. Physical matter is obligated to obey and obligated to respond to any stimulus that comes its way. Psyche is NOT!

The Ultimate Cause – immaterial psyche, or consciousness, or intelligence – designed and created the first particle of physical matter and brought order (efficient cause) to pre-existing spirit matter (the 'material cause' of the spirit realm or quantum realm), became the manufacturer behind the physical universe and the ordering (efficient cause) of physical matter into useful items and forms such as planets and stars, produced the various different consensus realities in the spirit realm or non-local realm (efficient causes of a spiritual nature), designed the blueprints, plans, and software behind all genomes and life forms and physical items on this planet (formal cause), and purposefully chose to set this whole plan into motion for its own reasons, purpose, and goals (final cause and teleology).

Ultimate Cause or Non-Local Consciousness is, well, the ultimate cause behind all of the other causes, whether we are talking about this physical 3D space-time realm or the non-local trans-dimensional quantum realm. Everything that has ever been organized, ordered, created, or brought into existence (including physical matter) was brought into existence by some kind of non-local ultimate cause or non-local consciousness, according to the Science of Quantum Mechanics.

According to Quantum Mechanics, psyche or consciousness or ultimate cause is non-local, non-physical, trans-dimensional, immaterial, supernatural, and spiritual in nature and origin. I'm going with the Science on this one, and not the materialistic or naturalistic philosophies. Materialism is based upon a denial of Reality and a refusal to look at contradictory evidence; whereas, this Ultimate Model of Reality is based upon a full embrace of the existential experiences of the whole of mankind. Every experience and truth reported by human beings throughout the whole of recorded history comes under the purview of this Ultimate Cause Personality Theory and this Ultimate Model of Reality, or at least that's its purpose, goal, and intent. ONLY Psyche can do life, experience, knowledge, and truth.

I eventually decided to take people at their word whenever they report a spiritual experience or a supernatural experience to me, which is something that the Materialists and the Scientific Naturalists refuse to do. I'm no longer an Atheist, and I'm no longer a Materialist. I am now willing to look for evidence of Psyche and Ultimate Cause. My eyes have been opened, and I have seen the light. My goal here is to include or to subsume every truth, every experience, and every reality that I can find into this Ultimate Model of Reality. It's existential and experiential. It's based upon the Lived Experiences of the human race.

This Ultimate Cause Model of Reality attempts to be all-inclusive and attempts to include everything that we humans have ever experienced or thought about, throughout the whole of human history. All my life, I have been looking for the true model of reality, a model which will explain everything and lay a foundation for everything which human beings (human psyches) have ever experienced, created, manufactured, or thought about. I believe that I have finally found what I was looking for, even though I had to create it for myself.

Our Experiences of Reality do and should take precedence over our philosophical speculation and our interpretations of any scientific evidence. In other words, our philosophical guesswork and our interpretations of scientific data should be forced to match with Reality or forced to match with the Lived Experiences or Psyche Experiences of the human race, rather than the other way around as the Materialists and Naturalists and Atheists do with their science and philosophy.

Materialism and Naturalism violate and deny the existence of the Lived Experiences or the Spiritual Experiences of the human psyche. That's Bad Philosophy, Bad Form, and Bad Science – the refusal to let eye-witness testimony into evidence, or the refusal to permit personal experiences into evidence. Materialism, Naturalism, and Atheism of any kind are based upon a refusal to look at the evidence. Materialism and Naturalism are Exclusionary Philosophies and are based upon a Denial of Reality. These exclusionary philosophies should be excluded from Science and Philosophy, especially if we want to experience Truth and Knowledge for ourselves.

Truth and Knowledge are based upon our Lived Experiences wherein we (our psyches) experience Reality directly for ourselves; and therefore, a Lived Experience Epistemology or a Psyche Epistemology is vastly superior to any epistemology that is based exclusively upon the Scientific Methods or philosophical guesswork. ONLY Psyche can do epistemology, ontology, lived experiences, truth, knowledge, science, philosophy, and reality. The rocks, which the Materialists and Naturalists idolize and worship, need not apply.

The Best Example Which I Currently Have of This Ultimate Reality

Empirical Knowledge, Experiential Knowledge, or Lived Experience is superior to scientific hypotheses, materialistic interpretations of scientific data, and philosophical guesswork.

A single Lived Experience is of more value to me than all of the philosophical speculation in the world combined. The materialistic and naturalistic philosophers tell us that psyche and spirit do not exist; however, the Lived Experiences of the human race tell us that the Materialists and Naturalists are wrong.

What was there in existence BEFORE the first particle of physical matter was designed, created, manufactured, and brought into existence? WE WERE! Our Non-Local Consciousness, Psyche, or Personality was there BEFORE the "Big Bang" or Prime Event was used to bring this physical universe and physical matter into existence. Everything that has a beginning has a Beginner, an Ultimate Cause, who set it in motion. It is illogical to think otherwise. Only something like Psyche or Non-Local Consciousness has always existed and will always exist. ONLY Psyche can act as the ultimate cause of physical matter. ONLY Psyche is infinite in nature and scope. Psyche resides in the realm of the infinite.

According to the Science of Quantum Mechanics, every particle of physical matter required an Observer, or a Non-local Consciousness, or an Ultimate Cause in order to bring it into existence. In other words, according to Quantum Mechanics or Spiritual Mechanics, every particle of physical matter had a beginning; and thus, had a Beginner or an Ultimate Cause behind its construction or its transformation from a quantum wave or a spiritual wave into a physical particle. That's what the Science of Quantum Mechanics is trying to tell us, which explains why the Materialists and Naturalists are unable to understand and accept Quantum Mechanics at face-value. Quantum Mechanics is Spiritual Mechanics, or the way that spirit matter really works.

I use William Buhlman's out-of-body experiences as Empirical Evidence and Living Proof that a Psyche Ontology is in fact the true reality of our existence. There is indeed a Fifth Cause or an **Ultimate Cause** behind ALL things which exist and behind ALL life forms; and, this **Ultimate Cause** is Psyche, or our Non-Local Immaterial Consciousness, or Intelligence.

William Buhlman Wrote

Journal Entry, October 2, 1982

I hear the buzzing, engine-like sounds and will myself out-of-body. [He left his physical body behind.] I step to the bedroom door and automatically request "Clarity now!" My vision improves and I step through the door, into the living room. Still feeling a little out of sync, I verbally repeat my request with more emphasis, "Clarity now!" I feel my awareness and vision snap into place.

My thoughts are clear, and I make a verbal demand, "I need to see the form I'm in now!" Instantly I feel an intense sensation of being drawn within myself. I'm suddenly different, weightless as though I'm floating in space. As I look forward I see a sparkling, bluish white form. For some reason, I seem to know that I'm looking at my nonphysical body from a different perspective. I stare in amazement at this form before me that shines and flows with energy and light. It looks like an energy mold created from a million tiny points of light; it radiates a bluish glow but appears to have a defined outer structure. The body of light before me is naked and is identical to my physical form. Even though my body looks firm, there is a noticeable energy motion and radiation present. I can see what appears to be an ocean of blue stars throughout my body. It's difficult to describe because the stars are stable, yet moving at the same time; the light and energy of my [spirit] body appear to change and flow almost like the waves of an ocean.

As I stare at the body of light, it hits me that I must be in another body. Yet I can't perceive any form or substance; I'm like a viewpoint in space without shape or form of any kind. [He, his immaterial viewpoint in space, is pure psyche or intelligence or consciousness.] As I reflect upon my new state of being, I feel a sensation of rapid motion and I'm instantly back within my physical body.

Lying still and reviewing my experience, I'm struck by an inescapable conclusion: I must possess multiple energy-bodies. The form I just experienced was noticeably lighter (less dense) than even my second nonphysical body. I realize that the traditional view of our possessing two bodies — a physical body and a spiritual body — is far too simplistic; we are much more complex than this. Just as there are multiple nonphysical energy dimensions within the universe, each of us must consist of multiple energy-bodies or vehicles of expression.

Now I seriously wonder just how many nonphysical bodies or forms this involves. I suspect that there must be one within each dimension of the universe and that all of these are interrelated and connected, just as the physical body is connected to its first nonphysical (spiritual) body.

Buhlman, William L. (1996). *Adventures Beyond the Body: How to Experience Out-of-Body Travel* (pp. 34-35). New York: HarperCollins. (HarperCollins permits quotes from their books in Critical Reviews and Promotional Reviews. I definitely promote this book any chance I get).

Notice that it's the immaterial viewpoint in space who does all of the thinking, observing, choosing, planning, feeling, and living – not the spirit body which he was looking at, and certainly not his physical body which he left behind in his bed. Only some kind of psyche or immaterial non-local consciousness can do teleology and life. Only some kind of psyche can do purposeful, deliberative, meaningful choosing, planning, decision-making, final causality, design and creation, and agentive moral action. It's that immaterial psyche, spark, or viewpoint in space who is alive and doing all the observing and thinking and choosing, not the spirit body and not the physical body.

William Buhlman's Psyche, Personality, Intelligence, or Non-Local Consciousness is the immaterial viewpoint floating in space looking at his most refined or inner-most Spirit Body or Non-Physical Body. It's this Immaterial Viewpoint, Intelligence, or Non-Local Consciousness which is in fact the **Ultimate Cause** of ALL things that exist, including the first particle of physical matter that was designed and created (material causes), the manufacturers who build or organize new things out of raw materials of both a spiritual and a physical nature (efficient causes), the (formal cause) designs and blueprints and computer programs and genomes, as well as the (final cause) purposes, goals, reasons, choices, intentions, affirmations, and ultimate decisions and destinations.

Personally, I don't believe in the parallel universe theory, wherein there are multiple versions of me doing different things. I personally believe that our single Psyche and single Spirit Body can manifest and function in all the multiple nonphysical energy dimensions within this universe. In other words, I believe that we have one spirit body and one psyche, but that these things can function in and interact with ALL the dimensions of this universe. They adjust as needed. Our Psyche and Spirit Body can even have a presence within and influence upon this Physical Dimension and our Physical Body.

The Psyche or Non-Local Consciousness is the **Ultimate Cause**, because the Psyche brings ALL of the other Causes into existence. The Psyche or Personality or Consciousness is the part of us that survives separation from the physical body, separation from the spirit body, physical death, and brain death according to the Empirical Science of Out-of-Body Experiences and Near-Death Experiences. I rely upon the Lived Experiences of the human race in order to establish the truth and validity of this Ultimate Model of Reality. I require evidence, lots and lots of evidence. I'm no longer willing to take things on blind faith, like I used to do when I was a Materialist and an Atheist.

I own literally hundreds of different books about Near-Death Experiences (NDEs), Shared-Death Experiences (SDEs), Out-of-Body Experiences (OBEs), Quantum Consciousness, Lived Experiences, and Spiritual Experiences (Spiritual Empiricism) including most of the books listed by Gibson in *They Saw Beyond Death: New Insights on Near-Death Experiences*; and, William Buhlman's book is the one I chose to quote first in support of my thesis or theory about an **Ultimate Cause**, a "fifth cause", a First Cause, or a Primal Cause. Buhlman was the first person to convincingly teach me that our Psyche is completely separate from our Spirit Body.

Psyche, personality, individuality, intelligence, awareness, intentionality, and non-local consciousness is the **ultimate cause**. I have my own personal library which I have been reading from and drawing upon, possibly as many as 7,000 different books, mostly digital in nature, but hundreds of them are hardback and paperback. Collectively, these books not only testify that psyche exists, but these books also testify that psyche must of necessity exist in order to successfully explain all of the different various phenomena which the human race has experienced and observed throughout the whole of human history.

Psyche is the ultimate causal agent; yet, I refused to look at any evidence for psyche and God while I was a Materialist and an Atheist, and thus I truly believed that there wasn't any evidence supporting the existence of God and psyche, and that there never would be. I was wrong. Self-deception works, and it works every time! However, the books I now have on hand and hundreds of YouTube videos have met their burden of proof and successfully convinced me that psyche and God do indeed exist; and, the evidence just keeps pouring in. I choose to believe that what the evidence is telling me is real and true; and, one of the things which these books and videos are telling me is that Materialism and Scientific Naturalism are false. (See Richards & Bergin; & Goswami). Ultimately, I chose to believe in the Lived Experiences of the human race rather than the philosophical guesswork of the Materialists, Naturalists, and Atheists.

Materialism is the pinnacle of self-deception and the worst form of prejudice. Materialists, Naturalists, and Atheists are anti-psyche, anti-belief, anti-trust, anti-Christ, anti-God, anti-science, anti-reality, anti-rationality, and anti-discovery. The hard-core dedicated Materialists and Naturalists never make any ground-breaking paradigm-shifting scientific discoveries because these people refuse to ask the questions which lead to new scientific discoveries. The non-local sciences, or the spiritual sciences, or the quantum sciences are the final frontier in Science; yet, the Materialists and Naturalists refuse to go there.

My personality theory is based upon **ultimate cause** or psyche. And, based upon the empirical evidence which I now have at hand, it is my theory and belief that personality or psyche is the part of us that survives bodily death and brain death according to the Empirical Sciences, the lived experiences, and the empirical evidence from NDEs, SDEs, OBEs, and Quantum Mechanics.

This Ultimate Cause Model of Reality is based upon the collective Lived Experiences of the human race. The ONLY way to KNOW the Truth is to live it and experience it for yourself, or to choose to trust someone who has.

Again, that immaterial and formless viewpoint in space, or non-local consciousness, or non-local psyche, or immaterial intelligence and awareness which Buhlman mentioned in his personal out-of-body experience is **THE ULTIMATE CAUSE** of everything that has been ordered, organized, or brought into existence including spiritual consensus realities, physical matter, stars, planets, genomes, life forms, this physical universe, and the physical objects which have been engineered and manufactured.

A spirit body localizes a Consciousness or Intelligence or Psyche there in the spirit realm or the quantum realm; and, a spirit body gives us an identity, a form, and a heritage. It is clear from Buhlman's OBE that his spirit body is in the image of the Gods, which means that his spirit body is one of the spirit children of the Gods. That's quite a heritage, to have all the potential of a God. (Richards and Bergin; Holy Bible; & Bishop, 1998).

As most of us already know, our physical body massively limits us to sub-light speeds. Because our physical body has physical mass, there is no way to accelerate it to the speed of light and beyond. Physical matter by definition in principle exists at sub-light-speed velocities – if you accelerate physical matter faster than the speed of light then it becomes spirit matter or a non-local quantum object, potentially infinite in frequency and scope and potentially instantaneous in velocity. Sub-light-speed limitations don't apply to the immaterial or the spiritual, such as our spirit bodies and our non-local consciousness. Spirit matter or quantum objects in their wave-like format exist at velocities greater than the speed of light, which is why we can't capture them nor detect them with our physical instruments. Quantum objects or spirit matter have to be converted into physical particles by conscious observation before we are then able to detect them with our physical instruments. This is what Quantum Mechanics or Spiritual Mechanics is trying to teach us.

Most of our physical bodies here on earth are mortal, although there are indications that the Gods who have proven themselves trustworthy have access to physical bodies that are immortal, trans-dimensional, and glorified in some way. Of course, a physical body is the best way to experience a physical reality and to go through this physical-schooling process in what it's like to have some real limitations. A physical reality is the ultimate consensus reality, and a mortal physical body is the ultimate experience in what it's like to have some real limitations.

Physical consensus realities do have their advantages, though. Here in this physical consensus reality, I can count on this paper being there on my flash drive and hard drive tomorrow when I turn on my computer and go looking for it. I can count on my land and some version of my house being there when I drive home from work today; and, I'm sure that my dog is going to be there wanting some attention and love. Due to all the limitations of a physical reality, a semi-sense of permanence, stability, control, and predictability comes from a physical reality or a physical consensus reality, which you don't necessarily get from a spirit reality or a quantum reality, even a spiritual consensus reality which was put together by a community of spirits. A physical reality is the ultimate consensus reality. Buhlman in his book was the first person to teach me about consensus realities and non-consensus realities.

I have observed that all of the unsolvable "problems" created by Materialism (such as the mind-brain problem, the nature vs. nurture problem, the binding problem, and the free-will vs. determinism problem) are instantly solved by getting rid of Materialism, Determinism, and Scientific Naturalism and replacing them with some type of teleology, quantum non-locality, final causality, agency, psyche, quantum consciousness, ultimate causality, and a Psyche Ontology. Materialism and Naturalism are archaic, primitive, useless, and worthless in comparison to all of this. You won't get any information about nonlocality or spirituality or non-consensus realities from a Materialist or an Atheist, because they have nothing like this to give you. That has been my scientific observation and my personal experience.

Now you know one of the reasons why I am no longer a Materialist and no longer an Atheist. I now own hundreds of different books about NDEs, OBEs, Quantum Consciousness, and other types of spiritual experiences. The empirical evidence, which I have on hand of Psyche as the Ultimate Cause is so overwhelming, that I no longer have blind faith enough to be a Materialist or an Atheist. It's too late for me now. I can't go back to my ignorance; and, why should I want to? Quantum Immateriality, Infinite Singularities, Quantum Consciousness, Non-local Consciousness, and Psyche are extremely interesting. Materialism is boring and very limited in comparison.

Astrophysicists have observed that during the Prime Event or "Big Bang", only 4% of this universe got converted into physical matter or "observed into existence" as physical matter. The other 96% of this universe remained spiritual, non-local, and quantum in nature as dark matter (spirit matter) and dark energy (the zero-point field of light or the Light of Christ). The Materialists and Naturalists are wrong about 96% of this universe. (Ross, 2008). So, why should we continue to believe them and their philosophical guesswork?

Choose the best and get rid of all the rest. That's what I decided to do.

Proof of Concept

This Proof of Concept was based upon Joseph Rychlak's books, class notes from Edwin Gantt, and the following article by Williams and Slife.

Slife, B. D. & Williams, R. N. (1995). Science and Human Behavior. In *What's Behind the Research? Discovering Hidden Assumptions in the Behavioral Sciences*, (pp. 167–204). Thousand Oaks, CA: SAGE Publications.

http://mypsyche.us/science/

Materialism is the philosophical belief that ONLY physical matter exists, and that spirit and psyche do not exist.

Naturalism is the philosophical belief that psyche, non-local consciousness, the spiritual, spirit matter, the immaterial, the supernatural, and the non-local realm or spirit realm do not exist.

I was being repeatedly told online that Science and the Scientific Methods cannot be used to prove anything. I was being chastised right and left for claiming that the Scientific Method had proven to me that God exists, or that it had proven to me that the Theory of Evolution is false. Of course, I wasn't telling these people what they wanted to hear, so they weren't willing to give me a hearing.

However, it used to bug me to no end whenever these people told me that I can't use the Scientific Method to prove things, because I was using the Scientific Method to prove things all the time. I'm a scientist; and, I remember thinking, "What good is it, if it can't be used to prove anything?"

Technically, these people were just as blind as I was whenever they made the claim that the Scientific Method can't be used to prove anything. When it comes to the Scientific Methods, it all comes down to what one is trying to prove! The Scientific Methods cannot be used to prove anything true, at least not directly.

However, in theory and in practice, the Scientific Methods can be used and ARE BEING USED to prove things false. And, I was falsifying things right and left at the time! Furthermore, through repeated falsification of various different scientific theories, we slowly arrive at the truth through a process of elimination. This is what Sherlock Holmes does.

How often have I said to you that when you have eliminated the impossible, whatever remains, *however improbable*, must be the truth? — Sherlock Holmes.

Sherlock Holmes slowly arrives at the truth by falsifying his theories and his imagined evidence. We can do the same with the Scientific Methods. The Scientific Method isn't completely worthless; but, it also isn't as powerful and convincing as I once thought it was. Nevertheless, it has its uses, as long as it is used for what it's good for and for what it was designed for.

Verification and Validation

First, let's discuss how Science can go wrong. We can't use Science and the Scientific Methods to verify and know the Truth. Scientific Methods cannot be used to produce Knowledge.

Due to the *affirming the consequent* logic fallacy, and the *jumping to conclusions* logic fallacy, and the *category error* logic fallacies which are built into the Scientific Methods, it is impossible to use Science and the Scientific Methods to know the truth and to prove the truth.

If your ultimate goal in life is to KNOW THE TRUTH and prove the truth, then Science and the Scientific Methods are in fact one of the worst ways for accomplishing that task.

This is how scientific verification works in practice.

The following logical argument outlines the basic approach that has been taken by scientists:

If Theory X is true, then we will observe Y.

We observe Y.

Therefore, Theory X is true.

This sort of thinking, however, reflects a logical fallacy called *affirming the consequent*. Here's a comparable example to demonstrate:

We hypothesize: If Sally's pet is a cat, it will have a tail.

We observe: Sally's pet has a tail.

We conclude: Therefore, Sally's pet is a cat.

We can easily see that this logic is fallacious. Just because we observe Y (a pet with a tail) that does not mean that our theory X (the pet is a cat) is true. After all, dogs and lizards have tails too. (Rychlak; Slife; Edwin Gantt; Mark My Words).

Yet, **this IS the Scientific Method**, and this is exactly how scientists use the Scientific Methods to demonstrate and prove the truth. The whole enterprise is based upon a logic fallacy or two.

Begging the question consists of making our conclusion one of the axiomatic premises, and effectively results in *jumping to conclusions*. The Materialists and Darwinists do this all the time, without realizing it. Now, let's run a science experiment using the Scientific Method, based upon a *jumping to conclusions* or *begging the question* logic fallacy.

Premise or Hypothesis: All cats have tails.

Observation or Test of Hypothesis: My pet has a tail.

Scientific Conclusion: My pet is a cat.

The premise seems to match with the conclusion; but, this scientist jumped to the wrong conclusion based upon his scientific observations or scientific evidence – unless of course his pet was really a cat, then he simply got lucky. But, his pet theory is in fact a dog, so he got this one wrong.

Let's try another one. This one combines *categorization errors* along with *jumping to conclusions*.

Premise or Hypothesis: Truth matches with reality; and, the fossil record is a part of reality.

Observation or Test of Hypothesis: Darwin's Tree of Life and the Theory of Evolution match with the fossil record.

Scientific Conclusion: Therefore, Darwin's Tree of Life and the Theory of Evolution are the Truth.

Most scientists can't see anything wrong with any of this, because **this IS the Scientific Method**, which they use each and every day to find and demonstrate the truth of their science. But, this scientific method and logic fallacy demonstrated here are based upon "jumping to conclusions" or "begging the question", along with "false equations" or "categorization errors". These people find some way to input their desired conclusion as an axiomatic indisputable premise; and, this is *begging the question*. This is what the Darwinists, Materialists, Naturalists, and Atheists do all the time; and, they don't even know it. They can't even see it or understand it, because they have been trained not to see it or understand it. They truly believe that their logic is sound, when it is not.

Jumping to conclusions and *affirming the consequent* is precisely what the Darwinists and Atheists DO each and every time they publicly declare that Science has proven the Theory of Evolution true.

Let's demonstrate their error first by *affirming the consequent*:

If the Theory of Evolution is true, the fossil record will match with Darwin's Tree of Life.

We observe that the fossil record matches with Darwin's Tree of Life.

Therefore, the Theory of Evolution and Darwinism are true.

This is *affirming the consequent*. This is how the Darwinists and Naturalists prove the theory of evolution to be true. They do it every day. You can go online, and you will find them online right now *affirming their conclusions* or *jumping to conclusions*.

Most of the scientists can't see anything wrong with this argument. They use it every day to prove to themselves that the Theory of Evolution is true.

Now let's demonstrate their logic error by *jumping to the conclusion* or *begging the question*, along with a *categorization error* or two:

Premise or Hypothesis: If our theory matches with and is based upon the fossil record, then our theory matches with reality and is true.

Observation or Test of Hypothesis: We observe that Darwin's Tree of Life and the Theory of Evolution match with the fossil record; and, we observe that the fossil record is truly there and really exists.

Scientific Conclusion: Therefore, the fossil record proves that Darwin's Tree of Life and the Theory of Evolution are true.

This is the argument which I see them using everyday online and in their books. They can't see anything wrong with it! I couldn't either, when I was a Materialist and an Atheist.

We can pull this one off multiple different ways, because these people are simply *jumping to the conclusion* that they want to affirm:

Premise or Hypothesis: All truth will match with the fossil record.

Observation or Test of Hypothesis: Darwinism, Darwin's Tree of Life, and the Theory of Evolution match with the fossil record.

Scientific Conclusion: Therefore, Darwinism, Darwin's Tree of Life, and the Theory of Evolution are true.

Can you see anything wrong with these arguments?

I couldn't, when I was a Materialist and an Atheist! I fell for it hook, line, and sinker!

These people are jumping to conclusions throughout the whole process, and they can't even see that it is so. Their conclusions require a leap of faith and a confounding of the evidence (premises), and their arguments contain hidden assumptions, which the Materialists and Atheists are unable to see or understand and are more than willing to provide. Confounded! These people mix up something with something else so that the individual elements become something different and something difficult to distinguish from one another. They take that leap of faith and equate everything with everything else, and then equate it with the truth as well.

There are *category errors* taking place here, wherein these people are equating their philosophical theories with the fossil record; but, what gives them the right to consider their philosophical explanations of the fossil record to be superior to other philosophical explanations which contradict their chosen point of view? There's no logical or rational reason why they should do so, except for the fact that they are prejudiced and biased.

Nevertheless, this IS the way that the Darwinists, Materialists, and Naturalists use Science and the Scientific Method to prove that the Theory of Evolution is true. They do it every day of their lives, because they are "scientists". They can't see anything wrong with their logic. They simply pick the conclusion they want, and then they make the evidence and their interpretation of the evidence fit their conclusion. There's nothing wrong with that! It's Science!

BUT, the whole enterprise is based upon severe logic fallacies, which the Materialists, Naturalists, Darwinists, Atheists, and scientists in general are unwilling and unable to see and understand. In other words, these people suffer from Confirmation Bias.

Using their logic, we can prove ANYTHING to be true.

Premise or Hypothesis: If our theory matches with and is based upon the fossil record, then our theory matches with reality and is true.

Observation or Test of Hypothesis: We observe that Intelligent Design Theory matches perfectly with the fossil record and matches much

better than the Theory of Evolution ever did; and, we observe that the fossil record is truly there and really exists.

Scientific Conclusion: Therefore, the fossil record proves beyond a shadow of a doubt that Intelligent Design Theory is real and true.

Using the logic of the Scientific Method, we can prove anything we want to be true. The Materialists and Naturalists have been doing that for thousands of years!

In other words, using the Scientific Methods, any conclusion we want can be proven to be true; and, any interpretation of the evidence that we want can be made to fit the scientific evidence. ONLY God knows which interpretation is the BEST interpretation or the RIGHT interpretation.

How's this possible?

It's due to all the different logic fallacies which are built into the Scientific Methods! The Scientific Methods are actually a very poor way of finding and knowing the truth; but, most scientists never make this realization throughout the whole of their careers. I never did; and, I considered myself to be a scientist. I went for 55 years BEFORE I learned anything about any of this! The Materialists and Atheists don't teach us about any of this stuff in college, for some strange reason. ONLY the Theists seem to care about the truth; and, most of them don't know where to find it.

Finally, the ultimate weakness of the Scientific Methods comes from the fact that every science experiment <u>begins</u> with a philosophical guess or hypothesis and <u>ends</u> with some kind of philosophical interpretation of the scientific evidence or the scientific data, which was produced by the science experiment. In other words, the Scientific Methods are based exclusively on philosophical speculation and a philosophical interpretation of the scientific data; and, Philosophy is typically considered to be the worst way or the most unreliable way of finding and knowing the truth, because with Philosophy various different logic fallacies, trickery, deception, self-deception, and sophistry abound and are at work. After all, Materialism, Naturalism, and Atheism are NOT science. They are philosophy, sophistry, self-deception, and metaphysics; and, these philosophies have taken over the whole of science.

When it comes to scientific data and scientific evidence, you can have ANY interpretation that you want; and, that's the fundamental weakness of the Scientific Methods.

Ironically, Darwin's Tree of Life does NOT match with the fossil record that we have on hand! The Materialists, Naturalists, and Darwinists simply affirm the consequent, jump to the conclusion, and assume that it does.

Affirming the consequent is a form of logical error which occurs when we uncritically accept as affirmed or confirmed the results of an "if-then" sequence of reasoning or scientific exploration. *Affirming the consequent* IS *jumping to the conclusion*.

This is why philosophers are wont to point out that for any given fact pattern which can be demonstrated or "discovered" empirically, an infinite number of theoretical explanations are possible. This is also why we say theories remain theories, even after they have been validated. All validating evidence can establish convincingly is the *negation* of a theoretical proposition. Having postulated an '*If A, then B*' sequence, when our researches *fail* to confirm this sequence, we can logically reject the theoretical relation originally postulated. [*Negating the consequent.*] Now this situation bothers philosophers. A datum accrued is not always knowledge gained, because sometimes genuine understanding is lacking. One can have a great deal of information and still suffer for lack of knowledge. (Rychlak, 1981, *A Philosophy of Science for Personality Theory*, pp. 81-82).

Imagine it! There are an infinite number of possible explanations or interpretations for every piece of validated scientific evidence or scientific data that we generate. Picking the explanation or interpretation you like the most and treating it as the God-given truth IS "affirming the consequent" or "affirming your chosen conclusion". This is what the scientists do all the time! We can have a ton of scientific evidence at hand, yet still suffer from a lack of knowledge and truth.

Once again, using the logic of the Scientific Method, we can prove anything we want to be true. The Materialists and Naturalists have been doing so for thousands of years!

The idea or "interpretation of the fossil record", which claims that Intelligent Beings or Intelligent Psyches terraformed this earth and seeded this planet in a systematic and progressive manner, matches infinitely better with the fossil record than Darwin's Tree of Life does, and is infinitely more plausible, logical, reasonable, and believable than claiming that physical matter designed and created the genomes and life forms on this earth as the Materialists, Naturalists, and Atheists claim.

But, the Naturalists, Materialists, and Atheists can't see this because they don't want to see it or understand it. Their ultimate goal is to reject it and to deny it. Anything other than Materialism and Naturalism and Atheism, they declare to be "NOT SCIENCE". I KNOW, because I have had them use that "NOT SCIENCE" argument on me – it's Not Materialism, and it's Not Naturalism, and it's Not Atheism; therefore, it's NOT SCIENCE – as a trump card against me every time I have interacted with them. It's petty, and it's unconvincing; but, it's their bread and butter, and their way of doing Science. Nevertheless, their science stinks; and, it's not science. It's philosophy.

I interacted with the Darwinists, Materialists, Skeptics, anti-Christs, and Atheists online for many years. At first, I was curious to know what I was. I think I was open-minded about it all. As my understanding began to grow and I started to understand what these people really believe, I found myself disagreeing with them at times. When one of my favorite friends turned against me and sicked his Flying Spaghetti Monster on me, I was hooked. I found myself diving into a study of the whole thing, studying it as much as I possibly could.

I started with Antony Flew, allegedly the world's most notorious atheist, and his book, *There Is a God: How the World's Most Notorious Atheist Changed His Mind*. I found out that he gave up his atheism because of Intelligent Design Theory. His reason for giving up his atheism was different than mine; and, I started to take Intelligent Design Theory seriously.

Eventually, I debated about the Intelligent Design Theory with a person online, and the ONLY argument he was ever able to produce against it is that "IT'S NOT SCIENCE" – meaning that it's not Atheism, not Naturalism, and not Materialism. It was then when I fully realized how weak these people's case really is. It was then that I realized that these peoples' philosophical arguments are not science. I began to learn exponentially ever since. These people jump directly to the conclusion that they want, and they exclude everything else that they don't want to hear about, accept, or believe. That's the way these people do science; but, that procedure is philosophy, not science!

In summary, when it comes to scientific data and scientific evidence, due to the philosophical nature of the whole enterprise, you can have ANY interpretation that you want for the results of a science experiment; and, that's the fundamental weakness of the Scientific Methods. Consequently, Science ends up being an extremely weak way of finding and knowing the truth, and no better than philosophy.

In contrast, God KNOWS how life came to be on this planet, because He was there, and He experienced it first-hand. He lived it, so He knows it. The truth is KNOWN by living it and experiencing it first-hand. Whenever the Biblical God Jesus Christ appears to mankind and claims to be the ONE (the Psyche) who designed and created this physical universe, this earth, and all of the life on this earth, I choose to believe Him because He was there and He knows how it was really done. The Materialists, Naturalists, Darwinists, and Atheists don't know and can't know – they are simply guessing, jumping to conclusions, and affirming the specific consequent or the specific conclusion which they desire most to affirm.

The Scientific Methods cannot be used to prove the truth and know the truth. If we want the truth, then we have to get at the truth through some other way or method. Does such a method exist? I say that it does. I KNOW that it does, because I have lived it and experienced it for myself. Through our Lived Experiences it is possible to find the truth, live the truth, experience the truth, and to KNOW THE TRUTH first-hand, which is something that cannot be accomplished with philosophical speculation, scientific methods, scientific experimentation, jumping to conclusions, affirming the consequent, categorization errors, and philosophical interpretations of the scientific data.

Scientific Methods aren't any good for finding and knowing the truth. Living the truth and experiencing the truth is an infinitely better way of KNOWING THE TRUTH.

Now, let's discuss what the Scientific Methods might be good for.

Falsification

In theory, Scientific Methods can be used to falsify theories or to prove theories false. The falsification approach is much more logically sound than the verification approach to the Scientific Methods. The falsification approach does not involve egregious logic fallacies, although it does involve various different types of philosophical interpretation, hypotheses, and philosophical assumptions!

The falsification approach is called, *negating the consequent*; and, it is logically sound, although it still suffers from philosophical interpretations and the fact that it's typically impossible to control ALL of the variables in a science experiment. For example, the human psyche is a hidden variable and an uncontrollable variable in EVERY science experiment, because ONLY the human psyche can do science experiments! The human psyche is mixed up in every science experiment and in the interpretation of the scientific data as well. That's one of the primary weaknesses of the Scientific Methods – personal interpretation of the scientific evidence.

There is a different argument structure that scientists have begun to use instead of verificationism. It is called the falsificationism approach to science. This approach is much more logically sound than the verificationism approach.

The argument structure for falsificationism, or *negating the consequent*, looks like this:

If Theory X is true, then we will observe Y.

We don't observe Y.

Therefore, Theory X is not true.

Rather than verifying a theory, this approach falsifies a theory. Unlike the verification argument, this argument is logically sound. The idea is that although we can't ever know (based on empirical data alone) when a theory is true, we can know (based on empirical data) when a theory is false.

In other words, we develop a theory, make predictions based on that theory, and if those predictions come true, rather than claim that we have confirmed our theory, we more humbly claim that our theory has not yet been proven false. We prove theories false by *negating the consequent*. Eventually, we arrive at the truth by a process of elimination. (Slife; Gantt; Mark My Words).

In effect, this is what Sherlock Holmes does in order to get at the truth of a matter. He centers in on the truth through a process of elimination, or a process of falsification.

How often have I said to you that when you have eliminated the impossible, whatever remains, *however improbable*, must be the truth? — Sherlock Holmes.

The Scientific Methods can be used to falsify theories and to prove things false. *Negating the consequent* is logically sound.

The following is how falsifying a theory works in practice. This example is a direct application of *negating the consequent*:

Scientific Definitions and Principles: By definition in principle, Materialism, Naturalism, and Darwinism are design and creation by physical matter, because the Materialists and Naturalists as a matter of principle by definition and fiat have chosen to believe that only physical matter exists and that the non-local spirit realm and psyche do not exist. For the Materialists and Naturalists, physical matter is all that there is; and therefore, physical matter is the only thing available that can design and create things for us. Technically, Materialism, Darwinism, and Naturalism ARE Creation by Rocks. It's important to get the fundamental and the correct definition for things!

Premise or Axiom or Law: The Scientific Methods can be used to falsify theories and to prove things false.

Scientific Hypothesis: If Materialism, Naturalism, and Darwinism are true, then we will observe the rocks and physical matter designing, creating, and manufacturing genomes and life forms from scratch.

Scientific Observations: Spontaneous generation and Abiogenesis have been proven false – Louis Pasteur proved Materialism, Darwinism, and Creation by Rocks false in 1859. We don't ever observe the rocks designing and creating anything. They can't.

Scientific Conclusion: Therefore, Materialism, Naturalism, and Darwinism are not true.

Simple truth!

Can you feel the truth of it?

I just falsified Materialism, Naturalism, and Darwinism! And, my argument is logically sound. It's also valid and true. This is precisely what I did the first time that I falsified the Theory of Evolution. It works, because it's true.

By getting the proper and the most fundamental definition for Materialism, Naturalism, and Darwinism, I have just successfully used the Scientific Method to falsify Materialism, Naturalism, and Darwinism. Said another way, I simply took the Materialists, Naturalists, and Darwinists at their word by assuming along with them that ONLY physical matter exists. Then based upon their assumption or scientific hypothesis, I proceeded to falsify their assumption or their hypothesis using the Scientific Method and observation, which is easy to do because raw physical matter has never been caught in the act of design and creation, and never will be.

Scientific Definition: Materialism, Naturalism, and Darwinism are design and creation by physical matter.

Scientific Hypothesis: If Materialism, Naturalism, and Darwinism are true, then we will observe the rocks and physical matter designing and creating genomes and life forms.

Scientific Observation: We have NEVER observed the rocks designing and creating anything; and, we NEVER will.

Scientific Conclusion: Therefore, Materialism, Naturalism, and Darwinism are false.

Materialism, Naturalism, and Darwinism have been falsified or proven false thousands of different ways. Consequently, if we want to KNOW the Truth, then we have to look someplace else besides Materialism, Naturalism, and Darwinism. This argument is logically sound, because the falsification of these scientific theories is logically sound.

According to the Materialists and Naturalists, raw physical matter was the ONLY thing in existence before the first genomes and the first life forms were designed and created; therefore, the raw physical matter or the rocks must have designed and created the first genomes and the first life forms on this planet. Such a materialistic and naturalistic claim VIOLATES the lived experiences, observations, and common sense of the human race, because we all KNOW for a fact that the rocks can't design and create anything at all. Our idols can't function in the role of Psyche or God. The rocks can't do formal cause and final cause.

In contrast, intelligent beings, human beings, human psyches, or psyche beings have been caught in the act of design and creation trillions of trillions of different times. Intelligent Design and Creation by Intelligent Psyches has been experienced and observed trillions of different times in trillions of different situations and ways. Follow the evidence! It ALL points to Psyche or Intelligence, because raw physical matter cannot design and create.

This is also an excellent Proof of God's Existence and an excellent proof of psyche's existence. God must of necessity exist in order to have done all of the design, creation, science, engineering, manufacturing, diversification, and systematic deployment that needed to be done, which the rocks and physical matter could NEVER have done, while the genomes and the life forms were being designed, created, and placed upon this earth.

This falsification approach to Science and the Scientific Methods is much more logically sound than the verification approach, because the falsification approach isn't based upon an egregious logic fallacy or two. In other words, my falsification of Materialism, Naturalism, and Darwinism IS logically sound! It also seems to be true, in that it actually matches with reality; whereas, Materialism and Naturalism don't.

Now, let's take this to its logical conclusion. I have observed that Materialism and Naturalism and even Atheism have been falsified trillions of trillions of different times in thousands upon thousands of different ways. There are so many different ways to falsify Darwinism, Naturalism, Atheism, Nihilism, and Materialism that it's impossible to count all the different ways that this has been done and can be done. Furthermore, the falsifications of these philosophical concepts continue to pour into the records of science, and into the records of the human race as well. In fact, every time you have a thought, or see a ray of light, or feel the press of gravity, or witness the effects of magnetism, or have an out-of-body experience, or have a spiritual experience, you have just falsified Materialism and Naturalism and Darwinism. It's easy to do! Thoughts and dreams are spiritual experiences, because there's no way possible to record our thoughts and dreams with our physical instruments. Our very thoughts and dreams – our Psyche Experiences – falsify Materialism and Naturalism and Darwinism.

In contrast, design and creation by Psyche or Intelligence – sometimes called Intelligent Design – has NEVER been falsified and never will be. In fact, Design and Creation by Psyche or Intelligence has been observed and verified and validated trillions upon trillions upon trillions of times, with an infinite number of more times to go. Follow the evidence. Think logically and rationally. The truth is staring us in the face; but, most of us refuse to see it.

This whole falsification process explains why the Scientists and Philosophers, whom I have been reading and studying the past couple of years, refuse to believe in Materialism, Naturalism, and Darwinism on a blind leap-of-faith that these things are axiomatically true, because these people too have used logic and the Scientific Methods to falsify Materialism, Naturalism, and Darwinism in one way or another. It's easy to do. Where there's a will, there's a way.

In contrast, it's literally impossible to falsify THE TRUTH using the Scientific Methods. The scientists will never be able to falsify Intelligent Design Theory, or design and creation by Intelligence, because it's true. Design and Creation by Intelligence or Psyche can be verified and replicated on demand! The truth never fails and is never falsified! ONLY Psyche – ONLY Intelligence – can design and create things, and do science. Psyche or Intelligence is the ONLY thing we have ever observed doing so. You can verify this right now for yourself, by choosing to design and create something for someone, because the rocks can't do it for you. It's elementary my dear friend.

See also:

Slife, B. D. & Williams, R. N. (1995). Science and Human Behavior. In *What's Behind the Research? Discovering Hidden Assumptions in the Behavioral Sciences*, (pp. 167–204). Thousand Oaks, CA: SAGE Publications.

http://mypsyche.us/science/

The Ultimate Alternative for Knowing the Truth

I think that the ultimate discovery of my Science Career came when I finally observed, realized, and accepted the fact that Lived Experience IS THE BEST WAY of finding and knowing the truth. In fact, it's the ONLY way to find and know the truth, because ONLY Psyche can do Science, Philosophy, Lived Experiences, finding, knowing, reality, existence, and truth. This Reality and Truth is what I call a Psyche Epistemology, which ends up being the BEST METHOD for finding and knowing the truth.

Lived Experience or Psyche Experience is the Ultimate Method for finding and KNOWING the TRUTH! Finding and knowing the truth is what Psyche does. Psyche or Intelligence is Light and Truth and Knowledge.

We KNOW from the Lived Experiences of the human race that Psyche or Non-Local Consciousness does indeed exist. (Buhlman; Durham; Storm; Alexander; Sharkey & McCormack; Neal; Van Lommel; Moody; Gibson; Long; Hinze; Ring; Rivas; Dispenza; Eccles; & Penfield). Psyche IS the fundamental unit of Reality. A Psyche Ontology is the Ultimate Ontology, because ONLY Psyche can do ontology and lived experiences. Consequently, a Psyche Epistemology is the Ultimate Epistemology, because ONLY Psyche can do epistemology.

There is a way to KNOW the TRUTH! We know the truth by living the truth and experiencing the truth for ourselves, or by choosing to trust someone who has.

Unlike the Scientific Methods, Philosophical Speculation, and the Philosophical Interpretation of Scientific Data, which technically can NEVER be used to find the truth, verify the truth, and prove the truth, we can KNOW THE TRUTH by living the truth and experiencing the truth directly for ourselves or by choosing to trust someone who has.

The Lived Experiences of the human race, both our physical experiences and our spiritual experiences, ARE the repository of Truth and Knowledge for the human race. Living the truth and experiencing the truth ARE the best and most efficient method for KNOWING the TRUTH.

Our Philosophical Speculation and our Scientific Interpretations have to match with the Lived Experiences of the human race; otherwise, our philosophies are wrong, and our scientific conclusions are incorrect. Our interpretations of scientific evidence have to match with Reality, or our interpretations are false! A materialistic, naturalistic, and Darwinian interpretation of scientific evidence is always false, because Materialism, Naturalism, and Darwinism don't match with the Lived Experiences of the human race and therefore don't match with Reality.

ONLY Psyche can have Lived Experiences; and therefore, ONLY Psyche can KNOW the TRUTH by directly experiencing, living, and remembering the truth for itself.

God, or God's Psyche, is KNOWN by direct experience and by what God chooses to do for us or chooses to reveal to us. God is KNOWN to us by doing for us the things which we could never have done for ourselves – such as by designing and creating our genomes, the first of our cells, and our physical bodies.

I KNOW from direct observation and experience that the rocks can't design, create, and manufacturing anything. Therefore, I KNOW that God must of necessity exist in order to have designed, created, engineered, and manufactured all of the genomes and life forms in the first place. I KNOW that God and God's Psyche must of necessity exist in order to have done all of the Science that needed to be done in order to bring life to this planet. I KNOW that ONLY Psyche can do Science.

I also KNOW from Quantum Mechanics that Psyche must exist in order to bring physical matter into existence – in order to convert spirit matter into physical matter – in order to convert non-locality into locality. In a sense, Science itself has proven to me that God exists; however, it was in fact my lived experiences and the lived experiences of the human race which proved to me that God does in fact exist. Unlike Philosophy and the Scientific Method which can never be used to prove the truth or know the truth, we can prove the truth and KNOW THE TRUTH through our Lived Experiences, by living the truth and experiencing the truth for ourselves, or by choosing to trust someone who has. Experiential Knowledge is vastly superior to Science, Scientific Methods, and the Philosophical Interpretation of Scientific Data.

An epistemology is a way of knowing the truth. A Psyche Epistemology is better than any Philosophical Epistemology, Naturalistic Epistemology, or Scientific Epistemology, because Lived Experiences are a better way of knowing the truth. In fact, Lived Experiences are the ONLY way to know the truth. Consequently, a Psyche Epistemology is the Ultimate Epistemology, because a Psyche Epistemology subsumes or includes ALL other epistemologies, because ONLY Psyche can do epistemology.

An ontology is a fundamental unit of reality. A Psyche Ontology observes and concludes that Psyche is the fundamental unit of reality. Psyche is axiomatic, foundational, fundamental, essential, necessary, self-evident, lawful, tautological, immanent, infinite, obvious, ultimately causal, and self-sustaining, because ONLY Psyche can do psyche. It's elementary my dear friend!

In other words, we come full circle by observing and accepting the fact that we KNOW from the Lived Experiences, Psychic Experiences, Spiritual Experiences, and Psyche Experiences of the human race that Psyche or Non-Local Consciousness does indeed exist.

ONLY Psyche can have Lived Experiences and remember these experiences in order to share them with others at a later date; and therefore, ONLY Psyche can KNOW the TRUTH by directly experiencing and living the truth for itself. A Psyche Epistemology is the Ultimate Epistemology. A Psyche Ontology is the Ultimate Ontology, because ONLY Psyche can do epistemology and ontology.

Lived Experience or Psyche Experience is the Ultimate Method for finding and KNOWING the TRUTH! In order to find the truth and KNOW the truth, I had to get rid of My Materialism, My Atheism, My Naturalism, and My Nihilism. We have to get rid of all the falsehoods in our lives, in order to find the truth and KNOW the truth. It's elementary my dear friend!

You, your psyche, will determine for yourself whether you believe any of this to be true or not. For you personally, your psyche is the ultimate arbiter of truth. You, your psyche, will decide if any of this is useful to you or not. You are your own punishment, and you are your own reward. I believe that I have met my burden of proof; but, you are free to decide differently if you want to.

My hope and my expectation is that if you have reached this far into my introduction to Psyche as the Ultimate Cause and to Psyche as the Ultimate Ontology, then you are starting to take me seriously. If not, please read it again. After reading it again, if you're still not getting it then it's probably not for you – your biases and prejudices are too strong and are preventing you from being able to see it and understand it. Try again some other time when you have changed your mind.

As you can probably tell, I love this stuff. I'm passionate about all of this, because I feel as if I have finally found the truth. Truth feels good and tastes good. We get goosebumps or we get animated and enlivened, whenever we speak the truth, write the truth, or are in the presence of truth, unless of course we have desensitized ourselves to the truth.

All of this is infinitely more interesting than My Materialism and My Atheism ever were!

This Ultimate Model of Reality is just a great deal of fun to think about, write about, and study!

Definitions Explaining Ultimate Cause

Truth Is Knowledge: Knowledge is truth, or at least it should be. It's kind of impossible to actually know the truth by choosing to believe in something that it false. Furthermore, knowledge of falsehoods doesn't do us any good, UNLESS we actually know that these things are false and know why they are false. Back when I was a Materialist, Nihilist, and Naturalist, I didn't even know what those terms actually meant. At the time, I didn't know what a Materialist, Naturalist, and Nihilist was. I simply was one. All I knew at the time was that I was an Atheist. My Atheism was based upon my lived experiences, wherein I believed at the time that God was completely absent from my life and that as a result of my experiences (or lack of experiences) I was convinced and believed that God does not exist. I was without God in the world. From my perspective during those days of my life, I seemed to know something that nobody else around me seemed to know – I knew that God doesn't exist. I had taken that leap of faith. Of course, I was wrong; but, I didn't know that, while I was right in the middle of it all. It was only after I – with God's help – had successfully gotten rid of My Materialism, My Nihilism, My Atheism, and My Skepticism that I was finally able to see and understand all of the different miraculous interventions which had taken place in my life to keep me alive, to keep me safe, to help me heal, to bring me back to sanity and sobriety, and to help me learn from my mistakes and the falsehoods that I had chosen to believe in. God had been helping me all throughout the whole process, sometimes in amazingly miraculous ways, but I couldn't see it at all when I was a Materialist and an Atheist and living right in the middle of it. Thank God He kept me alive long enough to see and understand the errors of my ways. God is merciful, gracious, forgiving, and kind; but, I wasn't able to see that when I believed that He was not.

We blind ourselves to the truth whenever we choose to believe in falsehoods. That's just the way it works. We KNOW the truth, the real TRUTH, by living it and experiencing it for ourselves. We know it, when we OWN it and take full responsibility for it. Nowadays, I even KNOW and understand how and why Materialism, Naturalism, Nihilism, Darwinism, and Atheism ARE false. I KNOW the truth about these things! It's amazing how that works! God was schooling me and teaching me by letting me experience these things for myself. I personally needed to know what it's like to be a Skeptic, Materialist, Nihilist, and Atheist – I needed to know and understand the hell that is typically associated with these worldviews, so that I could more fully appreciate and understand their opposite. The pain makes it real. The lessons are actually learned for real. Pain is our greatest teacher. It motivates us to learn. Through pain's ministrations, we OWN the knowledge that is received through the experience. We experience the knowledge. We experience the truth. We come to KNOW it. It becomes an integral part of us. I now have no desire whatsoever to go back to My Nihilism and My Atheism, because now I KNOW precisely what they are and what they did to me. It was hell. I can see that now; whereas, I couldn't see it before. I had to live it and experience it in order to KNOW it and understand it. Our lived experiences are our greatest and most effective teachers. I can see that now. I KNOW it, because I lived it.

Psyche Ontology or the Ultimate Ontology: With this Ultimate Model of Reality, I'm introducing a Psyche Ontology in which psyche is the fundamental unit of reality and the ultimate maker of meaning and reality. This Ultimate Cause Model of Reality is a Psyche Ontology. ALL the evidence led me to this truth. Brent Slife and friends expended a lot of effort developing a Relational Ontology in which relationships are the fundamental unit of reality; but, it quickly became clear to me that ONLY Psyche can do relationships and friendships, so it became obvious to me that Psyche is in fact the fundamental unit of Reality. Of course, I easily adopted this stance, because I have been developing and promoting Psyche as the Ultimate Cause for the past half year or so, in my theorizing and writing. I find it interesting that nobody else seems to have developed and presented a Psyche Ontology before. Everyone seems to be programmed and conditioned to overlook Psyche automatically; but, ever since I read Buhlman's book at the end of 2015, I have KNOWN that Psyche exists and have basically known what Psyche is. I have been looking at Psyche ever since!

Psyche Epistemology or the Ultimate Epistemology: Eventually, it became clear to me that I was also looking for the Ultimate Epistemology to go along with this Ultimate Model of Reality. George Kelly's Scientism Epistemology or Constructive Alternativism was my original preference, because I considered myself to be a scientist more than anything else. Eventually I found myself looking at Joseph Rychlak's focus upon Teleology and Final Cause, and some kind of Telic Epistemology. Then I developed Spiritual Empiricism as an epistemology, because that seemed to fit better with Psyche as the Ultimate Cause. Then I started reading Brent Slife and friends. In their writings, these people toyed with William James' Radical Empiricism (all of our experiences including our spiritual experiences), Levinas' Phenomenology (phenomena and events and lived experiences), and Hermeneutics (studying the interpretation and the meaning of experiences and events). Eventually it dawned on me that I was looking for a Psyche Epistemology, because ONLY Psyche can do epistemologies. ONLY Psyche can do science, constructs, relationality, spirituality, teleology, lived experiences, phenomenology, and hermeneutics. Therefore, I introduce my reader to a combined Teleological Epistemology, Hermeneutic Epistemology, Radical Empiricism Epistemology, Spiritual Empiricism Epistemology, Phenomenological Epistemology, and Traditional Empiricism Epistemology which is based upon the Lived Experiences of the human psyche and the Lived Experiences of the human race as a whole. This is a Psyche Epistemology – the ultimate way of knowing the truth. ONLY Psyche can do epistemology. ONLY Psyche can find and know the truth. The ultimate purpose of my books is to introduce my reader to a Psyche Ontology in which immortal psyche is the fundamental unit of Reality and therefore the Ultimate Reality. This leads naturally to a Psyche Epistemology, wherein the Lived Experiences of the human psyche become the ultimate best way of finding and knowing the truth. The truth is known by living it and experiencing it first-hand, whether we are talking about a spiritual reality or a physical reality. This is a Psyche Epistemology, the Ultimate Epistemology. In ALL of my theorizing, I kept adjusting this Ultimate Model of Reality until it matched with the Lived Experiences or the Psyche Experiences of the human race. I wanted to get this model as close to the Truth and Reality as I possibly could.

Radical Empiricism: This Ultimate Cause Model of Reality fully embraces what William James called Radical Empiricism, which means that this Ultimate Model of Reality subsumes or includes and encourages ALL epistemologies or all forms of knowing the truth, including philosophy, logic or rationalism, dialectical and demonstrative reasoning, physical empiricism, the Scientific Methodologies, scientific experimentation, qualitative methodologies, application or practice, quantum mechanics or spiritual mechanisms, lived experiences, hermeneutics, phenomenology, spiritual experiences of all kinds, theophanies or revelations of God, and revelations from God. Nothing is off-limits nor out-of-bounds, which is why I chose to call it the Ultimate Model of Reality. It is a Psyche Ontology – a sort of Holistic Ultimology. Its study is the whole of Reality, and not just a small part of reality. The Ultimate Model of Reality also subsumes a Psyche Epistemology, because ONLY Psyche can do epistemology. (James, W. (1912). *Essays in Radical Empiricism*. https://www.amazon.com/dp/B004TS16TI/).

Lived Experience: Truth is based upon Lived Experiences, and not philosophical abstractions! From the Wikipedia: Pragmatism rejects the idea that the function of thought is to describe, represent, or mirror reality. Instead, pragmatists consider **thought** [psyche] an instrument or tool for prediction, problem solving, and action. Pragmatists contend that most philosophical topics — such as the nature of knowledge, language, concepts, meaning, belief, and science — are all best viewed in terms of their practical uses and successes. The philosophy of pragmatism "emphasizes the practical application of ideas by acting on them to actually test them in **human experiences**". Pragmatism focuses on a "**changing universe**" rather than an unchanging one as the Idealists, Realists, and Thomists had claimed". (The preceding was taken from the Wikipedia.) The pragmatic focus of this Ultimate Cause Model of Reality is upon Lived Experiences, human experiences, or psyche experiences – the things which the human psyche actually experiences and acts upon, both while in the physical body and while separated from the physical body. Thought or Psyche is the fundamental unit of reality. Psyche as the Ultimate Cause is based upon Radical Empiricism, or Lived Experiences, or Phenomenology. ONLY Psyche can do life and lived experiences. Thought is a function of psyche, because we continue to think and perceive after our physical body is dead and our physical brain has gone offline, according to the Science of NDEs and OBEs. Every time you have a thought, you have had a spiritual experience. Psyche produces thoughts, and it remembers them. Psyche is a conserved quantum that is capable of making and storing conserved quantum information or memories. Psyche also remembers all of our choices and all of our experiences. Pragmatism is the Philosophy of Thoughtful Practice – practical uses, choices, lived experiences, spiritual empiricism or spiritual experiences, evidentiary pluralism, and successful acts. William James treats Pragmatism or Thought as an epistemology – a way of knowing the truth. For me, all of this started to point to a Psyche Epistemology! William James's Radical Empiricism or Lived Experience, Methodological Pluralism, and Pragmatism are briefly summarized in this article: http://brentslife.com/article/upload/evidence/EBPP%20-%20APA%20Taking%20Sides%20final.pdf. William James was a hundred years ahead of his time. The rest of the scientists are finally starting to catch up.

Phenomenology and Hermeneutics: Phenomenology is the philosophical and scientific study of Lived Experiences, events, and phenomena. ONLY Psyche can do life, consciousness, awareness, and lived experiences. Matter of any kind, which has no psyche residing within it, is incapable of experiencing life and remembering events. Psyche is Life. Psyche is our personality and our memories. Psyche is consciousness and awareness and personality. Psyche is the Ultimate Truth. Hermeneutics is the philosophical and scientific study of interpretations, explanations, and meaning. Teleology is the philosophical and scientific study of purpose, intentions, reasons, desires, wishes, final causality, and goals. Psychology is supposed to be the study of psyche. ONLY Psyche or Non-Local Consciousness can do hermeneutics and teleology and psychology. ONLY Psyche can do interpretation, meaning, purpose, reasoning, thinking, dreams, planning, and final causality. The rocks don't think, reason, nor dream unless they have some kind of living psyche residing within them. ONLY Psyche can do Phenomenology, or Lived Experiences, or Radical Empiricism.

Nonlocality: Although there are dozens of other observed and experienced scientific discoveries which support Psyche as the Ultimate Cause, an understanding and acceptance of Quantum Nonlocality, Quantum Entanglement, and Quantum Consciousness seems to be the most crucial and essential Science for understanding Psyche as the Ultimate Cause. Nonlocality means non-physical, or not located in our physical 3D space-time universe. Nonlocality by definition means that quantum objects (the wave form of quantum objects) and consciousness are trans-dimensional, non-physical and/or immaterial, and not located in our physical 3D space-time reality. Once a non-local trans-dimensional quantum object, or quantum wave, or light wave, or energy wave is consciously observed or receives the Word of Command, then its wave function collapses, and it becomes a physical particle, or a photon located here in our 3D physical space-time. Again, non-local means non-physical, or trans-dimensional and spiritual, in nature and origin. Quantum Entanglement has in effect proven Quantum Non-Locality to be real and true, because every science experiment regarding Quantum Entanglement has verified Quantum Entanglement to be real and true. Computer Science and the makers of computer chips are completely reliant upon Quantum Mechanics and limited by Quantum Mechanics (or spiritual mechanisms) in what they can do. They can only go so small before the thing ceases to be physical and starts to become spiritual and uncertain instead. The closer we get to the spiritual or the quantum, the more psyche's agency or choice comes into play, and the more unpredictable, unreliable, and uncertain things become. The physical realm is a highly reliable and highly predicable consensus reality, the ultimate consensus reality; but, a consensus reality is made up of non-consensus realities, quantum realities, or spiritual realities where agency, free-will, self-determination, uncertainty, and psyche reign supreme. I have stress-tested computer chips, and I have had them leak through or break past their physical limitations, and then become unpredictable – seeming to develop a mind of their own. Quantum Mechanics or Spiritual Mechanics is real stuff with real-world applications. Quantum Mechanics is verified and validated science, possibly the best-proven and most-used science that we currently have; but, Quantum Mechanics only makes sense from a spiritual or a non-local perspective. Quantum Mechanics when properly understood proves that Materialism is false (falsifies Materialism) and essentially demonstrates and "proves" that some kind of psyche, conscious observer, and non-local consciousness must of necessity exist. For most people, the Science of Quantum Mechanics and Quantum Non-Locality have proven beyond a shadow of a doubt that some kind of non-local or spiritual conscious observer is the ultimate cause behind each and every physical particle that has ever been brought into existence. Remember, it's possible to falsify theories or to prove them false using Science and the Scientific Methods; therefore, it's theoretically possible to center in on the truth through a process of elimination or falsification. In other words, Quantum Mechanics has always been verified. It has never been falsified. That reality gives us the impression that Quantum Mechanics has been proven true; and, Quantum Mechanics explains how Psyche and Spirit Matter really work. Again, Quantum Mechanics is the best proven and most-used Science that we currently have because it hasn't been falsified; and, Quantum Mechanics when properly understood proves through a process of elimination the existence of the non-local, the non-physical, the immaterial, and the spiritual, as well as "proving" that Psyche is necessary and must exist. In other words, Quantum Mechanics when properly understood falsifies Materialism and Naturalism and proves that they false; and through the process of elimination, indirectly "proves" that psyche or non-local consciousness must exist. That's the primary lesson of Quantum Mechanics or Spiritual Mechanics, because Quantum Mechanics is the science of how spirit matter and the spirit realm really work. Of course, we KNOW from the Lived Experiences of the human race that Psyche, Spirit Matter, the Spirit Realm, Spirit Bodies, and the Biblical God Jesus Christ exist and that Jesus rose from the dead; but, Quantum Mechanics becomes a Scientific Way of demonstrating that Psyche and Spirit Matter and God must exist. ALL TRUTH is mutually sustaining, because all truth will match with Reality and the Lived Experiences of the human psyches or the human race. We also learn from Quantum Mechanics that a wave-like quantum object or spirit matter is a completely different substance or essence than non-local consciousness, psyche, intelligence, awareness, life, or mind. Consciousness or Psyche chooses and acts; whereas, spirit matter or the wave form of quantum objects is acted upon by consciousness. Consciousness and spirit matter are two completely different things in the Non-Local Realm, or Quantum Realm, or Trans-Dimensional Realm, or Spirit Realm. (Van Lommel, 2010; Mark My Words, 2016, *Quantum Mechanics from a Non-Physical Spiritual Perspective*).

Psyche: Psyche is your spirit, soul, mind, consciousness, life, spark, individuality, intentionality, awareness, intelligence, personality, memories, immaterial entity, infinite singularity, living light, existence, long-term memory storage, and non-local life form. Although Psyche or Intelligence is often confounded or equated with spirit matter, spirits, and our spirit body, it has been observed that they are two completely different things. Psyche or non-local consciousness is an immaterial, limitless, infinite singularity or viewpoint in space. Psyche is your core being, your essence, your existence, and your life. Psyche or consciousness is the ultimate cause behind everything else that has ever been organized or brought into existence, including any organization of spirit matter and physical matter, and the existence of physical matter in the first place. Psyche is the eternal and the indestructible part of us. It has always existed and will always exist. ONLY Psyche can do lived experiences, life, memories, epistemology, truth, knowledge, and ontology. Psyche is where the memories of our lived experiences are stored; and, our memories go with us into the next life after our physical body and physical brain are dead and gone. (Buhlman, 1996, *Adventures Beyond the Body: How to Experience Out-of-Body Travel.*)

Psychology: Psychology is the study of the human psyche. Psychology is philosophy with a practical, useful, interesting, compelling, and realistic application. Psychology is also an attempt to study psyche scientifically through observation and empirical evidence. One of the goals of this Ultimate Cause Model of Reality is to put psyche back into Psychology. Psychology has been defined as the study of behavior and experience. The Materialists, Naturalists, and Atheists focus all of their attention on observable physical behavior; and, the rest of us study the Lived Experiences of human beings or human psyches. The one is infinitely more expansive than the other. I have learned to love the Study of Psyche. It's infinitely more interesting than My Materialism and My Atheism ever were!

Darwinism: The philosophical and religious belief taken on blind faith that physical matter designed and created the first functional genome on this planet and the first life form on this planet from scratch. In its purest and most basic form, Darwinism is Creation by Chance, Creation by Rocks, Abiogenesis, or Spontaneous Generation. Darwinism is another form of Materialism, and the foundation of Scientific Naturalism. The Theory of Evolution is Creation by Rocks. Darwinism or Creation by Rocks is false, because we all instinctively know that the rocks or physical matter cannot design and create anything at all. The rocks have never been caught in the act of doing so; and, they never will, because rocks can't design and program genomes and because rocks cannot create and manufacture cells and life forms. The evolutionary functions (or change functions) of natural selection and random mutations did not exist until AFTER some kind of Ultimate Cause or Ultimate Psyche designed and created the first genome and the first life form on this planet in the first place. This is common sense logic, which the philosophers behind Darwinism and Naturalism choose to completely ignore. Science itself has demonstrated that natural selection and random mutations cannot design and create anything at all, but the Materialists and Darwinists refuse to look at, study, and accept this evidence. Darwinism is fascinating to study, but only for purposes of debunking it and falsifying it. It's fun to see how many different ways we can falsify Darwinism and the Theory of Evolution. Remember, Scientific Methods can be used to falsify theories like the Theory of Evolution; and, it's a great deal of fun to do so and to learn how to do so. (Sanford, 2013; Sanford, 2014).

Materialism: To know what something truly is, you have to know what it is not. This Ultimate Cause Model of Reality is NOT Materialism. Ultimate cause or psyche is anti-Materialism, and vice versa. Scientific Naturalism is incompatible with Theism, Psyche, Final Cause, and Ultimate Cause. Materialism and Psyche are mutually exclusive – they both can't be true at the same time. We are forced to choose between them. Materialism let me down. Materialism almost killed me. Materialism is incapable of supporting a robust, believable, and all-inclusive Science. Materialism is the philosophical idea or religious metaphysical idea that spirit, psyche, the non-physical, the non-local, consciousness, and the immaterial do not exist. Materialism is the chosen belief that physical matter is all there is and that nothing else exists or matters. When applied to science, Materialism is often called Scientific Naturalism. Materialism is metaphysics, or a philosophical assumption, not a science. Materialism is a denial of empirical evidence and scientific evidence, and a refusal to look at that evidence. Materialism is the polar opposite of NDEs, SDEs, and OBEs; and, Naturalism is a denial that such Out-of-Body Experiences actually happen for real. There is no scientific evidence and can be no scientific evidence to support the claims or major premise of Materialism. Finding scientific support and evidentiary support for a Denialism, such as Materialism, is philosophically and logically impossible. Materialism is design and creation by rocks or by physical matter. Raw physical matter by definition in principle cannot design and create. ALL of the empirical evidence, observational evidence, and experiential evidence stands firmly against Materialism, because rocks or physical matter cannot design, create, nor manufacture anything at all. It is obvious to all of us, except for the atheists (atheists are only 3% of the population), that Materialism is false, doesn't match with reality, and doesn't hold water. Technically, Materialism is a denial of the existence of psyche, psychology, quantum non-locality, light, gravity, dark energy, dark matter, consciousness, spirit matter, forces, fields, time, magnetism, thoughts, dreams, space, and everything else that is non-physical which exists. Materialism is incompatible with the Science of Psychology, the study of psyche. Materialism is incompatible with Science, because Materialism is unscientific and doesn't match with reality. Materialism is the very definition of Bad Science and Pseudo-Science, because Materialism is in fact metaphysics or philosophy masquerading as science, but not science. (Richards & Bergin, 2005). A strong Personality Theory and a solid model of reality must match with Reality. Materialism and Scientific Naturalism do not match with reality, thus making them incomplete and unscientific models of reality. Materialism is an unsubstantiated model of reality that doesn't hold up under scrutiny. Darwinism is just another form of Materialism. They both reduce to Creation by Rocks or Creation by Chance, which are scientifically impossible and have never been observed in the wild. At their most fundamental, Darwinism and Materialism are Creation by Rocks or Creation by Physical Matter; but, we all know instinctively that the rocks cannot design and create anything at all, let alone fully functional genomes, computers, software, and complex life forms. Spontaneous generation, Creation by Rocks, Materialism, and Darwinism were proven false by Louis Pasteur in 1859, the same year that Charles Darwin published "On the Origin of Species". Furthermore, Quantum Mechanics when properly understood proves beyond a shadow of a doubt that Materialism and Naturalism are false. The Science of Quantum Mechanics trumps the philosophies of Materialism, Naturalism, and Darwinism proving Materialism of any type to be false. Once we have eliminated Materialism or Naturalism and its derivatives from our worldview, then we are obligated to try to put something much more viable and believable in its place. I suggest that psyche or ultimate cause should be the thing we use to replace the falsehoods generated by Materialism and Naturalism. (Van Lommel, 2010; Mark My Words, 2016, *Quantum Mechanics from a Non-Physical Spiritual Perspective*).

Buddhism: The Buddhists have chosen to believe that psyche or the soul does not exist and that our personality and memories cease to exist when we die. Such an idea is not compatible with Psyche as the Ultimate Cause. Buddhism is "spirituality" for Materialists, Naturalists, and Atheists. Buddhism is Atheism, although there is some debate as to whether Buddhism is Materialism and Naturalism, because Buddhism is internally inconsistent. Parts of Buddhism are deliberate non-sense, such as the koan. Materialism and Scientific Naturalism are incompatible with Theism, and therefore more amenable to something like Atheism or Buddhism. This Ultimate Cause Model of Reality is incompatible with Materialism, Scientific Naturalism, and Atheism; and therefore, incompatible with the fundamental philosophical beliefs of Buddhism. In contrast, the Hindus believe in atman or psyche; therefore, Hinduism would be supportive of and compatible with Psyche as the Ultimate Cause. Islam is theism; and, the Muslims believe in an afterlife and the continuation of the soul or psyche into the afterlife. So once again, Islam would be compatible with and supportive of Psyche as the Ultimate Cause. Christ and the Bible promote an afterlife for the psyche or the soul. Most of the world's religions and most of this world's population would be supportive of Psyche as the Ultimate Cause; but not the Materialists, Naturalists, Atheists, and Buddhists. The law of non-contradiction states that they both can't be right. Either psyche exists, or it does not. "The existence of psyche" and the "non-existence of psyche" can't be true at the same time, because that would be a violation of the law of non-contradiction. We are each forced to choose what it is that we want to believe. We are forced to choose between the two. Belief or faith is a decision or a choice, hopefully based upon lots of evidence. When choosing between psyche and the non-existence of psyche, I have chosen to go with the one that has the most evidentiary, scientific, empirical, observational, logical, and experiential support because there is no evidence and can be no evidence to support "the non-existence of psyche" or any other such Denialism like Materialism, or Naturalism, or Atheism, or Buddhism. ALL of the evidence is on the side of Psyche as the Ultimate Cause. I finally chose to follow the evidence, which any good scientist should do.

The Scientific Method: Although Scientific Methods can never be used to prove the truth, the Scientific Methods can be used to falsify theories or to prove theories false. Science and the traditional Scientific Methods prove Materialism and Naturalism false many different ways. First of all, there is no way possible to design a science experiment which demonstrates and proves that the non-physical or the spiritual does not exist. Instead, ALL of the science experiments which we have done over the centuries keep revealing to us the existence of the immaterial or the non-physical in some form or another – such as gravity, forces, fields, extra-dimensionality, trans-dimensionality, quantum non-locality, quantum entanglement, light, dark energy, dark matter, the zero-point field, magnetism, thoughts, dreams, space, time, non-local consciousness, and a whole host of other things which are immaterial and non-physical in origin and nature. In other words, Science and Common-Sense Logic as a whole prove Materialism, Naturalism, and Darwinism false. Remember, we can use the Scientific Methods to prove theories false or to falsify theories, and we have done so when it comes to Materialism, Naturalism, and Darwinism. Second of all, the various different forms of Materialism are in fact Creation by Rocks or Creation by Chance; and, there is no way to design and create a science experiment wherein we capture rocks or raw physical matter in the process of designing and creating genomes and life forms from scratch. It can't be done. Therefore, once again, we can't use the Scientific Method to demonstrate and prove that Materialism, Darwinism, and Naturalism are true. Instead, in 1859, Louis Pasteur used the Scientific Method to prove that spontaneous generation or Creation by Rocks is false. Therefore, the net effect of all of this is that the Scientific Method itself proves through Common Sense Logic, Experimentation, Procedural Evidence, and the Reality of the Situation that Materialism, Darwinism, and Naturalism are in principle false and patently non-starters. There's no way to use the Scientific Method to prove the philosophies of Materialism, Naturalism, and Darwinism true; consequently, the Scientific Method effectively proves Materialism, Darwinism, and Naturalism false. Furthermore, there is no evidence and can be no evidence to support the major premises of Materialism and Scientific Naturalism stating that psyche or spirit does not exist; whereas, there are thousands of books and videos and millions of personal experiences which provide empirical evidence supporting psyche and ultimate cause, in the form of NDEs, SDEs, OBEs, and other types of spiritual experiences or Spiritual Empiricism. If we choose to follow the evidence, we can easily prove to ourselves that Materialism and Scientific Naturalism are false. That's what happened to me. I used to be a Materialist and an Atheist, but then I finally started looking at all the evidence. Lastly, the Scientific Methods and Common Sense have already been used to prove (verify) Quantum Mechanics, Quantum Nonlocality, Quantum Entanglement, light, gravity, forces, fields, dark energy, and dark matter to be real and true, in that these things have always been verified and never been falsified; and, each one of these things proves Materialism and Scientific Naturalism false while at the same time being supportive of Psyche or Consciousness as the Ultimate Cause. Therefore, the Scientific Method has already been used to prove Materialism, Scientific Naturalism, and Creation by Rocks (the Theory of Evolution) false; but, the Materialists and the Atheists refuse to look at the evidence. Whenever we use the Scientific Method the way that it is supposed to be used, in an unbiased and unemotional and scientific manner, it proves beyond a shadow of a doubt that Materialism of any kind is false. Remember, we can indeed use the Scientific Methods to prove theories false. Such a procedure is logically sound.

Once we know that Materialism and Naturalism and Darwinism are false, then we find ourselves looking for a better and more believable Model of Reality which takes into account all of the evidence at our fingertips. I submit this Ultimate Cause Model of Reality or Psyche Ontology as a possible solution. (Mark My Words, 2016, *Using the Scientific Method: To Eliminate the Usual Suspects and to Prove the Truth*; Mark My Words, 2016, *The Scientific Method: Proves That the Theory of Evolution Is False*; Mark My Words, 2016, *The Theory of Evolution Proved to Me that God Exists: Why I Am No Longer an Atheist and Why I No Longer Believe in the Theory of Evolution*).

Material Cause: Material cause is one of Aristotle's four causes. Material cause is, traditionally, the materials or the physical matter from which different items are made here in this physical realm. This is the kind of cause that the Materialists have actually chosen to believe in. Material cause is the essence of a thing – what it is made of. In the spirit world, however, consensus realities and non-consensus realities are made up of spirit matter, which would become their material cause in that trans-dimensional realm, or non-local spirit realm. In the non-local realm, spirit matter and consciousness are the essence of that domain or the material cause of that domain. According to Quantum Mechanics, psyche or non-local consciousness is the material cause and efficient cause behind the conversion of spirit matter into physical matter through conscious observation. In a sense, spirit matter is the material cause of physical matter; whereas, psyche is the efficient cause of converting spirit matter into physical matter. In this physical 3D space-time realm, physical matter is the essence by which things are made; and therefore, their material cause. (See Rychlak. Anything from Joseph Rychlak will give a solid philosophical foundation into Aristotle's four causes, which are in fact the philosophical foundation for Psyche as the Ultimate Cause).

Efficient Cause: Efficient cause is one of Aristotle's four causes. The carpenter is the efficient cause of the table, cabinets, and chair. The engineer and manufacturer are the efficient cause of your computer. In our philosophy books, we also typically see efficient cause presented in physical terms as the motion of past events which combine to make up our current physical reality. Efficient cause is therefore typically presented as the cause-effect model of our physical reality. However, things are a bit different when we start applying efficient cause to the Non-Local Realm, Quantum Realm, or Spirit Realm. According to the Science of Quantum Mechanics, observation from a non-local consciousness is required to bring each particle of physical matter into existence. Observation from a psyche is the thing which transforms a quantum wave into a physical particle, and therefore is its efficient cause. Every physical particle was brought into existence by the observation or the will of a non-local consciousness, according to the Science of Quantum Mechanics. Psyche or non-local consciousness is the efficient cause of physical matter. (Van Lommel, 2010; Mark My Words, 2016, *Quantum Mechanics from a Non-Physical Spiritual Perspective*). In the spirit world or trans-dimensional realm, the psyche is the efficient cause of turning a non-consensus reality into a consensus reality, because it is the psyche who organizes the spirit matter according to its desire and will. That's the power of consciousness or psyche. (Buhlman, 1996). Rocks or physical matter cannot act in the role of carpenter, or manufacturer, or genetic engineer, or psyche, or designer, or creator, or ultimate cause. They have never been observed doing so. Rocks and raw physical matter cannot be the efficient cause of genomes, life-forms, meaningful and useful DNA, and computer programs despite what the Darwinists claim. In contrast, Psyche can act in the role of efficient cause. It has been observed doing so. Physical matter cannot choose to act. Physical matter was designed to react. Deleterious mutations outnumber beneficial mutations thousands to one, and at times even a million to one. Species will go extinct thousands of times over before one of them will get around to evolving something useful, beneficial, and productive. Evolution doesn't work. Mutation/selection cannot design and create genomes and life forms from scratch (Sanford, 2014). This means that only God knows which mutations are beneficial and where those beneficial mutations should be placed into the genome. If we had waited to evolve from some chimp-like ancestor, we and the chimpanzees would all be extinct by now. It has been estimated that it would take trillions of years for 10,000 chimpanzees to evolve into humans "naturally", and that is with God preventing ALL deleterious mutations and with God keeping that population of chimpanzees alive for trillions of years so that they can evolve into humans "naturally". It would take trillions of trillions of years for 10,000 chimpanzees to evolve into humans naturally without God's help, protection, and intervention. "Trillions of trillions" is synonymous with impossible. Mutation and selection cannot design and create (cannot do formal cause and final cause), nor can they do efficient causation (manufacturing and construction from blueprints). Evolution has NO psyche; therefore, evolution cannot do the conversion of spirit matter into physical matter (material cause). Evolution cannot do formal cause (design), nor final cause (purposeful telic creation). ONLY Psyche can do these kinds of things! Evolution is a process or a reaction within some types of efficient causality, but evolution of any kind can't do the manufacturing and engineering aspects of efficient cause. ONLY Psyche can do that. Yes, evolution, mutation, and natural selection do exist; but, just because these things exist doesn't make Creation by Rocks (the Theory of Evolution) true, as the Darwinists and Materialists claim. I have encountered many Darwinists and Materialists who simply believe, based upon the fact that evolution (change) happens or mutation exists, that the Theory of Evolution, or Darwinism, or Creation by Rocks has been proven true. These people are willing to make that leap of faith and eagerly do so; but, their blind faith in Darwinism is based upon a wide variety of logic fallacies and no empirical evidence whatsoever. In contrast, we ALL simply KNOW from experience that the rocks can't design and create anything at all. Knowledge of these realities gives us a clear understanding why Evolution or Creation by Rocks is insufficient, unscientific, and incapable of designing, creating, and manufacturing genomes, information, and life forms from scratch. Evolution of any kind can't do Psyche or Ultimate Cause. (Sanford, 2014; Sanford, 2013; Mark My Words, 2016, *The Theory of Evolution Proved to Me that God Exists*).

Formal Cause: Formal cause is one of Aristotle's four causes. The formal cause is the form, pattern, design, outline, or blueprint behind each and every physical object. In psychology and philosophy, a personality type, logical consistency, and lifestyle are considered to be formal causes. (Rychlak, 1981, p. 7). It is empirically obvious that some kind of psyche or mind is behind each and every design, pattern, software program, genome, and blueprint. Only the

Materialists deny this reality; and, it is obvious that the Materialists are wrong. The Materialists are always wrong, especially when it comes to the non-local sciences or the spiritual sciences. Rocks or physical matter cannot design and create – they have never been observed doing so. Rocks or raw physical matter cannot act in the role of formal cause. Spirit, psyche, or living consciousness can. Psyche, or consciousness, or intelligence has actually been observed doing so. In recent days, I like to think of formal cause as the Designer, the Genetic Engineer, the Architect, or the Programmer.

Final Cause: Final cause is one of Aristotle's four causes – the reason, end purpose, or goal why a line of behavior is being carried out or a physical object is under creation. Meaning, purpose, intentions, hopes, wishes, desires, expectations, reasons, aims, goals, choices, plans, teleologies, "for the sake of which", and even "just for the fun of it" are all some type of final cause or telic purpose. (Rychlak, 1981, p. 7 & p. 300). Psyche is the efficient cause of formal causes and final causes. In other words, ONLY Psyche can do formal cause and final cause. Psyche is the ultimate cause of formal cause and final cause. Rocks or physical matter cannot act in the role of final cause, because the rocks cannot make plans and goals. They have never been observed doing so. The rocks have no purpose in mind that we know of. The rocks can't choose and act, in any way that we humans would find noticeable. If rocks were to have a purpose, it would in fact be some kind of psyche who would give them that purpose. When was the last time you caught a rock thinking, reasoning, planning, and acting? Rocks or raw physical matter have never exhibited teleology or purpose. Only psyche or self-aware consciousness can function in the role of final cause. This reality in and of itself is enough to prove Materialism false, which is why the Materialists, Determinists, and Radical Behaviorists deny the existence of psyche, free-will, agency, teleology, and final cause. ONLY Psyche can do final cause or teleology.

Ultimate Cause: Psyche or non-local consciousness is the ultimate cause. The ultimate message from material cause, efficient cause, formal cause, and final cause is that behind every material cause or particle of physical matter there is a psyche or an ultimate cause, and that behind every efficient cause or manufacturer there is an ultimate cause, and that behind every formal cause or pattern or blueprint or design there is a designer or an ultimate cause, and that behind every final cause or goal or choice or reason for acting there is an agent, or a psyche, or an ultimate cause. There is an ultimate cause behind each and every one of Aristotle's four causes. Everything reduces to psyche or ultimate cause. Our whole existence and reality is in fact a Psyche Ontology, wherein Psyche is the fundamental unit of reality and existence, and the ultimate cause of everything. In contrast, the Materialists and Naturalists have chosen to believe that only physical matter exists; and, these people deny the existence of psyche, nonlocality, spirit bodies, and the spirit realm. Consequently, the Materialists and Naturalists in principle deliberately exclude formal cause, final cause, and ultimate cause from their models of reality because we all instinctively know that the rocks or physical matter can't do formal cause (design), final cause (purposeful goals or thoughtful reasoning or deliberate choice or telic creation), nor ultimate cause (psyche or living consciousness or lived experiences). In fact, rocks or inanimate physical matter can't do Science nor manufacturing either; consequently, the rocks can't even do the engineering type of efficient causality. Physical matter can't design and create physical matter. Once we fully understand causality and how it really works, then we just know that Materialism and Darwinism are false, because we know why they are impossible and false. Follow the logic and follow the evidence! That's what I finally chose to do. According to Quantum Mechanics, ultimate cause is the Observer, the non-local consciousness who organizes spirit matter into consensus realities and brings physical matter into existence. According to Psychology, ultimate cause is psyche or personality or intentionality. According to Philosophy, ultimate cause is the Agent and is synonymous with free-will, agency, morality, ethics, and choice. The ultimate cause is the person or personality who chooses the outcome of each event. According to metaphysics, ultimate cause or non-local consciousness has always existed and will always exist; there's something out there that has to have always existed, and Psyche is it. Ultimate cause is the primal construct and the Prime Mover. According to the Empirical Science of Near-Death Experiences, Shared-Death Experiences, and Out-of-Body Experiences, ultimate cause or psyche or personality is the part of us that survives separation from the physical body, separation from the spirit body, death of the physical body, and brain death. According to theology, ultimate cause is the Mind of God. Every time we choose to use a personal pronoun (with the exception of a mindless "it"), we witness ultimate cause or a mind of some kind in action. Psyche is the ultimate cause. Obviously, psyche is not rocks, nor physical matter, nor some kind of inanimate non-living object. Psyche is life! Psyche animates matter, both spirit matter and physical matter! Psyche is some kind of non-local trans-dimensional living consciousness, an Observer, according to the Science of Quantum Mechanics. These people call it Quantum Consciousness. Psyche is consciousness, awareness, intelligence, and life. Everything that I encounter testifies to the truthfulness, usefulness, and existence of Psyche and Ultimate Cause. Clearly, I'm going with the Science and the Empirical Evidence on this one, and not the Materialists and their mindless philosophy. The Materialists are simply guessing, and they guessed wrong. The ultimate cause or psyche is the efficient cause of formal causes and final causes, and any other type of spiritual cause or psychical cause as well. Ultimate cause is the efficient cause of physical matter, because Psyche or a Conscious Observer is needed to convert spirit matter into physical matter. Any way we choose to look at it, Psyche is the Ultimate Cause.

 Mechanists: Mechanism is another form of Materialism. Mechanists treat human beings (human psyches) as if they are robots or machines. The Mechanists purposefully limit themselves exclusively to material causes and efficient causes, which are often called mechanistic causes; and, these people formally reject formal causes, teleology, humanism, psyche, predication, agency, soul, mind, and final causes. In contrast, this Ultimate Cause Model of Reality is willing to include the whole of human experience into the mix, making for a richer and more robust Science and Model of Reality than what can be produced by Materialism, Determinism, Behaviorism, and the Mechanists.

 Entity Dualism: Dualism, as used in this Ultimate Model of Reality, is the philosophical belief and the Observational Knowledge that the spirit body and the physical body are two completely separate entities. Furthermore, psyche and its assigned spirit body are two completely separate and different entities, according to the Lived Experiences of the human race. Psyches control spirit bodies, and spirit bodies can be assigned to or housed within physical bodies. Psyche as the ultimate cause relies upon dualism, and it is based upon the observation or out-of-body experiences wherein the person or psyche observes that his spirit body is a completely separate and different entity than his physical body. Furthermore, the SOUL is a holistic unit comprised of the psyche, spirit body, and physical body united as one functional unity or entity. In other words, the Psyche controls the whole Unit, which we call a Soul.

Alas, classical dualists unfortunately tend to believe in a type of Deism, in which God's spirit cannot be functioning at the same time that natural laws are functioning. They can't wrap their minds around the integration of spirit matter and physical matter, nor comprehend the idea that the physical body is the temple for the human spirit to reside within. They refuse to accept the integration – the idea that a spirit or soul (psyche/spirit body) can control a physical body.

> Dualistic strategies [of the Deism variety] assign one realm or world to theistic assumptions (usually religion, values, and subjectivity) and another to naturalistic assumptions (science, facts, and objectivity). In a [classical or traditional] dualism the two types of worlds cannot, in principle, interact, so again, no integration here. Theism is the belief that God is active and involved in all aspects of the world. (Brent S. Melling, from class notes).

This Ultimate Cause Model of Reality is compatible with Theism, the idea that not only does God exist but also that God or the Ultimate Cause is active in this world and can intervene in this world at will. Thus, one could say that this Ultimate Model of Reality is based upon Theistic Dualism, because psyche has to be able to be active, intervene, and interact with this physical world in order to be an ultimate cause in this physical world. It's elementary my dear friend. We KNOW from the Lived Experiences of the human race that Psyche exists. Furthermore, the Non-Local Psyche has to exist in order to bring physical matter into existence, according to the Science of Quantum Mechanics. This Ultimate Model of Reality is a Psyche Ontology, wherein psyche is the fundamental unit of reality.

<u>Existentialism</u>: "I AM. I exist. What more needs to be said? Forget all the fighting about monism and dualism. I choose my own being and what I am becoming. Good enough!" The philosophy of existentialism can be made compatible with this Ultimate Cause Model of Reality, because existentialism tries to take into account the whole of human existence and experience. However, in my various models of reality, I now replace the existentialist "threat of non-being" or the existentialist concern about death, nihilism, ceasing-to-exist, and annihilation with the empirical evidence and the experiential hope of the survival of the psyche after bodily death as demonstrated and proven by NDEs, SDEs, and OBEs. Remember, we can KNOW the truth and prove the truth by experiencing and living the truth for ourselves, or by choosing to trust someone who has. Knowing the truth, experiencing the truth, IS proof of the truth! Existentialism is often parodied as nihilism, thus portraying existentialism as morbidly sad and depressing. But, my brand of hopeful existentialism is willing to include psyche and ultimate cause into the mix resulting in the continuation of the human psyche after bodily death and brain death. The existentialists are heavily into free-will, choice, freedom,

meaningfulness, teleology, final causality, and moral responsibility. These are in fact some of the logical conclusions of Psyche as the Ultimate Cause. There are theistic existentialists and atheistic existentialists. Obviously, this Ultimate Model of Reality would be most compatible with theism, because God does indeed qualify as an Ultimate Cause. In contrast, this Ultimate Cause Personality Theory would technically be incompatible with the atheistic and nihilistic forms of existentialism – the depressing forms of existentialism which people like to laugh at and joke about. Jokes about death are some of the funniest, but the psyche or non-local consciousness cannot die. Since psyche has no physical or material or corruptible component, psyche is eternal and immortal.

Spiritual Empiricism: Typically, whenever the word "empiricism" is used by the scientists and British Empiricists, they do in fact mean "physical empiricism" and try to limit their empiricism to physical evidence of some kind – British Empiricism is Traditional Empiricism. This Ultimate Cause Model of Reality not only fully embraces physical empiricism, but it also embraces spiritual empiricism, or non-local empiricism, or Radical Empiricism. The Materialists and British Empiricists are determined to limit and cripple their science; but, I am not. Been there and done that; and, I don't want to go back. Spiritual Empiricism is in fact spiritual experiences of any kind, including NDEs, SDEs, OBEs, and modern-day revelations from the Biblical God Jesus Christ. Evidence of non-locality and spirit has exploded exponentially since the year 2000 to the extent that the evidence is so overwhelming now that it is impossible to deny. There are now hundreds of different videos about NDEs and OBEs on YouTube, thousands of respondents to Dr. Long's online surveys about NDEs and seeing God while out-of-body (Long, 2016), and many different studies of hundreds of hospital patients who have had NDEs and OBEs (Gibson, 2006). Recent polling indicates that millions of people on this planet, as much as 10% of the human population has experienced some kind of OBE and/or NDE, and as much as 30% of this world's population has experienced a life-changing spiritual experience (See: http://mypsyche.us/how-common-are-ndes/). There's no question now that the NDE phenomenon is real and has happened to millions of different people on this planet – one would have to have the blind-faith and blind-loyalty of a Materialist or an Atheist in order to be able to deny all of the spiritual evidence or Spiritual Empiricism that we now have access to as a race. I couldn't do it anymore. I no longer have faith enough to be an Atheist or a Materialist. I now own hundreds of different books about NDEs, SDEs, OBEs, and other kinds of spiritual experiences. The evidence became too overwhelming to deny. I had to make a paradigm shift in my own thinking processes. I had to switch to a new and different model of reality. I can no longer straddle the fence like I tried to do for most of my life. Any kind of spiritual experience, psychic experience, spiritual evidence, trans-dimensional evidence, transpersonal evidence, and Quantum Non-locality is fair game within this Ultimate Cause Model of Reality. That's why I chose to call it the Ultimate Model of Reality, because it's not limited to Materialism and Naturalism.

Nihilism: Nihilism is the philosophical belief that "all we are is dust in the wind". Nihilism is another one of the Denialisms, a denial that there is an afterlife and a denial that life has any purpose or meaning. Denialisms, such as Nihilism, have to be taken on blind faith as being real and true, because the various

Denialisms contradict ALL of the empirical evidence that we have on hand as a race. From the Wikipedia, "Nihilism (from the Latin nihil, nothing) is a philosophical doctrine that suggests the lack of belief in one or more reputedly meaningful aspects of life. Most commonly, nihilism is presented in the form of *existential nihilism*, which argues that life is without objective meaning, purpose, or intrinsic value. Moral nihilists assert that morality does not inherently exist, and that any established moral values are abstractly contrived [man-made]. Nihilism can also take epistemological, ontological, or metaphysical forms – meaning that, in some aspect, [truth does not exist], knowledge is not possible, or that reality does not actually exist." (End of wiki quote). Nihilism is a type of relativism which claims that truth does not exist. Truth is knowledge, and knowledge is truth; and, Nihilism claims that truth and knowledge do not exist. Nihilism is a denial of reality. Nihilists deny the existence of an afterlife. The Atheists have chosen to believe in "Creation by Nothing", which takes a great deal of faith to believe to be true. The Nihilists believe in nothing. Nihilism leads a person to moral relativism, angst, depression, ennui, and Atheism. In contrast, this Ultimate Cause Model of Reality goes in pursuit of the True Reality of our existence, as revealed to us by empirical evidence, spiritual experiences, lived experiences, first-hand observations, scientific experimentation, NDEs, SDEs, OBEs, scriptural evidence or revealed evidence, and the like. This Psyche Ontology proves that Nihilism is false by falsifying Nihilism in many different ways.

Pursuing the Ultimate Psychotherapy: A study of Personality Theories will reveal that most, but not all, of the dozens of different Personality Theories and all of the major Personality Theories were based upon a psychotherapeutic treatment method of one sort or another. Since I'm a philosopher and a scientist, and not a psychotherapist, I'm not going to develop and promote a separate treatment method for this Ultimate Personality Theory. Instead, I prefer to adopt what has already been proven to work and has proven efficacious. Cognitive Therapy was designed by Atheists to treat and cure Atheists; and, it works for the treatment of depression. Cognitive Therapy will work for you whether you are an atheist or not. David Burns' book, "Feeling Good", is the ultimate self-help Cognitive Therapy book for treating depression; and, it worked for me. Other therapy techniques that seemed to be developed by Agnostics for Agnostics, which could be made compatible with the spirit and the feel of this Ultimate Cause Model of Reality, are William Glasser's Choice Therapy and Carl Rogers' Client-Centered Therapy. In my most recent round of psychotherapy, I found Mindfulness helpful and useful – learning to be present in the present, stop ruminating about the past which can't be changed, and stop fearing the future which isn't here yet. Live for the moment in the present moment! Depressed people live in the past and beat themselves up over their past mistakes; and, anxious people are constantly afraid of the future. Stop it! Learn to live in the present, here and now. Mindfulness is an Americanized adaptation from Buddhist Psychotherapy. Existential Psychotherapy as presented by Rollo May and Irvin Yalom in an article by the same name could be made compatible with the spirit of this Ultimate Cause Personality Theory. Existential Psychotherapy is supposed to be a useful supplement for any other type of psychotherapy, which one might choose to employ. Existential Psychotherapy emphasizes taking responsibility for one's own life and one's own future, a process which I found extremely helpful during my last round of psychotherapy. Existentialists believe in free will, choice, and moral responsibility which are all functions of Psyche. I went to a seminar by a clinical psychologist; and, he said that in his clinic the Biopsychosocial Model has become the standard model for therapeutic treatment – a holistic approach treating whatever currently ails the client. This Ultimate Cause Personality Theory is completely compatible with the Biopsychosocial Model of Reality, including the "psycho" or psyche part of the model. This Ultimate Cause Personality Theory would be amenable to an Eclectic Psychotherapy – pursuing whatever works to help and benefit the client. As already mentioned, I see Biopsychosocial Psychotherapy as the pinnacle of an Eclectic Psychotherapy – psychotherapy and treatment tailor-made to the needs of the client. The last time I got sick, dozens of different things went wrong; and in order to get better, dozens of different things had to go right. An eclectic approach and a holistic approach were the right way to go. Though I haven't experienced it personally, Theistic Psychotherapy as presented in the second chapter of *Casebook for a Spiritual Strategy in Counseling and Psychotherapy* and the fifth chapter of *Spiritual Strategy for Counseling and Psychotherapy* (2nd ed.) looks interesting and promising. Theistic Psychotherapy is something which I wish that I could have experienced or been involved with in my life. Even so, I can testify that bringing the healing power and the psychotherapeutic power of the Atonement of Christ (Christ's mercy, forgiveness, love, and grace) into my life has had massive life-changing and extremely positive psychotherapeutic and physical effects. As I see it, the Atonement of Christ is the Ultimate Psychotherapy. Christ can cure anything that ails you. The only question is whether He will do so, or whether He has other plans in mind for you. If I were forced to choose a single psychotherapy for this Ultimate Model of Reality, I would choose the Atonement of Christ, because it works and it works best, from all that I have experienced in my life so far.

<u>The Ultimate Epistemology</u>: While developing this book, I didn't even know what my ultimate goal was. I was just putting out commentary explaining how the various Personality Theorists led me to Psyche as the Ultimate Cause. In a class I took, I was called upon to develop a Personality Theory of my own, and I took it to heart, eventually developing this Ultimate Cause Model of Reality and a Psyche Ontology to go with it. In recent days, it dawned on me that I was in fact looking for the Ultimate Epistemology while developing my books. Epistemologies are various different ways of Knowing the Truth or coming to KNOW the Truth. During my research and commentary, I was in fact looking for the BEST Personality Theorist who had found the best way of uncovering and knowing the truth. Ironically, I didn't find a single one; so, I had to become that one, and develop the Ultimate Epistemology all on my own. It took half a year for me to figure out what I was doing, what I was looking for, and what is in fact the Ultimate Epistemology. The EVIDENCE kept pointing me to what I now call a Psyche Epistemology, wherein the Lived Experiences of the human race or the Lived Experiences of the human psyche becomes the BEST and the most efficient way of KNOWING the Truth. I simply kept looking at the EVIDENCE, which I personally found most convincing; and, ALL of that evidence was comprised of some kind of Psyche Experience or Spiritual Experience or Lived Experience; and then, it became obvious to me that our Lived Experiences are in fact the BEST, the most convincing, and the most

effective way we have of finding and knowing the Truth about our Reality. Lived Experience IS a Psyche Epistemology, because ONLY Psyche can do lived experience, truth, and knowledge. A Psyche Epistemology is based strongly upon the Phenomenological, Radically Empirical, and Hermeneutic Epistemologies. I also observed that a Psyche Epistemology subsumes ALL other epistemologies, because ONLY Psyche can do ALL of the other epistemologies. It was obvious and clear. I had finally found what I was searching for. A Psyche Epistemology is the Ultimate Epistemology, because ONLY Psyche can do epistemology and ONLY the human psyche can do epistemology to its fullest extent. It's elementary my dear friend. Scientism, Materialism, Naturalism, Nihilism, Darwinism, and Atheism ARE religions. The practitioners of those religions have chosen to believe that Scientific Methods and Scientific Experimentation ARE the best way of finding and knowing the truth. For the first fifty-five years of my life, I too was of that opinion. I relied heavily and exclusively on Science and Science Methods, because I truly believed that they were indeed the BEST way of finding and knowing the truth. Mine was a Scientism Epistemology. Even when I had successfully abandoned My Materialism and My Atheism, I continued to maintain my belief in Scientism. I knew of no other way. If you are going to tear something down, you are obligated to put something better in its place; and, for most of my life I had NOTHING better to put into the place of My Scientism. But now I do. Psyche is the Ultimate Epistemology and therefore the ultimate way of finding and knowing and living the truth. ONLY Psyche can do Science, Scientism, scientific experimentation, scientific interpretation, and scientific methodologies. A Psyche Epistemology subsumes ALL other epistemologies, because ONLY Psyche can do epistemology!

The Ultimate Personality Theory and Ultimate Ontology: While developing my books, I started by searching for the Ultimate Personality Theory. In one of my classes, I was assigned the task of developing my own Personality Theory. I eventually came to the conclusion that NONE of the Personality Theorists had the BEST personality theory; so, I decided to develop one of my own, which I thought was better than all the rest. Thus, this Ultimate Personality Theory was born. This Ultimate Personality Theory states that Psyche or Personality and our memories survive bodily death, separation from the spirit body, and brain death. Although I didn't know it at first, I was of the opinion that the Ultimate Personality Theory should in fact be based upon the Fundamental Unit of Reality. Over time, it became clear and obvious to me that Psyche is the Ultimate Causal Agent, and thus Psyche is the Fundamental Unit of Reality. As a result, this Psyche Ontology and this Psyche Personality Theory were born. I wanted to find and go with the BEST, and I believe that I have done so. I can't think of anything more fundamental and essential to our lives, our reality, and our existence than Psyche or our own personal Non-Local Consciousness. A Psyche Ontology is the Ultimate Ontology, because Psyche IS the fundamental unit of our own Personal Reality. Consequently, this Psyche Personality Theory is the Ultimate Personality Theory, because ONLY Psyche can do personhood and personality. Furthermore, we KNOW from the lived experiences of the human race that Psyche exists. Psyche or personality is the part of us that survives separation from our spirit body, separation from our physical body, and the death of our physical body. It's elementary my dear friend. It doesn't get more fundamental than Psyche. I now treat Psyche and Ultimate Cause as LAW, because it has become clear to me that Psyche is the Ultimate Ontology, that Psyche is axiomatic, and that Psyche IS the Hidden Assumption or the Invisible Variable in ALL of our theorizing, philosophizing, hypothesizing, experimentation, science, lived experiences, spiritual experiences, empiricism, and scientific interpretations. A Psyche Ontology subsumes ALL other ontologies, because ONLY Psyche can do ontology and metaphysics. This is the Ultimate Model of Reality, after all. In their many free articles, Brent Slife and friends present a compelling case for a Relational Ontology, wherein relationships and friendships are the fundamental unit of reality. That's all fine and well, but I noticed that Psyche is more fundamental than relationships, because ONLY Psyche can do relationships and friendships. This Psyche Ontology nicely subsumes or includes a Relational Ontology, because a Psyche Ontology subsumes or includes every other ontology because ONLY Psyche can do ontology. The goal was to find the fundamental unit of reality, and I believe that I have, because ONLY Psyche can do Reality and Lived Experiences and actually remember having done so. Go with the BEST and subsume all the rest – that's what I chose to do. I find this whole thing amazing. I don't see a Psyche Ontology anywhere in the literature. Everyone seems to have blinded themselves to the possibility. I didn't see it nor understand it either, until after it was revealed to me. It's so clear and obvious to me now, that I sometimes wonder how the whole world has been able to overlook it throughout the whole of human history. I sometimes wonder how I was able to overlook it for the first 55 years of my life; but then again, I wasn't looking for it, now was I? No seeking, then no finding! And, I never received any help or encouragement from my schoolteachers, either. A Psyche Ontology is definitely something that you will NEVER get from your materialistic, naturalistic, and atheistic college professors. Now is it?

The Ultimate Psychotherapy: All throughout my life, I have submitted to different types of Psychotherapy, because I really needed it at the time. While developing my books, I observed that ALL of the Personality Theorists developed their own unique brand of Psychotherapy, technique, and treatment for their clientele. As I put this Ultimate Personality Theory together, I suddenly found myself looking for the Ultimate Type of Psychotherapy. After examining my personal experiences and observing what worked BEST for me and what had the BEST and the most long-lasting benefits and effects, it became obvious and clear to me that the Atonement of Christ is the Ultimate Psychotherapy. There is nothing better or more effective! ONLY the Atonement of Christ can touch, influence, and change the Psyche directly; thus, it became obvious and clear to me that the Atonement of Jesus Christ IS the best and most effective form of Psyche-Therapy. It's elementary my dear friend. I went looking for the Ultimate Model of Reality, and I believe that I have found it.

The Ultimate Cause Model of Reality: This Ultimate Cause Model of Reality is a holistic model of reality, which means that it is automatically incompatible with the exclusionary models of reality that deny the existence of some aspect of Reality. The exclusionary models of reality that deny the existence of Psyche or Non-Local Consciousness cannot be made compatible with Psyche as the Ultimate Reality or Psyche as the Fundamental Reality of our existence. The exclusionary models of reality are Denialisms, which faithfully claim that some aspect of Reality does not exist. These Denialisms are philosophies such as Naturalism, Materialism, Scientism, Positivism, Physical Empiricism, Nihilism, and Atheism. These Denialisms, or exclusionary models of reality, deny the existence of some aspect of Reality. These exclusionary philosophies can NEVER be made compatible with holistic models of reality, because Denialisms always exclude some aspect of Reality. This is the Ultimate Model of Reality because it attempts to include and embrace the Whole of Reality. Nothing is excluded! In ALL of my theorizing, I kept adjusting this Ultimate Model of Reality until it matched with the Lived Experiences or Psyche Experiences of the human race. I wanted to get this model as close to the Truth and Reality as I possibly could.

A Lesson in Quantum Mechanics or Spiritual Mechanics

Quantum Mechanics explains the Supernatural – what it is and how it works.

As their mathematical equations approach infinity and break down, their equations are in fact approaching Psyche, who lives and works in the realm of the infinite. Whether we are talking Relativity, Calculus, Trigonometry, Geometry, Quantum Mechanics, or any other type of math, the Materialists, Naturalists, and Atheists freak out whenever their equations go infinite and break down on them; but, where else are their equations going to go? These people are thinking materialistically and spatially, rather than transdimensionally and psychically. Their naturalistic thinking greatly limits them.

Nevertheless, as their equations become infinite, their equations become Psyche, who IS infinite in every respect. Whether we are talking about the infinitely large, or the infinitely small, we are talking about Psyche, who is at home in the infinite and resides in the infinite. We are talking about God. (Please see the **$y = \tan(x)$** model in my book, *The Ultimate Model of Reality: Psyche Is the Ultimate Cause*, for a graphic explanation of how all of this works in practice.)

When it comes to infinity, it's all the same to Psyche, because Psyche IS infinite. Psyche is an infinite being. This Reality is so glaringly obvious, that I sometimes wonder how we all managed to overlook it for the whole of human history. Nevertheless, the Materialists and Naturalists have trained us not to see it. Their teachings are ingrained right into us and are hard to get rid of. I KNOW, because I used to be a Materialist and an Atheist.

The Materialists and Naturalists cannot deal with infinity nor handle the infinite.

Why?

It's because these people limit themselves to the physical; and, the physical is finite. The physical is finite! It's limited! As a result, these people and their science are limited and finite. It can't be helped, except by getting rid of their Materialism and Naturalism.

But, their math isn't lying to them. Everything moves towards the infinite. In other words, everything moves towards Psyche, because Psyche IS infinite in every respect.

When the math equations move towards the limit, or move towards an infinite sampling, or move towards infinity in general, they are in fact moving towards Truth, Knowledge, Reality, and Psyche. Anyone who truly understands Calculus understands this Reality. With an infinite sampling, the math moves towards the limit, and Psyche resides at the limit, because Psyche IS Truth, Knowledge, Existence, and Reality.

Naturalism and Materialism mess everything up, making them incomprehensible. This applies to Quantum Mechanics as well. Thanks to Materialism and Naturalism, there's a ton of confusion associated with Quantum Mechanics. Many mathematicians and scientists believe that Quantum Mechanics has been falsified because their equations break down or go infinite on them. However, they have NOT falsified Quantum Mechanics. Quantum Mechanics and the math are telling them the truth! Reality and the math do indeed move towards Psyche or the Infinite. What these people have indeed done is falsify the Finite! Remember, Materialism and Naturalism are Finite. In other words, these people have successfully falsified Materialism and Naturalism with their mathematical equations that go to infinity and break down.

It's all a matter of one's personally chosen interpretation of the math and the scientific data! There's more than one way to look at these things; but, the Materialists and Naturalists choose ONLY one way, a very finite and limited way, to look at science and reality. It holds them back. They can't deal with it or handle it, because it's Psyche and it's Infinite. As the math moves towards the limit or moves towards the infinite, it is in fact moving towards Psyche who lives and resides at the infinite. Materialism and Psyche are opposites. Materialism or physical matter is finite. Psyche IS infinite!

I'm passionate about Quantum Mechanics. Quantum Mechanics has been verified or validated dozens of different times in dozens of different ways. It has been estimated that a third of our economy in the United States is based upon Quantum Mechanics to one extent or another – its practical utility is undeniable. It has been claimed that Quantum Mechanics has never been falsified, and that no aspect of Quantum Mechanics has ever been falsified. Instead, Quantum Mechanics has been repeatedly verified through scientific experimentation. Quantum Mechanics is Spiritual Mechanics – the way that spirit matter really works. Quantum Mechanics IS the science behind consciousness, psyche, and near-death experiences! (Van Lommel, *Consciousness Beyond Life: The Science of the Near-Death Experience*).

Although Science and the Scientific Methods can't be used to prove the truth, Quantum Mechanics is probably the closest we have ever come to a Proven Science as a race, due to the fact that Quantum Mechanics has never been successfully falsified. Furthermore, Quantum Mechanics is the closest we have ever come to using Science to prove the existence of Psyche, Non-Locality, the Spirit Realm, and Non-Local Consciousness, because Quantum Mechanics teaches us that a Psyche or

Conscious Observer is necessary in order to bring physical matter into existence in the first place. Furthermore, Quantum Mechanics is Spiritual Mechanics, or the way that spirit matter works. Quantum Mechanics explains to us how spirit matter gets converted into physical matter through Conscious Observation or the Word of Command. (Van Lommel; and Mark My Words, 2016, *Quantum Mechanics from a Non-Physical Spiritual Perspective*.)

We KNOW from the Lived Experiences of the human race that Psyche exists. Quantum Mechanics lends evidentiary and scientific support to that KNOWLEDGE.

It had to be pointed out to me by my best friend, because I wasn't looking for it at the time, but there are now thousands of different Near-Death Experiences and Out-of-Body Experiences on YouTube just waiting for you to find them and watch them. Most of these verify the existence of our spirit body as a separate object distinct and different from our physical body – a spirit body, human in shape, comprised of some sort of light, photons, stars, or particles of spirit matter. Spirit matter is light, and light is spirit matter. The rest of these NDEs and OBEs on YouTube verify the existence of our Psyche or Intelligence as an immaterial, disembodied, third-person viewpoint in space – a spark of life, an infinite singularity, who is able to look at its assigned spirit body from a third-person vantage point outside of that spirit body and also from the third-eye perspective which is more of a first-person perspective.

One lady on YouTube said that when she passed away, she was floating up near the ceiling watching her physical body. Then she said that she went to look for herself; and, there was nothing there. She said, "I was just a spark on the ceiling". Spark is another word for Psyche, a viewpoint in space. Our spark is our life or our psyche. She had apparently separated from both her spirit body and her physical body. It can be done. (For another example see: Buhlman).

Psyche or non-local consciousness is some kind of matterless infinite singularity, quantum singularity, or unlimited singularity (not limited by physical laws, nor limited by spiritual laws). Psyche, or this immaterial and formless viewpoint in space which Buhlman mentioned, is the polar opposite of a physical 3D space-time universe – making psyche an infinite universe taking up no space whatsoever, or a universe in a viewpoint – infinite, non-physical, immaterial implosion having no size whatsoever.

Imagine it! No size and no dimension! Zero is a different type of infinity. Think of the other times when we might have heard of the concept of an "infinite singularity" being discussed and used in Science. I'm not the first person to come up with the idea, although I might be the first person to apply the idea to the immaterial, disembodied, and non-local Psyche.

Psyche is Intelligence, Light, Truth, and Knowledge. ONLY Psyche is capable of having and remembering lived experiences and knowing the truth. ONLY Psyche can do life, lived experiences, truth, and knowledge.

Psyche is infinite memory and information storage that takes up no space whatsoever. Psyche is our own personal truth – what we truly are at our very core. Psyche is the fundamental unit of reality. Psyche is Light – a Living Light – a Light which exists at an infinite frequency thus making it or transforming it into something that is completely different than light or spirit matter. It becomes consciousness and life.

Just as combining hydrogen gas with an oxygen gas transforms the mixture into a completely different substance, a liquid that we call water; taking light and ramping it up to an infinite frequency, infinite velocity, omnipresent, infinite potential, infinite capacity, and infinite implosion state completely transforms that light into a whole new unity or substance that we call Psyche, Consciousness, and Life. By taking on all those properties, Psyche becomes something completely different than it was before. It transforms into a whole other unique entity. It can see, without needing eyes. It can think without needing a brain. It can speak and hear without needing a mouth and ears. It can sense and feel without needing skin. It can live without needing sustenance. Psyche can move and interact psychically, telepathically, and telekinetically with its surroundings without needing a body to do so.

Psyche or living consciousness is a third type of thing, completely different than spirit matter and completely different than physical matter because non-local consciousness seems to be completely immaterial, both here in this physical realm and there in the spirit realm. That's what makes Quantum Consciousness or non-local consciousness so powerful, so unlimited, and so infinite and omnipresent and instantaneous in scope and range, because it doesn't have any material component at all limiting it and tying it down, neither a spirit matter component nor a physical matter component. Psyche has the potential to be omnipotent, omnipresent, infinite, and eternal or everlasting.

Pure Psyche or Pure Intelligence can function transpersonally or psychically instantaneously at infinite distances at an infinite velocity because it has no matter associated with it – no spirit matter and no physical matter. Psyche can actually be all-knowing and omnipresent. Psyche or Quantum Consciousness is a third type of thing because it is the only thing that can collapse a quantum wave thereby converting spirit matter into physical matter, according to the Science of Quantum Mechanics.

Matter of any kind, whether spirit matter or physical matter, provides limitations as well as new possibilities and new opportunities. Psyche or Quantum Consciousness is not a quantum object (matter), because consciousness has no spirit matter and no physical matter, although psyche or consciousness can be assigned to and attached to a spirit body and eventually to a physical body as well. Matter of any kind imposes limitations or boundaries or determinants, whether we are talking about spirit matter or physical matter; but, matter also opens up new and different opportunities.

Since psyche or non-local consciousness is completely immaterial, in its pure form and original form it is an infinite singularity with infinite possibilities and no limitations whatsoever. Quantum Consciousness is life – eternal life. Since psyche or consciousness is completely immaterial in nature and origin, it can neither be created nor destroyed. It just is. It's life. Non-local Consciousness is existence and life, intelligence and consciousness, light and truth in its pure and original form.

These are fascinating concepts which I was never able to get from my materialistic and atheistic friends and teachers because they had limited themselves exclusively to the study of physical matter.

Classical Physics explains the laws behind physical matter; and, Quantum Mechanics explains the laws associated with spirit matter. They are different, because spirit matter can theoretically exist and function at an infinite velocity and spirit matter has innate velocities which are at least faster than the speed of light; whereas, physical matter is spirit matter which has been reduced to sub-light speeds. Spirit matter is light; and light is spirit matter. Physicist David Bohm describes physical matter as frozen light, or light (spirit matter) which has been slowed down to sub-light speeds and thereby localized into our physical space-time realm. Conscious observation or the Word of Command is the thing that slows down spirit matter or light or an energy wave transforming it into physical matter or a photon thereby freezing it into physical matter or locality, and thereby localizing it into our physical 3D space-time universe.

We know that thoughts and dreams are spiritual experiences or quantum experiences, because our thoughts and dreams cannot be captured, measured, nor recorded by our physical instruments. If you have ever had a thought or have ever had a dream, then you have had a spiritual experience, a conscious experience, and a non-local experience. Even the Materialists and the Atheists have spiritual experiences, but they never recognize them as such because of their cognitive bias and chosen worldview which limits them to thinking of things only in physical terms.

Matter of any kind provides boundaries, determinants, and limitations along with a whole host of new and different opportunities and perspectives. In a similar manner, the philosophy of Materialism limits people to an extremely scaled-down and lobotomized form of science, a pseudo-science or semi-science of sorts. We have to be willing to break free from Materialism in order to make quantum leaps in our scientific understanding of Reality. Materialism holds us back and keeps us in elementary school. However, as part of Reality, we must also realize and acknowledge the fact that an understanding of the Physical Laws has supplied us with much of our technology and infrastructure; so, an understanding of the physical isn't completely worthless. A physical reality is the ultimate consensus reality; and, this is the Ultimate Model of Reality after all.

According to Quantum Mechanics, a **quantum object** has two different complementary states of existence, a spirit-like wave-like non-local infinite-velocity simultaneously-everywhere quantum wave state and a localized sub-light-speed finite physical particle state, after the quantum object has been observed by a living consciousness. Quantum objects can be either in spirit matter format or in physical matter format, but NOT in both formats at the same time according to the quantum law of complementarity. A quantum object manifests as either a spiritual non-local wave or as localized space-time particle, but not as both at the same time. Most scientists get this one wrong thinking that a quantum object is simultaneously wave-like and particle-like, which is not the case. A quantum object is in either one state or the other at any given point in time. (Van Lommel, 2010; & Bishop, 1998).

Interestingly enough, if spirit matter or a quantum wave can be converted into a physical particle through conscious observation or the Word of Command, then the reverse process is theoretically possible as well – converting a physical particle back into spirit matter or a quantum wave. We just have to learn how to do it; and, it ultimately won't be a physical process; it will be some kind of spiritual or conscious process instead – something ultimately caused by a Psyche. In other words, in order for it to happen, God has to allow it to happen or cause it to happen. God's Psyche holds the keys to these kinds of things. That's why He is God, and we are not.

Some people believe that there are quantum objects, such as electrons and subatomic particles, who pop back and forth between the wave-state and the physical-state at will. That would suggest that each electron has a spark of psyche residing within it, because ONLY Psyche can do agency and will. There's a lot more going on there than meets the eye.

Non-local consciousness or living psyche is a completely different animal and only has an immaterial format – it's not spirit matter, and it's not physical matter, which means that non-local consciousness is not a quantum object. Non-local consciousness is the thing which has the power and the ability to collapse a quantum wave function and slow down a quantum object, thus converting that quantum object from spirit matter into physical matter. Quantum objects were designed to be acted upon; whereas, Non-Local Consciousness or Intelligence is innately The Actor and The Observer – always has been and always will be.

In the Book of Abraham in the *Pearl of Great Price*, God tells us that He is more intelligent than all the rest of us combined, which is why He is God and the rest of us are not. God apparently has a complete omniscient and omnipotent understanding of non-local consciousness, psyche, infinite singularities, quantum mechanics, and classical physics, which we do not. God knows how to convert one of those infinite singularities or quantum singularities into a physical 3D space-time universe like ours. And when it is time to do so, I imagine that God also knows how to convert our physical universe back into a quantum singularity or an infinite singularity once again.

Each psyche or infinite singularity or point of consciousness has a different degree of light, glory, and intelligence than the one next to it. It's possible for there to be one of these infinite singularities inside each and every electron with plenty of room to spare. Physicists like David Bohm believe that each electron and each sub-atomic particle is conscious, alive, and aware of its surroundings and its existence, and that they can communicate with each other instantaneously across great distances. (Bishop, 1998). Therefore, it's theoretically possible that the consciousness or the infinite singularity associated with an electron could one day be upgraded to become the controlling consciousness of a spirit body or a physical body, or one day become a God. Each point of consciousness is capable of enlargement, or capable of a greater stewardship and responsibility.

Furthermore, if an electron is alive and aware of its surroundings, it's because it has a psyche of some sort within it. Raw physical matter and spirit matter by itself is incapable of having lived experiences and forming memories, as witnessed by Out-of-Body Travelers who have observed that their physical bodies stop having lived experiences and stop forming memories while their spirit body and psyche are separated from their physical body. While outside their physical body looking at their physical body, they can no longer feel their physical body's pain and their physical body doesn't remember that pain. It's only when they go back inside their physical body that the pain returns. Also, while outside their spirit body, their spirit body isn't making and storing memories either! Only the Naturalists, Materialists, and Atheists are incapable of wrapping their minds around concepts such as these!

Psyche or immaterial non-local consciousness appears to be a whole universe within an infinite, timeless, immaterial singularity with no physical limits and no spiritual limits whatsoever. Imagine what such a thing could design and create, having no limits and no limitations whatsoever. Such a thing could design and create universes, like our physical 3D space-time universe and comprehend the whole thing all at once. And, the Materialists and Darwinists want us to believe that we descended from pond scum, fish, monkeys, and apes. We humans are the children of the Gods – co-equal and co-eternal. That's what Quantum Mechanics and Buhlman's OBEs are trying to teach us, or at least that's the message I got from Quantum Mechanics and Buhlman's OBEs. (Buhlman; Van Lommel; Mark My Words, 2016, *Quantum Mechanics from a Non-Physical Spiritual Perspective*).

Again, Matter is a Quantum Object. A full understanding of both aspects of matter opens up a whole new world of possibilities and application. A Quantum Object, or matter, can either exist as spirit matter or as physical matter; but, it cannot exist in both states simultaneously, according to the Quantum Law of Complementarity. A Quantum Object is either in its spirit matter phase or in its physical matter phase; but, it can't be in both phases simultaneously.

We KNOW from Quantum Mechanics (or Spiritual Mechanics) that what the Bible calls the WORD – the Word of Command – and what the Quantum Physicists call a Conscious Observer or Non-Local Consciousness is necessary to convert spirit matter into physical matter. That's powerful and useful science there!

What else does this reality tell us?

It tells us that, if the Word of Command or God's Psyche can convert spirit matter into physical matter at will, then the same Word of Command should be able to convert physical matter back into spirit matter at will. Think about it! That's powerful science! It's much better than anything you will ever get from the Materialists and the Atheists.

It had to be pointed out to me by my best friend, because I wasn't looking for it at the time, there are now thousands of different Near-Death Experiences (NDEs) and Out-of-Body Experiences (OBEs) on YouTube just waiting for you to find them and watch them.

Many different people on YouTube have observed and stated that their spirit body can teleport from one location to another instantaneously. Their spirit body can levitate, and fly, and pass through ceilings and walls. They can simply walk, while in the spirit realm, if they want to. Their spirit body can be in two places at once. In non-consensus realms, their spirit body can take any form they like, and their psyche can conjure up any type of reality and surroundings that it wants. Their spirit body and/or psyche can hear the thoughts and prayers of others. Spirit beings communicate telepathically. The Angels have NO wings. Their spirit body can transform into trees and animals and other things; and, these people can spend time experiencing what it is like to be a bear, or an alien grey, a snake, or one of the blues or Nordics, or some such. One man on YouTube said that as he walked up a flight of stairs, he could see from the third-eye view that his spirit body began to dissolve beneath him. He said that by the time he got to the top of the stairs, his spirit body had transformed into an orb of light; and, he was apparently looking at this orb of light from a third-person perspective or psyche perspective. Others have said that their spirit bodies are comprised of millions of points of light – like an ocean of stars.

Physical matter is condensed light or frozen light – light which has been slowed down to below the speed of light. Physical matter is the ultimate consensus reality – highly reliable and stable and sure. Spirit matter is light; and, light is spirit matter. Spirit matter exists at velocities and frequencies faster than the speed of light. Spirit bodies can take upon themselves any shape they want. Spirit bodies are malleable, or they can manifest in different shapes and forms.

Most of the truly unidentified UFO's and aliens that people have seen are in fact spirits, who have transformed themselves into a different shape besides a human shape. UFO's function like spirit bodies, or spirit orbs, or quantum objects, or light. The alien spaceships which some people have seen and gone into are spirit realms or communities which have been transformed into alien spaceships by demons – Satan's followers – evil spirits – a group of human spirits who can take on any shape they desire and can create any kind of reality they desire, in an effort to deceive us and distract us from the truth. Demonic "spaceships" made out of spirit matter can travel much faster than the speed of light and teleport – or seem to. The spirit realm in the hands of evil spirits becomes a full holodeck experience, the matrix, and a complete star trek experience; and, it is infinitely more real than any of our science fiction movies can ever be. (Ross, Samples, & Clark).

Should we ever fall under or submit to Satan's power and influence, he and his demonic followers can run us through any holodeck experience of their choosing, just for the fun of it. While out-of-body, the human spirit can therefore have the full alien-autopsy experience under the ministration of demons who have made themselves look like aliens; and, the resulting spiritual experience will be as real as anything these people have ever experienced in their mortal lives or physical lives. Many demons or evil spirits have learned how to interact with and influence the physical realm telekinetically – with laser beams, pyrotechnics, UFO's, explosions, burns, levitation, cutting, slaps, punches, sound effects, dragon's breath, and the whole works. (Ross, Samples, & Clark).

People who use drugs, LSD, peyote, ayahuasca, alcohol, oxygen deprivation, witchcraft, sorcery, whirling, crystal balls, and hypnosis to force an NDE or OBE to happen often experience some really strange things, because they have unknowingly opened themselves up to Satan's influence and demonic influence. Satan is the master deceiver.

For example, under hypnosis we can form false memories and manufacture whole new realities, such as past lives that we allegedly lived as an ape, or a tree, or an ant, or Caesar, or Cleopatra, or an alien grey. We can also go and live in consensus realities in the spirit world, such as Atlantis or Lemuria and experience complete lifetimes of memories. We can contact aliens living on other worlds, or supposedly flying in spaceships through space, and then receive revelations from them. Or we can talk with snakes, bears, and other animals. They seem to be real aliens or animal spirits, but they are in fact Satan and his demonic followers giving us the holodeck experience of our choosing.

False memories, false lives, and manufactured lives and realities feel every bit as real and memorable as our own lives, if not more so. Most people who have OBEs and NDEs state that being Out-of-Body feels infinitely more real and clear and solid than being in a physical body. Anything becomes possible while out-of-body or in an altered state of consciousness; and, it will seem more real than our physical life when we are done. There's some really funny stuff that can happen as well.

This is one of the funniest NDE and OBE stories that I have ever watched:

https://www.youtube.com/watch?v=yjiSbpOJhZo

We learn from NDEs and OBEs – the lived experiences of the human race – how spirit matter and spirit bodies work in the non-local realm, quantum realm, or spirit realm. There are thousands upon thousands of these OBEs and NDEs online now. Most of them have gone online during the past five years. The evidence is impossible to deny, unless of course you are determined to do so. It's now April 2017 as I write this. God has started to reveal these things to us exponentially during the past five years or so.

We know from the Scriptures and from the Out-of-Body Travelers that a spirit body can travel much faster than the speed of light to remote destinations in other parts of this universe, like passing through a tunnel or a wormhole; yet, that same spirit body can come back to its physical body instantaneously – snap! We also know from the scriptures that Resurrected Beings can travel across this universe at the speed of thought, can levitate or stand in the air, and can walk through walls; yet, that same being can materialize and become fully physical, is able to eat and drink and be touched by mortal beings like us, when that Resurrected Being reaches its destination. The science fiction terms for these kinds of capabilities are called "teleportation," "levitation," and "phase-shifting".

Said another way, at the Word of Command, Resurrected Beings can phase-shift or transmute their physical bodies back into spirit matter, and then teleport at the speed of thought to anywhere in this universe. When this Resurrected Being reaches its destination, then at the Word of Command, this person can phase-shift back into being a physical body once again. If it's possible to convert spirit matter into physical matter through Conscious Observation or the Word of Command, then it should be equally possible to convert physical matter back into spirit matter by the same Word of Command. That's how the Resurrected Beings in the scriptures can do all of the different things that they can do, yet be fully and completely physical when they reach their final destination. Everything becomes possible once a person or psyche is granted full control over matter, or Quantum Objects.

If God were to give us humans the ability (or the permission) to phase-shift and teleport at will, we could travel to Alpha Centauri or Vega in a matter of seconds, and then rematerialize at our destination. Have you ever seen the movie *Contact*? I have watched and read NDEs and OBEs, wherein people (psyches) have connected with other humans in other parts of this galaxy and universe, while out-of-body and/or near-death. Through the Holy Spirit, the righteous spirits and mortal Saints of God can be in contact with each other, even though they reside on different sides of this universe.

I know people and I know of people who have phase-shifted and teleported – mortal beings like you and me. God intervened, decided to protect them, and decided to save them. My best friend has phase-shifted and teleported and slingshotted out of harm's way, many different times. He didn't actually consciously do it, though – God did it for him. These kinds of things can and do happen to us – the serendipitous and the miraculous – but they can't be replicated at will, because they are God's doing, not ours. God is KNOWN by what He does for us – things that we can't do for ourselves.

I have phase-shifted and teleported as well, because there is no other way to describe and explain what happened to me; but, it was nothing that I did. It was ALL God's doing and God's intervention. There is no other explanation for it.

There was no place to go when that car pulled out in front of me. I was going too fast and I was right on top of it. I hit the brakes. Time stopped. I was suddenly able to sense everything around me 360 degrees. It was like the whole of time had imploded into me, and I had an eternity of time to look into the mirrors, to look around, and to assess the situation. Time had simply stopped. Then, I saw a way out or I saw where I wanted to be. Snap! Time resumed and I was instantly in the middle of the lawn to the right and front of me, my car was completely stopped, and my car was dead or turned off. I had no memory of how I got there. I remember hitting the brakes and having time stop; and then snap, I just teleported or woke up at my destination twenty or thirty feet away with no memory of the intervening travel that took place and no knowledge of how the car had gotten stopped and turned off.

The guy who came running up to see if I was all right said that that was the coolest thing he had ever seen in his life, and that there was no way that I could have missed that other car, but I did.

There had been no contact, and there was no scratch or anything like that on my car. I was simply on that other car's side and bumper one instant and the next instant I was in the middle of the lawn completely stopped with the engine shut down, as if I had teleported or phased right through that other car. And it wasn't just me; it was my car as well that had phase-shifted and teleported. The only logical way that I can explain it is that God or the Spirit phase-shifted me or teleported me, in order to prevent me from merging with and becoming one with that other car.

The Materialists and Atheists will tell you that I imagined it all, or that it was some kind of hallucination, or that there must be some other logical and rational explanation. I KNOW, because I used to be a Materialist and an Atheist. I've talked myself out of it a few times before, so I know how it goes. I've also forgotten about the experience for long periods of time. People can become so separated from God and Spirit that they can no longer remember the things that they used to know. I KNOW this is true, because I have experienced this as well when I was addicted, a Nihilist, a Materialist, and an Atheist. These people can think what they want; but, they weren't there. I started the car, backed out into the street, and went off to work as if nothing had happened. There hadn't even been enough time for my heart to start racing or for the adrenaline to start flowing. The whole experience had been calm, peaceful, and serene, as if I had been in a bubble of protection or in the dream realm the whole time.

Ever since then, I have known (but haven't always wanted to admit or remember) that time can stop and that God can intervene in our lives and can phase-shift us, or teleport us, or miraculously save us if He wants to do so. I know others who have had similar experiences. These people have phased or teleported or slingshotted through no conscious will of their own. A Divine hand simply intervened in their lives. God is KNOWN by doing for us the kinds of things that we can't do for ourselves. Simply put, God is KNOWN by what He does for us. Jesus Christ has come and gotten many different people out of hell – just for the asking. God is KNOWN by living Him and experiencing Him. It all comes down to the psyche experiences, the spiritual experiences, the theophanies, and the lived experiences of the human race.

In conclusion, this Ultimate Model of Reality is the antithesis (the polar opposite) of Materialism and Scientific Naturalism. This Ultimate Cause Model of Reality is everything that Materialism and Naturalism are not, yet it subsumes an understanding of physical matter into the mix as well. This Psyche Ontology is a holistic model that's all-inclusive. That's why I decided to call it the Ultimate Model of Reality, because it attempts to include everything that human beings have ever experienced, studied, learned, thought about, researched, or observed throughout the whole of human history. I'm no longer limiting myself to physical matter. I have completely abandoned My Materialism, My Nihilism, and My Atheism. I'm now a Quantum Scientist to the fullest extent of the word! I'm a student of the Quantum Realm or the Spirit Realm.

What could be more exciting or interesting?

The Philosophical, Scientific, and Empirical Foundation Supporting Ultimate Cause

A strong Personality Theory and solid Model of Reality requires a strong philosophical base. It also requires strong empirical evidence.

Obviously, I have a lot of empirical evidence or KNOWLEDGE and many convincing reasons for choosing to believe that this Ultimate Cause Model of Reality is correct, useful, and true. William Buhlman's book is just a start. In ALL of my theorizing, I kept adjusting this Ultimate Model of Reality until it matched with the Lived Experiences or the Psyche Experiences of the human race, including Near-Death Experiences (NDEs), Out-of-Body Experiences (OBEs), Shared-Death Experiences (SDEs), and all the other types of Spiritual Experiences. I wanted to get this model as close to the Truth and Reality as I possibly could; so, I kept choosing to follow the evidence wherever it might lead me.

Because of his reliance on Aristotle's four causes, Joseph Rychlak's books and Logical Learning Theory provide a strong philosophical base for this Ultimate Cause Personality Theory and Ultimate Model of Reality. Logical Learning Theory (LLT) deliberately rejects Materialism and mechanistic exclusivity by including formal causes and final causes into all its theories, experiments, and models of reality. LLT is supportive of and compatible with this Ultimate Cause Model of Reality and is subsumed by this Ultimate Personality Theory.

I took instruction and inspiration from others as well, especially Brent D. Slife. Everything ever written and experienced by mankind supports Psyche as the Ultimate Cause, except for Naturalism and its derivatives such as Materialism which by definition in principle deny the existence of psyche, spirit, mind, consciousness, soul, and reality itself.

My personality theory is that psyche is personality and that psyche is the ultimate cause of everything that has ever been created, organized, and brought into existence from scratch. Furthermore, our psyche or personality or memories survive bodily death and brain death according to the KNOWLEDGE gained from NDEs, OBEs, SDEs, Radical Empiricism, Theophanies, and the Lived Experiences of the human race.

I submit the following books into evidence as proof of the truthfulness and usefulness of this **Ultimate Personality Theory** and this **Ultimate Cause Model of Reality**. As you can tell, I'm no longer an Atheist and no longer a Materialist, because I no longer refuse to look at evidence. I'm no longer afraid of the evidence for psyche and spirit and God, like I used to be. The preceding definitions and this **Ultimate Cause Theory of Personality** were informed strongly by the books in the following informal annotated bibliography:

Buhlman, W. L. (1996). *Adventures Beyond the Body: How to Experience Out-of-Body Travel*. New York: HarperCollins Publishing. (Out-of-Body Experiences). This book was my first exposure to someone who can replicate out-of-body experiences on demand. Buhlman has turned the whole thing into a Science, and he takes a non-denominational and secular approach to it all. He is non-religious and seems to be mildly against organized religion of any kind. This is a Science book, full of observations and empirical evidence and how-to suggestions, and not a religious book. I have encountered many other people who have this spiritual gift and can leave their physical body at will; but, this book remains my most favorite comprehensive introduction to the subject of OBEs. Best of all, Buhlman doesn't use drugs or hypnosis to go Astral – he does it naturally. The

Materialists, Atheists, Naturalists, and Mechanists refuse to look at this kind of evidence, because it proves that their philosophies are wrong. This book proves that psyche exists, and that psyche or personality survives separation from the physical body and bodily death. This book also proves that we humans have a spirit body that's in the same shape or form as our physical body. A careful reading of this book also proves that Psyche is something completely different than our spirit body and our physical body. We can KNOW the truth and prove the truth simply by living the truth and experiencing the truth first-hand for ourselves, or by choosing to trust someone who has. This book provides KNOWLEDGE and PROOF of the truth.

Gibson, A. S. (2006). *They Saw Beyond Death: New Insights on Near-Death Experiences*. Springville, UT: Horizon Publishers. (Near-Death Experiences). This book is an attempt to list and briefly discuss ALL of the books about Near-Death Experiences that were written by 2004. It's a good and semi-comprehensive introduction to the empirical or experiential Science of Near-Death Experiences. The Materialists, Atheists, Naturalists, and Mechanists refuse to look at this kind of evidence, because it proves that their philosophies are wrong. NDEs provide evidence in support of this Ultimate Cause Personality Theory, because it is the contention of this Ultimate Personality Theory that psyche or personality is the part of us that survives bodily death and brain death according to the empirical evidence from NDEs, OBEs, and SDEs.

Mark My Words. (2016). *Quantum Mechanics from a Non-Physical Spiritual Perspective*. Kindle. (Reference: https://www.amazon.com/dp/B01J023TGU). When it comes to Science, it's all about perspective or one's chosen interpretation of the available scientific data. The Materialists, Atheists, Naturalists, and Mechanists choose to interpret ALL scientific data from a physical perspective or a local perspective; and thus, they come up with completely different results and explanations of the scientific evidence than the people who choose to interpret all of the same scientific evidence and scientific data from a non-local, trans-dimensional, and spiritual perspective. Materialism provides a very limited view of Science and Reality, and an extremely scaled-down interpretation of scientific data. Quantum Mechanics becomes a completely different, and infinitely more comprehensive, Science when one looks at it from a spiritual or non-local perspective. Quantum Mechanics is better and makes more sense from a spiritual perspective than from a materialistic perspective. The Mechanists and Materialists are at a loss as to how to explain Quantum Mechanics from a physical perspective; but, the Spiritualists and Theists take to Quantum Mechanics like a fish to water, because Quantum Mechanics explains how spirit matter really works. Quantum Mechanics is in fact Spiritual Mechanics. Spirit is light; and, light is spirit, after all. The spirit realm is the quantum realm, the realm of light.

McTaggart, L. (2002). *The Field: The Quest for the Secret Force of the Universe*. New York: HarperCollins. (Quantum Mechanics, and the Immaterial or Non-Physical Zero-Point Field). This is an extremely popular book, and for good reason! In preparation for this book, Lynne McTaggart interviewed seventy-five different scientists, studying their scientific research which the Materialists and Naturalists ridicule, mock, censor, censure, and formally reject. In her book, *The*

Field, McTaggart wrote, "For a number of years, while researching *The Field* and subsequent work carried out in this area, I was patiently tutored in quantum physics by some seventy-five frontier scientists. I badgered, cajoled, demanded, and wheedled from each one of them countless hours, up to twenty interviews apiece, teasing out explanations, eventually wresting some crude translation for concepts that often exist for the physicist as pure mathematics. *What exactly is quantum coherence? Why does the Zero Point Field exist?* I would take their frequently incomprehensible answers and play them back via a metaphor until we could both agree on a lay approximation." (From the Preface to the 2008 Paperback Edition). McTaggart demanded to be taught the things which the Materialists and Naturalists refuse to study, accept, or acknowledge. There are many other books from Michael Talbot, Bernard Haisch, Gregg Braden, Rupert Sheldrake, and Amit Goswami (see the References for a few recommendations) who consider and study the hidden, immaterial, and unseen aspects of Quantum Mechanics which the Materialists and Atheists refuse to study and think about; but, *The Field* by Lynne McTaggart is considered to be the best of the bunch. Anything which deals with the Zero-Point Field of Light, the Quantum Sea of Light, the Light of Christ, Biophotonics, Dark Energy, Dark Matter, Spirit Matter, Trans-dimensionality, Parapsychology, Transpersonal Psychology, and Quantum Non-Locality will lend evidentiary support to psyche, non-local consciousness, and this Ultimate Cause Model of Reality. All one really needs to know is that these things exist, and that Materialism is false, in order to have all the support one needs for this Ultimate Model of Reality. Anything that explains why Materialism is wrong will generally be supportive of Psyche as the Ultimate Cause. Psyche and ultimate cause have always existed, just waiting for one of us to identify and label them as a causal agent in the overall scheme of reality. Alas, most of the pundits and theoreticians approach Quantum Mechanics and Consciousness from a physical perspective, because that's a no-brainer; but, that process ends up being completely worthless when we are trying to establish evidence for psyche, spirit, consciousness, or ultimate cause. The truly ingenious and creative people try to get a handle on Quantum Mechanics from a non-local or a spiritual perspective. These are the people McTaggart interviewed and reported on in her book. The non-local or the spiritual is by definition in principle non-physical, which means that it has to be inferred to exist or experienced in person first-hand because we will NEVER be able to measure it directly with our physical instruments. Spirit matter is light; and, light is spirit matter. We can't measure the waveform of light. We can only get a handle on it when it has slowed down, and become a physical photon, and left a mark. And, there are different types of light. According to Quantum Mechanics, only non-local immaterial consciousness has the ability to convert spirit matter into physical matter by collapsing the quantum wave function or the spiritual wave function thereby localizing that particular quantum object or spiritual object into our local 3D physical space-time universe. Spiritual objects or quantum objects only become detectable by our physical instruments after their quantum wave function has collapsed, they have slowed down, and they have become physical particles or photons instead of spiritual waves. That's what the Science of Quantum Mechanics is trying to teach us. If trying to establish evidence for psyche or ultimate cause, Michael Talbot's science books come highly recommended, because he was psychic and was surrounded by the supernatural while growing up. Michael Talbot, therefore, approached the Science of Quantum Mechanics from a psychical or a spiritual perspective which was rather unique at the time while he was alive and doing so. In the same vein, Dean Radin and Charles Tart approach Quantum Mechanics from the perspective of Parapsychology, Transpersonal Psychology, and psychical research. Rupert Sheldrake approaches the unseen and the non-local from a biological perspective, which gets the Atheists, Materialists, and Darwinists stirred up to no end directing a lot of persecution his way. Amit Goswami believes that consciousness, not matter, is the foundation of all existence, holds that the universe is self-aware, teaches that God is the source of "downward causation" (which is similar to first cause or ultimate cause), and claims that consciousness creates the physical world. Goswami uses Quantum Mechanics to establish and prove the existence of God. You know that's not going to sit well with the Atheists, Naturalists, and Materialists. But, the people who find evidence for the non-local, the paranormal, and the supernatural from Quantum Mechanics are geniuses, which is something the Materialists and Naturalists are not. The Materialists come across as lazy in comparison. The only thing the Materialists are interested in is limiting what we are permitted to study and research as scientists, psychologists, and theoreticians. The Materialists typically do their "science" by mocking and ridiculing the people who are doing the Real Science on the cutting-edge – the people whom Lynne Taggart interviews for this book. It doesn't take much ingenuity or creativity to ridicule and mock the people who are doing what you can't do. It's petty, actually. But, it's their modus operandi. Regarding his opponents and religious people, Atheistic Darwinist Richard Dawkins said at the Reason Rally, "Mock them! Ridicule them! In public! They need to be ridiculed with contempt!" Online, Richard Dawkins posted, "I lately started to think that we need to go further: go beyond humorous ridicule, sharpen our barbs to a point where they really hurt. I think we should probably abandon the irremediably religious precisely because that is what they are – irremediable. I am more interested in the fence-sitters who haven't really considered the question very long or very carefully. And I think that they are likely to be swayed by a display of naked contempt. Nobody likes to be laughed at. Nobody wants to be the butt of contempt." That's how Richard Dawkins does science, logic, and reason; and, he is not alone. Most of the Materialists and Atheists are this way – they do their science through mocking and ridicule rather than theorizing and research. While writing *The Field*, McTaggart went to the scientists (who were being mocked and ridiculed by the Atheists and Materialists) for her information and scientific understanding. An amazing, expansive, and popular science book was the result of her efforts. The Materialist's and Atheist's response to this book? "It's not science." That's their default response to everything that isn't Materialism, Naturalism, and Atheism.

    Moody, R., & Perry, P. (2010). *Glimpses of Eternity: Sharing a Loved One's Passage from This Life to the Next*. New York: Guideposts. (Shared-Death Experiences). Dr. Raymond Moody introduced the world to NDEs in his groundbreaking book *Life After Life*. Decades later, after having a Shared-Death Experience of his own, Raymond Moody introduces us to SDEs, which are solid and convincing proof that psyche, personality, perception, and memories survive separation from the physical body, because in an SDE a healthy living person accompanies the dead person on the first part of the dead person's afterlife journey and life review. The living person's brain isn't starved for oxygen, and thus can't be hallucinating as a result. SDEs are solid and convincing proof that psyche or personality survives brain death and bodily death, because the dead person is also there having the same out-of-body experience that the live people are having. And with some SDEs, there are two or more live people there with the dead person when the dead person passes on into the next life. In other words, there's independent confirmation of the same event from multiple different sources and perspectives. SDEs on YouTube are interesting to study and observe also.

Richards, P. S., & Bergin, A. E. (2005). *A Spiritual Strategy for Counseling and Psychotherapy* (2nd ed.). Washington DC: American Psychological Association. This book introduces a Theistic Spiritual Perspective, Spiritual Personality Theory, Theistic Personality Theory, and Theistic Psychotherapy, which means that this book is fully compatible with this Ultimate Cause Personality Theory. God fully qualifies as an Ultimate Cause, after all. Consequently, this book from Richards and Bergin provides a psychotherapeutic model, scientific evidence, clinical application, applied science, empirical evidence, and philosophical support for this Ultimate Cause Model of Reality. Any book that lends evidentiary support for theism, theophanies, revelations from God, and spiritual experiences will also lend evidence to this Ultimate Cause Model of Reality. The scriptures which the Biblical God Jesus Christ had a hand in writing and producing would be such a thing. For case studies supporting Theistic Psychotherapy, see the companion book, *Casebook for a Spiritual Strategy in Counseling and Psychotherapy* by Richards and Bergin. In contrast, any psychotherapy based exclusively on Materialism and Scientific Naturalism would technically be incompatible with Psyche and this Ultimate Cause Model of Reality. In these books, Richards and Bergin do an excellent job explaining what's wrong with Scientific Naturalism.

Ring, K., & Cooper S. (1999). *Mindsight: Near-Death and Out-of-Body Experiences in the Blind* (2nd ed.). Kearney, NB: Morris Publishing. (Near-Death Experiences and Out-of-Body Experiences). The people who were born blind can see while out-of-body during their NDEs and OBEs. This is solid evidence that psyche, memories, and personality survive separation from the physical body, bodily death, and brain death.

Rivas, T., Dirven, A., & Smit, R. H. (2016). *The Self Does Not Die: Verified Paranormal Phenomena from Near-Death Experiences*. Durham, NC: IANDS Publications. (Near-Death Experiences). This book turns NDEs into a verified and validated Science, by providing external confirmation that the NDEs and OBEs really took place. The truthfulness and usefulness of this Ultimate Cause Model of Reality hinges upon the fact that the human psyche survives bodily death and brain death according to the empirical sciences of NDEs, SDEs, and OBEs. If there were no empirical evidence for psyche surviving bodily death and separation from the physical body, then this Ultimate Cause Model of Reality would be no better than Materialism and Naturalism, which also have no evidence to support their primary premises. But, since there is plenty of evidence to support Psyche as the Ultimate Cause and no evidence to support Materialism's claim that psyche or spirit does not exist, this Ultimate Cause Model of Reality ends up being infinitely superior and a whole lot more complete, compelling, believable, and useful than Naturalism or

Materialism will ever be, as a model of reality. Materialism and Naturalism don't match with reality, so they make for a very poor model of reality. Something is said to be scientifically accurate if it matches with Reality. Since Materialism and Naturalism don't match with reality, they are by definition in principle unscientific – nothing but pure philosophy, sophistry, self-deception, and metaphysics.

Rychlak, J. F. (1970). The Human Person in Modern Psychological Science. *British Journal of Medical Psychology*, 43(3), 233–240. (Retrieve from: http://mypsyche.us/rychlak/). This article provides an excellent introduction to Joseph Rychlak and his use of Aristotle's four causes in psychological science. The rocks or physical matter cannot do science, teleology, predication, nor final cause; but, the human psyche or human person certainly can. Personality is psyche.

Rychlak, J. F. (1981a). *A Philosophy of Science for Personality Theory* (2nd ed.). Malabar, FL: Robert E. Krieger Publishing Company. Since one of my purposes in this essay is to introduce a new and unique Personality Theory, the Ultimate Personality Theory, this book from Rychlak serves as the perfect philosophical foundation for this Ultimate Cause Personality Theory, which I introduce in this essay. My Theory of Personality is that psyche is personality and that psyche is the part of us that survives bodily death and brain death, according to the Empirical Sciences of NDEs, SDEs, and OBEs. Under this Ultimate Personality Theory, psyche becomes the ultimate cause of everything that has ever been created, organized, and brought into existence from scratch, including genomes and life forms and physical matter. Any of Rychlak's books can be used to supply a philosophical introduction to Aristotle's four causes, which are necessary to understand, because ultimate cause is a fifth cause meant to explain the "who" behind each of Aristotle's four causes.

Rychlak, J. F. (1981b). *Introduction to Personality and Psychotherapy: A Theory-Construction Approach* (2nd ed.). Boston, MA: Houghton Mifflin Company. Since one of my purposes in this essay is to introduce a new and unique Personality Theory, the Ultimate Personality Theory, this book from Rychlak serves as the perfect introduction to Personality Theory and is a good foundation for this Ultimate Cause Personality Theory, which I introduce in this essay. I will take up this book in much greater detail in other books that I'm working on.

Rychlak, J. F. (1988). *The Psychology of Rigorous Humanism* (2nd ed.). New York: New York University Press. In this book, on pages 8 to 31, Joseph Rychlak goes through 104 philosophers and scientists who had a hand in laying the foundation for the Science of Psychology and Science in general; and, Rychlak identifies the philosophers and theoreticians who employed formal cause and final cause in their theories; and, those who did not. Fourteen of the 104 basically limited themselves to mechanistic causes, material cause and efficient cause. The other 90 slipped over into some kind of formal cause (design theory). This is a clear case where the minority seized control of Science and made the arbitrary rule that only material causes, and efficient causes, should be considered and allowed into evidence while doing Science. The Mechanists and Materialists are the people who lobotomized Science in an attempt to enforce their philosophical worldview or personal religion onto the rest of us. It's the atheistic, mechanistic, and materialistic philosophers and scientists, who reject teleology and final-cause, who have forced their way into controlling our public schools, scientific research labs, the tenure and peer-review process, and our college textbooks, so that we the paying public are prevented from ever learning about final-cause, teleology, psyche, mind, and soul. It worked. Most of us don't have a clue. The Materialists had to hijack Science and censor Science, because once a person understands and accepts teleology and final cause, then he or she just automatically knows why Materialism and Naturalism are false. Raw physical matter can't do teleology, design, formal cause, final cause, nor ultimate cause. Physical matter is an effect, a Quantum Effect; and therefore, physical matter can't be the ultimate cause for anything nor the first cause of anything. Of the 104 philosophers and scientists that Rychlak researched and documented, 58 of them employed some type of teleology or final cause in their theorizing. That's quite a coup for the Mechanists, Materialists, and Atheists; wherein, the minority clearly dominated and tried to exterminate the majority and pretty much succeeded in doing so where Science and Scientific Evidence are concerned. Rychlak's goal, as a Rigorous Humanist or Experimental Humanist, was to put teleology, predication, purpose, meaning, and final cause back into the Science of Psychology. The ultimate goal of this Ultimate Cause Model of Reality is to put psyche back into science, philosophy, clinical application, and psychology. It seems like the humane and right thing to do. These goals are in sync with each other. The goal is to get at the truths which the Materialists, Naturalists, and Atheists are trying to hide from us.

Rychlak, J. F. (1994). *Logical Learning Theory: A Human Teleology and Its Empirical Support*. Lincoln, NB: Nebraska University Press. This book from Rychlak contains empirical support or scientific evidence for a human teleology, or the human psyche's innate ability to function, predicate, and cognate in formal-cause and final-cause roles. The rocks and physical matter cannot do formal-cause design nor final-cause choice and creation; but, the human psyche can. I see this book as providing scientific evidence and experimental evidence for Psyche as the Ultimate Cause of everything that has ever been created, organized, and brought into existence. I submit this book into evidence as scientific support and empirical support for this Ultimate Cause Model of Reality, which I develop in this essay. For case studies supporting Logical Learning Theory, see, *Personality and Life-Style of Young Male Managers: A Logical Learning Theory Analysis*, by Joseph Rychlak. For a "Reader's Digest" version of Logical Learning Theory (LLT) and a simplified introduction to all of this see, *The Human Image in Postmodern America*, by Joseph Rychlak. Rychlak's book *Discovering Free Will and Personal Responsibility* is also a good introduction to his Logical Learning Theory, because ONLY Psyche can do discovery, free will, personality, and responsibility. I feel very lucky to have discovered Joseph Rychlak and his books, because Rychlak is one of the people that the Materialists, Mechanists, and Naturalists are trying desperately to keep hidden from us; and they have succeeded in doing so.

Rychlak, J. F. (1997). *In Defense of Human Consciousness*. Washington DC: American Psychological Association. I debated about whether to include this book in this list, because it is Rychlak's Logical Learning Theory yet again, but from the perspective of consciousness. Alas, as of 1997 Rychlak is actually very weak and lukewarm when it comes to psyche and non-local consciousness, leaving open what he apparently believes might be the possibility that the psyche or mind or consciousness is an epiphenomenon (a side-effect or a derivative) of the physical brain, an idea which is incompatible with the empirical evidence provided by NDEs, SDEs, and OBEs. Of course, the exponential explosion in books about NDEs and OBEs didn't start until about 2006; and, I didn't get onboard with the whole OBE thing until the end of 2015 after I read William Buhlman's book *Adventures Beyond the Body: How to Experience Out-of-Body Travel*; so, I can understand why Rychlak didn't mention NDEs in his books, because we were all in the same boat back then in the mid-1990's with limited NDE and OBE evidence at hand. In this book, *In Defense of Human Consciousness*, Rychlak employs *Logos* as the grounds for "psychic consciousness" making the two basically synonymous; yet also stating that Logos is not physical and that "psychic consciousness is not mysterious or spiritual". In other words, Rychlak implies that psyche (logos) is not physical and that it's not spiritual either. So, what is it? Well, Rychlak employs "psychic consciousness" as a philosophical concept, or a psychological construct, or a "mental" construct; and, he seems to leave it at that. In his books, Rychlak states that he is not going to take an official stand on the mind-brain problem, only stating that his Logical Learning Theory (LLT) can be made compatible with the convictions of those who have chosen to believe that the mind is a completely separate entity from the physical brain. Consequently, I turn to others for procedural evidence, validating evidence, and empirical evidence of psyche or mind being a non-local quantum phenomenon completely separate from the physical brain. I use Joseph Rychlak primarily for the philosophical, logical, and procedural foundation of this Ultimate Cause Model of Reality; but, his books do indeed provide experimental evidence supporting Humanism, Logos, Teleology, Predication, Final Causality, and this Ultimate Cause Personality Theory; and, only psyche or non-local consciousness can do final cause, choice, and predication. I turn to NDEs, SDEs, OBEs, revelations from the Biblical God Jesus Christ, and other spiritual experiences for empirical, experiential, and observational evidence supporting this Ultimate Cause Model of Reality. I turn to Pim van Lommel and his cutting-edge (2010) interpretation of Quantum Mechanics, from a spiritual perspective or an NDE perspective, for the primary scientific foundation of this Ultimate Cause Model of Reality. Gravity, dark matter, dark energy, forces, magnetism, and the zero-point field of light, as non-physical immaterial trans-dimensional spiritual phenomena, also lend evidentiary support to this Ultimate Model of Reality. The physical brain is a transceiver for the non-local living consciousness, who is in fact the broadcaster to the physical brain, the recorder of information coming from the physical brain, and the director of the physical brain. Psyche is the driver, and the physical brain is the machine that gets driven. The psyche or non-local consciousness and our memories survive bodily death and brain death, according to the empirical evidence from NDEs, SDEs, and OBEs. I finally chose to follow the evidence. Rychlak, for whatever reason, chose to straddle the fence and chose to leave the mind-brain problem unresolved in his theorizing; whereas with a couple of extra decades of empirical evidence to draw upon, I quickly noticed that the mind-brain problem is instantly and immediately solved once Materialism or Scientific Naturalism is eliminated from the equation and taken out of the picture with extreme prejudice. Along the same lines of inquiry, you might be interested in Rychlak's book, *Artificial Intelligence and Human Reason: A Teleological Critique*, which makes a solid case demonstrating that computers and artificial intelligence cannot do teleology nor final cause; but, the human psyche or "psychic consciousness" certainly can.

Sanford, J. (2014). *Genetic Entropy* (4th ed.). Cornell University: FMS Foundation. (Scientific Proof and Common-Sense Proof that All Types of Materialism or Naturalism Are False). This book provides convincing procedural evidence, mathematical modeling evidence, scientific evidence, and logical common-sense evidence demonstrating beyond a shadow of a doubt that Natural Selection and Random Mutations cannot design and create genomes and life forms from scratch as the Darwinists and Materialists claim. In fact, Natural Selection and Random Mutations didn't even exist at all, until after God or Psyche designed and created the first fully functional genome and life form on this planet, in the first place. A fully functional genome and life form had to be produced by some kind of Psyche or Consciousness or Intelligence, before Natural Selection, Random Mutations, and other Physical Processes could come into play; therefore, Mutation/Selection (Evolution) cannot be the origin of life on this planet. When fully understood, this book from Sanford provides proof of Psyche as the Ultimate Cause of everything that has ever been created, organized, and brought into existence from scratch, including physical matter, genomes, and life forms.

Van Lommel, P. (2010). *Consciousness Beyond Life: The Science of the Near-Death Experience*. New York: HarperCollins. (Quantum Mechanics from a Non-Local or Non-Physical Perspective). This just may be the most popular, most read, and most useful book about Quantum Mechanics and Near-Death Experiences that has ever been written. This book is Science, and it demonstrates clearly and conclusively that psyche or mind survives separation from the physical body and brain death. Quantum Mechanics is Spiritual Mechanics, the way that spirit matter really works. Classical Physics tells us how physical matter works; and, Quantum Mechanics tells us how spirit matter works. Quantum Non-Locality, Quantum Mechanics, and Spiritual Mechanics are the Science behind Psyche, Non-Local Consciousness, and Near-Death Experiences. I refer to this book almost constantly in all of my writings and theorizing.

Goswami, A. (2008). *God Is Not Dead: What Quantum Physics Tells Us about Our Origins and How We Should Live*. Charlottesville, VA: Hampton Roads. I completely designed and wrote this Ultimate Cause Model of Reality BEFORE reading any of Amit Goswami's books. After writing most of this essay, I started reading from *God is Not Dead*, and it quickly became apparent to me that Goswami's theories and ideas are compatible with Psyche as the Ultimate Cause. Of course, that shouldn't be too surprising, because anyone who blasts Materialism as heavily as Goswami does will end up by definition in principle being compatible with this Ultimate Cause Model of Reality. By the time Goswami gets done with Materialism, there's nothing left but a smoking crater. Ultimate Cause is the exact opposite of Materialism. Goswami mentions Near-Death Experiences (NDEs) in his book. Goswami makes his ideas compatible with Intelligent Design Theory. Goswami also tries to salvage the theory of evolution by describing evolution as a consciously directed process, which is something that I no longer feel the need to do. In my theorizing, I have observed that the various versions of evolution or Darwinism are synonymous with Materialism, which is synonymous with Creation by Rocks. Psyche or ultimate cause is the antithesis of Materialism and Darwinism; but technically, the Theory of Evolution can be salvaged if psyche or consciousness is employed to direct and do the evolution. Goswami defines God as "quantum consciousness", which is an idea that is completely compatible with Psyche and Ultimate Cause. Goswami explains evolution and creation as "downward causation" or top-down causation or psyche causation, which is completely compatible with **ultimate cause**, which I purposefully designed to be compatible with and explanatory of Aristotle's four causes in an attempt to give this Ultimate Cause Model of Reality a solid philosophical base. Goswami also fully embraces Rupert Sheldrake's *morphic resonance* and morphogenetic blueprints, which are spiritual blueprints or non-local quantum designs (formal causes) upon which life forms are built while developing in the womb, because there is not enough information in DNA to build a living organism from scratch, but instead just enough information in the DNA to keep a living organism alive and functioning after it has been built from its spiritual blueprint or morphogenetic blueprint. Cell differentiation, the arbitrary turning on and off of certain genes during the development of the fetus (just the right genes being turned on or off in each and every cell with one cell initially being completely different than the cell next to it or the cell from which it came) is not done by DNA but is instead done by this spiritual blueprint or quantum blueprint or the human spirit. Your spirit provided the design or blueprint from which your physical body was made, not your DNA. That's the lesson of epigenetics and morphogenetic blueprints. Goswami also repeatedly emphasizes that physical matter and physical machines like computers cannot do meaning and cannot process meaning, a reality which signifies that physical matter can't do final causality. Goswami covers all the bases necessary to make his unnamed model of reality compatible with this Ultimate Cause Model of Reality. Not being a Christian, Goswami has some strange ideas about Jesus Christ, though. Personally, I would adjust and tweak some of Goswami's ideas to make them match more fully with the modern-day revelations which we have received from the Biblical God Jesus Christ, because I prefer to get information about Jesus Christ straight from Jesus Christ himself rather than from Goswami's speculations about Jesus Christ; but, it's doable! With occasional adjustments, Goswami's ideas can be made compatible with the modern-day revelations from Jesus Christ as found in the *Book of Mormon: Another Testament of Jesus Christ*, the *Doctrine and Covenants*, and *Pearl of Great Price*. Where else is God going to reveal Himself to us besides the books He had a hand in writing and producing? In these books, the Biblical God Jesus Christ tells us that He created everything spiritually (quantumly), before it was organized physically. In other words, our physical bodies develop in the womb and our cells differentiate according to the spiritual blueprint or the quantum blueprint which the Biblical God designed and put together BEFORE this physical universe was called into existence by quantum consciousness, or psyche, or God. Amit Goswami has many other books about Quantum Consciousness, which I own, but haven't read yet. Based upon what I have read so far in *God Is Not Dead*, it looks promising. Goswami puts another nail into Materialism's coffin, which automatically lends support to this Ultimate Model of Reality and to Psyche as the Ultimate Cause of everything that has ever been brought into existence from scratch. Every truth will be compatible with every other truth; and, every truth will be incompatible with Materialism and Naturalism. That is my observation.

Bishop, B. G. (1998). *The LDS Gospel of Light*. USA: Ponce de Leon. (Consciousness, Energy, Waves, Light, Spirit, NDEs, Light of Christ, God is Light, and Quantum Mechanics). This is another book that I started reading after writing this essay; and, the whole book seems to be compatible with Psyche as the Ultimate Cause of our Reality. Of course, I already realized that this Ultimate Model of Reality would be compatible with the *Bible*; with the Gospel of Christ as presented to us in the *Bible*, the *Book of Mormon: Another Testament of Jesus Christ*, the *Doctrine and Covenants*, and the *Pearl of Great Price*; with the revelations of the Biblical God and the revelations from the Biblical God; and with Christ's Gospel of Light or the Light of Christ. Every truth lends evidentiary support to every other truth; and, every truth explains to us why Materialism and Naturalism are false. That has been my experience and my observation.

These were the main books which helped me to create this **Ultimate Cause Model of Reality** and that caused me to believe that this **Ultimate Personality Theory** is real and true.

This Ultimate Cause Model of Reality is an all-inclusive model of reality; and, its goal is to subsume all truth and every human experience into its ranks. It was actually difficult to pick the very best books to represent and support this Ultimate Cause Model of Reality, because everything in science, philosophy, and experience supports psyche as the ultimate cause, except for the books and people who deliberately limit themselves to Materialism, Mechanism, and Naturalism. Materialism and Naturalism are incompatible with the rest of Science, Philosophy, Religion, Theology, and Human Experience including this Ultimate Cause Model of Reality. Materialism and Naturalism are extremely limited and exclusive, so they can't be made compatible with the rest of Science, Philosophy, Applied Psychology, Empirical Evidence, and Lived Experiences because Naturalism and Materialism are based upon a refusal to look at evidence and a refusal to accept evidence. I'm not interested in excluding evidence from this Ultimate Model of Reality – been there and done that already, when I was an atheist and a materialist. I don't want to go back to My Materialism and My Atheism, because I have found something infinitely better to take their place.

Psyche is the ultimate cause. Joseph F. Rychlak provided the core philosophical foundation and the personality theory foundation for this Ultimate Cause Model of Reality. Richards and Bergin provided the psychological and psychotherapeutic foundation for this Ultimate Model of Reality. Sanford reveals the biggest error that has been made so far in the physical sciences, an error or falsehood that keeps on giving and keeps holding people back. One has to eliminate the falsehoods, deceptions, and lies before he or she can pursue the true reality or the ultimate reality of our existence.

Pim van Lommel, Lynne McTaggart, Amit Goswami, and Quantum Mechanics provide the scientific foundation for Psyche or Non-Local Consciousness as the Ultimate Cause. Hugh Ross and Astrophysics explains to us that only 4% of this universe got converted to physical matter during the Prime Event or "Big Bang", and that the other 96% of this universe is still in its original spiritual state of existence. Near-death Experiences (NDEs), Shared-Death Experiences (SDEs), Out-of-Body Experiences (OBEs), and Spiritual Experiences in general provide the empirical evidence, experiential evidence, and observational evidence needed to prove that this Ultimate Model of Reality is correct and true. The truth can be KNOWN and PROVEN by experiencing it and living it first-hand, or by choosing to trust someone who has. After having given it years of study, research, and thought, I build this Ultimate Cause Model of Reality on a solid foundation.

Every Spiritual Experience Supports Psyche as the Ultimate Cause

Empirical Knowledge, or Experiential Knowledge, or Lived Experience is superior to scientific hypotheses and philosophical guesswork.

Whenever I encounter people who are sharing with me some of the spiritual experiences, visions, psychic experiences, out-of-body experiences, and near-death experiences which they have had, I choose to believe them and take them at their word because I have had spiritual experiences of my own and know that the phenomenon is real and truly happens from time to time. Furthermore, I have friends who have seen and talked with the spirits of their dead relatives. I know of many people who have seen and talked with Jesus Christ while out-of-body. Quantum Mechanics has also taught me that the Non-Local or the Spiritual has to exist in order to explain all of the scientific evidence and empirical evidence associated with Quantum Mechanics. Psyche must exist, or physical matter would not exist.

The Materialists and Naturalists can't explain Quantum Mechanics in a way that makes logical sense, just like these people can't explain spiritual experiences in a way that is parsimonious and makes sense. You will never learn anything useful about the spiritual or the non-local from a Materialist or an Atheist. I have observed that once people have had spiritual experiences of their own, then they are no longer Materialists or Naturalists; and, once these people have seen God, then they are no longer Atheists.

The following YouTube Videos about OBEs and NDEs had an influence on my philosophical worldview and this Ultimate Cause Model of Reality; therefore, I submit them into evidence. These spiritual experiences and out-of-body experiences are sensational. Mine have been quite mellow in comparison. Nevertheless, my best friend is a prophet, seer, and revelator; and, he has had many different types of spiritual experiences, visions, out-of-body experiences, miraculous experiences, and near-death experiences and he has testified to me while watching these videos with me that these things are real and true and match perfectly well with his own Lived Experiences while out-of-body in the spirit realm. My friend told me that these OBEs and NDEs and Theophanies have all the signatures of authenticity. The ONLY way to KNOW the truth is to live it and experience for yourself, or to choose to trust someone who has. I trust my friend; and, I trust these people as well.

Let's start with Howard Storm. Howard Storm was an atheist, died, and went to hell; and, after finally calling upon Jesus Christ to save him, Jesus came and rescued Howard Storm from hell. Should you ever find yourself in hell, remember that Jesus Christ can get you out of there for the asking.

https://www.youtube.com/watch?v=UPj4wci_bcI

https://www.youtube.com/watch?v=Vm647n1360A

Ian McCormack was an atheist, went to hell, and was saved from hell by Jesus Christ. This is currently my most favorite version of Ian McCormack's NDE. There are many others:

https://www.youtube.com/watch?v=HbTAmN4m2lQ

Dr. Mary Neal had a very vigorous near-death experience, which kind of makes it hard to believe that she was hallucinating the whole time. I'm interested in the NDEs wherein the individual gets to see God.

https://www.youtube.com/watch?v=DX473dF7ChY

https://www.youtube.com/watch?v=ULsl92H-Noc

Each one of these people has written and/or published a book or two about their Near-Death Experience, which I own and have been reading from. Check the References for a list.

The Hell Experience of an Ex-Satanist, such as John Ramirez, can be informative and also provide empirical evidence for Psyche as the Ultimate Cause:

https://www.youtube.com/watch?v=I11L71PD3Lw

There are hundreds, if not thousands, more on YouTube right now. Anything from Raymond Moody or Eben Alexander proves fascinating. NDEs, SDEs, OBEs, and Spiritual Experiences provide empirical evidence that this Ultimate Cause Model of Reality is correct and true. Psyche is the Ultimate Cause. I could never find any supporting evidence for My Materialism and My Atheism; but, I have found reams of evidence for Psyche as the Ultimate Cause, and each day I come across more and more evidence which proves to me that Psyche exists. I finally chose to follow the evidence.

Remember, Empirical Knowledge, or Experiential Knowledge, or Lived Experience is superior to scientific hypotheses and philosophical guesswork. There's no substitute for KNOWING the truth, having lived it and experienced it for yourself.

The Ramifications of Psyche and Ultimate Cause

Human beings are composite entities having both an immaterial spiritual component and a physical body component. While here in mortality, a person is a spiritual being having a physical experience. In the afterlife, before the resurrection from the dead which Christ says is going to happen to all of us, a person is a spiritual being.

Personality is psyche; and, psyche is personality. Psyche is typically defined as one's spirit, soul, mind, consciousness, individuality, intelligence, awareness, spark, and life. Personality or psyche is the part of us that survives separation from our physical body, separation from our spirit body, death of our physical body, and brain death according to the empirical sciences of Near-Death Experiences (NDEs), Out-of-Body Experiences (OBEs), Shared-Death Experiences (SDEs), and other types of Spiritual Empiricism (spiritual experiences). Psyche is the Ultimate Cause of every contingent thing that has ever been brought into existence from scratch.

I define the Self as the Psyche, in Cartesian Dualism fashion, with the psyche or consciousness being an immaterial substance completely different than and separate from the physical brain. I rely upon Quantum Mechanics, Astrophysics, and Brain Stimulation in Neuroscience for the scientific evidence necessary to confirm the truthfulness and usefulness of this Ultimate Cause Model of Reality and the existence of Psyche. I rely upon NDEs, OBEs, and SDEs for the necessary empirical evidence. I also rely upon scriptural evidence or revelatory evidence for this Ultimate Cause Personality Theory.

Doctrine and Covenants 88: 15: "And the spirit and the body are the soul of man."

Doctrine and Covenants 93: 33-34: "For man is spirit. The elements are eternal, and spirit and element, inseparably connected, receive a fullness of joy; and when separated, man cannot receive a fullness of joy."

This is the Ultimate Model of Reality because it attempts to conform to all known truths and to the whole of human experience and knowledge.

What is truth, and how do we know it?

The truth is knowledge of things as they really are. If something is true, then it has always been true, and it will always be true. Even better, truth is Lived Experience! The best way to KNOW the truth is to live the truth and experience the truth for yourself, or to choose to trust someone who has.

There are at least three ways of knowing the truth.

1. The least effective and least reliable way of knowing the truth is philosophically or though logical common sense – sometimes called procedural evidence. Philosophy is guesswork or the making of hypotheses – a comparing of

ideas and points of view. Philosophy is used to explore the ideas and concepts that are unseen, intuitive, immaterial, and/or unknowable by our physical senses and physical instruments. Ideally, Philosophy is supposed to be the pursuit of the truth; but, many people (sophists and materialists) use Philosophy in an attempt to make their lies seem true, which is easy to do since Philosophy is an abstract mental activity. Philosophy is rife with deliberate deception, including self-deception. Philosophy is worthless if it is used to prove a lie true, as happens in the case of Materialism and Naturalism. Materialism, Naturalism, and Atheism are philosophy, because they have no evidence to support them and have to be taken on blind faith as being true. Materialism or Naturalism is metaphysics and religion, not science.

    2. A more effective way of knowing the truth (at least where physical reality is concerned) is the Scientific Method or Scientific Experimentation. However, the Scientific Method has an inbuilt logic fallacy called "affirming the consequent". What this means is that the final step of the Scientific Method calls for an interpretation of the scientific data or an explanation of the scientific evidence. This is where the flaws come into play with the Scientific Method, because the scientific data calls for a philosophical best explanation of the scientific evidence or a philosophical best interpretation of the scientific data. The Materialists and Naturalists simply guess wrong, and they interpret the scientific data incorrectly from an exclusively physical perspective. It requires faith to believe that one's chosen interpretation is correct, right, and true. It's very easy to provide the wrong interpretation or the worst possible explanation to scientific data and scientific evidence. The Materialists, Naturalists, and Darwinists do so all the time. The Materialists interpret everything in terms of Creation by Physical Matter or Creation by Rocks; yet, we all know that the rocks cannot design and create anything at all.

    3. The best way of knowing the truth is to experience it first-hand or first-person for yourself, or to choose to trust someone who has. I have taken to calling it The Art of Knowing, or Knowledge. Upon further research, I learned that most philosophers call it Lived Experience or Phenomenology. Knowing or certain knowledge trumps philosophical guesswork and scientific interpretation every time. Physical Empiricism and Spiritual Empiricism, experiencing the physical world and the spirit world directly for yourself, is the best way of knowing how these things really work. The people who have been to the spirit world, seen God, and talked with the angels of heaven during their Near-Death Experiences (NDEs), Out-of-Body Experiences (OBEs), Shared-Death Experiences (SDEs), and other types of Spiritual Empiricism (spiritual experiences) simply KNOW that God exists and that the human psyche survives separation from the physical body, bodily death, and brain death. My best friend has experienced many of these things directly, so he KNOWS that they are real and true. First-hand experience and first-person observation are the best way of knowing the truth. The scientists call this method of knowing the truth, Observation. Philosophers tend to call this method of knowing the truth, Experience or Empiricism, a branch of Epistemology; and, many people call this method of knowing the truth, Direct Revelation, Radical Empiricism, or Lived Experience. Knowledge of the truth, experiencing the truth and living the truth, is the best way of knowing the truth. Truth is knowledge.

Doctrine and Covenants 93: 24: "And truth is knowledge of things as they are, and as they were, and as they are to come." Truth is knowledge. John 8: 32: "And ye shall know the truth, and the truth shall make you free." Truth is synonymous with knowledge; and even though it's tautological, simply KNOWING THE TRUTH or experiencing the truth first-hand is in fact the BEST way of knowing the truth. It sets you free! It's elementary my dear reader.

Every time I pick up a book or article about Science, Quantum Mechanics, Quantum Objects, Quantum Entanglement, Quantum Nonlocality, Trans-Dimensionality, Action at a Distance, Forces, Fields, Magnetism, the Zero-Point Field, Conscious Observers, Particle Physics, Gravity, Dark Matter, Dark Energy, Light, Faster than Light Travel, Cosmological Constants, Intelligence, Thought, Dreams, Psychology, Philosophy, Psyche, Life, Consciousness, Mind-Over-Matter, the Placebo Effect, Time, Space-Time, Universal Constants, Physical Laws, Causality, Spirituality, Heaven, Hell, Revelations, Visions, NDEs, OBEs, SDEs, Spiritual Experiences, or God, I quickly realize that NONE of these things are possible if Materialism is true. Materialism or Naturalism precludes and excludes the existence of these kinds of things. If Materialism were really true, it would prevent these things from happening and make them impossible. You and I would not exist if Materialism or Naturalism were 100% true. Since these things have been experienced and thereby proven to exist, we KNOW that Materialism and Naturalism are false. Truth can be KNOWN by living it and experiencing it for ourselves. Truth falsifies lies such as Materialism, Naturalism, and Darwinism.

Materialism or Scientific Naturalism is a worthless and useless philosophy masquerading as Science; but, it's not Science. Materialism is nothing more than philosophy, pseudo-science, sophistry, and metaphysics. Philosophy, speculation, hypotheses, and the spreading of known falsehoods are the weakest way of knowing the truth, because they can actually prevent us from knowing the truth. All you really want is the truth, unless of course you are a Materialist, Naturalist, or Atheist – then any old lie will do. You'll even be eager to share that lie with someone else if you are a Materialist or an Atheist because you can make some good money if you do.

Ramifications of this Ultimate Cause Personality Theory

This Ultimate Model of Reality and Psyche Ontology is based heavily upon our Lived Experiences as a race, including our theophanies, spiritual experiences, visions, revelations from God, near-death experiences, and out-of-body experiences. ONLY Psyche can do Lived Experiences both in the spirit realm and here in this physical realm and actually remember those experiences when it's done. This chapter answers some of the questions that I was asked about this Ultimate Personality Theory or Psyche Personality Theory.

Psyche, or one's immaterial non-local trans-dimensional consciousness, is the basic structure of personality and self, because according to NDEs, OBEs, SDEs, and other types of Spiritual Empiricism our self, or psyche, or consciousness, or memories survive separation from the physical body, separation from our spirit body, death of the physical body, brain death, and resurrection from the dead.

Under this Ultimate Cause Personality Theory, going to hell is the ultimate cause of an abnormal personality, psychological illness, spiritual illness, or mental illness. It's possible to go to hell while here in mortality. I did, so I know of what I speak. There are many applicable definitions for hell. Hell is being completely alone, with no friends and no social support and no reason to live. Hell is being immobilized and trapped in anxiety, depression, hopelessness and fear year after year after year. Hell is having no purpose in life. Hell is being addicted to substances and thereby having no control over one's life and destination. Hell is constant never-ending anxiety and fear. Hell is being trapped in sin or addicted to sin, unable to get out. Sin is anything which prevents us from reaching our full potential. Sin is damnation and hell – being stopped in our progress, fulfillment, and actualization. Consequently, the Atonement of Jesus Christ and getting out of hell is the ultimate therapy for one's spirit, body, mind, and soul.

Psychological illnesses should be addressed or treated with a combination of the Atonement of Christ, repentance or change for the better, spiritual comfort and support through prayer, hope, judicious careful drug therapy if one's mental illness can be demonstrated to have a physical component, and Friendship Therapy or Social Interest. We are not afraid of our friends, and we look forward to seeing our friends. Making friends and supporting each other through the hard times is our primary reason for existence here in mortality. Try to make a new friend every day. The purpose of life is to make friends, including making friends with God.

What is the process of normal development under this Ultimate Personality Theory? The human psyche has multiple stages of development similar to a butterfly; but, the psyche is a bit more complex than an insect because psyche or consciousness has an immaterial, non-local, trans-dimensional, non-physical basis according to Quantum Mechanics and Lived Experiences.

In the beginning, psyche is pure immaterial consciousness. This stage is pre-egg, meaning that at this stage psyche is pure intelligence or pure thought, an unformed actuality. Eventually, psyche can be assigned to occupy spirit matter (a type of egg stage), and subsequently assigned to occupy physical matter (a type of larval or caterpillar stage capable of interacting directly with the physical world). In time, that larva or caterpillar will "die" and go into an underworld, or a cocoon, or a state of physical dormancy; yet, its psyche or mind or consciousness or individuality will go on. According to the scriptures which the Biblical God Jesus Christ had a hand in writing and producing, at some point in the future our psyche will be resurrected from the dead and emerge as a glorious immortal being (similar to the butterfly rising from the dead, the cocoon or chrysalis, after its heavenly transformation).

According to Jesus Christ, one of the main purposes of mortal life is to follow Him and to become like Him. Families are known to sup together. Christ invites us to become a part of His family.

Revelation 3: 20-22: "Behold, I stand at the door, and knock: if any man hear my voice, and open the door, I will come in to him, and will sup with him, and he with me. To him that overcometh will I grant to sit with me in my throne, even as I also overcame, and am set down with my Father in his throne. He that hath an ear, let him hear what the Spirit saith unto the churches."

During the normal process of development under this Ultimate Cause Model of Reality and this Ultimate Personality Theory, if successful in our development, we will eventually sup with Christ and sit down with Christ on His Father's throne.

Consequently, under this Ultimate Cause Model of Reality, the good life is ultimately defined as becoming like Christ and then sitting down with Him on His Father's throne. Meanwhile, here in mortality the ultimate goal is peace of mind.

Doctrine and Covenants 59: 23: "Learn that he who doeth the works of righteousness shall receive his reward, even peace in this world, and eternal life in the world to come."

The good life consists of peace in this world, and eternal life in the world to come. In other words, the good life consists of getting out of hell and staying out of hell. Likewise, psychological health implies having a sound mind, which is impossible if one is in hell.

2 Timothy 1: 7: "For God hath not given us the spirit of fear; but of power, and of love, and of a sound mind."

Within this Ultimate Cause Model of Reality and this Ultimate Personality Theory, psychological health consists of getting God's Spirit within us, so that His Spirit can fill us with hope, love, and a sound mind while at the same time vanquishing all anxiety, depression, addiction, sin, and fear. Psychotherapy consists of getting out of hell and into God's presence. Psychotherapy consists of finding the peace that surpasses all understanding. Psychotherapy is learning how to bring the full effects of the Atonement of Christ into our lives.

Remember! If it's motivated by guilt, force, anger, hatred, pride, profit, selfishness, competition, jealousy, or fear, then it's Satan's work which you are doing even if you are doing a good thing or trying to do a good thing. In order for the full blessings, happiness, joy, and peace to accrue, it must be motivated by friendship, charity, compassion, and love. We must learn to do things for the right reasons in order to achieve the best results both for ourselves and for others.

There is an element of force, fear, and intimidation in classical conditioning and operant conditioning. Conditioning works on the animals, but humans tend to rebel whenever someone tries to condition them or force them to comply. The human psyche is a choosing organism, an agent in the fullest sense of the word; and, the human psyche doesn't respond favorably to force, fear, guilt, hatred, and intimidation.

Each psyche is unique, a type of infinite singularity, originally without form or matter of any kind. This also means that every psyche or intelligence has its own unique set of traits, temperaments, likes, dislikes, and dispositions. Whenever presented with an opportunity or some kind of dialectical opposition, the psyche is the thing that chooses between the alternatives; and, the psyche tends to choose in conformity to its desires, likes, goals, and dispositions; yet, it can also choose to experiment and try-out alternative courses in an attempt to discern if its likes and dispositions and goals have changed.

Zion is meant to be a community affair. From an LDS Perspective, Paradise is the Celestial Kingdom or the City of Zion in the spirit world. The other part of the spirit world has been called by various names including purgatory, hell, Gehenna, the spirit prison, hades, Sheol, and outer darkness; and, it is probably composed of different degrees or different levels and types of existence – some people will be completely and utterly alone or self-absorbed, and others will have gathered into communities. Furthermore, the Latter-day Saints believe that our spirit body is literally the offspring of God the Father and Heavenly Mother, which means that all of us are related to each other as brothers and sisters.

Doctrine and Covenants 76: 22-24: "And now, after the many testimonies which have been given of him, this is the testimony, last of all, which we give of him: That he lives! For we saw him, even on the right hand of God; and we heard the voice bearing record that he is the Only Begotten of the Father — that by him, and through him, and of him, the worlds are and were created, and the inhabitants thereof are begotten sons and daughters unto God."

This scripture from the Biblical God Jesus Christ indicates that our spirit bodies are literally the sons and daughters of God the Father and Heavenly Mother, which means that in the spiritual plane of existence, we are in fact brothers and sisters. We are related to each other.

This Ultimate Cause Model of Reality is also contained within Abraham 3: 17-28:

17 There is nothing that the Lord thy God shall take in his heart to do but what he will do it.

18 He [God] made the greater star; as, also, if there be two spirits [psyches], and one shall be more intelligent than the other, yet these two spirits [psyches], notwithstanding one is more intelligent than the other, have no beginning; they existed before, they shall have no end, they shall exist after, for they are gnolaum, or eternal.

19 And the Lord said unto me: These two facts do exist, that there are two spirits [psyches], one being more intelligent than the other; there shall be another more intelligent than they; I am the Lord thy God, I am more intelligent than they all.

21 I dwell in the midst of them all [the intelligences or psyches, and their assigned spirit bodies]; I now, therefore, have come down unto thee [Abraham] to declare unto thee the works which my hands have made, wherein my wisdom excelleth them all, for I rule in the heavens above, and in the earth beneath, in all wisdom and prudence, over all the intelligences [psyches] thine eyes have seen from the beginning; I came down in the beginning in the midst of all the intelligences [psyches] thou hast seen.

22 Now the Lord had shown unto me, Abraham, the intelligences [psyches] that were organized [into spirit bodies] before the world was; and among all these there were many of the noble and great ones;

23 And God saw these souls [His spirit children] that they were good, and he stood in the midst of them, and he said: These I will make my rulers; for he stood among those that were spirits, and he saw that they were good; and he said unto me: Abraham, thou art one of them; thou wast chosen before thou wast born.

24 And there stood one among them that was like unto God [Jehovah or Jesus Christ], and he said unto those who were with him: We will go down, for there is space there, and we will take of these materials, and we will make an earth whereon these may dwell;

25 And we will prove them herewith, to see if they will do all things whatsoever the Lord their God shall command them;

26 And they who keep their first estate [their spiritual pre-mortal life] shall be added upon; and they who keep not their first estate [Satan and the demons] shall not have glory in the same kingdom with those who keep their first estate; and they who keep their second estate [mortal life or physical life] shall have glory added upon their heads for ever and ever.

27 And the Lord said: Whom shall I send? And one answered like unto the Son of Man: Here am I, send me. And another answered and said: Here am I, send me. And the Lord said: I will send the first.

28 And the second [Satan] was angry, and kept not his first estate; and, at that day, many followed after him.

These things only make sense in the light of psyche and ultimate cause. All of these things are silliness and foolishness from the perspective of Materialism and Naturalism. The perspective, worldview, philosophy of life, paradigm, or model of reality which we each choose to embrace makes all the difference in the world to the outcome and the results that we will be able to achieve.

Remember! If it's motivated by guilt, force, anger, hatred, pride, profit, selfishness, competition, materialism, naturalism, jealousy, or fear, then it's Satan's work that you are doing even if you are doing a good thing or trying to do a good thing. In order for the full blessings, happiness, joy, and peace to accrue, it must be motivated by friendship, charity, compassion, and love. We must learn to do things for the right reasons in order to achieve the best results both for ourselves and for others.

Remember, there is an element of force, fear, and intimidation in classical conditioning and operant conditioning. Conditioning works on the animals, but humans tend to rebel whenever someone tries to condition them or force them to comply. The human psyche is a choosing organism, an agent in the fullest sense of the word; and, the human psyche doesn't respond favorably to force, fear, guilt, hatred, and intimidation.

Conclusions Regarding this Ultimate Model of Reality

Consciousness or psyches "have no beginning; they existed before, they shall have no end, they shall exist after, for they are gnolaum, or eternal".

Our spirit bodies had a beginning, when our Heavenly Mother gave birth to our spirit bodies and our psyche or consciousness was assigned to our spirit body. It was at this point in time, during the birth of our spirit body, that our psyche was assigned a form, a gender, a heritage, a potential to become a God, and some spirit matter to occupy and control; and, we became the sons and daughters of God the Father.

God's Psyche is the Ultimate Cause of all other contingent realities that have ever been organized and brought into existence from scratch. The good life consists of keeping one's first estate, and then keeping one's second estate. The good life consists of getting out of hell and staying out of hell, and then getting into Paradise and the Kingdom of God instead. The good life consists of peace of mind in this world, and eternal life in God's presence as the ultimate end of our journey in the world to come. As I see it, this is the Ultimate Model of Reality because it conforms to the whole of human experience and knowledge.

Psyche Is the Ultimate Cause

The existence of **ultimate cause** is so obvious and so necessary that **we** tend to take it for granted and completely overlook it; however, every time that **you** use a personal pronoun in one of **your** sentences, **you** are in fact invoking and applying some type of ultimate cause. Whether expressly written or simply implied, ultimate cause is the subject of every sentence that **you** write or speak. Where **you** are concerned, **you** or **your** psyche is the ultimate cause of everything having to deal with **you**, including every sentence that **you** think about, read, or write. **You** wouldn't exist without it! Wherever **you** go, there it is. What do **you** think? Do **I** have a point or not? Only the personal pronoun "it" can be used by a psyche to refer to something that is not-psyche; but, it is always a psyche **who** writes sentences using the personal pronoun "it". The rocks or physical matter can't write books for **us** to study and learn from.

If you read, study, understand, and accept everything that has been written in the books and articles which I have mentioned so far in this essay, then you will have a good and solid understanding of ultimate cause, why it's necessary, why it must exist, and why it must be true.

Material Cause answers the question of **what** something is made of, its essence. Things can be made of consciousness, spirit matter, and/or physical matter. The lower ones can reside within the higher ones on the list –

consciousness can be housed in a spirit body, and a spirit body can be housed in a physical body. Efficient Cause and Formal Cause are an attempt to answer the question of **how** something was made or **how** something came into being or existence. Formal cause also explains **what** something is. Final Cause is an attempt to answer the question of **why** something was made or **why** something was brought into existence.

Ultimate Cause is always an attempt to answer the question of **who** wrote this sentence, or **who** brought order and structure to this physical object or that particular physical genome, or **who** did that particular material cause or efficient cause or formal cause or final cause, or **who** wrote that computer program or genome, or **who** designed and created and produced the first functional genome and the first physical life form, or **who** brought physical matter and this physical universe into existence.

This **Ultimate Cause Model of Reality** is an umbrella model that subsumes everything that stands in opposition to Materialism and Naturalism. **Psyche as the Ultimate Cause** is so obvious that after its existence dawned on me like a lightning bolt, I have found myself wondering ever since why nobody has ever thought of it before. I found myself wondering how I could have overlooked it for the first fifty-five years of my life. Psyche as the Ultimate Cause has been staring us in the face for the duration of human history, yet nobody seems to have noticed it. It's the core foundation of our Reality, but everyone seems to ignore it or take it for granted.

This Ultimate Model of Reality is the **Model of Lived Experience**. It is a Psyche Ontology. It subsumes or includes the whole of Reality and the whole of human experience, including the revelations of God and the revelations from God. I don't think it can get more expansive or all-inclusive than that. If it can, I'll have to let someone else figure out how.

I hope that what I have written to this point in this essay serves as a solid, useful, and constructive introduction to this **Psyche as Ultimate Cause Model of Reality**.

This essay, which I prepared for a class on Personality Theory that I was taking, serves as the introduction to a book I have written entitled, "The Ultimate Model of Reality: Psyche Is the Ultimate Cause". It will also be used as the introduction to my book, "Putting Psyche Back into Psychology: Restoring Science to Consciousness".

In the subsequent parts of those books, I discuss the various applications and ramifications of the **Ultimate Cause Model of Reality**, thereby hopefully bringing the items in the preceding paragraphs into sharper relief. It's all about finding evidence to support one's thesis, because models of reality such as Materialism and Naturalism are completely worthless because it's impossible to find any evidence to support their major premises which state that psyche, the non-physical, the immaterial, and quantum non-locality do not exist. Instead, ALL of the evidence we have on hand proves beyond a shadow of a doubt that Materialism and Scientific Naturalism are false, and that Psyche does indeed exist.

Within "Putting Psyche Back into Psychology: Restoring Science to Consciousness" and "The Ultimate Model of Reality: Psyche Is the Ultimate Cause", I explore many of the practical applications of this Ultimate Cause Model of Reality, as well as bring to light more of its evidentiary support. Ultimate cause is huge, with lots of evidence and logical common sense to support it, once a person is willing to start looking for that evidence. In contrast, I could never find any compelling or convincing evidence to support My Materialism and My Atheism. I eventually had to let them go. Materialism and its derivatives are the ultimate fiction; and, I was looking for something real and true.

Concluding Note: This part entitled, "THE ULTIMATE MODEL OF REALITY", is PART I of my books, "Putting Psyche Back into Psychology: Restoring Science to Consciousness" and "The Ultimate Model of Reality: Psyche Is the Ultimate Cause". Within those books, I eventually bring Doctorate of Theoretical and Philosophical Psychology, Brent D. Slife, into the mix by mentioning and/or discussing all of his free articles about Psychology and Philosophy for a more full and robust treatment of this Ultimate Model of Reality. Hold on to your hats! These books comprise my magnum opus – my primary contribution to philosophy, quantum realities, science, psychology, and the reality of our lived experiences as a race. If this topic interests you, then please look at those books for a continuation of this theme – Psyche is the Ultimate Cause. This first part is just a drop in the ocean compared to what I was able to accomplish with this theme in those books.

I hope you found this interesting and useful. I certainly did when it was first revealed to me. It came to me in a flash of insight; and, it has taken me months to put it into words. I discovered and realized that everything, all the evidence, points to Psyche as the Ultimate Cause. Only the things such as Materialism, Naturalism, and Atheism – things which have no evidence supporting them – fail to point to Psyche. These falsehoods fail to point to a lot of other truths and realities as well. It's in their nature to do so, because they are exclusionary philosophies and are based upon a refusal to look at evidence and based upon a denial of reality.

I eventually realized that if I wanted to get at the truth and KNOW the truth, then I had to get rid of My Atheism, My Materialism, My Nihilism, and any type of Naturalism including Darwinism. The truth cannot be built upon falsehoods such as these. That's the TRUTH, and I KNOW it!

Core Set of References

Alexander, E. (2012). *Proof of Heaven: A Neurosurgeon's Journey into the Afterlife*. New York: Simon & Schuster.

Alexander, E., & Moody, R. (2013). *Conversations Beyond Proof of Heaven*. Reference: https://www.amazon.com/dp/B00LYRYFCC/.

Bannister, D., & Fransella, F. (1986). *Inquiring Man: The Psychology of Personal Constructs* (3rd ed.). Dover, NH: Croom Helm.

Bishop, B. G. (1998). *The LDS Gospel of Light*. USA: Ponce de Leon.

Boeree, C. G. (2006). *Personality Theories*. Psychology Department: Shippensburg University.
(Retrieved from http://webspace.ship.edu/cgboer/perschapterspdf.html).

Braden, G. (2007). *The Divine Matrix: Bridging Time, Space, Miracles, and Belief*. Carlsbad, CA: Hay House.

Buhlman, W. L. (1996). *Adventures Beyond the Body: How to Experience Out-of-Body Travel*. New York: HarperCollins Publishing.

Denton, M. (1986). *Evolution: A Theory in Crisis*. Chevy Chase, MD: Adler & Adler.

Dispenza, J. (2014). *You Are the Placebo: Making Your Mind Matter*. USA: Hay House Inc.

Durham, E. (1998). *I Stand All Amazed: Love and Healing from Higher Realms*. Orem, UT: Granite Publishing and Distribution.

Eccles, J. C., & Popper, K. R. (1977). *The Self and Its Brain: An Argument for Interactionism*. New York: Routledge.

Eccles, J., & Robinson, D. N. (1984). *The Wonder of Being Human: Our Brain and Our Mind*. New York: The Free Press.

Eccles, J. (1985). *Mind and Brain: The Many-Faceted Problems*. New York: Paragon House Publishers.

Engler, B. (2009). *Personality Theories: An Introduction* (8th ed.). Boston, MA: Houghton Mifflin Harcourt Publishing Company.

Feser, E. (2009). *Aquinas (A Beginner's Guide)*. Oxford, England: Oneworld Publications.

Gibson, A. S. (2006). *They Saw Beyond Death: New Insights on Near-Death Experiences*. Springville, UT: Horizon Publishers.

Goswami, A. (2008). *God Is Not Dead: What Quantum Physics Tells Us about Our Origins and How We Should Live*. Charlottesville, VA: Hampton Roads.

Haisch, B. (2006). *The God Theory: Universes, Zero-Point Fields, and What's Behind It All*. San Francisco, CA: WeiserBooks.

Hinze, S. (1994, 1997). *Coming from the Light*. New York: Pocket Books.

Kalat, J. W. (2008). *Introduction to Psychology* (9th ed.). Belmont, CA: Wadsworth, Cengage Learning.

Kelly, E. F., Kelly, E. W., Crabtree, A., Grosso, M., & Gauld, A. (2007). *Irreducible Mind: Toward a Psychology for the 21st Century*. Plymouth, United Kingdom: Rowman and Littlefield.

Long, J., & Perry, P. (2016). *God and the Afterlife: The Groundbreaking New Evidence for God and Near-Death Experience*. New York: HarperOne.

Mark My Words. (2016). *Quantum Mechanics from a Non-Physical Spiritual Perspective*. Kindle. (Retrieve from: https://www.amazon.com/dp/B01J023TGU).

Mark My Words. (2016). *The Scientific Method: Proves That the Theory of Evolution Is False*. Kindle. (Retrieve from: https://www.amazon.com/dp/B01IAAIRT2).

Mark My Words. (2016). *The Theory of Evolution Proved to Me that God Exists: Why I Am No Longer an Atheist and Why I No Longer Believe in the Theory of Evolution*. Kindle. (Retrieve from: https://www.amazon.com/dp/B01HZYBZ7K).

Mark My Words. (2016). *Using the Scientific Method: To Eliminate the Usual Suspects and to Prove the Truth*. Kindle. (Retrieve from: https://www.amazon.com/dp/B01J6STHP0).

Marshall, P. D., Kelly, E. F., & Crabtree A. (Eds.). (2015). *Beyond Physicalism: Toward Reconciliation of Science and Spirituality*. London, United Kingdom: Rowman and Littlefield.

McTaggart, L. (2002). *The Field: The Quest for the Secret Force of the Universe*. New York: HarperCollins.

Meyer, S. C. (2010). *Signature in the Cell: DNA and the Evidence for Intelligent Design*. New York: HarperCollins.

Meyer, S. C. (2013). *Darwin's Doubt: The Explosive Origin of Animal Life and the Case for Intelligent Design*. New York: HarperCollins.

Moody, R. A. (1975, 2015). *Life After Life: The Bestselling Original Investigation That Revealed "Near-Death Experiences"*. New York: HarperCollins.

Moody, R., & Perry, P. (2010). *Glimpses of Eternity: Sharing a Loved One's Passage from This Life to the Next*. New York: Guideposts.

Neal, M. C. (2011). *To Heaven and Back: A Doctor's Extraordinary Account of Her Death, Heaven, Angels, and Life Again: A True Story*. Colorado Springs, CO: WaterBrook Press.

Penfield, W. (1978). *The Mystery of the Mind: A Critical Study of Consciousness and the Human Brain*. Princeton, NJ: Princeton University Press.

Radin, D. (2006). *Entangled Minds: Extrasensory Experiences in a Quantum Reality*. New York: Paraview Pocket Books.

Richards, P. S., & Bergin, A. E. (2004). *Casebook for a Spiritual Strategy in Counseling and Psychotherapy*. Washington DC: American Psychological Association.

Richards, P. S., & Bergin, A. E. (2005). *A Spiritual Strategy for Counseling and Psychotherapy* (2nd ed.). Washington DC: American Psychological Association.

Ring, K., & Cooper S. (1999). *Mindsight: Near-Death and Out-of-Body Experiences in the Blind* (2nd ed.). Kearney, NB: Morris Publishing.

Rivas, T., Dirven, A., & Smit, R. H. (2016). *The Self Does Not Die: Verified Paranormal Phenomena from Near-Death Experiences*. Durham, NC: IANDS Publications.

Ross, H. (1991). *The Fingerprint of God: Recent Scientific Discoveries Reveal the Unmistakable Identity of the Creator* (2nd ed.). Orange, CA: Promise Publishing Co.

Ross, H. (1996). *Beyond the Cosmos: The Extra-Dimensionality of God: What Recent Discoveries in Astronomy and Physics Reveal about the Nature of God*. Colorado Springs, CO: NavPress.

Ross, H. (2008). *Why the Universe Is the Way It Is*. Grand Rapids, MI: Baker Books.

Ross, H., Samples, K. R., & Clark, M. (2002). *Lights in the Sky & Little Green Men: A Rational Christian Look at UFOs and Extraterrestrials*. Colorado Springs, CO: NavPress.

Rychlak, J. F. (1979). *Discovering Free Will and Personal Responsibility*. New York: Oxford University Press.

Rychlak, J. F. (1981a). *A Philosophy of Science for Personality Theory* (2nd ed.). Malabar, FL: Robert E. Krieger Publishing Company.

Rychlak, J. F. (1981b). *Introduction to Personality and Psychotherapy: A Theory-Construction Approach* (2nd ed.). Boston, MA: Houghton Mifflin Company.

Rychlak, J. F. (1982). *Personality and Life-Style of Young Male Managers: A Logical Learning Theory Analysis*. New York: Academic Press.

Rychlak, J. F. (1988). *The Psychology of Rigorous Humanism* (2nd ed.). New York: New York University Press.

Rychlak, J. F. (1991). *Artificial Intelligence and Human Reason: A Teleological Critique*. New York: Colombia University Press.

Rychlak, J. F. (1994). *Logical Learning Theory: A Human Teleology and Its Empirical Support*. Lincoln, NE: Nebraska University Press.

Rychlak, J. F. (1997). *In Defense of Human Consciousness*. Washington DC: American Psychological Association.

Rychlak, J. F. (2003). *The Human Image in Postmodern America*. Washington DC: American Psychological Association.

Sanford, J. (2014). *Genetic Entropy* (4th ed.). Cornell University: FMS Foundation.

Sanford, J. C., Marks, R. J., Behe, M. J., Dembski, W. A., & Gordon, B. L. (Eds.). (2013). *Biological Information: New Perspectives*. Hackensack, NJ: World Scientific.

Sharkey, J. (2008). *A GLIMPSE OF ETERNITY: One man's story of life beyond death. Ian McCormack's Story*. Orewa, New Zealand: Arun Books.

Slife, B. D. & Williams, R. N. (1995). Science and Human Behavior. In *What's Behind the Research? Discovering Hidden Assumptions in the Behavioral Sciences*, (pp. 167–204). Thousand Oaks, CA: SAGE Publications.

Storm, H. (2000, 2005). *My Descent into Death: A Second Chance at Life*. USA: Random House.

Talbot, M. (1991, 2011). *The Holographic Universe: The Revolutionary Theory of Reality*. New York: HarperCollins.

Talbot, M. (1993). *Mysticism and the New Physics*. New York: Penguin Books.

Tart, C. T. (2009). *The End of Materialism: How Evidence of the Paranormal Is Bringing Science and Spirit Together*. Oakland, CA: New Harbinger Publications.

Van Lommel, P. (2010). *Consciousness Beyond Life: The Science of the Near-Death Experience*. New York: HarperCollins.

Wells, J. (2000. *Icons of Evolution: Science or Myth? Why Much of What We Teach About Evolution Is Wrong*. Washington, DC. Regnary.

PART VI — SCIENTISM

The Scientific Method Is Based Upon Affirming the Consequent

The Scientific Method is a good way for coming to know the truth; but, most people do not realize that the Scientific Method is not a foolproof way of knowing the truth nor is it the best way of knowing the truth.

Like me, most people are shocked when they first realize or are first taught that the Scientific Method is based upon a logic fallacy called "affirming the consequent".

It took an honest philosopher and scientist to reveal this truth to me, because many people like me have been brainwashed into believing that the Scientific Method is an infallible god. The people who are part of the religion called Scientism actually worship Science and treat Science as if it were God. These people pin all of their hope and faith on the Scientific Method, never once realizing that there is a fatal flaw at the heart of the Scientific Method which has the power to mess them up every time.

The Scientific Method typically runs through the following sequence in an attempt to arrive at the truth.

1) Form a HYPOTHESIS.

2) Select a Scientific Method or Scientific Methodology to TEST the hypothesis.

3) Run the Science Experiment; and then, Observe and Measure the RESULTS.

4) Find the BEST INTERPRETATION or the BEST EXPLANATION for the Scientific Data, the Scientific Evidence, the Scientific Observations, and the RESULTS of the Science Experiment.

The main flaw in the Scientific Method is found in the fourth and final step of the Scientific Method, where human error and human weakness comes into play by "affirming the consequent" or affirming the conclusion which the human wants the hypothesis to prove. In other words, the human being chooses the conclusion or the interpretation that he or she wants, never once realizing that there might in fact be a better explanation for the scientific data or a better interpretation of what the tested hypothesis and the chosen conclusion might in fact really truly mean. This happens all the time. Human beings make the data and the scientific evidence fit the conclusion they personally want, even though there are better explanations for the scientific data than the explanation or the interpretation which the person as the scientist has chosen to affirm.

One example: Materialism by definition is design and creation by physical matter, or Creation by Rocks. That's what Materialism, Darwinism, and Naturalism reduce to – Creation by Chance or Creation by Rocks. Millions of scientists across this world on a daily basis form hypotheses, run experiments and tests, get lots of results and scientific data, and then conclude that their hypotheses and scientific evidence have successfully proven Creation by Rocks, or Materialism, or Darwinism to be true – never once realizing that Intelligent Design, or Intelligent Manufacturing, or Creation by Psyche is in fact a far better, more logical, more realistic, and more parsimonious explanation of their scientific data than Creation by Rocks or Materialism.

Affirming or choosing Creation by Rocks as one's conclusion or as one's interpretation of the scientific data is the perfect example of the "affirming the consequent" logic fallacy, which the Scientific Method employs every time that the Scientific Method is used. Creation by Rocks is the "affirming the consequent" logic fallacy in action. The fourth step of the Scientific Method is unavoidable – conclusions have to be drawn, but there's no guarantee that one's chosen conclusions are correct. Creation by Rocks is an illogical and unsustainable conclusion; yet, it is THE CONCLUSION which millions of scientists choose on a daily basis.

Obviously, one's chosen consequent or one's chosen conclusion doesn't have to be wrong or false. These same scientists could have chosen Creation by Intelligent Designers or Creation by Psyche as their conclusion and greatly increased their chances that their chosen conclusion might in fact be real, right, correct, and true; but, most of the scientists will not and have not chosen Creation by Intelligent Beings as their conclusion or their interpretation of the scientific data because these people don't want to choose that conclusion even though it is in fact the more logical, more parsimonious, and most realistic conclusion that they could have chosen.

Ultimate Cause or Psyche will always be a better consequent, or a better conclusion, or a better interpretation of the scientific data than Creation by Rocks; but, Creation by Rocks will typically be the consequent which the majority of the scientists and that all of the Materialists and Naturalists will choose to affirm, because these people want to, not because it's true.

The thing we each need to realize in all of this is that it is always some kind of Psyche (or Ultimate Cause) who chooses the conclusion, or affirms the consequent, or chooses the interpretation of the scientific data that he or she personally desires the most; and, there is the flaw and the logic fallacy which is built into the Scientific Method right from the very beginning. A fallible psyche or fallible human being chooses the interpretation of the scientific data which he or she desires most never once realizing that there might in fact be another interpretation of the same scientific data which is in fact a better fit or a more realistic conclusion than the one they have chosen.

Design and Creation by Psyche or Creation by Intelligent Beings will always be a better interpretation of the scientific data than Creation by Rocks; but, Creation by Rocks has proven to be THE INTERPRETATION which the majority of the scientists have chosen to employ. Alas, the majority isn't always right. The majority can be right, but they aren't always right. And, therein is the flaw of the Scientific Method – the majority chooses the consequent or the conclusion they desire most and then unilaterally affirm that that consequent is true. It's happening right now even as I write this and even as you read this. Someone right now is affirming Creation by Rocks or Materialism as their consequent or their conclusion.

We human beings are ever-learning but never able to come to a knowledge of the truth, because we are constantly affirming consequents that can't possibly be true. Design and Creation by Rocks – how could that ever possibly be true? It can't. But, Creation by Rocks or Materialism is nonetheless affirmed by scientists across this world on a daily basis because it is the conclusion or the consequent which they desire to affirm the most. Interesting, is it not?

Can you see how an understanding of philosophy, psychology, and logic can greatly improve our understanding of Science and the Scientific Method? The Scientific Method has a HUGE human element or psyche element; and, human beings are known to make mistakes and are known to jump to conclusions. We humans often leap before we look. Creation by Rocks or Materialism can't possibly be true; yet, it is THE CONSEQUENT which most of the scientists choose to affirm for the duration of their careers.

> "This is why philosophers are wont to point out that for any given fact pattern which can be demonstrated or "discovered" empirically, an infinite number of theoretical explanations are possible. This is also why we say theories remain theories, even after they have been validated. All validating evidence can establish convincingly is the *negation* of a theoretical proposition." (Rychlak, *A Philosophy of Science for Personality Theory*, p. 81).

In other words, Science can be used to prove that Creation by Rocks is false by negating Creation by Rocks; but, Science cannot convincingly prove that Creation by Rocks is true. To conclude that Darwinism, Materialism, or Naturalism has been proven true by Science and the Scientific Method requires an act of faith or a leap of faith, which millions of scientists are willing to take. But, taking that leap of faith doesn't actually mean that these people have chosen the correct consequent or the right interpretation for their scientific data.

> "This is a good point at which to observe that the logic of empirical study is flawed – not fatally, but in a way that limits the certainty with which our explanation of empirically proven facts can be believed in. We never achieve logical necessity [certainty] in the proof garnered by a scientific experiment. This is because we always commit the logical error that Aristotle pointed out long ago, of *affirming the consequent* of an "If, then" line of argument. Another way of saying this is that the empirical findings act as a predicating meaning for our theory, but there are always going to be other theories that can take meaning from this data array as well." (Rychlak, *Artificial Intelligence and Human Reason: A Teleological Critique*, p. 33-34).

Drawing conclusions or interpreting the scientific data is an integral and essential part of the Scientific Method; but, this is also where the logic errors and the flaws are introduced into the process. I have observed that Creation by Rocks is never the best explanation that can be given to scientific evidence; yet, Creation by Rocks, or Materialism, is the explanation that is most-given to scientific evidence. Interesting, is it not?

I have also observed that there are infinitely better and more believable explanations for scientific evidence than Materialism, or Design and Creation by Rocks. Materialism, or Creation by Rocks, always provides a false interpretation or a false explanation to any data array or set of scientific evidence. That has been my observation, once I finally started looking at the empirical evidence and the logic associated with the Scientific Method. Materialism, Naturalism, and Atheism are based upon a refusal to look at contradictory evidence and a refusal to look at any other possible explanation for the scientific evidence. Once I started looking at the logic and the evidence, it was easy to see that Materialism and Naturalism are fatally flawed.

There's a lot of money that can be made telling the Atheists and the Materialists exactly what they want to hear; but, it's dishonest. Like Joseph Rychlak's books, my books go largely unnoticed, because I'm not telling the Materialists and the Atheists what they want to hear; but, I sleep well at night with a clear conscience knowing that I have finally found the truth that I have been searching for all of my life. This stuff is really cool, and it has set me free; but, most people will never see it because they don't want to see it. Such is life.

> "The fallacy of affirming-the-consequent stipulates the fact that it will always be possible for some other explanation to account for any empirically observed fact pattern. This loss of certainty in validation is not fatal for the Scientific Method, of course. It has not prevented scientists from curing polio or putting people on the moon. It merely alerts us to the fact that some conceptualizer [psyche] always has to make a decision as to which fundamental grounding [or interpretation or explanation] will be used in the sequence of theory formation and testing. The grounds are never 'out there' in the hard data but 'in here' as assumptive frameworks. If we who theorize appreciate that we will never attain certainty in validating our theories, we will be in a better position to see that alternative groundings that explain such empirical evidence can be complementary. To *complement* is to fill out or make up for what is lacking in any theoretical understanding of a subject." (Rychlak, *In Defense of Human Consciousness*, pp. 18-19).

Notice that there must be a conceptualizer, psyche, interpreter, theorizer, observer, decider, chooser, and assumer behind every theoretical hypothesis and the Scientific Method, or the science experiment will never take place. The grounds for doing science take place 'in here' in our psyche. Contrary to the claims of the Darwinists and the Materialists, you can't place the rocks or raw physical matter into the role of conceptualizer, psyche, theorizer, formal cause, final cause, or ultimate cause. The rocks won't go there and can't do that. Once again, logic and the Scientific Method have proven Materialism and Scientific Naturalism inadequate and false. The Scientific Method can definitely be used to prove things false, which has happened in the case of Materialism and Naturalism thousands of different ways. In fact, the falsification of Materialism and Naturalism is complementary, in that Naturalism and Materialism have been falsified thousands of different ways.

We have our fill of evidence demonstrating and proving what is lacking or false in Naturalism and Materialism.

> "Science can only work through a kind of negating procedure of falsifying claims put to nature by the theorist in question. As scientists, says Popper, we never really verify things but continually falsify – or fail to falsify – claims [theories, hypotheses, etc.] expressed by some investigator [recognizing, of course, that serendipitous findings occur as well]. This is why the scientist always restates his hypothesis into the null form. Ultimately the reason we must falsify has to do with the logical fallacy of 'affirming the consequent' of an 'If [antecedent] . . . then [consequent] . . .' proposition. We like to think our theory has necessarily been verified, but Popper teaches us that it has not. There will always be, in principle, other ways of accounting for the observed data [the facts] than our preferred theory." (Rychlak, *The Psychology of Rigorous Humanism*, pp. 181-182).

There will always be other ways of accounting for the scientific data than our preferred conclusion, Materialism.

Every scientist, myself included, loves to make the claim that Science has proven different concepts to us. In the past, I have made the claim that the Scientific Method proved to me that God exists, that the Theory of Evolution is false, and that Quantum Nonlocality or the Spirit Realm does indeed exist. I have also made the claim that the Scientific Method has eliminated falsehoods while at the same time pointing me to the truth or proving the truth. I have been called on it many times, but I still stand behind it. What good is Science and the Scientific Method if it can't be used to prove things?

Typically, though, the way that the Scientific Method proves the truth is by eliminating all of the associated falsehoods. The Scientific Method can be used thousands of different ways to falsify Materialism and Naturalism; and, it already has been used in that way thousands of different ways. The Scientific Method has in fact proven to me that Materialism and Naturalism are false by falsifying Materialism and Naturalism thousands of different ways. This also lends evidentiary support to the observation, claim, and conclusion that something other than Materialism must of necessity be true. Intelligence, Psyche, Intelligent Design, Creation by Psyche, and Ultimate Cause are the opposite of Materialism. Since the Scientific Method has proven to me that Materialism and Naturalism are false, the Scientific Method greatly increases my chances that Intelligent Design, Psyche, and Ultimate cause will be proven to be true.

The Scientific Method has proven to me that Quantum Mechanics, Quantum Entanglement, and Quantum Nonlocality are real and true. How? It's because Quantum Mechanics has never been falsified in any of the science experiments that have been run on it. In every science experiment that has been performed on Quantum Mechanics and Quantum Entanglement, Quantum Mechanics or Spiritual Mechanics has been verified as true. Consequently, I feel safe in claiming that the Scientific Method has proven to me that Quantum Nonlocality or the Spirit Realm is real and truly exists, because Quantum Nonlocality by definition means spiritual, or non-local, or non-physical, and because Quantum Nonlocality has been repeatedly verified and proven true.

Finally, I know that Intelligence, or Psyche and Ultimate Cause, exist because they are obvious and axiomatic. We all KNOW that Psyche or Intelligent Beings can design and create anything they set their minds to at will, because we have observed and experienced this Reality first-hand every single day of our lives. I don't need the Scientific Method to prove to me that Intelligent Beings exist, because if you can read this, then you already KNOW that Intelligent Beings exist. And since Psyche or Ultimate cause is synonymous with Intelligent Beings, whether looked at from a spiritual perspective or a materialistic perspective, I really don't need the Scientific Method to prove to me that Psyche and Intelligent Beings exist because I already KNOW that they exist. The Scientific Method is unnecessary if you already KNOW the truth.

KNOWING trumps the Scientific Method and philosophical speculation every time. Also, since I am willing to accept NDEs, SDEs, OBEs, and other spiritual experiences into evidence, I KNOW that Psyche or Ultimate Cause is non-local, non-physical, and spiritual in nature and origin. I KNOW that psyche or our personality is the part of us that survives bodily death and brain death. Since many people have seen and talked with the Biblical God Jesus Christ during their NDEs and OBEs, I know that He exists as well; and, I don't need the Scientific Method to try to convince me otherwise. I don't need the Scientific Method for any of these things, because I KNOW that they are real and true.

But, the failures and the inability of Materialism and Darwinism to account for the origin of the first physical genome and the first physical life form was in fact the first thing to prove to me scientifically that God must of necessity exist in order to do all of the design, science, and creation that the rocks, Materialism, and Naturalism could never have done. So in a very real sense, the Scientific Method and the falsification of Materialism and the Theory of Evolution proved to me that God must of necessity exist in order to do all of the different things that needed to be done which Materialism, the rocks, and the Theory of Evolution could never have done.

It's really not 100% accurate to make the claim that Science and the Scientific Method will never be able to prove the truth, because through the scientific process of eliminating every falsehood, we eventually find ourselves landing upon the truth as a last resort or a final default; and then, we find ourselves staring at and going with something which is axiomatic law, obvious, and 100% true and can never be falsified, or proven false, or proven not to exist – something like Psyche, Intelligence, Intelligent Design, and Ultimate Cause which we simply KNOW exists. I exist; therefore, I AM; and, I KNOW IT.

> "When we design and carry out a research experiment, it is easy to confuse the concrete empirical findings with the activity or the process that supposedly brought these findings about or made them happen. The logic of experimentation gives rise to the following problem: for any observed fact pattern there are, in principle, infinitely many possible explanations. This follows from the necessity that, in conducting research, all scientists are

constrained by the "*affirming the consequent*" fallacy. There will always be an alternative explanation possible for the observed fact pattern." (Rychlak, *Logical Learning Theory: A Human Teleology and Its Empirical Support*, pp. 3-4).

When the Naturalists and Materialists carry out their research experiments and choose to affirm Creation by Rocks or Materialism as their consequent, know that there are an infinitely many possible explanations for their science experiment and their scientific data, some of which might actually be true – such as using Intelligent Design, or Creation by Psyche, or Design and Creation by Intelligent Beings as the explanation for their science experiment and scientific data instead of using Creation by Rocks or Materialism as their explanation.

I have observed that Psyche, or Intelligence, or Ultimate Cause will always be a better explanation for research experiments than Materialism or Creation by Rocks. Would you agree with that observation, or not? Whether you agree or not, I have observed that Psyche or Intelligence can do research experiments; whereas, the rocks cannot. Consequently, science experiments done by psyche (or ultimate cause) are infinitely more plausible and believable than science experiments done by rocks. That's one of the things which the Scientific Method has taught me.

See how the Scientific Method and the Philosophy of Science can be used to get at the truth, often through a process of elimination? It's not as direct nor as immediate as Knowing or Knowledge, but it does have its uses.

Scientism

I had bought into all the hype, posturing, and promotion; and, I truly believed that Science and the Scientific Methods are the best way for finding and knowing the truth. I was a proponent of Scientism – the philosophical belief that scientific methods are the best, if not the only, way to find and know the truth. I had based my whole life on Scientism. For the first fifty-five years of my life, I considered myself to be first and foremost a scientist. I was dedicated to the system. I believed in it passionately.

I was in for a rude awakening!

Defining Scientism

The following definition and explanation of Scientism comes from class notes by Edwin Gantt:

> In the world at large, there is a troubling trend toward viewing science as not just a sturdy approach to answering some important questions about

the world, but rather as the only reliable source of truth, and as a way of making sense of the world that is superior to all others.

Many scientists unquestioningly adopt the position articulated by the British chemist Peter Atkins:

> Although some may snipe and others carp, there can be no denying the proposition that science is the best procedure yet discovered for exposing fundamental truths about the world. There appear to be no bounds to its competence.

Atkins has further stated, **"Science is the only path to understanding"**. [That's a very narrow and extremist point of view. That's a religious point of view. Atkins is posturing here.]

Scientism is a thoroughly unscientific approach to the world. This is because science, truly and properly understood, is inseparable from a deep epistemic humility—a genuine and thorough-going acknowledgement of the inherent limitations and fallibility of human understanding.

Genuine science, at the very least, does not permit making the sorts of sweeping metaphysical, theological, and moral claims to which the advocates of scientism are so frequently prone, since these sweeping metaphysical claims cannot be grounded in empirical experience.

For example, Atkins's claim that "science is the only path to understanding" is not some truth discovered by the methods of science and empirical observation. Rather, it is just a sweeping philosophical assertion.

Assuming that observed regularities are immutable or universal is entirely unjustified by empirical experience. We simply do not know and cannot know until we have made systematic observations of all of reality.

We cannot make broad, sweeping metaphysical claims about the world without leaving the realm of systematic empirical observation and entering the world of philosophical speculation.

(From class notes by Ed Gantt).

Scientists like Peter Atkins do an excellent job of explicating the dogmatic religious extremism of Scientism. Scientism is just another type of religious extremism – scientism is a worship of Scientific Methods. These people treat science as if it were God. Scientism is philosophical speculation.

These people don't realize that Science and Scientific Methods are extremely limited or bounded when it comes to the spiritual, non-physical, immaterial, and non-local Realities of our existence, because these people have deliberately blinded themselves to other realities. Science and Scientific Methods are not only bounded by physical limitations but are also extremely incompetent when it comes to the spiritual or the non-local Realities of our Lived Experiences.

When it comes to "psychology", the Behaviorists, Materialists, and Naturalists limit themselves exclusively to observable human behavior. For them, the word "observable" and the word "empiricism" are defined in terms of the five physical senses and therefore limited exclusively to the physical.

Therefore, whenever human behavior is caused by the human psyche or non-local consciousness, Scientific Methods are actually a very poor way to study human behavior, because there is no way for Scientific Methods and physical observations to study human behavior that's taking place in the spirit realm or the non-local realm.

Scientism is based exclusively on Naturalism and Materialism.

The Weaknesses of Naturalism

During our class, Ed Gantt outlined some of the weaknesses of Naturalism:

1. Observing regularities and patterns in the world does not itself warrant the belief that these "laws" cause events, or even serve as sufficient explanations for them.

2. The naturalistic worldview cannot be proven to be true by empirical evidence because it is really just a philosophical assumption.

3. Naturalism has not been a very successful approach in the social sciences — psychologists simply have not discovered the kinds of scientific laws they hoped to find.

4. The naturalistic worldview, by definition, precludes explanations that do not assume the sufficiency of scientific laws in accounting for the world.

5. Human beings are fundamentally different from the sorts of things that can be explained exclusively in terms of universal laws, natural processes, and physical reactions. For this reason, the assumptions of naturalism are simply inadequate in the social sciences and the behavioral sciences.

6. Merely observing regularities in the world is not itself reason to believe we have actually explained anything. We all know from personal experience that unsupported objects fall. Personal experience or lived experiences is how we come to know about immaterial and non-physical objects and subjects such as gravity.

These are undeniable regular and consistent patterns, and we give these patterns a name: *gravity*. However, to this day, nobody actually knows why things fall. We often say, "Things fall because of gravity." However, this is a logical fallacy, known as the *Nominalistic Fallacy*.

To commit the Nominalistic Fallacy is to make the mistake of assuming that merely because we have given something a name or described it in great depth we have, therefore, explained it. Things do not fall because of gravity; rather, gravity is simply the name we have given for the observed pattern of things falling.

      7. Naturalism is a philosophical perspective or worldview that assumes that everything that happens in this world can be explained in terms of physical processes and scientific laws. A naturalist, in this context, is someone who assumes that we do not need to invoke God, supernatural entities, or any religious mystery to fully explain events in the world. (Adapted from class notes by Ed Gantt).

Naturalism, Naturalism, Atheism, and Scientism are nothing more than biased philosophies and extremist dogmatic religions. The whole of Scientism, Naturalism, Materialism, and Darwinism are based exclusively on a wide variety of different logic fallacies.

I'm lucky to have found college professors and scientists who were willing to teach me about these kinds of things. That's why I dedicated this book to them!

Science and Human Behavior

Slife, B. D. & Williams, R. N. (1995). Science and Human Behavior. In *What's Behind the Research? Discovering Hidden Assumptions in the Behavioral Sciences*, (pp. 167–204). Thousand Oaks, CA: SAGE Publications.

    http://mypsyche.us/science/

The authors, contributors, and their disciples have been handing out this paper in class to the public at-large. They are not concerned about copyright. They want this thing spread about!

This article is excellent! This thing is paradigm-shifting and life-changing. Warning: This article is extremely painful if you are a promoter of Scientism, Naturalism, and Behaviorism.

I feel very lucky to have encountered this article and these ideas, because our materialistic, naturalistic, and atheistic college professors have no idea that such things actually exist; and, the few who do know refuse to teach us anything about them. I was exposed to these ideas a full twenty years after Slife and Williams presented them to the world – better late than never!

This article marked the END of my Scientism – my worship of the Scientific Method as an infallible god.

While reading this article, a lot of ideas came to mind and I wrote them as notes in the margins. Here's a sampling of what I came up with.

Science Is Based Upon Philosophical Assumptions and Philosophical Interpretations

Ironically, due to the fact that hypotheses, theorizing, and interpretation of scientific data are an integral part of the Scientific Method, Science and the Scientific Method ARE philosophy at their very core. We can't escape philosophy, or personal opinion and personal interpretations of the scientific data.

For example, this is currently my own personal interpretation of the scientific evidence; and, it tends to be quite different from the interpretation that is typically found in our college textbooks.

According to Quantum Mechanics, God or the Prime Observer organized this physical universe out of spirit matter, which was already in existence in great abundance. His Conscious Observation, His WORD, was necessary to transition, transfer, and transmute part of that pre-existing spirit matter into physical matter; and, He could pick what parts of that spirit matter to transform into physical matter and when to do so. His WORD, His Psyche, or His Light could command whole suns and whole galaxies into existence all at once.

According to Quantum Mechanics, ONLY Psyche can observe physical matter into existence or call physical matter into existence, which means that some kind of Psyche had to precede the creation or organization of the first particles of physical matter.

According to Astrophysics, 96% of this universe is still in its spirit matter and/or light and energy state of existence, as dark matter and dark energy respectively. Only 4% of this universe has been consciously observed into physical matter, or consciously called upon to be physical matter. The Big Bang was not an explosion. It was a birth, a transition, and the beginning of physical matter in this universe – a process that happened when God or the Prime Observer called that physical matter into existence from pre-existing spirit matter.

The Big Bang, or the beginning of space-time and the beginning of this physical universe happened; but, not in the way that the Materialists and Atheists typically visualize it. There was no explosion. There was NO physical matter until AFTER God's Psyche called, or WORDED, or observed the first physical matter into existence. It's interesting how the Biblical Prophets understood Quantum Mechanics and the power of the WORD millennia before our scientists did. In fact, there are many scientists and Quantum Mechanists who still have no idea how Quantum Mechanics really works or fits into the picture as a whole.

Likewise, there were no random mutations and no natural selection and no DNA on this planet until AFTER God's Psyche designed, organized, and created the first genomes and the first life

forms on this planet. Psyche is the only thing we know of that can convert spirit matter into physical matter. Psyche is the only thing we know of that can design, create, and do science. Psyche or Non-Local Consciousness, particularly God's Psyche, is in fact the fundamental unit of Reality.

There's more than one way to interpret the scientific data and the scientific evidence. In fact, there can be dozens and possibly even thousands of different ways to interpret a particular piece of scientific evidence. Furthermore, ONLY the Human Psyche is capable of doing a full-range of interpretation and meaning-making.

Defining Science

What is science?

Surprisingly, science is a philosophy, a worldview, which is why Scientism can be classified as a religion.

Science is just another philosophy attempting to determine what is true. Science is an advanced form of philosophy, or metaphysics.

"Science itself is based on a set of ideas – assumptions about what the world is like and how it should be studied. In the minds of many, science encompasses what we *know* to be true. We cannot have knowledge about what is not real or true."

Ironically, lived experiences end up becoming what a person or psyche KNOWS to be true. I learn best through comparison and contrast.

The strength and the weakness of Science is found in the fact that most scientists limit the discipline or philosophy purely to the physical, completely ignoring and discounting the other aspects of reality. That's good if the ultimate goal is technology that you can sell to others; but, it's really bad if your ultimate goal is to get a handle on the whole of our Reality and Existence.

Scientists, particularly Atheistic Scientists, base their philosophical assumptions on Materialism and Naturalism. By deliberately limiting their brand of science exclusively to what can be physically observed, these scientists deliberately limit the scope and the range of what they can study and discover. This philosophical focus on physical matter blinds them to everything else. In fact, it's extremely unscientific – especially if we can agree upon the fact that Science should really be about the pursuit of knowledge and not just about the promotion of Materialism, Naturalism, and Atheism!

The Atheists tend to promote themselves as the beacons of rationality. Rationality is synonymous with logic and interpretation. Unfortunately, there can be dozens, if not thousands, of different interpretations for the same piece of scientific data or scientific evidence. The Atheists, Materialists, and Naturalists can use their rationality to talk themselves out of the truth just as easily as we can use our rationality to talk ourselves into the truth. Rationality and logic have a blind spot, and it's called personal interpretation or personal preference.

Science was based upon the idea that "knowledge is gained through careful observation and skilled exercise of rationality".

Unfortunately, Science and the Scientific Method actually break down on two fronts. First of all, a hypothesis is a philosophical guess; and, it is therefore subjective and subject to a lot of personal bias and personal interpretation right up front even before the experiment is conducted and the results of the experiment gathered in. Second of all, the Scientific Method ends with a personal philosophical interpretation of the scientific data or scientific evidence. It's possible to interpret evidence and data incorrectly. In fact, I would venture to guess that most scientific evidence and data is interpreted incorrectly, especially if it's being interpreted from a materialistic and naturalistic worldview or philosophy.

Interpretation is subject to preferences, values, desires, biases, and pre-determined conclusions. And, the final step of Scientific Methods is to interpret the data or the results of the experiment, which places Science firmly on the grounds of philosophy once again, from whence it started in the first place. Unfortunately, a single observation can be interpreted a dozen different ways which only make sense to the interpreter but not necessarily to anyone else. Interpretation is a function of philosophy and psyche, and NOT observation or experimentation.

In fact, I have observed in recent years that the philosophical interpretation of scientific data which they call "Intelligent Design Theory" is infinitely more plausible, credible, and realistic than the philosophical interpretation of the same data which they call "The Theory of Evolution". I have lost my faith and trust in materialistic and naturalistic interpretations of scientific data and scientific evidence. My Materialism, My Nihilism, and My Atheism betrayed me and threw me to the wolves. They were no friends of mine, and they have become my mortal enemies.

"It is common in our culture to refer to the Scientific Method as if it were a single thing about which all people agree. [In fact, in the literature, most people simply call it *method*.] It should be pointed out, however, that there is no such thing as *the Scientific Method*. There are as many scientific methods as there are scientists doing research."

In fact, the authors of this article promote what they call Methodological Pluralism – the use of many different types of methods – in an attempt to get at the truth or what is often called *knowledge*.

The Purpose of Science

The goals of a science, which has been limited to the philosophical bounds of Materialism and Naturalism, is prediction and control – replication and reliability –

technology. The physical realm is the ultimate consensus reality after all! In a physical realm, I can trust that this essay will be on my hard drive and my flash drive tomorrow morning when I wake up. The same can't be said about spiritual realities which are created by psyche or mind – especially the non-consensus realities in the non-local realm or spirit realm.

We can't predict and control the human mind or the human psyche suggesting that it is part of the supernatural world rather than being a part of this physical world. Human beings cannot be programmed like computers. Human beings resist programming, and human beings can break their conditioning at will. In fact, any time you have a thought or a dream or make a choice, you are in fact having a spiritual experience. Your thoughts, dreams, memories, feelings, and choices show up in your life review after your physical body and physical brain are dead and gone, according to the empirical science of Near-Death Experiences.

There's a whole other Science of Lived Experiences, a Science of Choices and Actions, which the Materialists, Naturalists, Nihilists, Atheists, Determinists, and Behaviorists CHOOSE to ignore and reject, in a very unscientific manner.

"The test of whether we have, indeed, discerned the secrets [of nature] is whether we can exercise power and control over nature to bring about practical ends. This dual emphasis on careful observation in the pursuit of truth and practical control over the natural world is at the heart of our modern notion of science. Since the Enlightenment, science has been thought of as a method for testing ideas and opinions through observation. Observation takes place through the medium of sensory experience [defined as the five physical senses]. This emphasis on observation of the world as a way of attaining true knowledge is the essence of *empiricism*."

In the literature, empiricism is defined as physical empiricism or observations done by the five physical senses. These people have CHOSEN to completely ignore Spiritual Empiricism or the Lived Experiences of the human race.

Behavioral Sciences

Prediction and control are the hallmarks of a naturalistic and materialistic science, because the ultimate goal is technology. Unfortunately, we still can't force nature to produce new and unique genomes and life forms from scratch, according to our desires and our will. All we can do is mess around with what God has already created. In fact, Random Mutations and Natural Selection and Evolution didn't even exist until AFTER God designed and created all the genomes and life forms in the first place. Materialism and Naturalism are design and creation by rocks. God and God's Psyche must of necessity exist in order to have done all of the design, engineering, manufacturing, science, and creation that needed to be done which evolution and the rocks could NEVER have done and can NEVER do.

The reason why the physical sciences or natural sciences have so much prestige and success associated with them is because of all the wonderful and useful technology that has come as a result.

"Behavioral scientists hold out the hope that a technology for dealing with the problems of human behavior might be developed with much the same beneficial results as in the natural sciences."

The Behaviorists want to be able to predict and control human behavior, much the same way that we can predict and control robots and computers. But, the Materialists, Naturalists, and Atheists don't realize nor understand that physical technology will NEVER be able to touch, change, predict, nor control the immaterial, non-physical, non-local, human psyche! The human psyche has a life all its own. There really IS a ghost in that machine!

The scientists can want power, prediction, and control when it comes to the human psyche and human behavior, but human beings are not exclusively machines thanks to psyche!

In fact, it is ALWAYS some kind of Psyche or Ultimate Cause who designs, programs, engineers, builds, and creates the various machines in the first place. The Materialists, Naturalists, and Behaviorists don't get it, because they don't understand nor accept Psyche.

"Behaviorists have tried to explain human behavior in strictly naturalistic terms, but the attempt has not been wholly successful. The behavioral sciences have for the most part been unable to formulate universal laws that fully account for human behavior. Behavioral scientists have assumed a deterministic posture toward human behavior, but they have not been able to demonstrate the kind of control and prediction that have been achieved in the natural sciences."

This IS because Psyche is a supernatural or non-local phenomenon, and not a natural nor physical process. The universal laws associated with human behavior are in fact based upon a Psyche Ontology wherein psyche is the fundamental unit of reality; therefore, the Universal Law associated with human behavior IS free will, the pursuit of freedom, and agency because Psyche IS an independent agent and an independent entity completely separate from the physical body and the physical brain. Psyche continues to behave, and misbehave, long after the physical body is dead and gone, according the empirical evidence from Near-Death Experiences and other types of Spiritual Experiences. To study the human psyche and human behavior, we must study the Lived Experiences of the human race and not just the physical brain. It's elementary, my dear friend!

When it comes to human behavior or the human psyche's behavior, the Materialists, Naturalists, and Behaviorists are barking up the wrong tree!

The first common way of defining science is to consider science as the body of knowledge which explains the physical world and the natural sciences. Under this model for science, "scientific explanations for the most part involve rejecting supernatural explanations of phenomena in favor of naturalistic explanations. Scientific explanations are usually given in terms of matter or other naturalistic constructs [such as biology]. [As a result,] the behavioral sciences [and the psyche sciences] will never be 'scientific' in the same way as the natural sciences. The second common way of speaking about the nature of science is to say that science is primarily a *method* for studying phenomena." In other words, we could choose to broaden the horizons of science to include Phenomenology and formally state that science IS the study of the Lived Experiences of the human psyche. The Lived Experiences of the human race naturally include physical phenomena, but our Lived Experiences also include spiritual phenomena. This is the way that the behavioral sciences and psychology should go with the discipline; but, it's not going to happen while the Materialists, Naturalists, and Atheists are in control of these disciplines.

It is "difficult to measure many important aspects of human beings", particularly the spiritual aspects and the psyche aspects, "because we lack the instruments to do so. Furthermore, human beings in many settings actively resist strict control and prediction". Therefore, let's change the rules of the game so that we can study human beings, human choices, human actions, and human behavior as a whole. Let's make psychology what it was originally meant to be, a study of the human psyche!

"Remember that scientific methods are based on certain [materialistic and naturalistic] assumptions about the nature of truth and the world. If these assumptions are not true of the human world and of human phenomena, then the Scientific Methods based on these assumptions may not be appropriate for studying people [and the human psyche]". Therefore, we need to choose a different methodology if we truly want to do Psychology or a study of the human psyche! It's only logical!

Get rid of Materialism, Naturalism, Darwinism, Atheism, and Creation by Rocks as our major premises or primary assumptions, and then we can start to do Science for real! It's elementary my dear reader. By getting rid of all the falsehoods that we know about, we are then free to start a pursuit of the truth instead. That's what Science should really be about – getting rid of all the falsehoods that we know about so that only the truth remains. There's NO evidence to support the major premises of Materialism, Naturalism, and Atheism; and, there NEVER will be because there can't be. We should get rid of Scientism, and then replace it with some kind of Methodological Pluralism instead, which will continue to include scientific methods as one of the methods but develop and include other methods as well.

Materialism, Naturalism, Nihilism, Darwinism, Atheism, and Scientism are predatory religions. These things claim to be the ONLY source of truth, and therefore their proponents mock and ridicule everything else. These are exclusionary philosophies. These people are religious extremists and religious fanatics. The ONLY way that we will ever be free to study the inclusionary philosophies, the psyche sciences, and the holistic sciences is if we successfully EXCLUDE the exclusionary philosophies.

Knowledge

These are powerful ideas being presented in this article, but only if people are willing to understand them, accept them, and embrace them. This article is over twenty years old, so these changes and improvements have been slow in coming, because this is a message that you will never hear from the materialistic and naturalistic professors who control our public schools and academia. A person really has to dig in and study on the fringes if he or she truly wants to know what's wrong with Materialism, Naturalism, and Darwinism. I have been doing this for a few years now, and I'm finally starting to make some headway.

I KNOW how all this works, because I have lived it. I have experienced it for myself.

I KNOW what it is like to have one's physical body develop a mind of its own and a life of its own, and then become completely beyond my control. That's a special type of hell, which is no fun to live with and live through.

I KNOW that we can go to hell, while here in mortality while we are still in our physical body.

I KNOW what it is like to want to die and to want annihilation. I was hoping that I could die and cease to exist. I was no longer afraid of death. I wanted to die; and, I was hoping that my spirit and psyche would completely cease to exist when my physical body died.

I KNOW what it is like to have no friends. While I was attempting suicide, I remember thinking, "At least I will never have to come back here." I also remember thinking, "If the theists are right after all, at least in hell I'll finally have a chance to make some friends." Hell is being completely alone and having no friends.

I KNOW what it's like to finally reach a point in life where one is thinking, "If I end up in hell again, I'm going to try to help people and try to help them be happier and better." I KNOW what it's like to get outside oneself and to think about others and to think about helping others, every once in a while. I KNOW that there are better ways to live, other than the ways which we typically choose to live as mortals.

I KNOW what it's like to be addicted to crap; and, I KNOW what the months of psychosis are like while one is going through withdrawal.

I KNOW how important it is to keep God's Commandments to the best of our ability. After I got sober, one of the first things to come back to me was a deep and abiding faith or trust in the Commandments of God. I KNOW how good they are for us. I KNOW how happy and peaceful they can make us, by choosing to own them and live them. I KNOW what it's like to be completely at peace, thanks to the mercy and forgiveness and grace that comes from the Atonement of Christ.

I KNOW what it's like to be suckered into believing that philosophical concepts like Materialism, Naturalism, Nihilism, Scientism, Atheism, and Darwinism are real and true. I KNOW how gullible we can be as human beings, and how easily and eagerly we fall for obvious falsehoods such as Materialism and Naturalism and Atheism because we desperately want them to be true.

I KNOW what it's like to falsify philosophical theories such as Materialism, Naturalism, Darwinism, and Nihilism.

I KNOW these things, because I have experienced them.

Positivism and Realism

Slife, B. D. & Williams, R. N. (1995). Science and Human Behavior. In *What's Behind the Research? Discovering Hidden Assumptions in the Behavioral Sciences*, (pp. 167–204). Thousand Oaks, CA: SAGE Publications.

http://mypsyche.us/science/

Next, this article from Slife and Williams does a few pages on Positivism and Realism.

"Many people believe that science is self-correcting when it goes wrong."

It is NOT, because personal interpretation of the data and evidence is involved, when it comes to the Scientific Methods!

Each piece of scientific evidence or data can theoretically be interpreted an infinite number of different ways. Scientific data can be interpreted in ways that support the Biblical narrative and in ways that completely contradict and debunk the Biblical narrative. You can interpret the scientific data any way you want.

For example, I have observed that the scientific data makes a lot more logical sense if it's given an Intelligent Design interpretation than it does when it's given a materialistic, creation by rocks, Theory of Evolution interpretation. Psyche or Intelligence can design and create. The rocks can't. Psyche and Intelligence have been caught in the act of design and creation trillions of different times; whereas, the rocks NEVER have. This reality is so obvious and clear that I sometimes wonder how I was able to overlook it for the first fifty years of my life. Intelligent Design is an infinitely more plausible and believable interpretation of the scientific evidence than Creation by Rocks or the Theory of Evolution will ever be. But, I couldn't see it nor understand it when I was a Materialist and an Atheist. I was blind to those kinds of obvious realities; but, now I see.

I have tended towards Scientism and Realism most of my life. I wasn't going to believe in God unless I had Scientific Proof of God's Existence; and, I didn't think that such a thing was possible. It wasn't until the end of 2015 that Science started proving to me that God exists.

Why did it take so long?

It's because I wasn't looking for it and didn't believe it to be possible!

I had fallen for the Scientism and Materialism hook, line, and sinker!

Observation

"Both realists and positivists put great emphasis on observation" – a thing that they call empiricism.

When it comes to observation and empiricism, what these people really mean is physical observation of physical objects with one's physical senses and one's physical instruments. These people deliberately exclude spiritual experiences acquired by spiritual senses, even though the spiritual experiences are technically more powerful and real than anything acquired by the physical senses.

Science is based upon biased and prejudiced worldviews, assumptions, and interpretations of the data. There's really nothing objective about science. The whole thing is philosophical interpretation from beginning to end.

"Science is not as objective and free from cultural influences [and personal interpretations] as the realists and positivists have thought. The view of the world that each scientist has and shares with other scientists and non-scientists influences the way that science is done. This worldview (in Kuhn's terms, a *paradigm*) leads scientists to think about their science [and to interpret their science] in the way they do, even though they usually do not recognize their view as a worldview [or personal philosophy]. The way they formulate their questions, the methods they believe to be appropriate, and the sorts of explanations they hold to be acceptable are all influenced by the worldview in which they live. Scientific method itself is based on certain assumptions about the nature of the thing it studies."

The whole scientific enterprise is biased and subjective from bottom to top, completely colored and molded and controlled from beginning to end by a materialistic and naturalistic worldview. They stack the deck, pre-determine acceptable conclusions, and limit the conclusion to a materialistic explanation BEFORE they even start the science experiment. I have encountered people online who are dogmatically religious about this exclusionary process. Scientific method is based upon philosophical assumptions, and exclusionary philosophies such as Materialism, Naturalism, Darwinism, Scientism, and Atheism. That's its strength, but that's also its greatest weakness. Science and scientific methods are biased and prejudiced. And, that's not the only thing that's wrong with the Scientific Methods.

Scientism is based exclusively in Materialism and Naturalism.

Definition of Behaviorism

"Behavioral scientists have approached their subject matter – human beings – convinced of the importance of objectivity and scientific certainty, and intent on uncovering general or universal causal laws, formulating deterministic and mechanistic explanations of behavior, and try to develop technological solutions to human problems."

"It is important to consider whether human beings are really the sort of entities best described by mechanism, determinism, and efficient causality. If they are, then the traditional methods of science may be the most appropriate methods to employ. If they are not . . ." well, a whole mess of confusing interpretations and philosophical speculation is going to ensue.

Scientific method is limited to studying physical matter. It's a very good way to get at the truths about physical objects and physical realities. However, scientific method is a weak way of getting at the truth about the immaterial or the non-physical and non-observable realities of our existence. Furthermore, scientific methods seem to be absolutely worthless when trying to get at the truth about psyche, thoughts, dreams, and the afterlife. Scientific methods are not tuned for psyche!

Ironically, only the human psyche KNOWS how to use language, math, and science to make sense of things! Only the human psyche can do science! Yet, the proponents of Scientific Naturalism completely overlook this reality and choose to pretend that psyche does not exist.

Some Limitations and Weaknesses of Scientific Methods

"One of the main reasons science is so attractive is that it seems to hold the promise of truth. At least science appears to be a more trustworthy way of testing the truth of our ideas than other ways that have been developed. However, this perception of science is based upon" a wide variety of philosophical assumptions and logic fallacies. A logic fallacy is NOT the best platform upon which to try to build the truth; but, that's exactly what we see going on with the Scientific Methods and the *affirming the consequent* logic fallacy.

"Prior to the application or even the development of any scientific method, there is always an operative understanding of the truth", meaning that there are certain philosophical assumptions that are made. "It is this pre-understanding of the truth", a pre-selection of what is assumed to be true, "that makes it possible to frame any method at all. Without this understanding" or initial prejudice, "we could not formulate any scientific method because we would not know what the method should be like – or that we even need a method. This means that understandings of truth produce methods, rather than methods producing the truth. If this is the case, then we cannot be confident that properly and carefully applying scientific methods to the study of human behavior will get us to the truth." Rather, the whole enterprise could in fact lead us away from the ultimate truth.

Scientific Methods are NOT tuned for psyche! Scientific Methods are capable of dealing with a Psyche Ontology.

"Many important questions about human behavior can only be solved by careful theoretical work, not by the application of method. Scientific methods do not establish the truth of the matter", because they are limited exclusively to physical matter. "A related conclusion is that all methods, including scientific methods can only find the sorts of things that they are 'tuned for'. In the behavioral sciences, if our methods are not 'tuned' for human beings" and the human psyche, "then the method can miss what is true and important about human beings. In this way, scientific methods can act as blinders as much as they can reveal something important to us" that might be true.

The Affirming the Consequent Logic Fallacy

This article does the best treatment of *negating the consequent* and *affirming the consequent* that I have encountered so far.

I was shocked to learn that the whole Western Tradition has been built upon the *affirming the consequent* logic fallacy, which is built into the Scientific Methods and is unavoidable. I was shocked and stunned to learn that we can NEVER use the Scientific Methods to prove the truth. (Rychlak). I remember thinking, "What good is it, if we can't use it to prove the truth?"

This is the ONLY article that I have encountered so far that talks about *negating the consequent*, how it's logically and philosophically sound, and how scientific methods can be used to falsify hypotheses and theories that are physical in nature, and origin, and substance.

I had already been introduced to Karl Popper's idea that science and scientific methods should be used to falsify ideas; but, I had no idea why he had chosen to use science to falsify theories rather than using science to verify or prove theories. (Kalat). It's because *negating the consequent* is philosophically and logically sound; whereas, *affirming the consequent* is not.

I was also shocked to learn that *negating the consequent* isn't as effective nor as convincing as Popper made it out to be, due to the fact that there's theoretically an infinite number of variables to have to take into account and control while using scientific methods to falsify our theories and due to the fact that there are dozens of possible interpretations that can be given to each and every piece of evidence and scientific experiment which we come across.

Nevertheless, we can and we do use our scientific experiments, scientific methods, and *negating the consequent* to falsify our theories. I have used such procedures dozens of times myself to **falsify** Materialism, Naturalism, Nihilism, Darwinism, and Atheism to my satisfaction. The idea is that if we eliminate enough falsehoods using science and the Scientific Methods and logical common sense (procedural evidence), we eventually start centering in on the truth through a process of elimination. Slife and Williams mentioned this reality – we arrive at the truth through a process of eliminating the falsehoods. Science and the Scientific Methods can help us to do so.

Ironically, we can't even design and run scientific experiments on the major premises or primary assumptions of Materialism, Naturalism, Atheism, and even Darwinism (design and creation by rocks) which makes these philosophical concepts by definition, in principle, unscientific to begin with. We can, though, use scientific methods and logical common-sense methods and observational and experiential methods to **FALISFY** Darwinism, Materialism, Naturalism, and Atheism to our heart's content. ALL is not lost, because we can use scientific methods successfully to falsify these kinds of falsehoods. I still rely upon the Scientific Method to do so!

I spend some time discussing verification and falsification in the next chapters, so I won't go into the details here.

By pointing out all the weaknesses of the Scientific Method, this article forced me to face Reality. Scientific methods are NOT a sure way of knowing the truth. There are in fact better ways of knowing the truth than the Scientific Method! I never realized that before! My goal is to learn something new every day, and for me, this was a major breakthrough or epiphany. This article marked the END of My Scientism. You won't catch me worshipping the Scientific Method ever again. I have seen the light!

Instead, I now use the Scientific Method for what it is good for and what it was designed for – falsifying falsehoods such as Darwinism, Materialism, Naturalism, Nihilism, Atheism, Scientism, and Creation by Rocks. The Scientific Method, observation, empiricism, and logical common sense can be used to falsify ALL of the exclusionary philosophies that are based upon Naturalism and Materialism. Get rid of all the falsehoods and exclusionary philosophies that we know about, and slowly we should arrive at the truth through a process of elimination. Materialism, Naturalism, and Atheism are based upon a denial of Reality and a refusal to look at evidence, so they are worthless in the end. We have to get rid of them if we want to know the truth. It's elementary my dear friend! This reality is so clear and obvious to me now, that I sometimes wonder why I wasn't able to see it nor understand it for the first fifty years of my life. But, we had all been duped by the Materialists, Naturalists, Darwinists, and Atheists and had fallen for their deceptions and lies. Even the Theists that I knew in college back in 1980 had fallen for these kinds of things. We didn't know better, because we weren't taught better! We had all been fooled!

The good thing about Richard Dawkins and the other New Atheists is that they finally presented the weakness of their case to the world. Before their books came out, we took it on blind faith that these people knew what they were talking about. After their books came out, suddenly we knew that these people didn't have a clue about how science, philosophy, and reality truly work. Before they wrote their books, we thought their case was infinitely stronger than it really is. After they wrote their books, suddenly everyone could see that the Emperors have no clothes.

Dozens of books have been written explaining what's wrong with Dawkins' books, Dawkins' science, Dawkins' ideas, and Dawkins' philosophy. The books that have been written debunking Dawkins and the other New Atheists make infinitely more sense than anything that Dawkins and his friends ever wrote or said. Suddenly everyone could see that these New Atheists weren't so bright after all. In fact, I and my friends feel kind of embarrassed and sorry for them, because these New Atheists are so ignorant, uneducated, unscientific, closed-minded, prejudiced, and blind. We have watched their debates and were surprised how weak their case and their knowledge of science and philosophy really is. You would have to be a Materialist or an Atheist not to be able to see it.

Science, Logic, and Reason

Slife, B. D. & Williams, R. N. (1995). Science and Human Behavior. In *What's Behind the Research? Discovering Hidden Assumptions in the Behavioral Sciences*, (pp. 167–204). Thousand Oaks, CA: SAGE Publications.

http://mypsyche.us/science/

"The relation between science and logic can be traced to Aristotle, who invented formal logic as a way of doing science. The form of logical argument that Aristotle developed was called *syllogism*."

On pages 182 to 189, Slife and Williams discuss some of the logic fallacies upon which the Scientific Methods are based.

The conclusion I drew from all of this is that Scientific Methods are bad at getting at the truth, even the physical truths or "materialistic truths" that we are trying to get at. In other words, Scientific Methods can't be used to prove the truth. The first time that this reality was presented to me by Joseph Rychlak in his books, I remember thinking, "What good is it if it can't be used to prove the truth?"

Furthermore, before reading these pages in this article, it had already become clear to me that it is impossible to get at a Psyche Ontology and Psyche Truths using the traditional physicalist Scientific Methods.

I was extremely shocked and surprised to learn that Scientific Methods aren't even dependable and reliable when a person is trying to falsify theories by *negating the consequent*. I came away with the impression that Scientific Methods are good for nothing at all! That was definitely the END of my Scientism!

"Whatever scientific methods may be good for, they cannot be used to *verify* theories in the sense of affirming the consequent, or *falsify* theories, in the sense of negating the consequent. In other words, we cannot through scientific methods discover whether theories and hypotheses are true or false."

"It is in principle impossible for an experiment to establish or verify that a cause-and-effect relationship exists between any two variables in a

study, as Hume noted long ago. Just because the independent variable and dependent variable always seem to occur together does not mean that some other, as yet unidentified, variable could not always be present as the real cause."

YIKES! Well, shoot dang! I could feel my paradigms shifting underneath me! And, paradigm shifts can be painful!

Here we have people like me, scientists, who have built their whole worldview and placed all of their faith and trust in the Scientific Method. What an eye-opener and a shock that was to discover that Scientism – placing all of our faith and trust in scientific methods – is an unsustainable religion and an untenable epistemology!

Williams and Slife were the first scientists in my life to point out the fact that scientific methods can't even be relied upon to successfully falsify theories and hypotheses. They were the first people to tell me anything about *negating the consequent*, and its limitations. Scientism had been my epistemology ALL of my life! It looked like it was time for me to find a NEW epistemology – a new way of finding and knowing the truth. It suddenly made this Psyche Epistemology or a "Lived Experience Epistemology" look highly desirable, for the first time in my life.

In another article Richard Williams stated:

The truth of an event is very different from the truth of a proposition. The truth of propositions is established by reason and argument, the difficulty of which I have just described. The truth of events is established by witnesses. Once we know what is true, reason provides a wonderful tool for sorting out our obligations, anticipating consequences, and persuading others that what we know is true. Truth, I am convinced, can be rendered reasonable, but it does not arise from reason.

http://mypsyche.us/faith-vs-reason/

Witnessing events or lived experiences is a better way of knowing the truth than philosophical speculation or scientific interpretation of data. Truth and knowledge arise from lived experiences, or certain events in our lives.

The truth of Christianity does not rest on reason and logic, nor does it rest on scientific evidence. The truth of Christianity rests on the occurrence of certain events. The truth of Christianity rests upon the lived experiences of the human psyches or the human beings who were there at those events.

ONLY the human psyche can witness and then testify of events. In other words, only the human psyche can share its lived experiences with other human psyches. We humans are unique. We can share our lived experiences, including our spiritual experiences and out-of-body experiences, with each other in writing and through the spoken word.

I enter these Spiritual Experiences and Lived Experiences into evidence as proof of this concept:

http://www.markme.us/forums/forum/favorite-spiritual-experiences/

https://www.youtube.com/watch?v=UPj4wci_bcI

https://www.youtube.com/watch?v=HbTAmN4m2lQ

http://www.markme.us/forums/topic/appearances-of-our-resurrected-lord-jesus-christ/

Psyche and Lived Experiences Are Axiomatic

I now treat the **falsity** or **falsehood** of Materialism and Naturalism as axiomatic, which means that I now treat Psyche's existence as axiomatic and treat Psyche as the fundamental unit of Reality and Existence.

ONLY Psyche can do lived experiences, the highest form of knowledge and truth. ONLY the human psyche can share his or her lived experiences with other human psyches.

Scientism is no longer my chosen nor favorite epistemology. I have officially switched from a materialistic and naturalistic ontology in which physical matter is the fundamental unit of reality over to a Psyche Ontology and a Psyche Epistemology in which psyche is the fundament unit of Reality and lived experiences are the best way of finding and knowing the Truth. Psyche really is the Ultimate Cause! And this Ultimate Model of Reality was the result of all of this conversion, which has been taking place in my life during the past year or so.

I'm starting to look at things in new and different ways.

Unidentified Variables and Unknown Gods

Some things like Psyche cannot be falsified nor reliably verified, especially if one is relying exclusively on scientific methods and a materialistic epistemology. Things like forces, fields, light, spirit matter, non-locality, and psyche have to be experienced and observed directly in order to KNOW that they are real and truly exist. We can observe their effects on physical matter while in a physical environment; but, we have to switch to a non-local spiritual environment if we want to experience and observe these things directly. That's just the reality of the situation.

Only Psyche can do relationships, science, logic, language, math, experience, memories, interpretation, meaning, teleology, afterlife, life, friendship, charity, love, compassion, evaluation, experimentation, design, creation, programming, engineering, manufacturing, existence, and reality. These are all things that we can't get at directly with our scientific methods. Instead of trying to operationalize these things so that we can apply scientific methods to them, we NEED a new and better methodology if we truly want to get at these sorts of immaterial and non-physical things.

This article talks about unidentified variables in scientific research.

The ultimate unidentified variable in ALL scientific research is invariably PSYCHE, or Non-Local Consciousness. Psyche pervades the whole scientific enterprise from beginning to end – starting with philosophical hypotheses, moving into philosophical methodology, and ending with philosophical interpretations of the scientific data. The Scientific Method is pure philosophy from beginning to end; and, ONLY Psyche can do philosophy, hypothesizing, theorizing, methodology, experimentation, and hermeneutical meaning and interpretation of data.

If WE make something happen, it is in fact our Psyche who is making it happen or setting it up so that it can happen. This reality applies to scientific methods and everything else that WE make happen!

The Problem of Operationalization

In order to study concepts and realities such as forces, fields, light, consciousness, thoughts, dreams, and psyche, the scientists have to try to find some way to *operationalize* or *physicalize* these things so that their effects can be observed and measured.

On page 191, this article talks about the weaknesses and flaws that are introduced into science whenever we try to operationalize the immaterial and non-physical realities of our existence.

"A hallmark of traditional scientific method is observation. To perform a scientific test, we must be able to observe the phenomenon we are trying to study. This is not always an easy or straightforward thing to do, even in the natural sciences and physical sciences. For example, a clearly scientific construct like gravity is never observed directly. We can observe what we take to be the effects of gravity, but not gravity itself."

Just because a phenomenon can't be observed with our physical eyes and physical instruments doesn't mean that it doesn't exist. These types of things can be inferred to exist by the effect they have on physical objects and physical beings such as ourselves.

Gravity is a force or a field, and it is therefore like Psyche or Consciousness, something that is immaterial and non-physical in nature and origin. Because gravity is not physical in nature or origin, it can be omnipresent and instantaneously and simultaneously everywhere all at once. It is physicality or localization in 3D Space-Time which introduces limitations into our lives as mortal beings. Since gravity is not a part of our physical reality, it is NOT subject to the limitations of our physical existence. The same can be said of Psyche or Non-Local Consciousness. Physical matter imposes severe limitations on our Psyche and our Spirit Body; but, when our psyche and spirit body have separated from our physical body, psyche and spirit have no physical limitations. Physical matter cannot travel faster than the speed of light; but, Psyche and our Spirit Body and Gravity certainly can.

Gravity is not observable with our physical eyes and our physical instruments. It can only be experienced. In fact, it has to be experienced in order to know that it is real. This same reality applies to Psyche, Spirit Bodies, and Non-Local Consciousness. It has to be experienced in order to know that it is real. Thus, a Psyche Epistemology or an epistemology based upon Lived Experiences becomes the best and surest way of finding and KNOWING the Truth.

The moral of the story is that we SHOULD switch away from the Scientific Methods and over to knowledge and experience when it comes to human behavior and the human psyche! Lived experiences are an infinitely better way of getting at the truth and knowledge associated with Psyche and the Spirit Realm. Scientific methods can't even begin to touch these things! We need to pursue a Psyche Epistemology if we want to KNOW the truth about immaterial and non-physical realities such as forces, fields, light, psyche, consciousness, spirit matter, time, thoughts, dreams, and non-locality.

Materialists, Naturalists, Nihilists, and Atheists can't be objective about these kinds of immaterial things because these people have a prior dogmatic commitment to their chosen cause, chosen religion, and chosen worldview.

"In the behavioral sciences, however, it is open to debate whether people can in fact be manipulated and controlled by behavioral technology. There is wisdom in taking great care about how and why we would ever pursue such a project. There is also a good argument that the first and most important goal of science is not manipulation or control, but understanding."

Objectivity

Objectivity becomes an issue every time an interpretation or an explanation of the scientific data needs to be made.

Materialists, Naturalists, Nihilists, and Atheists can't be objective even when it comes to Science and the Scientific Methods because these people have a prior dogmatic commitment to their chosen cause, chosen religion, and chosen worldview, which completely colors their scientific interpretations.

"The very definitions and framing of a research question are shot through with traditions, history, expectations, values, biases, prejudices, and other subjective factors. It seems unlikely that at any stage the scientific research process is truly objective." Furthermore, Science can't make truth claims.

In contrast, we can KNOW what we have lived and experienced first-hand for ourselves.

Lived Experience becomes the best way of KNOWING the Truth, followed by trusting the witnesses who have had the lived experiences and therefore have KNOWLEDGE of the truth.

"What is important for the purposes of this book is that the characteristics of science, as they are often held up for the behavioral sciences to emulate, are fully exposed for thoughtful examination. Natural science methods may not be capable of serving the validational function that many behavioral scientists desire. The best that science can offer may be *one* way of viewing human behavior, without any special warrant for claiming that it is the only way or even the best way. If this is the case, then whether the behavioral sciences can or should be considered sciences, and just what it might mean to claim that they are sciences, remain open to question."

Science and Scientism are simply one way of looking at the world, and not necessarily the best or most effective way of looking at the world. This article suggests that we treat scientific methods as if they are one language, among many other possible languages.

Alternative Epistemologies

This article ends by discussing alternative epistemologies (ways of knowing the truth) and alternative methodologies (qualitative methods, human science methods, methodological pluralism, interview methods, and introspective methods).

"If we take the theoretical position that human beings are not simply natural objects, not fundamentally like mechanisms, and not determined by laws and forces the way natural objects are, then it is inappropriate to use the methods developed to study natural objects" to study human beings and the human psyche.

"Use of methods developed for natural objects will result in imposing this naturalistic theoretical outlook on human beings. Some who hold this perspective," namely that Scientism and a Naturalistic Ontology are not the best way to study the human psyche, "argue that adequate study of human beings requires a 'human science' – natural objects can be studied by the methods of natural science and Naturalism, but human beings require **human science methods**" and some kind of **Psyche Science**.

And here we finally have it!

"In this view, human scientists accept **lived experience** as the origin of understanding as well as the object to be understood. **Methods of study must be faithful to and grounded in lived experiences.** Methods developed for a detached study of natural objects will be inadequate."

Slife and Williams are talking about and suggesting a **Lived Experience Epistemology**, wherein lived experiences become the best way of finding and knowing the truth about Reality and our Existence. Although they don't say it in so many words, they are also talking about a **Psyche Ontology** and this Ultimate

Cause Model of Reality which I have been trying to develop in this book, wherein Psyche is the Ultimate Cause and Psyche is the fundamental unit of reality.

Slife and Williams then mention a few methods, typically qualitative methods, which are based upon this proposed **Lived Experience Epistemology**.

Qualitative Methods

"Many human science methods are grouped under the rubric ***qualitative methods***. The thrust of qualitative research methods is to reject the philosophical assumptions of the traditional scientific methods. Researchers avoid measurement and quantification, allowing subjects to describe their own behaviors and experiences in the language native to their experience. The analysis of the data is likewise carried out in conversational language rather than with statistics. The qualitative researcher is essentially involved in a project of careful questioning, describing, and interpreting. Many people who subscribe to **human science methods** argue that qualitative methods are superior to quantitative methods and ought to be the method of choice for all the behavioral sciences. Qualitative research methods are becoming increasingly accepted as a legitimate alternative to traditional empirical methods. Qualitative methods now form an important part of the literature of many behavioral science disciplines."

I never knew that! Imagine it, this article was written twenty years ago, and I'm just now hearing about qualitative methods for the first time in my life. The Materialists, Naturalists, and Atheists had successfully hidden these things from me for fifty years of my life. It's what they do, and they are very good at it.

Lived Experiences

The primary focus in the behavioral sciences is now upon **LIVED EXPERENCES**! Or, at least it should be!

Furthermore, I have observed that ONLY Psyche can do lived experiences, science, methodology, philosophy, theorizing, meaning, final causality, interpretation of data, ontology, epistemology, qualitative methods, psyche-study, and the human sciences. Psyche really is the Ultimate Cause and the Fundamental Unit of Reality, because only psyche can do these kinds of things!

"One possible organization of the methods groups them into three categories: ethnography, phenomenology, and studies of artifacts."

"Phenomenology has as its primary interest the study of the meaning of concrete human experiences." In other words, **Phenomenology is the study of Lived Experiences**. Phenomena are lived experiences and events which the human psyche encounters and experiences. This includes spiritual experiences as well as physical experiences. Nothing is off limits or out-of-bounds.

"To get at the meaning of experiences, phenomenological researchers rely heavily on interviews and other verbal or written accounts of experiences. The researchers then carefully analyze these accounts in order to understand not only individual private meanings of the experiences, but also what is general and illuminating in understanding the meaning of human experience in the wider context of people and situations."

The authors don't mention it in this article, but **Hermeneutics** has become the study of meaning and the interpretation of Lived Experiences.

Furthermore, **Teleology** is the study of final causality and the study of purpose, reasons, goals, and intentions.

It is my personal observation that the Human Psyche is the ONLY thing that can do phenomenology, psychology, science, psychotherapy, hermeneutics, teleology, epistemology, and ontology. The Human Psyche is the ONLY thing we know of directly that can actually do Philosophy and Science. How do we KNOW of Psyche directly? We KNOW of Psyche directly through Lived Experiences and a Lived Experience Epistemology. Lived Experience is the best way of finding and KNOWING the Truth. Lived Experiences cover everything, including our revelations from God, our spiritual experiences or Spiritual Empiricism, our physical experiences or traditional empiricism, our science experiments, our work and our construction, our creativity, our philosophical theories and ideas, and everything else we humans think about and do.

Memories of our Lived Experiences survive the death of our physical body and physical brain according to the Lived Experiences of OBErs and NDErs who have gone to the other side and KNOW what it's really like in the Spirit Realms, and have experienced a life review while in the spirit. According to the out-of-body travelers and NDErs, the human psyche continues to have Lived Experiences and memories of those Lived Experiences while their physical body and physical brain are either dead or offline.

Psychology really should have remained **The Study of Psyche**; but, the Materialists, Naturalists, Atheists, and Behaviorists got their way for a century. It's finally starting to come full circle and turn on them. But, they are still entrenched and not going anywhere.

Conclusion and Solution

When it comes to psyche, we humans are unique, in that we can actually share our Lived Experiences with other humans through speech and the written word. The animals can't do that, even though they do have psyches of their own. ONLY the human psyche can do language, writing, reading, and speaking in a way that we can share ALL of our Lived Experiences with each other, including our spiritual experiences and our revelations from God.

This is possible for the human psyche, because we humans are the children of the Gods. Both our spirit bodies and our physical bodies are descended from the Gods. According to the Bible, Adam was a son of God, and Eve was his wife; and, we are gods. Our surname is God. The human psyche is unique among all of God's creation, because we are the children of God the Father and a Heavenly Mother. Only the human psyche has the innate inborn ability to rebel against God, because our spirit bodies and physical bodies are descended from the Gods. We humans really did partake of the tree of knowledge. NO other animal in creation has done so.

It ALL points back to a **Psyche Ontology** and to Psyche as the Ultimate Cause of these Lived Experiences! Psyche really is the Ultimate Cause or the Fundamental Unit of our reality, existence, experiences, and lives. I believe that I have made my case and met my burden of proof. Of course, you as an individual unique psyche have every right to disagree. In your own personal life, you are judge, jury, executor, and executioner. By "you", I mean your psyche. Thanks to your psyche, you will still be you long after your physical body and physical brain are dead and gone. It ALL points back to Psyche.

PART VII — VERIFICATION VERSUS FALSIFICATION

Negating the consequent is logically valid and typically sound; but, *affirming the consequent* is not. The Scientific Method is based upon "affirming the consequent". I figured all of this out logically on my own before I knew the actual terminology for it, as demonstrated in my other books listed in the references under Mark My Words.

I'm not the first person to figure all of this out, though. For example, the following is adapted from class notes by Edwin Gantt:

Verificationism

Historically, scientists have imagined that if predictions (hypotheses) based on their theories came true, then their theories have been verified (in other words, proven to be true) by empirical evidence. That is, the best way to prove a theory true is to make a prediction and then make measurements. If whatever the scientists predict will happen does happen, then it is assumed that the theory that gave rise to the predictions is true.

It is important, however, to consider the logic of this approach. A prediction based on a theory usually takes this form: "If Theory X is true [antecedent], then we will observe Y [consequent]."

The following logical argument outlines the basic approach that has been taken by scientists:

If Theory X is true, then we will observe Y.

We observe Y.

Therefore, Theory X is true.

This sort of thinking, however, reflects a logical fallacy called *affirming the consequent*. Here's a comparable example to demonstrate:

If Sally's pet is a cat, it will have a tail.

Sally's pet has a tail.

Therefore, Sally's pet is a cat.

We can easily see that this logic is fallacious. Just because we observe Y (a pet with a tail) that does not mean that our theory X (the pet is a cat) is true. After all, dogs and lizards have tails too. (Taken from class notes by Edwin Gantt).

Falsificationism

There is a different argument structure that scientists have begun to use instead of verificationism. It is called the falsificationism approach to science. This approach is much more logically sound than the verificationism approach.

The argument structure for falsificationism looks like this:

If Theory X is true, then we will observe Y.

We don't observe Y.

Therefore, Theory X is not true.

Rather than verifying a theory, this approach falsifies a theory. Unlike the verification argument, this argument is logically sound. The idea is that although we can't ever know (based on empirical data alone) when a theory is true, we can know (based on empirical data) when a theory is false.

In other words, we develop a theory, make predictions based on that theory, and if those predictions come true, rather than claim that we have confirmed our theory, we more humbly claim that our theory has not yet been proven false. Eventually, we arrive at the truth by a process of elimination.

The problem with this approach, however, is that there is no way to actually falsify a theory with absolute certainty. Theories are a patchwork of assumptions, presuppositions, and ideas, and there is no way to know which of those are wrong if the predicted events don't occur in our experiments.

Also, more importantly, theories and worldviews guide our assumptions about what might even constitute falsifying evidence in the first place.

So, a theory that holds that God is not real or active should, in theory, be falsified by actual encounters with God. But, instead, such experiences are usually bracketed off [by the Materialists and Naturalists] as things yet to be fully explained by the theory, while the theory itself is still assumed to be true. (Taken from class notes by Edwin Gantt).

Scientific Methods are built upon and based upon the *affirming the consequent* logic fallacy. Joseph Rychlak mentions the *affirming the consequent* logic fallacy in most of his books (Rychlak), in a similar manner as has been presented here by Ed Gantt.

See also:

Slife, B. D. & Williams, R. N. (1995). Science and Human Behavior. In *What's Behind the Research? Discovering Hidden Assumptions in the Behavioral Sciences*, (pp. 167–204). Thousand Oaks, CA: SAGE Publications.

http://mypsyche.us/science/

Understanding the *affirming the consequent* logic fallacy basically put an END to my Scientism and my exclusive reliance upon Scientific Methods to discover the truth. Scientific Methods can't be used to prove the truth or to verify the truth, because of the logic fallacies and philosophical interpretations upon which Scientific Methods are based. Consequently, I'm now willing to look to and look for other epistemologies – other ways of finding and knowing the truth, in addition to the Scientific Methods. I don't reject Scientific Methods, because that would be stupid; but, I no longer rely upon Scientific Methods exclusively for finding the truth.

I still rely heavily on falsification, because falsification is logically sound and tends to be reliable!

The argument structure for falsification looks like this:

If the theory of evolution is true, we will observe the rocks and physical matter designing and creating genomes and life forms. We NEVER observe the rocks designing and creating anything; therefore, Creation by Rocks, Materialism, Naturalism, Darwinism, and the Theory of Evolution ARE FALSE.

Materialism, Nihilism, Atheism, Naturalism, Darwinism, Scientism, Spontaneous Generation, Abiogenesis, and Creation by Rocks have been **falsified** trillions of different times in thousands of different ways. I don't believe in these things anymore, because they have been **falsified** too many times.

I used Science and the Scientific Method to prove to myself that Materialism, Naturalism, and Darwinism are false. I finally chose to use the Scientific Methods for what they are truly good for, and that's **falsifying** materialistic ideas that are false. For that reason alone, I will always be grateful for the Scientific Method.

When it comes to Scientific Methods, I still believe in and trust the **falsification** process. Therefore, I'm still a scientist in the traditional sense of the word. Science itself and Scientific Methods have proven to me that Materialism, Naturalism, and Darwinism ARE FALSE. I still have a great deal of love and respect for the Scientific Methods, especially the **falsification process** that's associated with Scientific Methods.

I have observed that design and creation by rocks, or physical matter, IS *the* materialistic and naturalistic assumption or premise that is always FALSE! Only Psyche has the innate inborn ability to design and create! Materialism IS Creation by Rocks; and, the rocks can't design and create. The naturalistic worldview, by definition and fiat, excludes explanations and interpretations of the scientific evidence that do not assume a naturalistic worldview. The solution to this problem is to EXCLUDE Materialism, Naturalism, Nihilism, and Atheism from our scientific explanations and scientific interpretations of the data. Eliminate every known falsehood as the Scientific Method says we should, and we eventually end up with the truth through a process of elimination.

How often have I said to you that when you have eliminated the impossible, whatever remains, *however improbable*, must be the truth? — Sherlock Holmes

Eliminate Materialism, Naturalism, Nihilism, Atheism, and Creation by Rocks; and, we are left staring at the truth. It's obvious, and it's logical. It's elementary my dear reader!

Applying Falsification to Materialism

This is the process that I used in my books to falsify Materialism and the Theory of Evolution.

Whether the Materialists and Naturalists realize it or not, they believe that the rocks or physical matter designed and created and produced the first genomes and the first life forms on this planet. Materialism and Darwinism reduce to Creation by Rocks, or Spontaneous Generation, or Abiogenesis.

The argument structure for the falsification of the Theory of Evolution looks like this:

1. If the Theory of Evolution is true, then we will observe rocks and physical matter designing and creating and manufacturing genomes and life forms from scratch.

2. We don't observe rocks designing and creating genomes and life forms; and, we have never caught the rocks in the act of designing and creating life.

3. Therefore, the Theory of Evolution is not true.

This argument, which involves negating the consequent or falsifying the consequent is philosophically and logically valid and sound.

Now, let's run the Theory of Evolution through the verification process like the Materialists and Darwinists do.

1. If the Theory of Evolution is true, then we will observe progression in the fossil record.

2. We observe progression in the fossil record.

3. Therefore, the Theory of Evolution is true.

The Materialists, Naturalists, and Atheists can't see anything wrong with this argument, because they have blinded themselves into *affirming the consequent* and use this exact argument all the time to make their case. But now, let's run something else through this argument that is sure to get the Darwinists' attention.

1. If the Theory of Intelligent Design is true, then we will observe progression in the fossil record.

2. We observe progression in the fossil record.

3. Therefore, the Theory of Intelligent Design is true.

If the first argument is considered valid, then the second argument must also be considered valid. And, since Creation by Intelligent Design is an infinitely more solid and plausible explanation for the scientific evidence and the origin of life than Creation by Rocks, the second argument (Creation by Intelligent Design) should in fact be given priority over the first argument (Creation by Rocks or the Theory of Evolution).

In other words, ALL of the scientific evidence that is being produced by the Darwinists in an attempt to prove the Theory of Evolution true also ends up proving the Theory of Intelligent Design true as well. And, since Creation by Intelligence has been observed and caught in the act, and Creation by Rocks has not, Creation by Intelligent Design is an infinitely better interpretation of the scientific evidence than Creation by Rocks or the Theory of Evolution will ever be.

Creation by Rocks, Spontaneous Generation, Abiogenesis, Darwinism, Materialism, and Naturalism have been falsified thousands of different ways. In contrast, design and creation and manufacturing by Intelligent Beings has been observed, replicated, and verified trillions of different times over millennia of time. So, which theory and which interpretation of the evidence is most likely to be true – the one that has been falsified or the one that has been observed and verified?

It was in this manner that I was able to prove to myself why the Theory of Evolution and Materialism are false. In every sense of the word, the Scientific Method proved to me that Materialism and Darwinism are false, and that Intelligent Design Theory is much more likely to be true. Eventually, we arrive at the truth through a process of elimination. Most of us instinctively know that Creation by Rocks (Materialism) is impossible. By eliminating Creation by Rocks in all its different forms (Materialism, Darwinism, Naturalism, and the Theory of Evolution), we find ourselves zeroing in on the truth (Design and Creation by Intelligence or Psyche or Ultimate Cause). And for most of us, the truth is what we are after.

How often have I said to you that when you have eliminated the impossible, whatever remains, *however improbable*, must be the truth? — Sherlock Holmes

Reference Articles

The following articles lend evidentiary support for the concepts presented in this section:

1. Slife, B. D. & Williams, R. N. (1995). Science and Human Behavior. In *What's Behind the Research? Discovering Hidden Assumptions in the Behavioral Sciences* (pp. 167–204). Thousand Oaks, CA: SAGE Publications.

http://mypsyche.us/assumptions/

http://sk.sagepub.com/books/whats-behind-the-research/n6.xml

http://www.ldsphilosopher.com/blog_posts/the-logic-of-falsification/

The authors and contributors have been handing out this paper in class to the public.

This article was the end of my scientism – my erroneous belief that the Scientific Method is an infallible god.

For a year or two, I considered James Kalat's chapter 2 on the Scientific Method in his *Introduction to Psychology* to be the best and most informative chapter on the Scientific Method that I had ever read in my life (Kalat). This chapter from Slife and Williams about the limitations of the Scientific Method is just as good if not better than Kalat's chapter. But, I like them both.

It is impossible to use the Scientific Method to prove the truth. As this article states, it's also technically impossible to use the Scientific Method to falsify anything as well. However, you are much more likely to falsify something with the Scientific Method than you are going to be able to verify something with the Scientific Method. The Scientific Method isn't as infallible and God-like as it is cracked up to be. The Materialists, Naturalists, and Atheists place way too much unjustified blind faith in the Scientific Method, often making it their god.

There are better ways of knowing the truth than the Scientific Method.

2. Williams, R. N. (2000, February 1). Faith, Reason, Knowledge, and Truth. *BYU Speeches*.

https://speeches.byu.edu/talks/richard-n-williams_faith-reason-knowledge-and-truth/

http://mypsyche.us/faith-vs-reason/

This article demonstrates how we humans rely upon other ways of knowing the truth besides just the Scientific Method. Experience or knowledge trumps scientific speculation and scientific hypotheses every time.

Furthermore, scientists take a leap of faith every time they design and run a science experiment. They hope that their sciences experiments will produce predictable results and give them more control over their environment; but, their hopes aren't always realized. It takes just as much faith to run a science experiment as it does to believe that God is real and truly exists. Belief or faith is simply a matter of preference and choice. Faith and beliefs are chosen into existence by the human psyche.

We choose what we want to do with our faith, our hopes, and our beliefs. Some people choose to find God, who KNOWS the truth. Other people choose to run a science experiment hoping to get lucky and find the truth. Others choose to engage in reason and sophistry, trying to manufacture the truth in the process. The goal in each case is the find the truth, but there are different ways of going about this process – some more effective than others.

During long stretches of my life, I didn't choose wisely.

From this article: "Finally, there is a possibility of being both dark and despairing as well as beyond the pale of reason."

I lived and experienced this particular reality when I was suicidal, psychotic, and going through withdrawal. For the first year, it manifested itself primarily as anxiety and fear; but, it eventually devolved into apathy, pain, and suicidal ideation. I have had an experience or two during my fifty years of life, and one of my friends thinks I should share them with the world.

"I believe the anchor opposite faith is darkness, nihilism, despair — that state of the soul that comes from living 'without God in the world' (Ephesians 2: 12)."

I have been there and done that as well!

Desire to get out of the pit or desire to get out of hell was the beginning of my progress; and, I have made a great deal of progress during the past five years of my life.

Truth rests on phenomena, events, and lived experiences! Faith is trusting the witnesses who have had the lived experiences and thus have THE KNOWLEDGE which we seek. The lived experiences of witnesses IS the best way of knowing the truth!

PART VIII — LIVED EXPERIENCES

Lived Experiences ARE the pinnacle of knowledge and truth. There's NO better way for finding and knowing the TRUTH than Lived Experience or Direct Observation.

Science is supposed to be Observation, or Lived Experience, although for most people Science is treated as if it is nothing more than philosophy or sophistry, and they limit Science exclusively to our physical reality. In other words, these people use Science to try to deceive us and to deceive themselves as well; and, they are extremely successful at doing so. Self-deception works, and it works every time. That type of limiting, materialistic, and exclusive "science" really isn't Science – it is sophistry, metaphysics, and philosophy. Science based upon falsehoods such as Materialism and Naturalism are the very definition of BAD SCIENCE or pseudo-science.

In contrast, the TRUTH is KNOWN by living it, witnessing it, and experiencing it for yourself, or by choosing to trust someone who has done so for himself.

By Lived Experiences, we mean ALL of our psyche experiences, thoughts, dreams, choices, decisions, actions, spiritual experiences, conscious experiences, memories, out-of-body experiences (OBEs), mystical experiences, psychic experiences, physical experiences, science experiments, unexplainable "miraculous" events, healing, near-death experiences (NDEs), shared-death experiences (SDEs), theophanies, revelations of God, and revelations from God. Through Lived Experience, we can KNOW the TRUTH directly by experiencing it directly for ourselves.

We KNOW from the Lived Experiences of the human race that Psyche exists and that our Psyche or Intelligence is something completely different than our Spirit Body. We also KNOW from the Lived Experiences of the human race that our Human Psyche or Non-Local Consciousness has been assigned to a Spirit Body or united holistically with a Spirit Body, which looks like our Physical Body. What people call our "Spirit", or our "Ghost", is in fact our Psyche or Personality which has been united with a Spirit Body as a single functioning unit or whole. We KNOW from the Lived Experiences of the human race that Psyche can temporarily separate from its Spirit Body and look at its Spirit Body from an immaterial third-person viewpoint in space.

This is a Psyche Ontology. This is a holistic ontology or a holistic model of reality which takes into consideration ALL of our Lived Experiences as a race. The Human Psyche is unique in that it can record and then share its Lived Experiences with other human beings. The animals can't. This Psyche Ontology, wherein Psyche or Intelligence or Non-Local Consciousness is the fundamental unit of reality, is the Ultimate Ontology and the Ultimate Model of Reality. ONLY Psyche can do ontology, reality, and existence. Without Psyche or Non-Local Consciousness, there would be no existence.

A Psyche Epistemology, based upon our Lived Experiences or Direct Observations, is the ultimate way of finding and knowing the truth. A Psyche Epistemology is a much better way of knowing the truth than the Scientific Methods, scientific experimentation, or philosophical speculation which introduce elements of doubt, uncertainty, interpretation, and guesswork into the equation and which tend to limit us exclusively to the physical realm.

A Psyche Epistemology is all-inclusive; whereas, a Scientism Epistemology, or Materialism Epistemology, or Naturalism Epistemology are not. The Naturalistic Epistemologies and Scientific Epistemologies tend to be denialistic, exclusive, limited, materialistic, and restrictive – censoring what their adherents are permitted to study and learn about. Materialism, Atheism, and Naturalism really aren't Science or Observation – they are in fact Dogmatic Religions – the bad kind of blind-faith religions which the Materialists and Atheists claim to despise.

Materialism and Atheism of any kind are based upon a refusal to look at evidence, a denial of reality, and a rejection of the Lived Experiences or the Observations of the human race. A Naturalistic Epistemology is a highly restrictive and limited epistemology, which deliberately rejects most of the Lived Experiences or Observations of the human race. That's NOT science, because Science is supposed to be all about Observation, or Lived Experience, or Experiential Evidence! Naturalism and Materialism are religions and epistemologies, which have to be taken on blind faith as being true. They are NOT science.

In contrast, a Psyche Epistemology is the Ultimate Epistemology, because it's all-inclusive and based upon the Lived Experiences of the human race. Furthermore, ONLY Psyche can do knowledge and experience for REAL. We don't witness any of this from the rocks or physical reactions. These things don't learn from their experiences in any noticeable or detectable fashion. Furthermore, the rocks have no way to apply their knowledge if they have any, because technically they are not alive; whereas, Psyche is. Only Psyche can do observations, which means that only Psyche can do science!

From the Lived Experiences of the human race, we KNOW that our Psyche or Intelligence or Non-Local Consciousness is a completely different entity than our spirit body, and that the psyche/spirit body unit or combination is a completely different entity than our physical body. Psyche, spirit body, and physical body nest within each other holistically like Matryoshka dolls, from smallest to largest or from most refined to most coarse. It's really easy to understand if one chooses to do so. A child can understand it. Human beings are psyche beings. Human beings are intelligent beings. Being IS Psyche. Psyche IS Being and Becoming. Psyche is the sum total of our Lived Experiences. ONLY Psyche can have Lived Experiences and remember having done so. Raw matter can't, whether we are talking about physical matter or spirit matter.

We learn best through Observation and Experience – through the comparison and the contrast of our Lived Experiences as a race. Human beings or human psyches are unique in that we can record our Lived Experiences and then share our Lived Experiences with each other. The animals can't. For all we know, the animals see and talk with God every day; but, they have no way of telling us about it if they do.

I, personally, demanded Scientific Proof of God's Existence before I was going to let go and be willing to believe in God. Thankfully, God gave me what I was looking for and asking for; and over time, I found and developed a few Scientific Proofs of God's Existence that were totally convincing to me. After that, I simply KNEW that God exists, even though I had yet to come to KNOW God.

When it comes to Scientific Proof of God's Existence, the way this process works is through the falsification of Materialism, Naturalism, Nihilism, and Atheism; and then through a process of elimination, after falsifying all of these things, we arrive at the truth through a process of elimination. Remember, the Scientific Methods can't be used to prove anything true; but, the Scientific Methods can definitely be used to prove things false – such as Materialism, Naturalism, Darwinism, and Atheism.

Once you have successfully falsified Materialism, Naturalism, Darwinism, Nihilism, and Atheism there are only a few logical things which remain to explain the origin of genomes and life forms on this planet. In effect, ALL of the evidence ends up pointing to Psyche, Intelligence, and some kind of Advanced Alien Intelligence or God. After Naturalism and Darwinism have been **falsified**, then it becomes clear that this earth was terraformed and seeded.

Think about it logically. That's what I did. Once you have successfully eliminated ALL the falsehoods and lies, then you are left starting at the truth! It works. I KNOW, because that's what I did to get at the truth. Let me demonstrate the process.

What's the opposite of Materialism? Psyche, spirit, non-locality, the quantum realm, the transdimensional realm, and the spirit realm are the opposite of Materialism. Once you use the Scientific Methods to successfully **falsify** Materialism a trillion different ways, you are left staring at its opposite and are left staring at the truth, which is Psyche, Intelligence, Spirit, the Immaterial or Non-Physical, Forces, Fields, Light, Non-Locality, Non-Local Consciousness, Quantum Mechanics or Spiritual Mechanics, and the Transdimensional Realm.

What's the opposite of Naturalism? The supernatural, the spiritual, the non-local, spirit bodies, spirits, psyche, Theism, and God are the opposite of Naturalism. After you have used the Scientific Methods to **falsify** Naturalism a thousand different ways, through a process of elimination, you are left staring at the opposite of Naturalism as the only possible Truth that remains.

What's the opposite of Darwinism? Darwinism is reduced to design and creation by physical matter. Alas, the rocks can't design and create. They have NEVER been observed doing so. Once you have used the Scientific Methods to **falsify** Darwinism a dozen different ways, you are left staring at the truth that only Intelligent Beings or Intelligent Psyches can design and create. You are left staring at Darwinism's opposite – intelligent design and creation. Random Mutations and Natural Selection (Biological Evolution) did NOT exist until after God's Psyche designed and created the first genomes and life forms on this planet. God must of necessity exist in order to have designed and created the first genomes and life forms on this earth, because ONLY Psyche can design and create and do science. The various physical reactions or evolutionary processes could not have designed, programmed, engineered, field-tested, manufactured, and deployed themselves. ONLY Psyche or Intelligence can do these kinds of things, because only Psyche can do science, engineering, and manufacturing.

In a very real sense, for me personally, the repeated **falsification** of the Theory of Evolution became my first real and most convincing Scientific Proof of God's Existence. God must of necessity exist in order to have done ALL of the science and creation which the various types of evolution or "creation by rocks" could NEVER have done. Eventually, for me, that Reality became logical and obvious. If we successfully use the Scientific Methods to eliminate ALL the falsehoods, then ONLY the Truth remains.

What's the opposite of Nihilism? Knowledge of an afterlife and knowledge of some kind of meaning and purpose in our lives is the opposite of Nihilism. I, personally, used the Lived Experiences of the human race – NDEs, SDEs, OBEs, and other types of spiritual experiences – to prove that Nihilism is false. Direct Observations or Lived Experiences trump philosophical speculation and the Scientific Methods, or at least they should. We can and we should use our Lived Experiences to **falsify** the scientific theories and the philosophical ideas that don't match with Reality. Through Lived Experiences, we humans can go directly to KNOWING the TRUTH, which is something that we can't do with the Scientific Methods or philosophical guesswork. Science, ultimately, is supposed to be about Observation or Lived Experience or Psyche Experience, which is in fact the ONLY real way of finding and knowing the truth. Only Psyche can do Lived Experience, Observation, Knowledge, Reality, Existence, and Truth!

What's the opposite of Atheism? God's Psyche or God's Intelligence is the opposite of Atheism. Atheism is design and creation by chance, or design and creation by nothing. And, we all KNOW scientifically from observation and experience that chance and nothing can't design and create anything – at least when we are thinking rationally and logically we KNOW that this to be true. Once we have used the Scientific Methods and common sense to **falsify** Atheism in a variety of different ways, suddenly we find ourselves staring into the face of Intelligent Beings or the Gods as our ultimate designers and creators. In a very real sense, your Psyche could have been the designer and the creator of the frogs and the dogs and other types of critters, because ONLY Psyche can design and create and do science! In the next life, many of the Atheists, Materialists, and Naturalists will discover that they helped to design and create some of the life forms on this planet. That will be quite a surprise for them, won't it!

Once we have eliminated ALL of the falsehoods, then we are left staring at the truth! It's elementary my dear reader! That's how we use the Scientific Methods to "prove the truth" – by using the Scientific Methods to eliminate ALL of the falsehoods. Remember, we can and we should use the Scientific Methods to **falsify** all the theories and ideas which are in fact false. Once we have done so, once we have eliminated all the falsehoods, then we are in fact left with the truth.

What's this truth which we are left with after we have eliminated ALL the falsehoods?

THE TRUTH IS that ONLY Psyche or Intelligent Beings can design and create genomes and life forms, planets and universes, you and me. Best of all, the Scientific Methods can NEVER be used to falsify this truth, because the Scientific Methods or observations will always verify this truth. Our observations and lived experiences have PROVEN beyond a shadow of a doubt that **ONLY** Intelligent Psyches or Intelligent Beings can design and create and do manufacturing, engineering, and science. Scientific observation has verified this reality trillions of times, and it will verify this reality an infinite number of times more before we are done.

THE FUNDAMENTAL TRUTHS are always verified by our scientific methods and lived experiences, because this kind of truth is never falsified and can never be falsified. There is NO way to design and create a scientific method that will falsify Psyche's ability to design and create, because you would in fact VERIFY "design and creation by Psyche or Intelligent Beings" by trying to design and create such a scientific method in the first place. There's no way to use the Scientific Methods to falsify this TRUTH. Scientific Observation has verified "Creation by Psyche" or "Creation by Intelligent Beings" trillions of trillions of times. There's NO way to falsify something like that, because it is TRUE. We can ONLY use the Scientific Methods to falsify theories that are in fact false – things like Materialism, Naturalism, Darwinism, Nihilism, and Atheism.

Lived Experiences point us directly to a Psyche Ontology, the Ultimate Ontology, wherein psyche becomes the fundamental unit of reality. Only Psyche can design, create, and do manufacturing and science. ONLY Psyche or Non-Local Consciousness can convert spirit matter to physical matter, according to Quantum Mechanics. If it weren't for Psyche, this physical universe would not exist. Psyche is the Ultimate Cause of physical universes and physical matter. That's what Quantum Mechanics is trying to teach us. It doesn't get more basic or fundamental than that! Psyche or Non-Local Consciousness is the fundamental unit of Reality and Existence.

For me, personally, this whole thing was a paradigm shift that initially took place for me in the spring of 2016, because I used to be a Materialist, Nihilist, and Atheist and used to truly believe that it would always be impossible to prove the existence of Psyche, God, and Truth. I was wrong; but, I didn't know that at the time.

Let's use the Scientific Method for what it was designed for. Here's how falsifying a theory, or *negating the consequent*, works in principle:

Scientific Hypothesis: If Theory X is true, then we will observe Y.

Scientific Observations: We don't observe Y.

Scientific Conclusion: Therefore, Theory X is not true.

Here's how it works in practice:

Scientific Hypothesis: If the Theory of Evolution, Materialism, Naturalism, and Darwinism are true, then we will observe the rocks and physical reactions designing, creating, and manufacturing genomes and life forms from scratch.

Scientific Observations: We have NEVER observed the rocks and physical reactions designing and creating genomes and life forms; and, we NEVER will. They can't.

The Scientific Conclusion: Therefore, the Theory of Evolution, Materialism, Naturalism, and Darwinism are not true.

I just successfully falsified Materialism, Naturalism, Darwinism, and the Theory of Evolution; and, I used the Scientific Method to do so! Best of all, my argument is philosophically and logically sound. I have in fact falsified Materialism, Darwinism, Naturalism, and the Theory of Evolution for REAL! I have PROVEN them false! I successfully used the Scientific Methods for what they are good for – falsifying theories which are false. I wish I would have known how to do that forty years ago. It would have saved me a lot of confusion and grief.

After I had successfully used the Scientific Methods to falsify Materialism, Naturalism, Darwinism, and Atheism, I simply KNEW that God of necessity must exist in order to have done all of the different types of Science, which the rocks or physical matter could never have done. It was logical, and it was obvious.

I had effectively used the Scientific Method to prove to myself that God exists, by using the Scientific Methods to **falsify** everything that was the opposite of God, Non-Locality, Intelligence, and Psyche. Thereafter, the Lived Experiences of the human race were able to provide me with solid and convincing PROOF of God's Existence, although up to that point, I was always reluctant to accept Lived Experience into evidence until after the Scientific Method and Science had demonstrated to me that God must of necessity exist in order to have done all the Science which needed to be done in the first place.

Ironically, though, after God had provided me with a couple dozen different Scientific Proofs of His Existence, it still took me a year to figure out for myself that Lived Experience is a much better way for finding and knowing the truth than the Scientific Methods can ever be. For me, switching to Lived Experience is a recent paradigm shift in my way of thinking, which finally took place for me in April 2017. Before I saw the Light, I truly believed that the Scientific Method was the ONLY way for finding and knowing the truth. I was wrong, and it was holding me back!

Nowadays, I greatly value the Lived Experiences of the human race, and what they have taught me. I have slowly learned to value my Lived Experiences a lot more than I value my philosophical reasoning and the Scientific Methods. But for me, it wasn't always that way.

The TRUTH is KNOWN by observing it, living it, and experiencing it; or, by choosing to trust someone who has.

Through lived experience or direct observation, we human beings or human psyches can go directly to KNOWING the TRUTH without having to engage in philosophical speculation, scientific experimentation, and scientific methodology. Powerful! Is it not? Maybe you can sense why I found it so enticing and compelling, and why I was so eager to embrace it, once the idea of Lived Experience was finally presented to me and I accepted it as being real and true. Lived Experience is infinitely more powerful and convincing and real than the Scientific Methods can ever be. I never knew that before; but, now I do. Scientific Methods can't be used to prove the truth directly; but, Lived Experiences can! It doesn't get any better than that!

For example:

John 17: 3: And this is life eternal, that they might **know** thee the only true God, and Jesus Christ, whom thou hast sent.

John 8: 32: Ye shall **know the truth**, and the truth shall set you free from death, hell, ignorance, and sin.

Doctrine and Covenants 50: 25-27: "And again, verily I say unto you, and I say it that you may **know the truth**, that you may chase darkness from among you. He that is ordained of God and sent forth, the same is appointed to be the greatest, notwithstanding he is the least and the servant of all. Wherefore, he is possessor of all things; for all things are subject unto him, both in heaven and on the earth, the life and the light, the Spirit and the power, sent forth by the will of the Father through Jesus Christ, his Son."

John 14: 5-6:
5 Thomas saith unto him, Lord, we know not whither thou goest; and **how can we know the way**?
6 Jesus saith unto him, **I am the way, the truth, and the life**: no man cometh unto the Father, but by me.

Truth is knowledge. God KNOWS that Lived Experience is the best way of finding and knowing the truth. God is KNOWN by living Him and experiencing Him. Life, light, truth, and knowledge come to us from God the Father through His Son Jesus Christ.

Truth is knowledge. Truth is Lived Experience. We KNOW the things that we have lived and experienced! That's much more powerful than the Scientific Methods will ever be.

Mosiah 27: 36: Thus, they were instruments in the hands of God in bringing many **to the knowledge of the truth,** yea, **to the knowledge of their Redeemer.**

Ether 4: 12-13: "And whatsoever thing persuadeth men to do good is of me; for good cometh of none save it be of me. I am the same that leadeth men to all good; he that will not believe my words will not believe me — that I am; and he that will not believe me will not believe the Father who sent me. For behold, I am the Father, I am the light, and the life, and **the truth of the world**. Come unto me, O ye Gentiles, and I will show unto you the greater things, **the knowledge** which is hid up because of unbelief."

From the Lord's perspective, to be led astray means to be led away from the Church of Jesus Christ of Latter-day Saints, His Church, His priesthood power, His saving ordinances, His temple blessings, His Apostles, His people, His Saints, His Truth, and His Holy Spirit. That's what it truly means to be led astray; and, we KNOW from experience that the Prophets of God will never lead us astray, because they ALWAYS point us directly to the Church of Jesus Christ of Latter-day Saints, Christ's Own Church, Christ's Apostles, Christ's Saving Ordinances, Christ's priesthood power and priesthood blessings, and the Temple Ordinances.

This is how we discern truth from error. If it points us and leads us to the Church of Jesus Christ of Latter-day Saints, Christ's Church, His saving ordinances, His priesthood powers and keys and authority, His covenants, and His temple ordinances, then we KNOW that it is from God and Christ. If it points us to some other path or leads us down some other path, then we KNOW that it is from Satan and the evil spirits who follow him.

This is how I tell the difference between truth and error. It was simple, once I finally realized the truth of it, and extremely useful, powerful, and helpful. I KNOW, because I have lived it and experienced it for myself.

Truth is knowledge which is gained from our lived experiences.

1 Timothy 4: 1-3: "Now the Spirit speaketh expressly, that in the latter times some shall depart from the faith, giving heed to seducing spirits, and doctrines of devils; speaking lies in hypocrisy; having their conscience seared with a hot iron; forbidding to marry, and commanding to abstain from meats, which God hath created to be received with thanksgiving of them which believe and **know the truth.**"

Doctrine and Covenants 31: 2: Behold, you have had many afflictions because of your family; nevertheless, I will bless you and your family, yea, your little ones; and the day cometh that they will believe and **know the truth** and be one with you in my church.

Alma 24: 19: And thus we see that, when these Lamanites were brought to believe and to **know the truth**, they were firm, and would suffer even unto death rather than commit sin; and thus we see that they buried their weapons of war, for peace.

Moroni 10: 5: By the power of the Holy Ghost ye may **know the truth** of all things.

TRUTH is KNOWN by living it and experiencing it directly. God can reveal the truth to us, because God knows all truth, because God made reality or the truth. God IS the Truth.

Doctrine and Covenants 49: 2: Behold, I say unto you, that they desire to **know the truth** in part, but not all, for they are not right before me and must needs repent.

God can help us to overcome ALL things, except for a lack of desire. If we have NO desire to know the truth and to live the truth and to become the truth, then there's nothing that God can do for us. If we have no desire to repent or change for the better, God can't do anything to help us. God will never force the truth upon us. The truth and its associated knowledge have to be lived and owned by us, before it becomes useful and real to us. God knows this, which is why He will never force the truth upon us. We repent by turning to God and by choosing to embrace the truths which He freely gives us.

1 John 2: 3-4:

3 And hereby we do know that we **know** Him, if we keep His commandments.

4 He that saith, I know Him, and keepeth not His commandments, is a liar, and **the truth** is not in him.

The truth is known by living it and doing it. The truth is known by acting upon it and becoming it.

Helaman 15:7-8:

7 And behold, ye do **know** of yourselves, for **ye have witnessed it**, that as many of them as are brought to the **knowledge of the truth**, and to know of the wicked and abominable traditions of their fathers, and are led to believe the holy scriptures, yea, the prophecies of the holy prophets, which are written, which leadeth them to faith on the Lord, and unto repentance, which faith and repentance bringeth a change of heart unto them —

8 Therefore, as many as have come to this, ye **know** of yourselves are firm and steadfast in the faith, and in the thing wherewith they have been made free.

How did they know these things? By observing them! By living them and experiencing them! They witnessed it!

In order to get back to our Father in Heaven, we have to choose to follow someone who has already done so – Jesus Christ.

We KNOW the TRUTH by living the truth, doing the truth, keeping the truth, and becoming the truth. The Truth is KNOWN by observing it, witnessing it, living it, being it, and becoming it. The rebels and atheists don't have the truth in their lives, because they don't want it.

There are many more of these to be found in the Bible, the Book of Mormon: Another Testament of Jesus Christ, the Doctrine and Covenants, and the Pearl of Great Price. These are a sampling of a few of my favorites.

The following is probably the BEST discourse on **truth and knowledge** that has ever been given to us by the Biblical God Jesus Christ:

Doctrine and Covenants 93: 1-40:

1 Verily, thus saith the Lord: It shall come to pass that every soul who forsaketh his sins and cometh unto me, and calleth on my name, and obeyeth my voice, and keepeth my commandments, shall see my face and **know** that I am;

2 And that I am **the true light** that lighteth every man that cometh into the world;

3 And that I am in the Father, and the Father in me, and the Father and I are one —

4 The Father because he gave me of His fulness, and the Son because I was in the world and made flesh my tabernacle, and dwelt among the sons of men.

5 I was in the world and received of my Father, and the works of Him were plainly manifest.

6 And John saw and bore record of the fulness of my glory, and the fulness of John's record is hereafter to be revealed.

7 And he bore record, saying: I saw His glory, that He was in the beginning, before the world was;

8 Therefore, in the beginning the Word was, for He was the Word, even the messenger of salvation —

9 The light and the Redeemer of the world; **the Spirit of truth**, who came into the world, because the world was made by Him, and in Him was the life of men [physical body] and the light of men [psyche and spirit body].

10 The worlds were made by Him; men were made by Him; all things were made by Him, and through Him, and **of Him**.

11 And I, John, bear record that I beheld His glory, as the glory of the Only Begotten of the Father, **full of grace and truth**, even **the Spirit of truth**, which came and dwelt in the flesh, and dwelt among us.

12 And I, John, saw that He received not of the fulness at the first, but received grace for grace;

13 And He received not of the fulness at first, but continued from grace to grace, until He received a fulness;

14 And thus He was called the Son of God, because He received not of the fulness at the first.

15 And I, John, bear record, and lo, the heavens were opened, and the Holy Ghost descended upon Him in the form of a dove, and sat upon Him, and there came a voice out of heaven saying: This is my beloved Son.

16 And I, John, bear record that He received a fulness of the glory of the Father;

17 And He received all power, both in heaven and on earth, and the glory of the Father was with Him, for [the Father] dwelt in Him.

18 And it shall come to pass, that if you are faithful you shall receive the fulness of the record of John.

19 I give unto you these sayings that you may understand and **know** how to worship, and **know** what you worship, that you may come unto the Father in my name, and in due time receive of His fulness.

20 For if you keep my commandments you shall receive of His fulness, and be glorified in me as I am in the Father; therefore, I say unto you, you shall receive grace for grace.

21 And now, verily I say unto you, I was in the beginning with the Father, and am the Firstborn;

22 And all those who are begotten [reborn] through me are partakers of the glory of the same, and **are the church of the Firstborn**.

23 Ye were also in the beginning with the Father; that which is Spirit, even **the Spirit of truth** [sprit, psyche, truth is what we really are at our core being];

24 And **truth is knowledge** of things as they are, and as they were, and as they are to come;

25 And whatsoever is more or less than this is the spirit of that wicked one who was a liar from the beginning.

26 **The Spirit of truth is of God. I am the Spirit of truth**, and John bore record of me, saying: He received **a fulness of truth**, yea, **even of all truth**;

27 And no man receiveth a fulness unless he keepeth His commandments.

28 **He that keepeth His commandments receiveth truth and light, until he is glorified in truth and knoweth all things**.

29 Man was also in the beginning with God. Intelligence [psyche], or **the light of truth**, was not created or made, neither indeed can be.

30 **All truth is independent** in that sphere in which God has placed it, **to act for itself**, as all **intelligence** also; otherwise there is no existence. [Truth, psyche, intelligence is an independent entity, and the fundamental source and reality of our existence. Without psyche, there is no existence. Psyche acts for itself. Psyche is the ONLY thing that can. Psyche or intelligence is synonymous with the truth – the true reality of our existence. Psyche or intelligence is made of some kind of light. Intelligence or psyche is light and truth.]

31 Behold, here is the agency of man, and here is the condemnation of man; because that which was from the beginning is plainly manifest unto them, and they receive not the light.

32 And every man whose spirit receiveth not the light is under condemnation [and severely disadvantages himself, or damns himself, by stopping himself from finding and knowing the truth].

33 For man [human beings or human psyches] is spirit. [Each man or psyche has a spirit body]. The [physical] elements are eternal, and spirit [psyche and spirit body] and element [physical body], inseparably connected, receive a fulness of joy; [The physical elements are eternal, because they are made of a part of God's Psyche and are made from spirit matter, both of which are eternal in nature.]

34 And when separated [from his spirit body and his physical body], man [psyche] cannot receive a fulness of joy.

35 The elements [physical body] are the tabernacle of God; yea, man is the tabernacle of God, even temples; and whatsoever temple is defiled, God shall destroy that temple. [We destroy it ourselves, but God allows it to be destroyed and allows us to destroy it.]

36 The glory of God is **intelligence**, or, in other words, **light and truth**. [The glory of God is His Psyche or Non-Local Consciousness, or, in other words, light and truth. Intelligence or psyche is light and truth. Psyche is the true fundamental reality of our existence, and psyche is some type of light, a living light.]

37 Light and truth forsake that evil one. [Light and truth should reject the darkness].

38 Every **spirit of man** was innocent in the beginning; and God having redeemed man from the fall, men became again, in their [physical] infant state [as children], innocent before God.

39 And that wicked one cometh and taketh away light and truth [intelligence], through disobedience, from the children of men, and because of the tradition of their fathers. [Satan makes us stupid, and ignorant of the truth and the true reality of our existence.]

40 But I have commanded you to bring up your children in light and truth.

That's the best discourse on truth, knowledge, psyche, intelligence, spirit bodies, physical bodies, glory, and light that I have encountered so far in my life, and it came to us from the Biblical God Jesus Christ Himself – the pinnacle of truth and knowledge. It doesn't get any better than that.

I find it interesting whenever God says that the worlds were made "of Him", as He said above. Think about it! Only God would say such a thing. Only God could say such a thing.

Physicists have noticed that if they pour in a lot of energy, they can raise the electrons around an atom to a higher state or a higher level. Therefore, if we take a hydrogen atom, lower its electrons to a lower state or a lower level "creating" a hydrino or spirit matter or dark matter in the process, there should be a ton of energy released as a result.

What does this mean where God is concerned? It means that for every atom of physical matter that He brought into existence or commanded into existence, He had to pour some of His energy or His glory or His light or His life into it, in order to transform it from spirit matter into physical matter. Therefore, when God tells us that the worlds, the physical matter and the atoms, were created by Him, through Him, from Him, and of Him, He means it literally. There's a piece of God, a piece of

God's glory and light and consciousness, in each and every physical atom that He brings into existence. God KNOWS of what He speaks, because He lived it and experienced it. He gave of Himself to produce it! God had to pour Himself and His energy and His life into the creation of this physical universe in order to bring it into existence in the first place. God says as much here, and elsewhere.

<u>Doctrine and Covenants 88: 1-21</u>:

1 Verily, thus saith the Lord unto you who have assembled yourselves together to receive His will concerning you:

2 Behold, this is pleasing unto your Lord, and the angels rejoice over you; the alms of your prayers have come up into the ears of the Lord of Sabaoth, and are recorded in the book of the names of the sanctified, even them of the celestial world.

3 Wherefore, I now send upon you another Comforter, even upon you my friends, that it may abide in your hearts, even the Holy Spirit of promise; which other Comforter is the same that I promised unto my disciples, as is recorded in the testimony of John.

4 This Comforter is the promise which I give unto you of eternal life, even the glory of the celestial kingdom;

5 Which glory is that of the church of the Firstborn, even of God [the Father], the holiest of all, through Jesus Christ his Son!

6 He that ascended up on high, as also He descended below all things, in that He comprehended all things, that He might be **in all and through all things, the light of truth** [God is Reality, Truth, and Existence. That is the Truth of our existence. A bit of God's consciousness or life is in the whole of it];

7 Which **truth shineth! This is the light of Christ**. As also **He is in the sun**, and the light of the sun, and **the power thereof by which it was made**.

8 As also **He is in the moon**, and is the light of the moon, and the power thereof by which it was made;

9 As also the light of the stars, and **the power thereof by which they were made**;

10 And the earth also, and **the power thereof**, even the earth upon which you stand.

11 And the light which shineth, which giveth you light, **is through Him** who enlighteneth your eyes, which is the same light that quickeneth [enlivens] your understandings;

12 Which **light proceedeth forth from the presence of God to fill the immensity of space**; [God is speaking of dark energy here, the zero-

point field, the quantum sea of light, or the light of Christ. Dark energy or the light of Christ counteracts gravity and causes our universe to stretch, spread, and expand. Dark energy or the light of Christ also keeps the atoms from imploding and becoming spirit matter once again. Spirit matter is Dark Matter. Dark energy, the light of Christ, the Word of Command, or Conscious Observation is what keeps the electrons in their orbits – creating and then sustaining physical atoms. Dark energy or the light of Christ is what keeps this universe from imploding back into the infinite singularity from whence it came! Imagine it! It ALL comes from God. God had to pour Himself, His energy and light and life, into this physical universe in order to convert parts of it, 4.6% of it, to physical matter and to organize that physical matter into planets and stars, and to cause this universe to expand so that we would have space. They say that an atom is 99.999% empty space; and, it is God who is creating that space through His light, power, glory, life, command, and will. God's light and life are needed to keep those electrons in orbit, because positive attracts negative! An atom should implode, but it doesn't. God's light and life and psyche is preventing each atom from imploding! God's light and life are needed to keep this universe expanding and growing. God's light and life are needed, in order to choose what parts of that primal spirit matter to convert into physical matter, stars, planets, gas clouds, and galaxies. Galaxies come into existence whole. Galaxies aren't grown. They spring into existence as a unit or a whole, and then start to wind up. That's what we observe. If God didn't intervene in just the right spots to create stars, planets, and galaxies, then this universe would be nothing but one huge homogeneous ball of gas. Because according to Big Bang Theory every atom is supposedly moving away from every other atom, there could be no planets, stars, and galaxies without God's intervention and direction – choosing which parts of this universe to convert to physical matter and which parts of this universe to leave as dark matter or spirit matter. I think God gave this whole thing some thought and influence. God brought this whole thing to life. There is no other way to account for it logically.]

13 **The light which is in all things, which giveth life to all things, which is the law by which all things are governed, even the power of God who sitteth upon His throne, who is in the bosom of eternity, who is in the midst of all things.**

14 Now, verily I say unto you, that through the redemption which is made for you is brought to pass the resurrection from the dead.

15 And the **spirit** and the body are the soul of man. [The **spirit** or ghost is comprised of our psyche and our spirit body as a united whole. The Gods use the terms "intelligence" and "spirit" interchangeably because from their point of view, the unification of psyche and spirit body is a done deal. The soul of man is comprised of our physical body and our **spirit** or ghost].

16 And the resurrection from the dead is the redemption of the soul [spirit and physical body].

17 And the redemption of the soul is through him that quickeneth [enlivens] **all thing**s, in whose bosom it is decreed that the poor and the meek of the earth shall inherit it.

18 Therefore, it must needs be sanctified from all unrighteousness, that it may be prepared for the celestial glory;

19 For after it hath filled the measure of its creation, it shall be crowned with glory, even with **the presence of God the Father**;

20 That bodies who are of the celestial kingdom may possess it forever and ever; for, for this intent was it made and created, and for this intent are they sanctified.

21 And they who are not sanctified through the law which I have given unto you, even the law of Christ, must inherit another kingdom, even that of a terrestrial kingdom, or that of a telestial kingdom.

God's Light, Life, and Glory is within each atom in our physical bodies. God's programming skills are manifest in our genomes; and therefore, God's signature is written on every living cell in our physical body within the software that is our genome. God has filled us with Himself and written Himself upon us. It doesn't get more personal than that. Think about it if you can. God KNOWS the truth because He is the truth and because He makes the truth and becomes the truth. God KNOWS reality, because God makes reality. Reality is made up of God and came from God.

As physical mortal beings, we don't think about these things nor have much of an appreciation for these things. But God does, because He had to give of Himself in order to bring physical matter into existence in the first place. God had to make a sacrifice of Himself in order to bring this physical universe into existence and in order to bring us to life.

The Truth is KNOWN by living it and experiencing it! This reality applies as much to God as it does to us.

You come to have faith in the truth by living it and acting upon it. Over time, that faith transforms into knowledge, because it is based upon the truth. Truth is knowledge; and, knowledge is truth, because it is impossible to know the truth based upon falsehoods and lies. Lived Experience or Observation is the pinnacle of knowledge and truth.

I find it interesting how a change in philosophical perspective makes things like Psyche and God's Psyche useful, feasible, realistic, credible, and even necessary. Everything points to Psyche or Intelligence, except for the exclusionary philosophies and sciences which deliberately and knowingly exclude Psyche or Non-Local Consciousness from consideration. But, even they point us to Psyche by trying to steer us away from Psyche.

I have learned to trust the Lived Experiences and Observations of the human race and the prophets of God, because Lived Experiences are our ONLY way of finding and knowing the truth. You can't find and know the truth through the Scientific Methods, because the Scientific Methods are based upon a wide variety of logic fallacies and philosophical assumptions which prevent the Scientific Methods from being able to prove the truth.

The TRUTH is KNOWN by observing it, living it, witnessing it, and experiencing it for ourselves – not through philosophical speculation or the philosophical interpretation of scientific evidence. Lived Experience IS the BEST way of finding and knowing THE TRUTH.

Doctrine and Covenants 76: 22-24

22 And now, after the many testimonies which have been given of him, this is the testimony, last of all, which we give of him: That he lives!

23 For we saw him, even on the right hand of God; and we heard the voice bearing record that he is the Only Begotten of the Father;

24 That by him, and through him, and of him, the worlds are and were created, and the inhabitants thereof are begotten sons and daughters unto God.

Due to the inherent weaknesses of the Scientific Methods, Lived Experience or Observation is technically the ONLY way of finding and knowing the truth. It's elementary, my dear friend!

I think that this is the greatest scientific discovery of my career as a scientist, theoretician, psychologist, observer, and philosopher. Lived Experience or Observation is the pinnacle of knowledge and truth. We KNOW the Truth by living it and experiencing it for ourselves, or by choosing to trust someone who has.

Doctrine and Covenants 46: 13-14: "To some it is given by the Holy Ghost to **know** that Jesus Christ is the Son of God, and that He was crucified for the sins of the world. To others it is given to believe on their words, that they also might have eternal life if they continue faithful."

Experience or knowledge is a better way of knowing the truth than philosophical speculation or scientific experimentation. For some, knowledge and truth are revealed to them by God. They are eyewitnesses. Others choose to believe on their words. I have tended to be in the latter camp, whenever I haven't been deceiving myself, although the other kind of knowledge has been coming online for me in recent years because I chose to open myself up to it and chose to ask God for it. I have a testimony of asking, knocking, and seeking because I have observed that it leads to receiving and knowing of The Truth, who is God.

The following scriptures sum up my life, during the past five years of my life:

Doctrine and Covenants 88: 118: "And as all have not faith, seek ye diligently and teach one another words of wisdom; yea, seek ye out of the best books words of wisdom; seek learning, even by study and also by faith."

Doctrine and Covenants 46: 13, 14: "To some it is given by the Holy Ghost to know that Jesus Christ is the Son of God, and that He was crucified for the sins of the world. To others it is given to believe on their words, that they also might have eternal life if they continue faithful."

I didn't have a lot of faith, but I definitely know how to study and do research. I'm not a prophet or seer or clairvoyant, but I have learned to believe and trust the people who are. I have learned to trust and believe their Lived Experiences.

The purpose of life is to make friends, and to seek The Truth and find The Truth. I have been doing a little bit of both ever since I got sober and gave up My Materialism, My Nihilism, My Scientism, and My Atheism.

When it comes to the therapeutic benefits of a Relational Ontology and the truthfulness of a Psyche Ontology, and when it comes to the powerful effects of the Atonement of Christ for physical healing and for psyche-therapy, I KNOW these things to be real and true, because I have lived them and experienced them for myself. It doesn't get any better than that!

Faith in the unseen becomes KNOWLEDGE of the unseen through Lived Experience. I KNOW that the non-physical or the non-local exists, because I have experienced parts of it for myself and because I have chosen to trust those who have experienced all of it for themselves. When it comes to the non-local or the non-physical or the spiritual, I have chosen to believe in and trust the Lived Experiences of the human race, because our collective Lived Experiences or Psyche Experiences PROVE that these non-local realities do in fact exist.

The Biblical God Jesus Christ really isn't hiding from us. Thousands, if not millions, have experienced Him directly through their Lived Experiences. However, there are in fact millions if not billions of Materialists, Naturalists, Nihilists, Darwinists, Behaviorists, and Atheists who are trying to hide from the Biblical God Jesus Christ and believe that they are having some success in doing so. I used to be one of them, so I KNOW. I lived it and experienced it.

PART IX — MY SCIENTIFIC DISCOVERIES

1. One of my greatest scientific observations and scientific discoveries came when I finally realized that Materialism, Naturalism, and Darwinism have been falsified trillions of different times in thousands of different ways. That was a major conceptual breakthrough for me, because I used to be a Materialist, Nihilist, and Atheist.

Materialism, Naturalism, and Darwinism are BEST defined as "Design and Creation by Physical Matter". These people literally believe and teach that the rocks – physical reactions – designed, programmed, created, engineered, field-tested, manufactured, and deployed ALL of the different genomes and life forms on this planet. That's what these people really believe, and that's what these people teach in all of our public classrooms.

Ironically, Louis Pasteur falsified Materialism, Naturalism, Darwinism, and the Theory of Evolution in 1859 by demonstrating that the rocks – raw physical matter – cannot design and create. That was the same year that Charles Darwin published, "On the Origin of Species". Louis Pasteur proved the Theory of Evolution or Creation by Rocks false the same year that Charles Darwin introduced his theory to the world. Remember, Science and the Scientific Methods can be used to prove theories false. It's called falsifying a theory; and, that's what the Scientific Methods do, or are supposed to do.

Technically, the Scientific Methods can't be used to prove anything true. Due to a wide variety of different logic fallacies which are built directly into the Scientific Methods, the Scientific Methods cannot be used to prove a theory true, as the Materialists, Naturalists, and Darwinists try to do.

Instead, we can use the Scientific Methods to falsify theories. Using the Scientific Methods to prove our theories false is philosophically and logically sound. In other words, if you falsify a theory by *negating the consequent*, you have in fact proven that theory false.

Here's how it works in principle:

Scientific Hypothesis: If Theory X is true, then we will observe Y.

Scientific Observations: We don't observe Y.

Scientific Conclusion: Therefore, Theory X is not true.

Here's how it works in practice:

Scientific Hypothesis: If the Theory of Evolution, Materialism, Naturalism, and Darwinism are true, then we will observe the rocks and physical reactions designing, creating, and manufacturing genomes and life forms from scratch.

Scientific Observations: We have NEVER observed the rocks and physical reactions designing and creating genomes and life forms; and, we NEVER will. They can't.

The Scientific Conclusion: Therefore, the Theory of Evolution, Materialism, Naturalism, and Darwinism are not true.

I just successfully falsified Materialism, Naturalism, Darwinism, and the Theory of Evolution; and, I used the Scientific Method to do so! Best of all, my argument is philosophically and logically sound. I have in fact falsified Materialism, Darwinism, Naturalism, and the Theory of Evolution for REAL! I just used the Scientific Method to PROVE these things false. I successfully used the Scientific Methods for what they are good for – falsifying theories that are false. I wish I would have known how to do that forty years ago. It would have saved me a lot of confusion and grief.

In contrast, there is NO way to falsify theories that are true, such as "Design and Creation by Intelligent Beings or by Intelligent Psyches". Instead, true theories are continuously verified over and over again. "Design and Creation by Psyche or by Intelligence" has been observed and verified and experienced trillions of trillions of different times and ways, with an infinite number of more to go. See how that works? That's Science and the Scientific Methods in action for REAL, doing what they are best at!

(See: Slife, B. D. & Williams, R. N. (1995). Science and Human Behavior. In *What's Behind the Research? Discovering Hidden Assumptions in the Behavioral Sciences*, (pp. 167–204). Thousand Oaks, CA: SAGE Publications.

http://mypsyche.us/wp-content/uploads/2017/04/Science.pdf).

This has been and is one of my greatest scientific discoveries and scientific observations of all time, during the whole of my science career. I have observed that there are literally thousands of different ways to falsify Creation by Rocks – or Materialism, Naturalism, Darwinism, and the Theory of Evolution; and, they are all philosophically and logically sound. Thanks to the Scientific Methods and scientific observations, we KNOW that Creation by Rocks, Materialism, Naturalism, and Darwinism are false, because we KNOW why they are false – rocks and physical reactions cannot design and create. It's elementary my dear friend.

There's NO way that Evolution (Mutation and Selection) could have designed, programmed, engineered, created, and manufactured the first genomes and life forms, because Evolution didn't even exist until AFTER God had designed, created, and deployed the first genomes and life forms in the first place.

I have observed that the Scientific Methods or Observation have been used thousands of different ways to falsify Materialism, Naturalism, Darwinism, and even Atheism. Atheism is Creation by Nothing, or Creation by Chance. The Atheists really truly believe that Nobody and Nothing designed and created everything. Nobody, Nothing, and Chance are the holy trinity of Atheism. These people believe in those false gods or idols with a passion. Materialism, Naturalism, and Atheism are in fact our modern-day form of idolatry; and, these people are idolaters. These people worship the rocks and physical reactions, as if these things were God.

Ironically, God must of necessity exist in order to have done ALL of the science that needed to be done, which the rocks or raw physical matter could never have done. Only Psyche can design and create; and, God's Psyche is the only one we know of who was there at the time and could have done the job.

Finally, as a capstone, the Biblical God Jesus Christ has told us repeatedly in the Bible, Book of Mormon: Another Testament of Jesus Christ, the Doctrine and Covenants, and the Pearl of Great Price that HE designed and organized the heavens, this earth, and all of the life forms on this earth. The Biblical God confessed to doing the job.

For me personally, falsifying Materialism and Darwinism became my first really convincing Scientific Proof of God's Existence. After falsifying Materialism, Naturalism, and Darwinism, I simply KNEW that God exists. I was finally willing to follow the evidence, wherever it might lead me.

(See: Mark My Words. (2016). *The Theory of Evolution Proved to Me that God Exists: Why I Am No Longer an Atheist and Why I No Longer Believe in the Theory of Evolution*. Kindle. Retrieve from: https://www.amazon.com/dp/B01HZYBZ7K).

(See also: Mark My Words. (2016). *The Scientific Method: Proves That the Theory of Evolution Is False*. Kindle. Retrieve from: https://www.amazon.com/dp/B01IAAIRT2).

2. My greatest scientific observation and scientific discovery is the realization that Lived Experience IS the BEST way of finding and knowing the truth. Lived Experiences or Psyche Experiences are the best way of finding the truth and knowing the truth in EVERY realm of existence, including this physical realm.

Due to the fact that the Scientific Methods can't be used to find the truth and prove the truth directly, and due to the fact that the Scientific Methods are typically restricted to the physical realm or the local realm, and due to the fact that there are a wide variety of logic fallacies built into the Scientific Methods, the Scientific Methods are in fact a much weaker source of knowledge and truth than Lived Experiences are. Lived Experiences or Psyche Experiences are in fact an infinitely better way of finding and knowing the Truth than science and the Scientific Methods can ever be. I wish I would have known that forty years ago. It would have saved me a lot of frustration and grief.

My greatest scientific discovery of all time is that the Lived Experiences of the human race are in fact a much better way of finding and knowing the truth than Science and the Scientific Methods. For me, that was a major epiphany and scientific breakthrough, because I used to be a Materialist, Nihilist, and Atheist.

Lived Experience is extremely powerful. Lived Experience is Science, Observation, Knowledge, and Truth in their purest form. Science is supposed to be

Observation and Knowledge and the pursuit of the Truth. In other words, Science is supposed to be Lived Experience. The Materialists, Naturalists, Behaviorists, and Atheists did the world a great disservice when they hijacked and stole Science and limited Science exclusively to physical matter. One of my greatest scientific discoveries was to finally realize how wrong these people really are.

I KNOW how it goes, though, because I allowed My Scientism, My Nihilism, My Materialism, and My Atheism to blind me to Lived Experience as a source of evidence and to blind me as to how powerful and useful Lived Experiences really are as a source of knowledge and truth. But, that's what we Materialists and Atheists do. We refuse to allow Lived Experiences into evidence. Materialism and Atheism of any kind is based upon a refusal to look at evidence.

I truly believed that Science and the Scientific Methods were the ONLY way for finding and knowing the truth. I WAS WRONG! In fact, the Scientific Methods can't be used to prove the truth, which means that Lived Experience is in reality the ONLY way to find and KNOW the truth directly. The truth is KNOWN by living it and experiencing it for yourself, or by choosing to trust someone who has.

I never fully realized until April 2017 that Lived Experience or Direct Observation is a much better way of finding and knowing the truth than the Scientific Methods and philosophical guesswork will ever be; but, I'm seeing it now and understanding it now.

We KNOW from the Lived Experiences of the human race that the Biblical God Jesus Christ does indeed exist and truly rose from the dead – although it's nice to have Science pointing us to the same truths, in its roundabout way.

(See: *Bible, Book of Mormon: Another Testament of Jesus Christ, Doctrine and Covenants,* and *Pearl of Great Price* – available for free at:

https://www.lds.org/scriptures/bible?lang=eng).

3. During the Prime Event, or what most scientists call the Big Bang, only 4.6% of this universe was converted to physical matter. According to Astrophysics, the other 95.4% of this universe remained Non-Local or Spiritual as dark matter (spirit matter) and dark energy (the Light of Christ or the zero-point field of light).

The Materialists, Naturalists, Darwinists, and Atheists are WRONG about 95.4% of this universe! Imagine it! 95.4% of this universe is still spiritual or non-local. Doesn't it make you wonder what else the Materialists might be wrong about?

(See: Ross, H. (2008). *Why the Universe Is the Way It Is.* Grand Rapids, MI: Baker Books.)

There are indications that a lot of energy has to be poured into spirit matter in order to convert spirit matter to physical matter. In other words, God poured a tiny bit of His own Psyche, Light, Intelligence, Power, Consciousness, Energy, or Life into each physical atom that God brought into existence or commanded into existence. ONLY God could have done such a thing. We human beings certainly don't know how to do that kind of Science!

Because of this KNOWLEDGE, I'm now willing to admit Lived Experiences into evidence.

(See: https://www.lds.org/scriptures/dc-testament/dc/93?lang=eng).

(See: https://www.lds.org/scriptures/dc-testament/dc/88?lang=eng).

(See: https://www.lds.org/scriptures/dc-testament/dc/76?lang=eng).

(See: https://www.lds.org/scriptures/dc-testament/dc/130?lang=eng).

4. One of my greatest scientific discoveries came when I first realized that Quantum Mechanics is in fact Spiritual Mechanics, or the way that spirit matter really works. When I first realized that Psyche or a Non-Local Conscious Observer and the Word of Command are needed to convert spirit matter into physical matter, then suddenly the whole of Quantum Mechanics became clear to me.

God must of necessity exist in order to have provided the necessary Word of Command or Conscious Observation, which Quantum Mechanics tells us must take place in order to convert spirit matter into physical matter. The Bible actually tells us that Jesus Christ is the WORD or the Word of Command who brought physical matter into existence in the first place. If God's Psyche didn't exist, then there would be NO physical matter.

For me personally, Quantum Mechanics became my first positive Scientific Verification of God's Existence. The scientists keep verifying Quantum Mechanics or Spiritual Mechanics over and over again; so, it must be the truth, because it's impossible to use the Scientific Methods to falsify the truth. Quantum Mechanics or Spiritual Mechanics has never been falsified, although the materialistic interpretations of Quantum Mechanics have been falsified trillions of times.

After realizing that God's Psyche must of necessity exist in order to convert spirit matter into physical matter, I simply KNEW that God exists.

(See: Van Lommel, P. (2010). *Consciousness Beyond Life: The Science of the Near-Death Experience*. New York: HarperCollins.)

(See: Goswami, A. (2008). *God Is Not Dead: What Quantum Physics Tells Us about Our Origins and How We Should Live*. Charlottesville, VA: Hampton Roads.)

(See also: Mark My Words. (2016). *Quantum Mechanics from a Non-Physical Spiritual Perspective*. Kindle. Retrieve from: https://www.amazon.com/dp/B01J023TGU).

The existence of Light falsifies Materialism and Naturalism, and it ends up pointing us directly to God and God's Psyche, because psyches or conserved quanta are made from energy or light. The glory of God is intelligence, or light and truth.

(See: Mark My Words. (2017). *God Is in the Light: God is light, and in Him is no darkness at all*. Retrieve from: https://www.amazon.com/dp/B07168S37N).

5. Some of the scientists have been smart enough and open-minded enough to observe that a genome is a radically advanced software program, that DNA is in fact like a radically advanced hard drive or memory storage device, and that a living cell is a radically advanced piece of computer hardware or nanotechnology. We, who have made this realization, also realize that ONLY Psyche or Intelligence is capable of designing, creating, engineering, and manufacturing hard drives and hardware; and, ONLY Intelligent Beings or Psyche Beings can do computer programming or the creation of software and genomes.

God must of necessity exist in order to have designed, engineered, manufactured, and created all of the DNA molecules, proteins, and living cells on this planet. These things don't put themselves together. They have never been observed doing so! In fact, Louis Pasteur used Science in 1859 to prove that these things can't put themselves together, by falsifying Spontaneous Generation or Materialism and Creation by Rocks.

Furthermore, God must of necessity exist in order to have written or programmed each and every genome that is encoded as software in the DNA of each life form on this planet. ONLY Psyche or intelligent beings could have done such a thing. We have NEVER caught the rocks in the act of designing, programming, engineering, and manufacturing genomes and life forms from scratch; and, we never will. They can't!

A Genome IS God's Signature!

What's most interesting about all of this is that God wrote His signature in each one of your cells. Your genome IS God's signature. Each genome on this planet is God's signature. God has written Himself upon you and within every life form on this planet. The rocks or raw physical matter can't write genomes or computer software, but Psyche certainly can. God's Psyche must of necessity exist, or you wouldn't have a genome. God's Psyche must exist, or you wouldn't have a physical body. That's what Science is trying to teach us!

(See: Mark My Words. (2016). *Using the Scientific Method: To Eliminate the Usual Suspects and to Prove the Truth*. Kindle. Retrieve from: https://www.amazon.com/dp/B01J6STHP0).

6. In many of his books, Hugh Ross uses Cosmic Fine-Tuning as Scientific Proof of God's Existence. The whole thing made logical sense to me, because ONLY Psyche can convert or transmute spirit matter into physical matter. ONLY Psyche can organize physical matter into useful and productive forms such as planets, stars, galaxies, genomes, and life forms. ONLY Psyche can do Cosmic Fine-Tuning. God's Psyche must of necessity exist, in order to have done all of the cosmic fine-tuning and science that needed to be done; otherwise, you and I would not be here right now in this physical realm. Science has made it obvious and clear to me that

it must be so. It's obvious that the rocks or raw physical matter can't do Cosmic Fine-Tuning. That kind of process requires an active, living, agentic, conscious, intelligent Being or Psyche, whom many of us tend to call God or the Ultimate Scientist.

(See: http://www.reasons.org/articles/fine-tuning-for-life-in-the-universe).

7. Many different scientists taught me that Random Mutations and Natural Selection (Evolution) cannot design and create genomes and life forms. They met their burden of proof and demonstrated to me that it must be so. Once we have eliminated all of the falsehoods such as Materialism, Naturalism, Atheism, Nihilism, and Darwinism, then we are left staring at THE TRUTH, which is that ONLY Psyche can design, program, engineer, manufacture, create, and do science.

By eliminating all of the falsehoods or pseudo-sciences, it becomes obvious that God's Psyche must of necessity exist in order to have DONE all of the Science, which evolution and the rocks could NEVER have done.

(See: Wells, J. (2000). *Icons of Evolution: Science or Myth? Why Much of What We Teach About Evolution Is Wrong*. Washington, DC. Regnary.)

(See: Sanford, J. (2014). *Genetic Entropy* (4th ed.). Cornell University: FMS Foundation.)

(See: Sanford, J. C., Marks, R. J., Behe, M. J., Dembski, W. A., & Gordon, B. L. (Eds.). (2013). *Biological Information: New Perspectives*. Hackensack, NJ: World Scientific.)

(See: Meyer, S. C. (2010). *Signature in the Cell: DNA and the Evidence for Intelligent Design*. New York: HarperCollins.)

(See: Meyer, S. C. (2013). *Darwin's Doubt: The Explosive Origin of Animal Life and the Case for Intelligent Design*. New York: HarperCollins.)

(See: Mark My Words. (2016). *The Scientific Method: Proves That the Theory of Evolution Is False*. Kindle. Retrieve from: https://www.amazon.com/dp/B01IAAIRT2).

(See: Mark My Words. (2016). *The Theory of Evolution Proved to Me that God Exists: Why I Am No Longer an Atheist and Why I No Longer Believe in the Theory of Evolution*. Kindle. Retrieve from: https://www.amazon.com/dp/B01HZYBZ7K).

8. Neuroscientists and brain stimulation have demonstrated that Psyche or Mind is a completely different entity than the physical brain. Any Science which points us to Psyche and Mind and Quantum Nonlocality is going to end up being vastly superior to the watered-down stuff that we get from the Materialists, Naturalists, and Atheists.

(See: Penfield, W. (1978). *The Mystery of the Mind: A Critical Study of Consciousness and the Human Brain*. Princeton, NJ: Princeton University Press.)

(See: Eccles, J. C., & Popper, K. R. (1977). *The Self and Its Brain: An Argument for Interactionism*. New York: Routledge.)

(See: Eccles, J., & Robinson, D. N. (1984). *The Wonder of Being Human: Our Brain and Our Mind*. New York: The Free Press.)

(See: Eccles, J. (1985). *Mind and Brain: The Many-Faceted Problems*. New York: Paragon House Publishers.)

These scientists solved the mind-brain problem over thirty years ago; but, the Materialists and Atheists refuse to read the memo. Materialism and Atheism of any kind are based upon a refusal to look at evidence. I KNOW, because I used to be a Materialist and an Atheist, until I started looking at the evidence.

Once you KNOW that Psyche or Mind exists, then you are suddenly free to go looking for the Mind of God.

9. Any Science that demonstrates Mind over Matter is going to point us directly to Psyche and establish the fact that Psyche is something completely different than matter, whether we are talking about spirit matter or physical matter.

The Placebo Effect is scientific proof of Psyche's existence.

(See: Dispenza, J. (2014). *You Are the Placebo: Making Your Mind Matter*. USA: Hay House Inc.)

(See also: McTaggart, L. (2002). *The Field: The Quest for the Secret Force of the Universe*. New York: HarperCollins.)

A complete understanding of Quantum Mechanics explains how God's Psyche was able to convert or transmute spirit matter into physical matter at the locations in this universe where He wanted there to be physical matter. It explains how the galaxies came into existence all at once!

If the Big Bang Theory were 100% true, then every particle in this universe should be moving away from every other particle in this universe; and, there would be NO planets, stars, and galaxies! This universe should be nothing more than one huge homogeneous ball of gas. It required God's Psyche in order to transmute spirit matter into physical matter at the current LOCATION of the planets, stars, and galaxies. God's Psyche is the ONLY way to explain the existence of planets, stars, and galaxies, because they shouldn't exist at all if the Big Bang Theory as typically presented to us were 100% true.

The Big Bang Theory "proves" that this physical universe had a beginning. That's the part of the Big Bang Theory that I truly believe in. However, the Big Bang Theory FAILS to explain how the stars, planets, and galaxies came together;

and, for an explanation of that reality we have to turn to God's Psyche and Mind-over-Matter.

A complete understanding of Mind over Matter ends up explaining how God's Psyche was able to organize planets, stars, galaxies, genomes, and life forms from raw unorganized physical matter. These are all things which the rocks or raw physical matter could NEVER have done!

For me personally, John Pratt's sacred calendars provided Scientific Proof of God's Existence and scientific proof of mind over matter at a cosmic scale:

http://www.johnpratt.com/items/docs/lds/dates.html

http://www.johnpratt.com/items/docs/article_nums.html

That scientist certainly had a vision, which would ONLY have been possible by choosing to believe in God's existence in the first place.

(See: https://www.lds.org/scriptures/pgp/abr?lang=eng).

You particularly want to look at Abraham 4: 10, 12, 18, 21, and 25; where the Gods watched and waited to make sure that the physical matter obeyed them, during the organization and terraforming of this earth.

The Gods are Scientists, not magicians! There is NO such thing as Creation Ex Nihilo or magic.

(See also: Mark My Words. (2017). *I Am Not a Creationist: So What Am I?*

https://www.amazon.com/dp/B071XTM8XY).

10. The book, *Adventures Beyond the Body: How to Experience Out-of-Body Travel*, by William Buhlman was a game-changer for me. Here was my first encounter with someone who can go out-of-body basically "at will". He turned the whole experience into a Science. I'm talking about a REAL SCIENCE of direct observation, experimentation, and lived experience – NONE of that garbage where they deliberately restrict their "science" to physical matter.

This book about Out-of-Body Experiences (OBEs) provided me with some of my very first evidentiary PROOF that Materialism and Naturalism are false. Buhlman ran science experiments and tried things out while out-of-body on the astral plane or in the spirit realm. He did REAL SCIENCE, doing direct observation of the spirit realm and direct experimentation with the spirit realm, while out-of-body. I thought this whole thing was really cool and fascinating.

With Lived Experiences, you go directly to KNOWLEDGE, TRUTH, and PROOF. Lived Experiences are vastly superior to the Scientific Methods for getting at truth and proof and knowledge. Technically, you can't use the Scientific Methods to prove anything due to the wide variety of logic fallacies and biases which are built into the Scientific Methods. But, you can definitely use Lived Experiences or Direct Observations as knowledge and proof of the truth.

This was the first book to help me see what Science should really be all about – Observation and Lived Experience, which document KNOWLEDGE and PROVE the TRUTH.

Buhlman helped to set me up, so that I was ready to take the NDErs and their out-of-body experiences seriously.

There are at least 13 million documented cases of Near-Death Experiences (NDEs) in the record of Lived Experience for the human race. The Materialists and Naturalists have chosen to deny and reject every one. Denial of evidence and rejection of evidence is NOT science! But, this is exactly the way that the Materialists and Atheists do science, by denying and rejecting evidence or Lived Experience.

The discovery of all those different NDEs and OBEs is in fact one of my greatest scientific discoveries!

The best way to KNOW God is to live Him and experience Him directly, or to choose to trust someone who has. Science and Scientific Methods can point us to God as the BEST EXPLANATION for the scientific evidence. However, through Lived Experience we can KNOW the truth and KNOW God directly if we want to.

That's very powerful Science there!

(See: Buhlman, W. L. (1996). *Adventures Beyond the Body: How to Experience Out-of-Body Travel*. New York: HarperCollins Publishing.)

(See: Durham, E. (1998). *I Stand All Amazed: Love and Healing from Higher Realms*. Orem, UT: Granite Publishing and Distribution.)

(See also: Storm, H. (2000, 2005). *My Descent into Death: A Second Chance at Life*. USA: Random House.)

Conclusion: All of this was a long time in coming, fifty-five years, but I finally got there in the end. Still, I wish I would have known about all of this stuff forty years ago. It would have saved me a lot of time and grief.

ALL of my greatest scientific discoveries and scientific observations have pointed me directly to God, because ONLY God's Psyche could have done ALL of the science which needed to be done at the time that these various things came into existence. This is obvious and clear to me now; but, it wasn't when I was a Materialist, Nihilist, and Atheist. Go with the BEST and get rid of all the rest. That's what I finally decided to do; and, you can too. All of this came open to my view, once I got rid of My Materialism, My Scientism, My Nihilism, and My Atheism.

If you really want to do Science, you start by getting rid of the pseudo-sciences, such as Darwinism, Materialism, and Naturalism. Then you switch over to Observation or Lived Experience for your scientific evidence and your Science. Through Lived Experiences, including spiritual experiences, you can go directly to KNOWING the TRUTH which is something that you can't do with the Scientific Methods and philosophical speculation. I never realized how powerful and revelatory Lived Experience can be, until after I got rid of My Materialism, My Atheism, My Nihilism, and My Scientism.

I never realized how weak and insubstantial the Scientific Methods and science really are until AFTER Joseph Rychlak, Brent Slife, and Edwin Gantt pointed out all the different logic fallacies that are built into the Scientific Methods. These are concepts which you will never get from your materialistic and atheistic college professors, because these people want you believing that science and the Scientific Methods are infallible.

There was a time in my life when I truly believed that Science and the Scientific Methods are the ONLY way of finding and knowing the truth. Boy, was I wrong! Over fifty years of being wrong! I'd been duped! Due to all the different logic fallacies that are built into the Scientific Methods, we can't use the Scientific Methods to prove the truth and to know the truth. If we want to KNOW the TRUTH, then we have to switch over to Lived Experience or Direct Observation – or choose to trust someone who has had the kinds of Lived Experiences that we wish we would have had. I KNOW of what I speak, because I have lived it and experienced it, on both sides of the fence.

A Philosophy of Science for Personality Theory or Psyche Theory

INDIVIDUAL QUANTUM OBJECTS

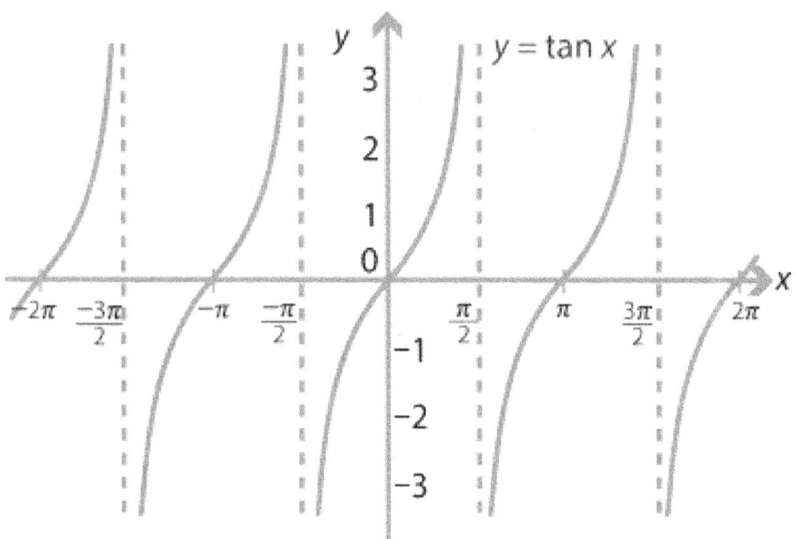

A Psyche Ontology is based upon the idea that Psyche, or Quantum Non-Local Consciousness, is the fundamental unit of reality. If there really is such a thing as a point particle or an infinite singularity, Psyche is it. Psyche or consciousness is the "elementary particle" upon which everything else is based. Psyche determines or chooses everything else.

I use **y = tan(x)** to mathematically model my Psyche Ontology or my Ultimate Cause Model of Reality, quantum objects, spirit matter, physical matter, spirit bodies, physical bodies, universes, infinite singularities, psyche, and non-local consciousness. Each wave on the graph is a different Quantum Object, or a different particle of matter, or a different universe.

This model extends all the way from the infinitely small to the infinitely large; and, I was extremely satisfied with its explanatory power when it was revealed to me.

A single Universe, or Quantum Object, or Particle of Matter, and/or its lifeline exists in the domain between the -π/2 and π/2 asymptotes. Matter or a Quantum Object has some limitations. Matter in any format, whether spiritual or physical, reacts. There's a time lag associated with Matter or a Quantum Object, because it waits for and then reacts to Psyche's Word of Command.

When a psyche or consciousness enters a non-consensus reality, that spiritual reality conforms itself to that psyche's demands and commands. In contrast, a physical reality is the ultimate consensus reality. It doesn't reshape itself automatically to the demands of our psyche. Instead, we have to work it in order to reshape it.

Again, a single Universe, or Quantum Object, or Particle of Matter (and its size or the amount of space it takes up) exists in the domain between the $-\pi/2$ and $\pi/2$ asymptotes. That Quantum Object, or matter, or particle, or wave, or material is spiritual and takes up little space when it is below (0, 0) on the graph; and, that same Quantum Object or matter is physical and can theoretically take up infinite space and become infinite in size when it is above (0, 0) on the graph above – although it's prevented from reaching infinite size and infinite velocity by the restraints of physical law, on the physical side of the equation.

Psyche resides ON the $-\pi/2$ and $\pi/2$ asymptotes. Think about it. That means that Psyche or Non-Local Consciousness has NO size and takes up NO space, yet is simultaneously infinite in size, range, scope, and presence. It's a paradox, but it's Reality. It's one eternal round! Psyche goes down the rabbit hole and climbs the stairway to heaven simultaneously, because Psyche exists at the infinite and IS infinite.

Psyche or Intelligence is something completely different from a spirit body; but, they can be combined together into a functional whole. The spirit body and psyche combination are typically called a Spirit, Ghost, or Soul.

At the $-\pi/2$ asymptote, Psyche or Quantum Non-Local Consciousness is an infinite singularity or point particle which has NO physical size and takes up NO physical space whatsoever; yet, simultaneously at the $\pi/2$ asymptote, Psyche is capable of being omnipresent and omniscient. Psyche resides in the realm of the infinite. Psyche is infinite and instantaneous. In its native format, there are no speed limits placed upon Psyche. Psyche is capable of being instantaneously and simultaneously everywhere. Psyche is infinitely small; but, it is also simultaneously capable of infinite omnipresence and infinite omniscience.

Psyche is viewed or seen as a spark or a pinpoint of light. Psyche is experienced as an immaterial viewpoint in space. Psyche or Intelligence is a whole other animal besides matter or dust; yet, both are comprised of different types and/or different frequencies of quantum waves.

Quantum waves are also thoughts and memories. They survive the death of our physical brain according to the observational evidence obtained from Near-Death Experiences (NDEs), Out-of-Body Experiences (OBEs), Shared-Death Experiences (SDEs), and our after-death Life Reviews. A Psyche is capable of transmitting, receiving, and storing quantum waves, which are thoughts and memories.

In contrast, Quantum Objects, particles, or matter have two states of existence according to the Quantum Law of Complementarity – a non-local spiritual state and a localized physical state. According to the Quantum Law of Complementarity, a Quantum Object can be either spirit matter or physical matter, but it can't be in both states simultaneously. A choice has to be made by Someone Psyche to determine which of the two states it will be in. Psyche ACTS. Quantum Objects REACT to psyche's demands and intervention.

According to the Quantum Law of Superposition, a Quantum Object or Particle of Matter can be phase-shifted, dimension-shifted, or frequency-shifted along the domain between the $-\pi/2$ and $\pi/2$ asymptotes. Matter is finite and reacts to psyche's demands, particularly to God's Psyche who provided matter with Laws, Order, Structure, Restrictions, and Purpose.

Thanks to the Quantum Law of Superposition, a psyche, a spirit body, and a physical body can occupy the same space at the same time because they are out of phase with each other, exist at different frequencies, and therefore exist in different dimensions, but in the very same space at the exact same time.

A Quantum Object or particle of matter can be transitioned, or phase-shifted and moved, all along the x-axis between the $-\pi/2$ and $\pi/2$ asymptotes. Out of body travelers have observed that on the spiritual side, their spirit body can phase-shift or go into different dimensions at will. Being in a different phase or at a different frequency, their spirit body can walk through physical walls and doors; and, they can levitate and rise into air through physical ceilings and roofs.

However, on the physical side, the physical laws that God put into place prevents us from phase-shifting our physical bodies at will and then walking through walls and doors. God retains the capability of phase-shifting physical matter for himself and apparently doesn't share that capability with mortals. The same thing applies to the teleportation or the quantum tunneling of physical matter. God put the physical laws into place in order to restrain it and contain it so that your physical body doesn't quantum tunnel away on you one atom at a time. God put the sub-light speed limitations and entropy into place on physical matter in order to create the Ultimate Consensus Reality for us to live and experience.

In a physical reality, I can depend on my house, my car, this paper, and my dog being there when I go looking for them tomorrow. A physical reality is the ultimate consensus reality because it is dependable, reliable, and predictable. It provides an excellent school-ground upon which to learn how to interact with other psyches in a social manner.

Physical matter is still capable of phase-shifting and teleportation – changing frequencies – but the whole thing is under God's control. God teleported me and my car to safety one time. It wasn't anything I did. It was something that God did for me to save me. My best friend has experienced phase-shifting. An elk passed through his truck un-phased. Again, it wasn't anything that my friend did. It was something that God did for him in order to save him, or the elk.

Quantum Objects – spirit matter and physical matter – reside in the realm of the finite. Psyche, Intelligence, Consciousness, or Life is the Master of the Infinite!

This IS the Ultimate Model of Reality and the Grand Unified Theory of Everything; and, it's really quite simple to visualize and understand once a person chooses to do so.

Mark My Words

—

Source

The Ultimate Model of Reality: Psyche Is the Ultimate Cause

https://www.amazon.com/dp/B071NC9JK6

A Psyche Ontology – The Grand Unified Theory of Everything

Rychlak, J. F. (1981a). *A Philosophy of Science for Personality Theory* (2nd ed.). Malabar, FL: Robert E. Krieger Publishing Company.

I used this book from Joseph Rychlak to point me to a Philosophy of Psyche or a Philosophy of Personality; and, this book was instrumental in helping me to develop and present my Psyche Ontology or Ultimate Model of Reality. Every truth points to every other truth. By observing what was missing from all the different models of reality or personality theories, this book from Joseph Rychlak and its discussion of Aristotle's four physical causes pointed me directly to Psyche as the Ultimate Cause, because physical causes CAN'T DO psyche, non-local consciousness, spirituality, intelligence, science, philosophy, teleology, and life.

Ever since a Psyche Ontology was revealed to me, I have wondered why nobody has ever thought of it before. Why didn't God reveal it to someone thousands of years ago? Why do I see no mention of a Psyche Ontology in any of the literature?

The only answer I have is that people weren't wanting a Psyche Ontology; and therefore, people weren't looking for a Psyche Ontology. God gives us what we want most; and, most people don't want an ontology that points them to God's Psyche or that reduces to God's Psyche. Depending upon how much you are resisting all of this, you might be one of them.

Nevertheless, this Psyche Ontology or Ultimate Model of Reality or Ultimate Paradigm ends up being the Grand Unified Theory of Everything, because it explains everything and subsumes the whole of existence, reality, knowledge, lived experience, and truth.

The scientists and mathematicians are NEVER going to be able to unite the Theory of Relativity with Quantum Mechanics directly, because Classical Physics and the Theory of Relativity explain the LIMITATIONS of physical matter whereas Quantum Mechanics or Spiritual Mechanics or Non-Local Mechanics explains to us that the quantum or the spiritual or the non-local has NO physical limitations. The scientists will NEVER be able to explain Quantum Mechanics or Non-Local Mechanics in materialistic or physical terms, because Quantum Mechanics is a non-local, non-physical, transdimensional, spiritual phenomenon. Quantum Mechanics explains how spirit matter works, and how spirit matter becomes physical matter by taking upon itself some additional limitations.

To unify Classical Physics with Quantum Mechanics we have to provide a mathematical model, like **y = tan(x)**, to model how Matter or Quantum Objects REALLY work. We have to explain how spirit matter becomes physical matter! When we do so, this Psyche Ontology or Ultimate Model of Reality becomes the Grand Unified Theory of Reality, which everyone has been looking for.

When Matter or a Quantum Object is below (0, 0) and approaching infinite velocities, infinite frequencies, and zero size, that particle of Matter or that

Quantum Object can function instantaneously and simultaneously with NO distance and speed limitations whatsoever, because the stuff is spirit matter and NOT physical matter. The only real limitation which spirit matter has is that it reacts and therefore it lags just slightly behind Psyche's Word of Command, where the timing is concerned.

When Matter or a Quantum Object is above (0, 0) and approaching infinite mass and infinite inertia and zero velocity, that particle of physical matter is in fact approaching LIMITATIONS which it can't get beyond. The ONLY way that that particle of physical matter or that Quantum Object can get beyond those limitations is to transform back into spirit matter where it will have no such limitations.

It is my observation that Seraphim, Translated Beings, Resurrected Beings, God the Father, and Jesus Christ can convert or transmute their physical bodies into spirit matter at will, and then convert that spirit matter back into physical matter when they reach their chosen destination. That's how these people do what they do. While their bodies are spirit matter, these people can travel across the universe instantaneously at the speed of thought. When these people reach their chosen destination, they then convert their bodies back into physical matter. It's a Quantum Jump Drive or a Blink Drive, and it is infinitely faster than a warp drive which tries to keep everything in a physical matter format.

Warp drive will NEVER be possible because it's impossible to push physical matter faster than the speed of light. However, when we temporarily convert physical matter back into spirit matter, then we are dealing with Quantum Leaps and a Blink Drive or a Quantum Jump Drive, which has NO physical limitations! It's teleportation and it works. It has also been EXPERIENCED and OBSERVED. It solves everything that we have been looking for.

The ONLY problem is that God or God's Psyche is in control of it all, and He gifts this teleportation ability only to His righteous followers whom He has selected for the gift or the experience. Teleportation or Blink Drives are something that we mortal beings will never gain control of, because God is in control of this gift or this "technology"; and, it is God who decides whom He is going to teleport or blink from one location to another. ONLY God's Psyche controls this spiritual gift. This is an ability or a gift that is ONLY God's to give.

I and my car have teleported. I blinked or I jumped; BUT, it was nothing that I consciously did or consciously chose. God did it for me. God is the ONLY one who can do these kinds of things for us. If we are going to blink or jump out of harm's way or walk through fire unscathed, this is something that God is going to have to choose to do for us, because it is something that we mortal fallen beings can't do for ourselves. God is KNOWN by what He chooses to do for us; and, God does for us the things that we can't do for ourselves.

This is a Psyche Ontology, wherein Psyche is the fundamental unit of reality and God's Psyche is the Ultimate Psyche or the Ultimate Cause. This IS the Ultimate Model of Reality and the Grand Unified Theory of Everything. A Psyche Ontology subsumes ALL of the other ontologies, because ONLY Psyche can do ontology, reality, truth, knowledge, life, transmutation of matter, teleportation, lived experience, memories, and existence. This really IS the Ultimate Model of Reality, which is why nobody has ever thought of it before.

Evolution Is Entropy

Web Page: https://evolution-is-entropy.com/evolution-is-entropy/

The Materialists, Naturalists, Darwinists, Nihilists, Behaviorists, and Atheists axiomatically define entropy as "disorder".

They call this falsehood or lie the Second Law of Thermodynamics, which axiomatically and erroneously states that the total amount of entropy or disorder in our universe is constantly increasing and that it can never decrease and go to zero. This falsehood predicts that we shouldn't be here, yet here we are. Furthermore, according to the equations for heat and entropy, as mass goes to zero, entropy goes to zero along with it. Entropy can and does go to zero whenever mass or heat goes to zero. Therefore, the second law of thermodynamics is false.

There is NO CORRELATION between entropy and disorder. Disorder doesn't cause heat, and disorder doesn't eliminate heat either. Disorder is the wrong definition for entropy; yet, that definition persists and is dominant anyway because it is false. It is its falsehood that makes it preferable to everything else, because above everything else the Materialists and Naturalists are trying to trick us and deceive us just as they have been tricked and deceived. The corollary to the Second Law of Thermodynamics states that entropy, chance, or disorder can design and create anything, if given enough time to do so. The Theory of Evolution has the same exact corollary as the Second Law of Thermodynamics. They are the same thing in the end. They are the same deception and lie.

These people have chosen to believe that entropy or disorder can design and create things if given enough time to do so. For the Materialists, Naturalists, and Darwinists, both the theory of evolution and the second law of thermodynamics are Creation by Entropy, Creation by Disorder, or Creation by Chance. They are the same thing. They are magic! The Atheists are superstitious because they believe in Creation Ex Nihilo or Creation by Magic.

Once I realized and understood what the Materialists, Naturalists, and Darwinists are teaching and preaching, I immediately realized that in their worldview, evolution is entropy – the false definition for entropy. In other words, according to the Darwinists, evolution is Creation by Entropy, Creation by Random Disorder, or Creation by Chance. Evolution works like magic.

There are a couple of things that you need to know about evolution, or creation by chance. First of all, extinction, genetic deterioration, or genetic entropy is the dominant law in biology – NOT spontaneous generation, abiogenesis, or chemical evolution. Evolution doesn't work as advertised. Evolution cannot design and create. Evolution or genetic mutations can only maim, kill, and destroy. Evolution produces extinction – NOT creation and life. The theory of evolution, spontaneous generation, chemical evolution, or abiogenesis was falsified in 1859 by Louis Pasteur, and it has been false ever since. Second of all, the Primary Axiom of Science and Statistics axiomatically states that chance is not causation. In other words, chance cannot do choice, correlation, design, causation, or creation. Chance and causation are mutually exclusive, which means that they falsify each other.

Once chance starts doing causation, then it is no longer chance but has become some type of deliberate causation or intelligent design and creation instead. The Primary Axiom of Science and Statistics falsifies the theory of evolution and the second law of thermodynamics.

The theory of evolution is just as false as the second law of thermodynamics because they are the same thing in the end. They both are design and creation by chance, disorder, chaos, death, or entropy. Evolution is entropy – the false definition for entropy. Therefore, evolution or creation by chance is just as false as creation by entropy or creation by disorder, because they are the same thing after all. They are nothing but superstition or magic, because chance cannot design and create things at will, no matter how much time it is given to do so. The "magic" or quantum mechanics exists at the quantum level, NOT the physical level. At the physical level, creation by chance, the theory of evolution, and the second law of thermodynamics are physically impossible. This is what we have actually experienced and observed.

Now pay attention!

Despite the FACT that I have and will have debunked and falsified the Theory of Evolution thousands of times in hundreds of different ways, don't ever once let me convince you that there is NO such thing as evolution or random mutations.

Evolution is REAL, very REAL. Evolution or random mutation is entropy or the second law of thermodynamics. It's REAL. Entropy is very REAL at the physical level; and, it works as advertised at the physical level. Entropy produces death and extinction.

One of my all-time most significant scientific discoveries came to me when I first realized that the Theory of Evolution is based upon entropy.

Obviously, I'm not the first person to discover entropy; but, I seem to be the first person on the planet to realize that Materialism, Naturalism, Darwinism, Nihilism, Atheism, Determinism, Physicalism, Behaviorism, Scientism, Classical Physics, Chemical Evolution, Macro-Evolution, Natural Selection, Random Mutations, and the Theory of Evolution are based exclusively on entropy; and, I seem to be the first person to realize what that truly means for science in general.

You see, these people have chosen to believe that physical matter and entropy are the ONLY thing that exists. They based ALL of their science exclusively on entropy.

Are you starting to see the significance of this reality? Can you see the problem?

Well, ask yourself, "What is entropy?"

Entropy is corruption, disease, disorder, chaos, death, and extinction; and, the Theory of Evolution is based exclusively on entropy. Are you starting to see the problem? It's a BIG ONE! This is HUGE!

Evolution of any type is entropy; and, evolution of any type is based upon entropy. Entropy is death and extinction. When was the last time that you caught **death** in the act of designing, creating, and producing new and unique life forms from scratch? It has never happened, and it will never happen. Entropy or evolution prevents it from happening.

Entropy literally prevents the Theory of Evolution from becoming true. Evolution produces entropy. The different types of evolution produce death and extinction. Entropy prevents the different types of evolution from designing, creating, and producing something new. Entropy cannot design and create, which means that evolution of any kind cannot design and create. Entropy can only destroy, which means that the different types of evolution can only destroy. This reality is so obvious, that I sometimes wonder why nobody has ever thought of it before.

Life, design, creation, organization, order, information, intelligence, proteins, genes, genomes, computer programs, hardware, cars, buildings, computers, phones, televisions, physical matter, and physical life forms REQUIRE an infusion of Syntropy in order to come into existence. We KNOW that Syntropy must exist, or there wouldn't be any entropy. We KNOW that Syntropy must exist, or there wouldn't be any physical matter. We KNOW that Syntropy must exist, or there wouldn't be any life – the whole thing would be death, or chaos, or entropy. We KNOW that Syntropy must exist, or we wouldn't be here to think about it.

The Theory of Evolution can't work as advertised. It's prevented from doing so by entropy, or death and extinction. Evolution is entropy; and, evolution prevents the Theory of Evolution from becoming true.

The Fruits of Evolution or Entropy

Whenever the Materialists, Naturalists, Darwinists, Nihilists, Behaviorists, and Atheists start talking about evolution, they get most everything wrong because evolution or entropy cannot design and create. However, these people do indeed get one thing perfectly right. Evolution, or random mutation, or entropy is indeed the CAUSE of ALL of our heritable diseases, developmental diseases, and heritable mental illnesses.

Remember, the Theory of Evolution is FALSE because random mutations or entropy cannot design and create genes, proteins, and life forms. However, evolution or random mutation or entropy is very REAL; and, it can indeed destroy genes, proteins, and life forms. Do you see how that works? It's important to understand.

The Theory of Evolution is a fictional story that they made up out of thin air. There is NO empirical evidence supporting any version of it. In fact, ALL of the empirical evidence and experimental evidence that we have on hand as a race FALSIFIES the different versions of the Theory of Evolution and VERIFIES Quantum Mechanics instead. Quantum Mechanics is Supernatural. Quantum Mechanics is the Priesthood Power of God. Quantum Mechanics and Psyche are Pure Syntropy. There is NO entropy in the spirit realm, the non-local realm, the quantum realm, or the transdimensional realm. Psyche, Spirit Matter, and Quantum Mechanics ARE Syntropy.

The very existence of something like the Orthodox Interpretation of Quantum Mechanics from Henry P. Stapp FALSIFIES Materialism, Naturalism, and the various versions of the Theory of Evolution. Quantum Mechanics FALSIFIES Materialism, Naturalism, and their derivatives. Fictional stories like the Theory of Evolution cannot stand in the light of truth.

The Theory of Evolution is the very pinnacle of fictional ad hoc just-so story telling; and, *ad hoc just-so stories* are logic fallacies. The Theory of Evolution is fictional because it never happened – none of it happened! The chemical evolution of proteins and genes from atoms is physically impossible. Abiogenesis, spontaneous generation, and the various different forms of macro-evolution are physically impossible. They are prevented from happening by entropy.

The Theory of Evolution is science fiction. The fictional nature of the story becomes most egregious whenever they try to guesstimate how many millions of years it took for evolution to do something for us, because evolution or entropy can't do anything for us – not ever. They are making the Theory of Evolution up as they go along. It's a fictional story, and nothing more.

Since the whole Theory of Evolution is nothing but a fictional story, you can successfully and rightfully make up fictional stories of your own to debunk it. That's the way fiction works!

Scientific Inference

Comparative Psychology and Evolutionary Psychology are based upon Darwinism and the Theory of Evolution, which means that they too are nothing more than *fictional ad hoc just-so stories* that these people have made up out of thin air.

In fact, in his book *Biopsychology*, John Pinel tells us as much when he tells us that his "evolutionary perspective" is based upon *scientific inferences*. *Scientific inferences* are fictional ad hoc just-so stories. *Scientific inferences* are logic fallacies. The whole of their evolutionary perspective, evolutionary psychology, and comparative psychology is based upon *scientific inferences* or stories that they have manufactured out of thin air. Making up stories is what makes being a scientist fun, according to John Pinel.

From *Biopsychology* page 13, John Pinel writes:

> **Scientific inference is the fundamental method of biopsychology and of most other sciences – it is what makes being a scientist fun. This section provides further insight into the nature of**

biopsychology by defining, illustrating, and discussing scientific inference.

The Scientific Method is a system for finding things out by careful observation, but many of the processes studied by scientists cannot be observed. For example, scientists use empirical (observational) methods to study ice ages, gravity, evaporation, electricity, and nuclear fission – none of which can be directly observed; their effects can be observed, but the processes themselves cannot. Biopsychology is no different from the other sciences in this respect. One of its main goals is to characterize, through empirical methods, the unobservable processes by which the nervous system controls behavior.

The empirical method that biopsychologists and other scientists use to study the unobservable is called <u>scientific inference</u>. Scientists carefully measure key events they can observe and then use these measures as a basis for logically inferring the nature of events that they cannot observe.

Like a detective carefully gathering clues from which to recreate an unwitnessed crime, a biopsychologist carefully gathers relevant measures of behavior and neural activity from which to infer the nature of the neural processes that regulate behavior.

The fact that the neural mechanisms of behavior cannot be directly observed and must be studied through scientific inference is what makes biopsychological research such a challenge – and as I said before, so much fun. (*Biopsychology*, p. 13.)

Pinel, J. (2014). *Biopsychology* (9th ed.). New York: Pearson.

John Pinel is trying to lay a scientific foundation for his belief in Evolution, Materialism, Naturalism, and Darwinism; and, he is using *scientific inferences* to do so. He wrote this section of his book as empirical proof that the Theory of Evolution is true. *Scientific inferences* are how the Materialists, Naturalists, and Darwinists prove that the Theory of Evolution is true. So, what are *scientific inferences*?

He defines *scientific inference* as an empirical method or a scientific method. Then he uses *scientific inference* to prove that the Theory of Evolution is true. Because he has never studied the Philosophy of Science, he has NO idea how faulty, fallacious, and weak *scientific inferences* really are. He has NO idea that *scientific inferences* are logic fallacies. Instead, he literally treats *scientific inferences* as empirical evidence, so that he can prove that the Theory of Evolution is true.

Notice carefully how he defines Evolution, Materialism, Naturalism, and Darwinism as events that we cannot observe. These are events that have NEVER been observed. So, we are going to use *scientific inferences* to manufacture "empirical evidence" for these events that cannot be observed and that have never been observed. *Scientific inference* is a logic fallacy; yet, he erroneously calls it an empirical method. They ALL do that in one way or another, without even realizing that they are using logic fallacies to prove that Evolution, Materialism, Naturalism, and Darwinism are true. These people tack on the word "scientific" to make it seem like science, but it is nothing more than "inference" or personal interpretation masquerading as empirical fact. These people are *begging the question* and *jumping to conclusions*.

Inference is the *affirming the consequent* logic fallacy in action. If you ever study the Philosophy of Science and get a professor like Joseph Rychlak who knows what he is talking about, he will teach you that for every single piece of scientific evidence, there are literally an infinite number of interpretations or inferences that can be given to that single piece of scientific evidence or that single event; and, many of them will seem plausible or believable, but only one of them can be true.

Scientific inference is the *affirming the consequent* logic fallacy in action.

Scientific inference is the *jumping to conclusions* logic fallacy in action.

Scientific inference involves *begging the question* or *circular reasoning*, which are also logic fallacies. *Scientific inference* is a logic fallacy; but, John Pinel is choosing to use *scientific inferences* as empirical evidence and as an empirical method or a scientific method. It's a logic fallacy to do so; but, he doesn't know that. He's never been taught that *scientific inference* is a logic fallacy. Scientists typically don't study the Philosophy of Science in college and grad school; and if they do, they are taught the philosophy of science by Materialists, Naturalists, Darwinists, and Atheists who don't know what they are talking about.

When it comes to the origin of life, there are an infinite number of hypotheses or inferences that are being used to explain the origin of life on this planet. In fact, the Intelligent Design Theory is infinitely more plausible and believable than the Theory of Evolution because intelligent design has actually been OBSERVED trillions of trillions of times, whereas the evolution of genes and proteins from atoms has NEVER been observed and NEVER will be because it's physically impossible. Evolution of any kind has NEVER been observed doing spontaneous generation, abiogenesis, macro-evolution, or the creation of new life forms from scratch because it's physically impossible for them to do so.

Intelligent Design Theory is an infinitely more plausible, believable, credible, and parsimonious *scientific inference* than the Theory of Evolution is because the evolution of atoms into genes, proteins, eyes, brains, and life forms is physically impossible. If you are going to use logic fallacies and are determined to use *scientific inferences* as empirical evidence, then you owe it to yourself to go with the BEST explanation or the BEST *scientific inference* – one that has actually been OBSERVED and caught in the ACT – Intelligent Design Theory. Get rid of the ones that have NEVER been observed such as Materialism, Naturalism, Darwinism, Nihilism, Determinism, and Atheism.

The whole Theory of Evolution is based upon *scientific inferences* or *fictional ad hoc stories* that these people have manufactured out of thin air. There's nothing empirical about it! There is NO empirical evidence demonstrating the creative powers of entropy, evolution, or random mutations. Chemical evolution of proteins and genes from atoms is physically impossible thanks to entropy or the second law of thermodynamics. It can't happen, which means that it never happened. Evolution is entropy, or the second law of thermodynamics. It can't design and create. It's physically impossible.

Furthermore, evolution (genetic change), random mutations, and natural selection didn't even exist until AFTER God designed and created the proteins, genes, genomes, brains, eyes, and life forms in the first place. It's physically impossible for something that doesn't even exist yet to design, program, engineer, manufacture, and create proteins and genes out of thin air. Spontaneous generation or abiogenesis is physically impossible. Entropy or the second law of thermodynamics prevents it from happening.

Evolution is entropy, which means that it can't design and create anything. Evolution or entropy can only deteriorate and destroy things. It can't design and create. It's physically impossible for evolution or entropy to design and create proteins, genes, genomes, brains, eyes, and life forms. That's just the way it is, because evolution of any kind is entropy.

It took me years, even decades, to discover that evolution or random mutation is entropy or the second law of thermodynamics. That discovery also came with a powerful gift. I discovered Syntropy! Since evolution or entropy cannot design and create, some type of Syntropy must have done the job.

I finally realized that Quantum Mechanics is the exact opposite of Materialism, Naturalism, Darwinism, Classical Physics, and Entropy. Quantum Mechanics and Psyche are Pure Syntropy. Quantum Mechanics is the Power of God or the Priesthood Power of God. Psyche or Quantum Non-Local Consciousness is the only thing that can control Quantum Mechanics at the quantum level or the psyche level. Physical matter and entropy can't touch nor control the quantum level. Only the smaller can dwell within and control the larger. Psyche is Syntropy. Quantum Mechanics is Syntropy.

There is NO aging or entropy in the Quantum Realm, Psyche Realm, Spirit Realm, or Transdimensional Realm. Transdimensional means non-physical and non-local – not located in our physical 3D space-time realm. Everything in the Quantum Realm is Pure Syntropy. It is endless, timeless, eternal, and everlasting because there is NO entropy which means that nothing ages, gets old, or dies in the Non-Local Realm or the Syntropy Realm.

The Gods create physical matter by infusing a particle of spirit matter with space-time and the ability to acquire entropy. The Gods create a particle of physical matter by taking a particle of spirit matter, filling it full of space, slowing it down to sub-light speeds, and making it subject to entropy or the passage of time. According to the theory of relativity, the particles of spirit matter existing at velocities faster than the speed of light experience NO passage of time, meaning that they do not age and are not subject to entropy. Entropy is a function of time or an aging process. Spirit matter and physical matter are the same thing – they are quantum objects. However, spirit matter is pure syntropy; whereas, physical matter has been slowed down by being infused with space-time and made subject to entropy or the passage of time.

What is the difference between a hypothesis and a theory?

A hypothesis is an idea. A theory is a hypothesis that has observational evidence supporting it. It really should be called the "hypothesis of evolution" because there is NO observational evidence supporting macro-evolution, chemical evolution, abiogenesis, and spontaneous generation; and, there NEVER will be. Thanks to entropy, random diffusion, or the second law of thermodynamics, it's physically impossible for atoms to spontaneously generate into functional genes, proteins, genomes, eyes, brains, and life forms. It can't be done, which means that it wasn't done.

A hypothesis is a prediction. A hypothesis is a testable proposition.

> **In their quest for insight, social psychologists propose *theories* that organize their observations and imply testable *hypotheses* and practical predictions.**
>
> **A theory is an integrated set of principles that explain and predict observed events.**
>
> **To a scientist, facts and theories are** [the same difference]. **Facts are agreed upon statements about what we observe. Theories are ideas that summarize and explain facts.** (*Social Psychology*, p. 17.)

What are the Theory of Evolution's hypotheses or predictions?

According to the science fiction that I have watched on television – *Star Trek*, *Babylon 5*, and *Earth: Final Conflict* – a million years from now human beings are going to evolve into energy beings. There's a serious problem with that prediction. There has been life on this earth for billions of years, and NONE of it has ever evolved into an energy being. It's not going to happen because it's physically impossible. It won't be done because it can't be done.

The theory of evolution makes NO realistic or useful predictions. It just catalogues what happened in the past according to the fossil record, and then it forces a personal interpretation or a scientific inference onto the fossil record after-the-fact. *Scientific inference* is a logic fallacy. *Personal interpretations* introduce a wide variety of logic fallacies into science. The theory of evolution is based upon a wide variety of scientific inferences, logic fallacies, personal interpretations, and wishful thinking.

Macro-evolution, chemical evolution, abiogenesis, and spontaneous generation have NEVER been observed because they are physically impossible thanks to entropy, random diffusion, and the second law of thermodynamics.

Technically, the theory of evolution makes NO testable predictions because ALL of the observational evidence, empirical evidence, experimental evidence, and experiential evidence FALSIFIES the theory of evolution.

The FACT is that the spontaneous generation of atoms into functional genes, proteins, genomes, eyes, brains, and life forms is physically impossible. Since macro-evolution, spontaneous generation, chemical evolution, and abiogenesis are physically impossible, that means they never happened because they couldn't happen.

The FACT is that the major premises or primary hidden assumptions of Materialism, Naturalism, Darwinism, Nihilism, Behaviorism, Determinism, and Atheism – claiming that the quantum or the supernatural does not exist – are FALSIFIED due to a complete lack of observational evidence or a complete lack of supporting evidence. Furthermore, they are FALSIFIED by observational evidence and experimental evidence. The verified and proven existence of Quantum Mechanics and Action at a Distance FALSIFIES Materialism, Naturalism, and their derivatives.

The primary assumptions or hidden premises associated with Naturalism and Darwinism are NOT testable.

There are NO observed events and NO observed facts associated with chemical evolution, design and creation by random mutations, abiogenesis, spontaneous generation, macro-evolution, or design and creation by genes, proteins, RNA, and amino acids. Stand-alone atoms cannot spontaneously generate into functional genes, proteins, genomes, eyes, brains, and physical bodies because spontaneous generation is physically impossible thanks to entropy or random diffusion.

There are NO observed facts associated with creation by evolution, which means that the "theory of evolution" is in fact a falsified hypothesis and NOT a verified theory.

The theory of evolution is based exclusively upon correlation, and NOT observation. There is a huge difference between the two. Correlation does NOT prove causation. The Darwinists have carefully correlated their hypotheses with the fossil record. The Darwinists have deliberately correlated the fossil record with Darwin's Tree of Life. However, NOBODY except for the Biblical God Jesus Christ actually observed the production of the fossil record. Our fossil record could have been produced in many different ways; and with hindsight, all of these different ways or hypotheses can be made to correlate with the fossil record. Only God knows which one of those ways is the actual way by which He produced the fossil record.

Observation trumps correlation, or at least it should. However, when it comes to the theory of evolution, the Materialists, Naturalists, Darwinists, and Atheists make their correlations trump ALL the observational evidence that falsifies their pre-chosen beliefs. They cheat in order to make their case.

ORIGIN: Probability of a Single Protein Forming by Chance

https://www.youtube.com/watch?v=W1_KEVaCyaA

http://www.originthefilm.com/mathematics.php

http://science-2-0.com/wp-content/uploads/2018/03/THE-MATHEMATICS-OF-ORIGIN.pdf

https://www.youtube.com/watch?v=cQoQgTqj3pU

https://www.youtube.com/watch?v=_zQXgJ-dXM4

http://bio-complexity.org/ojs/index.php/main/article/view/BIO-C.2010.1/BIO-C.2010.1

http://science-2-0.com/wp-content/uploads/2018/03/Case-Against-Darwinian-Origin.pdf

http://science-2-0.com/wp-content/uploads/2018/03/Chemical-Evolution-Is-Impossible.zip

http://science-2-0.com/wp-content/uploads/2018/04/Origin-Of-Life.zip

Once you have eliminated everything that is FALSE and everything that is IMPOSSIBLE, then ONLY the Truth remains. It's elementary.

The theory of evolution has been tested and falsified. There's NO practical application for creation by rocks, chemical evolution, or the theory of evolution because spontaneous generation is physically impossible. There are NO observations when it comes to the theory of evolution. The theory of evolution makes NO testable and verifiable predictions because ALL of observed evidence falsifies Materialism, Naturalism, Nihilism, Atheism, and the Theory of Evolution.

Materialism, Naturalism, Darwinism, Nihilism, Behaviorism, Determinism, and Atheism are FALSIFIED due to a complete lack of observational evidence or a complete lack of supporting evidence for their hidden assumptions or major premises which claim that the quantum or the supernatural does not exist. In contrast, it is said that Quantum Mechanics or Action at a Distance is the most-verified, best-proven, and most-used science that we currently have.

Quantum Mechanics and Quantum Neuroscience

On page 25 of *Biopsychology*, John Pinel quotes another worshipper of the theory of evolution:

> **Evolution is both a beautiful concept and an important one, more crucial nowadays to human welfare, to medical science, and to our understanding of the world than ever before. It's also deeply persuasive – a theory you can take to the bank. The supporting**

evidence is abundant, various, ever increasing, and easily available in museums, popular books, textbooks, and a mountainous accumulation of scientific studies. No one needs to, and no one should, accept evolution merely as a matter of faith (Quammen, 2004, p. 8).

Ironically, the mountains of evidence supporting the theory of evolution are correlational – designed and manufactured after-the-fact to fit the fossil record. NONE of that evidence is observational. Correlation is not causation. Instead, ALL of the observational evidence and experiential evidence that we have on hand as a race FALSIFIES Materialism, Naturalism, Darwinism, Nihilism, Behaviorism, Scientism, Determinism, and Atheism. Correlation cannot be used to prove causation. All of these falsified philosophies have to be taken on blind faith as being true because there is NO observational evidence supporting them. The theory of evolution is bankrupt. There's no logical reason to believe in the theory of evolution, except for the fact that everybody else has been deceived by it and has chosen to believe in it.

I talk about all of this in great detail in my book, *Quantum Neuroscience: The Answer to Life, the Universe, and Everything*. If a comparison between Evolution and Quantum Mechanics interests you, I recommend you take a look at that book:

https://www.amazon.com/dp/B079Z6QQQB

The book, *Quantum Neuroscience: The Answer to Life, the Universe, and Everything*, makes a detailed comparison between Neuroscience and Quantum Neuroscience, which means that it makes a detailed comparison between Classical Physics and Quantum Mechanics.

When properly understood, Quantum Mechanics FALSIFIES Classical Physics, Materialism, Naturalism, Darwinism, Nihilism, Scientism, Behaviorism, Determinism, and even Atheism. Quantum Mechanics is Supernatural. Quantum Mechanics or Syntropy is the exact opposite of Classical Physics, Entropy, Random Mutations, Materialism, Naturalism, Darwinism, and the Theory of Evolution.

Quantum Mechanics is a proven and verified science. Quantum Mechanics is the best-proven and most-used science that we have. In contrast, the Theory of Evolution has NO empirical evidence supporting it. In fact, ALL of the empirical evidence that we have on hand as a race, including Quantum Mechanics, FALSIFIES Materialism, Naturalism, Darwinism, and the Theory of Evolution. Do you see how that works? It's important to understand.

Quantum Mechanics, Spirit Matter, and Quantum Non-Local Consciousness (Psyche or Intelligence) are PURE SYNTROPY. The Syntropy has to exist somewhere someplace somehow because, according to the Law of Entropy and the Second Law of Thermodynamics, the physical Multiverse should have burned out and suffered heat death an eternity or two ago; and, there should be NO more physical universes anywhere, but here we are, nonetheless. The very existence of this physical universe – it's beginning full of syntropy and its ongoing existence billions of years later – is positive proof that Someone Psyche knows how to do syntropy or Someone Psyche is syntropy. Quantum Mechanics is syntropy or the

Priesthood Power of God. God's Psyche knows how to do syntropy or Quantum Mechanics; otherwise, this physical universe would not exist.

This is what Quantum Mechanics or Transdimensional Physics is trying to teach us. Quantum Mechanics, Spirit Matter, and Psyche are pure syntropy. Evolution, Random Mutations, Physical Matter, and Classical Physics are entropy.

Quantum Mechanics is an infinitely better theory than Materialism, Naturalism, Darwinism, Nihilism, Atheism, Classical Physics, and the Theory of Evolution. Old theories are supposed to fall by the wayside whenever a better theory is proposed to account for the findings. Intelligently designed quantum machinery is an infinitely better theory for the origin of life than the theory of evolution because Quantum Mechanics and Intelligence have been experienced, observed, replicated, verified, and proven to be real and true through a preponderance of the evidence. Design and creation by entropy or evolution has NOT. Instead, design and creation by entropy or evolution has been falsified. It's physically impossible.

Evolution Is Entropy

Random mutations are entropy.

It's physically impossible to produce a functional protein by throwing dice. It can't be done, which means that it wasn't done.

Furthermore, designing, engineering, and making a matching gene to go along with a functional protein requires deliberation, planning, engineering, and intelligence. It can't happen through random chance or luck. It's physically impossible. The production of a protein, a matching gene, as well as a functional genome requires a programmer, an engineer, a designer, a fine-tuner, and Syntropy of the highest order. Someone Psyche has to infuse syntropy or order into the equation in order to make it happen because entropy naturally prevents it from happening. Your genome is God's Signature.

Random chance produces disorder, disorganization, chaos, cancer, death, extinction, and entropy.

Random mutations are the mechanism of change when it comes to evolution, genetic change, genetic drift, or genetic entropy – not natural selection. Natural selection doesn't touch our genes and proteins. Natural selection is worthless when it comes time to change our genes and proteins.

Natural selection is also a product of chance or a function of luck. The ultimate product of natural selection is death and extinction. Death and extinction are entropy. The physical mechanism that we call natural selection causes entropy, death, and extinction. Natural selection cannot design and create because natural selection doesn't touch our genes and proteins.

Since the very beginning of the theory of evolution, Darwinists and Evolutionists have erroneously stated that natural selection is the causative agent behind genetic change – the origin of species by means of natural selection. They are wrong. They don't even understand their own theory. Selection or sexual activity determines whether our genes get passed on to the next generation, or not; but, natural selection or selection of any kind does NOT change our genes. Selection or sexual activity does NOT touch our genes!

Natural selection or survival of the fittest doesn't do anything except result in death and extinction. Death and extinction are entropy, not syntropy, design, and creation. Natural selection is NOT a mechanism of change. Natural selection doesn't do anything. Natural selection is death, extinction, and entropy. Natural selection doesn't touch our genes. It can't. It's physically impossible. There's no invisible person called "Natural Selection" who is reaching into our genes and making them come alive.

The majority of our heritable mutations come from the genetic recombination process that takes place during the production of our gametes (egg and sperm). Genetic recombination is the process that shuffles our genes and introduces errors or mutations into the system; and, genetic recombination or the production of gametes takes place BEFORE any type of selection or sexual activity. Remember, natural selection doesn't touch our genes. God is the person who designed and created our proteins and the matching genes to go along with them, which means that God is the one who designed the physical process that we call genetic recombination. Genetic recombination is the primary mechanism of change driving evolution, random mutations, and genetic drift. Random mutations are entropy, an integral part of physical matter.

Combine random mutations with natural selection and we end up with genetic entropy – the devolution of our genome to the eventual point where our species goes extinct. It all ends in death or entropy. Technically, the "theory of evolution" is the origin of species by means of genetic entropy, random mutations, and extinction, which is physically impossible. Natural selection has nothing to do with the origin of species. Death and extinction are the natural result of evolution or genetic change.

[See: Sanford, J. (2014). *Genetic Entropy* (4th ed.). Cornell University: FMS Foundation.]

Evolution or Entropy Puts on the Brakes

Evolution of any kind cannot design and create. It's physically impossible. It's prevented from happening by entropy.

Chemical evolution, abiogenesis, spontaneous generation, or macro-evolution is physically impossible. Entropy prevents these things from happening, thereby making them physically impossible. That's the way it really is in the natural world.

The chemical evolution of functional genes and proteins is physically impossible. It can't happen, which means that it didn't happen.

ORIGIN: Probability of a Single Protein Forming by Chance.

https://www.youtube.com/watch?v=W1_KEVaCyaA

I find all of this fascinating. I used to be a Materialist, Naturalist, Nihilist, and Atheist; so, I find it fascinating to observe how wrong I really was in my chosen conclusions at the time. But, I'm not alone. Millions have fallen for the deceptions and the lies. Nevertheless, we have had the truth all the way along, if we were willing to look for it, find it, and accept it.

Spontaneous generation, abiogenesis, chemical evolution, or macro-evolution was FALSIFIED in 1859 by Louis Pasteur – the very same year that Charles Darwin published "On the Origin of Species". We've known that the Theory of Evolution is false from the very beginning; but as a race, we chose to ignore the evidence.

In fact, evolution (genetic change), random mutations, and natural selection didn't even exist until after God designed and created the proteins, the matching genes, the genomes, the eyes, the brains, and the physical bodies in the first place. It's so obvious that I sometimes wonder how I managed to overlook it for the first fifty years of my life; but, I wasn't looking for it, so I wasn't able to see it. Evolution is entropy. Each species has been de-evolving or degenerating ever since God created its genome, thanks to evolution or entropy.

Entropy prevents a genome from spontaneously generating out of thin air. Spontaneous Generation or Chemical Evolution was FALSIFIED in 1859 by Louis Pasteur, which means that Materialism, Naturalism, Darwinism, Abiogenesis, Macro-Evolution, and the Theory of Evolution were FALSIFIED at the same time in 1859. Isn't it fascinating to observe that we KNEW that the theory of evolution is false from the very beginning; but, our scientists chose to go along with it anyway? The truth is that genomes don't spontaneously generate out of thin air. They NEVER have, and they NEVER will. The same can be said of proteins.

Your genome is programming code. Your genome is God's Signature. I was a computer programmer for a decade. I know for a fact that programming code doesn't spontaneously generate out of thin air. That's physically impossible. Entropy or evolution cannot do computer programming. You cannot do computer programming or genome programming by throwing dice, shuffling the deck, random mutations, or random chance. I also know what happens whenever there is a bug in the code. The results are often fatal; and, if the program manages to survive, the results from a bug in the code are less than optimal. Disease, deformity, cancer, pain, suffering, death, and extinction are the natural result of entropy, or random mutations, or evolution.

Finally, take note that entropy or random mutations prevent a genetically compatible Mr. and Mrs. Mutant from coming into existence at the same time in the same place. Thanks to entropy or the randomness of mutations, it's physically impossible for chimp-like ancestors to evolve into chimpanzees and humans, because it's physically impossible to produce the requisite Mr. and Mrs. Mutant each and every time that one is needed, so as to make a sexually reproducing species evolve naturally into some other species. Macro-evolution of this type is prevented from happening by entropy and genetics when it comes to a sexually reproducing species.

Macro-evolution (lizards in a cage giving birth to rabbits) doesn't exist. Evolution of any kind is blocked and prevented from happening among sexually reproducing species by their genetics. Our genomes were deliberately designed to prevent macro-evolution from happening. Sexually reproducing species can't evolve into other types of animals. We have to be of the same species in order for males and females to produce viable offspring. The only place where evolution even remotely applies is when it comes to viruses and bacteria, which don't produce sexually. Viruses and bacterial were designed to mutate. Even then, after their genetic changes, they are still bacteria or viruses. They don't evolve into sexually reproducing animals. Macro-evolution is science fiction. It has NEVER been experienced NOR observed in the lab or in the wild. When macro-evolution goes down, the theory of evolution goes down with it; and, we are left studying the mutations that take place in viruses and bacteria.

Remember, the theory of evolution was made to fit the fossil record after-the-fact in hindsight, and it's constantly adjusted every time another fossil is dug up. The theory of evolution is reactive and not predictive. The theory of evolution is correlational, and not observational. Correlation cannot be used to prove causation. The missing links really are missing. There are NO observations supporting any aspect or any version of the theory of evolution. Design and creation by evolution has NEVER been caught in the act. There are infinitely better and more plausible explanations for the origin of life than random chance or blind luck. The only way to make the theory of evolution work as advertised is to get God to intervene and force it to work; and, God has infinitely better and infinitely faster ways of designing and creating life than random mutations and natural selection.

The theory of evolution is wishful thinking in action. The theory of evolution is the product of Confirmation Bias, which is the psychological tendency to search only for information that confirms one's preconceptions or one's pre-chosen conclusions. These people automatically reject and dismiss anything that falsifies the Theory of Evolution, Materialism, and Naturalism. They cheat, in order to make their case. They employ a wide variety of logic fallacies to make their case for the theory of evolution.

The Summary

So, what have we learned?

Evolution is entropy. Entropy cannot design and create. These people fail to meet their burden of proof, because the preponderance of the evidence falsifies Materialism, Naturalism, Darwinism, Nihilism, Atheism, and the Theory of Evolution.

A hypothesis is an idea. A theory is a hypothesis that has observational evidence supporting it. It really should be called the "hypothesis of evolution" because there is NO observational evidence supporting macro-evolution, chemical evolution, abiogenesis, and spontaneous generation; and, there NEVER will be. Thanks to entropy, random diffusion, or the second law of thermodynamics, it's physically impossible for atoms to spontaneously generate into functional genes, proteins, genomes, eyes, brains, and life forms. It can't be done, which means it wasn't done.

Isn't it fascinating how we can find and know the truth simply by thinking about things critically, logically, impartially, and rationally?

I don't make my living by preaching and teaching evolution, so I'm free to see through all of its deceptions and lies. I'm not motivated by confirmation bias where the theory of evolution is concerned, so I'm free to allow ALL of the evidence into evidence and free to pursue a preponderance of the evidence.

I have observed that the preponderance of the evidence falsifies Materialism, Naturalism, Nihilism, Atheism, and the Theory of Evolution. I have also observed that there are NO observations wherein any aspect of the theory of evolution has been caught in the act. Evolution of new life-forms from pre-existing life-forms has never been observed, which means that the theory of evolution has never been verified. In contrast, design and creation by Intelligent Beings or Intelligent Psyches has been observed, verified, and experienced trillions of trillions of times. So, which explanation for the origin of life is true – the one that has been falsified or the one that has been verified?

Repeated falsification proves that a theory is false. Constant verification implies that a theory is true.

Entropy FALSIFIES the claims of Materialism, Naturalism, Darwinism, and the Theory of Evolution, because evolution is entropy, and entropy cannot design and create anything. Entropy can only degrade and destroy. Entropy cannot design, program, fine-tune, field-test, manufacture, engineer, and create, which means that evolution of any kind cannot design, program, fine-tune, and create. Evolution is entropy.

So, where did life come from since it obviously didn't come from evolution or entropy?

Life came from Syntropy.

Life is syntropy. Psyche is syntropy. Intelligence is syntropy. Physical matter originated from syntropy or spirit matter. Quantum Mechanics is syntropy. The Priesthood Power of God is Syntropy. Quantum Mechanisms are the Priesthood Power of God. Programming code or genomes are the result of syntropy. The arrow of progression in the fossil record, or the ever-increasing complexity of life in the fossil record, is the result of syntropy. WE KNOW this is so because entropy or evolution would prevent it from happening. God is Syntropy.

Life came from Syntropy, not entropy or evolution. Evolution or entropy cannot do life. Evolution or entropy can only do disorder, disease, chaos, cancer, death, and extinction. It takes an infinite amount of blind faith to believe that entropy or evolution can design and create, when ALL of the physical evidence we have on hand is telling us that entropy or evolution can only do disorder, disease, deformity, death, and extinction. The Darwinists and Evolutionists say that evolution is a beautiful concept, but there's nothing beautiful about cancer, disease, death, and extinction. Cancer, disease, deformity, death, and extinction are the result of evolution or random mutations.

Ironically, the very existence of entropy or evolution is Scientific Proof of God's Necessity; and therefore, Scientific Proof of God's Existence. Evolution is entropy, which means that evolution of any kind cannot design and create anything. The chemical evolution of proteins, and the matching genes to go along with them, NEVER happened because entropy prevents it from happening. The development of new species through random mutations and natural selection never happened because entropy prevents it from happening. Consequently, we have to look someplace else besides entropy or evolution for the origin of life on this planet. What's the opposite of entropy or evolution? Syntropy is the opposite of entropy. Syntropy is order and organization. Syntropy is intelligence. God is Syntropy.

It's elementary.

God has to exist, because Someone Psyche had to be there in the first place to wind up the clock or wind up this physical universe with Syntropy in the first place. We couldn't have all of that subsequent entropy without in initial infusion of syntropy. The initial syntropy within this physical universe had to come from someplace, and it came from God and the quantum realm. There's no other logical explanation for its origin. The Syntropy had to come from the syntropy realm, or quantum realm, or psyche realm, or spirit realm because it clearly doesn't exist here in this physical realm. The order, organization, or syntropy couldn't have come from evolution because evolution is chaos and entropy. The order, organization, programming, and structure within 3D proteins, genes, eyes, brains, and genomes had to come from Someone Psyche or Someone Syntropy.

This physical realm and classical physics are based upon entropy. Entropy cannot do design and creation. The psyche realm, quantum realm, and Quantum Mechanics are based upon syntropy, order, and intelligence. Syntropy, order, organization, psyche, and intelligence have the innate ability or the inherent capability of doing design and creation.

Remember, evolution of any kind is entropy; and, evolution of any kind is prevented from happening by entropy. The production of functional genes and proteins from atoms is prevented from happening by entropy. Evolution or entropy cannot produce order or syntropy. It's physically impossible. Therefore, we are looking for a non-physical explanation for the origin of life or the origin of syntropy.

Order and Organization are Syntropy. Life is Syntropy. Psyche is Syntropy. Intelligence is Syntropy. Quantum Mechanisms are Syntropy. Functional genomes and functional proteins are the result of Syntropy. Your genome is God's Signature. God is Syntropy.

Quantum Mechanics is Syntropy, which means that quantum mechanisms are supernatural in nature and origin. Quantum Mechanics is the Priesthood Power of God. Quantum Mechanics is observed, proven, and verified science. It's time that we start using Quantum Mechanics to explain how things really work. The very existence of Quantum Mechanics, Syntropy, Transdimensional Mechanisms, or Supernatural Mechanisms FALSIFIES Materialism, Naturalism, Darwinism, Nihilism, the Theory of Evolution, and even Atheism.

Materialism, Naturalism, Darwinism, Entropy, and Classical Physics completely lack explanatory power when it comes time to explain what the Human Psyche and Nature's Psyche are doing in the quantum realm and how they work in the quantum realm or the syntropy realm. For that explanation, we need quantum mechanisms, supernatural mechanisms, or psyche mechanisms.

Quantum Mechanics is our best-proven, most-verified, and most-used science that we currently have. It's time that we start using Quantum Mechanics or Transdimensional Physics to explain what's happening in the quantum realm, syntropy realm, or psyche realm.

If your interpretation of Quantum Mechanics cannot explain what the Human Psyche and Nature's Psyche are doing at the quantum level in order to get things done for us at the physical level, then your interpretation of Quantum Mechanics is worthless because it's based upon Materialism, Naturalism, Nihilism, and Classical Physics. Naturalism and Classical Physics lack explanatory power when it comes to the psyche realm or the quantum realm because Materialism and Naturalism deny the existence of psyche or syntropy.

You can't use something that is incomplete or false to demonstrate and verify the truth. Materialism, Naturalism, Darwinism, Nihilism, Behaviorism, Determinism, Atheism, and Classical Physics cannot be used to verify the truthfulness and usefulness of Quantum Mechanics, Syntropy, and Psyche. Quantum Mechanics is supernatural in nature and origin, which means that Naturalism and Quantum Mechanics are mutually exclusive. If the one is true, then the other is automatically false.

You are going to have to decide for yourself which one is true, and which one is false; but, I KNOW for myself which one the preponderance of the evidence verifies and which one the preponderance of the evidence falsifies; and, that's good enough for my needs. All I ever really wanted to know is the truth, and now I do.

Mark My Words

—

Source Material

1. ***Scientific Proof of God's Existence: Finding God Where the Atheists Refuse to Look for Him***.

https://www.amazon.com/dp/B07B26CRHX

2. Myers, D. G. (2010). *Social Psychology* (10th ed.). New York: McGraw-Hill.

3. Pinel, J. (2014). *Biopsychology* (9th ed.). New York: Pearson.

4. Sanford, J. (2014). *Genetic Entropy* (4th ed.). Cornell University: FMS Foundation.

References

1. **Quantum Neuroscience: The Answer to Life, the Universe, and Everything**.

https://www.amazon.com/dp/B079Z6QQQB

2. **NATURE vs. NURTURE vs. NIRVANA: An Introduction to Reality**

https://www.amazon.com/dp/B01JWRCSVA

https://www.amazon.com/dp/1521132615

3. **BioPsychoSocial: Including Psyche or Light into our Theoretical Models**

https://www.amazon.com/dp/B0713NDHVW

4. **Science 2.0: I Upgraded My Science**.

https://www.amazon.com/dp/B0771K6WTX

Using the Scientific Method to Falsify Theories

Web Page: https://philosophy-of-science.com/falsifying-theories/

Ask yourself, "What hidden assumptions do the Materialists, Naturalists, Darwinists, Nihilists, and Atheists make at the beginning of each of their scientific arguments or scientific proofs, every time they try to do science?" A lot can be learned by choosing to study people's hidden assumptions.

Ask yourself how these people convince themselves that the Theory of Evolution is true, because they successfully do so.

Ask yourself what logic fallacies these people use to verify and prove that the Theory of Evolution is true.

The Scientific Method as the Materialists, Naturalists, Darwinists, Nihilists, Behaviorists, Determinists, and Atheists typically use the Scientific Method is based upon the *begging the question* and the *affirming the consequent* logic fallacies. These people rely upon these logic fallacies in order to make their case for the Theory of Evolution; therefore, if you are a real scientist, it's extremely important to know what these logic fallacies are and how they work.

This is Philosophy of Science 101.

This topic is important to know about and understand. Science is supposed to be all about falsifying and eliminating everything that is false. Science is supposed to be about observation and experience. Traditional science is not about these things. Traditional science, based upon Materialism and Naturalism, is all about trying to trick us and deceive us into believing that science is Materialism and Naturalism. It's a hoax and a scam that millions of us have fallen for.

Begging the Question

Begging the question is the process of using your pre-chosen conclusion as your hidden assumption, major premise, or primary assumption in all of your scientific arguments, as evidence or proof that your pre-chosen conclusion is true. These people literally start every science experiment with the pre-chosen conclusion that the Theory of Evolution is true; and then, these people use the Truthfulness of Evolution as their first primary assumption or major premise, as evidentiary proof that the Theory of Evolution is true. It's *circular reasoning*, another logic fallacy. Their pre-chosen conclusion is used axiomatically as the hidden assumption or first premise in every scientific argument they make and every science experiment that they do. That's the way these people do science. It's posturing, smoke and mirrors, and deception. It's cheating.

Begging the question consists of making your conclusion one of the axiomatic premises, and effectively results in *jumping to conclusions*. The Materialists and Darwinists do this all the time, without realizing it.

Affirming the Consequent

Let's discuss how Science can go wrong. We can't use Science and the Scientific Methods to verify and know the Truth. Scientific Methods cannot be used to produce Knowledge.

Due to the *affirming the consequent* logic fallacy, and the *jumping to conclusions* logic fallacy, and the *category error* logic fallacies which are built into the Scientific Methods, it is impossible to use Science and the Scientific Methods to know the truth and to prove the truth.

If your ultimate goal in life is to KNOW THE TRUTH and prove the truth, then Science and the Scientific Methods are in fact one of the worst ways for accomplishing that task.

This is how scientific verification works in practice.

The following logical argument outlines the basic approach that has been taken by traditional scientists throughout the history of science:

Scientific Hypothesis: If Theory X is true, then we will observe Y.

Scientific Observations: We observe Y.

Scientific Conclusion: Therefore, Theory X is true.

This sort of thinking, however, reflects a logical fallacy called *affirming the consequent*. Here's a comparable example to demonstrate:

We hypothesize: If Sally's pet is a cat, it will have a tail.

We observe: Sally's pet has a tail.

We conclude: Therefore, Sally's pet is a cat.

We can easily see that this logic is fallacious. Just because we observe Y (a pet with a tail) that does not mean that our theory X (the pet is a cat) is true. After all, dogs, lizards, birds, and mice have tails too.

Yet, **this IS the Scientific Method**, and this is exactly how traditional scientists use the Scientific Methods to demonstrate and prove the truth. The whole enterprise is based upon a logic fallacy or two.

By *affirming the consequent*, you can prove anything to be true – anything – including Materialism, Naturalism, Darwinism, Nihilism, Atheism, and the Theory of Evolution. By *affirming the consequent*, you can prove that Sally's pet is a cat, even though it is in fact a bird. By *affirming the consequent*, you can prove that the theory of evolution is true, even though it is in fact a dog. Do you see how that works?

We hypothesize: If the Theory of Evolution is true, it will match with the fossil record and Darwin's tree of life.

We observe that the Theory of Evolution matches with the fossil record and Darwin's tree of life.

We conclude: Therefore, the Theory of Evolution is true.

This is how they prove that the theory of evolution is true. In fact, I see it on almost every page of my college textbooks where the Evolutionists are simply amazed at how miraculously all that evidence supports the truthfulness and predictions of the Theory of Evolution. These people never realize that they are *affirming the consequent*. They never realize that Darwin's tree of life and the theory of evolution were carefully designed and tailor-made to match with any and all evidence – past, present, and future. But, their arguments are fallacious, and they don't even know it.

Why is this argument fallacious?

It's because some version of Darwin's tree of life can also be made to match with Intelligent Design Theory or Christianity; and, the different types of Christianity and Islam can be made to match the fossil record as well. The whole thing is based upon *circular reasoning*, *begging the question*, or *affirming the consequent*. It's a logic fallacy.

Affirming the consequent is how the Materialists, Naturalists, Darwinists, Nihilists, Behaviorists, Determinists, and Atheists use the Scientific Method to prove that the Theory of Evolution is true. They cheat; and, most of them don't even realize that they are doing so, because they have never studied the Philosophy of Science.

Let's try it and see how it works in practice.

Jumping to conclusions and *affirming the consequent* is precisely what the Darwinists and Atheists DO each and every time they publicly declare that Science has proven the Theory of Evolution true.

Let's demonstrate their error by *affirming the consequent*:

If the Theory of Evolution is true, the fossil record will match with Darwin's Tree of Life.

We observe that the fossil record matches with Darwin's Tree of Life.

Therefore, the Theory of Evolution and Darwinism are true.

This is *affirming the consequent*. This is how the Darwinists and Naturalists prove that the Theory of Evolution is true. They do it every day. You can go online, and you will find them online right now *affirming their conclusions* or *jumping to conclusions*.

Most of the scientists can't see anything wrong with this argument. They use it every day to prove to themselves that the Theory of Evolution is true. They've been taught to do so in school by their science professors.

The problem with this argument, besides the fact that it is *affirming the consequent*, comes from the fact that Intelligent Design Theory, the Spaghetti Monster Theory, and even Christianity match with the fossil record, or can be made to match with the fossil record. Anything can be made to match with the fossil record and the future fossils that we dig out of the ground, including the Theory of Evolution and Darwin's Tree of Life.

Online, we see the Darwinists, Evolutionists, Materialists, Naturalists, and Atheists telling us every day that the fossil record is real and that the fossil record proves that the Theory of Evolution is true. "I've seen the fossils myself. I've seen evolution in action," they say. They are *affirming the consequent* and don't even realize that they are doing so. They have convinced themselves that the fossil record proves that the Theory of Evolution is true. Self-deception works, and it works every time.

Begging the question and *affirming the consequent* are two of the ways by which these people convince themselves that the Theory of Evolution is true. It works, because it was designed to work that way.

The few scientists, who fully understand that *affirming the consequent* can be used and is used to prove that anything is true, will correctly state that the Scientific Methods cannot be used to prove anything true. But, these same people do not realize that that truth and reality applies directly to the Theory of Evolution as well. When it comes to the Theory of Evolution, the very same people employ a *special pleading,* attempting to grant Darwinism, Materialism, and Naturalism immunity from the rules of science. They cheat and grant Darwinism an exemption and make it an exception. The *special pleading* logic fallacy is alive and well where Materialism, Naturalism, Darwinism, Nihilism, and Atheism are concerned. They cheat. Their first premise is allowed to be an Axiom or a Given Truth, which doesn't require any evidence, verification, or proof of its truthfulness; and, their first premise is invariably their pre-chosen conclusion. It's called *begging the question*, and it's a logic fallacy.

Scientific Hypothesis: If the Theory of Evolution is true, we will observe that the Theory of Evolution is true.

Scientific Observation: We observe that the Theory of Evolution is true.

Scientific Conclusion: Therefore, the Theory of Evolution is obviously true.

This is precisely the scientific argument that I watched a couple of the New Atheists using the other day. They were totally convinced that the Theory of Evolution is true. Go figure! They take a blind leap of faith and go straight to their pre-chosen conclusion. They use their pre-chosen conclusion as proof that their pre-chosen conclusion is true.

You can't lose when it comes to *begging the question* and *affirming the consequent*. It is "heads I win, tails you lose". It makes you a winner every time! But, that's exactly the way that these people do science, by *affirming the consequent* with every word that they speak.

Proving the Theory of Evolution True

Due to the *affirming the consequent* logic fallacy, and the *jumping to conclusions* logic fallacy, and the *category error* logic fallacies which are built into the traditional scientific methods, it is impossible to use Science and the Scientific Methods to know the truth and to prove the truth. This is Philosophy of Science 101; and, almost everyone is completely oblivious to how often they use *affirming the consequent* to make their case and to prove their point. Scientists and PhDs are particularly vulnerable to this logic fallacy because this is the way they were taught to do science while they were in school.

This is how *affirming the consequent* works in practice. The following logical argument is the traditional Scientific Method in action. The following logical argument outlines the basic approach that has been taken by scientists to prove that the Theory of Evolution is true:

Scientific Hypothesis: If Theory X is true, then we will observe Y.

Scientific Observations: We observe Y.

Scientific Conclusion: Therefore, Theory X is true.

This IS the Scientific Method in action! This is the way that the Scientific Method is traditionally used to do science. This sort of thinking, however, reflects a logic fallacy called *affirming the consequent*.

Now, let's do as the Materialists, Naturalists, Darwinists, and Atheists do – let's use *affirming the consequent* to prove that the Theory of Evolution is true.

If the Theory of Evolution is true, then we will observe natural selection.

We observe natural selection.

Therefore, the Theory of Evolution is true.

We see the Materialists, Naturalists, Darwinists, and Atheists everyday online and within their research papers using natural selection to prove that the theory of evolution is true; and, these people don't stop there.

If the Theory of Evolution is true, then we will observe random mutations.

We observe random mutations.

Therefore, the Theory of Evolution is true.

We see the Materialists, Naturalists, Darwinists, and Atheists everyday online and within their research papers using random mutations to prove that the theory of evolution is true.

These people literally use the existence of natural selection and random mutations as scientific proof that Evolution happens and as scientific proof that the Theory of Evolution is true. That's how these people prove that the Theory of Evolution is true; and, millions have fallen for the deception or the ruse. They can't see anything wrong with it.

If Chemical Evolution is true, then we will observe genes and proteins.

We observe genes and proteins.

Therefore, Chemical Evolution is true, and the Theory of Evolution is true.

We see the Materialists, Naturalists, Darwinists, and Atheists everyday online and within their research papers using Chemical Evolution and the existence of genes and proteins as proof that the Theory of Evolution is true. Clearly, evolution made those genes and proteins, because nothing else could have. That's *begging the question* and *affirming the consequent*.

These people literally use the existence of genes and proteins as scientific proof that Chemical Evolution happened and as scientific proof that the Theory of Evolution is true.

These people are using the *affirming the consequent* logic fallacy to prove that the Theory of Evolution is true; and, the unwary, the uneducated, and the scientists themselves don't even realize that they are doing so. We are being tricked and deceived, and we don't even know it.

We see these arguments being used every day online; and, nobody is aware of the logic fallacy that is contained within them.

I have seen this argument hundreds of times, and so have you.

Definition: Evolution is natural selection and random mutations. Evolution is genetic change.

Obviously, we observe natural selection, random mutations, and genetic change every day on this planet.

Therefore, the Theory of Evolution is obviously true; and, natural selection, random mutations, and genetic change obviously designed and created all of the life forms on this planet.

By *affirming the consequent*, these people *jump to the conclusion* that the Theory of Evolution is true without even realizing that they are doing so. It makes logical sense to them. They simply know that the Theory of Evolution is true, and never once do they question it. Many of them promote it with a blind-faith and zeal that rivals anything that can be found from an Evangelical Christian or a Muslim Fanatic.

> **Of course, the Theory of Evolution is true because natural selection and random mutations have been observed.**

These people don't see anything wrong with their argument. In their minds, they have proven that the Theory of Evolution is true.

> **Of course, Evolution is true. It has been observed. Random mutations and natural selection have been observed. Of course, the Theory of Evolution is true. It's obvious. It's been observed.**

These people *affirm the consequent* and *jump straight to the conclusion* that the Theory of Evolution is true. This is how they do science, and they can't see anything wrong with it.

The following is one of my favorites. I see it being used all the time by PhD scientists online, in their research papers, and everywhere else.

> **Scientific Hypothesis: If Materialism, Naturalism, Darwinism, Nihilism, Atheism, and the Theory of Evolution are true, then we will observe entropy.**
>
> **Scientific Observation: We observe entropy all around us all the time.**
>
> **Scientific Conclusion: Therefore, Materialism, Naturalism, Darwinism, Nihilism, Atheism, and the Theory of Evolution are true.**

This IS the Scientific Method in action – the way that the Scientific Method is traditionally used to do science. Most people can't see anything wrong with it; but, it's based upon a logic fallacy called "affirming the consequent".

There's nothing wrong with *affirming the consequent*; is there? As long as it gets us the truth, it should be okay; shouldn't it?

Well, let's demonstrate what's wrong with *affirming the consequent* and *jumping to conclusions*.

> **If the Intelligent Design Theory is true, then we will observe natural selection.**
>
> **We observe natural selection.**
>
> **Therefore, the Intelligent Design Theory is true, and God clearly designed living organisms to be influenced by natural selection and sexual selection.**

Notice how I first *affirmed the consequent*, and then I *jumped to the conclusion* that I wanted to be true.

By *affirming the consequent* and *jumping to conclusions*, we can prove anything to be true.

> **If the Christianity is true, then we will observe random mutations.**
>
> **We observe random mutations.**
>
> **Therefore, Christianity is true, and God clearly designed and created our genomes so that random mutations would be possible and would be observed.**

Notice how I first *affirmed the consequent*, and then I *jumped to the conclusion* that I wanted to be true.

By *affirming the consequent* and *jumping to conclusions*, we can prove anything to be true; and, that's why *affirming the consequent* and *jumping to conclusions* are logic fallacies. By using logic fallacies, we can prove anything to be true.

> **If Materialism and Naturalism are true, then we will observe physical matter.**
>
> **We observe physical matter.**
>
> **Therefore, Materialism and Naturalism are true.**

By *affirming the consequent*, Materialists, Naturalists, and Scientists prove that Materialism and Naturalism are true. This is the traditional Scientific Method in action. You see them online each and every day within their scientific arguments and research papers proving that Materialism, Naturalism, and the Theory of Evolution are true by *affirming the consequent* and *jumping to conclusions*. That's the way these people do science. We observe them doing so every day and for centuries at a time.

It's easy to prove anything true by *affirming the consequent*. These people do it all the time, and they don't even know it.

> **Scientific Hypothesis: If Darwinism and the Theory of Evolution are true, then we will observe DNA, genes, genomes, proteins, eyes, brains, and living life forms.**
>
> **Scientific Observations: We observe DNA, genes, genomes, proteins, eyes, brains, and living life forms.**
>
> **Scientific Conclusion: Therefore, Darwinism and the Theory of Evolution are true; and, Evolution clearly designed and created our DNA, genes, genomes, proteins, eyes, brains, and physical bodies.**

Notice how I first *affirmed the consequent*, and then I *jumped to the conclusion* that I wanted to be true. It works every time, even though it is a logic fallacy.

We see the New Atheists using this argument all the time to make their case and prove that the Theory of Evolution is true. These people are *affirming the consequent* and *jumping to conclusions*, and they don't even know it.

I see these very same arguments in almost ALL of the college textbooks that I own and online every day. By *affirming the consequent* and *jumping to conclusions*, these scientists prove that the Theory of Evolution is true. They are completely oblivious and unaware of the logic fallacies that they used to make their case. They were taught in college that *affirming the consequent* is the way that they are supposed to use the Scientific Method to do science. They were taught wrong; but, most of them don't even know it.

Falsifying Theories by Negating the Consequent

Now, let's use the Scientific Method for what it was designed for. Let's falsify some theories. Unlike *affirming the consequent*, *negating the consequent* or *falsifying a theory* is philosophically and logically sound. Here's how *falsifying a theory*, or *negating the consequent*, works in principle using the Scientific Method:

Scientific Hypothesis: If Theory X is true, then we will observe Y.

Scientific Observations: We don't observe Y.

Scientific Conclusion: Therefore, Theory X is false and has been falsified by the Scientific Method.

Here's how *negating the consequent* or *falsifying a theory* works in practice:

Scientific Hypothesis: If the Theory of Evolution, Materialism, Naturalism, and Darwinism are true, then we will observe the rocks and physical reactions designing, creating, and manufacturing genomes and life forms from scratch. If the Theory of Evolution is true, we will observe chemical evolution, abiogenesis, and spontaneous generation all around us all the time in real time.

Scientific Observations: We have NEVER observed the rocks and physical reactions designing and creating genomes and life forms; and, we NEVER will. They can't. The chemical evolution, abiogenesis, or spontaneous generation of proteins and genes from atoms is physically impossible and has NEVER been observed.

The Scientific Conclusion: Therefore, the Theory of Evolution, Materialism, Naturalism, and Darwinism are false and have been falsified by the Scientific Method or Scientific Observations.

I just successfully falsified Materialism, Naturalism, Darwinism, and the Theory of Evolution; and, I used the Scientific Method to do so! Best of all, my argument is philosophically and logically sound. I have in fact **falsified** Materialism, Darwinism, Naturalism, Creation by Rocks, and the Theory of Evolution for REAL! I have PROVEN them false! It's that simple! I successfully used the Scientific Methods for what they are good for – falsifying theories which are false. I wish I would have known how to do that forty years ago. It would have saved me a lot of confusion, frustration, time, and grief.

Materialism, Naturalism, Atheism, and their derivatives are falsified by a complete lack of observational evidence supporting their Major Premises or Hidden Assumptions. That's what falsifying a theory means – eliminating every premise, theory, and conclusion that has NEVER been experienced nor observed.

Now, let's run Atheism through the Scientific Method by *negating the consequent* to falsify it:

Scientific Hypothesis: If Atheism is true, we will observe "nothing" designing, creating, and manufacturing everything, including genomes and life forms. It would be chaos, but that's exactly what we should be observing if Atheism were true.

Scientific Observations: Obviously, we have NEVER observed "nothing" designing and creating anything; and, we NEVER will. It's patently absurd.

The Scientific Conclusion: Therefore, Patent Absurdity or Atheism is false and has been falsified by the Scientific Method and Scientific Observations.

I just falsified Atheism for REAL; and, it's philosophically and logically sound. I have in fact falsified Atheism, using the Scientific Method and Scientific Observations to do so!

Materialism, Naturalism, Atheism, and their derivatives are falsified by a complete lack of observational evidence supporting their Major Premises or Hidden Assumptions. That's what falsifying a theory means – eliminating every premise, theory, and conclusion that has NEVER been experienced nor observed. Materialism, Naturalism, Darwinism, and Atheism have been falsified or proven false. There's no way to observe that something does not exist. There's no way to observe that the non-physical does not exist. There's no way to observe that the supernatural or the quantum does not exist. There's no way to observe that an after-life does not exist. There's no way to observe that life has no purpose and no meaning. There's no way to observe that God does not exist. There's NO way to use *affirming the consequent* to prove the truth.

However, you can definitely use *negating the consequent* to prove that your favorite theories are false. Such a course of action is philosophically valid and scientifically sound.

Remember, there's only one way to use the Scientific Method "to prove the truth" and that is to use *negating the consequent* and the Scientific Methods to eliminate everything that is false so that ONLY the truth or ONLY the observed remains. It's fascinating to observe what remains after eliminating everything that you know to be false or everything that has been proven false, such as Materialism, Naturalism, Darwinism, Nihilism, Atheism, and their derivatives.

Falsifying the Theory of Evolution

Now, let's use the Scientific Method for what it was designed for. Let's falsify some of people's favorite theories. *Negating the consequent* or *falsifying a theory* is philosophically valid and logically sound. *Negating the consequent* is NOT a logic fallacy. *Negating the consequent* is the way that Science should be done and the way that the Scientific Methods should be used.

Sherlock Holmes taught me how this works in principle or practice.

> **How often have I said to you that when you have eliminated the impossible** [and the false], **whatever remains, however improbable, must be the truth? — Sherlock Holmes.**

Sherlock Holmes and the Philosophy of Science taught me how to use the Scientific Method to falsify the Theory of Evolution, Materialism, Naturalism, and Atheism.

Here's how *falsifying a theory*, or *negating the consequent*, works in principle using the Scientific Method:

Scientific Hypothesis: If Theory X is true, then we will observe Y.

Scientific Observations: We don't observe Y.

Scientific Conclusion: Therefore, Theory X is false and has been falsified by the Scientific Method.

This scientific argument is philosophically valid and logically sound. It works. Theory X has been proven false by the Scientific Method. Theory X has been falsified by a complete lack of observational evidence supporting the hypothesis. That's how you falsify a theory! It's dependable. It works.

One of the primary tenets of Science 2.0 is the observation that if we successfully eliminate everything that is false or everything that has been falsified, then ONLY the truth will remain. Under Science 2.0, the truth is found by eliminating everything that is false, everything that is impossible, and everything that has been falsified.

Observe that *negating the consequent* or *falsifying a theory* requires careful consideration of what one is really trying to prove. In other words, *negating the consequent* requires a great deal more intelligence and precision than *affirming the consequent*.

Let's use *negating the consequent* and the Scientific Method to falsify a theory or two. This can be a great deal of fun to do.

Scientific Hypothesis: If Chemical Evolution is true, then we will observe atoms spontaneously generating into DNA, genes, genomes, and proteins.

> **Scientific Observations:** We have NEVER observed atoms spontaneously generating into DNA, genes, genomes, and proteins. It's physically impossible. It's prevented from happening by entropy or the second law of thermodynamics.
>
> **Scientific Conclusion:** Therefore, Chemical Evolution is false and has been falsified by the Scientific Methods. In other words, Chemical Evolution is falsified by a complete lack of observational support. Chemical Evolution or spontaneous generation has NEVER been observed.

Thanks to entropy or the second law of thermodynamics, it's physically impossible for atoms to spontaneously generate into functional proteins and the matching genes to go along with those proteins. Chemical Evolution is physically impossible. It's prevented from happening by entropy. That's the REAL science. Chemical Evolution has been falsified by a complete lack of supporting evidence. Since Chemical Evolution is false and impossible, the Theory of Evolution is false and impossible.

Let's try another one by using the *negating the consequent* version of the Scientific Method:

> **Scientific Hypothesis:** If Macro-Evolution is true, then we will observe cats occasionally giving birth to dogs, and we will observe the occasional ape giving birth to humans.
>
> **Scientific Observations:** We have NEVER observed one species giving birth to a completely different species. It's physically impossible. It's prevented from happening by genetics and random mutations.
>
> **Scientific Conclusion:** Therefore, Macro-Evolution is false and has been falsified by the Scientific Method. In other words, Macro-Evolution is falsified by a complete lack of observational support.

Thanks to entropy or the second law of thermodynamics, it's physically impossible for random mutations to produce a functionally compatible or genetically compatible Mr. and Mrs. Mutant at the very same time in the very same place. It can't be done, which means that it wasn't done. Thanks to random mutations or entropy, it's physically impossible for chimp-like ancestors to give birth to genetically compatible male and female chimpanzees, or genetically compatible male and female human beings. It can't be done, which means that it wasn't done. That's the REAL science.

Macro-Evolution is false because it is prevented from happening by entropy, random mutations, and genetics. Macro-Evolution has NEVER been experienced nor observed; and, it NEVER will be, because it is physically impossible.

> **Scientific Hypothesis:** If the Theory of Evolution is true, then we will observe natural selection and random mutations producing new and unique life-forms in real-time. It should be happening here and now, and not in some distant unseen past.

> **Scientific Observations:** We have NEVER observed one species giving birth to a completely different species. It's physically impossible. It's prevented from happening by genetics and random mutations.
>
> **Scientific Conclusion:** Therefore, the Theory of Evolution is false and has been falsified by the Scientific Method. In other words, the Theory of Evolution is falsified by the fact that random mutations and natural selection have NEVER been caught in the act of designing and creating new and unique life-forms from scratch.

These are powerful observations, which just happen to be true. The Theory of Evolution is falsified by a complete lack of observational support for its major claim or primary premise, which states that natural selection and random mutations designed and created ALL of the life forms on this planet. In truth, evolution (genetic change), random mutations, and natural selection didn't even exist until AFTER God designed, created, and manufactured the genes, proteins, genomes, eyes, brains, and physical life forms in the first place. Your genome is God's Signature. This is Logic 101; and, it falsifies the Theory of Evolution. Evolution, natural selection, and random mutations didn't exist and couldn't exist until after Someone Psyche or Someone Intelligent designed and created the genes and the genomes in the first place.

Furthermore, random mutations and natural selection cannot do computer programming or genetic programming. A genome is programming code and computer hardware. Thanks to entropy, it's physically impossible for random mutations to design, program, engineer, create, and manufacture functional genomes from scratch. Random mutations are entropy, NOT designers and creators.

The false is falsified by the truth. The false is falsified by observation and common sense; whereas, the truth is repeatedly experienced and observed by Someone Psyche somewhere sometime.

Let's falsify a couple other theories by using the *negating the consequent* version of the Scientific Method:

> **Scientific Hypothesis:** If the claims of Materialism and Naturalism are true, then we will observe atoms spontaneously generating into DNA, genes, genomes, and proteins.
>
> **Scientific Observations:** We have NEVER observed atoms spontaneously generating into DNA, genes, genomes, and proteins. In fact, spontaneous generation was falsified in 1859 by Louis Pasteur.
>
> **Scientific Conclusion:** Therefore, Materialism and Naturalism are false and have been falsified by the Scientific Method. In other words, Materialism and Naturalism are falsified by a complete lack of observational evidence supporting their hidden assumptions or major premises which claim that the non-physical, the supernatural, or the

quantum does not exist. It's physically impossible to observe that something does not exist.

The Materialists and Naturalists *jump to the conclusion* that only physical matter exists and therefore atoms must spontaneously generate into DNA, genes, proteins, genomes, and life forms. They are wrong. Spontaneous Generation, Abiogenesis, Materialism, Naturalism, Darwinism, and the Theory of Evolution were FALSIFIED by Louis Pasteur in 1859, the very same year that Charles Darwin published "On the Origin of Species". We knew the truth from the very beginning but chose to ignore it.

Once we have successfully falsified Materialism and Naturalism, then the Theory of Evolution and Atheism go down with them as fruit from the poisoned tree. Atheism – the idea that Nothing designed and created everything – is patently absurd. Atheism is creation ex nihilo – creation from nothing by nothing. Atheism or creation ex nihilo is obviously false to any rationally thinking person who comes across it. "Nothing" cannot design and create anything.

Jumping to conclusions, the Materialists, Naturalists, Darwinists, Nihilists, and Atheists define science as Materialism, Naturalism, Darwinism, Nihilism, and Atheism. That's the way these people do science. It's *begging the question* – using their pre-chosen conclusion as scientific evidence that their pre-chosen conclusion is true. It's *circular reasoning*.

The Materialists and Naturalists define science as Materialism and Naturalism, thereby making Materialism and Naturalism axiomatically true and thereby proving that Materialism and Naturalism are true. That's *begging the question*. By *affirming the consequent*, *begging the question*, and *jumping to conclusions*, you can prove anything to be true – even Materialism, Naturalism, Atheism, and the Theory of Evolution. It works, because logic fallacies can be used to prove anything true. It's called sophistry. Self-deception works, and it works every time.

In contrast, *negating the consequent* or *falsifying a theory* involves careful examination of the observational evidence and careful examination of the hidden conclusions that these people are jumping to.

I wish I would have known about *negating the consequent* or *falsifying theories* forty years ago when I was first in college; but, these are concepts that you will NEVER learn about from the Materialists, Naturalists, and Atheists who control and run our public schools. This type of information is banned from our public schools and science labs because it falsifies Materialism, Naturalism, and Darwinism. This type of information is forbidden in our public schools. The purpose of our public schools is to indoctrinate you in Materialism, Naturalism, Darwinism, Scientism, Nihilism, Behaviorism, Determinism, Classical Physics, and Atheism. Yet, these falsifiable and falsified philosophies or religions are in fact FALSIFIED by *negating the consequent*. The truth can be found through a process of elimination by falsifying and eliminating from science anything that is false.

This is the way that science should be done – by falsifying and eliminating theories. This is the way that science should have been done but wasn't. Instead, by defining science as Materialism, Naturalism, Darwinism, Nihilism, and Atheism, falsified theories are retained as the core foundation of the physical sciences and classical physics. In other words, these people cheat and use logic fallacies to make their case and prove that Materialism, Naturalism, and the Theory of Evolution are true.

When it comes to Science and my understanding of the Scientific Methods, the eight months I spent studying the Philosophy of Science, *affirming the consequent*, and *negating the consequent* paid the most dividends in the end. It is powerful stuff and very interesting too. By using the *negating the consequent* version of the Scientific Method, I have successfully FALSIFIED Materialism, Naturalism, Darwinism, Atheism, and the Theory of Evolution many different times in many different ways.

In order to find and know the truth, we must successfully eliminate everything that is false, starting with Materialism, Naturalism, Darwinism, Nihilism, Atheism, and their derivatives.

Negating the consequent or *falsifying theories* was really cool once I discovered how it truly works. Now that I KNOW how *negating the consequent* works and how it can be used to prove theories false, there's NO going back to *affirming the consequent* and the falsehoods and lies that it is being used to produce. Cool, huh?

Using the Scientific Method to falsify Materialism, Naturalism, Darwinism, Atheism, and the Theory of Evolution just might be the greatest scientific discovery that I have made during my lifetime. I certainly find it powerful and useful. I wish I would have known how to falsify these false philosophies or false religions forty years ago when I was first in college. It would have saved me a ton of time, confusion, and grief. All you want is the truth, because falsehoods and falsified theories don't do anyone any good, unless you are making a living off the lies.

What do you think?

Is my discovery of the difference between *affirming the consequent* and *negating the consequent* one of my greatest scientific discoveries of all time; or, am I deluding myself? Self-deception works, and it works every time. Am I deceiving myself; or, do you think I have a point?

Obviously, I'm not the first person to discover *affirming the consequent* and *negating the consequent*; but, as far as I know, I am the first person to apply them directly to Materialism, Naturalism, Darwinism, Nihilism, Atheism, and the Theory of Evolution.

Ever since I successfully falsified Materialism, Naturalism, and their derivatives, I HAVE KNOWN that they are false because I HAVE KNOWN why they are false. That knowledge is particularly valuable to me because I used to be a Materialist, Naturalist, Nihilist, and Atheist. In my ignorance, I couldn't see anything wrong with the arguments and scientific proofs that were being presented to me by these people. I fell for it – hook, line, and sinker.

But now, I have seen the light.

Science 2.0 Is the Way that Science Should Have Been Done

Science 2.0 deliberately allows ALL of the evidence into evidence, and the goal is to pursue a preponderance of the evidence. Like the forensic sciences, Science 2.0 relies upon a Burden of Proof Methodology when it comes to the non-physical and the non-replicable; and, a Burden of Proof Methodology is dependent upon a preponderance of the evidence just like in a court of law. The goal is to get beyond a reasonable doubt.

The cool thing about Science 2.0 is that it allows into evidence EVERYTHING that we have observed and experienced as a race concerning the physical realm, as well as EVERYTHING that we have lived and experienced and observed in respect to the quantum realm or the supernatural realm.

Science 2.0 defines "science" as observation and experience. Under Science 2.0, nothing is off-limits or out-of-bounds. ALL of the evidence is allowed into evidence, as science should be done. When it comes to the quantum realm, or the supernatural realm, or the transdimensional realm, or the spirit world, Science 2.0 relies upon the observed, verified, and proven evidence from Quantum Mechanics, or Transdimensional Physics, or Supernatural Physics to explain what's happening in the quantum realm. It works, if allowed to work.

Science 2.0 demands that we find an Interpretation of Quantum Mechanics that explains in detail what the Human Psyche and Nature's Psyche are doing at the quantum level in order to get things done for us at the physical level. I found two such Interpretations for Quantum Mechanics, and I discuss them in detail in my books about Quantum Mechanics:

"Quantum Neuroscience: The Answer to Life, the Universe, and Everything".

"Quantum Mechanics from a Non-Physical Spiritual Perspective".

How often have I said to you that when you have eliminated the impossible [and the false], **whatever remains, however improbable, must be the truth? — Sherlock Holmes.**

This is Science 2.0 in action. The goal of Science 2.0 is to eliminate the impossible, the false, and the falsified while at the same time keeping and promoting the experienced and the observed.

If we successfully eliminate everything that is false, then only the TRUTH will remain. It's fascinating to observe what remains after Materialism, Naturalism, Darwinism, Nihilism, Behaviorism, Determinism, Scientism, Atheism, and the Theory of Evolution have been FALSIFIED and ELIMINATED from science.

The truth remains.

The VERIFIED, the OBSERVED, and the EXPERIENCED remain. The eye-witness evidence, the experiential evidence, the observational evidence, and the empirical evidence remain. Science is observation, or it should be. Science 2.0

defines science as Observation and Experience. The TRUTH is repeatedly verified, repeatedly experienced, and repeatedly observed by everyone who is willing to observe and learn. Science 2.0 allows ALL of the evidence into evidence and then pursues a preponderance of the evidence. Science 2.0 uses a Burden of Proof Methodology, which is based upon a preponderance of the observed evidence. Science 2.0 is how science should have been done but wasn't.

According to Science 2.0, if we successfully eliminate everything that is false, then ONLY the TRUTH will remain. According to Science 2.0, the BEST and FASTEST way to find and know the truth is to observe it, live it, and experience it for yourself, or to choose to trust someone who has. According to Science 2.0, science is observation, and NOT philosophical speculation and wishful thinking. According to Science 2.0, Materialism, Naturalism, Darwinism, Nihilism, Atheism, and the Theory of Evolution are FALSIFIED due to a complete lack of observational evidence supporting their hidden assumptions or major premises which claim that the supernatural or the quantum does not exist.

Quantum Mechanics is Supernatural Mechanics – it's non-physical in nature and origin. Quantum Mechanisms or Psychic Mechanisms in their natural state or their original non-local non-physical state are NOT limited by physical laws nor physical restrictions. Quantum Mechanisms are Syntropy, which means that they are not subject to entropy in their original primal state. In a physical state, however, Quantum Mechanisms, Supernatural Mechanisms, or Psyche Mechanisms are dampened, limited, and restrained by physical matter and entropy. This is the way things really work, at both the quantum level and the physical level.

Science 2.0 chooses to give precedence to quantum mechanisms or supernatural mechanisms, because they have been observed, experienced, and verified. Likewise, Science 2.0 eliminates Materialism, Naturalism, and Darwinism from science because they are falsifiable and have been falsified due to a complete lack of observational evidence supporting their hidden assumptions and hidden conclusions which claim that the supernatural or the quantum does not exist.

Under Science 2.0, the psyche level or the quantum level or the supernatural level remains, because it has been OBSERVED, EXPERIENCED, and repeatedly VERIFIED by lived experiences and scientific experimentation. Quantum Mechanics is our best-proven, most-verified, and most-used science that we have. Quantum Mechanics is Supernatural Mechanics. Likewise, the physical level remains under Science 2.0, because the physical level has been OBSERVED, EXPERIENCED, and repeatedly VERIFIED by lived experiences and scientific experimentation.

In contrast, Materialism and Naturalism are falsified by Science 2.0, because the non-existence of the supernatural or the non-existence of the quantum has NEVER been observed, experienced, nor verified. Materialism, Naturalism, and their derivatives are FALSIFIED by observational evidence to the contrary. The observation and experience of Quantum Mechanisms, or Psychic Mechanisms, or Supernatural Mechanisms FALSIFIES Materialism, Naturalism, and their derivatives.

Under Science 2.0, design and creation by Intelligent Beings remains because it has been OBSERVED, EXPERIENCED, and repeatedly VERIFIED by lived experiences and scientific experimentation. Science 2.0 retains what is TRUE because it retains what has been observed, experienced, lived, and remembered. Likewise, Science 2.0 consistently eliminates everything that is false and has been falsified like Materialism, Naturalism, Darwinism, Nihilism, Scientism, Behaviorism, Determinism, Atheism, and the Theory of Evolution.

Science 2.0 is the way that science should have been done but wasn't.

Mark My Words

—

Source

This was adapted from: *Using the Scientific Method to Eliminate the Usual Suspects and to Prove the Truth*.

https://www.amazon.com/dp/B01J6STHP0

https://www.amazon.com/dp/1521133581

References

Rychlak, J. F. (1981). *A Philosophy of Science for Personality Theory* (2nd ed.). Malabar, FL: Robert E. Krieger Publishing Company.

Slife, B. D. & Williams, R. N. (1995). Science and Human Behavior. In *What's Behind the Research? Discovering Hidden Assumptions in the Behavioral Sciences*, (pp. 167–204). Thousand Oaks, CA: SAGE Publications.

http://mypsyche.us/science/

Science 2.0: I Upgraded My Science.

https://www.amazon.com/dp/B0771K6WTX

Quantum Neuroscience: The Answer to Life, the Universe, and Everything.

https://www.amazon.com/dp/B079Z6QQQB

Quantum Mechanics from a Non-Physical Spiritual Perspective.

https://www.amazon.com/dp/B01J023TGU

https://www.amazon.com/dp/1521132380

Using the Scientific Method to Eliminate the Usual Suspects and to Prove the Truth.

https://www.amazon.com/dp/B01J6STHP0

https://www.amazon.com/dp/1521133581

Analysis of Falsification

What did we just learn from this exercise?

We learned that the Scientific Methods can be used to prove theories false or to falsify theories. Technically, according to the laws of logic, the Scientific Method cannot be used to prove anything true, due to the *affirming the consequent* logic fallacy; however, the Scientific Methods can definitely be used to falsify theories and thereby prove that they are false. Falsifying a theory means proving that a theory is false. The Scientific Methods are used to prove theories false; and, it works. In contrast, *affirming the consequent* can be used to trick us and deceive us. Self-deception works, and it works every time.

What else did we learn from this exercise in falsification?

We learned that a lack of observation FALSIFIES theories. That's what we really learned, isn't it? We learned that a lack of observation or a lack of observational evidence is the foundational hallmark of Materialism, Naturalism, Darwinism, Nihilism, and Atheism. These philosophical theories are FALSE because there is NO observational evidence supporting their major premises or primary assumptions which claim that the quantum, or the supernatural, or the non-physical does not exist.

We learned that there is no evidence and can be no evidence for the claim that Psyche does not exist. We learned that there is no evidence and can be no evidence for the claim that the Non-Local or the Non-Physical does not exist. We learned that there is no evidence and can be no evidence supporting the materialistic and naturalistic claim that the quantum level or the psyche level does not exist. We learned that ALL of the observational evidence and experimental evidence that we have on hand as a race FALSIFIES Materialism, Naturalism, Darwinism, Nihilism, Behaviorism, Scientism, Determinism, Physical Reductionism, and Atheism.

We learned that science begins by allowing ALL of the evidence into evidence. We learned that whenever science is done right, by *negating the consequent* or by falsifying and eliminating theories that are false, we are pointed to the truth as a result. We learned that the truth is discovered by falsifying and eliminating Materialism, Naturalism, Darwinism, Atheism, and their derivatives. We will learn that it can be extremely fascinating and informative to study what remains after everything that is false has been deliberately eliminated.

We learned that if you successfully eliminate everything that is false, then only the truth will remain. If you successfully eliminate everything that has NEVER been experienced nor observed by anyone, then ONLY the truth will remain. If you successfully allow ALL of the evidence into evidence, then ONLY the truth will remain. We learned that science is personal experience and observation.

We learned that we don't have to run any science experiments in order to find and know the truth. We can go straight to KNOWING the TRUTH simply by observing it, experiencing it, and living it for ourselves. We learned that the BEST and the fastest way to find and know the truth is to live it and experience it for yourself, or to choose to trust someone who has. We learned that the second-best way to find and know the truth is to eliminate everything that is false so that only the truth remains. We learned that the discovery of truth starts by eliminating Materialism, Naturalism, Darwinism, Nihilism, Atheism, and their derivatives.

Powerful, is it not?

We learned that there is NO empirical evidence and will NEVER be any observational evidence supporting the major premises or primary assumptions of Materialism, Naturalism, Darwinism, and Atheism. We learned that the Scientific Method FALSIFIES Materialism, Naturalism, Darwinism, and Atheism. We learned that there is NO evidentiary support for Materialism, Naturalism, Darwinism, Atheism, and the Theory of Evolution. There are NO observations and can NEVER be any observations supporting the claims of Materialism, Naturalism, Darwinism, and Atheism, because these things have to be taken on blind faith as being real and true, because they can't be experienced nor lived nor learned from.

We learned that there is NO phenomenological support or observational support for Materialism, Naturalism, Darwinism, and their derivatives. These things are pure hypothesis, pure philosophy, pure fiction, pure speculation, pure dogma, pure religion, pure pseudo-science, and the very pinnacle of wishful thinking.

We learned that Materialism, Naturalism, Darwinism, Nihilism, Atheism, and the Theory of Evolution are FALSE because there is no observational evidence supporting them. We learned that science is OBSERVATION. We learned that you can get a lot of mileage out of the truth, especially when that truth is experienced and observed.

We learned that most of us have been brainwashed and conditioned in our public schools not to be able to see, understand, nor accept these realities and truths. You will NEVER learn anything about any of this from your materialistic, naturalistic, and atheistic professors and teachers in our public schools because this kind of information has been deliberately banned, blocked, censored, and eliminated from the curriculum. The purpose of our public schools is to indoctrinate us into Materialism, Naturalism, Darwinism, Nihilism, Scientism, Behaviorism, Determinism, and Atheism.

Fascinating, is it not?

It explains what has been done to us. It answers a lot of unasked questions, doesn't it? Instantly, the first fifty years of my life made sense to me, as it finally dawned on me how I had been tricked and deceived all of my life. Ye shall know the truth, and the truth shall set you free. My hope is to set a lot of people free, as I have been set free.

Ask yourself, "What hidden assumptions do the Materialists, Naturalists, Darwinists, Nihilists, and Atheists make at the beginning of each of their scientific arguments or scientific proofs, every time they try to do science?"

These people always start with the hidden assumption that Psyche, Spirit Matter, Quantum Non-Local Consciousness, Syntropy, God, Action at a Distance,

Quantum Non-Locality, and Quantum Mechanisms do NOT exist. *Begging the question*, these people always start with these hidden assumptions; and therefore, these people always come to the very same conclusions – there is NO such thing as Syntropy, God, Quantum Non-Locality, Action at a Distance, Psyche, Choice, Spirit Matter, Consciousness, or Eternal Life. Garbage in, then garbage out. These hidden assumptions are inserted at the beginning of every argument that these people make, whether they realize it or not. Consequently, these people always come to the same conclusions. It's unavoidable. If you start your science with false assumptions, you are going to end up with false conclusions when you are done. Simple. Logical. True.

There's no way possible to observe that Psyche and Syntropy do not exist. It's physically impossible to prove that Psyche and Syntropy do not exist. There's no way to observe that spirit matter does not exist. There's no way to observe that God does not exist. There is no way to observe that the quantum realm, the transdimensional realm, the spirit realm, or the non-local realm does not exist. In fact, Materialism, Naturalism, Darwinism, Nihilism, and Atheism completely fall apart and fail each and every time someone sees, talks with, and OBSERVES our Resurrected Lord Jesus Christ. Observation destroys Materialism, Naturalism, Darwinism, and their derivatives.

Each time we OBSERVE or experience Syntropy, God, Action at a Distance, Resurrected Beings, Consciousness, Quantum Mechanics, Psyche or Choice, Thoughts and Memories, and the Spirit World, those observations and experiences FALSIFY Materialism, Naturalism, Darwinism, Nihilism, Behaviorism, Scientism, and Atheism. That's the way evidence works! Evidence falsifies everything that is false. Science is observation, or it should be. Observation falsifies Materialism, Naturalism, and their derivatives. Observations should be given precedence and priority over philosophical speculations and wishful thinking, especially when it comes to science.

The very existence of Quantum Mechanics proves that the Materialists and Naturalists are wrong. Quantum Mechanics is Syntropy. The very existence of Quantum Mechanics, Syntropy, Action at a Distance, Quantum Non-Locality, and Psyche or Quantum Non-Local Consciousness FALSIFIES Materialism, Naturalism, Darwinism, and their derivatives.

Quantum Mechanics is very difficult to falsify, due to the fact that it exists at the quantum level or the psyche level. However, Quantum Mechanics is constantly and repeatedly VERIFIED each and every time we get smart enough as a race to design an experiment that will indeed verify it or use it. Quantum Mechanics is verified by the effects that it has on physical matter. Syntropy, Quantum Mechanics, Quantum Non-Local Consciousness, Intelligent Intervention, Action at a Distance, or Transdimensional Physics is very REAL; and, Quantum Mechanics is in fact our most-verified, best-proven, and most-used science on the planet right now.

WE KNOW that Materialism, Naturalism, Darwinism, Nihilism, Behaviorism, Scientism, and Atheism are FALSE, because there is NO observational evidence supporting their hidden assumptions, major premises, or primary assumptions. Materialism, Naturalism, Darwinism, and their derivatives are unscientific because there is NO observational evidence supporting their major premises or primary assumptions. That's the way science works; or at least, that's the way science is supposed to work.

Anything that has been successfully falsified should be eliminated from science as being unscientific. If you want to find and know the truth, then you must eliminate everything that is false and everything that has been falsified, starting with Materialism, Naturalism, Darwinism, Nihilism, Behaviorism, Scientism, Atheism, and the Theory of Evolution. There is no other way to find and know the truth. The truth is what remains after everything that is false has been eliminated.

Remember, entropy degenerates genomes. It doesn't improve and enhance them. It can't. It's physically impossible. Entropy cannot do order, organization, design, programming, engineering, fine-tuning, field-testing, and manufacturing. Entropy makes chemical evolution, abiogenesis, spontaneous generation, and macro-evolution physically impossible. Random mutations are entropy. Natural selection is a dead-end, death, entropy, or the extinction of species. These things destroy and end genomes. They do NOT enhance and improve genomes. Evolution of any kind is entropy. It can't design and create.

Begging the Question

One of my most important and significant discoveries in Science and Philosophy is the discovery that Materialism, Naturalism, Darwinism, Nihilism, Atheism, and their derivatives are based exclusively on *Begging the Question*, which is a logic fallacy. Materialism, Naturalism, and Atheism have been based upon the *Begging the Question* logic fallacy since the beginning of recorded history.

So, what is *Begging the Question*?

Begging the Question is a logic fallacy wherein they take their pre-chosen conclusion and insert their pre-chosen conclusion as evidence and proof that their pre-chosen conclusion is true. In other words, these people are using their pre-chosen conclusions as their Major Premises or Hidden Assumptions in every scientific argument, scientific interpretation, scientific inference, and scientific proof that they make.

So, what are these Hidden Assumptions or Pre-Chosen Conclusions that these people are using to make their case and prove that they are right? They have been doing so for thousands of years. Wouldn't you want to know?

Materialism is based upon the pre-chosen conclusion that the non-physical does not exist. Therefore, these people (often unknowingly) insert the **non-existence of the non-physical** as their first Major Premise or Hidden Assumption; and thereafter, these people ONLY consider physical matter, and they completely and deliberately exclude and refuse to do research into all the different invisible, non-physical forces, fields, and entities which have been experienced, observed, and proven to exist.

Naturalism is based upon the pre-chosen conclusion that the supernatural or the quantum does not exist. Naturalism is also based upon the pre-chosen conclusion that Psyche, Spirit Matter, and Quantum Non-Local Consciousness do not exist. These people deny the existence of Intelligence, Psyche, or the Life Force. It's philosophically impossible to verify, or to prove, or even to falsify such a pre-chosen conclusion; but, they don't care. These people deny the existence of Supernatural Mechanisms or Quantum Mechanisms; and therefore, these people insert the **non-existence of the supernatural** or the **non-existence of the quantum** as their first Major Premise or Hidden Assumption in every scientific argument, scientific proof, scientific interpretation, and scientific inference that they make. Consequently, these people ONLY consider nature or classical physics, and they completely and deliberately exclude the quantum, the supernatural, the psyche, or Syntropy from consideration when it comes to their science and their philosophy.

Nihilism is the philosophical belief that there is no after-life for the Psyche or the Soul, and therefore no such thing as a Quantum Realm, Syntropy Realm, Psyche Realm, or Spirit Realm. These people choose to believe that there is no such thing as the Non-Local Realm, Non-Physical Realm, or Transdimensional Realm. These people use the **non-existence of psyche**, and the **non-existence of spirit matter**, and the **non-existence of syntropy** as scientific proof that these things do not exist. They *beg the question*. These people insert the non-existence of these things into every scientific argument, scientific proof, scientific interpretation, and scientific inference that they make as evidence or proof that their pre-chosen conclusions are true. These people use the non-existence of these things as evidence or proof that these things do not exist. It's illogical and irrational; but, that's the nature of logic fallacies such as *begging the question*. Most of them don't even realize that they are *begging the question* and *affirming the consequent*. They do it automatically because they have been trained to do so in college by their college professors who were also Materialists, Naturalists, Darwinists, Nihilists, and Atheists.

Atheism is based upon the pre-chosen conclusion that God or God's Psyche does not exist. It's philosophically impossible to verify, or to prove, or even to falsify such a pre-chosen conclusion. Atheism is the design and creation of something by nothing. Atheism is irrational, illogical, and patently absurd. There's no logic to it whatsoever. These people deny the existence of a Maker, even when it comes to things that were obviously made, such as proteins and genomes; consequently, these people insert the **non-existence of God** or the **non-existence of Makers** as their first Major Premise or Hidden Assumption into every scientific argument, scientific proof, scientific interpretation, and scientific inference that they make. As a result of this pre-chosen conclusion, these people convince themselves that proteins, genes, genomes, eyes, brains, and life forms spontaneously generate out of thin air all the time. Such a belief or claim has NEVER been experienced nor observed. In fact, Louis Pasteur falsified spontaneous generation in 1859, the very same year that Charles Darwin published "On the Origin of Species". It's ironic, but Louis Pasteur falsified Spontaneous Generation, Abiogenesis, Creation Ex Nihilo, Materialism, Naturalism, Darwinism, Chemical

Evolution, Macro-Evolution, and the Theory of Evolution in 1859; and, the Naturalists have been denying that fact ever since.

Darwinism is a carefully designed combination of Materialism, Naturalism, Nihilism, and Atheism. Darwinism or the Theory of Evolution is based upon the **non-existence of God**, the **non-existence of the non-physical**, the **non-existence of the supernatural**, the **non-existence of psyche**, and the **non-existence of the quantum realm** or transdimensional realm. These people quietly insert the non-existence of these things as their Major Premise or Hidden Assumption into every science argument, science proof, interpretation of data, and scientific inference that they make. These people use their pre-chosen conclusions as evidence and proof that their pre-chosen conclusions are true.

This is how these people do Science and Philosophy, by excluding a priori anything and everything that falsifies Materialism, Naturalism, Darwinism, Nihilism, and Atheism. They cheat. They *beg the question*! They declare their pre-chosen conclusions to be Axioms, Laws, or Given Truths, and then insert those Axioms or Given Truths (their pre-chosen conclusions) as their Major Premise into every scientific argument, scientific proof, scientific interpretation, and scientific inference that they make. Most of them don't even realize that they are doing so; but, they are. It's what they do. It's what they have been trained to do in college.

These people also make a *special pleading* and adjust the rules of science and philosophy by axiomatically declaring that their First Premise in any argument that they make is allowed to be a Given Truth or an Axiom, and therefore doesn't have to be seen, verified, nor proven to be true. That's how they get around the rules of logic and the rules of science, by unilaterally declaring that their First Premise in any argument that they make is allowed to be an Axiom or a Given Truth. That's how these people trick us and deceive us. That's how they trick and deceive themselves. They cheat. They *beg the question* and *affirm the consequent*. They use *circular reasoning* to make their case and prove their point.

Because I understand the *begging the question* and the *affirming the consequent* logic fallacies, whenever I do philosophical proofs and scientific arguments, I demand that ALL of my premises as well as my conclusion be experienced and observed by Someone Psyche somewhere sometime, or I consider my philosophical arguments and scientific proofs to be worthless. I upgraded my science to Science 2.0. I don't allow Axioms or Given Truths to be used as the Premises in any of my scientific arguments or philosophical proofs. Everything – the premises and the conclusion – has to be based upon some kind of observational evidence, experiential evidence, or empirical evidence; or, I'm liable to be deceived. Using their pre-chosen conclusions as their First Premise, Given Truth, or Axiom is precisely how the Materialists, Naturalists, Darwinists, Nihilists, and Atheists deceive themselves. Self-deception works, and it works every time.

By using their pre-chosen conclusions as evidence or as proof that their pre-chosen conclusions are true, these people easily and convincingly prove to themselves and to others that their pre-chosen conclusions are indeed true. You can't miss by *begging the question*! By *begging the question* and *affirming the consequent*, you can prove anything to be true, including Materialism, Naturalism,

Darwinism, Nihilism, and Atheism. *Begging the question* is cheating. It's a logic fallacy. But, these people don't care, because *begging the question* and *affirming the consequent* proves that their theories, philosophies, and ideas are true; and, that's all they really care about – winning the argument. And, these people have won millions if not billions over to their side.

When it comes to Materialism, Naturalism, Darwinism, and Atheism, the whole thing is a sham or a scam; and, most people don't even know it. Fascinating, is it not? Well, I think so. I used to be a Materialist, Naturalist, Nihilist, and Atheist. I'd been tricked and deceived just like everyone else; and, I didn't even know it. Philosophy (or religion and blind faith) has a definite impact in our lives, especially when it comes to Science. By using philosophy and *begging the question*, we can prove anything to be true.

This is the benefit of studying the Philosophy of Science and the various different logic fallacies that the Materialists and Naturalists use to trick us and deceive us. You can actually figure out how they are tricking you and deceiving you. You can also take steps to correct the problem.

It's time that we upgrade our science to Science 2.0.

Begging the question, traditional science is defined as Materialism, Naturalism, Darwinism, Behaviorism, Scientism, Determinism, Physical Reductionism, Nihilism, and Atheism. It's time for us to re-define Science as observation and experience. Let's get rid of the philosophical speculation and the *begging the question* logic fallacy, and switch over to observational evidence, experiential evidence, eye-witness evidence, and empirical evidence for our proof. Science 2.0 allows ALL of the evidence into evidence and pursues a preponderance of the evidence. Let's demand that every premise and every conclusion be experienced and observed by Someone Psyche somewhere sometime – or consider our scientific proofs and philosophical arguments to be falsified if there is no observational evidence supporting one of our premises or our pre-chosen conclusion. The different types of evolution have NEVER been caught in the act of design and creation; therefore, they are false and have been successfully falsified. They should be eliminated from consideration. That's the way science should be done but isn't.

Remember, Materialism, Naturalism, Darwinism, Nihilism, and Atheism are FALSIFIED by a complete lack of observational evidence supporting their Major Premises or Hidden Assumptions. These things should have been falsified and eliminated long ago, but they weren't because their proponents cheat and *beg the question* in order to make their case and prove their point. These falsified philosophies should have been eliminated from science, but they weren't. That's the way science should have worked but didn't. Materialism, Naturalism, Darwinism, Nihilism, and Atheism should be eliminated from science; but, they aren't because they have been declared axiomatically true and are then used as Axioms or Given Truths as the First Premise in every scientific argument, scientific inference, scientific interpretation, and scientific proof that these people make.

By *begging the question* and *affirming the consequent*, these people successfully prove that the Theory of Evolution, Materialism, Naturalism, Nihilism, and Atheism are true; and, millions if not billions have fallen for the ruse because they don't understand the Philosophy of Science and how things really work.

If you tear something down, you are obligated to try to build something better in its place, which I have also tried to do.

Mission Statement

After successfully using the Scientific Method and Logic to falsify Materialism, Naturalism, Darwinism, and Atheism, then I felt a compelling need to re-envision and re-do Science and Philosophy in order to bring them into line with the millions of Near-Death Experiences (NDEs) and Out-of-Body Experiences (OBEs) which are on record and that have been experienced and observed first-hand by real human beings or human psyches. In order to do so, I had to switch Science and Philosophy from a platform of speculation and wishful thinking based upon Materialism and Naturalism over to a new paradigm based upon the observations and verified proof that have been obtained from Quantum Mechanics or Supernatural Mechanics. In other words, I had to bring the quantum or the supernatural or the non-physical into Science and Philosophy in order to upgrade them, fix them, and bring them into the modern age.

As a result, Science 2.0 and the Ultimate Model of Reality were born and brought into existence.

Mark My Words

—

The Fruits of These Truths

Science 2.0 was the result of these science discoveries. If you study this essay carefully, you will notice that Science 2.0 follows naturally from *negating the consequent* and using the Scientific Method the way it was meant to be used. The Scientific Method was supposed to be used to falsify and eliminate everything that is false; but, it is seldom used that way by the Materialists, Naturalists, and Darwinists because the Scientific Method and *negating the consequent* falsifies Materialism, Naturalism, and their derivatives. Science 2.0 is discussed in much greater detail in the following book.

Science 2.0: I Upgraded My Science.

https://www.amazon.com/dp/B0771K6WTX

The Ultimate Model of Reality and Ultimate Cause were also the result of using the Scientific Method and Logic to successfully falsify Materialism, Naturalism, Darwinism, and Atheism. The purpose of the Ultimate Model of Reality, a Psyche Ontology, a Psyche Epistemology, and Ultimate Cause is to add a necessary Fifth Cause to Aristotle's Four Causes which were originally based upon Materialism, Naturalism, and Atheism. In other words, I added Syntropy, Psyche, Non-Locality, Quantum Mechanics, the Non-Physical, or the Supernatural to Aristotle's Four Causes so as to bring Philosophy and the Philosophy of Science into the modern age or the quantum age. The Ultimate Model of Reality is discussed in much greater detail in the following books.

The Ultimate Model of Reality: Psyche Is the Ultimate Cause

https://www.amazon.com/dp/B071NC9JK6

Putting Psyche Back into Psychology: Restoring Science to Consciousness

https://www.amazon.com/dp/B071NC987S

BioPsychoSocial: Including Psyche or Light into our Theoretical Models

https://www.amazon.com/dp/B0713NDHVW

NATURE vs. NURTURE vs. NIRVANA: An Introduction to Reality

https://www.amazon.com/dp/B01JWRCSVA

https://www.amazon.com/dp/1521132615

Source

Science 2.0: I Upgraded My Science.

https://www.amazon.com/dp/B0771K6WTX

Using the Scientific Method to Eliminate the Usual Suspects and to Prove the Truth.

https://www.amazon.com/dp/B01J6STHP0

https://www.amazon.com/dp/1521133581

Reference Materials

Using the Scientific Method to Eliminate the Usual Suspects and to Prove the Truth.

https://www.amazon.com/dp/B01J6STHP0

https://www.amazon.com/dp/1521133581

The Theory of Evolution Proved to Me that God Exists: Why I Am No Longer an Atheist and Why I No Longer Believe in the Theory of Evolution.

https://www.amazon.com/dp/B01HZYBZ7K

https://www.amazon.com/dp/1521131228

The Scientific Method Proves That the Theory of Evolution Is False.

https://www.amazon.com/dp/B01IAAIRT2

https://www.amazon.com/dp/1521133611

Summary Of: The Theory of Evolution Proves that God Exists.

https://www.amazon.com/dp/B01GQCWED6

https://www.amazon.com/dp/1521130485

Quantum Mechanics from a Non-Physical Spiritual Perspective.

https://www.amazon.com/dp/B01J023TGU

https://www.amazon.com/dp/1521132380

Quantum Neuroscience: The Answer to Life, the Universe, and Everything.

https://www.amazon.com/dp/B079Z6QQQB

Rychlak, J. F. (1981). *A Philosophy of Science for Personality Theory* (2nd ed.). Malabar, FL: Robert E. Krieger Publishing Company.

Slife, B. D. & Williams, R. N. (1995). Science and Human Behavior. In *What's Behind the Research? Discovering Hidden Assumptions in the Behavioral Sciences*, (pp. 167–204). Thousand Oaks, CA: SAGE Publications.

http://mypsyche.us/science/

—

What did we learn from this?

We learned that the dude wrote some books, and he wouldn't mind if you were to take some time to buy them and read them.

Obviously Made

A computer, a computer program, a calculator, a battery, a book, a watch, a car, a bridge, a table, a chair, a skyscraper, a house, a phone, a tablet, an airplane, a jet, and even a baby was obviously made. Likewise, a genome was obviously made. They had a beginning which means that they had some kind of Beginner or Maker who brought them into existence.

ALL of these things were obviously made.

NONE of these things has ever been observed spontaneously generating out of thin air ex nihilo, because they can't. Entropy or the second law of thermodynamics prevents spontaneous generation from happening. Furthermore, spontaneous generation was falsified in 1859 by Louis Pasteur. Spontaneous generation is physically impossible. That's the real science behind all of this.

Things that were obviously made obviously had a Maker who made them. This truth is unavoidable, because there is NO such thing as abiogenesis, spontaneous generation, creation ex nihilo, chemical evolution, or macro-evolution. They are physically impossible. They are prevented from happening by entropy or the second law of thermodynamics.

ALL of the things on the list were obviously made.

However, a genome is different than the others in a very obvious and unique way.

Can you tell what that is?

The first things on the list were obviously man-made. However, a genome obviously was NOT man-made. A genome and the proteins which that genome produces obviously pre-date man or human beings, which means that a genome and proteins obviously were NOT man-made. So, who made them? It's obvious that they were made, because such things have NEVER been observed spontaneously generating out of thin air ex nihilo.

Observation or Premise: Anything that was obviously made obviously had a Maker or Creator who made it.

Observation or Premise: A genome was obviously made.

Logical Conclusion: Therefore, a genome obviously has a Maker or a Creator who designed it, programmed it, engineered it, field-tested it, fine-tuned it, made it, manufactured it, and deployed it.

This syllogism is both a Philosophical Proof of God's Existence and a Scientific Proof of God's Existence. If its premises are true, then its conclusion must be true as well.

Genomes were obviously made; so, who made them?

That is the million-dollar question and the 800-pound gorilla, is it not?

There's only one logical answer to this question because genomes cannot spontaneously generate out of thin air as the Atheists and Darwinists claim.

Your genome is God's Signature, and God's Signature is written on every cell in your body. Simple. Logical. Parsimonious. True. Q.E.D.

We shouldn't be embarrassed by the truth. Science or observation makes the truth obvious, or it should.

—

Natural Selection Is a Blind Watchmaker

Richard Dawkins is the best that the Naturalists, Darwinists, and Atheists have been able to produce. In one of his best-selling books, Dawkins claims that natural selection is a blind watchmaker. How many blind watchmakers have you encountered during your life? They don't exist. Likewise, design and creation by natural selection does not exist. It's science fiction.

> **All appearances to the contrary, the only watchmaker in nature is the blind forces of physics, albeit deployed in very special way. A true watchmaker has foresight: he designs his cogs springs, and plans their interconnections, with a future purpose in his mind's eye. Natural selection, the blind, unconscious, automatic process which Darwin discovered, and which we now know is the explanation for the existence and apparently purposeful form of all life, has no purpose in mind. It has no mind and no mind's eye. It does not plan for the future. It has no vision, no foresight, no sight at all. If it can be said to play the role of watchmaker in nature, it is the *blind* watchmaker.**
>
> **Mutation is random; natural selection is the very *opposite* of random.**
>
> **We have seen that living things are too improbable and too beautifully 'designed' to have come into existence by chance.**
>
> **The essence of life is statistical improbability on a colossal scale.**
>
> **— Richard Dawkins**
>
> *The Blind Watchmaker* (1986), 5, 41, 51, 317.

Richard Dawkins claims that a blind watchmaker, called natural selection, made your proteins, genes, genomes, eyes, brain, and physical body. Richard Dawkins is talking about "spontaneous generation by natural selection". These claims are so obviously false that it's simply amazing to find people who actually believe that they are true.

In fact, these quotes are based upon a *category error* logic fallacy, wherein Richard Dawkins places natural selection into the category of "Invisible Gods Who

Can Design and Create Anything They Set Their Minds To". Richard Dawkins just elevated natural selection to the status of Godhood by turning natural selection into a Designer and a Creator, but one that is running around blind and half-cocked.

It's obvious that genomes required vision, foresight, and some kind of purpose in mind. Dawkins even says as much in many of his statements. Genomes don't just spontaneously generate out of thin air. And, if you take Dawkins' explanation of natural selection in this quote seriously, you simply KNOW that natural selection could never have designed and created a genome from scratch or from atoms. It's physically impossible. Chemical Evolution and Macro-Evolution of any kind are prevented from happening by entropy or the second law of thermodynamics.

Furthermore, natural selection did not exist and could not exist until after God designed, programmed, engineered, fine-tuned, field-tested, created, manufactured, and deployed the genomes in the first place. This reality is so obviously true that Dawkins' claims don't make any logical sense to the rational mind.

Dawkins portrays natural selection as if it were some kind of magical God. Dawkins describes natural selection is if it were spontaneous generation, creation ex nihilo, or magic. These things don't exist. They have been falsified by science – meaning that they have been falsified by observation and experience.

In fact, natural selection doesn't really exist either. Natural selection is defined into existence by the scientists, but natural selection doesn't exist as a person, entity, psyche, intelligence, field, or force that can actually do something constructive for us. Natural selection is a function of entropy or work; and, entropy cannot design and create. Dawkins portrays natural selection as some type of spontaneous generation. His analogy is flawed, because there is no such thing as spontaneous generation or abiogenesis.

Dawkins is wrong. Natural selection is purely random – purely a function of chance. When you get selected against and die as a result, that's just as much a function of chance as anything else in this universe. The fittest are typically the first to die or the first to get selected against through war, risk-taking, infection, bad luck, and accidents of every kind. The timid, the meek, and the restrained really do inherit the earth.

There's no such thing as "spontaneous generation by natural selection". In fact, there's no such thing as spontaneous generation. Spontaneous generation, abiogenesis, or macro-evolution was falsified in 1859 by Louis Pasteur – the very same year that Charles Darwin published "On the Origin of Species". We have KNOWN from the very beginning of the theory that the Theory of Evolution, Creation by Natural Selection, Abiogenesis, Chemical Evolution, Materialism, and Naturalism are false because they are physically impossible and prevented from happening by entropy or the second law of thermodynamics.

Mankind, human beings, and blind watchmakers obviously did NOT make the genomes, so who did?

Well, WE KNOW that it wasn't natural selection because natural selection cannot design and create. Natural selection doesn't touch our genes. It can't. It's physically impossible. Dawkins doesn't even understand his own theory. The "agent of change" in the theory of evolution is the random mutations which take place during the genetic recombination that is used to produce our gametes – egg and sperm. Natural selection doesn't touch our genes. Natural selection results in entropy, death, and extinction. That's it!

Natural selection doesn't do anything to our genes or for our genes. Natural selection doesn't do anything except wait for us to die. Even sexual selection doesn't do anything for you after you have been conceived. Natural selection is not the panacea or magic potion that Dawkins claims that it is. Richard Dawkins' claims are obviously false to anyone who actually understands the Theory of Evolution, Entropy, Natural Selection, Random Mutations, and Science.

Furthermore, Richard Dawkins' blind watchmaker model is obviously self-refuting. A blind watch maker is an intelligent being; and, intelligent beings obviously qualify as Makers or Creators or Gods. Just because he is blind, it doesn't satisfy the demands of Richard Dawkins' claim that proteins, genomes, and life forms just spontaneously generated out of thin air by natural selection. Furthermore, a blind watchmaker (or natural selection) is definitely NOT going to be making a watch or genome without some help from some other intelligent beings. An intelligent watch maker (blind or otherwise) actually falsifies and refutes Dawkins' claim that our proteins, genes, genomes, eyes, brains, and physical bodies spontaneously generated out of thin air by natural selection.

A watch or genome is NOT going to spontaneously generate out of thin air. Entropy or the second law of thermodynamics prevents it from doing so. Entropy prevents Chemical Evolution of any kind from happening in the first place. That's the real science behind all of this.

A watch of any kind was obviously made. Likewise, a genome of any kind was obviously made. These things do NOT spontaneously generate out of thin air by natural selection or by any other means. They have NEVER been observed doing so, because they can't. It's physically impossible. Entropy or the second law of thermodynamics prevents watches and genomes from spontaneously generating into existence. Therefore, Richard Dawkins' claims are falsified by science, entropy, random diffusion, and the second law of thermodynamics. His claims are also falsified by logical common sense.

Yes, a blind watchmaker can theoretically make watches because he is an intelligent being. NO, a gene, protein, genome, eye, brain, or physical body cannot spontaneously generate into existence. They are prevented from doing so by entropy. Spontaneous generation, abiogenesis, or macro-evolution of any kind is physically impossible. It can't happen, which means that it didn't happen.

Most people can quickly see where Richard Dawkins goes wrong, because he is not a good philosopher, and he's not much of a scientist either. Most of what he promotes was falsified decades ago.

—

It's Obvious that Genomes Were Made

Even Richard Dawkins seems to agree that genomes were obviously made. So, what's his hang-up? His problem is that he doesn't want the Biblical God to be his Maker. Anything but God!

https://www.youtube.com/watch?v=GlZtEjtlirc

https://evolution-is-entropy.com/wp-content/uploads/2018/05/Richard-Dawkins.zip

In the following quote, Richard Dawkins states in a round-about way that genomes were obviously made. It would take from here to infinity for a genome to self-assemble by sheer higgledy-piggledy luck. I believe it, because it is true.

> **It is grindingly, creakingly, crashingly obvious that, if Darwinism were really a theory of chance, it couldn't work. You don't need to be a mathematician or physicist to calculate that an eye or a hemoglobin molecule would take from here to infinity to self-assemble by sheer higgledy-piggledy luck. Far from being a difficulty peculiar to Darwinism, the astronomic improbability of eyes and knees, enzymes, and elbow joints and all the other living wonders is precisely the problem that any theory of life must solve, and that Darwinism uniquely does solve. It solves it by breaking the improbability up into small, manageable parts, smearing out the luck needed, going round the back of Mount Improbable and crawling up the gentle slopes, inch by million-year inch. Only God would essay the mad task of leaping up the precipice in a single bound.**
>
> **— Richard Dawkins**
>
> *Climbing Mount Improbable* (1996), 67-8.

Dawkins is trying to make a case for Evolution or Darwinism, but he inevitably ends up refuting it, and debunking it, and falsifying it with truths hidden among the deceptive lies. A genome would take from here to infinity to self-assemble by sheer higgledy-piggledy luck. That is the truth; and, natural selection is nothing but luck – bad luck at that. It's never your lucky day when you get selected against and die.

ALL of Richard Dawkins' books are self-refuting and self-defeating. They contain within them their own seeds of destruction.

Why?

It's because his books were obviously made, which means that they obviously had a Maker, Creator, or God who made them.

In our own limited way, human beings are Gods. We are Creators. Adam was a son of God. That means that both his spirit body and his physical body were descended from the Gods. We human beings are Gods. We are descended from the Gods. Both our spirit body and our physical body are descended from the Gods. We inherited the ability to design and create from the Gods, our ancestors.

In this quote, Dawkins actually explains why Darwinism can't work and doesn't work – because it's a theory of chance or a theory of spontaneous generation. Smeared-out luck is still luck. It is still chance. In fact, it becomes less likely the more smeared-out it is. Infinite luck is even more impossible and improbable than one-time luck. Infinite luck or smeared-out luck has to get lucky each inch by million-year inch, instead of getting lucky only once. That's impossible! It's physically impossible for genetically compatible Mr. and Mrs. Mutants to be born at the same place at the same time year after year for millions of years. Genetics, random mutations, and entropy prevent it from happening! It can't be done, which means that it wasn't done. We don't have chimp-like ancestors because that's physically impossible. Both genetics and random mutations prevented it from happening.

Dawkins is right. ONLY God could have taken the task of designing and creating a genome in a single bound; and, the Gods have admitted that they are scientists and took their time getting our physical world, physical genome, and physical body just right for our existence. Even the Gods didn't do it in a single bound; but, they did do it. The Gods are scientists, not magicians; and, Richard Dawkins repeatedly portrays Darwinism as magic. The contrast couldn't be more obvious.

The Gods watched until they were obeyed.

https://evolution-is-entropy.com/wp-content/uploads/2018/05/The-Gods-Watched-Until-They-Were-Obeyed.pdf

While using God to make his case, Richard Dawkins actually falsifies his case. Dawkins actually admits that ONLY God could do "life" in a single bound; and, we all know instinctively that a genome has to be fully in place and fully functional in the first place, or an organism is going to die and go extinct. Life had to be done in a single bound, because it couldn't have been done over millions of years – one gene at a time. A gene by itself is worthless. It doesn't do anything. It requires all 20,000 of them to make a genome. They all have to be there from the very beginning, or an organism dies and goes extinct.

Likewise, a million apes pounding away on a typewriter could NEVER produce Dawkins' books in a trillion zillion years. First of all, they wouldn't know how to load the paper into the machine. Second of all, that typewriter, that paper, and those apes were obviously made, which means that they each had an intelligent Maker – a fact that falsifies Dawkins' claims that natural selection made them. Natural selection didn't make the typewriter and the paper, and it certainly didn't make the apes.

http://www.blogos.org/thinkabout/infinite-monkey-theorem.html

https://evolution-is-entropy.com/wp-content/uploads/2018/05/Infinite-Monkey-Theorem.pdf

Dawkins' arguments are self-defeating, because Dawkins' deliberately ignores and excludes the Maker, Creator, or God behind each one of them. Richard Dawkins' arguments and claims are non-starters. Everything that he talks about was obviously made. The blind watchmaker was obviously made. His books and concepts were obviously made. Even natural selection is a made-up concept concocted out of thin air by the scientists or intelligent psyches who thought up the idea.

There's NO getting around it – anything that was obviously made obviously had a Maker, Creator, or God who made it. That's just the way it is. Proteins and the matching genes to go along with them don't spontaneously generate out of thin air. That's physically impossible. It's prevented from happening by entropy or the second law of thermodynamics. Proteins and genes were obviously made.

Richard Dawkins' examples and claims are self-defeating. They are defeated by science, logic, common sense, entropy, and reality itself. They are defeated by observation and experience. Science is observation and experience, not spontaneous generation. Spontaneous generation, chemical evolution, or macro-evolution was falsified by Science, by experience, and by observation.

I buy into the science, and not Dawkins' wishful thinking and confirmation bias.

—

Belief in Darwinism Requires an Infinite Amount of Blind Faith

I have repeatedly observed that belief in Darwinism, Materialism, Naturalism, Nihilism, and Atheism require an infinite amount of ignorance and blind-faith in order to believe in them because their Hidden Assumptions or Major premises cannot be experienced nor observed and have to be taken on blind-faith as being true.

Faith is belief without evidence and reason; coincidentally that's also the definition of delusion.

Do not indoctrinate your children. Teach them how to think for themselves, how to evaluate evidence, and how to disagree with you.

— Richard Dawkins

I love to read Dawkins. He always puts a smile on my face. Practically everything he says is self-defeating in one way or another.

Dawkins debunks himself and falsifies himself as he goes along. Dawkins defeats himself with practically everything he says.

I also love the hypocrisy. He should practice what he preaches, but he's incapable of seeing and understanding the contradictions.

Teach your children how to think for themselves, how to evaluate the evidence, and how to falsify and debunk Richard Dawkins. Teach your children to allow ALL of the evidence into evidence and not just the selective sub-set that Richard Dawkins allows into evidence.

Ironically, Dawkins' beliefs or blind faith are without evidence and run contrary to reason, which according to Dawkins means that they are delusional. That always puts a smile on my face every time I think of it. Everything that Dawkins says is based upon blind-faith and runs contrary to the scientific evidence and observational evidence that we have on hand as a race. Gotta love it! Self-deception works, and it works every time – especially when it comes to PhD scientists and authors. Remember, Dawkins' beliefs or blind faith are contradicted by the scientific evidence or empirical evidence which has been experienced and observed.

We should be thankful to Richard Dawkins for giving us a nearly endless stream of material to falsify and debunk. It's good for us. It forces us to think. It teaches us how to evaluate the evidence and how to disagree with and falsify Richard Dawkins.

Dawkins' books are NOT best-sellers because they are true. They are best-sellers because they are telling their readers what they want to hear – namely that God does not exist. People will pay a lot of money to be told that God does not exist because that's what they want to hear.

In contrast, my books will never be best-sellers, because I'm not telling the Materialists, Naturalists, Darwinists, and Atheists what they want to hear. I'm telling them the truth. These people aren't going to pay money to be told that God exists and why He must exist. And, the Christians and Muslims already know that God exists, so technically they don't need my books and won't be buying my books either. That's just the way it is.

I make and produce these books for my own benefit, so that I have a way of remembering what I have learned. While in a physical package, memory is such a fickle thing – one day you wake up and it is gone. But, it's not going to disappear on me and be gone if I can get it into a book.

—

The Philosophy of Science Falsifies Darwinism

I'm big on the Philosophy of Science. I actually used the Philosophy of Science and *negating the consequent* to falsify Materialism, Naturalism, Darwinism, and Atheism in many of my books.

I'm a generalist. I'm good at everything and master of nothing. The breadth and depth of my education is quite amazing, actually. My problem is that I get bored easily unless I'm learning something new.

I'm a scientist, psychologist, theoretician, logician, philosopher, accountant, mathematician, statistician, and computer scientist.

When I first watched Richard Dawkins debate others, I was surprised when I first realized that I KNOW a lot more science, theology, psychology, theory, logic, history, and philosophy than Richard Dawkins does. I was embarrassed for him. I felt sorry for him. He was making a fool of himself and didn't even know it. His ignorance was simply amazing. He has no breadth. His atheism has stunted his growth. He stopped learning long ago.

I was expecting Dawkins to have something significant or convincing to say, but he didn't. His science is over fifty years old and out-of-date. It expired long ago. And, Dawkins has absolutely no knowledge of philosophy, logic, and the Philosophy of Science. His arguments didn't work, because they were easily falsified as he went along; yet, all the time he actually thought that he was winning the debate. Go figure!

Philosopher of science, Karl Popper, wrote: "Testable theories are scientific, but those that are 'untestable' are not."

In *Unended Quest*, Popper declared, "I have come to the conclusion that Darwinism is not a testable scientific theory, but a metaphysical research programme, a possible framework for testable scientific theories."

Karl Popper is right.

I came to the same conclusion myself multiple times in many different ways.

Darwinism is a faulty and falsified religion. Evolution and natural selection are man-made gods.

Darwinism, Materialism, Naturalism, Nihilism, Atheism, and the Theory of Evolution are NOT testable scientific theories. They are philosophy, religion, or metaphysics because they require a ton of blind faith in order to believe in them and because ALL of the observational evidence and experiential evidence falsifies them.

Materialism, Naturalism, and Darwinism are NOT science. They are not testable, and there is NO evidence and can be NO evidence supporting their Hidden Assumptions or Major Premises which claim that the quantum or the supernatural does not exist. ALL of the evidence that we have on hand as a race tells us clearly and conclusively that the quantum or the supernatural does in fact exist in spades. Quantum Mechanics or Supernatural Mechanics or Transdimensional Physics is our best-proven, most-verified, and most-used science that we have. WE KNOW that it exists because it has been experienced and observed. WE KNOW that it exists by the effects that it has on physical matter.

The very existence of Syntropy or Quantum Mechanics or Intelligence falsifies Materialism, Naturalism, Darwinism, Nihilism, and Atheism. The proven, verified, and experienced existence of magnetism, gravity, the strong nuclear force, the weak nuclear force, psyche, dark energy, and dark matter or spirit matter FALSIFIES Materialism, Naturalism, Darwinism, Nihilism, and even Atheism.

The false is falsified by the truth; and, the truth is repeatedly experienced and observed. Science is observation and experience, not philosophical speculation and wishful thinking.

Remember, Darwinism is metaphysics or religion – not science. It requires an infinite amount of blind-faith and ignorance in order to believe in it. The same applies to Materialism, Naturalism, Nihilism, and Atheism. These are metaphysics or religion, and not science. That is what I have experienced and observed. I used to be a Materialist, Naturalist, Nihilist, and Atheist until ALL of the scientific evidence and observational evidence convinced me that I was wrong.

In my psychology courses in college, each author paid homage to Darwinism and the Theory of Evolution in their texts. Most of the things that they said were obviously false. It was so obvious that it was painful at times. It's unfortunate that they chose to call it evolutionary psychology and an evolutionary perspective, because evolution is in dispute. Something that is in dispute shouldn't be a part of science, especially a metaphysical philosophy such as Darwinism.

Evolutionary psychology is actually comparative psychology, because they are comparing animals with humans. They should have called it Comparative Psychology and not evolutionary psychology. Evolution is in dispute. Comparisons are not. Furthermore, they should have called it a Genetic Perspective rather than an evolutionary perspective for the very same reason. Evolution is in dispute. Genes and genomes are not. The genes and genomes have been experienced and observed; whereas, the different types of evolution have not. The different types of evolution – including random mutations and natural selection – have NEVER been caught in the act of designing and creating new unique proteins, genes, genomes, and life-forms from scratch because they can't. It's physically impossible.

Philosophies like Materialism, Naturalism, Darwinism, Nihilism, and Atheism cannot do science. They will never be caught in the act of doing science, design, creation, engineering, field-testing, fine-tuning, manufacturing, distribution, or anything else because they can't. This is so obviously true, that everyone has completely overlooked its significance. Only Psyche, Intelligence, or Syntropy can do design and creation; and, there's no intelligence anywhere within the Theory of Evolution. Richard Dawkins even says as much in the different statements that he has made.

Evolution is entropy. The different types of evolution are based upon entropy. Materialism, Naturalism, Darwinism, Nihilism, Behaviorism, Scientism, Determinism, Classical Physics, Physical Reductionism, and Atheism are based exclusively on entropy. Evolution is prevented from happening by entropy. Entropy or evolution cannot design and create. There is no intelligence in evolution or entropy.

Fascinating, is it not, how the truth rises to the top if you choose to let it do so?

—

Computer Science Falsifies Darwinism

Bill Gates — DNA is like a computer program but far, far more advanced than any software ever created.

A genome is like a computer program. It was obviously made.

Those of us who are Computer Scientists recognize a genome for what it truly is. A genome is both hardware and software. It's a computer program, a radically advanced four-dimensional computer program. A genome is also precision hardware. A genome is both software and a machine.

https://creation.com/four-dimensional-genome

https://evolution-is-entropy.com/wp-content/uploads/2018/05/Four-Dimensional-Genome.pdf

I was a computer programmer for a decade in the 1980's. I KNOW for a fact from experience and observation that elegant code does NOT spontaneously generate out of thin air. Computer programs don't spontaneously generate from nothing or natural selection, as Richard Dawkins claims. Elegant code is designed, planned, and made. It requires a Maker, Creator, or God. Computer programs were obviously made, which means that they obviously have a Maker who made them.

Likewise, a genome is obviously elegant programming code. A genome is also hardware. Programming code and hardware don't spontaneously generate into existence automatically. A million monkeys pounding away on a keyboard won't produce them either, because that keyboard and those monkeys were obviously made by some kind of Maker, Creator, or God – and they still can't produce a genome from scratch. A million monkeys and a keyboard do NOT qualify as an adequate example of spontaneous generation, chemical evolution, or macro-evolution because they were obviously made. Anything that was obviously made obviously had a Maker, Creator, or God who made it. That's logical common sense.

A complex computer program such as a genome was obviously made, which means that it obviously has a Maker who made it.

—

Scientific Proof of God's Existence

I found this philosophical proof of God's existence and Scientific Proof of God's Existence convincing ever since it was first revealed to me. Genomes are what convinced me that God exists. God must of necessity exist in order to have done ALL of the science and made ALL the genomes which chemical evolution, macro-evolution, natural selection, and random mutations NEVER could have done or made. To me personally, this reality is obviously true. Genomes were obviously made. Ever since that revelation or epiphany, I have known that God exists because I have known why He must exist. There's no denying it because I believe the science and KNOW that the premises are true.

Let's take another look at this Proof of God.

First Observation or Premise: Anything that was obviously made obviously had a Maker or Creator who made it. This is Logic 101.

Second Observation or Premise: A genome was obviously made. Such a thing doesn't just spontaneously generate out of thin air. It took some planning, programming, science, and manufacturing to get the job done.

Conclusion: Therefore, a genome obviously has a Maker or a Creator who designed it, programmed it, engineered it, field-tested it, fine-tuned it, made it, manufactured it, and then deployed it.

The first premise is obviously true. If you don't believe that the first premise is true, then you aren't living in the same dimension or the same universe as the rest of us. Anything that was obviously made obviously had a Maker or Creator who made it. This claim is obviously true.

Consequently, only the second premise is in dispute. Obviously, if you don't believe that genomes were made by Someone Intelligent or Someone Psyche, then you are not going to believe the conclusion either. However, most of us are convinced that genomes were obviously made. Even Richard Dawkins said as much, without saying as much.

You don't need to be a mathematician or physicist to calculate that an eye or a hemoglobin molecule [or a genome] **would take from here to infinity to self-assemble by sheer higgledy-piggledy luck. — Richard Dawkins**

Even to Dawkins, it's clear that genomes were obviously made – made by natural selection. The thing he doesn't realize or understand is that natural selection doesn't touch our genes, which means that natural selection cannot design and create genomes. Natural selection doesn't have any teeth. It doesn't touch our genes.

Dawkins is right, it would take from here to infinity for genomes to self-assemble by sheer higgledy-piggledy luck; but, Dawkins is wrong every time he claims or suggests that natural selection did it because natural selection can't touch our genes. Natural selection doesn't do anything for us or against us, before or after the selection has been made. If we are talking about natural selection or survival of the fittest, we can only die once. If we are talking about sexual selection, we can only be conceived once. Either way, selection doesn't touch our genes. Selection doesn't change our genes. Selection of any kind doesn't have anything to do with our genes.

It's the random mutations that affect our genes; and, random mutations are entropy. Every scientist knows, or should know, that entropy cannot design and create anything. Instead, entropy leads us towards disease, disability, mental illness, cancer, death, and extinction. That is what has been experienced and observed. Entropy is death and extinction. Natural selection produces entropy; and, random mutations are entropy. Entropy cannot design and create. It's physically impossible.

Self-assembly of genes, proteins, genomes, eyes, brains, and life forms from atoms is impossible. Spontaneous generation is impossible. Design and creation by random mutations and natural selection is impossible. Chemical evolution is impossible. Macro-evolution of any kind is impossible. Self-assembly of genomes from scratch is impossible. It can't happen, which means that it didn't happen. Genomes were obviously made. There's no denying it. It's obvious.

The mathematicians have indeed run the numbers as Dawkins suggested; and, they have proven beyond a shadow of a doubt that it is physically impossible for proteins and their matching genes to spontaneously generate or self-assemble by sheer luck – even if given an infinite amount of time to do so. Dawkins is right, it would take from here to infinity for genomes to self-assemble by sheer higgledy-piggledy luck. It can't happen, which means that it didn't happen. It's physically impossible. Self-assembly of genomes from atoms is prevented from happening by entropy, random diffusion, or the second law of thermodynamics. Chaos or entropy cannot design and create. Chance or blind luck cannot design and create. Chance or blind luck is also a function of entropy. Natural selection produces death or entropy; and, entropy or natural selection or death cannot design and create as Richard Dawkins says that it does.

The Probability of a Protein Forming by Chance
https://www.youtube.com/watch?v=W1_KEVaCyaA
https://evolution-is-entropy.com/wp-content/uploads/2018/05/Probability-of-a-Protein-Forming-by-Chance.zip

The Probability of Making a Protein
https://www.youtube.com/watch?v=cQoQgTqj3pU
https://evolution-is-entropy.com/wp-content/uploads/2018/05/Probability-of-making-a-protein.zip

Evidence for Creation by Outside Intervention
https://www.youtube.com/watch?v=ci4s75al1Rw
https://evolution-is-entropy.com/wp-content/uploads/2018/05/Evidence-for-Creation-by-Outside-Intervention-01.zip
https://evolution-is-entropy.com/wp-content/uploads/2018/05/Evidence-for-Creation-by-Outside-Intervention-02.zip
https://evolution-is-entropy.com/wp-content/uploads/2018/05/Evidence-for-Creation-by-Outside-Intervention-03.zip

These videos are worth owning, watching, and keeping. I archived them so that they can't disappear on me while I'm alive.

Observation or Premise: Anything that was obviously made obviously had a Maker or Creator who made it.

Observation or Premise: A genome was obviously made.

Logical Conclusion: Therefore, a genome obviously has a Maker or a Creator who designed it, programmed it, engineered it, field-tested it, fine-tuned it, made it, manufactured it, and deployed it.

These observations convinced me that God does in fact exist, because I'm a scientist and I believe in the scientific evidence that has been experienced and observed. In other words, I KNOW that the premises are true; therefore, I KNOW that the conclusion MUST be true as well.

Some type of Syntropy, Psyche, or God MUST exist; or, physical matter, entropy, genomes, proteins, and this physical universe would NOT exist. It's elementary, and it's obvious.

Consequently, since WE KNOW for a fact that God does indeed exist, the next task is to figure out who He is. I choose to go with our resurrected Lord Jesus Christ, because He has been EXPERIENCED and OBSERVED both in the flesh and during our Near-Death Experiences (NDEs), after He rose from the dead. I choose to go with the ONE who has been experienced and observed in real life by real people like you and me. Go with the true and living God – the one who isn't a man-made God, like evolution or natural selection.

Evolution and natural selection are man-made gods, imbued by definition with all of the power and foresight of a God. But, you don't want to go with any of the man-made gods because they are nothing but fiction. A man-made god isn't going to come and get you out of hell. You want to go with the True and Living God, Jesus Christ, because He has been experienced and observed.

https://www.youtube.com/results?search_query=NDE+Jesus

https://www.youtube.com/results?search_query=NDE+Atheist

Thousands of people have seen Jesus during their Near-Death Experiences (NDEs). Check out the videos! Jesus Christ even came and got some of the atheists out of hell during their NDE. Remember, should you ever find yourself in hell, Jesus Christ can get you out of there just for the asking. Jesus Christ could even get Richard Dawkins out of hell if Richard Dawkins were to ask. That's good enough for me. That's the God that I want to go with – the one who can make good on His promises and get me out of a jam.

Go with the best and get rid of all the rest. That's what I have decided to do, and you can too.

—

Genomes Were Obviously Made

Genomes were obviously made, which means that they obviously had a Maker, Creator, or God who made them.

A genome is Scientific Proof of God's Existence. It was the first convincing Scientific Proof of God's Existence that I encountered in science. It's so obvious to me that the genomes were made, that it was obvious to me that some kind of Maker, Creator, or God made them. The truth of it was undeniable. Genomes are God's Signature.

In contrast, it was also obvious to me that spontaneous generation, creation ex nihilo, abiogenesis, natural selection, random mutations, or macro-evolution could never have made or produced a genome. It's physically impossible. It's prevented from happening by entropy. It can't happen, which means that it didn't happen.

Chemical Evolution of any kind is prevented from happening by entropy or random diffusion. That's the real science behind all of this. It's obvious that it requires some type of Syntropy, Intelligence, or Psyche in order to be able to design and create anything – including physical matter and genomes.

Furthermore, it was obvious to me that evolution (genetic change), random mutations, and natural selection did not exist until after God designed and created the physical matter, proteins, genes, genomes, eyes, brains, and life forms in the first place. Random mutations and natural selection obviously could never have produced our genomes because they obviously didn't exist until after our genomes were made. Evolution didn't exist until after God made the genomes.

Proteins and the matching genes to go along with them were obviously made, which means that they obviously had a Maker, Creator, or God who made them.

In fact, physical matter, proteins, genes, genomes, eyes, brains, and physical bodies were obviously made, which means that they obviously had a Maker, Creator, or God who made them. This is Logic 101 and positive proof that God does in fact exist.

It's obvious to me that natural selection, random mutations, chemical evolution, and macro-evolution are man-made gods and are typically portrayed as if they were God; but, they can't design and create. Since we KNOW for a fact that God must exist, who is the True and Living God that does exist? Jesus Christ is the only one who qualifies because He has been experienced and observed after He rose from the dead. Jesus Christ has the necessary credentials, qualifications, capabilities, and pedigree. It doesn't get more God-like than rising from the dead or showing up and getting an atheist out of hell.

So, do you want the truth, or do you want convenient fables and science fiction? The choice is yours. That's not a choice that anybody else can make for you. Where your choices and beliefs are concerned, you are their Maker, Creator, and God. God definitely exists. In your own limited way, you are a God; and, you definitely exist. Get used to it! You'd just as well, because that's where we are going as a race if we don't destroy ourselves first.

Mark My Words

—

Source

The Scientific Method Proves That the Theory of Evolution Is False

https://www.amazon.com/dp/B01IAAIRT2

https://www.amazon.com/dp/1521133611

My Philosophical Proof of God's Existence

Web Page: https://philosophy-of-science.com/my-philosophical-proof-of-god/

I've been working on this Philosophical Proof of God's Existence for three years now, because it wasn't obvious at the beginning how to proceed.

An argument is *valid* if the conclusion necessarily follows from the premises. An argument is *sound* if it is valid and the premises are true.

I have developed a philosophical proof of God's existence – or two – that I find logically compelling, philosophically valid, and scientifically sound. Let's start with the best and most believable conventional philosophical proof of God's existence, the Kalam Cosmological Argument.

Kalam Cosmological Argument

Let's start with William Lane Craig's version of the Kalam Cosmological Argument.

Premises:

Whatever begins to exist has a cause;

The universe began to exist;

Conclusion:

The universe has a cause.

https://en.wikipedia.org/wiki/Kalam_cosmological_argument

https://philosophy-of-science.com/wp-content/uploads/2018/04/Kalam-Cosmological-Argument.pdf

It begs the question, "Which universe began?" And, that's the main problem with the Kalam Cosmological Argument. The term "universe" is too vague and needs to be more carefully defined because we know of some universes that had NO beginning. Nevertheless, the Kalam Cosmological Argument is a good place to start, when it comes to a philosophical proof of God's existence.

This argument or proof is a syllogism. If the premises are true, then the conclusion has to be true. This is a properly constructed syllogism; therefore, it is a *valid* philosophical argument. An argument is *valid* if the conclusion necessarily follows from the premises.

This argument is also believed by many people to be a *sound* argument. Remember, an argument is *sound* if it is valid and the premises are true. This argument is considered TRUE by many people because it is philosophically *valid* and seems to be scientifically or observationally *sound*.

I find this philosophical proof of God's existence logically consistent, philosophically *valid*, and somewhat *sound*. It works in part, because its Premises are based somewhat upon observed truths. However, it can be greatly clarified and improved upon, because I can think of many exceptions to Craig's second premise.

Craig's Kalam Cosmological Argument FAILS to convince me in the end, because I can think of many exceptions to his second premise. There are some universes that had NO beginning; and therefore, they have NO cause. They are uncaused.

Proofs Demand Perfectly Sound Premises

You have got to get your premises perfect – perfectly true – or the proof doesn't work because it isn't *sound*. Craig's Kalam Cosmological Argument is a *valid* or well-constructed argument; but, it isn't *sound* – it isn't true. Craig can't see it nor understand it because he is using a falsehood as one of his hidden premises – namely creation ex nihilo. Falsehoods negate or falsify proofs.

Genesis 1:1-2:

1 In the beginning God created the heaven and the earth.

2 And the earth was without form, and void; and darkness was upon the face of the deep. And the Spirit of God moved upon the face of the waters.

IN THE BEGINNING OF WHAT?

It's extremely important to figure out and KNOW for sure which beginning they are actually talking about. You see, verse one of Genesis is also supposed to be a philosophical proof of God's existence; but, it only works if you know which beginning they are talking about.

Clearly God existed BEFORE this beginning, or He wouldn't have been able to begin it. Verse two of Genesis actually tells us that God was in a spirit form or a spiritual body during the formation of our physical earth.

Clearly, whatever substance God organized our heavens and our earth from had to exist BEFORE the beginning of our heavens and our earth because creation ex nihilo is impossible – meaning that it is scientifically and logically *unsound*. Creation ex nihilo is philosophically and scientifically absurd. It didn't happen because it can't happen. Not even God can create something from nothing. This is what science and logic tells us is true. Something from nothing leaves nothing.

Every aspect of your philosophical proof has to be perfect, perfectly sound or perfectly true, and perfectly understood; or, your philosophical proof isn't going to work and isn't going to be convincing.

SO, IN THE BEGINNING OF WHAT?

In the beginning of our physical universe!

The majority of the scientists are Big Bang proponents, which means that the majority of the scientists really truly believe and KNOW that our physical universe as well as physical matter had a beginning of some kind, which means that our physical universe had a Maker, or Creator, or Organizer, or Ultimate Cause. It's unavoidable. Every beginning has a Cause. If there was a Big Bang, somebody pushed the button! Clearly, our physical universe began, which means that it has a Maker or Creator of some sort.

Once we successfully identify "the beginning" as the Beginning of our Physical Universe, then the first verse of Genesis becomes a successful and believable and convincing Kalam Cosmological Argument because we KNOW that all the premises are true. We scientists KNOW for a fact that our physical universe and physical matter had a beginning of some sort, which means that our physical universe had a Beginner or Maker who made it. It's undeniable, since we KNOW that the premises are true.

Then we also KNOW for a fact that whatever substance God used to make or to construct this physical universe, it wasn't physical matter! It had to be some kind of Dark Matter or Spirit Matter because physical matter didn't exist yet in this universe. We also KNOW that whoever and whatever God is, He pre-dated or pre-existed the beginning of this physical universe. Now, we have a platform of truth and knowledge upon which we can build.

My Adjustments to the Kalam Cosmological Argument

An argument is *valid* if the conclusion necessarily follows from the premises; and, an argument is *sound* if it is valid and the premises are true. Remember, the premises have to be true in order for the argument to be *sound*.

Even though Craig's argument is philosophically valid and scientifically or observationally sound, at least superficially, I found that I had to adjust it a bit so as to make it specifically clear what I'm talking about and what I'm trying to prove. Craig's argument is too vague and universal for my tastes, which allows falsehoods, faulty interpretations, confusion, logic fallacies, and unscientific ideas to creep into it too easily.

So, I chose to adjust it in the following manner and tighten it up a bit.

Premises or Observations:

Whatever begins to exist has a Beginner, Maker, or an Ultimate Cause;

Our physical universe began to exist;

Conclusion:

Therefore, our physical universe has a Beginner, Maker, or an Ultimate Cause.

Now this thing is bullet-proof. The premises are TRUE and universally agreed upon, which means that the conclusion has to be TRUE as well.

These adjustments are important and essential, because I'm trying to bring the premises and conclusion into line with what has been experienced and observed in real life by real people. I'm trying to make my argument scientifically valid and sound, not just philosophically so. A philosophical proof that proves a lie to be true is of no value to anyone. I want this thing solid and sound when I'm done with it so that it actually does prove that some kind of God exists. In other words, I was able to think of exceptions to Craig's second premise that actually falsified and ruined his Kalam Cosmological Argument for me so that it no longer worked as a proof of God's existence. The Premises of a proof have to be empirical, obviously true, and bullet-proof in order for the proof to be efficacious, convincing, real, and true.

Whenever I develop a philosophical proof, I try to switch over to observational evidence, experiential proof, verified reality, scientific evidence, and logical common sense as quickly as I can in order to prevent myself from being tricked and deceived. Self-deception works, and it works every time.

It's the **physical universe** that we are talking about here because the non-physical universe, quantum universe, non-local universe, psyche universe, syntropy universe, transdimensional universe, transcendent universe, or spiritual universe had NO beginning; and therefore, it will have no end. It's the **physical universe** that had a beginning, not the transdimensional universe. This is crucial to get straight and get right.

Craig's argument talks about "the universe" – that's way too vague.

Why?

Well, the Multi-Verse, or the Non-Physical Universe, or the Chaos Construct, or the Quantum Transdimensional Transcendent Universe, or the Spirit Realm had NO beginning, which technically makes Craig's premise #2 false to begin with when he says that "our universe began," because the Primal Universe or Original Universe had NO beginning and therefore had NO cause. However, it has been observed and agreed upon by the scientists in general that our physical universe definitely had a beginning; therefore, our physical universe definitely had to have had an Ultimate Cause or a Person who organized it and brought it into existence. It's our **physical universe** that began, NOT "the universe".

Furthermore, the Beginner or Ultimate Cause has to be a person, or a living entity, or a Psyche, or an intelligent being of some kind because we KNOW that raw matter, or dead matter, or inert matter, or chaotic matter, or entropy cannot spontaneously generate into anything whatsoever. There's no such thing as spontaneous generation or creation ex nihilo where physical matter is concerned.

Psyche is the Ultimate Cause; and, Psyche is the ultimate causal agent. Matter cannot design and create. There is NO such thing as spontaneous generation or creation ex nihilo. Spontaneous generation or creation ex nihilo has been FALSIFIED by the Scientific Method. Such a concept as Creation Ex Nihilo is philosophically, logically, and scientifically unsound. It doesn't make any logical sense. Something from nothing is illogical. It can't be done, which means that it wasn't done. Not even God can do the impossible. Not even God can do creation ex nihilo!

Our physical universe came from something that already existed, and our physical universe was organized by Someone Psyche or Someone Spiritual who also already existed.

Creation Ex Nihilo and Atheism are kissing cousins. They are of the same kind. You could even say that they are siblings.

Creation Ex Nihilo is the creation or manufacture of something from nothing. Creation Ex Nihilo has NEVER been experienced nor observed; and, it never will be, because it's impossible. Creation Ex Nihilo is illogical and patently absurd. It's magic. It won't happen, because it can't happen. It never happened because it can't happen.

Atheism is creation or manufacture of something by NOTHING. Technically, Atheism has NEVER been experienced nor observed; and, it never will be, because it's impossible. Atheism is illogical and patently absurd. It won't be verified, because it can never be verified. NOTHING will never be caught in the act of manufacturing something.

Likewise, spontaneous generation, macro-evolution, chemical evolution, and abiogenesis are a type of Atheism – the creation or manufacture of something by NOTHING. It can't be done, which means that it wasn't done. These types of "evolution" are physically impossible and prevented from happening by entropy or the second law of thermodynamics. They can't happen, which means that they didn't happen.

Most people are content to simply provide a *valid* philosophical proof; but when I'm doing proofs, I demand that the Premises be based upon scientific proof, actual observations, and real-life experiences. In other words, I demand that the Premises be scientifically and observationally *sound*. It makes for a better and an infinitely more convincing proof. The *soundness* of a proof is everything, in my humble opinion.

Ultimately, Craig's version of the Kalam Cosmological Argument isn't *sound* because I can think of many exceptions to his second Premise. My second premise doesn't allow for any exceptions because I tightened it up and made it specific.

My version of the Kalam Cosmological Argument does indeed PROVE that some kind of God exists – the God who designed, created, and produced our physical universe.

Why?

My proof works and is convincing because MOST of the scientists in the world are in agreement that our physical universe began in some sort of Big Bang; and, this proof should be convincing to the Christians and Muslims because their God and scriptures tell us that He created the heavens and the earth – namely our physical universe. Physical matter and physical universes have a beginning, which means that they have a Beginner, Maker, or some sort of Ultimate Cause. Big Bangs have a beginning, which means that they have somebody who pushed the button and made them go bang.

God's Psyche didn't have a beginning, because it has always existed. The spirit matter or dark matter from which God organized this physical universe has always existed. But, practically everyone is in complete agreement that this physical universe, physical matter, and our physical earth had a beginning, which means that the Beginner or Ultimate Cause was some type of Psyche, Spirit, or God BEFORE the beginning or organization of physical universes and physical matter.

My adjustment to the Kalam Cosmological Argument is already philosophically *valid* and scientifically *sound*; so, it already works as a philosophical proof of God's existence; but, let's formalize it and expand it just a bit, so that I can comment on it some more.

My Philosophical Proof of God's Existence

Remember, an argument is *valid* if the conclusion necessarily follows from the premises; and, an argument is *sound* if it is valid and the premises are true.

Premise and Observation: Anything that has a beginning has a Beginner or an Ultimate Cause – a Designer, Creator, and Manufacturer who brought it into existence. This is Logic 101. It's inherently logical and sound. It has been experienced and <u>observed</u>.

Observations: It has been <u>observed</u> that our physical universe had a beginning, which means that it had a Beginner or an Ultimate Cause – a Designer and Creator who brought it into existence. That means that physical matter had a beginning, which means that it had a Beginner or Ultimate Cause – a Designer, Creator, and Manufacturer who organized it or brought it into existence.

Conclusion: When it comes to the organization and beginning of our physical universe and physical matter, by definition, in principle or practice, that Beginner, Designer, Manufacturer, Maker, and Creator has to be a God.

This is a logical conclusion since the rest of us here on this physical earth aren't in the habit of making physical matter and physical universes from scratch. As physical beings, we can't touch the sub-atomic or the quantum or the spiritual. Only a transcendent and transdimensional God would have such capabilities. Such a God would have to exist BEFORE the organization of this physical universe from spirit matter, chaotic matter, dark matter, or primal matter. Such a God would have to be transcendent without any physical limitations whatsoever – an omnipotent God who is the master of transdimensional physics or quantum mechanics. Such a God would have to have sufficient knowledge and power for such an endeavor.

The FACT that our physical universe had to have had a Beginner or an Ultimate Cause tells us some important things about the nature and capability the Being, the Psyche, the Spirit, or the God who designed and organized our physical universe. The Person or God who organized our physical universe had to pre-date our physical universe; and, the stuff that He used to form or organize our physical universe also had to pre-date our physical universe. This is Logic 101. First things first!

Technically, according to the rules associated with arguments and proofs, the first premise is allowed to be a Given Truth, Axiom, or a freebie; but, I demand that the first premise be scientifically proven, or observed and experienced by Someone Psyche sometime somewhere. All of the hidden premises and freebies that philosophers use is what gets us into trouble in the first place. I want all of my premises and my conclusion to be proven and verified in one way or another; otherwise, my philosophical proofs and scientific arguments are worthless in my humble opinion – as worthless as the philosophical proofs and scientific arguments being used to support and promote Materialism, Naturalism, Darwinism, Nihilism, and Atheism.

I demand that ALL of the premises be observed, verified, experienced, and proven to be real and true. ONLY then is the conclusion guaranteed to be true. A valid proof is worthless, if it proves a lie to be true. The premises and the conclusion must be *sound*! Consequently, I demand that ALL of the premises and even the conclusion be observed, verified, experienced, and proven to be real and true in Real Life by Real People. ONLY then can you be sure that you have indeed found the truth.

Notice that in this Philosophical Proof of God's Existence, I used OBSERVATIONS as my Premises so that I could be absolutely sure that my conclusion is true. My Premises are not as concise as in the previous examples, but I KNOW that they are true, which is all that really matters to me in the end.

The Materialists, Naturalists, and Theists are simply satisfied to win the argument. But, winning the argument isn't enough for me. I want to have the truth and KNOW the truth when I am done. I'm a scientist. I want to know the truth, not just win the argument.

Try this one! It's got teeth and claws.

Observation: Anything that was obviously made obviously had a Maker or Creator who made it.

Observation: A genome was obviously made.

Logical Conclusion: Therefore, a genome obviously has a Maker or a Creator who designed it, programmed it, engineered it, field-tested it, fine-tuned it, made it, manufactured it, and deployed it.

My philosophical proofs of God's existence really do PROVE that some kind of God exists, because my philosophical proofs of God's existence are in fact Scientific Proofs of God's Existence or OBSERVED Proofs of God's Existence. There's a huge difference there; but, only for those who are actually looking for such a thing. Like I said, MOST philosophers are simply satisfied to provide a *valid* argument and could care less if their Premises are scientifically and observationally *sound*. MOST philosophers and scientists are satisfied with a valid argument and a valid conclusion, and they could care less if their Conclusion has actually been experienced and observed in real life by real people.

I want EVERY PART of my philosophical proofs and scientific arguments to be experienced and observed; or, my philosophical proofs, scientific arguments, and chosen conclusions end up being completely worthless to me. I even want my Conclusions to have been experienced and observed by Someone Psyche, sometime somewhere. Only then can I KNOW that I have finally found the truth.

Remember, an argument is *valid* if the conclusion necessarily follows from the premises; and, an argument is *sound* if it is valid and the premises are true.

While doing philosophical proofs, theological proofs, and scientific proofs, I DEMAND that ALL of the Premises and the Conclusion be *sound*, which means that I demand that all of the Premises and the Conclusion be experienced and observed by Someone Psyche somewhere sometime. That's one of the reasons why Islam FAILS for me, because nobody has ever seen Allah; however, thousands, if not millions, of different people have seen and experienced our resurrected Lord Jesus Christ either in the flesh or during their Near-Death Experiences. Jesus Christ is the being of light and love whom people encounter and experience after they have died, and their brain is clinically dead. Jesus Christ is REAL, and He truly exists, because ALL of the observational evidence is telling us that it is so.

It's the observational evidence that I find convincing, and NOT the philosophical proofs. The philosophical proofs simply give us a way to structure, compound, and then multiply the effects of the observational evidence. It's the observational evidence, knowledge, and truth that I'm after, and NOT victory over my foes. Victory is hollow if what you win ends up being worthless and false.

Commentary on My Philosophical Proof of God

Every beginning has a Person or a Psyche who caused it to happen. Psyche is the Ultimate Cause; and, Psyche is the ultimate causal agent.

See: ***The Ultimate Model of Reality: Psyche Is the Ultimate Cause***

https://www.amazon.com/dp/B071NC9JK6

Remember, every beginning has a Person, Intelligence, or a Psyche who caused it to happen or brought it into existence. This is Logic 101.

My Philosophical Proof is a philosophical proof of God's existence that actually makes logical sense to me and works for me. I find it compelling and convincing. It's based upon the Kalam Cosmological Argument, which I also find convincing and believable, as long as it is worded in a manner such that the Premises are based upon observational evidence and lived experiences. It's the Observational Evidence that I find convincing, not necessarily the philosophical proof.

https://en.wikipedia.org/wiki/Kalam_cosmological_argument

https://philosophy-of-science.com/wp-content/uploads/2018/04/Kalam-Cosmological-Argument.pdf

With Observed and Proven Premises in place, I found my version of the Kalam Cosmological Argument convincing; but, just because it works for me is no guarantee that it will work for you.

As with any philosophical proof, if you don't find the premises, arguments, and observations compelling, credible, and convincing, then you won't find the conclusion convincing either. That's the weakness of philosophical proofs. Observation and experience can easily falsify them; and, a lack of personal observational experience can easily make that person a victim of the philosophical proofs that are faulty and false. I seldom find any philosophical proof convincing because they are all subject to legerdemain or trickery. Most of them have NO observational evidence supporting them.

When it comes to philosophical proofs, their conclusion is also subject to *personal interpretation*, which introduces a wide variety of logic fallacies into the mix. Philosophical proofs are worthless if their Premises have NEVER been experienced nor observed. Their conclusions and the interpretations of their conclusions are worthless too if they have NEVER been experienced nor observed. That's why Materialism, Naturalism, Nihilism, and Atheism are worthless, because their hidden assumptions or major premises have NEVER been experienced and observed, nor can they be experienced and observed. Materialism, Naturalism, and their derivatives are FALSIFIED automatically due to a complete lack of observational evidence supporting their hidden assumptions or major premises which claim that the quantum, or the non-local, or the non-physical does not exist.

I was never really convinced by philosophical proofs. Most people aren't. There's too much self-deception going on when it comes to philosophy. Materialism and Naturalism are philosophy or religion, not science. They have to be taken on blind faith as being true. In contrast, whether I'm talking about science or philosophy, the thing that I do find compelling, convincing, and believable is Evidence, Observation, and Experience. Science is observation, or it should be. I've always found a Scientific Proof of God's Existence infinitely more believable and convincing than a philosophical proof of God's existence, because the science is based upon observation and experience.

Furthermore, a complete lack of observational experience or verified proof can easily falsify a premise, a philosophical proof, a hypothesis, or even a theory. That's what happened to Materialism, Naturalism, Darwinism, Nihilism, Behaviorism, Determinism, Scientism, Atheism, Macro-Evolution, and therefore the Theory of Evolution. <u>A complete lack of observational evidence</u> supporting their major premises, primary assumptions, or hidden assumptions completely FALSIFIES them.

It's the observations and experiences that I find convincing, not the philosophical proofs or philosophical sophistry. The wonderful thing about the Kalam Cosmological Proof of God's Existence, especially my modified version, is that it's backed by tons of observational evidence and it's easily transformed into a Scientific Proof of God's Existence simply by finding some observed evidence and experiential evidence to insert as your premises.

For example:

Premises or Observations:

Anything that has been fine-tuned has a Fine-Tuner;

Our physical universe has clearly been fine-tuned;

Conclusion:

Therefore, our physical universe has a Fine-Tuner.

This is a powerful syllogism that works as an excellent philosophical proof of God's Existence because it is based exclusively upon observational evidence and logical common sense. This argument is philosophically and logically SOUND because its premises have been experienced and observed by real-life human beings.

This argument is solidified by the scientific observation that spontaneous generation is physically impossible, thanks to entropy or the second law of thermodynamics. There's no such thing as spontaneous generation, abiogenesis, chemical evolution, creation ex nihilo, or macro-evolution thanks to entropy or the second law of thermodynamics which prevents these types of things from happening in the wild.

This Fine-Tuner Argument is as much a Scientific Proof of God's Existence as it is a philosophical proof of God's existence. The evidence of fine-tuning is all around us and impossible to deny. Fine-tuning is one of the most convincing Scientific Proofs of God's Existence.

Every instance or example of cosmic fine-tuning, mechanical fine-tuning, and biological fine-tuning is a miniature scientific proof of God's existence.

How do we know?

We scientists KNOW because living cells, genomes, genes, proteins, eyes, brains, and life forms have <u>NEVER been observed</u> spontaneously generating from atoms out of thin air. They can't because it's physically impossible for them to do so thanks to entropy, random diffusion, or the second law of thermodynamics.

For me personally, I found all of those precision-tuned Cosmological Constants or Physical Constants to be one of the most convincing Scientific Proofs of God's Existence that I have ever come across, especially since we scientists KNOW for a fact that such precision fine-tuning doesn't spontaneously generate out of thin air, because of entropy or the second law of thermodynamics. Likewise, we scientists KNOW for a fact that the exquisite and precise programming found in our genomes doesn't just spontaneously generate out of thin air either, here in this physical realm – once again thanks to entropy or the second law of thermodynamics. Your genome is God's Signature.

Again, for your convenience.

Observation: Anything that was obviously made obviously had a Maker or Creator who made it.

Observation: A genome was obviously made.

Logical Conclusion: Therefore, a genome obviously has a Maker or a Creator who designed it, programmed it, engineered it, field-tested it, fine-tuned it, made it, manufactured it, and deployed it.

My Genomic Argument is a variation on the Fine-Tuning Argument. They both are excellent Scientific Proofs of God's Existence, which also makes them some of the very best philosophical proofs of God's existence around because they are real and true, having been experienced and observed.

The BEST listing and documentation of God's precision fine-tuning is found on Reasons to Believe, by Hugh Ross, and is freely available to the public. Ross presents a link to this list in the appendix of his book, *Why the Universe Is the Way It Is*, and in the appendix of a couple of his other books. I archived it on a couple of my websites because it kept disappearing or moving around on me, and I could never find it again.

<u>Fine-Tuning Is Scientific Proof of God's Existence</u>

https://philosophy-of-science.com/wp-content/uploads/2018/04/compendium_part1.pdf

https://philosophy-of-science.com/wp-content/uploads/2018/04/compendium_part2.pdf

https://philosophy-of-science.com/wp-content/uploads/2018/04/compendium_Part3_ver2.pdf

https://philosophy-of-science.com/wp-content/uploads/2018/04/compendium_Part4_ver2.pdf

Ross, H. (2008). *Why the Universe Is the Way It Is*. Grand Rapids, MI: Baker Books.

For me personally, the first truly convincing Scientific Proof of God's Existence came to me on the day when I first realized that entropy or the second law of thermodynamics prevents chemical evolution, random mutations, and natural selection from spontaneously generating proteins, genes, genomes, eyes, brains, and life forms out of thin air from scratch. The theory of evolution is spontaneous generation, and spontaneous generation is physically impossible thanks to entropy or the second law of thermodynamics. Functional information-rich genomes, functional proteins, and life forms do not and cannot spontaneously generate from atoms. They just can't. It's physically impossible. It was then that I realized that God must exist in order to have done all of the Science and Fine-Tuning and Programming which chemical evolution, random mutations, and natural selection could NEVER have done. On that day, I simply KNEW that God exists, because I KNEW why He must exist.

Ironically, Louis Pasteur FALSIFIED spontaneous generation (and therefore Materialism, Naturalism, Darwinism, Atheism, Chemical Evolution, Macro-Evolution, Creation Ex Nihilo, and the Theory of Evolution) in 1859 – the very same year that Charles Darwin published "On the Origin of Species". We have KNOWN since the very beginning of the theory of evolution that spontaneous generation, abiogenesis, macro-evolution, chemical evolution, creation ex nihilo, or the theory of evolution is FALSE; but, most scientists have deliberately chosen to ignore that evidence and pursue wishful thinking instead. *Wishful thinking*, or *confirmation bias*, or *blind-faith* is a logic fallacy – one of the logic fallacies upon which Materialism, Naturalism, and Darwinism are based.

Remember, your philosophical arguments and syllogisms are NOT sound if their premises have NEVER been experienced nor observed. They may be logically valid, but they are NOT sound. That's how the Materialists, Naturalists, and Darwinists trick us and deceive us. They give us syllogisms and philosophical arguments that are logically valid, but totally unsound due to a complete lack of observational evidence supporting their chosen premises or hidden assumptions. Consequently, because of faulty premises, these people also *jump to conclusions* that are philosophically, logically, and scientifically impossible and unsound. *Jumping to conclusions* or *begging the question* is also a logic fallacy. Materialism, Naturalism, Darwinism, and Atheism are unscientific because their premises are scientifically unsound, having never been experienced nor observed.

Materialism, Naturalism, Darwinism, and their derivatives are based exclusively upon entropy; and, entropy by definition in principle cannot do Fine-Tuning and Programming. Therefore, God or Syntropy must of necessity exist in order to have done ALL of the Fine-Tuning and Programming that needed to be done. Simple. Logical. Parsimonious. TRUE!

Notice that whenever I present a scientific argument or a philosophical proof, I try to switch over to observed science or proven science for my Premises, as quickly as I can. It's the Observations and Experiences and Science that are convincing, and NOT the philosophical proof or syllogism. Remember, Materialism, Naturalism, Darwinism, Nihilism, Behaviorism, Determinism, Scientism, and Atheism have NO observed and NO proven Premises. Consequently, they FAIL before they even get started. There's no way to support them with science or observation because there is none, where their major premises or hidden assumptions are concerned.

Discussion of fine-tuning and the physical constants.

https://en.wikipedia.org/wiki/Fine-tuned_Universe

https://philosophy-of-science.com/wp-content/uploads/2018/04/Fine-tuned-Universe.pdf

https://en.wikipedia.org/wiki/Dimensionless_physical_constant

https://philosophy-of-science.com/wp-content/uploads/2018/04/Dimensionless-physical-constant.pdf

https://en.wikipedia.org/wiki/Physical_constant

https://philosophy-of-science.com/wp-content/uploads/2018/04/Physical-constant.pdf

https://philosophy-of-science.com/wp-content/uploads/2018/04/Physical-constant.pdf

Remember, the second law of thermodynamics, random mutations, death (natural selection), random diffusion, or entropy cannot do Physical Constants, Precision Fine-Tuning, and Programming. It's physically impossible. It can't be done, which means that it wasn't done.

Now, try this one:

Premises or Observations:

1. We have observed that everything that has a beginning has some kind of Creator or Maker who made it.

2. We have observed that our physical universe had a beginning. Consequently, we have observed that physical matter had a beginning.

Conclusion:

3. Therefore, it is logical to conclude that our physical universe and the physical matter within it had some kind of Creator or Maker who made them, organized them, or brought them into existence.

This argument has teeth because its premises have been experienced and observed. This argument is philosophically and logically SOUND because its premises have been experienced and observed. This Creator Argument or Maker Argument is as much a Scientific Proof of God's Existence as it is a philosophical proof of God's existence.

We have caught intelligent beings in the ACT of design, creation, manufacturing, and production zillions of times in trillions of different ways. Intelligent Design or Intelligent Creation is an OBSERVED, verified, and proven science. We KNOW that it is REAL because it has been EXPERIENCED and OBSERVED. It has been caught in the act. There's NO philosophical speculation, guesswork, or wishful thinking going on here where the observation of intelligent beings (or intelligent psyches) in action is concerned.

I find these Philosophical Proofs of God's Existence equally as compelling as the others because their premises have been experienced and observed in real life by real people. If you want the truth, then go with what has been experienced and observed, because the philosophical speculation, wishful thinking, and blind faith of the Materialists and Naturalist are worthless in the end.

Notice once again that when it comes to My Philosophical Proofs of God's Existence, I try to turn them into a Scientific Proof of God's Existence as quickly as I can. In other words, I try to turn them into an OBSERVED Proof of God's Existence. When it comes to science and proof, observation is where the tires really hit the pavement. Observed Proof of God's Existence is the most convincing and believable proof of God's existence. That's just the way it is.

In contrast, I also found the complete lack of observational evidence supporting their major premises or hidden assumptions equally as convincing when it came time to falsify theories and ideas such as Materialism, Naturalism, Darwinism, Nihilism, Behaviorism, Determinism, and Atheism. These philosophies or hypotheses are FALSIFIED by a complete lack of observational evidence supporting their major premises or hidden assumptions which state that the Non-Local, the Quantum, or the Non-Physical does not exist. The very existence and the verified existence of Quantum Mechanics and Action at a Distance FALSIFIES Materialism, Naturalism, Darwinism, Nihilism, Atheism, Classical Physics, and their derivatives.

Introducing Science 2.0

As a result of these observations, I upgraded my science to Science 2.0.

Science 2.0: I Upgraded My Science

https://www.amazon.com/dp/B0771K6WTX

Science 2.0 allows ALL of the evidence into evidence, and then it pursues a preponderance of that evidence. Evidence or observation is the only thing that has value to us. Philosophical arguments and philosophical proofs are absolutely worthless if the Premises that are employed are false and have been falsified, as has happened in the case of Materialism, Naturalism, Darwinism, and their derivatives.

When it comes to the non-physical sciences, Science 2.0 uses a Burden of Proof Methodology that is based upon a preponderance of the observational evidence. Under Science 2.0, observation and experience of any kind take precedence over philosophical speculation, guesswork, hypothesis, wishful thinking, scientific inferences, and confirmation biases. Science 2.0 is the way that science should have always been done but wasn't.

Science 2.0 is a new and better way of doing science that is based upon observational evidence, eye-witness evidence, and experiential evidence.

Under Science 2.0, the BEST way to find and know the truth is to live it, experience it, and observe it for yourself, or to choose to trust someone who has. The second-best way to find and know the truth is to use the Scientific Methods to falsify and eliminate everything that is false such as Materialism, Naturalism, Darwinism, Atheism, and their derivatives. If you successfully eliminate and remove everything that is false, then ONLY the truth will remain. It's fascinating to observe what remains after you have eliminated everything that is false and everything that has been falsified.

Remember, a syllogism can be logically valid but totally unsound. If the Premises of your argument, syllogism, or logic proof are NOT backed by observational evidence, then your philosophical argument is unsound. That's yet another serious problem with Materialism, Naturalism, Darwinism, Nihilism, Atheism, Behaviorism, Determinism, Scientism, Atheism, and the Theory of Evolution. There is NO observational evidence supporting their major premises or hidden assumptions, and there can NEVER be any observational evidence supporting their primary assumptions or major premises; therefore, the arguments produced by these philosophies, religions, dogmas, or pseudo-sciences can be forced to be valid, but they will NEVER be philosophically or logically sound. In other words, they will always be false. These falsified philosophies are unscientific and unsound because there can never be ANY observational evidence supporting their hidden assumptions or major premises which claim that the non-physical or the quantum does not exist.

The moral of the story is that when it comes to your philosophical proofs, be sure to use observed evidence, verified scientific evidence, experienced evidence, veridical evidence, and empirical evidence as your Premises if you want to have the greatest chance of your conclusions and subsequent interpretations being true.

My Philosophical Proofs of God's Existence have teeth and claws because their premises are OBSERVED evidence and have actually been experienced in real life. My Philosophical Proofs of God's Existence are both valid and sound because they are in fact Scientific Proofs of God's Existence, meaning that their premises are backed by tons of observational evidence and experience. We KNOW that the experienced and the observed are TRUE because they have been experienced and observed. We KNOW that Materialism, Naturalism, Darwinism, Nihilism, and Atheism are false because their major premises or hidden assumptions have NEVER been experienced nor observed and can't be experienced or observed.

We KNOW that God exists, because the Philosophical Arguments and Scientific Arguments that I have presented here PROVE that He exists. Once I knew that God exists, then I realized that I still have a long way to go before I get to know God. I used to be a Materialist, Naturalist, Nihilist, and Atheist; so, I still have a lot of work to do and a lot to accomplish where knowing God is concerned.

Go Out of Your Way to Get the Right Interpretation

Remember, philosophy has NO value whatsoever if it doesn't match with what has been experienced and observed. Notice how I tried to switch over from logic to what has actually been experienced and observed, before drawing any conclusions. The observations are essential. Science is observation. A philosophical proof has to match with reality, or it's worthless.

It's also important to draw the correct conclusion.

Online, when I provided a scientific proof of God's existence, one of the readers asked me which of all the man-made gods we should believe in.

Therefore, in harmony with his question, I ask, "Which of all the man-made gods is the one who was the Beginner, Designer, Manufacturer, Maker, and Creator of this physical universe and the physical matter in this universe?"

I chose to go with the True and Living God, Jesus Christ, because He has been experienced and observed after He rose from the dead. Thousands have seen Him and touched Him in the flesh; and, thousands have seen Him and embraced Him during their Near-Death Experiences. Should you ever find yourself in hell, remember that Jesus Christ can get you out of there just for the asking.

We KNOW that the Biblical God Jesus Christ exists, because He has been experienced and observed by thousands after He rose from the dead. Jesus Christ claims to be the God who organized the heavens and this earth, as well as all the life forms on this earth.

Notice once again, that when it comes to God's Existence, I try to switch over to observation and experience as quickly as I can, because philosophical speculation and wishful thinking are absolutely worthless in the end.

I define "science" as observation and experience. I have trained myself to switch away from philosophical speculation, wishful thinking, hypothesis, confirmation biases, and sophistry over to observation, experience, and empirical evidence as quickly as I can in order to prevent myself from being deceived. I encourage my readers to do the same.

A Proof or Argument Has to Be Logically Sound

This is most important thing to know and understand about philosophical proofs and scientific arguments – they MUST BE *valid* and *sound*. In other words, their Premises MUST BE true; otherwise, the proof, argument, and conclusion are absolutely worthless and liable to deceive.

An argument is *valid* if the conclusion necessarily follows from the premises. An argument is *sound* if it is valid and the premises are true.

Remember, a philosophy and a philosophical proof are worthless if their Premises are false. Likewise, a scientific argument is worthless if its Premises are false. That's precisely what's wrong with Materialism, Naturalism, Darwinism, Nihilism, Atheism, and their derivatives. Their hidden premises are false and can never be true.

If the premises of your argument don't add up to the conclusion, then your proof is worthless no matter how valid it might seem to be. If your premises have NEVER been experienced nor observed in the wild, then your conclusion is nothing but fiction and is philosophically unsound. Your chosen conclusion is automatically falsified by a complete lack of observational evidence. It's called falsifying a theory!

In order to be logically sound, your Premises have to be experienced, observed, and verified. In order for your Conclusion to be logically sound, your Premises have to be logically sound, and your Conclusion also has to be experienced, observed, and verified in real life by real individuals. Missing evidence cannot be used as evidence. The missing links really are missing, so get used to it.

Remember, a God who has never been experienced nor observed is completely worthless to us. Evolution is a man-made god who has never been experienced, observed, nor caught in the act. A premise or a conclusion has to be experienced and observed in order for it to be philosophically and logically sound. A premise and a conclusion that has never been experienced nor observed is completely worthless to us. If the Biblical God hadn't revealed Himself to us, He would have remained forever unknown. In any philosophical proof, religious argument, or scientific argument, the Premises absolutely MUST BE true, or those arguments and proofs are totally worthless to us. That's just the way things work.

If your Premises completely lack confirming evidence or verified evidence, then your conclusion is automatically false. Design and creation by Chemical Evolution, Spontaneous Generation, Abiogenesis, or Macro-Evolution have NEVER been experienced nor observed, which means that these concepts or theories are automatically false. We KNOW that these things are prevented from happening by entropy or the second law of thermodynamics.

Technically, random mutations and natural selection have NEVER been caught in the act of design and creation, which means that these theories or concepts are automatically false. In other words, random mutations and natural selection cannot design and create. Random Mutations are entropy. Natural Selection ultimately results in death and extinction, which means that natural selection is also a type of entropy. Evolution is entropy. By definition, in principle, Materialism, Naturalism, Darwinism, Nihilism, Atheism, Classical Physics, and the Theory of Evolution are based upon entropy. Entropy cannot design and create anything. So, who did?

The Premises behind Materialism, Naturalism, Darwinism, and their derivatives are false and have been falsified by science and by observation, which means that the Conclusions that these people make are also false and have already been falsified. Materialism, Naturalism, and Darwinism are FALSIFIED by a complete lack of confirmed evidence, verified evidence, or observed evidence supporting their major premises and hidden assumptions, which claim that the quantum or the supernatural does not exist. A claim that something or someone does not exist cannot be verified, which means that it can NEVER be proven true; therefore, such a Premise is assumed to be false and will always be false until someone finds a way to prove it true, which they NEVER will. The Premises behind Materialism, Naturalism, Darwinism, Nihilism, Behaviorism, Determinism, Physical Reductionism, Scientism, Atheism, and their derivatives will always be false because it's impossible to prove them to be true.

In contrast, the verified, observed, and proven existence of Quantum Mechanisms, Action at a Distance, or Supernatural Mechanisms FALSIFIES the major premises of Materialism and Naturalism which claim that the quantum or the supernatural does not exist. The very existence of Quantum Mechanics or Transdimensional Physics FALSIFIES Materialism, Naturalism, and their derivatives such as Nihilism, Darwinism, and Atheism.

I have observed and experienced the FACT that anything that has a beginning has a Beginner or an Ultimate Cause, who is the person who designed, created, manufactured, and produced that thing. This reality has always held true.

A Beginner or Creator or Maker is the person who brings a thing into existence in the first place. This is a FACT that I find fully compelling, believable, and incontrovertible whether we are talking about science, philosophy, logic, religion, reality, or existence in general. I find it convincing, because it has been experienced and observed. We have observed people or intelligent beings bringing things into existence or choosing things into existence zillions of times in trillions of different ways.

Likewise, I find the existence of the Biblical God Jesus Christ compelling and believable because He has been experienced and observed by thousands, if not millions, of different people after He died and rose from the dead. Science is Observation, or it should be. These people have seen Him and touched Him both here on the physical plane as well as on the spiritual, quantum, or transdimensional plane.

Every time that Jesus Christ is seen, experienced, touched, embraced, and observed whether in the flesh, in visions and theophanies, or during our Near-Death Experiences (NDEs), those experiences and observations are Scientific Proof of His Existence. Science is observation, or it should be.

https://www.youtube.com/watch?v=UPj4wci_bcI

https://www.youtube.com/watch?v=Vm647n1360A

https://www.lds.org/scriptures/bofm/3-ne/11?lang=eng

https://www.lds.org/scriptures/pgp/js-h/1?lang=eng

In order for an argument or a proof to be logically sound, the premises have to be true, which means that the premises have to be experienced and observed by Someone Psyche. A philosophical proof that has NO observational evidence supporting it is completely worthless in the end. That's why Materialism, Naturalism, Darwinism, Nihilism, Behaviorism, Determinism, Atheism, and the Theory of Evolution are completely worthless. They don't have and can't have any observational evidence supporting their major premises or hidden assumptions which claim that the quantum or the supernatural does not exist. In fact, the observed and proven existence of Quantum Mechanics and Action at a Distance FALSIFIES Materialism, Naturalism, and their derivatives.

It took me three years to develop a Philosophical Proof of God's Existence that I found completely and totally believable and true.

Why did it take so long?

I realized in hindsight that it took me so long to develop such a proof because I was looking for Premises that I KNOW to be true. I was looking for Scientific Proof of God's Existence, not just philosophical proof of God's existence. I was looking for OBSERVED Proof of God's Existence or EXPERIENCED Proof of God's Existence. It's the observations and experiences that I find convincing, NOT the philosophical proofs. It's the empirical evidence that I find convincing, NOT the philosophical arguments.

I don't trust philosophical proofs, because MOST of them have hidden premises that I KNOW to be false. Materialism, Naturalism, Darwinism, Nihilism, Behaviorism, Determinism, Atheism, and their derivatives ARE philosophical proofs or philosophical arguments. I don't believe them to be true because they have hidden premises that are demonstrably and empirically false. As a race, WE have FALSIFIED Materialism, Naturalism, Darwinism, Atheism, the Theory of Evolution, and their derivatives trillions of times in thousands of different ways. Falsified theories cannot be used as convincing Premises, especially when you KNOW that they are false and why they are false.

If you successfully eliminate everything that is false, then ONLY the truth will remain. It's fascinating to study and observe what remains after you have successfully falsified and eliminated Materialism, Naturalism, Darwinism, Nihilism, Atheism, and their derivatives. The VERIFIED and the OBSERVED remain. The Truth remains. Suddenly you find yourself looking at TRUE and VERIFIED Premises, as well as philosophically sound, scientifically observed, verified, experienced, and proven Conclusions. Your Conclusions ARE TRUE because they have been experienced and observed, and because their Premises are true and philosophically sound. Your Premises ARE TRUE because they too have been experienced and observed in real life.

Do you see how that works?

It works because it is TRUE.

In the end, all you really want is the truth.

Observational Reality and Common-Sense Logic

Anything that has a beginning has a Beginner or a Creator.

In contrast, God's Psyche and your psyche have always existed and will always exist. Your psyche and God's Psyche have no beginning, which means that they will have no end. This also means that your psyche or intelligence has NO Beginner, Designer, Creator, Manufacturer, Maker, or Ultimate Cause who brought it into existence in the first place. You have always existed, and you will always exist.

Things that have a beginning have a Beginner or a Creator, the person or psyche who brought them into existence. Physical matter, physical universes, physical genomes, and physical life forms have a beginning, which means that they have a Designer, Creator, and Maker who brought them into existence. In contrast, the things that have always existed, such as Psyche and Syntropy and Quantum Mechanics, have NO Creator because they have always existed.

Do you see how that works? This is Logic 101.

The things that have always existed, such as Psyche and Syntropy, have NO Creator. They are without a beginning of days or an end of years. Psyche is Syntropy. Quantum Mechanics or Transdimensional Physics is Syntropy. Syntropy is a type of unity, wholeness, completeness, or perfection. The Atonement of Christ is Syntropy. Syntropy means "without a beginning of days or an end of years". Syntropy means eternal, everlasting, and infinite. The Priesthood Power of God is Syntropy. Quantum Mechanics is the Priesthood Power of God. Syntropy had no beginning, and it will have no end. Syntropy, or Psyche, or Intelligence, or Consciousness has always existed. Syntropy is an organizing force that counteracts the effects of entropy. Syntropy has to exist or entropy wouldn't exist. Our physical universe had to receive an initial infusion of Syntropy, or none of that subsequent entropy would have been possible.

Do you see how that works?

Its explanatory power is through the roof and without limit!

I just provided the answer to life, the universe, and everything. That's the explanatory power of Syntropy, Psyche, and Quantum Mechanics.

Critiquing Craig's Cosmological Argument

William Lane Craig and others have declared him to be the master of the Kalam Cosmological Argument.

https://en.wikipedia.org/wiki/Kalam_cosmological_argument

Notice that William Lane Craig makes an argument that is philosophically valid and semi-sound; but then, he *jumps to a conclusion* at the end that is philosophically illogical and scientifically unsound. Craig is not going to critique his own argument, because he is too close to it and too emotionally invested in it to be able to see what might be wrong with it.

William Lane Craig states the Kalam cosmological argument as a brief syllogism, most commonly rendered as follows:

Whatever begins to exist has a cause;

The universe began to exist;

Therefore:

The universe has a cause.

From the conclusion of the initial syllogism, he appends a further premise and conclusion based upon ontological analysis of the properties of the cause:

The universe has a cause;

If the universe has a cause, then an uncaused, personal Creator of the universe exists who *sans* **the universe is beginningless, changeless, immaterial, timeless, spaceless and enormously powerful;**

Therefore:

An uncaused, personal Creator of the universe exists, who *sans* **the universe is beginningless, changeless, immaterial, timeless, spaceless and enormously powerful.**

Referring to the implications of Classical Theism that follow from this argument, Craig writes:

"... transcending the entire universe there exists a cause which brought the universe into being ex nihilo **... our whole universe was caused to exist by something beyond it and greater than it. For it is no secret that one of the most important conceptions of what theists mean by 'God' is Creator of heaven and earth."**

https://en.wikipedia.org/wiki/Kalam_cosmological_argument

William Lane Craig deliberately left his Kalam cosmological argument vague so that he could sneak a falsehood in at the end.

The only thing Craig gets wrong is that he *jumps to the conclusion* that God brought our universe into being **ex nihilo**. *Jumping to conclusions* or *begging the question* is a logic fallacy. In his conclusion, **ex nihilo** is an *unjustified add-on* or a *special pleading*, which are logic fallacies. In his promotion of Creation Ex Nihilo, Craig is *assuming facts not in evidence*.

There's NO such thing as Creation Ex Nihilo. Technically, Creation Ex Nihilo is Atheism – design and creation from nothing by nothing. It's impossible to create something from nothing. It's also impossible for nothing to design and create something. Even God can't do the impossible. Atheism is also impossible, because it's impossible for nothing to create something from nothing. Creation Ex Nihilo is philosophically and logically unsound because it doesn't make logical sense and because it's the result of *jumping to conclusions*, which is a logic fallacy. Creation Ex Nihilo or spontaneous generation is also scientifically unsound because it has been falsified by the Scientific Method.

Clearly, God existed BEFORE He organized this physical universe and brought it into existence. The very existence of God, the pre-existence of God, FALSIFIES creation ex nihilo. God did not spontaneously spring into existence from nothing. Creation ex nihilo does not and cannot apply to God, nor can it apply to the dark matter or spirit matter from which God organized this physical universe. God's Psyche has always existed and will always exist. The dark matter or spirit matter has always existed and will always exist. Creation Ex Nihilo or spontaneous generation is false and has been falsified. Creation Ex Nihilo is unnecessary and unjustified, so let's delete it. If you successfully eliminate everything that is false, then only the truth will remain.

Therefore, Craig's concluding statement should read:

"Transcending the entire universe there exists a cause which brought the universe into being. Our whole universe was caused to exist by something beyond it and greater than it. For it is no secret that one of the most important conceptions of what theists mean by 'God' is Creator of heaven and earth."

As long as we get rid of the logic fallacy and scientific falsehood that Craig tacks on at the end of his argument, I find the Kalam Cosmological Argument philosophically, logically, and scientifically valid and sound.

William Lane Craig teaches us by example that a philosophical proof can be valid, the premises can be sound, and the conclusion absolutely true; but, the person can still draw false conclusions and produce faulty and invalid interpretations from that True Conclusion or Proven Conclusion. Interpretation of the evidence or scientific inference is where the Scientific Method always falls down and dies whenever the Scientific Method is used to prove that a lie is true.

Remember, even William Lane Craig falls down and FAILS whenever he switches away from observational evidence over to *wishful thinking* and starts *jumping to conclusions* rather than relying upon observational evidence to make his case. Creation Ex Nihilo has NEVER been experienced nor observed; and, it never will be, because it's impossible. It's the observational evidence that we find convincing, and NOT the philosophical arguments. Philosophical arguments are worthless if their premises are not backed-up by observational evidence, verified evidence, and experiential evidence. Even a valid and sound conclusion from a philosophical argument is worthless, if it has NO observational evidence supporting it.

We find the existence of the Biblical God Jesus Christ compelling, believable, and convincing ONLY because it's backed-up by observational evidence and experiential evidence. Philosophical proof of God's existence is worthless if the premises are NOT backed up by observational evidence, verified evidence, scientific evidence, experiential evidence, eye-witness evidence, and empirical evidence. A conclusion is also worthless if it has NO observational evidence supporting it.

I was able to tighten up my version of the Kalam Cosmological Argument and get even more specific with it because I wasn't trying to insert a falsehood at the end. I was only interested in finding the observed and verified truth. Creation ex nihilo has never been experienced nor observed. It doesn't exist. It's impossible. Even God cannot do the impossible. Since I wasn't trying to support a falsehood with my philosophical argument, I was able to go directly to the Observed and Experienced Truth of the matter.

The theory of evolution is spontaneous generation; and, spontaneous generation is a type of Atheism or Creation Ex Nihilo – creation of something from nothing or by nothing. Something from nothing or something by nothing is impossible. It didn't happen because it can't happen.

If you accept the premises of the Kalam Cosmological Argument as being true, then they PROVE that our physical universe had a Creator, Beginner, or Ultimate Cause; however, they do NOT prove the veracity of creation ex nihilo. In fact, ALL of the observational evidence and experiential evidence that we have on hand as a race FALSIFIES creation ex nihilo, spontaneous generation, or the theory of evolution. Furthermore, it's ONLY our physical universe that had a beginning. The quantum universe, or the supernatural universe, or the spiritual universe, or the non-physical universe, or the non-local universe has always existed and will always exist. It has NO Creator or Ultimate Cause because it has always existed.

The Appearance of Design

Biology is the study of complicated things that have the appearance of having been designed with a purpose. – Richard Dawkins.

They have the appearance of having been designed because they were designed. There's NO other logical explanation for their origin because entropy or physical matter or natural selection cannot design and create anything. It's NEVER been caught in the act of doing so. It can't, which means that it didn't.

Spontaneous generation, creation by entropy, macro-evolution, abiogenesis, creation ex nihilo, chemical evolution, OR the theory of evolution was FALSIFIED in

1859 by Louis Pasteur. Biology appears to have been designed because it was designed.

Now try this philosophical proof of God's existence. It's an extremely powerful syllogism, because it is also a Scientific Proof of God's Existence or an OBSERVED Proof of God's Existence.

> **First Observation or Premise: Anything that was obviously made obviously had a Maker or Creator who made it. This is Logic 101.**
>
> **Second Observation or Premise: A genome was obviously made. Such a thing doesn't just spontaneously generate out of thin air. It took some planning, programming, science, and manufacturing to get the job done.**
>
> **Conclusion: Therefore, a genome obviously has a Maker or a Creator who designed it, programmed it, engineered it, field-tested it, fine-tuned it, made it, manufactured it, and then deployed it.**

These observations convinced me that God does in fact exist, because I'm a scientist and I believe in the scientific evidence that has been experienced and observed. In other words, I KNOW that the premises are true; therefore, I KNOW that the conclusion MUST be true as well.

Some type of Syntropy, Psyche, or God MUST exist; or, physical matter, entropy, genomes, proteins, and this physical universe would NOT exist. It's elementary.

Consequently, since WE KNOW for a fact that God does indeed exist, the next task is to figure out who He is. I choose to go with our resurrected Lord Jesus Christ, because He has been EXPERIENCED and OBSERVED both in the flesh and during our Near-Death Experiences (NDEs), after He rose from the dead. I choose to go with the ONE who has been experienced and observed in real life by real people like you and me.

Mark My Words

Conclusions

The Beginner or the Ultimate Cause of this physical universe has to pre-date the beginning of this physical universe, and so does the stuff of which He was made and the stuff by which He made. This is Logic 101.

God organized the heavens and the earth from already pre-existing matter. God transformed some of that pre-existing spirit matter or dark matter into physical matter; and therefore, our physical universe had a physical beginning, even though the spirit matter or dark matter from which it was made has always existed. We KNOW that the dark matter or spirit matter exists by the effect that it has on ordinary physical matter. It has to exist, or our observations and measurements don't make any logical sense.

Do you see how that works?

Everything makes sense to me once we get rid of the logic fallacies and switch over to observations and measurements instead. Then suddenly, everything is philosophically, logically, and scientifically valid and sound.

Verified means that it has been experienced and observed. Falsified means that it has NEVER been experienced nor observed. Materialism, Naturalism, Darwinism, Nihilism, Behaviorism, Determinism, Creation Ex Nihilo, and Atheism have been FALSIFIED by the fact that their major premises or hidden assumptions have never been experienced nor observed and can't be. Materialism, Naturalism, Darwinism, Nihilism, and Atheism are FALSIFIED by the things that have been experienced and observed. They are FALSIFIED by Science.

The lesson in all of this is to switch away from philosophical speculation and sophistry over to observational evidence, experiential evidence, eye-witness evidence, empirical evidence, veridical evidence, scientific evidence, and phenomenological evidence as soon as you possibly can.

The truth has been repeatedly verified, experienced, and observed. The falsified and the false have NEVER been experienced nor observed, which is why they have been falsified. The false is falsified by the truth; and, the truth has been experienced and observed. That's just the way things work. That's the way things should work in science; and, that's definitely the way that things should work in religion, philosophy, and logic as well. Go with the experienced and the observed, and get rid of the wishful thinking, scientific inferences, blind-faith, and confirmation biases as quickly as you can. Go with the best and get rid of all the rest.

If you successfully eliminate everything that is false, then ONLY the truth will remain.

Mark My Words

—

Source

1. ***Syntropy in Defense of Quantum Mechanics: The Answer to Life, the Universe, and Everything***

https://www.amazon.com/dp/B07BPT3W8R/

2. Web Page:

https://philosophy-of-science.com/my-philosophical-proof-of-god/

3. The Official Website:

https://philosophy-of-science.com/

Source for the Original One

Some might prefer the tighter version with less commentary; but, it's still pretty long. With this essay, I wanted to explain why I now believe in God's existence. It's the culmination of years of research, study, and thought. I'm not a prophet, so I don't know the Biblical God in person; but, I am a scientist, scholar, philosopher, logician, and theoretician, so I'm able to think things through carefully, logically, and rationally in order to reach a reasonable conclusion on the matter.

1. ***The Ultimate Model of Reality: Psyche Is the Ultimate Cause***

    https://www.amazon.com/dp/B071NC9JK6

2. ***Science 2.0: I Upgraded My Science***

    https://www.amazon.com/dp/B0771K6WTX

3. Web Page:

    https://philosophy-of-science.com/original-one/

References

1. Slife, B. D. & Williams, R. N. (1995). Science and Human Behavior. In *What's Behind the Research? Discovering Hidden Assumptions in the Behavioral Sciences*, (pp. 167–204). Thousand Oaks, CA: SAGE Publications.

    https://mypsyche.us/wp-content/uploads/2017/04/Science.pdf

    https://philosophy-of-science.com/wp-content/uploads/2018/04/Science.pdf

2. Gantt, E. (2014). Logical Arguments. In *Psychology 353 – LDS Perspectives in Psychology*, (pp. 8-11). Provo, UT: Brigham Young University.

    https://philosophy-of-science.com/wp-content/uploads/2018/04/Logical-Arguments.pdf

3. Gantt, E. (2014). Leveling the Playing Field – Why Science is Not a Trump Card. In *Psychology 353 – LDS Perspectives in Psychology*, (pp. 50-58). Provo, UT: Brigham Young University.

    https://philosophy-of-science.com/wp-content/uploads/2018/04/Verification-vs-Falsification.pdf

4. Rychlak, J. F. (1981a). *A Philosophy of Science for Personality Theory* (2nd ed.). Malabar, FL: Robert E. Krieger Publishing Company.

5. Ross, H. (2008). *Why the Universe Is the Way It Is*. Grand Rapids, MI: Baker Books.

6. **My Amazon Page:**

   https://amazon.com/author/science

7. ***Science 2.0: I Upgraded My Science***

   https://www.amazon.com/dp/B0771K6WTX

Quantum Mechanics Is Spiritual Mechanics

The goal in all of this is to hook people up and give them a Model of Reality that actually matches with reality. I want to help people find the truth. I put a lot of effort into this so that I would have a Model of Reality that actually matches with the Scientific Evidence and the Observational Evidence that we have on hand as a race.

In contrast, the goal of the Materialists, Naturalists, Darwinists, Nihilists, Behaviorists, Determinists, Physical Reductionists, Atheists, and Classical Physicists is to convince people that certain things DO NOT EXIST. That goal is totally worthless. That's NOT science. That's religion and dogma. It takes an infinite amount of blind faith to believe that something DOES NOT EXIST. Furthermore, it's philosophically and scientifically impossible to prove that something does not exist. It can't be done, which means that they will NEVER be able to prove that Materialism, Naturalism, Darwinism, and Atheism are true.

These people are tilting windmills, and they don't even know it. These people are fighting imaginary enemies which are nothing more than a figment of their imagination. They are also promoting pseudo-sciences that are also nothing more than wishful thinking which they have manufactured within their minds out of thin air. There's nothing tangible within Materialism and Naturalism that we can get our hands on. The whole thing is purely a Grand Illusion. It doesn't match with Reality, which makes it unscientific to begin with.

When it comes to Scientific Naturalism, Darwinism, and Physicalism, these pseudo-sciences are based exclusively on entropy. Entropy is death, which means that these pseudo-sciences are literally a Dead-End to begin with. There's NO life there! Entropy cannot produce life. Entropy cannot produce order. Entropy can only produce death. Entropy IS death. Evolution is entropy. Evolution is death. The theory of evolution is Creation by Death. Please explain to me how that's supposed to work.

One of my greatest and most useful scientific discoveries is that Quantum Mechanics is Spiritual Mechanics or Supernatural Mechanics. The proven and verified existence of Action at a Distance falsifies Materialism, Naturalism, Darwinism, Nihilism, Behaviorism, Atheism, Classical Physics, and their derivatives. All you really want is the truth, unless you are trying to deceive yourself.

Interpretations of Quantum Mechanics and Science

I used to be a Physicalist, Naturalist, Nihilist, and Atheist. I wasted a lot of time and effort learning concepts that are demonstrably false.

My Atheism was a very dark time in my life. If I can help to prevent even one person from going through what I went through, then it will have all been worth it!

What was the net effect of My Atheism?

It dumbed me down. It made me ignorant. It made me dumb and blind even though I still had a soul or a mind.

I knew that God does not exist.

Atheism makes you believe that you know more than you really do. Atheism makes you ignorant of all the different things that you need to learn and know about. Atheism made me ignorant of Quantum Mechanics or Supernatural Mechanics. Atheism made me ignorant of Syntropy. I'd never heard of Syntropy. I didn't know what it was. Atheism of any kind is based upon a refusal to look at evidence. Atheism makes you ignorant. Atheism makes you think that you are smart when you are not. You become arrogant and condescending towards others. That's the fruits of Atheism, Materialism, Naturalism, Nihilism, and Darwinism. You become dumber than what you used to be. Some people have called it the "dumbing down of America".

That's why it took me over 55 years to figure out that evolution is entropy. The theory of evolution is based upon entropy. That's why it doesn't work as advertised. It can't! Entropy can't design, create, and manufacture anything! Entropy is death. Entropy can only destroy; and think about it, the theory of evolution is based exclusively on entropy. The theory of evolution is Creation by Death. The theory of evolution is Creation by Entropy. The theory of evolution was false to begin with; but, I couldn't see that nor understand that because I had been educated and trained in Entropy, Materialism, Naturalism, Nihilism, Darwinism, Atheism, and Classical Physics. It made me dumb and blind.

The Materialists and Naturalists deliberately remove the quantum non-local sciences from consideration. These people are actively trying to hide Syntropy, Psyche, Quantum Mechanics, and Supernatural Mechanisms from us. The various different types of macro-evolution, or spontaneous generation, or chemical evolution have NEVER been observed in the wild. They have NEVER been experienced nor observed because they are physically impossible. They are prevented from happening by random diffusion or entropy. The Naturalists and Darwinists are trying to hide these scientific truths from us.

The Physicalists, Naturalists, and Atheists ONLY permit correlational research and experimental research into evidence. They ONLY permit physical evidence into evidence. They only permit the correlational method and the experimental method. They deliberately BAN all other sources of information and knowledge. However, there is also a Burden of Proof Methodology that is used in archeology and the forensic sciences. It is based upon the preponderance of the evidence. We have to use a Burden of Proof Methodology when going after spiritual, supernatural, and quantum realities and forces because we can't use our physical instruments to detect and record these things.

Correlations indicate a relationship, but that relationship is not necessarily one of cause and effect. Correlational research allows us to *predict,* but it cannot tell us whether changing one variable (such as social status) will *cause* changes in another (such as health).

> The correlation-causation confusion is behind much muddled thinking in popular psychology.
>
> The great strength of correlational research is that it tends to occur in real-world settings where we can examine factors such as race, gender, and social status (factors that we cannot manipulate in the laboratory). Its great disadvantage lies in the ambiguity of the results. This point is so important that even if it fails to impress people the first 25 times they hear it, it is worth repeating a twenty-sixth time: *Knowing that two variables change together (correlate) enables us to predict one when we know the other, but correlation does not specify cause and effect.* (*Social Psychology*, pp. 20-21.)

One of my greatest and most useful scientific discoveries came when I first realized that correlation-causation confusion is behind much of the muddled thinking that's associated with Darwinism and the theory of evolution. Materialism, Naturalism, Darwinism, Nihilism, Atheism, Behaviorism, Determinism, Physical Reductionism, and the Theory of Evolution are purely correlational! There's NO evidence backing their claims that that Quantum or the Supernatural does not exist; and, there never will be. These falsified philosophies or falsified religions are purely correlational. They have NO evidence supporting them. They have to be taken on blind faith as being real and true. In fact, the verified and proven existence of Action at a Distance or Quantum Mechanics falsifies Materialism, Naturalism, Darwinism, and their derivatives. ALL the EVIDENCE falsifies Physicalism, Naturalism, and Darwinism. The theory of evolution is based exclusively on correlation-causation confusion. Fascinating, is it not?

Scientific experimentation has its weaknesses too. Sometimes laboratory research and science experiments fail to generalize to real-world settings. Manufacturing amino acids in a beaker in a simulated laboratory setting has absolutely NOTHING to do with getting entropy and physical matter to spontaneously generate into amino acids in the wild.

Experimental Research: Searching for Cause and Effect

> The difficulty of discerning cause and effect among naturally correlated events prompts most social psychologists to create laboratory simulations of everyday processes whenever this is feasible and ethical.
>
> We need to be cautious, however, in generalizing from laboratory to life.
>
> ### Generalizing from Laboratory to Life
>
> As the research on children, television, and violence illustrates, social psychology mixes everyday experience and laboratory analysis. Throughout this book we will do the same by drawing our data mostly from the laboratory and our illustrations mostly from life. Social psychology displays a healthy interplay between laboratory research and everyday life. Hunches gained from everyday

experience often inspire laboratory research, which deepens our understanding of our experience.

This interplay appears in the children's television experiment. What people saw in everyday life suggested correlational research, which led to experimental research. Although the laboratory uncovers basic dynamics of human existence, it is still a simplified, controlled reality. It tells us what effect to expect of variable *X*, all other things being equal — which in real life they never are! Moreover, as you will see, the participants in many experiments are college students. Although that may help you identify with them, college students are hardly a random sample of all humanity. Would we get similar results with people of different ages, educational levels, and cultures? That is always an open question.

Nevertheless, we can distinguish between the *content* of people's thinking and acting (their attitudes, for example) and the *process* by which they think and act (for example, *how* attitudes affect actions and vice versa). The content varies more from culture to culture than does the process. People from various cultures may hold different opinions yet form them in similar ways. (*Social Psychology*, pp. 24, 28, 29.)

The ONLY way to distinguish between Psyche's thoughts and the observed physical actions is to accept the fact that both exist and that both are REAL. The Psyche's choices do indeed correlate with a person's physical behavior or physical actions; but if we want a fullness of the truth, we must accept the independent evidence or experiential evidence that verifies and proves the existence of Psyche because we are not going to get any of that from physical matter and our physical instruments. We must treat the observations and experiences associated with Near-Death Experiences and Out-of-Body Experiences as Scientific Evidence and allow them into evidence, if we want to find and know the truth about Psyche Mechanics or Quantum Mechanics; otherwise, we choose to remain in ignorance. That's just the way it is.

The laboratory is NOT a real-world setting most of the time. The variables in our lives aren't typically controlled, constrained, and manipulated in an artificial manner. Furthermore, a physical laboratory is slow to reveal the existence of Psyche or Quantum Mechanics. It can be done if the scientist is smart enough and creative enough; but, direct observation and experience is a better way for finding and knowing the truth about Psyche, Quantum Mechanics, and Syntropy.

Thought is a function of the Human Psyche. Thoughts and memories are quantum waves.

How do we know?

We know because our thoughts and memories survive the death of our physical brain. We know because our thoughts and memories can't be detected nor recorded by our physical instruments.

If a thought or memory survives the death of your physical brain, then it's truly a memory in every sense of the word, is it not?

This is just a small portion of what I learned while studying Social Psychology. Social Psychology is the scientific study of the Human Psyche and its interaction with all those other psyches. Social Psychology is valid on both sides of the veil – both here in this physical reality and afterwards in the spirit world. This is radically advanced science when it is no longer being restrained and limited by Naturalism, Darwinism, and Atheism. Social Psychology has whole chapters on the Self or the Psyche. When it comes time to define and explain the Self or the Mind, Psyche or Quantum Non-Local Consciousness ends up being the BEST explanation. That is what I have experienced and observed. Science is observation and experience after all.

In contrast, Atheism is the making of something by "Nothing". Atheism is patently absurd. Creation ex nihilo or spontaneous generation is the making of something from "Nothing". Creation ex nihilo or spontaneous generation is a type of Atheism, and it's just as absurd as Atheism. Creation ex nihilo assumes facts not in evidence. In fact, it's contradicted by logic and by evidence. The same can be said of Atheism, Physicalism, Darwinism, Nihilism, and Classical Physics. They are contradicted or falsified by logic and by evidence. They are falsified by Quantum Mechanics or Supernatural Mechanisms.

Do you want the truth, or do you want the science fiction? You can't have both. The one is true, and the other has been falsified by the truth. The one is true, and the other has been falsified by observation and experience. The false is falsified by the truth; and, the truth is repeatedly experienced and observed. Science is observation and experience – NOT Physicalism and Naturalism. Technically, the "science" behind Materialism, Naturalism, Darwinism, Nihilism, Atheism, and their derivatives has NEVER been experienced nor observed.

There are as many different interpretations of Quantum Mechanics as there are scientists, from what I can tell.

https://en.wikipedia.org/wiki/Interpretations_of_quantum_mechanics

https://en.wikipedia.org/wiki/Minority_interpretations_of_quantum_mechanics

Quantum Mechanics or Supernatural Mechanics is our most-used, most-verified, and best-proven Science that we have; but, nobody seems to know what it means.

I have experienced and observed the fact that interpretations of Quantum Mechanics that are based upon Physical Matter, Materialism, Naturalism, Darwinism, Nihilism, Atheism, and Classical Physics completely lack explanatory power and are therefore absolutely worthless.

Quantum Mechanics or Transdimensional Physics ONLY makes sense from a non-local, non-physical, transdimensional, spiritual perspective.

Mark My Words

Source Material

Quantum Mechanics from a Non-Physical Spiritual Perspective

https://www.amazon.com/dp/B01J023TGU

https://www.amazon.com/dp/1521132380

Myers, D. G. (2010). *Social Psychology* (10th ed.). New York: McGraw-Hill.

—

Entropy or the Passage of Time

According to the theory of relativity, anything traveling faster than the speed of light experiences NO passage of time. Time STOPS at the speed of light or faster.

Do you fully understand what this means?

It means that everything existing or traveling at frequencies faster than the speed of light is Syntropy. It doesn't age because there is NO passage of time. In other words, there is NO entropy. Entropy is a function of time. ONLY physical matter experiences entropy or is subject to entropy. Everything else is Syntropy.

Light is Syntropy. There is NO age limit on the Light that has been traveling for 13.2 billion years to get here from the other galaxies in this universe. There's NO entropy in Light. It doesn't age, get old, and die. From its perspective, it arrives the very moment it launches, even though from our perspective it took 13.2 billion years to get here. In fact, it has been theorized that Light already knows where it is going to land 13.2 billion light years away, or it doesn't launch in the first place.

Likewise, Psyche or Intelligence is Syntropy. It is ageless and timeless. There is NO entropy in Psyche, Intelligence, or Quantum Non-Local Consciousness. It is eternal and everlasting, without a beginning of days or an end of years. It is Syntropy. It doesn't bleed; so, there's no way to kill it or end it.

We will NEVER be able to accelerate or push physical matter faster than the speed of light because then it would no longer be physical matter, and it would no longer be subject to entropy or the passage of time. It would be spirit matter or dark matter instead. It would be Syntropy. It would be spiritual matter.

Physical matter has been caught in the act of Quantum Tunneling; so, it still retains that inherent capability. If you stop the passage of time or temporarily remove the entropy from physical matter, then you can teleport it or quantum tunnel it anywhere you want in the universe instantly. During the quantum tunneling process, it experiences NO passage of time. It doesn't age.

Those of us who have experienced Quantum Tunneling or Teleportation KNOW that time STOPS when you Quantum Tunnel. I felt time stop. It was like I had passed out. And then, I was just instantly somewhere else instead, with NO knowledge or memory of how I got there. I and the car I was driving had just jumped, blinked, or teleported. It was nothing that I did, though. It was something that God did for me in order to save me.

The science is REAL, and it's powerful; and, it's totally consistent, coherent, and harmonious with itself. It ALL makes sense. The Truth matches perfectly with all other truths. The Truth has been experienced and observed.

Quantum Tunneling is REAL. It has been experienced and observed. The only drawback or disappointment when it comes to Quantum Tunneling is that ONLY God has access to it, or ONLY God can do it and trigger it at will. ONLY God holds all the KEYS to Quantum Mechanics or Supernatural Mechanics. Mortal fallen beings like us aren't given direct access to these things. God resides within Syntropy. God is Syntropy. Whereas, we mortal fallen beings are stuck within or placed within entropy and physical matter. We age and die. When we die, our psyche or soul is released back into Syntropy.

God imposes entropy, the aging process, or the passage of time onto physical matter in order to keep the physical matter within your physical body from teleporting or quantum tunneling away from you at will. God keeps the physical matter within your body, house, computer, and car from dissolving into thin air by quantum tunneling to the other side of the universe at will. Physical matter still retains that capability; but, God prevents physical matter from quantum tunneling by making it subject to entropy or the passage of time. God deliberately slows the physical matter down to sub-light speeds so that we can live it, experience it, observe it, and remember having done so.

The explanatory power of this Science is through the roof. By choosing to allow Psyche, Syntropy, and Quantum Mechanics in to play, we can literally explain everything that comes our way.

Mark My Words

—

Quantum Mechanics Is Mind Over Matter

I used to be a Physicalist, Naturalist, Nihilist, and Atheist.

One of my greatest scientific discoveries came when I first realized that Quantum Mechanics is in fact Spiritual Mechanics, or the way that spirit matter really works. When I first realized that Psyche or a Non-Local Conscious Observer and the Word of Command are needed to convert spirit matter into physical matter, then suddenly the whole of Quantum Mechanics became clear to me.

God must of necessity exist in order to have provided the necessary Word of Command or Conscious Observation, which Quantum Mechanics tells us must take place in order to convert spirit matter into physical matter. The Bible actually tells us that Jesus Christ is the WORD or the Word of Command who brought physical matter into existence in the first place. If God's Psyche didn't exist, then there would be NO physical matter.

For me personally, Quantum Mechanics became my first positive Scientific Verification of God's Existence. The scientists keep verifying Quantum Mechanics or Spiritual Mechanics over and over again; so, it must be the truth, because it's impossible to use the Scientific Methods to falsify the truth. Quantum Mechanics or Spiritual Mechanics has never been falsified, although the materialistic and naturalistic interpretations of Quantum Mechanics have been falsified trillions of times.

After realizing that God's Psyche must of necessity exist in order to convert spirit matter into physical matter, I simply KNEW that God exists.

Here's a list of some of the books and articles that confirmed or verified these scientific observations for me.

See: Van Lommel, P. (2010). *Consciousness Beyond Life: The Science of the Near-Death Experience*. New York: HarperCollins.

http://avalonlibrary.net/ebooks/Pim%20van%20Lommel%20-%20Consciousness%20Beyond%20Life.pdf

https://science-2-0.com/wp-content/uploads/2018/05/Consciousness-Beyond-Life.pdf

https://quantum-neuroscience.com/wp-content/uploads/2018/05/Consciousness-Beyond-Life.pdf

See: Goswami, A. (2008). *God Is Not Dead: What Quantum Physics Tells Us about Our Origins and How We Should Live*. Charlottesville, VA: Hampton Roads.

See also: Mark My Words. (2016). *Quantum Mechanics from a Non-Physical Spiritual Perspective*. Kindle. Retrieve from:

https://www.amazon.com/dp/B01J023TGU

The existence of Light falsifies Materialism and Naturalism, and it ends up pointing us directly to God and God's Psyche.

See: Mark My Words. (2017). *God Is in the Light: God is light, and in Him is no darkness at all*. Retrieve from:

https://www.amazon.com/dp/B07168S37N

Quantum Mechanics is Supernatural Mechanics or Supernatural Mechanisms. Quantum Mechanics or Transdimensional Physics IS Action at a Distance; and,

Action at a Distance falsifies the claims of Classical Physics or Naturalism which claim that Action at a Distance is impossible and does not exist.

I find it fascinating to observe that the Real Science – the experienced and verified observations of the human race – repeatedly and steadily FALSIFY the claims of Materialism, Naturalism, Darwinism, Nihilism, and Atheism. EVERYTHING that we have experienced and observed as a race FALSIFIES Materialism, Naturalism, and their derivatives; whereas, there is NO evidence and can be NO evidence supporting the hidden assumptions or major premises of Materialism and Naturalism which claim that the quantum, or the supernatural, or action at a distance DOES NOT EXIST.

Every time we observe, verify, and catch Quantum Mechanisms or Action at a Distance in the act, those scientific observations and experiences FALSIFY Materialism, Naturalism, Nihilism, and their derivatives such as Darwinism and Atheism.

If you successfully eliminate everything that is false, then ONLY the truth will remain. If you successfully eliminate Materialism, Naturalism, Darwinism, Nihilism, Atheism, and their derivatives, then ONLY the truth remains.

Physical matter or entropy cannot explain everything that comes our way. Classical Physics, Physicalism, or Naturalism cannot explain everything that we encounter. Materialism, Naturalism, Darwinism, Nihilism, Behaviorism, Determinism, Physical Reductionism, Atheism, and Classical Physics LACK explanatory power when it comes to Action at a Distance, Psyche, Syntropy, and Quantum Mechanics.

Here are some of the articles and books that convinced me that the Materialists and Naturalists are wrong in their assumptions.

Is Your Brain Really Necessary?

https://quantum-neuroscience.com/wp-content/uploads/2018/05/Is-Your-Brain-Really-Necessary.pdf

A brain really isn't needed for thoughts and memories to take place. Thoughts and memories are quantum waves. We KNOW this is so because it is physically impossible to record our thoughts and memories with our physical devices or physical machines. Thoughts, memories, or quantum waves are NOT a product or epiphenom of physical matter.

Mind Time

https://zodml.org/sites/default/files/%5BBenjamin_Libet%2C_Professor_Stephen_M._Kosslyn%5D_Min.pdf

https://quantum-neuroscience.com/wp-content/uploads/2018/05/Mind-Time.pdf

https://science-2-0.com/wp-content/uploads/2018/05/Mind-Time.pdf

We continue to think, learn, and remember long after our physical brain is dead and gone according to the scientific evidence that has been obtained from Near-Death Experiences (NDEs), Shared-Death Experiences (SDEs), Out-of-Body Experiences (OBEs), and our after-death Life Reviews. Thoughts and memories are a product of the Human Psyche.

The Mind and the Brain: Neuroplasticity and the Power of Mental Force - Jeffrey M. Schwartz, Sharon Begley (2003):

http://publicism.info/psychology/mind/

https://science-2-0.com/wp-content/uploads/2018/05/The-Mind-and-the-Brain.zip

https://quantum-neuroscience.com/wp-content/uploads/2018/05/The-Mind-and-the-Brain.zip

Quantum Mechanics NEVER made any sense to me while I was a Materialist, Naturalist, Nihilist, and Atheist. Eventually I learned that Quantum Mechanics or Action at a Distance is a non-local, non-physical, immaterial, spiritual, transdimensional phenomenon.

When I finally realized that Quantum Mechanics or Supernatural Mechanics is in fact how spirit matter and psyche work at the quantum level or psyche level, then suddenly it became clear to me what Spiritual Mechanics or Quantum Mechanics is and how it works. Quantum Mechanics only makes sense from a Spiritual Perspective. Classical Physics, Materialism, Naturalism, Darwinism, Nihilism, Behaviorism, Determinism, Physical Reductionism, and Atheism CANNOT explain how Quantum Mechanics or Syntropy works, nor can they explain what Quantum Mechanics or Supernatural Mechanics is.

The Spiritual Brain

https://epdf.tips/the-spiritual-brain-a-neuroscientists-case-for-the-existence-of-the-soul4a910f6e71b0354edde057767fea620d82703.html

https://science-2-0.com/wp-content/uploads/2018/05/The-Spiritual-Brain.pdf

https://quantum-neuroscience.com/wp-content/uploads/2018/05/The-Spiritual-Brain.pdf

These are a few books that you must read and understand if you have a desire to move your Science into the Quantum Age.

These are some of the Science Books that the Materialists, Naturalists, Nihilists, and Atheists are trying to convince you do not exist. They don't want you to find this material because it FALSIFIES Materialism, Naturalism, Nihilism, Atheism, and even Darwinism or the Theory of Evolution.

Statistics seem to suggest that the majority of scientists and philosophers have chosen to believe in and promote Scientific Naturalism, Physicalism, Darwinism, Nihilism, Atheism, and Classical Physics. NO amount of evidence will ever convince them that they are wrong until they are actually willing to look at the evidence and accept it as being real and true.

But, we don't need to convince the scientists that they are wrong. We – the public at-large – can skip right past them and figure out for ourselves why they are wrong and how they are wrong. That's what I eventually chose to do. So long as you have found the truth, it doesn't really matter what the scientists choose to believe; now does it? All they really need is classical physics and the physical sciences to do their work of saving world – or destroying the world. God holds ALL the KEYS to Quantum Mechanics in any case; and, God doesn't give us mere mortals direct access nor direct control over Quantum Mechanics anyway. There's nothing that the scientists can do to get at Quantum Mechanics or Supernatural Mechanics and use that Science against us or against God. Most of them don't need Quantum Mechanics to do their work; so, just let them remain in the dark if they want to.

The beauty of Psyche, Syntropy, and Quantum Mechanics is that you can literally explain everything that happens to you and everything that comes your way once you choose to allow them into play.

Mark My Words

—

The Self Does Not Die

The Self or the Psyche does not die. It can't die.

Why?

This reality and truth just might be one of the most important scientific discoveries that a person can make during his or her scientific career.

The theory of relativity teaches us that at the speed of light or faster than the speed of light TIME STOPS. There is NO passage of time at the speed of light or faster than the speed of light.

Do you know what that means?

Well, being a Materialist, Naturalist, and Classical Physicist, Albert Einstein erroneously concluded that nothing exists faster than the speed of light. He was wrong!

Tachyons or spirit matter exist at frequencies faster than the speed of light.

In a book entitled *Physics of the Impossible* by Michio Kaku, we find the following description for Tachyons.

> **Tachyons live in a strange world where everything travels faster than light. As tachyons lose energy [mass], they travel faster, which violates common sense. In fact, if they lose all energy [or**

mass], they travel at infinite velocity. As tachyons gain energy [or mass], however, they slow down until they reach the speed of light. (p. 280).

That's the exact SAME definition for Spirit Matter that I have been giving in all of my books for the past couple of years.

Tachyons ARE Spirit Matter!

Tachyons make perfect sense to me, because I'm no longer a Materialist, Naturalist, Nihilist, and Atheist. At velocities slower than the speed of light, objects take on mass or energy, which slows them down. All of that energy is needed to create the space between the nucleus of the atom and its electrons. Some type of force or field is creating that space-time between the nucleus and the electrons. It has been observed that one has to infuse a lot of energy into an atom in order to raise the electrons to a higher state or a higher orbit. The net result of this slow-down process is that these objects also take on entropy and are forced to experience the passage of time or the aging process. They become subject to space-time. In contrast, if objects lose ALL of their mass, then they are capable of infinite velocities and they experience NO passage of time or NO entropy! Entropy is a function of time.

There's the truth of the matter!

There's the answer to life, the universe, and everything. It's been there all along just waiting for someone to see it, discover it, and declare it.

A photon traveling at the speed of light experiences NO passage of time. From its perspective, it arrives at its destination instantly the very moment it launches. From our perspective, existing at velocities much slower than the speed of light, it took that photon 13.2 billion years to reach us. From its perspective, it took NO time at all.

At the speed of light, TIME STOPS. There is NO passage of time. There is NO entropy. At the speed of light or faster, everything is SYNTROPY, meaning that everything is eternal and everlasting.

The Psyche, Self, Mind, Soul, Intelligence, or Quantum Non-Local Consciousness exists at frequencies faster than the speed of light, which means that it is SYNTROPY. It is eternal and everlasting, without a beginning of days or an end of years.

This just might be one of my most significant and most useful scientific discoveries of all time. It fits and makes sense to me. It's predicted by the theory of relativity.

Entropy is death.

Why is entropy death?

It's because entropy involves the passage of time. Things age and die at velocities slower than the speed of light.

In contrast, SYNTROPY is eternal life. The Psyche is eternal and everlasting because it experiences NO passage of time. It doesn't age. It has NO entropy. That pretty much explains life, the universe, and everything – now doesn't it?

What's the answer to life, the universe, and everything?

Well, it's not 42.

The answer to life the universe, and everything is SYNTROPY or PSYCHE or INTELLIGENCE. The question is, "What is eternal and everlasting, without a beginning of days or an end of years?" The answer is Psyche or Syntropy. SYNTROPY is the fundamental basis for life, the universe, and everything!

Quantum Mechanics is a type of Syntropy. It's eternal and everlasting.

These realities and truths have been experienced and observed. Science IS observation and experience. Science or observation is an infinitely better way for getting at the truth than just to say that psyche or syntropy does not exist, as the Materialists and Naturalists do. Saying that psyche or syntropy DOES NOT EXIST is a cop out. It's the way that these people avoid doing the science and the observations that need to be done in order to establish the truth of the matter. It's an evasion or a dodge, NOT science. These people are avoiding the truth or hiding from the truth rather than seeking for the truth. Scientists are supposed to search for and find the truth; but, these people refuse to do so.

The Self Does Not Die: Verified Paranormal Phenomena from Near-Death Experiences

> https://www.amazon.com/dp/0997560800
>
> https://www.amazon.com/dp/B01LYMHXCO

Physicalism, Naturalism, Darwinism, Nihilism, and Atheism are based upon a refusal to look at evidence. I KNOW because I used to be a Materialist, Naturalist, Nihilist, and Atheist. We Naturalists and Atheists have to castrate and lobotomize science in order to prevent people from falsifying our theories and our ideas because ALL of the scientific evidence or observational evidence that we have on hand as a race falsifies the claims of Materialism, Naturalism, Darwinism, Nihilism, and Atheism which state that such falsifying evidence does not exist.

Science itself convinced me that Materialism, Naturalism, Darwinism, Nihilism, and Atheism are FALSE. I believe the Science – the observations and the experiences of real people – and not the philosophical wishful thinking of the Materialists, Naturalists, Darwinists, Nihilists, and Atheists.

Mark My Words

—

Orthodox Interpretation of Quantum Mechanics

The Orthodox Interpretation of Quantum Mechanics by Henry P. Stapp was the FIND of a lifetime. Henry P. Stapp's work is important to know about and understand – everything else pales in comparison. This is the information and the scientific evidence that the Materialists, Naturalists, Darwinists, Nihilists, and Atheists are trying to hide from you; and, they are very good at what they do. I KNOW because I used to be a Materialist, Naturalist, Nihilist, and Atheist. These people are actively banning, blocking, and destroying this information about Syntropy and Quantum Mechanics because it falsifies Physicalism, Naturalism, Darwinism, and Nihilism.

Remember, Materialism, Naturalism, Darwinism, Nihilism, Behaviorism, Determinism, Physical Reductionism, and Atheism are based exclusively on Classical Physics and Entropy. Scientific Naturalism or Physicalism IS Classical Physics. Quantum Mechanics falsifies Classical Physics, or Materialism and Naturalism.

Stapp seems to be giving everything away for free on his website.

https://sites.google.com/a/lbl.gov/stappfiles/

http://www-physics.lbl.gov/~stapp/

He seems to be much more interested in getting his message out to the world than he is in maintaining the copyright for his books.

The Orthodox Interpretation explains what the Human Psyche and Nature's Psyche are doing at the quantum level in order to get things done for us at the physical level.

If your interpretation of Quantum Mechanics can't explain what the Human Psyche and Nature's Psyche are doing at the quantum level in order to get things done for you at the physical level, then your interpretation of Quantum Mechanics is absolutely worthless.

Quantum Mechanics makes absolutely NO sense whatsoever whenever it is reduced to Classical Physics, Materialism, Naturalism, Physicalism, and Nihilism. Most of the interpretations of Quantum Mechanics found online and within our college textbooks have indeed been reduced to Materialism, Naturalism, and Classical Physics; and thereby, they have completely lost their explanatory power and their scientific value. Materialism and Naturalism neuter and destroy everything that comes their way, including Quantum Mechanics.

To demonstrate some of these truths, first I will quote from:

NEUROSCIENCE, ATOMIC PHYSICS, AND THE HUMAN PERSON

http://www-physics.lbl.gov/~stapp/2nd.doc

https://quantum-neuroscience.com/wp-content/uploads/2018/05/Neuroscience-and-the-Human-Person.pdf

http://www-physics.lbl.gov/~stapp/phys.txt

The Quantum Approach.

Classical physics is *an approximation* to a more accurate theory - called quantum mechanics - and quantum mechanics makes mind efficacious. Quantum mechanics *explains* the causal effects of mental intentions upon physical systems: it *explains* how your mental effort can produce the brain events that cause your bodily actions. Thus, quantum theory converts science's picture of you from that of *a mechanical automaton* to that of *a mindful human person*. Quantum theory also shows, explicitly, how the approximation that reduces quantum theory to classical physics completely eliminates all effects of your conscious thoughts upon your brain and body. Hence, from a physics point of view, trying to understand the mind-brain connection by going to the classical approximation is absurd: it amounts to trying to understand something in an approximation that eliminates the effect you are trying to study.

Quantum mechanics arose during the twentieth century. Scientists discovered, empirically, that the principles of classical physics were not correct. Moreover, they were wrong in ways that no minor tinkering could ever fix. The *basic principles* of classical physics were thus replaced by *new basic principles* that account uniformly both for all the successes of the older classical theory and also for all the newer data that is incompatible with the classical principles.

Physical theory was turned inside out.

The most profound alteration of the fundamental principles was to bring the consciousness of human beings into the basic structure of the physical theory. In fact, the whole *conception of what science is* was turned inside out. The core idea of classical physics was to describe the "world out there," with no reference to "our thoughts in here." But the core idea of quantum mechanics is to describe *our activities as knowledge-seeking and knowledge-using agents*. Thus, quantum theory involves, basically, not just what is "out there," but also what is "in here," namely "our knowledge." Consciousness is thus introduced into contemporary orthodox physical theory, not as something *whose existence needs to be explained,* but as rather as something whose detailed structure and detailed connection to brain activities needs to be *further* explicated.

Science must bridge the psycho-physical divide.

The basic philosophical shift in quantum theory is the *explicit* recognition that science is about *what we can know*. It is fine to have a beautiful and elegant mathematical theory about an imagined "*really existing physical world out there*" that meets a lot of intellectually satisfying criteria.

But the essential demand of science is that the theoretical constructs be tied to the experiences of the human scientists who devise ways of testing the theory, and of the human engineers and technicians who both participate in these tests, and eventually put the theory to work. So, the structure of a proper physical theory must involve not only the part describing the behavior of the not-directly-experienced theoretically postulated entities, expressed in some appropriate symbolic language, but also a part describing the human experiences that are involved in these tests and applications, expressed in the language that we actually use to describe such experiences to ourselves and each other. Finally, we need some "bridge laws" that specify the connection between the concepts described in these two different languages.

Classical physics met these requirements in a rather trivial kind of way, with the relevant experiences of the human participants being taken to be direct apprehensions of various gross behaviors of large-scale properties of big objects composed of huge numbers of the tiny atomic-scale parts. And these apprehensions were taken to be *passive*: they had no effect on the behaviors of the systems being studied. But the physicists who were examining the behaviors of systems that depend sensitively upon the behaviors of their tiny atomic-scale components found themselves forced to go to a less trivial theoretical arrangement, in which the human agents were no longer passive observers, but were *active participants* in ways that contradicted, and were impossible to comprehend within, the general framework of classical physics, *even when the only features of the physically described world that the human beings observed were large-scale properties of measuring devices.*

The two-way quantum psycho-physical bridge.

The sensitivity of the behavior of the devices to the behavior of some tiny atomic-scale particles propagates in such a way that the acts of observation by the human observers of *large-scale properties of the devices* could no longer be regarded as passive: these acts were assigned a crucial *selective* action. Thus, the core structure of the basic general physical theory became transformed in a profound way: the connection between physical behavior and human knowledge was changed from a one-way bridge to a mathematically specified two-way interaction that involves *selections* performed by conscious minds.

This profound change in the principles is encapsulated in Niels Bohr dictum that "in the great drama of existence we ourselves are both actors and spectators." (Bohr, 1963: 15 & 1958: 81) The emphasis here is on "actors": in classical physics we, and in particular our minds, were mere spectators.

This revision must be expected to have important ramifications in neuroscience, because the issue of the connection between mind (the psychologically described aspects of a human being) and brain/body (the physically described aspects of that person) has recently become a matter of central concern in neuroscience.

The Copenhagen formulation.

The original formulation of quantum theory was created mainly at an Institute in Copenhagen directed by Niels Bohr and is called "The Copenhagen Interpretation." Due to the profound strangeness of the conception of nature entailed by the new mathematics, the Copenhagen strategy was to refrain from making ordinary ontological claims, but to take, instead, a fundamentally pragmatic stance. Thus, the theory was formulated *basically* as a set of practical rules for how scientists should go about their tasks of acquiring knowledge, and then using this knowledge in practical ways. Speculations about "what the world out there – apart from our knowledge of it - is really like" were regarded as "metaphysics," and hence outside *real* science.

Copenhagen quantum theory is about the relationships between human agents (called *participants* by John Wheeler) and the systems that they act upon. In order to achieve this conceptualization, the Copenhagen formulation separates the physical universe into two parts, which are described in two different languages. One part is the observing human agent and his measuring devices. That part is described in mental terms - in terms of our instructions to colleagues about how to set up the devices, and our reports of what we then learn. The other part of nature is *the system that the agent is acting upon*. That part is described in physical terms - in terms of mathematical properties assigned to tiny space-time regions.

Von Neumann's Process II.

The great mathematician and logician John von Neumann formulated Copenhagen quantum theory in a rigorous way.

Von Neumann identified two very different processes that enter into the quantum theoretical description of the evolution of a physical system. He called them Process I and Process II (Von Neumann, 1955: 418). Process II is the analog in quantum theory of the process in classical physics that takes the state of a system at one time to its state at a later time. This Process II, like its classical analog, is *local* and *deterministic*. However, Process II by itself is not the whole story: it generates physical worlds that do not agree with human experiences. For example, if Process II were the *only* process in nature then the quantum state of the moon would represent a structure smeared out over large part of the sky.

Process I: A dynamical psycho-physical bridge.

To tie the quantum mathematics to human experience in a rationally coherent and mathematically specified way quantum theory introduces *another process*, which Von Neumann calls Process I. It is a *selection* process that is tied to conscious experience, and it is not determined by the micro-local deterministic Process II. It is a selection made by an agent about how he or she will act or attend.

Any physical theory must, in order to be complete, specify how the elements of the theory are connected to human experience. In classical physics this connection is part of a *metaphysical* superstructure: it is not part of the core dynamical description. But in quantum theory this connection of the mathematically described physical state to conscious experiences is part of the essential dynamical structure. And this connecting process is not passive: it does not represent a mere *witnessing* of a physical feature of nature by a passive mind. Rather, the process is active: it injects into the physical state of the system being acted upon properties that depend upon the intention of the observing agent.

Quantum theory is built upon the practical concept of intentional actions by agents. Each such action is expected or intended to produce an experiential response or feedback. For example, a scientist might act to place a Geiger counter near a radioactive source and expect to see the counter either "fire" during a certain time interval or not "fire" during that interval. The experienced response, "Yes" or "No", to the question "Does the counter fire during the specified interval?" specifies one bit of information. Quantum theory is thus an information-based theory built upon the knowledge-acquiring actions of agents, and the knowledge that these agents thereby acquire.

Probing actions of this kind are performed not only by scientists. Every healthy and alert infant is engaged in making willful efforts that produce experiential feedbacks, and he or she soon begins to form expectations about what sorts of feedbacks are likely to follow from some particular kind of effort. Thus, both empirical science and normal human life are based on paired realities of this action-response kind, and our physical and psychological theories are both basically attempts to understand these linked realities within a rational conceptual framework.

The basic building blocks of quantum theory are, then, a set of intentional actions by agents, and for each such action an associated collection of possible "Yes" feedbacks, which are the possible responses that the agent can judge to be in conformity to the criteria associated with that intentional act. For example, the agent is assumed to be able to make the judgment "Yes" the Geiger counter clicked or "No" the Geiger counter did not click. And he must be able to report. "Yes" the counter is in the specified place, or "No" it is not there. Science would be difficult to pursue if scientists could make no such judgments about what they were experiencing.

All known physical theories involve idealizations of one kind or another. In quantum theory the main idealization is not that every object is made up of miniature planet-like objects. It is rather that there are agents that perform intentional acts each of which can result in a feedback that may conform to a certain criterion associated with that act. One bit of information is introduced into the world in which that agent lives, according to whether the feedback conforms or does not conform to that criterion. Thus, knowing whether the counter clicked or not places the agent on one or the other of two alternative possible separate branches of the course of world history.

These remarks reveal the enormous difference between classical physics and quantum physics. In classical physics the elemental ingredients are tiny invisible bits of matter that are idealized miniaturized versions of the planets that we see in the heavens, and that move in ways unaffected by our consciousness, whereas in quantum physics the elemental ingredients are intentional actions by agents, the feedbacks arising from these actions, and the effects of our actions on the physical systems that our actions act upon.

Consideration of the character of these differences makes it plausible that quantum theory may be able to provide the foundation of a scientific theory of the human person that is better able than classical physics to integrate the physical and psychological aspects of his nature. For quantum theory describes the effects of a person's intentional actions upon the physical world, whereas classical physics systematically leaves these effects out.

An intentional action by a human agent is partly an intention, described in psychological terms, and partly a physical action, described in physical terms. The feedback also is partly psychological and partly physical. In quantum theory these diverse aspects are all represented by logically connected elements in the mathematical structure that emerged from the seminal discovery of Heisenberg. That discovery was that in order to get the quantum generalization of a classical theory one must formulate the theory in terms of *actions*. A key difference between *numbers* and *actions* is that if A and B are two actions then AB represents the action obtained by performing the action A upon the action B. If A and B are actions then, generally, AB is different from BA: the order in which actions are performed matters.

The intentional actions of agents are represented mathematically in Heisenberg's space of actions. Here is how it works.

Each intentional action depends, of course, on the intention of the agent, and upon the state of the system upon which this action acts. Each of these two aspects of nature is represented within Heisenberg's space of actions by an action.

The idea that a "state" should be represented by an "action" may sound odd, but Heisenberg's key idea was to replace what classical physics took to be a "being" by a "doing." I shall denote the action that represents the state being acted upon by the symbol S.

An intentional act is an action that is intended to produce a feedback of a certain conceived or imagined kind. Of course, no intentional act is sure-fire: one's intentions may not be fulfilled. Hence the intentional action puts in play a process that will lead either to a confirmatory feedback "Yes," the intention is realized, or to the result "No", the "Yes" response failed to occur.

The effect of this intentional mental act is represented mathematically by an equation that is one of the key equations of quantum theory. This equation represents, within the quantum mathematics, the effect of the

Process I mental action upon the quantum state S of the system being acted upon. The equation is:

$$S \to S' = PSP + (1-P)S(1-P).$$

This formula exhibits the important fact that this Process I action changes the state S of the system being acted upon into a new state S', which is a *sum* of two parts.

The first part, PSP, represents the possibility in which the experiential feedback called "Yes" appears, and the second part, $(1-P)S(1-P)$, represents the alternative possibility "No", this feedback does not appear. Thus, the intention of the action and the associated experiential feedback are tied into the mathematics that describes the dynamics of the physical system being acted upon.

The action P is important. It represents an action upon the system that is being acted upon by the agent, and it depends on *the intention of the agent*. The action represented by the symbol P, acting both on the right and on the left of S, is the action of eliminating from the state S all parts of S except the "Yes" part. That particular retained part is determined by the intentional choice of the agent. The action of $(1-P)$, acting both on the right and on the left of S, is, analogously, to eliminate from S all parts of S except the "No" parts.

The projection operator P is required to satisfy $P = PP$. This implies $P(1-P) = (1-P)P = 0$, which says that the sequence of these two actions, P and $(1-P)$, in either order, leave nothing.

Thus, the action P is an action in the space in which the physical system is represented, and it reduces to zero all components that correspond to the "No" response, but it leaves intact the components corresponding to the "Yes" response to the intentional action. The action of $(1-P)$ is the analogous action with "Yes" and "No" interchanged. The action of P is the representation of an intentional mental action upon a physically described system.

Notice that Process I produces the *sum* of the two alternative possible feedbacks, not just one or the other. Since the feedback must either be "Yes" or "No = Not-Yes," one might think that Process I, which *keeps* both the "Yes" and the "No" parts, would do nothing. But that is not correct! This is a key point. It can be verified by noticing that S can be written as a sum of four parts, only two of which survive the Process I action:

$$S = PSP + (1-P)S(1-P) + PS(1-P) + (1-P)SP.$$

This formula is a strict identity. The dedicated reader can easily confirm it by collecting the contributions of the four occurring terms PSP, PS, SP, and S, and verifying that all terms but S cancel out. This identity shows that the state S can be expressed as a sum of four parts, *two of which are eliminated by Process I.*

But this means that Process I has a *nontrivial effect* upon the state being acted upon: it eliminates the two terms that correspond neither to the appearance of a "Yes" feedback nor to the failure of the "Yes" feedback to appear.

That is the *first key point*: quantum theory has a specific dynamical process, Process I, which specifies the effect upon a physically described system of an *intentional act* by a conscious agent.

Free Choices.

The second key point is this: the agent's choices are "free choices," *in the specific sense specified below.*

Orthodox quantum theory is formulated in a realistic and practical way. It is structured around the activities of human agents, who are considered able to freely elect to probe nature in any one of many possible ways. Bohr emphasized the freedom of the experimenters in passages such as:

"The freedom of experimentation, presupposed in classical physics, is of course retained and corresponds to the free choice of experimental arrangement for which the mathematical structure of the quantum mechanical formalism offers the appropriate latitude." (Bohr, 1958: 73)

This freedom of action stems from the fact that in the original Copenhagen formulation of quantum theory the human experimenter is considered to stand outside the system to which the quantum laws are applied. Those quantum laws are the only precise laws of nature recognized by that theory. Thus, according to the Copenhagen philosophy, *there are no presently known laws that govern the choices* made by the agent/experimenter/observer/participant about how the observed system is to be probed. This choice is, *in this very specific sense*, a "free choice." It is not ruled out that some deeper theory will eventually provide a causal explanation of this "choice."

Probabilities.

The predictions of quantum theory are generally statistical: only the *probabilities* that the agent will experience each of the alternative possible feedbacks are specified. Which of these alternative possible feedbacks will actually occur in response to a Process I action is not determined by quantum theory.

The formula for the probability that the agent will experience the feedback 'Yes' is Tr PSP/Tr S, where the symbol Tr represents the trace operation. This trace operation means that the actions act in a cyclic fashion, so that the rightmost action acts back around upon the leftmost action. Thus, for example, Tr ABC=Tr CAB =Tr BCA. The product ABC represents the result of letting A act upon B, and then letting that product AB act upon C. But what does C act upon? Taking the trace of ABC means specifying that C acts back around on A.

An important property of a trace is that the trace of any of the sequences of actions that we consider must always give a positive number or zero. Thus, this trace operation is what ties the actions, as represented in the mathematics, to measurable numbers.

[The trace operation, and in fact the operation of multiplying together any two operators, is the quantum analog of the classical process of integrating over all of "phase space," giving equal a prior weighting to equal volumes of phase space. Thus, the trace operation is in effect a statistical sum over all of the "loose ends" that are not fixed in the expression upon which the trace operation acts.]

Von Neumann's psycho-physical theory of the conscious brain.

The Copenhagen approach separates the world into two parts: "The Observer" which includes the mind, brain, and body of the personal observer together with his measuring devices; and "The System" that this observer is acting upon. "The Observer" is described in psychological terms, whereas "The System" is described in physical/mathematical space-time terms.

This procedure works very well in practice. However, it seems apparent that the body and brain of the human agent, and his devices, are parts of the physical universe. Hence a complete theory ought to be able to include our bodies and brains in the physically described part of the theory. On the other hand, the structure of the theory depends critically also upon the features that are represented in Process I, and that are described in mentalistic language as intentional actions and experiential feedbacks.

Von Neumann showed that it was possible, without significantly disturbing the predictions of the theory, to shift the bodies and brains of the agents, along with their measuring devices, into the physical world, *while retaining and ascribing to the mind of the agent, those mentalistically described properties of the agents that are essential to the structure of the theory.* The system acted upon by the mind is the brain. Thus, in this von Neumann re-formulation the Process I action is an action of mind upon brain. Hence von Neumann's re-formulation provides us with the core of a science-based dynamical theory of the conscious brain.

It is worthwhile to reflect for a moment on the ontological aspects of Von Neumann quantum theory. Von Neumann himself, being a clear-thinking mathematician, said very little about ontology. But he called the mentalistically described aspect of the agent "his abstract 'ego'." (Von Neumann, 1955: 421). This phrasing tends to conjure up the idea of a disembodied entity, standing somehow apart from the body/brain. But another possibility is that consciousness is an *emergent property* of the body-brain. Notice that some of the problems that occur in trying to defend the idea of emergence within the framework of classical physical theory disappear when one accepts the validity of quantum theory. For one thing, one no longer has to defend against the charge that the emergent property, consciousness, has no "genuine" causal efficacy, because anything it does is done already by the physically described process, independently of whether the psychologically described aspect emerges of not. In quantum theory the causal efficacy of our thoughts is no illusion: it's the real thing!

Another difficulty with "emergence" in a classical physics context is in understanding how the motion of a set of miniature planet-like objects, careening through space, can *be a painful experience.* But within the quantum framework the basic physical structure, namely the quantum state, is essentially knowledge or information imbedded in space-time. Hence there is no intrinsic problem with the idea that a sudden increment in a person's knowledge should be represented by a sudden jump in the quantum state of his brain. The identification of conscious actions with physical actions is no longer problematic. This is because the old idea of "matter" has been eradicated and replaced by a mathematical representation of an information-based psycho-physical reality.

In this connection, Heisenberg remarked:

"The conception of the objective reality of the elementary particles has thus evaporated not into the cloud of some obscure new reality concept, but into the transparent clarity of a mathematics that represents no longer the behavior of the particle but rather our knowledge of this behavior." (Heisenberg, 1958).

Conservation of Causality.

The question arises: How can the effect of a psychologically described action be injected into the dynamics of a physically described system without upsetting the causal structure of the latter.

The answer is this: Physicists have discovered an important and unexpected property of nature. It pertains to observable phenomena that depend upon microscopic properties that are *in principle inaccessible to observation.* In such a situation we are *in principle* unable, due to the lack of crucial micro-data, to give a complete causal description of the observable phenomena. However, our principled inability to give a complete causal account of the psychologically described phenomena, due to this inherent gap in the micro-data, can be partially offset by introducing into the theory, *instead of the inaccessible micro-data,* the *psychologically described selection of an action* made upon the system by an agent.

Thus, the loss of causal determination at the microlevel, due to the limitations imposed by Heisenberg's uncertainty principle, allows an alternative (statistical) causal account to be achieved by replacing the inaccessible micro-data by empirically available and controllable data about human selections of actions!

This feature discovered in atomic science should be equally importance in neuroscience. That is because the basic problem in neuroscience is essentially the same as the one in atomic physics. In both cases the problem is to provide a causal account of connections between experiences that depend sensitively upon micro-properties that are in principle inaccessible. But quantum theory shows how the principled loss of information at the microlevel can be partially offset by using, instead, the controllable and reportable variables of the intentional actions of human beings. Nature left open a causal gap for us to occupy.

The Quantum Brain.

The quantum state of a human brain is, of course, a very complex thing. But its main features can be understood by considering first a classical conception of the brain, and then folding in some key features that arise already in the case of the quantum state of a single particle, or object, or degree of freedom.

Source

http://www-physics.lbl.gov/~stapp/2nd.doc

https://quantum-neuroscience.com/wp-content/uploads/2018/05/Neuroscience-and-the-Human-Person.pdf

http://www-physics.lbl.gov/~stapp/phys.txt

https://docslide.com.br/download/link/stapp-mind-matter-and-quantum-mechanics

https://quantum-neuroscience.com/wp-content/uploads/2018/05/Mind-Matter-and-Quantum-Mechanics.pdf

https://science-2-0.com/wp-content/uploads/2018/05/Mind-Matter-and-Quantum-Mechanics.pdf

Stapp, H. P. (2009). *MIND, MATTER, AND QUANTUM MECHANICS* (3rd ed.). Berlin: Springer.

Process 1 consists of the choices that are made by the Human Psyche.

Process 2 happens when Nature's Psyche collapses the necessary wave functions and/or fires the specific neurons needed to make the Human Psyche's choices a physical reality.

Orthodox Quantum Mechanics explains what the Human Psyche and Nature's Psyche are doing at the quantum level in order to get things done for us at the physical level. This is good stuff! It's thousands of years ahead of anything that we have gotten from the Materialists, Naturalists, Darwinists, Nihilists, and Atheists.

If your interpretation of Quantum Mechanics can't explain what Nature's Psyche and the Human Psyche are doing at the quantum level or the psyche level, then your interpretation of Quantum Mechanics is absolutely worthless when it comes time to figure out how Quantum Mechanics or Action at a Distance really works.

Quantum Mechanics explains how your Psyche, Intelligence, Spirit, Soul, Spark, Mind, or Quantum Non-Local Consciousness controls your physical body and your physical brain. That's the purpose of Quantum Mechanics. In other words, explaining Psyche or Non-Local Consciousness scientifically is what Quantum Mechanics is good for. Quantum Mechanics or Supernatural Mechanics completes science, finishes science, and perfects science. Science is incomplete without Psyche or Syntropy.

For me personally, this is one of the most important, most significant, and most powerful scientific discoveries of my science career. It's essential. It's foundational. It changes everything!

If your interpretation of Quantum Mechanics can't explain how the Human Psyche is controlling its physical brain and physical body, then your interpretation of Quantum Mechanics is worthless. That is what I have experienced and observed.

The Human Psyche makes the decisions and the choices. Nature's Psyche collapses the wave functions and/or turns on the specific neurons needed to make the Human Psyche's choices a physical reality. We KNOW that it must be Nature's Psyche (or God's Psyche) who is collapsing the wave functions and firing the neurons because the Human Psyche isn't consciously aware of any of these quantum mechanical processes.

By choosing to allow Psyche, Syntropy, and Quantum Mechanics in to play, we can literally explain everything that comes our way. That is what I have experienced and observed.

I used to be a Materialist, Naturalist, Nihilist, and Atheist. Quantum Mechanics never made any sense to me until after I got rid of My Materialism, My Naturalism, My Nihilism, and My Atheism. Materialism, Naturalism, Darwinism, Nihilism, and Atheism claim that Psyche, Syntropy, and Supernatural Mechanisms do not exist; but, Quantum Mechanics IS Action at a Distance or Supernatural Mechanisms. Materialism, Naturalism, and their derivatives are FALSIFIED by the verified and proven existence of Action at a Distance or Quantum Mechanics.

Materialism, Naturalism, Darwinism, Nihilism, Behaviorism, Determinism, Physical Reductionism, Atheism, Entropy, and Classical Physics CANNOT EXPLAIN how the Human Psyche and Nature's Psyche are interacting with each other at the quantum level in order to get things done for us at the physical level. Therefore, these "sciences" are worthless, incomplete, and completely lack explanatory power when it comes to the Human Psyche, Nature's Psyche, Quantum Mechanics, Action at a Distance, Invisible Non-Physical Forces and Fields, and Supernatural Mechanisms.

Materialism and Naturalism have NO EXPLANATION for invisible non-physical forces and fields such as Psyche, Gravity, Magnetism, Nuclear Forces, Radio Waves, Microwaves, X-Rays, Dark Energy, Quantum Waves, Thoughts, and our after-death Memories which have been experienced, observed, and caught in the act.

It's time for the world at-large to upgrade science to Quantum Mechanics or Supernatural Mechanics. Once you choose to allow Psyche, Syntropy, and Quantum Mechanics in to play, instantly you can explain everything that comes your way.

Mark My Words

—

The Mindful Universe

The Orthodox Interpretation of Quantum Mechanics by Henry P. Stapp was the FIND of a lifetime. Henry P. Stapp's work is important to know about and understand – everything else pales in comparison. This is the information and the scientific evidence that the Materialists, Naturalists, Darwinists, Nihilists, and Atheists are trying to hide from you; and, they are very good at what they do. These people are actively banning, blocking, and destroying this information because it falsifies Physicalism, Naturalism, Darwinism, and Nihilism.

Stapp seems to be giving everything away for free on his website.

https://sites.google.com/a/lbl.gov/stappfiles/

http://www-physics.lbl.gov/~stapp/

He seems to be much more interested in getting his message out to the world than he is in maintaining the copyright for his books.

Stapp, H. P. (2011). *Mindful Universe: Quantum Mechanics and the Participating Observer*. Heidelberg: Springer.

https://epdf.tips/mindful-universe-quantum-mechanics-and-the-participating-observer-second-edition.html

https://science-2-0.com/wp-content/uploads/2018/05/Quantum-Mechanics-and-the-Participaing-Observer.pdf

https://quantum-neuroscience.com/wp-content/uploads/2018/05/Mindful-Universe.pdf

—

Snippets from the *Mindful Universe*:

http://www-physics.lbl.gov/~stapp/Key.doc

https://quantum-neuroscience.com/wp-content/uploads/2018/05/Mindful-Universe-Key.doc

http://www-physics.lbl.gov/~stapp/Book.doc

https://quantum-neuroscience.com/wp-content/uploads/2018/05/Mindful-Universe-Booklet.doc

http://www-physics.lbl.gov/~stapp/Chapt3.txt

http://www-physics.lbl.gov/~stapp/Chap4-6.txt

—

Quote from the *Mindful Universe*:

The World of Actions

Werner Heisenberg was, from a technical point of view, the principal founder of quantum theory. He discovered in 1925 the completely amazing and wholly unprecedented solution to the puzzle: the quantities that classical physical theory was based upon, and which were thought to be numbers, must be treated not as numbers but as actions! Ordinary numbers, such as 2 and 3, have the property that the product of any two of them does not depend on the order of the factors: 2 times 3 is the same as 3 times 2. But Heisenberg discovered that one could get the correct answers out of the old classical laws if one decreed that certain numbers that occur in classical physics as the magnitudes of certain physical properties of a material system are not ordinary numbers. Rather, they must be treated as *actions* having the property that the order in which they act matters!

This 'solution' may sound absurd or insane. But mathematicians had already discovered that logically consistent generalizations of ordinary mathematics exist in which numbers are replaced by 'actions' having the property that the order in which they are applied matters. The ordinary numbers that we use for everyday purposes like buying a loaf of bread or paying taxes are just a very special case from among a broad set of rationally coherent mathematical possibilities. In this simplest case, A times B happens to be the same as B times A. But there is no logical reason why Nature should not exploit one of the more general cases: there is no compelling reason why our physical theories must be based exclusively on ordinary numbers rather than on actions. The theory based on Heisenberg's discovery exploits the more general logical possibility. It is called quantum mechanics, or quantum theory.

The difference between quantum mechanics and classical mechanics is specified by Planck's constant, which is a tiny number on the scale of human actions. Thus, this tweaking of laws of physics might seem to be a bit of mathematical minutia that could scarcely have any great bearing on the fundamental nature of the universe, or of our role within it. But replacing *numbers* by *actions* upsets the whole apple cart. It produced a seismic shift in our ideas about both the nature of reality, and the nature of our relationship to the reality that envelops and sustains us. The aspects of nature represented by the theory are converted from elements of *being* to elements of *doing*. The effect of this change is profound: it replaces the world of *material substances* by a world populated by *actions*, and by *potentialities* for the occurrence of the various possible observed feedbacks from these actions. Thus, this switch from 'being' to 'action' allows – and according to orthodox quantum theory demands – a draconian shift in the very subject matter of physical theory, from an imagined universe consisting of causally self-sufficient mindless matter, to a universe populated by allowed possible physical actions and possible experienced feedbacks from such actions. A purported theory of matter alone is converted into a theory of the relationship between matter and mind.

What is this momentous change introduced by Heisenberg?

In classical physics the center point of each physical object has, at each instant of time, a well-defined location, which can be specified by giving its three coordinates (x, y, z) relative to some coordinate system. For example, the location of (the center point of) a spider dangling in a room can be specified by letting z be its distance from the floor, and letting x and y be its distances from two intersecting walls. Similarly, the velocity of that dangling spider, as she drops to the floor, blown by a gust of wind, can be specified by giving the rates of change of these three coordinates (x, y, z). If each of these three *rates of change*, which together specify the velocity, are multiplied by the *weight* (= mass) of the spider, then one gets three numbers, say (p, q, r), that define the *momentum* of the spider. In classical physics one uses the set of three numbers denoted by (x, y, z) to represent the position of the center point of an object, and the set of three numbers labeled by (p, q, r) to represent the momentum of that object. These six numbers are just ordinary numbers that obey the commutative property of multiplication that we all, hopefully, learned in third grade: $x * p$ equals $p * x$, where $*$ means multiply.

The six-dimensional space of all possible values $(x, y, z; p, q, r)$ is called *phase space*: it is the space of all possible instantaneous 'states' of the particle.

Heisenberg's analysis showed that in order to make the formulas of classical physics work in general, $x * p$ must be different from $p * x$. He found that the difference between these two products must be Planck's constant. (Actually, the difference is Planck's constant divided by 2π and multiplied by the imaginary unit i, which is a number such that i times i is minus one.) Thus, modern quantum theory was born by recognizing, or declaring, that the symbols used in classical physical theory to represent ordinary numbers actually represent actions such that their ordering in a sequence of actions matters. The procedure of creating the mathematical structure of quantum mechanics from that of classical physics, by replacing numbers by corresponding actions, is called 'quantization'.

The idea of replacing the numbers that specify where a particle is, and how fast it is moving, by mathematical quantities that violate the simple laws of arithmetic may strike you – if this is the first you've heard about it – as a giant step in the wrong direction. You might mutter that scientists should try to make things simpler, rather than abandoning one of the things we really know for sure, namely that the order in which one multiplies factors does not matter. But against that intuition one must recognize that this change works beautifully in practice: all of the tested predictions of quantum mechanics are borne out, and these include predictions that are correct to the incredible accuracy of one part in a hundred million. There must be something very, very right about this replacement of numbers by actions.

In classical physical theory each elementary particle is asserted to have at each instant of time a definite location, defined by a set of three numbers (x, y, z), and definite momentum, defined by a set of three numbers (p, q, r). In quantum theory one generally considers systems of many particles, but insofar as one can consider one particle alone the state of that particle at any instant of time would be represented by a *cloud of pairs of numbers*, with one pair of numbers (called a complex number) assigned to each point in three-dimensional (position) space. Someone might choose to perform a phenomenologically (i.e., experimentally/experientially) described probing action on this 'particle'. In quantum mechanics each such possible probing action turns out to have an associated set of distinct *experientially distinguishable* possible outcomes. The cloud of numbers *taken as a whole* determines the probability for the appearance of each of the alternative possible outcomes of that chosen probing action. The theory thus gives specified rules for computing the probabilities for each of the distinct alternative possible empirically described feedbacks from each of the alternative possible experimental probing actions that the human experimenter might chose to perform, but no rules that specify which probing action he or she will choose.

In classical physical theory when one descends from the macroscopic world of visible objects to the microscopic world of their elementary constituents one arrives at a world containing the 'solid, massy, hard, impenetrable moveable particles' that Newton spoke of. But in quantum theory one arrives instead at clouds, or quantum smears, of numbers that *taken as a whole* have empirical meaning in terms of probabilities of alternative possible experiences.

Briefly stated, the orthodox formulation of quantum theory (see Appendix D) asserts that, in order to connect adequately the mathematically described state of a physical system to human experience, there must be an abrupt *intervention* in the otherwise smoothly evolving mathematically described state of that system.

According to the orthodox formulation, these interventions are probing actions *instigated by human agents who are able to 'freely' choose which one, from among various alternative possible probing actions, they will perform*. The physically describable effect of the chosen probing action is to separate (partition) the prior physical state of the system being probed in some particular way into a set of component parts. Each physically described part corresponds to one *perceivable* outcome from the set of distinct alternative possible perceivable outcomes of that particular probing action.

If such a probing action is performed, then *one* of its allowed perceivable feedbacks will appear in the stream of consciousness of the observer, and the mathematically described state of the probed system will then jump abruptly from the form it had prior to the intervention to the partitioned portion of that state that corresponds to the observed feedback. *This means that, according to orthodox contemporary physical theory, the 'free' choices of probing actions made by agents enter importantly into the course of the ensuing psychologically and physically described events. Here the word 'free' means, however, merely that the choice is not determined by the (currently) known laws of physics; not that the choice has no cause at all in the full psychophysical structure of reality. Presumably the choice has some cause or reason – it is unreasonable that it should simply pop out of nothing at all – but the existing theory gives no reason to believe that this cause must be determined exclusively by the physically described aspects of the psychophysically described nature alone.*

If one sets Planck's constant equal to zero in the quantum mechanical equations, then one recovers (the fundamentally incorrect) classical mechanics. Thus, classical physics is *an approximation* to quantum physics. It is the approximation in which Planck's constant, wherever it appears, is replaced by zero. In this approximation the quantum smearing does not occur – each cloud is reduced to a point – and one recovers classical physics, along with the physical determinism (the causal closure of the physical) entailed by classical physics.

In the classical approximation there is no need for, *and indeed no room for*, any effect of any probing action. The *uncertainty* – arising from the non-zero size of the quantum cloud – that in the unapproximated theory needs to be resolved by the intervention of some particular probing action is already reduced to zero by the replacement of Planck's constant by zero. Thus, all effects upon the physically/mathematically described aspects of nature's process that are instigated by the actions 'freely' chosen by agents are eliminated by the classical approximation. *Consequently*, any attempt to understand or explain within the framework of classical physics the physical effects of consciousness is irrational, *because the classical approximation eliminates the effect one is trying to study*.

Intentional Actions and Experienced Feedbacks

The concept of intentional actions by agents is of central importance. Each such action is intended to produce an experiential feedback. For example, a scientist might act to place a Geiger counter near a radioactive source, with the intention to see the counter either 'fire', or 'not fire', during a certain time interval. The experienced response, 'Yes' or 'No', to the query 'Does the counter fire?' specifies one bit of information. The basic move in quantum theory is to shift, *fundamentally*, from the airy plane of high-level abstractions, such as the unseen precise trajectories of invisible elementary material particles, to the nitty-gritty realities of consciously chosen intentional actions and their experienced feedbacks, and to the theoretical specification of the mathematical procedures that allow us successfully to predict relationships among these empirical realities.

Probing actions of this kind are performed not only by scientists. Every healthy and alert infant is engaged in making willful efforts that produce experiential feedbacks, and he or she soon begins to form expectations about what sorts of feedbacks are likely to follow from some particular kind of felt effort. Thus, both empirical science and normal human life are based on paired realities of this action–response kind, and our physical and psychological theories are both basically attempts to understand these linked realities within a rational conceptual framework.

A purposeful action by a human agent has two aspects. One aspect is his conscious intention, which is described in psychological terms. The other aspect is the linked physical action, which is described in physical terms; i.e., in terms of mathematical entities assigned to space-time points. For successful living the physically described action should be a *functional counterpart* of the conscious intention: after sufficient empirical honing by effective learning processes the physically described aspect of the felt intentional act should have a tendency to produce the intended experiential feedback.

John von Neumann, in his seminal book, *Mathematical Foundations of Quantum Mechanics*, calls by the name 'process 1' the basic probing action that partitions a potential continuum of physically described possibilities into a (countable) set of empirically recognizable alternative possibilities. I shall retain that terminology. Von Neumann calls the orderly mechanically controlled evolution that occurs between interventions by name 'process 2'. This process is the one controlled by the Schrödinger equation. The numbering, 1 and 2, emphasizes the important fact that the conceptual framework of orthodox quantum theory requires *first* an acquisition of knowledge, and *second*, a mathematically described propagation of a representation of this acquired knowledge to some later time at which a further inquiry is made.

There are two other associated processes that need to be recognized. The first of these is the process that *selects the outcome*, 'Yes' or 'No', of the probing action. Dirac calls this intervention a "choice on the part of nature", and it is subject, according to quantum theory, to statistical rules specified by the theory. I call by the name 'process 3' this statistically specified choice of the *outcome* of the action selected by the prior process 1 probing action.

Finally, in connection with each process 1 action, there is, presumably, some process that is not described by contemporary quantum theory, but that determines what the so-called 'free choice' of the experimenter will actually be. This choice *seems to us* to arise, at least in part, from conscious reasons and valuations, and it is certainly strongly influenced by the state of the brain of the experimenter. I have previously called this selection process by the name 'process 4'; but will use here the more apt name 'process zero', because this process must precede von Neumann's process 1. It is the absence from orthodox quantum theory of any description on the workings of process zero that constitutes the causal gap in contemporary orthodox physical theory. It is this 'latitude' offered by the quantum formalism, in connection with the "freedom of experimentation" (Bohr 1958, p. 73), that blocks the causal closure of the physical, and thereby releases human actions from the immediate bondage of the physically described aspects of reality.

Cloudlike Forms

The quantum state of a single elementary particle can be visualized, roughly, as a continuous cloud of (complex) numbers, one assigned to every point in three-dimensional space. This cloud of numbers evolves in time and, taken as a whole, it determines, at each instant, for each allowed process 1 action, an associated set of alternative possible experiential outcomes or feedbacks, and the 'probability of finding (i.e., experiencing)' that particular outcome.

Heisenberg's uncertainty principle specifies that if one squeezes this spatial cloud – the spatial region in which the numbers are nonzero – into a sufficiently small region, it will violently explode outward when the constricting force is removed.

https://epdf.tips/mindful-universe-quantum-mechanics-and-the-participating-observer.html

https://books.google.com/books?id=yVpyM3dJOqYC&pg=PA24#v=onepage&q&f=false

Stapp, H. P. (2011). *Mindful Universe: Quantum Mechanics and the Participating Observer* (2nd ed.). Berlin: Springer-Verlag.

If any of this is true, and I believe that it is, then it completely falsifies Radical Behaviorism and Hard Determinism. It also falsifies Materialism, Physicalism, Naturalism, and Nihilism. The false is falsified by the truth; and, the truth is repeatedly experienced and observed.

The Materialists, Naturalists, and Classical Physicists are right in that there is no such thing as free will or choice at the physical level. Physical matter doesn't make choices. Physical processes such as random mutations, natural selection, evolution, and entropy don't make choices. Your physical brain doesn't make choices because it can't make choices! Physical matter can't do choice. Physical matter has no choice in the matter.

But, these people don't realize why they are right because they don't realize that choice is a function of Psyche and that every choice takes place at the quantum level or the psyche level instead of the physical level. Instead, these people deny the existence of choice and the existence of Psyche because their naturalistic model for reality can't explain psyche nor choice; consequently, these people erroneously choose to state that psyche or choice does not exist, and that's as far as their science can take them.

Materialism and Naturalism completely lack explanatory power when it comes to Psyche, Syntropy, Non-Locality, Action at a Distance, Quantum Mechanics, and Choice.

Mark My Words

—

On the Nature of Things

"On the Nature of Things: Human Presence in the World of Atoms" by Henry P. Stapp.

http://www-physics.lbl.gov/~stapp/NOT72C.pdf

https://quantum-neuroscience.com/wp-content/uploads/2018/05/On-Nature-of-Things.pdf

Here's a free book from Henry P. Stapp that he is handing out to the public.

The important thing is to get this information out to the public so that they can take advantage of it. Most of the scientists have officially rejected this information, so you won't get any of this information from them. You will have to look someplace else besides our college classrooms and public schools if you want to find this information because it's being banned from our public schools by the Materialists, Naturalists, Darwinists, and Atheists.

—

Quantum Physics in Neuroscience and Psychology

This article was my first introduction to Henry P. Stapp and the Orthodox Interpretation of Quantum Mechanics. I mention and discuss it elsewhere, so I won't do so here.

Schwartz, J. M., Stapp, H. P., & Beauregard, M. (2004). *Quantum Physics in Neuroscience and Psychology: A Neurophysical Model of Mind-Brain Interaction*. Published Online: Phil. Trans. R. Soc. B.

http://www-physics.lbl.gov/~stapp/PTRS.doc

https://quantum-neuroscience.com/wp-content/uploads/2018/05/PTRS.zip

http://www-physics.lbl.gov/~stapp/PTRS.pdf

http://mypsyche.us/wp-content/uploads/2017/10/PTRS.pdf

https://www.researchgate.net/publication/7613549_Quantum_physics_in_neuroscience_and_psychology_A_neurophysical_model_of_mind-brain_interaction

http://escholarship.org/uc/item/4w8665vk

This paper summarizes everything I have been looking for during the past fifty-five years of my life. That's a long time to wander in darkness looking for the truth; but luckily, our generation finally has the truth, if we know where to find it and recognize it as true when we do find it.

This paper and Henry P. Stapp's papers and books taught me what Stapp calls the **Orthodox Interpretation of Quantum Mechanics**, which explains the interplay and the scientific interface between mind and matter.

Examine these two websites, from Henry P. Stapp:

https://sites.google.com/a/lbl.gov/stappfiles/

http://www-physics.lbl.gov/~stapp/

They explain the Orthodox Interpretation of Quantum Mechanics in great detail. You could spend a lifetime studying them, and still have things to learn.

Mark My Words

—

The Grand Designer

In the book, *The Grand Designer: Discovering the Quantum Mind Matrix of the Universe*, Graham Smetham gathers together some of the very best from Henry P. Stapp into a very useful and interesting section entitled, "Henry Stapp's Mindful Universe and the Psycho-Physical Bridge".

This section is available on Google Books – just page down and up to populate all the relevant pages.

https://books.google.com/books?id=Z0FOAgAAQBAJ&pg=PA158

I also archived it at this link:

https://quantum-neuroscience.com/wp-content/uploads/2018/05/The-Grand-Designer.zip

I quote from Graham Smetham:

Henry Stapp outlines his version of von Neumann's analysis as the following three stages of the quantum experimental process:

Process 1: The 'free choice' of the experimental setup. Heisenberg called this phase "a choice on the part of the 'observer' constructing the measuring instruments and reading their recording". This choice is "not controlled by any known physical process, statistical or otherwise, but appears to be influenced by understandings and conscious intentions." Whilst this process was originally delineated as a phase within the experimental setting, Stapp indicates that such 'free choice' of 'probing actions' is a part of the general human condition:

> Probing actions of this kind are performed not only by scientists. Every healthy and alert infant is engaged in making willful efforts that produce experiential feedbacks, and he/she soon begins to form expectations about what sorts of feedbacks are likely to follow from some particular kind of effort. Thus, both empirical science and normal human life are based on paired realities of this action-response kind.

The hugely significant point in this 'process 1' 'free choice' is that it poses a question to which 'reality' [or Nature's Psyche] can feedback a 'yes' or a 'no', and the fact that the choice of the question is free means that the 'free choice' actually determines the nature of the possible feedbacks. Thus the 'free choices' determine the nature of the experienced reality:

> The process is active: it injects into the physical state of the system being acted upon the properties that depend upon the intentional chosen action of the observing agent.

Stapp calls this process 1 'a dynamical psychophysical bridge.'

Process 2: The deterministic quantum evolution of the potentialities within the Schrödinger wave function, this is the mathematical description of the development of the probabilities associated the potentialities within the quantum realm.

Process 3: This is what Paul Dirac called a 'choice on the part of nature.' It is the yes or no feedback from the experimental setup – yes, reality is this way or no, reality is not this way; Stapp indicates that complex questions can be reduced to yes-no choices. This delineation of the experimental process, which starts in the classical realm, then evolves within the quantum realm, and finally gives its answer within the classical level again, is forced upon us because we are limited to perceptions and language within the classical realm. This, of course, is a fundamental aspect of our embodied existential situation which was often commented upon by the early quantum physicists. In other words, we can only ask questions and receive answers in the experiential classical domain whilst the processes which determine the nature of the answer take place in the 'ultimate" quantum domain.

The Grand Designer: Discovering the Quantum Mind Matrix of the Universe. Graham Smetham.

This three-step process is the BRIDGE between Psyche and physical reality. It's called a psychophysical bridge. The Orthodox Interpretation of Quantum Mechanics explains how the Human Psyche and Nature's Psyche work together at the quantum level in order to get things done for us at the physical level.

This is the answer to life, the universe, and everything.

Process 1 – Choices are made by the Human Psyche. The Human Psyche chooses what it wants to do with is physical body and physical brain.

Process 2 – Nature's Psyche collapses the wave functions and/or turns on the necessary neurons in order to make the Human Psyche's choices a physical reality. We know that it is Nature's Psyche or "Reality" who is controlling all of this at the quantum level because the Human Psyche isn't consciously aware of any of this quantum mechanical stuff. The firing of neurons and the collapsing of wave functions happen outside our conscious awareness in the quantum realm under the command and control of Nature's Psyche.

Process 3 – Nature's Psyche responds to the Human Psyche's requests by determining what is possible and what is impossible, based upon the physical laws and the quantum mechanical rules which God has set into place. God's Psyche controls and restricts our access to quantum mechanics, and God's Psyche made the physical laws. Nature's Psyche enforces God's will both at the physical level and the quantum level. God and Nature can read our mind.

God holds ALL the KEYS to both the physical laws and the quantum mechanisms. God is Syntropy.

Try as he might, a fallen mortal physical being can't do anything that God, quantum mechanics, and the physical laws won't let him do. God created and enforces the physical laws in order to prevent us from using quantum mechanisms to destroy ourselves and our physical existence. God created the physical laws to prevent the physical matter in our bodies from quantum tunneling away from us at will. The physical laws make physical life possible.

The physical laws force order, structure, restrictions, and limitations into spirit matter or dark matter thereby slowing it down and converting it into physical matter. Entropy or physical law slows down physical matter to sub-light speeds, makes physical matter subject to time or an aging process, localizes physical matter in space, and prevents the physical matter from quantum tunneling away from us at will. The result is the Ultimate Consensus Reality, a physical reality. It is highly reliable, dependable, controllable, predictable, and stable. In a physical reality, I can actually count on this paper being on my hard drive and cloud drive tomorrow when I go looking for it.

Scientific Observation: Anything that began obviously has Someone Psyche or Someone Intelligent who caused it to begin. There's no such thing as spontaneous generation or creation ex nihilo.

Scientific Observation: Physical matter, physical laws, and entropy began.

Scientific Conclusion: Therefore, physical matter, physical laws, and entropy have Someone Psyche or Someone Intelligent who caused them to begin or made them begin.

It is also logical to conclude that physical matter has Someone Psyche or Someone Intelligent who is currently forcing it or making it obey the physical laws, including the second law of thermodynamics.

The physical laws are NOT an inherent part of spirit matter or dark matter – they ONLY apply to physical matter. Someone Psyche or Someone Intelligent is making the physical laws and entropy apply to physical matter. God's Psyche forced the physical laws and entropy (an aging process) into a small portion of the spirit matter or dark matter thereby converting that spirit matter into physical matter. God deliberately slowed down the spirit matter converting it into physical matter and subjecting it to the passage of time (entropy) so that we can live it, experience it, depend on it, control it, learn from it, and remember having done so.

Remember, at velocities or speeds faster than the Speed of Light, TIME STOPS – there is NO entropy or passage of time. Light doesn't age because it doesn't experience the passage of time, meaning that it has NO entropy. Light is Syntropy. Furthermore, Someone Psyche is forcing the wave functions to collapse into physical realities. This is what Quantum Mechanics is trying to tell us and teach us.

This IS Science at the cutting-edge.

If you consider yourself to be a scientist, then it's extremely important for you to find, read, and understand this material, so that you can bring yourself into the modern age where science is concerned.

My ultimate desire and goal is to set people free so that they can find the truth and know the truth so that they don't have to go through a lifetime of darkness like I did.

Quantum Mechanics or Syntropy is thousands of years ahead of what the Materialists, Naturalists, Darwinists, Atheists, and Classical Physicists have been able to produce.

Mark My Words

—

Synaptogenesis Is Applied Quantum Mechanics

Synaptogenesis, synaptic mapping, synapse pruning, and synapse elimination are dynamic and alive.

Who or what is overseeing and controlling this process? There's nowhere near enough information in our genome to code for synaptogenesis and synaptic pruning. Our genes code for proteins, NOT synapses. Our synaptic mapping is taking place someplace else besides our genes.

Who or what is mapping the synapses and deciding what each synapse means or what each synapse does? A synapse has no meaning at the physical level. The synapses can't communicate with each other at the physical level. Technically, neurons can't communicate with each other at the physical level.

Who or what is using this synaptic map or synaptic network at the quantum level to get things done for us at the physical level? In other words, who is doing this synaptic mapping, and then using these quantum maps of physical functionality at the quantum level or the psyche level in order to get things done for us at the physical level?

It can't be the Human Psyche because the Human Psyche isn't consciously aware of any of these quantum processes. It can't be our genes because our genomes have nowhere near enough information storage capacity within them to code for and control synaptogenesis, synaptic mapping, and synaptic pruning.

We are in fact looking at Nature's Psyche and/or God's Psyche for an answer to the source and the cause of Synaptogenesis, Synaptic Mapping, Synaptic Meaning, Synaptic Purpose, and Synaptic Pruning.

I'm a scientist. I'm also a generalist. I'm good at everything, master of nothing. Among many other things, I have been a computer scientist all of my life. I understand physical storage capacity.

Our genome is capable of storing 725 megabytes to 750 megabytes of information. In order to successfully MAP or NETWORK one quadrillion synapses – the estimated number of synapses in a newborn child – assign a 3D address to each, and then assign a meaning or purpose to each synapse, as well as regulate its synaptogenesis and any synaptic pruning, do REMAPPING or UPDATING or RE-NETWORKING of what remains, while keeping it all straight in one's mind would require petabytes of information storage capacity to accomplish. It's physically impossible to shove petabytes of information into our puny 750-megabyte genome. It can't be done, which means that it isn't being done.

Furthermore, genes ONLY code for proteins. There is enough memory storage capacity within our genome to code for all the different proteins that are used to make up our physical body; but, where is the information that's needed to organize all those different proteins into 3D machines and 3D cells being stored? Where are all the 3D blueprints being stored? These millions of different 3D blueprints telling the proteins how to assemble themselves into working machines can't be stored within our genome because there's simply not enough memory storage capacity within our physical genome to store all of that information. That, too, would require petabytes of information storage that simply isn't possible within our 750-megabyte genome.

Who, within a living cell, decides that that particular cell needs a new protein? Who sends the transcription enzymes to the right location in the genome for the specific gene needed to code for that protein? How is that information transferred throughout the cell? Once the mRNA molecule has been made and is on its way to a ribosome, who tells the activation enzymes the correct amino acid to join to the correct codon during the production of transfer RNA (tRNA), so that each tRNA molecule arrives at the ribosome in the correct sequence and at the right time for protein synthesis and the translation process? How is all of this information being transmitted from atom to atom, and molecule to molecule? How would you get two physical atoms to communicate with each other and coordinate with each other? How would you get two or more proteins to communicate with each other and coordinate with each other so as to assemble into much larger nanomachines at the right time in the correct 3D location?

https://en.wikipedia.org/wiki/Protein_biosynthesis

These are atoms and molecules that we are talking about here. Who is telling these different atoms and molecules where to go during the construction of amino acids, during the construction of mRNA and tRNA, and during the construction of proteins? Who tells these atoms and molecules where to go? Who tells the different enzymes which molecules to make? Who tells the transcription enzyme which protein is needed and which gene to read? Who is telling the different proteins where to go and what type of nanomachine to become when they get there?

If given the assignment, how would you make the atoms and the molecules communicate with each other and share information with each other at a distance? Remember, there's NO wires connecting the different atoms together at a distance. Communication between the atoms and molecules takes place wirelessly at a distance.

There's only one logical answer to all of this.

Have you seen it yet?

Telepathy or quantum waves are Wi-Fi at the quantum level or the psyche level. Quantum waves are the answer to life, the universe, and everything. Psyche or Intelligence produces, transmits, receives, and stores quantum waves.

The Materialists, Naturalists, Darwinists, Nihilists, and Atheists assure us that physical atoms are the ONLY thing that exists. They are either right, or they are wrong. Which is it?

Proteins are like tinker toys. The genes determine which shape of tinker toy to produce; but, who or what decides how the different tinker toys should be assembled so as to construct different structures or machines? There's NOT enough memory storage capacity within a genome to store all the different 3D blueprints for all the different gadgets, structures, and nanomachines that can be made with those tinker toys. The genome only has enough memory storage capacity to code for the construction of the individual tinker toys. So, where is all of that other information being stored for all those different 3D blueprints? How is all that information being transmitted from atom to atom, molecule to molecule, and protein to protein?

Who is transmitting all of this information from atom to atom? What makes these atoms and molecules move from one location to their target location? Who is telling these atoms and molecules where their target is located? How is all of this information being transmitted wirelessly from atom to atom, and molecule to molecule?

Where is all of this COMMAND and CONTROL information being stored, and how is it being transmitted throughout the cell to the different atoms and molecules within the cell?

It can't be our genes that are storing and transmitting petabytes of this invisible command and control information. The genes are one of the targets of that information but can't be the source of that information. Our physical genes can't store and transmit petabytes of information at the physical level through physical mechanisms. There's no mechanism in place for doing so at the physical level.

So, what are we left with since the source of this information can't be our genes?

The MAPPING and NETWORKING information for Synaptogenesis and synaptic organization, as well as the meaning or purpose of each synapse cannot be stored within our genome. Furthermore, all the different blueprints for all the different 3D nano-machines and 3D cells can't be stored within our genome. Command and control information, target location addressing, and assignments for trillions of different cells and quadrillions of different molecules can't be stored within our genome. So, where is all this information being stored because it clearly must exist and must be getting stored someplace? How is this invisible information being transmitted wirelessly from atom to atom and molecule to molecule since it can't be transmitted by the atoms and molecules themselves at the physical level? It can't be our physical genes doing all of this information storage and information exchange because our genes are just another molecule.

So, once again, what are we left with?

Well, think about it logically. Physical information storage or physical memory storage is limited by the size of an atom. Physical atoms take up space. It has been estimated that a genome is about as densely packed as information can be stored in a physical format. It doesn't get much more optimal than that at the physical level. But, we NEED millions or billions of times more memory storage capacity in the same amount of space that atoms occupy in order to achieve petabytes to exabytes of memory storage capacity within something the size of a physical brain.

The ONLY solution to this conundrum is to go sub-atomic or quantum. You have to go sub-atomic or invisible in order to achieve information storage densities and information storage capacities greater than what's possible with atoms at the physical level. Physical matter greatly limits information storage density and therefore information storage capacity because physical matter takes up space. Those limitations disappear if you go sub-atomic or quantum.

If you were to choose to store information as quantum waves, you could theoretically store exabytes of information in all of that "empty space" between the nucleus of an atom and its orbiting electron shells. I mean, ask yourself what type of "invisible information" is holding those electrons in orbit in the first place. Positive and negative attract, so there should be NO physical space within an atom; yet, there is. Why? Who is forcing the electrons to orbit the nucleus rather than merging with the protons in the nucleus? There has to be some kind of invisible force and intelligent force doing so; otherwise, physical matter wouldn't exist.

Entropy is death. Syntropy is eternal life. Syntropy is eternal and everlasting. There has to be some type of Syntropy or Psyche in existence, or all of that subsequent entropy and physical matter wouldn't have been possible in the first place.

Telepathy or quantum waves are Wi-Fi at the quantum level or the psyche level. Quantum waves or telepathy could be used to transmit information from atom to atom. Quantum waves could also be used to store information invisibly within atoms. Quantum waves or telekinesis could move an atom from one location to a different target location. There is NO answer at the physical level; but, there are an endless number of different answers to be found at the quantum level or the psyche level.

I have hypothesized that Psyche or Quantum Non-Local Consciousness is an infinite singularity. The math points to the existence of such a thing. If I'm right, then that means that Psyche is capable of an infinite amount of memory storage capacity within something that has NO size and takes up NO space whatsoever. Now, that's infinitely dense information storage capacity that we are talking about there within an infinite singularity or Psyche. That's infinitely more information storage capacity than what's possible at the physical level.

Mark My Words

—

Source

Quantum Mechanics from a Non-Physical Spiritual Perspective

    https://www.amazon.com/dp/B01J023TGU

    https://www.amazon.com/dp/1521132380

NATURE vs. NURTURE vs. NIRVANA: An Introduction to Reality

    https://www.amazon.com/dp/B01JWRCSVA

    https://www.amazon.com/dp/1521132615

Reference Material

Quantum Neuroscience: The Answer to Life, the Universe, and Everything

      https://www.amazon.com/dp/B079Z6QQQB

The Ultimate Model of Reality: Psyche Is the Ultimate Cause

https://www.amazon.com/dp/B071NC9JK6

PART X – EVOLUTION IS ENTROPY

The Materialists, Naturalists, Darwinists, Nihilists, Behaviorists, and Atheists axiomatically define entropy as "disorder".

They call this falsehood or lie the Second Law of Thermodynamics, which axiomatically and erroneously states that the total amount of entropy or disorder in our universe is constantly increasing and that it can never decrease and go to zero. This falsehood predicts that we shouldn't be here, yet here we are. Furthermore, according to the equations for heat and entropy, as mass goes to zero, entropy goes to zero along with it. Entropy can and does go to zero whenever mass or heat goes to zero. Therefore, the second law of thermodynamics is false.

There is NO CORRELATION between entropy and disorder. Disorder doesn't cause heat, and disorder doesn't eliminate heat either. Disorder is the wrong definition for entropy; yet, that definition persists and is dominant anyway because it is false. It is its falsehood that makes it preferable to everything else, because above everything else the Materialists and Naturalists are trying to trick us and deceive us just as they have been tricked and deceived. The corollary to the Second Law of Thermodynamics states that entropy, chance, or disorder can design and create anything, if given enough time to do so. The Theory of Evolution has the same exact corollary as the Second Law of Thermodynamics. They are the same thing in the end. They are the same deception and lie.

These people have chosen to believe that entropy or disorder can design and create things if given enough time to do so. For the Materialists, Naturalists, and Darwinists, both the theory of evolution and the second law of thermodynamics are Creation by Entropy, Creation by Disorder, or Creation by Chance. They are the same thing. They are magic! The Atheists are superstitious because they believe in Creation Ex Nihilo or Creation by Magic.

Once I realized and understood what the Materialists, Naturalists, and Darwinists are teaching and preaching, I immediately realized that in their worldview, evolution is entropy – the false definition for entropy. In other words, according to the Darwinists, evolution is Creation by Entropy, Creation by Random Disorder, or Creation by Chance. Evolution works like magic.

There are a couple of things that you need to know about evolution, or creation by chance. First of all, extinction, genetic deterioration, or genetic entropy is the dominant law in biology – NOT spontaneous generation, abiogenesis, or chemical evolution. Evolution doesn't work as advertised. Evolution cannot design and create. Evolution or genetic mutations can only maim, kill, and destroy. Evolution produces extinction – NOT creation and life. The theory of evolution, spontaneous generation, chemical evolution, or abiogenesis was falsified in 1859 by Louis Pasteur, and it has been false ever since. Second of all, the Primary Axiom of Science and Statistics axiomatically states that chance is not causation. In other words, chance cannot do choice, correlation, design, causation, or creation. Chance and causation are mutually exclusive, which means that they falsify each other. Once chance starts doing causation, then it is no longer chance but has become some type of deliberate causation or intelligent design and creation instead. The Primary Axiom of Science and Statistics falsifies the theory of evolution and the second law of thermodynamics.

The theory of evolution is just as false as the second law of thermodynamics because they are the same thing in the end. They both are design and creation by chance, disorder, chaos, death, or entropy. Evolution is entropy – the false definition for entropy. Therefore, evolution or creation by chance is just as false as creation by entropy or creation by disorder, because they are the same thing after all. They are nothing but superstition or magic, because chance cannot design and create things at will, no matter how much time it is given to do so. The "magic" or quantum mechanics exists at the quantum level, NOT the physical level. At the physical level, creation by chance, the theory of evolution, and the second law of thermodynamics are physically impossible. This is what we have actually experienced and observed.

There are literally dozens of different, contradictory, mutually exclusive, and self-defeating definitions for "entropy" on our books and in our "science". They can't all be right because they contradict each other and falsify each other.

The second law of thermodynamics erroneously claims that the total amount of entropy or disorder in this universe is constantly increasing and that it can never decrease and go to zero.

The very fact that we exist is Scientific Proof that this claim is false. If this claim were true, then planets, stars, and galaxies wouldn't exist; and, we wouldn't exist.

The second law of thermodynamics axiomatically defines entropy as "disorder". This is a false and falsified definition for entropy, heat, and thermodynamics, because there is NO CORRELATION whatsoever between disorder and thermodynamics. The two don't have anything to do with each other. Disorder doesn't make heat, and disorder doesn't eliminate heat either.

The second law of thermodynamics is typically portrayed as Creation by Chaos, or Creation by Random Disorder, or Creation by Entropy, or Creation by Random Chance.

The theory of evolution is also typically portrayed as Creation by Chaos, or Creation by Random Disorder, or Creation by Entropy, or Creation by Random Chance.

The theory of evolution ends up being the same exact false and falsified philosophy of science that the second law of thermodynamics is. They are the same falsehood in the end. They are both Creation by Entropy, Creation by Disorder, or Creation by Chance. They both are the same philosophy of science or interpretation of science in the end.

Evolution of any kind is Creation by Random Disorder, or Creation by Entropy, or Creation by Chaos, or Creation by Chance. EVOLUTION IS ENTROPY – the false definition for entropy. Therefore, evolution defined as entropy, and entropy defined as disorder or evolution, are false.

The true definitions for entropy can actually be used to falsify the theory of evolution and the second law of thermodynamics; therefore, it is obvious that the true definitions for entropy have absolutely nothing to do with "entropy defined as disorder" or the false definitions for entropy. Evolution defined as entropy, creation by entropy, creation by random disorder, creation by chaos, creation by death, or creation by chance IS FALSIFIED by the Scientific Methods and the true definitions for entropy. Evolution defined as entropy, and entropy defined as evolution, are falsified by the Truths in Science.

Once you KNOW that Creation by Entropy or Creation by Evolution is false, then you are finally free to start looking for the Truths in Science instead.

Mark My Words

The Dedicated Website

The Materialists, Naturalists, Darwinists, Nihilists, Behaviorists, and Atheists axiomatically define entropy as "disorder".

These people also have chosen to believe that entropy or disorder can design and create things if given enough time to do so. For the Materialists, Naturalists, and Darwinists, both the theory of evolution and the second law of thermodynamics are Creation by Entropy, Creation by Disorder, or Creation by Chance.

Once I realized and understood what the Materialists, Naturalists, and Darwinists are teaching and preaching, I immediately realized that in their worldview, evolution is entropy. In other words, evolution is Creation by Entropy, Creation by Random Disorder, or Creation by Chance.

I have dedicated a website to the "Evolution Is Entropy" concept; therefore, I won't duplicate all of that here. I will just link to it instead, so that you can find it.

Evolution Is Entropy

https://evolution-is-entropy.com/

    The Key to Entropy

    https://evolution-is-entropy.com/the-key-to-entropy/

    Entropy Defined

    https://evolution-is-entropy.com/entropy-defined/

    Falsifying the Second Law of Thermodynamics

    https://evolution-is-entropy.com/falsify-2nd-law/

Evolution Is Entropy

https://evolution-is-entropy.com/evolution-is-entropy/

    Affirming the Consequent

    https://evolution-is-entropy.com/affirming-the-consequent/

    Defining Syntropy

    https://evolution-is-entropy.com/defining-syntropy/

    Symbolizing Syntropy

    https://evolution-is-entropy.com/symbolizing-syntropy/

    Syntropy Is Order and Organization

    https://evolution-is-entropy.com/syntropy-is/

Fixing Science with Quantum Mechanics

https://evolution-is-entropy.com/fixing-science/

A New More Realistic Model for Entropy

https://evolution-is-entropy.com/a-new-more-realistic-model-for-entropy/

Evolution Falsified

https://evolution-is-entropy.com/evolution-falsified/

What Do Genes Code For?

https://evolution-is-entropy.com/what-do-genes-code-for/

The Scientific Method Is Based Upon Affirming the Consequent

https://evolution-is-entropy.com/affirming-the-consequent/

Using the Scientific Method to Falsify Theories

https://evolution-is-entropy.com/falsifying-theories/

Genetic Entropy

https://evolution-is-entropy.com/genetic-entropy/

Entropy Falsifies Evolution

https://evolution-is-entropy.com/entropy-falsifies-evolution/

No Explanatory Power

https://evolution-is-entropy.com/2018/05/20/no-explanatory-power/

I'm Not a Creationist

https://evolution-is-entropy.com/im-not-a-creationist/

The Altruistic Gene

https://evolution-is-entropy.com/the-altruistic-gene/

Papers – Dozens of Contradictory Definitions for Entropy

https://evolution-is-entropy.com/papers/

Evolution Made Your Brain

https://evolution-is-entropy.com/evolution-made-your-brain/

One science article after another states, "Over millions of years of time under the unrelenting pressure of natural selection on your genes, evolution made your brain." That means that for millions of years, we didn't have a brain, yet we were still somehow magically the fittest creatures on this planet anyway. This claim is nothing but science fiction.

Comparing the Evolutionary Perspective with the Cultural Perspective

https://evolution-is-entropy.com/comparing-perspectives/

Reflections on Evolutionary Psychology

https://evolution-is-entropy.com/reflections-on-evolutionary-psychology/

The Brain Is MAPPED

https://evolution-is-entropy.com/the-brain-is-mapped/

Obviously Made

https://evolution-is-entropy.com/obviously-made/

Proof of God's Existence

https://evolution-is-entropy.com/proof-of-god/

Convincing Proof of God's Existence

https://evolution-is-entropy.com/convincing-proof-of-gods-existence/

My Philosophical Proof of God's Existence

https://evolution-is-entropy.com/2018/05/02/my-proof-of-god/

BLOG

https://evolution-is-entropy.com/blog/

Introduction: Evolution Is Entropy

https://evolution-is-entropy.com/2018/04/29/intro-evolution-is-entropy/

Kin Selection

https://evolution-is-entropy.com/2018/05/18/kin-selection/

No Explanatory Power

https://evolution-is-entropy.com/2018/05/20/no-explanatory-power/

The Logic or Thinking behind Naturalism and Its Derivatives Is Flawed

https://evolution-is-entropy.com/2018/05/18/naturalism-is-logically-flawed/

Physical Matter or Entropy Began

https://evolution-is-entropy.com/2018/05/23/physical-matter-or-entropy-began/

The Books

https://evolution-is-entropy.com/the-books/

Contact

https://evolution-is-entropy.com/contact/

Enjoy!

PART XI — FIXING OR REPAIRING THE SECOND LAW OF THERMODYNAMICS

It's obvious that the second law of thermodynamics is false.

The second law of thermodynamics states that the total amount of entropy or disorder in our universe is constantly increasing and that it can never decrease or go to zero.

This is NOT what we experience and observe! Is it?

Of course not! We don't observe the proton decay that the second law of thermodynamics predicts. We don't observe an ever-encroaching gray goo coming in at us from all sides, as the second law of thermodynamics predicts. We don't observe the ever-increasing disorder and chaos that the second law of thermodynamics predicts. We don't even observe the heat death that the second law of thermodynamics predicts. We don't observe anything that the second law predicts.

WHY?

The obvious reason is that the second law of thermodynamics is false.

The less obvious reason is that some type of Syntropy or Exergy or Conservation is at work within our universe at a fundamental level.

The second law of thermodynamics is obviously false. The very fact that you exist and are reading this right now is Scientific Proof that the second law of thermodynamics is false. The second law predicts that you shouldn't be here, yet here you are! Science is observation, experience, and knowledge. WE KNOW that the second law of thermodynamics is false because you exist, and you are reading this right now.

The Scientific Method IS observation and experience. We OBSERVE that you are here, therefore, WE KNOW that the second law of thermodynamics is false. This is the Scientific Method in action! Your very existence falsifies the second law of thermodynamics or PROVES that it is false. That's how we do science!

Since the second law of thermodynamics is false, some type of Syntropy, Exergy, or Conservation MUST EXIST and be at work within our universe right now.

So, what is it? Where is it? It has to be all around us and within us, because ALL that we ever experience and observe in this part of the multiverse is constantly conserved Order and Organization. We can see it all around us, therefore, WE KNOW that it exists.

This Syntropy or Exergy is found in many different forms. Energy is infinitely malleable. According to the Law of Psyche, each psyche that exists has a certain amount of energy that's under its control, and that controlling psyche can form or transform the energy under its control into anything that it wants that energy to be, anytime and anywhere that it CHOOSES to do so. This is what we actually experience and observe.

Syntropy or Exergy is comprised of energy, and energy is always conserved. Energy is Syntropy!

Furthermore, ALL of the different perpetual motion machines that exist at the quantum level in the Quantum Realm or Spirit World are pure Syntropy. They are constantly being conserved! Ever since the Quantum Fields were designed by the Gods and made by Nature's Psyche or the Controlling Psyches within Nature, the Quantum Fields have been conserved. The Quantum Fields are Syntropy. The Quantum Fields are massless, entropyless, non-physical, exergic, syntropic perpetual motion machines. They will never get old, and they will never die. The Quantum Fields are perfect Order and Organization at the quantum level. The second law of thermodynamics will NEVER be true as long as the Quantum Fields exist.

According to the Quantum Law of Thermodynamics, there is NO thermodynamics, NO heat, NO mass, NO resistance to acceleration, and therefore NO entropy at the quantum level in the Quantum Realm or Spirit World. The whole thing is Pure Syntropy. The second law of thermodynamics will NEVER be true as long as the Conservation of Energy remains true.

The first law of thermodynamics and the second law of thermodynamics are mutually exclusive. They falsify each other. That means that if we can demonstrate that one of them is true, then the other one is automatically false. So, which one is true, and which one is false? The one that has actually been experienced and observed IS THE ONE that's Real and True.

Both the theory of evolution and the second law of thermodynamics ARE "creation by chance", "creation by random disorder or chaos", "creation by blind luck", "causation by chance", or the Null Hypothesis. Evolution is entropy. Evolution is Creation by Chance or Creation by Random Disorder. Evolution is the "creation by random disorder" or the "creation by chance" definition for entropy. The theory of evolution and the second law of thermodynamics are the same thing in the end. They are Materialism, Naturalism, Darwinism, Nihilism, Atheism, Creation by Chance, Creation Ex Nihilo, Chance Causation, or the Null Hypothesis which is why they are automatically false, because WE KNOW that chance cannot do causation.

It's obvious that the second law of thermodynamics is false. It's not so obvious that the theory of evolution is false until we realize that the theory of evolution is also Creation by Chance, Causation by Chance, or the Null Hypothesis. Once we realize that the second law of thermodynamics and the theory of evolution are the same thing – the same falsehood – then WE JUST KNOW that they are false because WE KNOW why they are false. They have never been experienced nor observed in a lab or in the wild.

Creation by Chance, Causation by Chance, Abiogenesis, Spontaneous Generation, Macro-Evolution, Creation by Death or Natural Selection, and Chemical Evolution have NEVER been experienced NOR observed because they are impossible. In fact, Louis Pasteur FALSIFIED Spontaneous Generation, Abiogenesis, Naturalism, Darwinism, and the Theory of Evolution in 1859 – the very same year that Charles Darwin published "On the Origin of Species by Means of Natural Selection". We've known from the very beginning that the Theory of Evolution is false, but collectively as a race, we have chosen to ignore the evidence because we want to believe that the Theory of Evolution is true instead.

Likewise, WE KNOW that the second of thermodynamics is false, but we have chosen collectively to ignore the evidence that falsifies it because we want it to be true.

True heat death will NEVER be possible as long as the Quantum Fields exist. The second law of thermodynamics will NEVER be true as long as the Quantum Fields exist. Heat death or absolute zero temperature IS maximum efficiency at the quantum level. Photons, quantum waves, and superconductors function at maximum efficiency at or near absolute zero in what our scientists erroneously call "heat death". Heat death at the physical level represents maximum efficiency at the quantum level.

We will never have true heat death as long as the Quantum Fields exist. Thanks to the Conservation of Energy and Psyche, the Conservation of Quantum Information within Psyche, the LAW of Psyche, the massless entropyless conserved Quantum Fields, and the Perpetual Motion Cycle $E = mc^2$, Nature's Psyche or the Controlling Psyches within nature can make NEW mass, NEW heat, NEW resistance to acceleration, and NEW heat storage capacity (entropy) anytime and anywhere that they CHOOSE to do so. And, there's nothing that we can do to stop them.

According to Quantum Field Theory, particles are born, and particles die. In other words, particles or quantum waves or photons are MADE by Nature's Psyche; and then later, Nature's Psyche transforms those massless and entropyless omnipresent quantum waves or photons INTO mass, heat, resistance to acceleration, OR mass's heat storage capacity which is entropy. Anytime a quantum wave or photon CHOOSES TO STOP, it transforms its infinite acceleration and omnipresent energy wave INTO mass, heat, resistance to acceleration, or heat storage capacity (entropy).

This is what we actually experience and observe, is it not?

This is the Perpetual Motion Cycle $E = mc^2$ in action!

Then after Nature's Psyche or the Controlling Psyches within Nature have made the planets, stars, and galaxies, WE OBSERVE Nature's Psyche actively transforming the mass, heat, resistance to acceleration, and entropy (mass's heat storage capacity) within our Sun INTO massless, heatless, chargeless, entropyless, non-physical, infinite acceleration quantum waves or photons.

This is what we actually experience and observe, is it not?

This is the Perpetual Motion Cycle $E = mc^2$ in action!

WE KNOW THAT THIS IS TRUE because it's constantly being experienced and observed!

The second law of thermodynamics will NEVER be true as long as the Perpetual Motion Cycle exists. The second law of thermodynamics will NEVER be true so long as the Quantum Fields exist. You would have to destroy the Conservation of Energy and Psyche, the Conservation of Quantum Information within Psyche, the massless and entropyless Quantum Fields, and the Perpetual Motion Cycle in order to restore everything to the chaos from whence it originally was made, in order to make the second law of thermodynamics true. The second law of thermodynamics hasn't been true, ever since the Gods designed the Quantum Fields and Nature's Psyche made the Quantum Fields. True disorder or chaos hasn't existed in our part of the multiverse ever since the Quantum Fields were designed and made.

The false is falsified by the truth, and the truth is constantly and repeatedly experienced and observed.

Likewise, anything that is obviously made obviously has a Maker who made it. Physical matter, quantum waves, and the quantum fields were obviously made; therefore, they obviously have some type of Maker who made them. Who or what could make quantum waves and quantum fields? Who or what could make physical matter from quantum waves or quantum fields? It's definitely not going to be something physical.

Our physicists are in agreement that the LAW of Quantum Information Conservation has to be true, or Quantum Field Theory and Quantum Mechanics cannot be true. That means that we MUST find or identify some type of massless, entropyless, syntropic, exergic, conserved, quantum machine or quantum mechanism that is innately capable of making, transmitting, receiving, processing, and storing Quantum Waves and Quantum Information. Psyche is that machine. It's the ONLY one that has been experienced and observed. Psyche, or intelligence, or quantum consciousness MUST EXIST, or the Quantum Law of Information Conservation could never be true.

Remember, the second law of thermodynamics will NEVER be true as long as the LAW of Quantum Information Conservation remains true, which means that the second law of thermodynamics will NEVER be true so long as the Conservation of Energy and Psyche as well as the Conservation of Quantum Information within Psyche remain true.

Since the second law of thermodynamics is obviously false, it's time for us to find some replacements for the second law of thermodynamics that are obviously true. That's the purpose of this particular group of essays. Let's go find some replacements for the second law of thermodynamics that are actually true.

It's easy to tear things down. It takes a bit more thought and work to build them correctly in the first place.

We are off to a good start here, though.

Sometime in the distant future, when our scientists catch up to Reality and realize that the second law of thermodynamics is false, collectively they will vote to replace the second law of thermodynamics with Quantum Field Theory, the

Perpetual Motion Cycle E = mc², the Quantum Law of Thermodynamics, the Conservation of Energy and Psyche, and the Conservation of Quantum Information within Psyche. ALL of these things falsify the second law of thermodynamics and the "disorder" definition for Entropy; therefore, they would be the perfect replacement for the falsified second law of thermodynamics because they have actually been experienced and observed.

So, what we really NEED to do here in this group of essays is to find a TRUE definition for Entropy to replace the falsified definitions for entropy that we get from the second law of thermodynamics. I've found a couple of them, that have actually been experienced and observed. They are precisely what we are looking for, now that WE KNOW why the second law of thermodynamics is false and why it needs to be replaced.

Mark My Words

The Key to Entropy

Our scientists have the wrong definition for entropy. They erroneously define entropy as disorder. Heat or thermodynamics has zero correlation with disorder. Disorder doesn't cause heat, and disorder doesn't eliminate heat. There's no observed correlation between the two. According to the equation for heat, heat actually has a great deal of order and organization within it, and our scientists don't even know it. Likewise, the absolute zero temperature space all around us and all through us also has a great deal of order and organization within it, and our scientists don't even know it.

Thanks to the quantum fields that were designed by the Gods and made by the Controlling Psyches within Nature, there is perfect order and organization at the quantum level in the transdimensional realm or the conserved realm at absolute zero heat or temperature. Disorder doesn't cause thermodynamics, and it doesn't eliminate thermodynamics either. There's no correlation between the two.

Conserved means entropyless or no-entropy. Conserved means syntropic or exergic. Conserved means perpetual motion. Conserved means eternal and everlasting. Therefore, entropy means "not conserved". In other words, entropy is not conserved. Entropy is an aging process that is built exclusively into physical matter and nothing else. This is the Ultimate Law of Thermodynamics. It defines entropy in terms of what is and is not conserved.

https://ultimate-law-of-thermodynamics.com/

This is science that our scientists haven't discovered yet, because they have already rejected it without even giving it a hearing or knowing that it exists.

This is the key to entropy:

Study the following equation for heat (Q). There is something important here that they completely overlooked, because they aren't searching for it and don't want it to be true because it falsifies their second law of thermodynamics.

$Q = mc\Delta T$

Thermodynamics is the transfer of heat from a hot system to a cold system. According to thermodynamics and the equation for heat (Q), anything that has NO mass ($m = 0$) has NO heat storage capacity ($mc = 0$), which means that the massless has NO heat ($Q = 0$) and is natively at absolute zero instead ($T = 0$). Everything goes to zero when mass goes to zero!

Remember that, it's important!

Now, let's look at the official heat-based equation for entropy, and see what we can learn.

$S = Q/\Delta T$

The heat-based equation for entropy (S) is based upon heat (Q). When heat goes to zero ($Q \to 0$), entropy goes to zero ($S \to 0$)! It's that simple! This is the

KEY to entropy and thermodynamics, and it has been completely overlooked by the scientists on our earth. They have never looked at this nor thought about it even once during their entire lives. It's there in plain sight and obvious, but they have already rejected it, choosing to go with materialism and the second law of thermodynamics instead. Their materialism and naturalism prevent them from discovering this simple truth.

Now, let's combine the two equations and solve the equation for entropy (**S**).

S = mcΔT/ΔT

So, what do we get?

We get:

S = mc

This is science that our scientists haven't discovered yet, and will never discover, because they have already rejected it and chosen to go with materialism and their erroneous second law of thermodynamics instead.

Entropy is mass's heat storage capacity (**S = mc**). Entropy (**S**) is mass (**m**) times a mass's specific heat capacity (**c**). No mass (**m = 0**), then no entropy (**S = 0**). It really is that simple. No mass (**m = 0**), then no heat (**Q = 0**), according to the equation for heat. No heat storage capacity (**mc = 0**), then no heat (**Q = 0**) and no entropy (**S = 0**), according to the equations for heat and entropy.

The equations for heat and entropy falsify the second law of thermodynamics which states that entropy can never go to zero and cease to exist. The second law erroneously states that the total amount of entropy or disorder in the universe is constantly increasing and can never decrease and go to zero. This is false. The second law of thermodynamics erroneously states that entropy, disorder, and physical matter are conserved. It's wrong. Entropy is not conserved. Disorder is not conserved. Mass is not conserved! Entropy and mass come and go in a Perpetual Motion Cycle (**E = mc²**) as Nature's Psyche sees fit. Entropy means "not conserved". The truth is internally consistent and self-supportive, while at the same time completely falsifying the second law of thermodynamics as well as materialism and naturalism.

The Ultimate Law of Thermodynamics states that entropy is not conserved, mass is not conserved, disorder is not conserved, heat is not conserved, resistance to acceleration is not conserved, and heat storage capacity is not conserved. It also states that energy's form is not conserved. ONLY psyche and energy are truly conserved. Psyche is a quantum computer that can form and transform information and energy making them into anything that this psyche or intelligence wants the energy and information under its control to become.

https://ultimate-law-of-thermodynamics.com/

Mass and entropy are purely physical phenomena and only exist at the physical level. They completely cease to exist at the quantum level. This is what we actually experience and observe. It's there to be found within the equations for heat and entropy. This is the Quantum Law of Thermodynamics, which states that there is no heat or thermodynamics at the massless level, entropyless level, heatless level, quantum level, or conserved level. Heat is a function of mass. Entropy is also a function of mass. No mass or resistance to acceleration, then no heat and no entropy. It really is that simple.

According to the combined equations for heat and entropy (**S = mc**), entropy, mass, heat storage capacity, and resistance to acceleration do not exist in the quantum realm, transdimensional realm, syntropic realm, absolute zero temperature realm, or conserved realm. In other words, the perpetual motion machines that we call the "quantum fields" are completely massless, heatless, entropyless, and conserved. This is the Quantum Law of Thermodynamics, which uses the equations for heat and entropy to tell us that heat and entropy don't exist at the quantum level in the massless realm, or the quantum realm, or the heatless realm, or the syntropic realm, or the exergic realm, or the conserved realm where there is no resistance to acceleration.

https://quantum-law-of-thermodynamics.com/

Our scientists have already rejected all of this and have chosen to go with their second law of thermodynamics instead, which erroneously states that the total amount of entropy or disorder in our universe is always increasing and can never go to zero. This is not what we experience and observe, therefore, it is obviously false. WE KNOW it is false because their own equations for heat and entropy falsify their second law of thermodynamics, replacing it with the truth instead.

Quantum Field Theory and the Perpetual Motion Cycle also falsify the second law of thermodynamics, and our scientists don't even know it.

Nature's Psyche is the thing that made and is now running the quantum fields. The quantum fields are the medium through which quantum waves travel and psyches interact.

Nature's Psyche makes quantum waves or photons at will from mass, heat, resistance to acceleration, or entropy; and then later, when that same psyche chooses to stop the quantum wave or photon it has made, that psyche transforms the infinite acceleration omnipresent quantum wave it has made INTO some type of localized mass, heat, resistance to acceleration, or entropy (mass's heat storage capacity) instead. This is the Perpetual Motion Cycle (**E = mc^2**). It has always existed, and it will always exist, because the underlying energy and psyche are always conserved. The quantum fields are syntropic and conserved by Nature's Psyche; and, the Perpetual Motion Cycle is Syntropy, or the conservation of energy and psyche. Our scientists will never discover Syntropy or the Perpetual Motion Cycle because they have already rejected these things.

A physical atom is one specific form of energy, and it is made from a choreographed combination of different energies or information, by a bunch of different psyches or intelligences. There's a lot of information or energy within a physical atom, and it's controlled by a lot of different psyches or intelligences.

Nature's Psyche makes the quantum waves or photons from mass, heat, entropy, or resistance to acceleration; and then later, the same psyche or intelligence within the quantum wave chooses to stop that wave or chooses to collapse the wave, thereby transforming that quantum wave or photon INTO some type of mass, heat, entropy, or resistance to acceleration – whatever it chooses for that energy or information to become. Then later, Nature's Psyche transforms some of that mass, heat, or entropy back into massless, heatless, chargeless, and entropyless infinite acceleration quantum waves or photons instead within our stars. This is the Perpetual Motion Cycle and Syntropy; and, our scientists deny its existence.

According to the Law of Psyche, each psyche or intelligence has a certain amount of energy that's under its control, and that controlling psyche can form or transform the energy under its control into anything that it wants that energy to be, anytime and anywhere that it chooses to do so.

Quantum information can be formed and transformed, while the underlying energy and psyche are always conserved. This is what we have actually experienced and observed. Energy's form, or quantum information, or memory is never conserved unless some psyche or intelligence chooses to conserve it, chooses to remember it, or chooses to store it somewhere within its memory banks at the quantum level. Quantum information is processed and stored within Psyche or Intelligence. Energy and information are synonymous with each other. Information is transmitted within energy. Energy or information can be transformed anytime that its controlling psyche chooses to transform it. All the while, the psyche, quantum information processor, or quantum information storage device is conserved and preserved. This is the way things really work at the quantum level.

Nature's Psyche makes the quantum waves or quantum information in the first place from mass or energy, and then Nature's Psyche later collapses or processes that quantum wave in some fashion transforming that energy into something else instead, including information, resistance to acceleration, mass, mass's heat storage capacity (entropy), or possibly entropic physical matter or a physical atom. Nature's Psyche can then take any atom of physical matter and transform it instantly into massless and entropyless and heatless quantum waves or photons as is currently happening within our suns. Then when that photon reaches your skin, the psyche within it can choose to stop the photon and transform that infinite acceleration quantum wave INTO heat, mass, or resistance to acceleration instead. Nature's Psyche could even choose to transform quantum waves or photons back into physical atoms if it wants to do so. This is the Perpetual Motion Cycle.

Our scientists haven't discovered any of this, and they never will, because they have already rejected it and decided that it does not exist. Denying the existence of something will never help you to discover it. You can't make scientific discoveries by denying their existence. You can't find and know the Truths in Science by denying their existence. This is logical common sense.

Our scientists deny the existence of the Perpetual Motion Cycle. Our scientists deny the existence of Syntropy. Our scientists deny the existence of consciousness, intelligence, life force, or psyche. Most of our scientists deny the existence of the quantum fields, because they are invisible, massless, heatless, intangible, entropyless, and non-physical. Our scientists deny the existence of Quantum Waves, Quantum Wave Processors, Quantum Information Processors, and Quantum Information Storage Devices. Yet something has to exist at the quantum level that is capable of making, transmitting, receiving, processing, analyzing, choosing among, and storing quantum waves and quantum information. Psyche is that Quantum Computer, Quantum Wave Processor, Quantum Information Storage Device, Infinite Singularity, Ultimate Point Particle, and Quantum Machine; and, our scientists don't even know it because they have already rejected it and denied its existence.

Thanks to their erroneous and falsified second law of thermodynamics, our scientists haven't yet discovered Syntropy, Psyche, or the Perpetual Motion Cycle. They don't know what any of this is, how it works, why it is important, or that it even exists. Instead, our scientists have erroneously concluded in advance that these things do not exist. Therefore, our scientists aren't looking for these things and will never find them. They don't want to find them. They don't want these things to be true.

Our scientists can't have the Truths in Science by constantly denying their existence or by constantly rejecting them. Scientific discoveries cannot be made by denying their existence. Our scientists will never discover these things because they don't want to.

This is the greatest science ever discovered, but our scientists have already rejected it, choosing to go with materialism, naturalism, creation by chance, and their second law of thermodynamics instead.

We each receive precisely what we are willing to accept. Our scientists don't want the truth. They prefer their deceptions and lies instead.

Mark My Words

PART XII – THE OBVIOUS LIES AND HOLES IN OUR SCIENCE

My goal is ambitious. I want to identify and fix everything that is wrong with "science" and our philosophies of science.

In these essays, I identify the denialistic philosophies of science, and then I use the Scientific Method and negating the consequent to falsify them. This process does indeed fix everything that is wrong with science and the philosophy of science. Then I point people to the Truths in Science instead, which are the things that have actually been experienced and observed and verified.

Whenever you deny the existence of anything, you are immediately on shaky ground because you automatically have no observational evidence to support your chosen belief. Whenever you deny the existence of anything, you immediately enter into the realm of religion, blind faith, and dogma. There will never be and can never be any observational evidence to support your chosen belief or your chosen claim that something specific does not exist. You cannot provide evidence for the non-existence of something. It's philosophically and logically impossible.

Behaviorism is one of the denialistic philosophies of science. The denialistic philosophies deny the existence of the evidence that falsifies them. Radical behaviorism formally rejects and at times denies the existence of choice, free will, consciousness, unconsciousness, mind, soul, intelligence, cognitions, thoughts, or psyche. Therefore, the proven and verified existence of any one of these things falsifies radical behaviorism or hard determinism. It's that simple! The philosophical soundness and validity of the denialistic philosophies is constantly in doubt, because the denialistic philosophies are easily falsified with observational evidence to the contrary. Behaviorism claims that consciousness and unconsciousness do not exist; therefore, the observation of consciousness and unconsciousness falsifies behaviorism. It's that simple.

Radical behaviorism or hard determinism predicts that we should be observing the "non-existence of choice" everywhere on this planet. According to radical behaviorism, there should be NO observable difference between animate objects and inanimate objects. In other words, according to radical behaviorism, there is NO such thing as free will, agency, volition, or choice. Therefore, the obvious fact that each one of us is making choices each waking second of our lives, as well as the obvious fact that our legal system holds us responsible for our choices, is scientific proof that radical behaviorism and hard determinism are false.

The false is falsified by the truth; and, the truth is repeatedly or constantly experienced and observed. If you successfully identify and eliminate everything that is false, then ONLY the truth will remain; and, the truth that remains will consist of everything that has ever been experienced or observed. This is Logic 101.

The behaviorism of Hull and Skinner would not allow nor admit concepts such as expectancies, cognitive maps, mental events, choice, cognition, thoughts, selective attention, free will, intentions, purpose, and teleology. The thing I don't like about the radical behaviorists or hard determinists is that they deny the existence of things that obviously exist – things that have actually been experienced and observed.

I hate deceptions and lies; and, I hate being lied to. Some people are fine with being tricked and deceived. Some people are fine with deleting or hiding the evidence that falsifies their beliefs, but I'm not fine with it. The denialistic philosophies of science are always lies because they invariably deny the existence of something that obviously exists. Then begging the question, these people literally use their "denial of evidence" or their ignorance as scientific proof that they are right; and, these people always deny the existence of everything that falsifies their beliefs. In other words, they lie, and they cheat in order to "meet" their burden of proof. I don't like their deceptions and their lies. That is my right.

There is a lot of truth in the denialistic philosophies of science, but they get you by inserting the single lie and treating that lie as if it were the truth. The lie comes in the form of "the denial of the existence of something" that obviously exists. Their lies are hidden within their hidden premises or their hidden conclusion. These people have to hide their premises, because their hidden premises are obviously false. In other words, when it comes to the denialistic philosophies of science, their hidden premises are always a deceptive lie. These people use *bait and switch* as their methodology.

The truth is the bait, but then comes the switch which is the lie. Physical matter obviously exists – that's the bait. It's obviously true. But then the materialists and naturalists immediately switch to the lie and claim that physical matter is the ONLY thing that exists or that EVERYTHING is made from physical matter. They lure you in with the bait or the truth, and then once they have you on the hook, then they switch to the lies and try to get you to swallow those a well. That's their modus operandi.

The Materialists, Naturalists, Darwinists, Nihilists, Behaviorists, and Atheists are united in their claim that the non-physical or the transdimensional does not exist. That's the lie! Quantum waves, magnetic waves, radio waves, microwaves, x-rays, light, photons, space, time, and quantum fields are obviously massless, entropyless, transdimensional, supernatural, and non-physical; therefore, materialism, naturalism, and their derivatives such as the theory of evolution and the second law of thermodynamics are obviously false. The non-physical does indeed exist. The massless and the entropyless do indeed exist. We KNOW that the non-physical or the transdimensional or the quantum exists, because it has been experienced and observed. We have actually seen the light! Therefore, we KNOW that materialism and naturalism are false.

The denialistic philosophies of science deny the existence of the evidence that falsifies them. The Materialists, Naturalists, Darwinists, Nihilists, Behaviorists, Hard Determinists, Atomists, Physical Reductionists, and Atheists deny the existence of the evidence that falsifies their beliefs. These people teach, preach, and believe that their theories, philosophies, and ideas cannot be falsified by the Scientific Method nor by observational evidence to the contrary. They are wrong.

The hidden premises or hidden axioms within these denialistic theories, philosophies, and ideas are easily falsified with observational evidence, which means that they are easily falsified by the Scientific Method.

Observation is a scientific method; and, observation easily falsifies the hidden premises within the denialistic philosophies of science. That's why these people have to hide their premises and their conclusions, because their hidden premises are obviously false. This is one of my greatest scientific discoveries of all time. Just identify their hidden premise, falsify it with observational evidence, and that's the end of that specific denialistic philosophy. It's that simple.

The denialistic philosophers and atheistic philosophers are like the flat-earthers, because they deny the existence of the evidence that proves that they are wrong. These are the people who denied the existence of the bacteria and viruses because they couldn't see these things with their physical eyes. Flat-earthers deny the existence of the evidence that proves that our earth is round. Likewise, young-earth creationists and young-universe creationists deny the existence of the evidence that proves that our earth and our universe is billions of years old.

The denialistic philosophies exist within every religion or creed; and, there's no sense denying it. The atheists deny the existence of designers and creators. The atheists deny the existence of intelligence or psyche. A psyche is a conserved quantum or a conserved packet of energy; and, the atheists deny the existence of conserved quanta. Atheism is creation ex nihilo or creation by magic – the creation of something from nothing by nothing. The design and creation of something from nothing by nothing is patently absurd. Alas, the Catholics and the Protestants also believe in creation ex nihilo or creation by magic, but with a twist – they believe in the creation of something from nothing by God who also happens to be nothing. These people don't believe in the God and the Creator who has actually been experienced, observed, and verified – any more than the atheists do.

I know how it goes, because I used to be a materialist, naturalist, nihilist, and atheist until Science, the Scientific Method, and the Scientific Evidence convinced me that I was wrong. I have changed now. I'm no longer denying the existence of anything anymore. Instead, I choose to believe in everything that has ever been experienced, observed, or verified. I now allow all of the evidence into evidence, which is why I can now explain everything that has ever been experienced or observed – what it is and how it works.

The denialistic philosophies of science are always false, because they are based upon hidden premises that are obviously false and are easily falsified with observational evidence to the contrary.

The denialistic philosophies of science are demonstrable lies. They were deliberately designed to trick you and deceive you. The denialistic philosophies were designed to convince you that the truth does not exist. The denialistic philosophies of science are evil because they were designed to hide the truth from you and prevent you from finding and knowing the truth. That's the very definition of evil. The people behind the denialistic philosophies are the people who make and love a lie. They are ever-learning but unable to find and know the truth.

The Materialists, Naturalists, Darwinists, Nihilists, Behaviorists, and Atheists have periodically voiced their disgust at my visceral and naked dislike of their deceptions, lies, falsified axioms, and hidden premises. These people have learned to love their lies; and, I have learned to hate their lies. To each their own. I hate being lied to. I immediately rejected their beliefs once I realized that they are lying to us and trying to trick us and deceive us.

However, not everything is a lie. There's nothing wrong with Materialism, Naturalism, Behaviorism, or Darwinism until they start denying the existence of things that obviously exist. These philosophies of science or interpretations of science are perfectly fine until they start lying to us. Physical matter obviously exists. Nature obviously exists. Behavior obviously exists, including chosen behaviors, conditioned behaviors, and reflexes. The fossil record obviously exists. There's nothing wrong with any of these obvious truths. Where the Materialists, Naturalists, and Atheists go wrong is when they start denying the existence of things that obviously exist. The non-physical, quantum, or transdimensional obviously exists. The supernatural or the quantum mechanical obviously exists. Chosen behaviors and choice obviously exist. Designers and creators obviously exist. Intelligence and consciousness obviously exist. Massless and entropyless quantum waves, photons, and quantum fields obviously exist. There's no sense denying it because their existence is obvious.

Intelligence, consciousness, conserved quanta, or conserved psyches obviously exist. They have to exist; otherwise, there would be no place to conserve quantum information, quantum laws, quantum order, quantum blueprints, and physical laws. Psyche is a conserved quantum computer, a conserved quantum wave processor, a conserved quantum information processor, and a conserved quantum information storage device. Psyche or syntropy has to exist; otherwise, quantum information, order, organization, standardization, quantum laws, physical laws, and physical matter would not exist. There has to be a conserved quantum machine that is capable of making, transmitting, receiving, processing, analyzing, transforming, and storing quantum waves or quantum information; otherwise, conserved quantum information, quantum laws, physical laws, and quantum waves would not exist. Conserved quanta or conserved psyches are the ultimate source of order, organization, structure, law, syntropy, quantum waves, the quantum fields, and physical matter. This is obviously true because we obviously exist.

The very fact that we exist is scientific proof that the second law of thermodynamics is false, because the second law predicts that nothing should exist and that everything should be nothing but random chaos or complete disorder. The second law of thermodynamics as well as creation by chance or the theory of evolution are just another form of materialism, which predicts that the non-physical or the transdimensional or the quantum does not exist.

All of this is an integral part of the Philosophy of Science. As far as I know, I'm the premier expert in the Philosophy of Science on the planet right now (2020). I have taken the Philosophy of Science further than anyone has ever taken it. I took it seriously and developed it to its fullest extent possible. The Denialistic Philosophies of Science is one of my unique discoveries in science and the philosophy of science, as well as the obvious fact that the hidden premises in the denialistic philosophies are easily falsified with observational evidence. I'm the only one on the planet using scientific observations, the scientific method, and negating the consequent to falsify the denialistic philosophies of science such as Materialism, Naturalism, Darwinism, Nihilism, Creation by Chance, Behaviorism, Hard Determinism, Atomism, Physical Reductionism, Creation Ex Nihilo, the Theory of Evolution, and the Second Law of Thermodynamics. I seem to be the only person on the planet who knows how to use negating the consequent to falsify materialism, naturalism, and their derivatives.

I suspect that my discoveries in science, statistics, and philosophy will go undiscovered and unnoticed for the duration of my life, because they have already been rejected by the materialists, naturalists, and atheists who control our public schools and publication system. These people deny the existence of all the evidence that falsifies their beliefs. In other words, these people deny the existence of the truths that falsify their lies. That's what they do. That's how they operate. That's how they keep themselves in business. They censor, ban, delete, deny, vilify, and reject everything that falsifies their obvious deceptions and lies.

I chose to separate myself from the materialists, naturalists, nihilists, and atheists. I'm no longer one of them. I chose to falsify their deceptions and their lies instead. I chose to pursue the truth. As part of my enhancement to the Philosophy of Science, I'm also the first person on the planet to discover the Primary Axiom of Science and Statistics. I made a lot of interesting scientific discoveries once I decided to abandon the deceptions and the lies.

Our scientists don't know what the Primary Axiom of Science and Statistics is. They have never thought about anything like it. Instead, our scientists have embraced its exact opposite. The Primary Axiom of Science and Statistics states that "chance is not causation". Once chance starts doing causation, then it is no longer chance. It has become some type of deliberate choice or intelligent intervention instead. Chance cannot do causation. Chance and causation are mutually exclusive. The one precludes the other. This is Science and Statistics that our scientists have never thought about before.

These essays should prove informative and instructional for those who want to know the Truths in Science. Sometimes a deliberate course correction is necessary; otherwise, we just continue with the deceptions and the lies that our society tells us is the truth. I'm actively promoting Truth in Science, because the lies can kill you, whether you realize it or not.

Truth is always better than the lies, unless of course you are one of those who is making his living and his money from the deceptions and the lies. If you are one of those who made his tenure and his living and his millions from the deceptions and the lies in philosophy and "science", then you are going to find this section of this book hard to swallow. It's going to hurt all the way down, and all the way through as well.

I think the rest of you will find this New Science and New Statistics refreshing and informative, because it is true, and it will set you free. Science is worthless if it's a demonstrable lie, or at least that's the way I see it now that I'm no longer a materialist, naturalist, nihilist, and atheist. I chose to go with the truth instead. I chose to go with the things that have actually been experienced and observed.

There's no secret knowledge here, just obvious truths.

Rather than deleting evidence or adjusting the evidence to match my pre-chosen beliefs like I did when I was a materialist and naturalist and atheist; I am now actively and deliberately adjusting my beliefs in order to make them match with the abundant evidence that we have on hand as a race.

I'm no longer hiding from evidence. I'm no longer denying the existence of evidence. I'm no longer afraid of finding evidence that falsifies my pre-chosen beliefs. In fact, I welcome it and actively look for it, because it invariably leads me to the truth. Think about it logically and rationally, instead of emotionally, and you will soon see that falsifying my personal beliefs with observational experiences is bound to lead me to the truth. We human beings find the truth through the process of identifying and eliminating everything that is false. The truth is what remains after the falsehoods have been eliminated, and the truths that remain end up being everything that has ever been experienced or observed.

Consequently, I have gotten into the habit of changing and fixing my beliefs every day. I'm a real scientist now, rather than a destroyer of evidence like I was before, when I was a materialist, naturalist, nihilist, and atheist.

Instead of hiding and deleting and denying evidence as the materialists and naturalists do, I now actively and deliberately adjust my beliefs to bring them into line and make them match with the things that have actually been experienced and observed. Instead of adjusting the evidence and making it match with my beliefs, I now adjust my beliefs and make my beliefs match with the evidence. This is a new and better way of doing Science that has never been discovered nor used before now.

Mark My Words

SOCIOLOGY IN THE NEWS

After thousands of years of ignorance, our world has finally started to discover Transdimensional Physics or Non-Physical Physics. Our scientists don't understand it – what it is and how it works. And most of our scientists will continue to reject it and ridicule it, because they don't want it to be true. But, it exists, nonetheless. Their ignorance doesn't magically make it cease to exist simply because they don't know about it yet.

Let me show you a stunning and simple example of what they are missing.

If we choose to define psychology as "the scientific study of the human psyche", then sociology becomes "the scientific study of all those other psyches that currently surround us." Then sociology becomes an eternal and everlasting science, because we are going to be interacting with other psyches or other intelligences for the rest of eternity – long after our physical brain is dead and gone.

Using this upgraded and enhanced definition for the word "sociology", one of the most newsworthy events in sociology is the ongoing verification of the Quantum Zeno Effect.

The researchers demonstrated that they were able to suppress quantum tunneling merely by observing the atoms. This [is the] **so-called "Quantum Zeno effect"** (Steele, 2015).

The Quantum Zeno Effect demonstrates clearly and conclusively the existence of Transpersonal Psychology or psyche-to-psyche psychology. The psyches or the intelligences within the atoms can literally read your mind at a distance; and, they know that you are looking at them or concentrating on them.

According to the Law of Psyche, each psyche or intelligence has a certain amount of energy that's under its control, and that controlling psyche can form or transform the energy under its control into anything that it wants that energy to be, anytime and anywhere that it chooses to do so. Psyche has control over energy at the quantum level or the non-physical level. Psyches communicate with each other through quantum waves or thoughts. Quantum waves or thoughts are WiFi at the quantum level. Quantum waves are transmitted through the massless and entropyless medium that we call the quantum fields.

It is obvious that the human psyche has control over the atoms, molecules, and cells within its assigned physical body, just as it is now obvious that the controlling psyches within a single physical atom have control over the energies and quanta that comprise that physical atom.

The psyches or intelligences who control a physical atom literally freeze an atom into place when you look at that atom or concentrate on that atom, in anticipation that you might require that atom to do something specific for you. That reality or truth is sociology at the quantum level in the transdimensional realm or the non-physical realm. Those psyches or intelligences within that single physical atom are literally socializing with your human psyche transpersonally or telepathically at a distance. That reality or truth is scientific evidence or scientific proof of transdimensional physics, non-physical physics, or transpersonal psychology.

This is the kind of science that the behaviorists, materialists, and naturalists have chosen to ridicule and mock because they have chosen in advance not to believe it and not to accept it. Nevertheless, the verified and proven existence of the Quantum Zeno Effect is additional scientific evidence that quantum mechanics is true and that the different types of behaviorism, materialism, naturalism, and nihilism are false because they claim that psyche or non-local consciousness does not exist, when in fact psyche or intelligence has been proven to exist.

Thanks to the ongoing verification of the Quantum Zeno Effect, we now have scientific proof that sociology or psyche-to-psyche interaction at a distance takes place at the quantum level in the transdimensional realm, spirit world, non-physical realm, or quantum realm. Sociology is now a quantum science.

D&C 93: 29: Man was also in the beginning with God. Intelligence, or the light of truth, was not created or made, neither indeed can be.

Our intelligence, spark, psyche, or non-local consciousness is co-eternal with God's psyche or intelligence. Like energy, psyche or intelligence is conserved. It cannot be made, and it cannot be destroyed. It has always existed, and it will always exist. God knows what he is talking about. Since psyches or intelligences are eternal, everlasting, and conserved according to God, that means that the same sociality or psyche-to-psyche interaction that we have here can continue onward into the eternities in the next life, whether we have a spirit body or not.

When we choose to define sociology as the scientific study of all the psyches that surround us, instantly sociology becomes an eternal and everlasting science. This is news! But unfortunately for Science, this is news that most of our scientists and psychologists continue to reject because they haven't heard about it nor accepted it as being real and true quite yet.

You might even resist it because you have been trained and conditioned and brainwashed in our public schools not to believe it, for your entire life; but, just because you don't believe it doesn't mean that it isn't true. Science has proven it true through scientific experimentation, which means that the Quantum Zeno Effect, Transpersonal Psychology, or quantum sociality among psyches is true, whether we want it to be true or not. That's news for most people!

Reference

Steele, B. (2015). *'Zeno effect' verified – atoms won't move while you watch.* Cornell University. Retrieved from https://phys.org/news/2015-10-zeno-effect-verifiedatoms-wont.html

SIGNS OF PSYCHE

The evidence for the non-physical is all around us and all through us. It shines on us every day, makes our hair stand on end, and we can feel the gravity of it; but, the Materialists, Naturalists, Nihilists, Darwinists, Behaviorists, and Atheists automatically deny its existence and pretend that the non-physical or the transdimensional does not exist.

If any of this Science or Knowledge is true, then there should be abundant scientific evidence on our world to support it.

Let me provide an example that should hold true on any world where God's children reside:

Back in 1975 and 1977, Neurosurgeon Wilder Penfield and Neurophysiologist John Eccles demonstrated clearly and conclusively that Psyche or Non-Local Consciousness is completely separate from the physical brain.

http://www.simpletoremember.com/articles/a/neurology-and-the-soul/

1) Penfield, W. (1975). *The Mystery of the Mind*. New Jersey: Princeton University Press.

2) Popper, K. R., & Eccles, J. C. (1977). *The Self and Its Brain*. New York: Springer-Verlag.

I found this quote useful:

Dr. Penfield had started his brain research with the explicit intention of disproving the existence of the soul, but after conducting experiments like the above one for forty years, he came to an unambiguous conclusion: "The brain is a computer, but it is programmed by something outside of itself."

https://www.thespiritualscientist.com/2011/10/proof-of-souls-existence/

I also found this quote useful:

To those who consider a discussion of souls and spirit to be wishful thinking or worse, consider the conclusions of Wilder Penfield, M.D., widely considered to be the father of neurosurgery. Like some of his medical colleagues, Penfield thought for many years that there was no consciousness independent of the brain. After fifty years of research, he changed his mind.

In his last book, *The Mystery of the Mind*, Penfield stated: "I came to take seriously, even to believe, that the consciousness of man, the mind, is NOT something to be reduced to brain mechanisms. Where did the mind – call it the spirit if you like – come from? Who

can say? It exists. We humans are lofty creations, eternal souls, and timeless spiritual beings. This view is difficult for some to fathom."

http://recursed.blogspot.com/2016/05/yes-your-brain-certainly-is-computer.html

The cognitive psychologists treat the human brain as a computer. It has been a useful analogy for me, since I am a computer scientist. Since the human brain is a computer, who designed and made that computer? Computers don't design and make themselves. Who builds that computer to design specifications each time one is made? Computers don't manufacture, engineer, and field-test themselves. Who mapped the synapses and neurons to perform specific functions? Quantum maps require some kind of Map Maker and Map User. Every computer has a user, so who is the user for the human brain? Who programmed that computer? Computers have to be programmed before they can be taught to program themselves. Who turns the computer on, and who puts the computer into sleep mode? The thing with a computer is that you can pull the plug anytime you want, and it ceases to function. So, who plugged the thing in, in the first place, and who pulls the plug when they are done with it?

The obvious answer to all of these questions is some type of Psyche, Soul, Intelligence, or Consciousness.

We have KNOWN for forty years that Psyche is a separate entity from the Physical Brain; but, the Materialists, Naturalists, and Atheists who control our public schools and media outlets have prevented this SCIENTIFIC EVIDENCE from becoming common knowledge. That's the way these people do science – by suppressing, ridiculing, censoring, and banning Scientific Evidence. They cheat, and millions of people have fallen for the deception and the ruse.

When it comes to Penfield and Eccles, the Materialists and Naturalists deliberately leave their scientific discovery of the Psyche unmentioned and undocumented. Materialism, Naturalism, and Atheism of any kind are based upon a refusal to look at evidence. That's how these people do science – by picking and choosing which scientific evidence they will accept and which scientific evidence they will reject and try to hide from us. They do the same thing with the Scientific Evidence that comes from Near-Death Experiences (NDEs), Shared-Death Experiences (SDEs), Out-of-Body Experiences (OBEs), Psyche Experiences, Revelations of the Biblical God, and Revelations from the Biblical God. These materialistic and atheistic people try to ridicule, mock, censor, and block this Non-Local or Non-Physical Scientific Evidence from becoming common knowledge.

The truth is there to be found; but, the Materialists, Naturalists, Darwinists, Nihilists, and Atheists are deliberately and even desperately trying to hide those truths from the rest of us; and, these people have had a great deal of success in doing so.

The Materialists and Naturalists deliberately suppress any and all evidence or proof of Non-Local Consciousness and Psyche.

Why?

It's because the proven existence of Psyche FALSIFIES Materialism, Naturalism, and their derivatives such as Darwinism, Nihilism, and Atheism. Each verification of Psyche falsifies Materialism and Naturalism; and, Psyche or Intelligence has been verified trillions of trillions of times.

Each time you have a thought or a dream, you verify the existence of Psyche, because people continue to think and dream long after their physical body and physical brain are dead and gone, according to the Empirical Evidence or Scientific Evidence or Observational Evidence from NDEs, SDEs, OBEs, and other types of Spiritual or Non-Local Experiences. We KNOW that our thoughts and dreams are Non-Local Experiences or Spiritual Experiences, because the contents of our thoughts and dreams cannot be captured nor recorded by our physical instruments.

Each time we use Science to prove the existence of Psyche, we indirectly prove the existence of God's Psyche, because God's Psyche is the Original Scientist from which ALL other Science derives. If it weren't for God's Psyche, then there would be NO Science! This Reality and Truth becomes Scientific Proof of God's Existence.

A materialist, naturalist, nihilist, atheist, or behaviorist won't find any of this convincing, because they have programmed themselves to automatically reject it. But, I found it convincing, which explains in part why I am no longer a materialist, naturalist, nihilist, and atheist.

Open-minded scientists on our world are coming to the same conclusion – psyche, mind, soul, intelligence, choice, quantum non-locality, quantum mechanisms, quantum consciousness, quantum waves, and the massless and entropyless quantum fields do in fact exist.

The following also hits the nail on the head:

While in the quantum realm or spirit world, psyche is experienced firsthand as being an immaterial viewpoint in space. While in the transdimensional realm or non-physical realm, psyche or intelligence is observed at the quantum level or the subatomic level as being a pinpoint of light, a photon, an infinite singularity, a conserved quantum, or the ultimate point particle. Psyche is the fundamental unit of reality and existence

(See: https://psyche-ontology.com/psyche-experienced-and-observed/).

Neurosurgeon Wilder Penfield was the first medical doctor, surgeon, and scientist to prove scientifically that psyche, mind, or consciousness is completely separate from one's physical brain. During brain surgery while stimulating people's physical brain, he soon discovered that the person or the human psyche was someone completely different than their physical brain.

Thanks to Penfield, we have known since 1975 that one's psyche or intelligence is a completely different entity or unit than one's physical brain.

As one of its primary axioms or primary truth claims, behaviorism or hard determinism makes the "scientific claim" that psyche, intelligence, choice, mind, quantum consciousness, or soul does not exist. Therefore, any verified example of mind-over-matter falsifies that claim. Wilder Penfield provided us with verified evidence of mind-over-matter.

The placebo effect is also mind-over-matter, and our pharmacological industry takes the placebo effect very seriously, because it is proven and verified science that faith, belief, prayers, and expectations can heal the physical body and can also enter into a science experiment as a confound or as an independent variable, and override any drugs that might be in the system. Anyone who has suffered from the hell of substance-induced psychosis, withdrawal symptoms, and addiction knows that it is psyche's choice to go cold turkey, quit the drugs, and suffer the subsequent months of psychosis or withdrawal symptoms that is in fact the deciding factor, and not one's addicted biology, not one's addicted brain, and not one's society or medical doctors or spouse who want you to keep taking the drugs.

Your biology definitely wants you to keep taking the drugs. Your psychiatrists, spouse, and medical doctors definitely want you to keep taking the drugs. The choice to quit is done by the human psyche, and then the hell really begins, because the system was designed by God to be mind-over-matter; but, whenever it becomes matter-over-mind as in the case of addiction, withdrawal, and substance-induced psychosis, then the person experiences the full wrath of hell having absolutely no control over his physical body. The physical body literally becomes a prison and a torture chamber for the psyche or the soul. In contrast, God and the Atonement of Christ are mind-over-matter. They can intervene and set the prisoner free.

The human psyche can survive just fine without a physical brain and has been observed to function infinitely more efficiently without a physical brain; but once a human psyche gets "trapped" within a physical body or "married" to a physical body, soon the human psyche and the human brain become symbiotes. Whatever happens to the physical body and physical brain happens directly to the human psyche as well. It's all made from energy – both the physical atoms and the human psyche – therefore they have no problem whatsoever interacting with each other at the quantum level. They can effectively become one at the quantum level. The human psyche, the human spirit body, and the human's physical body can become one united Energy Being at the quantum level. This is what we learn from Quantum Mechanics or Transdimensional Physics, which is proven and verified science.

The materialists, naturalists, nihilists, and atheists *automatically reject, hide, and ignore* these verified science experiments in mind-over-matter or the placebo effect because these science experiments falsify their pre-chosen belief that psyche, or mind, or consciousness, or choice does not exist. *Selective amnesia* is a logic fallacy. *Rejecting and dismissing a science experiment*, because you personally don't like the results, is a logic fallacy. This behavioristic and naturalistic practice, of *rejecting the scientific observations that they don't like*, contributed to the demise of behaviorism and a rejection of the behaviorists within psychology.

Hiding evidence, destroying evidence, and *rejecting evidence* are logic fallacies. *Rejecting verified science experiments*, because they didn't like the evidence or didn't like the conclusions, is one of the logic fallacies that helped to contribute to the demise of behaviorism and is now contributing to the demise of naturalism. *Rejecting evidence* is a demonstration of bad faith. The behaviorists want you to take their science experiments seriously, but they refuse to take your science experiments seriously if your science experiments demonstrate that the behaviorists or materialists are wrong.

By definition, observations and experiences are scientific evidence.

Selectively rejecting the evidence that falsifies your pre-chosen beliefs is a logic fallacy. *Refusing to allow evidence into evidence* is a logic fallacy. *Automatically rejecting the evidence*, that you don't like, is a logic fallacy. *Refusing to look at evidence* is a logic fallacy. *Denying the existence of evidence* is a logic fallacy. *Hiding and destroying evidence* is a logic fallacy. *Refusing to believe verified evidence or obvious evidence* is a logic fallacy. *Refusing to believe obvious truths* is a logic fallacy. *Using one's ignorance as scientific evidence* is a logic fallacy. *Assuming that evidence does not exist* is a logic fallacy. *Rejecting specific science experiments* because you don't like their conclusions, or the results, is a logic fallacy. *Rejecting obvious truths* is a logic fallacy. *Self-deception* is a logic fallacy. Nobody is immune. Self-deception works and it works every time.

These are some of the logic fallacies upon which materialism, naturalism, nihilism, atheism, behaviorism, hard determinism, and their derivatives are based.

You have to be willing to falsify the lies and eliminate the lies, if you are ever going to find and know the truth. Scientific discoveries are not made by denying their existence.

The atheists are deathly afraid of consciousness, psyche, mind, soul, or God. Materialists, naturalists, nihilists, atheists, and behaviorists deliberately hide and destroy any scientific evidence or observational evidence that falsifies their beliefs; but, they can't destroy the massless, entropyless, transdimensional, and non-physical photons or quantum waves. It's called "seeing the light" for a reason. The materialists, naturalists, nihilists, and atheists can't destroy the massless, entropyless, chargeless, non-physical, immaterial quantum waves, photons, or light. The very existence of light falsifies their beliefs, and they don't even know it, because they aren't looking for it and don't want to discover it. The materialists and naturalists won't ever discover any of this Science, because they don't want to find it and have convinced themselves that it doesn't exist. No seeking, then no finding. It's that simple.

A materialist, naturalist, nihilist, atheist, or behaviorist won't find any of this convincing, because they have programmed themselves to automatically reject it. They will see it as aggressive, inflammatory, offensive, and vindictive; and then, they will use that as an excuse to ignore it and reject it. But, I found it convincing, which explains in part why I am no longer a materialist, naturalist, nihilist, and atheist.

If we want to learn anything, we have to learn to let other people say what they have to say, and we have to learn to let the evidence say what it wants to say as well. Learning to listen is the ultimate sign of wisdom; whereas, denying the existence of evidence is the ultimate sign of dogmatic ignorance.

Mark My Words

FALSIFIED PHILOSOPHIES OF SCIENCE

Transdimensional means non-physical; therefore, transdimensional physics, or non-local mechanics, or energy mechanics, or quantum mechanics is the scientific study of non-physical physics.

Quantum fields, quantum waves, magnetic waves, radio waves, microwaves, x-rays, gravity waves, light waves, and photons are obviously massless, entropyless, and non-physical. Dark matter, phase-shifted matter, or spirit matter is considered to be non-physical or transdimensional in that it is by definition invisible and intangible. In other words, we can't get our hands on it, and we can't see it with our physical eyes. The non-physical or the transdimensional obviously exists, because these various different quantum phenomena or transdimensional phenomena have been experienced and observed; therefore, we know that they exist.

The truth is really simple to understand, because it is obviously true or self-evident. We know that the light exists because we have seen the light. We know that massless and entropyless quantum waves or photons exist, because we have seen them and felt them. The verified and proven existence of light falsifies materialism, naturalism, nihilism, and their derivatives which claim that the light or the truth does not exist.

A philosophy of science is an interpretation of science. A philosophy of science is a personal belief system. These philosophies of science or interpretations of science are chosen into existence by the human psyche or human intelligence, in every realm or dimension of existence. Remember, the human psyche chooses things into existence, both at the quantum level and at the physical level. A specific psyche is identified by the choices that it has made. A psyche or an intelligence is a conserved quantum or a conserved packet of energy. We know that these things exist because they have been experienced and observed.

The denialistic philosophies of science are the dogmatic and restrictive personal belief systems that deny the existence of one thing or another by decree, by fiat, by peer review, or by axiom. The denialistic philosophies of science are automatically false because there can never be any scientific evidence or observational evidence to support their truth claims or hidden premises, because by definition and by decree that supporting evidence does not exist.

Denialism, radicalism, dogmatism, or extremism tends to destroy itself over time. A science or a philosophy of science can't last if it is self-falsifying or self-defeating.

Philosophies of science or interpretations of science – such as materialism, naturalism, nihilism, atheism, structuralism, scientism, and behaviorism – that are deliberately designed to be narrow and restrictive can't possibly fill everyone's needs. They are destined to fail in the end. Censorship, or telling people what they can't study, is not the proper way to do science if you want to discover everything that is to be found within science. That's just the way it is.

Denying the existence of something you don't like or don't want to believe in is not the proper way to do science; and, it's definitely not the best nor the most efficient way to make new and interesting scientific discoveries. Denying the existence of Quantum Mechanics or Non-Physical Mechanisms won't help you make new and interesting discoveries in Non-Physical Physics, Transdimensional Physics, Quantum Field Theory, or Quantum Mechanics. Denying the existence of God won't help you to find and know God.

Every philosophy of science that denies the existence of one thing or another, as one of its primary axioms, is destined to fail in the end. You can't make scientific discoveries by denying their existence. You can't find and know the Truths in Science by denying their existence. This is logical common sense. We can't find and know the Truths in Science if we are constantly denying their existence.

The Materialists, Naturalists, Darwinists, Nihilists, Behaviorists, and Atheists define science axiomatically as Materialism, Naturalism, Darwinism (Creation by Chance), Nihilism, Behaviorism, and Atheism (Creation Ex Nihilo) so that their "science" will always be true. By defining science as Materialism and Naturalism, it becomes impossible to use science to falsify Materialism and Naturalism. Tricky, huh? They stack the deck and load the dice so that they can't lose. It's a scam. It's a ruse. It's a con. We let them get away with it because we refuse to develop a better definition for Science, because these people have given themselves a monopoly on "science". They cheat so that they will always win. Then they deny the existence of any evidence that falsifies their beliefs. That's how these people prove that they are right – by denying the existence of anything that proves that they are wrong. In other words, they cheat.

Materialism or physicalism or mechanism is a philosophy of science or interpretation of science that denies the existence of the transdimensional, or the non-physical, or the quantum. Therefore, non-physical physics, quantum mechanics, quantum field theory, and transdimensional physics are completely out of reach for the materialists. These people can't comprehend nor wrap their minds around the non-physical. They prevent themselves from doing so. They refuse to go there.

The second law of thermodynamics is a form of materialism that denies the existence of the non-physical, syntropy, or the perpetually conserved. The second law of thermodynamics denies the existence of the massless and entropyless quantum fields, quantum waves, and photons. The second law denies the existence of Syntropy and the Perpetual Motion Cycle. The second law of thermodynamics is obviously false, because it is nothing more than materialism. None of the things that the second law of thermodynamics predicts has ever been experienced or observed. The second law predicts that we shouldn't be here, yet here we are, which means that there is something wrong with the second law of thermodynamics. The very fact that we exist is scientific proof that the second law of thermodynamics is false. Quantum Field Theory falsifies the second law of thermodynamics, and our scientists don't even know it, because they aren't looking for it.

Naturalism is a philosophy of science or interpretation of science that denies the existence of the supernatural, quantum mechanics, superpowers or miracles, non-physical physics, psyche, quantum consciousness, quantum phase-shifting, other phases or dimensions, the quantum zeno effect, quantum entanglement, quantum tunneling, quantum non-locality, quantum omnipresence, quantum information, conserved quanta, quantum wave processors, quantum information processors, and God. The Naturalists will never discover any of this by denying its existence.

The Materialists and Naturalists will never discover Syntropy or the Perpetual Motion Cycle because they have already concluded that it does not exist. You can't make scientific discoveries by denying their existence!

Nihilism is a philosophy of science that denies the existence of the transdimensional realms, or the other phases or dimensions of existence. Nihilism claims that we had no pre-mortal existence and that we cease to exist when our physical brain dies. Nihilism denies that our lives have any purpose or meaning. Nihilism denies teleology or purpose. Nihilism denies the existence of Syntropy, the Perpetual Motion Cycle, Quantum Phase-Shifting, Quantum Tunneling, the After-Death Life Review, the perpetual motion machines that we call the Quantum Fields, as well as Conserved Quanta, Quantum Wave Processors, Quantum Information Processors, and Quantum Information Storage Devices known as Psyches or Intelligences. The Nihilists will never discover any of this by denying its existence.

Atheism, or hard determinism (behaviorism), or creation ex nihilo, or creation by chance is a philosophy of science that denies the existence of intelligence, choice, consciousness, life force, free will, agency, creators, psyche, quantum consciousness, universal consciousness, conserved quanta, and God.

The theory of evolution is just another form of atheism, creation by chance, or creation ex nihilo. The theory of evolution is a philosophy of science that denies the existence of intelligence, choice, creators, psyches, and consciousness. The theory of evolution, abiogenesis, chemical evolution, macro-evolution, or spontaneous generation is obviously false because it is physically impossible and has never been experienced nor observed in the lab or in the wild.

If a phenomenon, such as creation by chance or creation by random disorder, has never been experienced nor observed by anyone, not even God, then it can't possibly be real and true. This is Science 101. Creation by Random Disorder or Creation by Entropy is physically impossible. Entropy means "not conserved". The conserved cannot be designed and created by something that is not conserved! Entropy and disorder cannot do design and creation. By definition, chance cannot do causation, design, correlation, choice, or creation. It has never been observed doing so. Once it starts doing so, then it is no longer chance. The theory of evolution can't possibly be real or true because it is Creation by Chance, Creation by Entropy, or Creation by Random Disorder, which are impossible and have never been experienced nor observed.

Anything that is obviously made obviously has a Maker who made it. Genomes were obviously designed and made; therefore, genomes and life forms obviously have a Maker who made them. Rather than denying the existence of intelligent creators as Darwinism does, this is a positive statement of fact that has actually been experienced and observed. Your genome is God's Signature, and it is written on every cell in your body.

The denialistic and restrictive "sciences" are worthless. You will never discover anything new by denying its existence. It's that simple. Instead, we discover everything in this universe by observing it, experiencing it, and verifying it. There's more to life and existence than just the physical.

The flaw within all of these different faulty and falsified philosophies of science is that they are based upon the pre-chosen belief that the transdimensional, or the non-physical, or the quantum does not exist. It is impossible to establish through a preponderance of the evidence that the non-physical does not exist, especially since the vast majority of our universe is obviously non-physical, transdimensional, or quantum in nature or origin – being comprised of massless and entropyless space or quantum fields. Photons or quantum waves are obviously massless, chargeless, entropyless, and non-physical. The non-physical surrounds us, and permeates us, and fills the whole of space. Materialism of any kind is obviously false.

Since warm-blooded physical beings obviously freeze to death at absolute zero heat or absolute zero temperature, the materialists and naturalists have become overly fixated on their pre-chosen belief that our whole universe is going to end in heat death. They never once realize that there is no such thing as heat death. "Heat death" or absolute zero is steroids at the quantum level. Everything is in perpetual motion at the quantum level at the absolute zero state that the materialists call "heat death". It has been observed that massless, entropyless, and ageless psyches, quantum fields, quantum waves, photons, and superconductors function best and most efficiently at "heat death" or absolute zero. Absolute zero temperature is when these things are most alive and most efficient. This is what we have actually experienced and observed.

At "heat death" or absolute zero temperature, the quantum level remains perfectly organized, ordered, syntropic, perpetual, everlasting, and conserved. It's the physical level that introduces inefficiencies into the system such as mass, mass's heat storage capacity, resistance to acceleration, an aging process, and entropy. Entropy is mass's heat storage capacity, the aging process, and resistance to acceleration. Entropy has no correlation with heat death or disorder, because there is no entropy, disorder, or heat death at the quantum level. Entropy is purely a product of mass and mass's heat storage capacity. No mass, then no heat and no entropy. Entropy and heat cease to exist whenever mass ceases to exist, and then we switch over to the perpetual motion realm, or the transdimensional quantum realm, or the conserved realm, or the syntropic realm instead.

There is nothing within Science that supports the pre-chosen conclusions or the primary axioms of materialism, naturalism, nihilism, atheism, behaviorism, hard determinism, non-existence of the non-physical, heat death, creation ex nihilo, or creation by chance. In fact, Science falsifies these things through a vast preponderance of the evidence. Therefore, it is faulty, irrational, and illogical to choose to believe in these things.

The false is falsified by the truth; and, the truth is repeatedly and constantly experienced and observed by Someone Psyche or Someone Intelligent somewhere sometime somehow.

There are complete science disciplines that were automatically rejected by the Materialists, Naturalists, Darwinists, Nihilists, Behaviorists, and Atheists before they even got started. Materialism, Naturalism, and Darwinism were designed to prevent us from discovering the transdimensional or the non-physical. These faulty and falsified philosophies of science or interpretations of science are not good for Science, new scientific discoveries, new breakthroughs in knowledge, and the discovery of new Truths in Science.

The Materialists, Naturalists, Darwinists, Nihilists, Behaviorists, and Atheists deny the existence of the transdimensional or the non-physical. Consequently, these denialistic philosophies of science have no explanatory power when it comes to the transdimensional or the non-physical. Therefore, these people refuse to study and do science in this topic. These people deliberately delete, censor, ban, and destroy any scientific evidence that falsifies their beliefs. They pretend that it doesn't exist. This is Bad Faith and Bad Science; and, they don't even know it, because they refuse to think about it or study it.

Denying the existence of something you don't like or don't want to believe in is not the proper way to do science; and, it's definitely not the best nor the most efficient way to make new and interesting scientific discoveries. Denying the existence of Quantum Mechanics or Non-Physical Mechanisms won't help you make new and interesting discoveries in Non-Physical Physics, Transdimensional Physics, Quantum Field Theory, or Quantum Mechanics. Denying the existence of God won't help you to find and know God. Every philosophy of science that denies the existence of one thing or another, as one of its primary axioms, is destined to fail in the end. You can't make scientific discoveries by denying their existence. This is logical common sense.

I'm a scientist, so when it comes to psyche or mind, I want to explain it scientifically, not deny its existence. I want to examine what has actually been experienced and observed. I learned to love the scientific study or observational study of the people who have died, went to some type of afterlife, and observed first-hand what it is like being there in that transdimensional realm or non-physical realm.

According to quantum mechanics and quantum field theory, the most verified or best proven science on this planet, it is a conscious choice on the part of Nature or Nature's Psyche that makes the quantum waves and then later collapses the wave function. This has actually been experienced and observed. It's been caught in the act.

Without this conscious choice on the part of Nature or Nature's Psyche, nothing would exist, including physical matter. There would be no quantum waves, no conserved quantum information, and no collapsed wave functions. Without the conscious choice that collapses the wave function, physical matter would not exist, order and organization would not exist, we would not exist, and there would be nothing but chaos. This is how quantum mechanics put lie to or falsified the primary axiom of hard determinism or radical behaviorism which makes the scientific claim that choice, free will, or agency does not exist. In other words, radical behaviorism and hard determinism were falsified by observational evidence to the contrary. Any time we observe a choice being made, that observation falsifies radical behaviorism or hard determinism.

By definition, there is no evidence and can be no evidence for the "science" that does not exist. The denialistic "sciences" such as Materialism, Naturalism, Darwinism, Nihilism, Behaviorism, Atheism, and their derivatives are automatically false because there can be no evidence to support their primary truth claims or primary axioms which claim that something does not exist. There can't be evidence for the things that by definition or by axiom do not exist; therefore, there can never be any supporting evidence for Materialism, Naturalism, Nihilism, Behaviorism, Atheism, and their derivatives which claim that specific things do not exist. There is no way to support their claims with evidence that does not exist.

Consequently, these denialistic philosophies are not science because they have to be taken on blind faith as being real and true, because there will never be any evidence to support them. They are philosophies of science or interpretations of science, but they can't be science because they have no evidence to support them. True Science is based upon observational evidence, verified evidence, and experiential evidence. There can never be any evidence for a "science" that claims that something does not exist, because by definition and by axiom, the evidence does not exist. Consequently, Materialism, Naturalism, Darwinism, Nihilism, Behaviorism, and Atheism are self-defeating because the evidence supporting them does not exist.

The Materialists, Naturalists, Darwinists, Nihilists, Behaviorists, and Atheists deny the existence of everything that proves that they are wrong. That's how these people do "science", by denying the existence of everything that falsifies their beliefs.

These people literally deny the existence of the science experiments that prove that they are wrong, and these people refuse to allow these science experiments into evidence. We know this is so, because we have caught them in the act. That's how these people are identified or reveal themselves to us – they deny the existence of the evidence that proves that they are wrong and refuse to allow it into evidence.

Is it not so?

You can't make scientific discoveries by denying their existence! It's fascinating to observe how these people lie to themselves and deceive themselves so that they can always be right.

Assuming that evidence does not exist and *refusing to allow evidence into evidence* are the logic fallacies upon which Materialism, Naturalism, Darwinism, Nihilism, Behaviorism, and any form of Atheism are based. In contrast, we use scientific evidence, observational evidence, experiential evidence, replicated evidence, and verified evidence to falsify or put lie to Materialism, Naturalism, Behaviorism, and their derivatives.

How do we discover that the invisible or the non-physical exists? We study the invisible objects such as quanta, psyches, black holes, quantum waves, quantum fields, action at a distance, non-locality, quantum entanglement, quantum information, magnetism, microwaves, x-rays, gravity, wind, and radio waves by observing the effect they have on their surroundings! That's how we detect and study the non-physical, or the transdimensional, or the invisible. We observe the effect that they have on their surroundings. A specific psyche or intelligence is identified by the choices it has made. The results of those choices are observable.

A trial means to try things – to prove them true or to prove them false. A science experiment is a trial. What are we trying to prove by running an experiment? We are trying to prove whether our belief system, or theory, is true or false. A belief system, philosophy, theory, or "science" that is demonstrably false should have no value to us, but it does if we are emotionally invested in it. If we are emotionally invested in it, then we can't let it go, even though it is obviously false.

The things that have been experienced and observed end up being the things that are real and true. Observations are scientific evidence.

Lack of observational evidence supporting the primary axioms of materialism, naturalism, nihilism, and atheism helped to contribute to the demise of materialism, naturalism, nihilism, and atheism. Since Darwinism, hard determinism, and behaviorism are derivatives of materialism, naturalism, nihilism, and atheism, behaviorism and hard determinism went down with materialism and naturalism as fruit from the poisoned tree.

Falsifying the denialistic philosophies of science is easy to do, and it can be a fascinating exercise in the use of the Philosophy of Science. We falsify these things by observational evidence to the contrary.

In order for this to work, we first have to get our definitions and theories straight.

In science, falsification means that we use science experiments or actual first-hand observations to prove that specific theories, ideas, philosophies, belief systems, or hypotheses are false. By negating the consequent and by first-hand observations, we can indeed prove that specific theories and belief systems are false. According to the philosophy of science, this is called falsifying a theory or negating the consequent; and, it is philosophically valid and logically sound.

Here's how *falsifying a theory*, or *negating the consequent*, works in principle using the Scientific Method:

Scientific Hypothesis: If Theory X is true, then we will observe Y.

Scientific Observations: We don't observe Y.

Scientific Conclusion: Therefore, Theory X is false and has been falsified by the Scientific Method.

Here's how *negating the consequent* or *falsifying a theory* works in practice:

Scientific Hypothesis: If the Theory of Evolution, Materialism, Naturalism, and Darwinism are true, then we will observe the rocks and physical reactions designing, creating, and manufacturing genomes and life forms from scratch. If the Theory of Evolution is true, we will observe chemical evolution, abiogenesis, and spontaneous generation all around us all the time in real time.

Scientific Observations: We have NEVER observed the rocks and physical reactions designing and creating genomes and life forms; and, we NEVER will. They can't. The chemical evolution, abiogenesis, or spontaneous generation of proteins and genes from atoms is physically impossible and has NEVER been observed.

The Scientific Conclusion: Therefore, the Theory of Evolution, Materialism, Naturalism, and Darwinism are false and have been falsified by the Scientific Method or Scientific Observations.

The complete lack of observational evidence supporting a specific truth claim falsifies that specific axiom or truth claim. It is that easy to falsify the denialistic philosophies of science because by definition, by axiom, and by decree the evidence supporting the denialistic philosophies of science does not exist. The denialistic philosophies of science deny the existence of one thing or another. They deny the existence of the evidence that is needed to prove that they are right, so that they can allegedly never be proven wrong.

Materialism, Naturalism, Darwinism, Nihilism, Atheism, and their derivatives are automatically false, because they deny the existence of the evidence that is needed to support them and verify them. They have to be taken on blind faith as being real and true because the evidence supporting their truth claims does not exist and will never exist. Materialism, Naturalism, and their derivatives are self-defeating because they deny the existence of the evidence that is needed to verify them and prove that they are true. The denialistic philosophies of science will never be true because there will never be any evidence to support them, because these people have already denied its existence.

Darwinism, materialism, physicalism, mechanism, naturalism, nihilism, behaviorism, hard determinism, and atheism are belief systems, philosophies of science, or interpretations of science that are united in their claim or their belief that the non-physical does not exist, that psyche or mind or consciousness or intelligence does not exist, that choice does not exist, and that God does not exist.

Therefore, observation of anything non-physical, observation of psyche or the spark of life, observation of choices being made, or observation of God literally falsify these specific belief systems which claim that God, choice, psyche, and the non-physical do not exist. That's the way science, the scientific methods, and science experiments work. We prove theories, hypotheses, and belief systems false by observing the existence of their opposite.

This is a science experiment that anyone can perform. Just find something non-physical (such as space, time, or photons) and convince yourself that it really is non-physical, and immediately you have successfully falsified or put lie to materialism, naturalism, nihilism, and behaviorism which claim as one of their primary axioms that the non-physical does not exist.

Just observe a dog making a choice between two desirable competing options (you can just see and feel their little psyche grinding away trying to make the choice, sense their frustration and anxiety and confusion and indecision and excitement, and then observe the very moment when the choice is made) or watch your male dog deliberately reject your commands, and immediately you have falsified materialism, naturalism, nihilism, hard determinism, and radical behaviorism which claim as one of their primary axioms that choice or free will does not exist.

It really is that simple to falsify a theory, a philosophy, or a belief system that is obviously false. I know that these are false because I know why they are false.

When it comes to materialism, naturalism, nihilism, and behaviorism, we just identified and falsified a couple of their primary axioms or primary truth claims with observational evidence. We have in fact successfully falsified materialism, naturalism, behaviorism, and their derivatives.

Is that cool, scary, or what?

Each person will see it differently, depending upon their personal agenda, personal desires, and personally chosen belief system. Some people will be unable to accept it and refuse to accept it because it goes against or falsifies everything that they have chosen to believe in. I have had college professors with PhDs tell me that they can't understand this, can't see the relevance, can't see the significance of it, find it illogical or deceptive or irrational, label it as pseudoscience or myth, or can't accept it. They prefer and choose the lie or the deception instead. It's a choice.

The human psyche (and animal psyche) is a quantum machine or subatomic machine that makes choices between competing options at every level of reality and existence and in every realm of reality and existence. The materialists, naturalists, nihilists, behaviorists, and atheists can't and won't accept this obvious truth, that the non-physical does in fact exist.

I have had college professors with PhDs tell me that I didn't falsify their beliefs – I only think that I did. They have told me that it is impossible to prove things false. Have you heard that one from them? I have. I had one college professor tell me that "falsified" is an erroneous term and that I shouldn't be using it in a scientific research paper. I have some of them tell me that my observations are irrelevant. These people have already rejected the scientific evidence and the science experiments that prove that they are wrong, and they refuse to allow that evidence into evidence. These people deny the existence of the evidence that falsifies their beliefs.

I have had college professors with PhDs tell me that I can't falsify their beliefs and that there is no evidence that will ever falsify their beliefs – that the evidence for falsification does not exist. These people deny the existence of falsification and negating the consequent. That's how they prevent their theories and ideas from being falsified. Self-deception works, and it works every time, especially when it comes to college professors who have PhDs in self-deception.

We have in fact falsified or put lie to materialism, naturalism, nihilism, and behaviorism by demonstrating clearly and conclusively that the non-physical does indeed exist and that choice does in fact exist.

So, what else did the Materialists and Naturalists get wrong?

Remember, when it comes to the denialistic philosophies of science, the evidence supporting them does not exist. They are not science, because there is no evidence to support them. They are dogmatic and extremist religions instead, that have to be taken on blind faith as being real and true. Materialism, naturalism, and their derivatives are pseudoscience.

Pseudoscience is "science" that has no observational support. Materialism and naturalism are pseudoscience, philosophies of science, and falsified religions which are masquerading as science, because there is no observational support for their hidden premises.

So, how do the Materialists and Naturalists get around these obvious truths?

These people define Science axiomatically as Materialism, Naturalism, Darwinism, Nihilism, Behaviorism, and Atheism so that their "science" will always be true and so that their "science" can never be falsified. Tricky, huh? They cheat. That's how they do it. That's how they do science. They cheat.

Darwinism, materialism, physicalism, naturalism, mechanism, nihilism, behaviorism, hard determinism, and atheism are united in their claim or belief that the spiritual or the non-physical does not exist. One of my all-time greatest scientific discoveries came on the day that I first realized that everything in this universe is spiritual and non-physical except for an entropic physical atom; and technically, even physical atoms aren't really physical or solid either. Physical atoms are made from a choreographed dance of non-physical or spiritual quanta, forces, fields, space, quantum fields, and quantum waves. The physical is made from the spiritual or the non-physical.

Materialism, Physicalism, Naturalism, Darwinism, Nihilism, Behaviorism, and Atheism deliberately censor and delete and hide this scientific evidence, so that we don't see it and can't find it. These philosophies of science or interpretations of science are dogmatic, narrow, limited, and restrictive. They have a censorship arm that they call "peer review". They automatically and deliberately hide and delete everything that falsifies their pre-chosen conclusions. That's how they protect themselves and keep themselves in existence – by hiding and deleting the scientific evidence that they don't like or don't want to believe in. These people delete anything and everything that falsifies their beliefs and their pre-chosen conclusions. *Refusing to allow evidence into evidence* is a logic fallacy.

The fastest and most effective way to falsify materialism, naturalism, nihilism, and behaviorism is with observational evidence of the non-physical. According to the astrophysicists, WMAP, and NASA, over 95% of our universe is intangible, spiritual, and non-physical, which means that the materialists and naturalists are wrong about 95% of our universe. That's huge! If they are wrong about that, what else are they wrong about? Only 4.6 percent of our universe is made from physical matter. The rest of our universe is made from something intangible, invisible, and non-physical, such as dark matter (spirit matter or phase-shifted matter) and dark energy.

Quantum Mechanics and Quantum Field Theory falsify the materialistic and naturalistic claim that the non-physical does not exist. The quantum fields are obviously non-physical or immaterial. A photon or a quantum wave is obviously massless, entropyless, and non-physical. So is magnetism, radio waves, space, time, gravity, x-rays, microwaves, and dark energy. According to the astrophysicists, over 95% of our physical universe is non-physical or intangible, being made from dark matter, or spirit matter, and dark energy.

The non-physical does not reduce to the physical. It works the other way around. The physical reduces to the non-physical or the transdimensional. A physical atom is made from non-physical quanta and non-physical forces and fields. A physical atom really isn't physical.

Discovery of the non-physical or the transdimensional contributes to the demise of materialism, naturalism, behaviorism, and their derivatives. Since the demise of behaviorism, complete science books have been written on these topics. Quantum Mechanics and Quantum Field Theory are hard science and verified science; and, they literally falsify materialism, naturalism, behaviorism, and their derivatives such as atheism and Darwinism which erroneously claim that the non-physical does not exist.

Structuralism, functionalism, behaviorism, experimental psychology, and scientific psychology were based upon Darwinism to one extent or another; therefore, any scientific evidence that falsifies Darwinism, creation by chance, or the theory of evolution helps to falsify behaviorism and its derivatives. Newly discovered truths in science along with brand new science experiments in quantum mechanics helped to contribute to the demise of behaviorism by demonstrating that the non-physical does in fact exist and by demonstrating that psyche, mind, consciousness, or intelligence does in fact exist.

The Quantum Zeno Effect and the Law of Quantum Information Conservation within Psyche falsify the behavioristic claim that psyche or mind does not exist. Falsification is the correct term, because science, science experiments, and the scientific methods were designed to falsify theories, and they work as advertised. Falsifying a theory does indeed prove that it is false. That's what falsification means according to the philosophy of science. The Quantum Zeno Effect does indeed falsify the behavioristic claim that mind or psyche does not exist. We have to start somewhere, and these scientists chose to start with quantum mechanics, because quantum mechanisms have been experienced, observed, and repeatedly verified through science experiments.

Psyches are conserved quanta. Conserved quanta have to exist; otherwise, conserved quantum information, or order and organization, cannot exist. Conserved quanta have to exist; otherwise, physical atoms and physical laws would not exist. Psyche is a conserved photon. Transformable photons obviously exist. When a transformable photon or quantum wave chooses to stop, it transforms itself into mass, resistance to acceleration, heat, or entropy which is mass's heat storage capacity. If transformable photons exist, then why can't conserved photons exist?

The materialists, naturalists, nihilists, and atheists automatically reject conserved quanta and these verified science experiments because these experiments and observations falsify their pre-chosen belief that psyche, or mind, or consciousness, or choice does not exist. Hiding evidence, destroying evidence, and rejecting evidence are logic fallacies. Rejecting verified science experiments, because they don't like the evidence or don't like the results, is one of the logic fallacies that helped to contribute to the demise of behaviorism as the dominant school in psychology. One can't deny the existence of evidence forever and get away with it, because it eventually becomes obvious that the evidence does in fact exist. It has become obvious that the non-physical or the transdimensional does in fact exist. It has to exist; otherwise, physical matter would not exist. A physical atom is made from a bunch of non-physical quanta, psyches, forces, and fields. If the non-physical did not exist as the materialists and naturalists claim, then physical matter wouldn't exist.

The obvious falsehoods in behaviorism contributed to its demise. The behaviorists' continued insistence on rejecting obvious truths led to a rejection of the behaviorists and the eventual demise of behaviorism among the scientists, medical doctors, psychiatrists, and psychologists who knew better.

It has been demonstrated through scientific experimentation that the human psyche is the thing that causes neuroplasticity to take place within the physical brain. If the human psyche makes no effort to use its damaged physiology and damaged physical brain, then no neuroplasticity takes place, and nothing gets "re-wired". Since the demise of behaviorism, the placebo effect, the quantum zeno effect, and other mind-over-matter effects such as neuroplasticity have been identified and verified through scientific experimentation.

Many of us have observed that we need a new definition for science if we want to find and know the truths in science that are waiting to be discovered. If we define "science" axiomatically as materialism, naturalism, nihilism, atheism, and behaviorism, then "science" will never be able to prove the existence of God, psyche, mind, consciousness, or an afterlife, because these philosophies of science were deliberately created to prevent us from using "science" to prove the existence of God, psyche, consciousness, and an afterlife. We actually need a new definition for Science in order to use Science to prove that God, psyche, consciousness, choice, and an afterlife exist.

Science needs to be redefined as observation, experience, knowledge, and truth. Then anything that has been experienced and observed will qualify as Science.

This is the new cutting edge of Science and the Philosophy of Science that actually requires a new definition for Science, where Science is defined axiomatically as phenomenology, observation, and the lived experiences of the human race. Once we redefine Science as observation and experience, then instantly any experience of the afterlife becomes scientific evidence or scientific proof of an afterlife; and, any in-person experience or observation of psyche or God becomes scientific evidence or scientific proof of the existence of psyche or God. The new upgraded Science and definition for Science turns Science into observation, experience, knowledge, truth, and phenomenology or the lived experiences of the human race.

Behaviorism was designed to remove psyche, intelligence, consciousness, or mind from the science of psychology. Therefore, any attempt within science and the science of psychology to reintroduce psyche back into psychology contributes to the demise of behaviorism. Any time that psyche, intelligence, consciousness, choice, or mind-over-matter is experienced or observed, it falsifies the primary purpose of behaviorism, materialism, naturalism, nihilism, and atheism.

Science is supposed to be observation and experience – or knowledge and truth. Therefore, observational evidence of psyche or experiential evidence of psyche is the best, most effective, and most convincing way to falsify behaviorism. This essay is an attempt to replace behaviorism with observations, phenomenology or the lived experiences of the human race, quantum mechanics, and verified science.

When I was a materialist, naturalist, nihilist, and atheist, I had convinced myself that there is no evidence of an afterlife and no evidence of God's existence. I was wrong, and I didn't know it. I was using my ignorance as scientific proof that there is no evidence for God's existence and no evidence for an afterlife. Such a logic fallacy is called *begging the question* or *jumping to conclusions*. I had no idea that I was using logic fallacies as scientific evidence, but I was. Every materialist, naturalist, nihilist, and atheist does; and, they don't even know it.

On this earth, the internet and bookstores are now being flooded with evidence that falsifies Materialism, Naturalism, Darwinism, Nihilism, and Atheism. Ironically, that evidence has always existed, but the Materialists and Naturalists never go looking for it.

As a materialist, naturalist, nihilist, and atheist, I was completely unaware that this evidence falsifying or putting lie to the primary axioms of behaviorism and materialism even existed. My atheistic college professors claimed that "there is no such flood of evidence". We were wrong, and we didn't even know it. My atheistic college professors also assume that "there can never logically be scientific proof or observational proof" that materialism, naturalism, nihilism, behaviorism, and atheism are false. Once again, we were wrong.

Assuming that your theory or idea cannot be falsified is a logic fallacy. The Materialists, Naturalists, Darwinists, Nihilists, Behaviorists, and Atheists teach, preach, and believe that there is no evidence and will never be any evidence that falsifies their beliefs and their theories. Every atheistic college professor that I have encountered teaches and believes that his or her atheism, materialism, naturalism, and nihilism are infallible, unassailable, and can't be falsified by the scientific methods. These people are wrong, and they don't even know it. The scientific methods falsify or put lie to the theories and ideas that are false, including materialism, naturalism, nihilism, behaviorism, and atheism. Observation of their opposite falsifies these things.

Remember, *assuming that one's theory or belief system is infallible and unassailable* is a logic fallacy. It is also the definition for Bad Science.

Observation of the effects of massless, entropyless, and non-physical quantum waves, radio waves, microwaves, magnetic waves, and x-rays falsifies the materialistic axiom that states that the non-physical does not exist. Psyches are conserved quanta or conserved packets of energy. Psyches or particles or quanta communicate with each other at the quantum level through quantum waves. Quanta and quantum waves obviously exist. Radio waves, microwaves, and magnetic waves are quantum waves. These waves transmit massless and entropyless non-physical quantum information through invisible, transdimensional, and non-physical quantum fields. The non-physical or transdimensional obviously exists, therefore, materialism is obviously false.

Observation of the effects of action at a distance, such as invisible and intangible and non-physical magnetism, microwaves, x-rays, and other types of quantum waves falsifies the naturalistic claim that the supernatural, or the quantum, or the transdimensional does not exist. Action at a distance or the different types of quantum waves that have been experienced and observed obviously exist, therefore, naturalism or the claim that the supernatural does not exist is obviously false.

Observation of choices being made falsifies hard determinism or radical behaviorism which claims that choice, free will, and agency do not exist. Anytime we observe an animal making a choice between two competing alternatives, hard determinism and radical behaviorism are falsified or proven false. It really is that simple. These philosophies of science make truth claims, and they are falsified by observational evidence to the contrary.

Atheists deny the existence of intelligence, psyche, consciousness, design, purpose, teleology, choice, and creators. The atheists claim that everything happens by chance. Atheism is creation by chance and creation ex nihilo. The Primary Axiom of Science and Statistics falsifies this atheistic claim. The Primary Axiom of Science and Statistics states that chance cannot do causation, correlation, design, choice, or creation. Chance is not causation. Chance and causation are mutually exclusive. Chance cannot do causation. Once it starts doing so, then it is no longer chance but has become some type of deliberate choice or intelligent causation instead. This is what has actually been experienced and observed.

Nihilists deny the existence of psyche, intelligence, consciousness, spirit matter, soul, a pre-mortal life, and an afterlife. Therefore, any observational experience of an afterlife, including shared-death experiences and after-death life reviews and after-death encounters with God, falsifies or puts lie to this nihilistic claim. If you want to find and know the truth, you have to be willing to allow the evidence to lead you wherever it wants to take you. The Nihilists refuse to do so. These people *automatically reject all of the evidence* that falsifies their beliefs.

By 2015, the evidence for the existence of psyche, an afterlife, and God had hit critical mass on the internet and in the bookstores. People can no longer delude themselves into believing that evidence for psyche, an afterlife, or God does not exist.

In this essay, I mentioned a few of the scientific discoveries that helped to contribute to the demise of behaviorism as the dominant school in psychology. Many science books provide verified scientific evidence that psyche or mind exists separate from its physical brain, and that psyche, personality, and one's memories survive the death of one's physical brain.

You will have to find these books, read them, and decide for yourself if they met their burden of proof or not; but, that was the purpose for which they were written. They were written specifically to falsify or put lie to materialism, naturalism, nihilism, and behaviorism; and, if you are willing to allow the evidence into evidence, they do precisely what they were intended to do. Remember, *refusing to allow evidence* into evidence is a logic fallacy. There was a time in my life when I refused to look at the evidence that I didn't like, so I know how it goes. I used to be a materialist, naturalist, nihilist, and atheist because I *refused to look at the evidence* that falsified my personally chosen beliefs.

Materialism, Naturalism, Darwinism, Nihilism, Atheism, and their derivatives are philosophies of science or interpretations of science. The most fascinating, instructive, and interesting activity within the Philosophy of Science is to identify and list the logic fallacies that we are using to trick ourselves and deceive ourselves.

While doing research on this topic, I discovered a few new logic fallacies that I was guilty of committing while I was a materialist, naturalist, nihilist, and atheist. *Assuming that evidence does not exist* is a logic fallacy, because it is impossible for a single human being to know everything that there is to know. *Refusing to look at evidence* is a logic fallacy. *Rejecting evidence* is a logic fallacy. These logic fallacies are otherwise known as the *head-in-the-sand* logic fallacy. All of the evidence should be allowed into evidence, and that includes science experiments, observational evidence, and phenomenology which is the lived experiences of the human race as a whole. *Deliberately misinterpreting evidence* and *watering-down evidence* is a logic fallacy.

Evidence is dynamic and alive, because it is real and truly exists. Evidence should be allowed to tell us what it wants to tell us, and we should allow the evidence to take us wherever it wants to take us. On this planet, evidence is discovered by a real and breathing human being.

Thinking that you are always right and that everyone else is wrong is a logic fallacy. This is also known as the *professor's logic fallacy*, the *PhD logic fallacy*, *credentialism*, or an *appeal to authority*. *Labeling science experiments as pseudoscience* because they falsify your pre-chosen beliefs is a logic fallacy. *Assuming that nothing can be falsified* is a logic fallacy. Negating the consequent, science experiments, and the scientific methods were designed to falsify our theories. *Labeling science experiments as pseudoscience* because you disagree with the evidence or disagree with the conclusions of the science experiment is a logic fallacy. It is bad science and bad form to *label the conclusions and the evidence* that falsify your pre-chosen beliefs as pseudoscience.

Scientific inferences are logic fallacies. It is a logic fallacy to treat scientific inferences, assumptions, premises, or hypotheses as evidence and as proof that your pre-chosen conclusions are true. This logic fallacy is also known as *begging the question* or *jumping to conclusions*. *Using your ignorance as evidence or proof that you are right* is a logic fallacy. *Automatically rejecting* new science, new science experiments, and new knowledge is a logic fallacy. This one is also known as *automatically rejecting paradigm shifts*.

Behaviorism, Materialism, Naturalism, Darwinism, Nihilism, Atheism (Creation Ex Nihilo), the Theory of Evolution (Creation by Chance), and the Second Law of Thermodynamics (Creation by Entropy or Random Chaos) are propped up by these logic fallacies, and by even more that we haven't listed here. The fact that behaviorism was based upon logic fallacies helped to contribute to its demise.

If psyche, psychology, and personality continue to exist after a person's physical brain is dead and gone, then psyche really is an integral part of the Science of Psychology both now and in our afterlife. Consequently, the fact that behaviorism officially rejected the existence of psyche, mind, intelligence, consciousness, or soul helped to contribute to its demise as the dominant school in psychology. Furthermore, the fact that behaviorism rejected the existence of psyche or mind makes behaviorism a bad fit for any type of psychology or science that is now trying to reintroduce the quantized psyche or consciousness back into psychology, quantum mechanics, and science.

We have each had atheistic college professors, who automatically reject this kind of science, and who write in response to this kind of information: "There is no scientific evidence for this claim. What you have provided thus far is more rant than essay. It is tendentious without explicit support and often highly misleading. There is much at its core that is false and exposes a regrettable reliance on pseudoscience and myth."

According to the atheistic college professors in this universe, I'm lying to you, and I have been lying to you all throughout this essay. That's how they deal with this kind of information – by discounting it and dismissing it.

They *label it as pseudoscience and myth* so that they can discount it and dismiss it. That's the *labeling* logic fallacy. These people *label the scientific experiments and scientific evidence* that falsify their beliefs as pseudoscience and myth. That's how they discount it and dismiss it. That's how they get the evidence that falsifies their beliefs deleted. An atheistic college professor *won't allow the evidence that falsifies his beliefs into evidence*. Instead, he will say that these comments don't make sense to him or that these comments are irrelevant. In other words, he discounts them and dismisses them so that he doesn't have to deal with them. That's how these people do "science" – by *ignoring, deleting, and refusing to understand everything* that falsifies their pre-chosen conclusions or beliefs.

Notice how they *jump to the conclusion* that there is no scientific evidence that can falsify materialism, naturalism, nihilism, behaviorism, and atheism. This is *begging the question* or *using one's ignorance as scientific evidence and proof* that the scientific evidence falsifying one's beliefs does not exist. *Assuming that your beliefs cannot be falsified* is a logic fallacy. *Assuming that scientific evidence does not exist* is a logic fallacy. *Automatically rejecting and denying the existence of the science experiments and the observations* that falsify or put lie to your beliefs is a logic fallacy. *Assuming that your belief system can never be falsified* is a logic fallacy. These people assume and actively teach that their *beliefs cannot be falsified*, that there is *no scientific evidence* that falsifies their beliefs, and that *there will never be any observational evidence* that falsifies or puts lie to their beliefs.

Self-deception works, and it works every time. *Self-deception* is a logic fallacy. A PhD in Materialism, Naturalism, Darwinism, Nihilism, Atheism or one of their derivatives is a PhD in self-deception. If you are not an atheist by the time you graduate from college, then your college professor has failed in his primary objective. These people have convinced themselves that there *will never be any scientific evidence* and *can never be any scientific evidence* that will falsify their beliefs. *Convincing yourself that evidence does not exist*, and that *evidence will never exist*, are logic fallacies. *Denying the existence of evidence that obviously exists and is obviously true* is a logic fallacy. *Assuming that your philosophical belief system is infallible and unassailable* is a logic fallacy. *Assuming that science experiments and scientific observations and the scientific methods cannot be used to falsify your personal beliefs* is a logic fallacy. These people are deceiving themselves, and they don't even know it. *Self-deception of any kind* is a logic fallacy.

These are just a few of the logic fallacies that the Materialists, Physicalists, Naturalists, Darwinists, Nihilists, Reductionists, Behaviorists, and Atheists use to make their case and to prove that they are right. These people deny the existence of everything that proves that they are wrong; and, they get away with it because they have axiomatically and tautologically defined "science" as Materialism, Naturalism, Darwinism, Nihilism, Behaviorism, and Atheism so that their theories and ideas will always be true and so that their belief systems can never be falsified. In other words, they cheat and lie and deceive themselves in order to prove that they are correct; and, they have a PhD that says that they are officially granted the right to do so. Everything they touch is blessed and sanctified by Materialism, Naturalism, and Atheism; or, they automatically reject it and pretend that it doesn't exist. That's how these people do science. The fact that they have been caught doing so has contributed to their demise within every human society and culture who is open to the Truths in Science.

The false is always falsified by the truth; and, the truth is constantly and repeatedly experienced and observe by Someone Intelligent or Someone Psyche somewhere somehow sometime.

Remember, the denialistic philosophies of science are automatically false because by decree or by axiom the evidence supporting their primary truth claims or primary axioms does not exist. The denialistic philosophies of science or "does not exist" philosophies of science will never be true because the evidence supporting them does not exist and will never exist. The denialistic philosophies of science have to be taken on blind faith as being real and true because there is no evidence to support them. Remember, when it comes to the denialistic philosophies of science, the evidence supporting them does not exist.

Mark My Words

NEGATING THE CONSEQUENT

Breakthroughs in the Philosophy of Science contributed and contribute to the demise of materialism, naturalism, and behaviorism.

The denialistic philosophies of science are the dogmatic and restrictive personal belief systems that deny the existence of one thing or another by decree, by fiat, by peer review, or by axiom. Darwinism, materialism, naturalism, physicalism, nihilism, behaviorism, hard determinism, and atheism deny the existence of one thing or another; therefore, these are some of the denialistic philosophies of science or denialistic interpretations of science. The denialistic philosophies of science are automatically false because there can never be any scientific evidence or observational evidence to support their truth claims, hidden premises, or primary axioms which claim that something specific does not exist, because by definition that supporting evidence does not exist. Supporting evidence for the denialistic philosophies of science is defined out of existence by axiom, by decree, and by peer review. A denialistic philosophy of science has to be taken on blind faith as being real and true.

In science, falsification means that we use science experiments, scientific data, or actual first-hand observations to prove that specific theories, ideas, philosophies, belief systems, or hypotheses are false. By negating the consequent and by first-hand observations, we can indeed prove that specific theories and belief systems are false. According to the philosophy of science, this is called falsifying a theory or negating the consequent; and, it is philosophically valid and logically sound.

The negating the consequent version of the Scientific Method is the most power, most reliable, and most credible version of the Scientific Method because it can be used to prove that our philosophies, theories, hypotheses, and belief systems are false.

Here's how *falsifying a theory*, or *negating the consequent*, works in principle using the Scientific Method:

Scientific Hypothesis: If Theory X is true, then we will observe Y.

Scientific Observations: We don't observe Y.

Scientific Conclusion: Therefore, Theory X is false and has been falsified by the Scientific Method and by negating the consequent.

This version of the Scientific Method is philosophically valid and logically sound. Here's how *negating the consequent* or *falsifying a theory* works in practice:

Scientific Hypothesis: If the theory of evolution, materialism, naturalism, or Darwinism is true, then we will observe the rocks and physical reactions designing, creating, and manufacturing genomes

and life forms from scratch. If the theory of evolution is true, we will observe chemical evolution, abiogenesis, and spontaneous generation all around us all the time in real time.

Scientific Observations: We have never observed the rocks and physical reactions designing and creating genomes and life forms; and, we never will. They can't. The chemical evolution, abiogenesis, or spontaneous generation of proteins and genes from atoms is physically impossible and statistically impossible and has never been observed. In fact, Louis Pasteur falsified spontaneous generation, abiogenesis, chemical evolution, and the theory of evolution in 1859; and, they have been false ever since.

The Scientific Conclusion: Spontaneous generation, chemical evolution, and abiogenesis have never been caught in the act. Therefore, the theory of evolution, materialism, naturalism, and Darwinism are false and have been falsified by the scientific method because these phenomena have never been experienced nor observed in the lab or in the wild. They are false because they cannot be verified with scientific evidence nor replicated with science experiments.

We do not observe any of the things that the theory of evolution predicts that we should be observing; therefore, the theory of evolution is false.

We don't observe macro-evolution. If the theory of evolution were true, we should observe the occasional rat couple in the zoo produce the occasional cat. If macro-evolution were true, we should observe the occasional pair of domesticated cats give birth to a lion or a tiger. If macro-evolution were true, we should observe the occasional chimpanzee couple in the zoo give birth to a human being. If macro-evolution were true, we should observe the occasional human couple give birth to an X-man, a God, a Being of Light, or whatever else is supposed to be next on the evolutionary ladder. We don't observe any of the things that the theory of evolution predicts that we should be observing; therefore, the theory of evolution can't possibly be true.

When negating the consequent, the fact that there is no observational evidence supporting the hypothesis falsifies the hypothesis. It's that simple. The fact that there is no evidence supporting the "non-existence of the immaterial" falsifies a hypothesis or a theory which states that the non-physical does not exist. The fact that there is no evidence and can be no evidence to support the claim that "God does not exist" falsifies that specific premise, axiom, hypothesis, or truth claim leaving it open for verification instead.

By negating the consequent, the truth is eventually found through the process of eliminating everything that is false. If you successfully eliminate everything that is false, then only the truth will remain; and, the truth that remains will consist of everything that has ever been experienced or observed. This is Philosophy of Science 101.

The denialistic philosophies of science are automatically false because there can never be any observational evidence supporting their hidden premises, hypotheses, primary axioms, or truth claims, because these people deny the existence of that evidence whether they realize it or not. The other way the denialistic philosophies are falsified is through observational evidence that proves that they are false. The false is falsified by the truth, and the truth is repeatedly and constantly experienced and observed by someone, somewhere, sometime.

The complete lack of observational evidence supporting a specific truth claim falsifies that specific axiom or truth claim. It is that easy to falsify the denialistic philosophies of science because by definition, by axiom, and by decree the evidence supporting the denialistic philosophies of science does not exist. The denialistic philosophies of science deny the existence of one thing or another. They also deny the existence of the evidence that is needed to prove that they are right, so that they can allegedly never be proven wrong.

Darwinism, materialism, naturalism, hard determinism, nihilism, behaviorism, atheism, and their derivatives are automatically false, because they deny the existence of the evidence that is needed to support them and verify them. They have to be taken on blind faith as being real and true because the evidence supporting their truth claims, or primary axioms, or hidden premises does not exist and will never exist. Materialism, naturalism, behaviorism, and their derivatives are self-defeating because they deny the existence of the evidence that is needed to verify them and prove that they are true. The denialistic philosophies of science will never be true because there will never be any evidence to support them, because these people have already denied its existence. These people sink their own boat, and they don't even know it.

Darwinism, materialism, physicalism, mechanism, naturalism, nihilism, behaviorism, hard determinism, and atheism are belief systems, philosophies of science, or interpretations of science that are united in their claim or their belief that the non-physical does not exist, that psyche or mind or consciousness or intelligence does not exist, that choice does not exist, and that God does not exist. These are the denialistic philosophies of science because by axiom, by peer review, and by decree they deny the existence of one thing or another.

Therefore, observation of anything non-physical, observation of psyche or the spark of life, observation of choices being made, or observation of God literally falsify these specific belief systems which claim that God, choice, psyche, and the immaterial do not exist. We falsify the denialistic philosophies of science by observing the scientific evidence that puts lie to their primary axioms, hidden premises, or primary truth claims.

This is a science experiment that anyone can perform. Just find something non-physical or intangible or immaterial (such as space, time, radio waves, quantum waves, magnetic waves, gravity waves, dark energy, quantum fields, light waves, or photons) and convince yourself that it really is non-physical or immaterial, and immediately you have successfully falsified or put lie to materialism, naturalism, nihilism, and behaviorism which claim as one of their primary axioms that the non-physical or the immaterial does not exist. It really is that simple to falsify the denialistic philosophies. Just observe the scientific evidence that falsifies them, and they have indeed been falsified.

One can also observe that the denialistic philosophies have no evidence to support their primary axioms, hidden premises, or scientific truth claims; and once again, that will falsify them by negating the consequent.

> **Scientific Hypothesis: If materialism, naturalism, nihilism, or behaviorism is true, then we will observe that everything is made from physical atoms and that the immaterial or the non-physical does not exist.**
>
> **Scientific Observations: This is not what we observe. Even physical atoms aren't made from physical atoms. Physical atoms are made from a collection of non-physical quanta, forces, and fields by Nature's Psyche or Nature's Intelligence, who collapses the wave function thereby making mass or physical particles. Astrophysicists have observed that over 95% of our universe is in fact massless, entropyless, intangible, invisible, immaterial, non-baryonic, and non-physical. Space, time, quantum waves, quantum fields, radio waves, magnetic waves, microwaves, x-rays, light waves, and photons are obviously massless, entropyless, immaterial, intangible, and non-physical. Practically everything that exists falsifies materialism, naturalism, and their derivatives such as behaviorism and atheism.**
>
> **Scientific Conclusion: Therefore, materialism, naturalism, and their derivatives such as behaviorism are false and have been successfully falsified by the "negating the consequent" version of the Scientific Method, as well as by direct observational evidence that falsifies them.**

Here we have successfully used the Scientific Method to falsify materialism, naturalism, nihilism, and their derivatives such as behaviorism and atheism. It is really simple to use the negating the consequent version of the Scientific Method to falsify the hidden premises of the denialistic philosophies of science, such as materialism, naturalism, nihilism, behaviorism, and atheism. Without exception, each denialistic philosophy of science has a hidden premise, primary axiom, or scientific truth claim that is obviously false and has no evidence to support it.

We don't observe the things that materialism and naturalism predict that we should be observing, therefore, materialism, naturalism, and their derivatives are false. We don't observe the "non-existence of the immaterial" or the "non-existence of the non-physical" as materialism and naturalism predict that we should be observing. Therefore, we know that materialism and naturalism are false because we are not observing the things that materialism and naturalism predict that we should be observing.

> **Scientific Hypothesis: If radical behaviorism or hard determinism is true, then we will observe that nothing on this planet is capable of making a choice. There will be no difference between**

the inanimate objects and the animate objects if radical behaviorism or hard determinism is true.

Scientific Observations: We don't observe the complete absence of choice. We are constantly making choices every single day of our lives, and our legal system holds us accountable for the choices that we have made.

Scientific Conclusion: Therefore, radical behaviorism and hard determinism are false and have been successfully falsified by the "negating the consequent" version of the Scientific Method.

Here we have successfully used the Scientific Method to falsify radical behaviorism and hard determinism. It really is that simple to falsify the denialistic philosophies of science. Just look at and accept the evidence that proves that they are false, and immediately, they have been proven false! The negating the consequent version of the Scientific Method falsifies the hidden premises or primary axioms of behaviorism, materialism, naturalism, and their derivatives with observational evidence to the contrary. The Scientific Method really does prove that the denialistic philosophies of science are false, if we choose to let it do so.

In this particular case, we do not observe the thing that radical behaviorism or hard determinism predicts that we should be observing; therefore, radical behaviorism or hard determinism is false. We don't observe the "non-existence of choice"; therefore, radical behaviorism or hard determinism can't possibly be true. Every animate object or living object that we observe is capable of making choices; therefore, radical behaviorism or hard determinism is obviously false. It's that simple.

Mark My Words

NEGATING THE CONSEQUENT 2.0

Let's take this thing up a notch.

The negating the consequent version of the Scientific Method is the most power, most reliable, and most credible version of the Scientific Method because it can be used to prove that our philosophies, theories, hypotheses, and belief systems are false.

Here's how *falsifying a theory*, or *negating the consequent*, works in principle using the Scientific Method:

Scientific Hypothesis: If Theory X is true, then we will observe Y.

Scientific Observations: We don't observe Y.

Scientific Conclusion: Therefore, Theory X is false and has been falsified by the Scientific Method and by negating the consequent.

This version of the Scientific Method is philosophically valid and logically sound. It holds up under scrutiny. Negating the consequent really can take a false idea and prove that it is in fact false, by demonstrating that the false idea has no observational evidence to support it. In other words, we won't observe any of the things that a false idea predicts that we should be observing, if that idea is indeed false.

The denialistic philosophies of science are easy to falsify because they deny the existence of the evidence that is needed to verify them. In other words, the hidden premises of the denialistic philosophies cannot be verified with observational evidence. Instead, the denialistic philosophies have to be taken on blind faith alone, as being real and true.

—

Materialism, Naturalism, Nihilism, and Atheism are the most common denialistic philosophies of science. Each one of these denies the existence of one thing or another.

On this earth, we also have scientism, solipsism, behaviorism, hard determinism, reductionism, relativism, the theory of evolution, creation by chance, creation by entropy, creation by disorder, creation ex nihilo, and the second law of thermodynamics which are denialistic philosophies of science. Each one of these denies the existence of one thing or another.

The denialistic philosophies of science are automatically false because they deny the existence of the evidence that is needed to verify them and prove that they are true. The denialistic philosophies are unable to meet their burden of proof, because they have no evidence to support their hidden premises or primary axioms.

Every denialistic philosophy of science has a hidden premise or a primary axiom that has to be taken on blind faith as being real and true because there is no observational evidence and no scientific evidence to support it. Therefore, identify the hidden premise or the hidden axiom, falsify it with observational evidence, and that is the end of that specific denialistic philosophy of science.

By axiom, by law, by peer review, and by decree, the denialistic philosophies of science deny the existence of the evidence that falsifies them. *Denying the existence of evidence* is a logic fallacy. *Deliberating rejecting the evidence* that you personally don't like or that falsifies your personal beliefs is a logic fallacy. *Refusing to allow evidence* into evidence is a logic fallacy. *Assuming that evidence falsifying your beliefs does not exist* is a logic fallacy. *Begging the question* or *jumping to conclusions* is a logic fallacy. These are just a few of the logic fallacies upon which materialism, naturalism, behaviorism, and atheism are based.

The denialistic philosophies of science are easy to falsify because they deny the existence of the evidence that is needed to verify them. In other words, the hidden premises of the denialistic philosophies cannot be verified with observational evidence. Instead, the denialistic philosophies have to be taken on blind faith alone, as being real and true.

—

MATERIALISM

Materialism is the philosophy of science that denies the existence of the non-physical. These people deny the existence of the quantum, the transdimensional, the supernatural, the syntropic, the conserved, and the non-physical. These people deny the existence of the massless and entropyless photons, quantum waves, conserved quanta, quantum information, and quantum fields. According to the materialists, the quantum, the spiritual, the transdimensional, the immaterial, or the non-physical does not exist. Violating the Law of Non-Contradiction, these people deliberately and erroneously define energy as "mass" or "physical matter", so that their "science" will always be true. These people are deliberately sloppy with their terms and definitions so that they can maintain their existence.

The hidden premise or primary axiom within Materialism or Physicalism is the scientific truth claim that the immaterial or the non-physical does not exist. This is the hidden assumption upon which Materialism is based. Therefore, find and observe something non-physical such as radio waves, magnetic waves, photons, or dark energy, and you have successfully falsified Materialism. It's that simple!

Scientific Theory: According to materialism, the non-physical, or the transdimensional, or the quantum does not exist.

Scientific Hypothesis: If materialism is true, then we will observe that the non-physical does not exist.

Scientific Observations: We don't observe the non-existence of the non-physical. Everything in this universe, except for a physical atom, is non-physical; and, even a physical atom is made from a bunch of non-physical quanta, forces, and fields. The photons, quantum waves, and quantum fields are obviously massless, entropyless, and non-physical. Quanta or psyches are by definition non-physical or immaterial. Quantum information, thoughts, or memories are non-physical, especially the ones that survive the death of your physical brain. Space and time are non-physical. The immaterial, intangible, or non-physical obviously exists.

Scientific Conclusion: Therefore, materialism is obviously false and has been falsified by the Scientific Method, by observational evidence to the contrary, and by negating the consequent.

We don't observe what materialism predicts that we should be observing; therefore, materialism is obviously false. We don't observe the non-existence of the non-physical, therefore, materialism can't possibly be true.

Cool!

Is it not?

It's definitely cool if you are a scientist or a philosopher, rather than a dogmatic nutcase, who is denying the existence of everything that exists.

—

SOLIPSISM

Solipsism is the philosophy of science that denies the existence of physical matter and physical reality.

Solipsism's premise or primary axiom is not hidden. Solipsism makes the scientific truth claim that physical matter does not exist. Solipsism states that physical matter is nothing but a figment of the imagination. This idea is just as stupid as the idea that conserved quanta or conserved psyches do not exist. Psyche is a conserved quantum or a conserved packet of energy. Clearly, conserved quanta have to exist; otherwise, there would be no such thing as conserved quantum information or conserved memories. Clearly, quanta have to exist, or there would be no such thing as physical matter! Psyches and quanta have to exist because physical matter and a physical atom is made from non-physical psyches and quanta held together by a bunch of different non-physical forces and fields. Likewise, physical matter obviously exists, therefore, solipsism is obviously false.

Scientific Theory: According to solipsism, the physical matter does not exist.

> **Scientific Hypothesis:** If solipsism is true, then we will observe that physical matter and physical atoms do not exist.
>
> **Scientific Observations:** We don't observe the non-existence of physical matter. Scientists use electron microscopes to "see" physical atoms. Physical atoms clearly exist. IBM spelled "IBM" with 35 xenon atoms. Physical atoms clearly exist.
>
> **Scientific Conclusion:** Therefore, solipsism is obviously false and has been falsified by the Scientific Method, by observational evidence, and by negating the consequent.

We don't observe what solipsism predicts that we should be observing; therefore, solipsism is obviously false. We don't observe the non-existence of the physical, therefore, solipsism can't possibly be true.

Observe that every denialistic philosophy is automatically false, because there is NO evidence to support their hidden premises or primary axioms.

—

FALSIFYING THE MAJOR DENIALISTIC PHILOSOPHIES

The same scientific evidence that is used to falsify solipsism can also be used to falsify the main denialistic philosophies.

Darwinism, materialism, physicalism, naturalism, nihilism, atomism, behaviorism, hard determinism, atheism, and their derivatives are UNITED in their claim that ONLY physical matter exists. These denialistic philosophies of science claim that physical matter is the fundamental unit of reality and existence, and that there is nothing smaller than a physical atom. They each deny the existence of something smaller than a physical atom, because they each claim that ONLY physical matter exists.

The Obvious Law of Physics states that the smaller dwells within and controls the larger. The electron microscopes that are used to see and push around atoms are producing quantum waves, quanta, electrons, or packets of energy that are smaller and more elementary than a physical atom. Therefore, the primary scientific truth claims of materialism, naturalism, and their derivatives are proven false once again.

> **Scientific Hypothesis:** If materialism and naturalism are true, then we will observe that there is nothing smaller or more fundamental than a physical atom.
>
> **Scientific Observations:** We don't observe that physical atoms are the fundamental unit of reality and existence. Non-physical quanta, including non-physical conserved quanta (psyches), seem to be the fundamental unit reality and existence, not physical matter.

Once again, we don't observe what materialism and naturalism predict that we should be observing.

Scientific Conclusion: Therefore, materialism and naturalism are obviously false and have been falsified by the Scientific Method and by negating the consequent.

We don't observe what materialism and naturalism predict that we should be observing; therefore, materialism, naturalism, and their derivatives are obviously false. We don't observe the non-existence of the non-physical; therefore, materialism and naturalism can't possibly be true. We don't observe that physical atoms are the fundamental unit of reality; therefore, materialism and naturalism are false.

Observe that every denialistic philosophy is automatically false, because there is NO evidence to support their hidden premises or primary axioms. The evidence, that is needed to prove that the denialistic philosophies are true, is denied out of existence.

Remember, materialism, naturalism, nihilism, atheism, and their derivatives unitedly predict that we should be observing the non-existence of the non-physical. As one of their hidden premises, materialism, naturalism, and their derivatives state that the non-physical or the immaterial does not exist. Nobody on this planet has ever observed the non-existence of the non-physical. In fact, everything in this universe is non-physical or transdimensional, except for physical atoms; and, physical atoms aren't really physical or solid either. Physical atoms are made from a bunch of non-physical quanta and psyches that are held together by a bunch of different non-physical forces and fields. Therefore, materialism, naturalism, nihilism, and atheism are false and have been successfully falsified by the scientific method.

—

NATURALISM

Naturalism is the philosophy of science that denies the existence of the quantum, or the supernatural, or the spiritual.

The false is falsified by the truth; and, the truth ends up being everything that has ever been experienced or observed.

The hidden premise or primary axiom within Naturalism is the scientific truth claim that the supernatural, transdimensional, spiritual, or quantum does not exist. This is the hidden assumption upon which Naturalism is based. Therefore, find and observe something transdimensional or supernatural such as a photon or a quantum wave moving at the speed-of-light or faster (tachyon), and you have successfully falsified Naturalism. Action at a distance falsifies Naturalism. The verified and proven existence of the quantum fields falsifies Naturalism.

Scientific Theory: According to naturalism, quantum mechanisms do not exist.

Scientific Hypothesis: If naturalism is true, then we will observe that photons, quantum waves, quantum fields, quantum tunneling, action at a distance, quantum mechanics, and the quantum zeno effect do not exist.

Scientific Observations: We don't observe the non-existence of the transdimensional. We don't observe the non-existence of non-locality. We don't observe the non-existence of quantum mechanics or non-physical physics. The psyches or conserved quanta, non-local consciousness, photons, quantum waves, and quantum fields are obviously supernatural, transdimensional, and non-physical. The supernatural or transdimensional obviously exists, because light waves or photons obviously exist. Everything at the quantum level is supernatural in nature and origin.

Scientific Conclusion: Therefore, naturalism is false and has been falsified by the Scientific Method, by observational evidence, and by negating the consequent.

We don't observe what naturalism predicts that we should be observing; therefore, naturalism is obviously false. We don't observe the non-existence of the supernatural or the transdimensional; therefore, naturalism can't possibly be true. Action at a distance, transdimensional physics, quantum mechanics, and non-physical physics are supernatural and therefore falsify naturalism and its derivatives.

Quod erat demonstrandum!

Naturalism also predicts that God does not exist. Therefore, the thousands of different times that our resurrected Lord Jesus Christ has been experienced and observed in the flesh and in the spirit world falsifies Naturalism.

https://www.youtube.com/results?search_query=NDE+Jesus

https://www.youtube.com/watch?v=UPj4wci_bcI

https://www.youtube.com/results?search_query=NDE+Howard+Storm

The false is falsified by observational evidence to the contrary. Observations are scientific evidence. The thousands of times that the Biblical God has been experienced and observed after He rose from the dead is scientific proof that Naturalism and Atheism are false. That's why God appears to some of us – so that we know that He exists.

—

ATHEISM

Atheism is the philosophy of science that denies the existence of intelligent designers and intelligent creators. The atheists make CHANCE their Designer, Creator, and God.

The hidden premise or primary axiom within Atheism is the scientific truth claim that Creators, Intelligent Designers, Conserved Quanta, and Quantized Psyches do not exist. This is the hidden assumption upon which Atheism is based. Atheism is creation ex nihilo or creation by chance. Atheism denies the existence of Intelligence, Creators, and Designers. Therefore, catch a Creator, Intelligent Designer, or Intelligent Psyche in the act of design and creation, you have successfully falsified Atheism. Of course, all you have to do is to see God, touch God, talk with God, and walk with God, and you have also falsified Atheism.

Scientific Theory: According to atheism, designers, creators, intelligent beings, conserved quanta (psyches), and conserved quantum information within conserved psyches do not exist.

Scientific Hypothesis: If atheism is true, then we will observe that creators and designers do not exist. If atheism is true, then we will observe that everything comes into existence by chance alone.

Scientific Observations: This is not what we observe! We don't observe the non-existence of creators and intelligent designers. Anything that is obviously designed and made obviously has a Designer and a Maker who designed it and made it. We have never caught chance in the act of designing and creating proteins, genes, eyes, brains, genomes, and life forms.

Scientific Conclusion: Therefore, atheism is false and has been falsified by the Scientific Method and by negating the consequent.

We don't observe what atheism predicts that we should be observing; therefore, atheism is obviously false. We don't observe the non-existence of creators and intelligent designers; therefore, atheism can't possibly be true.

And there you have it.

—

NIHILISM

Nihilism is the philosophy of science that denies the existence of quantum conservation and the laws of conservation. According to Nihilism, conserved quanta, conserved physical laws, conserved physical constants, and other types of conserved quantum information do not exist. The second law of thermodynamics is a type of nihilism. The second law of thermodynamics denies the existence of the Perpetual Motion Cycle, Syntropy, Conserved Quanta, Conserved Quantum Information, as well as the Conservation of Energy and Psyche.

The hidden premise or primary axiom within Nihilism is the scientific truth claim that our pre-mortal life and our afterlife do not exist. Nihilism states that your conserved psyche, or conserved intelligence, or conserved quantum information does not exist. Nihilism denies the existence of the conservation of energy. Nihilism denies the existence of quanta or psyches. Nihilism denies the existence of quantum information or memories.

These are some of the hidden assumptions upon which Nihilism is based. Nihilism is the philosophical belief that our spirit, or psyche, or quantum, or intelligence, or energy ceases to exist when we die. Nihilism denies the existence of the conservation of energy, psyche, mind, soul, quantum information, and consciousness. Therefore, establish the fact that quantum information has to exist and has to be conserved, and you have successfully falsified Nihilism. Demonstrate that the conservation of energy or the conservation of quantum information is real and true, and you have falsified Nihilism. Of course, all you have to do is step out of your physical body into the transdimensional realm or spirit world, have an after-death life review, see and talk with dead relatives, see and talk with God, and you have also falsified Nihilism.

Scientific Theory: According to nihilism, conserved quanta (psyches), the perpetual motion cycle, syntropy, conserved quantum information within psyches, and the conservation of energy and psyche do not exist. According to nihilism, conservation does not exist.

Scientific Hypothesis: If nihilism is true, then we will observe that the conservation of energy and the conservation of quantum information do not exist.

Scientific Observations: This is not what we observe! Our physicists have proven to themselves that the Law of Quantum Information Conservation within Quanta or Psyches has to be true to one extent or another, or quantum mechanics and quantum field theory can't possibly be true. Quantum information within psyches or intelligences or quanta has to exist, or it would be impossible for there to be quantum laws and quantum blueprints and quantum standards by which a physical atom is designed and made. Without this quantum information and these quantum blueprints, each physical atom would be incompatible with every other physical atom, assuming that it even existed in the first place. Some type of conserved quantum wave processor, quantum information processor, and quantum information storage device has to exist, or quantum waves and quantum information couldn't possibly exist, and the Law of Quantum Information Conservation within Psyches or Quanta couldn't possibly be true. Psyche or intelligence has been experienced and observed at the quantum level. Psyche, or that pinpoint of light, or quantum, or packet of energy is that conserved quantum computer who makes, transmits, receives, processes, and stores quantum waves and quantum information. Psyches or quanta

have to exist, or quantum waves and quantum information and physical matter would not exist and could not exist.

Scientific Conclusion: Therefore, nihilism is false and has been falsified by the Scientific Method, by negating the consequent, by common sense, and by observational evidence to the contrary.

We don't observe what nihilism predicts that we should be observing; therefore, nihilism is obviously false. We don't observe the "non-existence of conserved energy" or the "non-existence of conserved information"; therefore, nihilism can't possibly be true. We don't observe complete and utter chaos, therefore nihilism can't possibly be true. We don't observe the non-existence of quanta or psyches, therefore, nihilism is false.

Something is being conserved even after the death of our physical bodies. Our personal energy is being conserved. The conservation of energy falsifies nihilism. Energy contains information! Psyche is a quantum or a packet of energy. Psyche and its information or energy are conserved or preserved, even when our physical body dies.

Quanta or psyches are packets of energy that contain, process, transmit, and receive quantum information. Consequently, the observed existence of memories, that survive the death of your physical brain and show up in your after-death life review, also falsifies Nihilism and its derivatives. The memories, thoughts, and personality that survive the death of your physical brain obviously are not stored within your physical brain. They are clearly being stored within your energy, psyche, or quantum instead.

This one was a bit more complex, but the process was the same. Negating the consequent is philosophically valid and logically sound. It works. It eliminates the things that are obviously false and should be eliminated.

Conservation of energy, conservation of psyche, conservation of quanta, and conservation of quantum information falsify Nihilism, because they survive the annihilation of one's physical body and physical brain.

Brutal, is it not?

—

INTRODUCING VERIFICATION

Some people will start panicking by now and close the book because they can't handle the thought of their favorite beliefs, deceptions, and lies getting flushed down the toilet like this. They are probably thinking that if you can do this to materialism, naturalism, nihilism, and atheism, then you can do it to anything. Well, that's not quite so.

You can't negate the consequent when it comes to the things that have actually been experienced and observed, because the things that have been experienced and observed have actually been verified and have proven themselves to be real and true.

Scientific Hypothesis: If Theory X is true, then we will observe Y.

Scientific Observations: We keep observing Y over and over and over again. Anytime, we go looking for Y, we find Y. Theory X keeps getting verified over and over again by all the different Scientific Methods and observational methods that we have at our disposal.

Scientific Conclusion: Therefore, there is a high probability that Theory X is in fact true because we keep observing Y every time that we go looking for Y. Yes, it is possible that our own personal interpretation of Theory X is wrong, and that Y doesn't really mean X; but, there is no doubt in our minds that Y truly exists because it has been experienced and observed.

You see, negating the consequent only works with the theories and hypotheses that have no supporting evidence. If we have observational evidence, that fact trumps theory, speculation, hypothesis, and wishful thinking every time.

There is a reason why the creators of the various different denialistic philosophies go out of their way to keep their hidden premises or primary axioms hidden. When it comes to the denialistic philosophies of science, their hidden premises or primary axioms or scientific truth claims are obviously false and obviously have no evidence to support them. That's why their hidden premises have to be hidden. Once you have identified the hidden premise, then it is immediately obvious that it is false.

When it comes to the denialistic philosophies of science and the false philosophies of science, all you have to do is to identify their hidden premise, notice that their hidden premise is obviously false or falsify their hidden premise with observational evidence, and that's the end of that specific denialistic philosophy of science. That interpretation of science or philosophy of science is obviously false because it has been falsified by observational evidence.

In contrast, you can't use observational evidence to falsify observational evidence, because that observational evidence is in fact verifying that observational evidence. Do you see how that works?

Verification is something completely different than falsification. We can successfully and often easily falsify the things that have never been experienced nor observed; but, it quickly becomes impossible to falsify the things that one keeps observing over and over again from one day to the next. The prophets of God who keep seeing God and keep talking with God, day after day after day, have a hard time denying His existence because they KNOW that He exists. Do you see how that works? It is important to get this straight, when a person is trying to do Science, Truth, or Knowledge.

Instead of defining "science" as Materialism, Naturalism, Darwinism, Nihilism, Behaviorism, and Atheism as the Materialists, Naturalists, Darwinists, Nihilists, Behaviorists, and Atheists do – so that their "science" will always be true; philosophers of science and scientists should define Science as observation, experience, replication, verification, knowledge, and truth. Experiencing and observing a phenomenon over and over again is how we come to know that it is real and truly exists.

We can see from the negating the consequent version of the Scientific Method that the people behind the denialistic philosophies of science are trying to trick us and deceive us with their hidden premises or primary axioms, which can never possibly be true. These people want us to take a blind leap of faith into nothing, just as they have done.

—

The denialistic philosophies of science are easy to falsify because they deny the existence of the evidence that is needed to verify them. In other words, the hidden premises of the denialistic philosophies cannot be verified with observational evidence. Instead, the denialistic philosophies have to be taken on blind faith alone, as being real and true.

By axiom, by law, by peer review, and by decree, the denialistic philosophies of science deny the existence of the evidence that falsifies them. *Denying the existence of evidence* is a logic fallacy. *Deliberating rejecting the evidence* that you personally don't like or that falsifies your personal beliefs is a logic fallacy.

On this planet, the scientists have set up a Censorship Arm that they call "peer review" consisting of PhDs who have been called, authorized, and set apart to bless and sanctify the lies, so that their deceptions and lies remain pure and undefiled and untouchable.

An essay like this one wouldn't make it past "peer review" because it falsifies their religion.

So, are you ready to finish this? It's gonna hurt if you are not ready for it. If you are ready, then let's burn these things down and be done with them. This can be a lot of fun, if you are ready for it.

—

BEHAVIORISM OR HARD DETERMINISM

Radical behaviorism or hard determinism is the philosophy of science that denies the existence of choice, free will, agency, or volition.

The hidden premise or primary axiom within radical behaviorism or hard determinism is the scientific truth claim that choice, free will, or agency does not exist. This is the hidden assumption upon which radical behaviorism or hard determinism is based. Therefore, find and observe some animate object or living object making a choice between two desirable competing options, and you have successfully falsified behaviorism or hard determinism.

Scientific Theory: According to radical behaviorism and hard determinism, choice, free will, agency, volition, psyches, and conserved quanta do not exist. According to radical behaviorism and hard determinism, conserved choices within conserved quanta do not exist.

Scientific Hypothesis: If radical behaviorism or hard determinism is true, then we will observe that nothing on this planet is capable of making a choice. There will be no difference between the inanimate objects and the animate objects if behaviorism or hard determinism is true.

Scientific Observations: We don't observe the complete absence of choice. We are constantly making choices every single day of our lives, and our legal system holds us accountable for the choices that we have made.

Scientific Conclusion: Therefore, radical behaviorism and hard determinism are false and have been successfully falsified by the Scientific Method and by negating the consequent.

We don't observe what radical behaviorism or hard determinism predicts that we should be observing; therefore, behaviorism is obviously false. We don't observe the non-existence of choice, therefore, radical behaviorism or hard determinism can't possibly be true.

Hard determinism or behaviorism is the easiest one to falsify because it is obviously false. The only thing standing in your way is that you have to be willing to choose to falsify it. If you are not willing to make that choice, then you are going to have a harder time convincing yourself that choice does in fact exist. Nevertheless, you are still choosing not to accept the obvious truth of the evidence; therefore, you are still making a choice, and choice does in fact exist.

—

CREATION BY CHANCE

Creation by Chance or Causation by Chance is the philosophy of science that denies the existence of Designers, Creators, Psyches, Intelligences, and Gods. The theory of evolution is Creation by Chance. The second law of thermodynamics is Creation by Chance or Creation by Random Disorder. Creation by Chance, as well as Creation Ex Nihilo, are different forms of atheism. They deny the existence of designers and creators, and instead, they make the claim that CHANCE designed and created everything that exists in this universe.

The hidden premise or primary axiom within Creation by Chance is the scientific truth claim that chance can design and create anything if given enough time to do so. This is the hidden assumption upon which Creation by Chance is based. Therefore, tossing coins, stirring soup, shuffling a deck of cards, hurricanes, and tornadoes should be producing new and interesting life forms right and left if Creation by Chance is true. They have definitely had enough time to do so. Spontaneous generation should be true if Creation by Chance, Creation by Random Disorder, Creation by Entropy, or Creation Ex Nihilo is true.

Creation by Chance and Creation Ex Nihilo are different types of Atheism. They are magic! Therefore, the other hidden assumption behind Creation by Chance is the primary axiom or scientific truth claim which states that Creators, Designers, Gods, and Intelligent People do not exist. We have already dealt with this one elsewhere, so we won't deal with it here, because it is obviously false.

Scientific Axiom or Law: The Primary Axiom of Science and Statistics states that chance cannot do choice, correlation, design, causation, nor creation. Chance cannot do causation. Chance is not causation. Chance and causation are mutually exclusive.

Scientific Theory: According to creation by chance, designers, creators, choosers, and causation do not exist. According to creation by chance, everything in this universe was designed and created by chance alone.

Scientific Hypothesis: If creation by chance is true, then we will observe chance, or abiogenesis, or chemical evolution, or spontaneous generation, or random disorder producing new and interesting life forms all around us all the time. We should observe the dust bunnies under our beds and the mud in our ponds coming alive all the time all around us, if creation by chance is true. We should observe hurricanes and tornadoes making functional and useful cars, computers, homes, and airplanes from scratch all the time all around us, if creation by chance is true. A new unique computer with new unique software and a new unique operating system should be blown onto your doorstep from time to time if creation by chance is true. The dishes should occasionally wash themselves if creation by chance is true. Macro-evolution (reptiles giving birth to birds) would actually be experienced and observed from time to time, if creation by chance were true.

Scientific Observations: We don't observe any of the things that creation by chance predicts that we should be observing. Furthermore, the primary axiom of science and statistics falsifies creation by chance. Chance cannot do causation. Chance is not causation. Chance and causation are mutually exclusive. Creation by chance is physically impossible. Creation by chance or spontaneous generation was falsified on this planet in 1859 by Louis Pasteur.

Scientific Conclusion: Therefore, creation by chance is false and has been successfully falsified by the Scientific Methods and by negating the consequent.

We don't observe any of the things that Creation by Chance predicts that we should be observing; therefore, Creation by Chance is obviously false. We don't observe chance designing and creating new and interesting things all around us all the time, because chance cannot do design and creation. Once chance starts doing causation, then it is no longer chance, but has become some type of deliberate choice, intentional design, sentient causation, or intelligent creation instead.

Remember, games of chance require an intelligent assist in order to exist. Games of chance do not and cannot design and run themselves. Someone psyche or someone intelligent has to make the coins, toss the coins, count the coins, and determine if they are all heads or all tails. Without an intelligent assist, chance does absolutely nothing.

—

For some people, this can be a fascinating exercise in philosophy and science – watching the different lies in "science" get identified and falsified one-by-one. Others will be mortified by this process, as their favorite lies are exposed and then debunked.

The ultimate goal is to get to where you can do this for yourself to every "truth claim" or "axiom" or "law" that comes your way.

The True Version of the Scientific Method:

Scientific Hypothesis: If Theory X is true, then we will observe Y.

Scientific Observations: We don't observe Y.

Scientific Conclusion: Therefore, Theory X is false and has been falsified by the Scientific Method and by negating the consequent.

As a scientist and a philosopher, your ultimate goal should be to get to the point where you can use the True Version of the Scientific Method anytime that you need to do so, to falsify all of the hidden premises and denialistic philosophies that come your way.

Most of the scientists on this planet can't see nor understand any of this; and, they don't realize that "science", "knowledge", "axioms", and "laws" are worthless if they are in fact demonstrably false or deliberate lies.

Remember, the denialistic philosophies of science deny the existence of the evidence that falsifies them. *Denying the existence of evidence* is a logic fallacy. *Deliberating rejecting the evidence* that you personally don't like or that falsifies your personal beliefs is a logic fallacy.

The denialistic philosophies of science are easy to falsify because they deny the existence of the evidence that is needed to verify them. In other words, the hidden premises of the denialistic philosophies cannot be verified with observational evidence. Instead, the denialistic philosophies have to be taken on blind faith alone, as being real and true.

—

RELATIVISM

Relativism is the philosophy of science that denies the existence of truth. According to the relativists, truth does not exist.

The hidden premise or primary axiom within Relativism is the scientific truth claim that states that "the truth does not exist". This is the hidden assumption upon which Relativism is based. Therefore, find and observe something that is absolutely true or always true, and you have successfully falsified Relativism.

By using logic, Relativism is self-defeating. If Relativism is true, and truth does not exist, then Relativism can't possibly be true; and therefore, Relativism has been falsified, and Relativism is false.

Quod erat demonstrandum!

However, let's play around with this as scientists, and see what else we might be able to learn from it.

Scientific Theory: According to relativism, truth does not exist.

Scientific Hypothesis: If relativism is true, then we will observe that there is no such thing as conservation of energy, conservation of psyche or life force, or conservation of quantum information within psyche and energy. Conservation means that the observed phenomenon remains absolutely true all the time. The laws of conservation within physics conserve truth, because they conserve information within quanta or psyches. If relativism is true, then we will observe that there is no such thing as memories, because there is absolutely no truth to be remembered.

Scientific Observations: Psyche or intelligence has been observed to be eternal, and everlasting, and conserved. Energy has also been observed to be eternal, everlasting, and conserved. Energy is absolutely true all of the time, because energy is conserved. Quantum information conserved within psyche or energy or a quantum has to be true, or quantum mechanics and quantum field theory can't possibly be true. These three – conservation of energy, conservation of psyche or intelligence, and conservation of quantum information within psyches or quanta – have to be absolutely true all the time, or we wouldn't exist.

Scientific Conclusion: Therefore, relativism is false and has been falsified by the Scientific Method and by negating the consequent.

We don't observe what relativism predicts that we should be observing; therefore, relativism is obviously false. We don't observe the non-existence of truth, therefore, relativism can't possibly be true.

—

SCIENTISM

Scientism is the philosophy of science that denies the existence of truth outside of science or separate from science.

Scientism is the philosophical belief that science or the scientific method is the ONLY way to find and know the truth.

The hidden purpose behind Scientism is to convince you that it is impossible to see, find, and know God. The hidden corollary or hidden truth claim within Scientism is the claim that there will NEVER be any scientific proof of God's existence.

The hidden premise or primary axiom within Scientism is the scientific truth claim that "science" defined as materialism, naturalism, nihilism, and atheism will NEVER be able to prove the existence of God because it can't be used to prove the existence of anything. This is the hidden assumption upon which Scientism is based. Therefore, find and observe better and faster ways for finding and knowing the truth, such as first-hand observation and experience, and you have successfully falsified Scientism.

Through observation and experience, a human being can go directly to knowing the truth, and he or she doesn't even have to design and run a science experiment in order to do so. Observation and experience are vastly superior to science experiments and the scientific methods; and, they are a lot more cost effective too.

Of course, the other way to falsify Scientism is to use science and the scientific methods to develop convincing Scientific Proofs of God's Existence. Since the hidden purpose of Scientism is to convince you that science will never be able to prove that God exists, using science to prove that God exists will falsify Scientism.

Currently, our scientists have defined "science" as materialism, naturalism, nihilism, and atheism so that "science" cannot be used to prove that God exists. Consequently, redefining Science as observation, experience, verification, replication, truth, and knowledge will also falsify Scientism or put lie to Scientism. All you have to do is to see God, touch God, and talk with God – or choose to trust someone who has done so – and you have successfully falsified Scientism.

Scientific Theory: According to scientism, truth does not exist outside of science or separate from science.

Scientific Hypothesis: If scientism is true, then we will observe that God does not exist.

Scientific Observations: It is impossible to observe that God does not exist. If you can go into every dimension, alternative reality, phase, level, time, universe, and space in order to see for yourself that God does not exist and then take us with you so that we can see for ourselves that God does not exist, then you are a God, and God does indeed exist.

Scientific Conclusion: Therefore, scientism is false and has been falsified by the Scientific Method and by negating the consequent.

We can't observe what scientism predicts that we should be observing; therefore, scientism is obviously false. We can't observe the non-existence of God, therefore, scientism can't possibly be true. The only thing we can do is to observe God, see God, experience God, touch God, and verify that God does in fact exist.

We can directly observe and experience the truth and immediately know that it is true; therefore, scientism's claim – that materialism, naturalism, and the affirming the consequent version of the scientific method is the ONLY way to find and know the truth – is obviously false.

We can also develop convincing Scientific Proofs of God's Existence that will put lie to the hidden premises within scientism.

Remember, by axiom, by law, by peer review, and by decree, the denialistic philosophies of science deny the existence of the evidence that falsifies them. *Denying the existence of evidence* is a logic fallacy. *Deliberating rejecting the evidence* that you personally don't like or that falsifies your personal beliefs is a logic fallacy.

The denialistic philosophies of science are easy to falsify because they deny the existence of the evidence that is needed to verify them. In other words, the hidden premises of the denialistic philosophies cannot be verified with observational evidence. Instead, the denialistic philosophies have to be taken on blind faith alone, as being real and true.

—

PHYSICAL REDUCTIONISM

Physical Reductionism or Atomism is the philosophy of science that denies the existence of anything smaller than a physical atom. Atomism is a type of materialism.

Physical Reductionism or Atomism is the scientific belief that everything reduces to a physical atom and that there is nothing smaller than a physical atom. Atomism is the scientific belief that physical matter or a physical atom is the fundamental unit of reality and existence.

Atomic theory, quantum mechanics, and quantum field theory falsify this belief. In this particular case, we can use Hard Science to falsify "science" or the scientific theories that are false.

The hidden premise or primary axiom within Reductionism is that nothing smaller than a physical atom exists. This is the hidden assumption upon which Physical Reductionism or Atomism is based. Therefore, find and observe something smaller than an atom such as a psyche, photon, or quantum, and you have successfully falsified Atomism.

Scientific Theory: According to atomism and physical reductionism, anything smaller than a physical atom does not exist.

Scientific Hypothesis: If physical reductionism is true, then we will observe that photons, quantum waves, quantum fields, and psyches or conserved quanta do not exist.

Scientific Observations: We don't observe the non-existence of quanta, or psyches, or packets of energy. The psyches or quanta, packets of energy, photons, quantum waves, and quantum fields that we see or experience all around us and all through us obviously exist. It is obvious that a physical atom is made from a bunch of different non-physical quanta held together by a bunch of different non-physical forces and fields.

Scientific Conclusion: Therefore, physical reductionism or atomism is obviously false and has been falsified by the Scientific Method and by negating the consequent.

We don't observe what atomism predicts that we should be observing; therefore, physical reductionism is obviously false. We don't observe the non-existence of quanta or packets of energy, therefore, atomism or physical reductionism or materialism can't possibly be true.

That's just the way it is!

A quantum is a psyche or a packet of energy. A psyche makes, transmits, receives, processes, and stores quantum waves and quantum information. Such a thing has actually been experienced and observed. Experiencing and observing massless quanta or psyches, massless and entropyless quantum waves or photons, and massless and entropyless quantum fields falsifies atomism and nihilism, which state that these things do not exist.

—

CREATION EX NIHILO

Creation Ex Nihilo is the philosophy of science that denies the existence of conserved quanta (psyches), conserved quantum information, quantum consciousness, quantum designers, and quantum creators.

The hidden premise or primary axiom within Creation Ex Nihilo is the scientific truth claim that "nothing" can design and create things at will from nothing. This is the hidden assumption upon which Creation Ex Nihilo and Atheism are based. Creation Ex Nihilo and Atheism are the same thing. They each deny the existence of Creators and Intelligent Designers. Creation Ex Nihilo is the scientific theory that is based upon nothing, which means that it has nothing in terms of evidence to support it. Creation Ex Nihilo or Atheism is patently absurd and obviously false.

Scientific Theory: According to creation ex nihilo, quantum designers and transdimensional creators do not exist.

Scientific Hypothesis: If creation ex nihilo is true, then we will observe cars, computers, buildings, and life forms springing into existence from nothing in a vacuum all around us all the time.

Scientific Observations: We don't observe anything like what creation ex nihilo predicts that we should be observing. We don't observe magic taking place all around us all the time.

Scientific Conclusion: Therefore, creation ex nihilo is false and has been falsified by the Scientific Method and by negating the consequent.

We don't observe what creation ex nihilo predicts that we should be observing; therefore, creation ex nihilo is obviously false. We don't observe the non-existence of creators and intelligent designers; therefore, creation ex nihilo or atheism can't possibly be true.

There you have it.

—

THE THEORY OF EVOLUTION

The Theory of Evolution is every denialistic philosophy of science rolled into one.

The Theory of Evolution is Creation by Chance. The Theory of Evolution is Materialism and Naturalism. The Theory of Evolution is Creation by Entropy or Creation by Random Disorder. The Theory of Evolution is Creation by Death or Creation by Natural Selection. Natural selection doesn't touch our genes. Natural selection simply sits around and waits for us to die. The Theory of Evolution really is Creation by Death. The Theory of Evolution is spontaneous generation or abiogenesis. The Theory of Evolution is Atheism and predicts that designers and creators do not exist. The Theory of Evolution is Creation Ex Nihilo or Magic.

The Theory of Evolution is practically every denialistic philosophy of science combined into one. The Theory of Evolution is comprised of everything that we have used the Scientific Method and negating the consequent to falsify. The Theory of Evolution is comprised of almost all of the different hidden premises within all of the different denialistic philosophies of science that we have ever encountered. The Theory of Evolution is the king of denialistic philosophies.

The hidden premise or primary axiom within the Theory of Evolution is the scientific truth claim that designers and creators do not exist, and that chance can design and create things at will if given enough time to do so. These are a couple of the hidden assumptions upon which the Theory of Evolution is based. In fact, the Theory of Evolution is based upon every hidden premise found in every denialistic philosophy of science. Therefore, falsify any denialistic philosophy or any hidden premise that you can find, and you have successfully falsified the Theory of Evolution, Chemical Evolution, Macro-Evolution, Creation by Chance, Creation by Entropy, Creation by Disorder, Creation by Magic, and Creation by Death.

Scientific Hypothesis: If the theory of evolution is true, then we will observe that intelligent designers, conserved quantum information, transdimensional non-physical quanta, conserved quanta or psyches, and creators do not exist.

Scientific Observations: We don't observe any of the things that the theory of evolution predicts that we should be observing. We don't observe creation by death. We don't observe creation by chance. We don't observe creation by entropy or random disorder. We don't observe the non-existence of designers and creators. We don't observe chemical evolution, abiogenesis, spontaneous generation, or macro-evolution in the lab or in the wild.

Scientific Conclusion: Therefore, the theory of evolution is false and has been falsified by the Scientific Method and by negating the consequent.

We don't observe what theory of evolution predicts that we should be observing; therefore, the theory of evolution or naturalism is obviously false. We don't observe the non-existence of intelligent designers; therefore, the theory of evolution can't possibly be true.

Quod erat demonstrandum!

Identify the hidden premises or the primary axioms of the theory of evolution, falsify them, and that's the end of the theory of evolution. May it rest in pieces.

—

THE SECOND LAW OF THERMODYNAMICS

The second law of thermodynamics is the philosophy of science or interpretation of science that denies the existence of conservation. The second law of thermodynamics was created to falsify the first law of thermodynamics or the Conservation of Energy and Psyche. Psyches are conserved quanta, and the second law of thermodynamics denies the existence of conserved quanta, the conservation of energy and psyche, as well as existence of the quantum fields and the perpetual motion cycle.

The second law of thermodynamics is a type of materialism, naturalism, nihilism, and creation by chance.

This planet's second law of thermodynamics states that the total amount of entropy or disorder in this universe is constantly increasing and that the total amount of entropy or disorder can never decrease or go to zero. The second law of thermodynamics is conservation of disorder, or conservation of entropy.

The equations for heat and entropy falsify this claim. We see from the equations for heat and entropy that whenever mass goes to zero, entropy also goes to zero. Heat, thermodynamics, and entropy are functions of mass or resistance to acceleration. No mass, then no heat and no entropy. No mass, then no thermodynamics. When mass goes to zero, entropy or heat or thermodynamics goes to zero along with it. Mass has heat storage capacity. No mass, then no heat; and, no heat or no mass, then no entropy. That's what the equation for heat and entropy are trying to tell us. There is no mass, heat, thermodynamics, or entropy at the quantum level among the massless, entropyless, and non-physical quantum waves and quantum fields. Mass, heat, thermodynamics, resistance to acceleration, mass's heat storage capacity, or entropy is purely a physical phenomenon.

Disorder is the wrong definition for thermodynamics or entropy. There is NO correlation whatsoever between heat and disorder. Disorder doesn't make heat, and disorder doesn't eliminate heat either. Every part of the second law of thermodynamics is a demonstrable lie.

The second law of thermodynamics predicts that we shouldn't be here, yet here we are. The second law of thermodynamics predicts that we should see nothing but disorder and chaos all around us; but instead, we observe nothing but order and organization all throughout our universe. The second law of thermodynamics predicts that the massless and entropyless and heatless quantum waves, photons, and quantum fields should not exist.

The second law of thermodynamics is just another form of Materialism, Naturalism, Creation by Chance, Creation by Random Disorder, Creation by Entropy, Creation by Death, Creation Ex Nihilo, and Atheism.

The hidden premise or primary axiom within the Second Law of Thermodynamics is the scientific truth claim that the transdimensional, spiritual, or quantum does not exist. The Second Law of Thermodynamics states as one of its hidden premises that the conservation of energy, as well as the conservation of quantum information within psyche, does not exist. The Second Law of Thermodynamics states that the Perpetual Motion Cycle does not exist. The Second Law of Thermodynamics states that the massless and entropyless quantum waves, photons, and quantum fields do not exist. These are some of the hidden assumptions upon which the Second Law of Thermodynamics is based. Therefore, find and observe something like the massless and entropyless Quantum Fields, the Perpetual Motion Cycle, Conservation of Energy, Exergy, or Syntropy, and you have successfully falsified the Second Law of Thermodynamics.

Scientific Theory: According to the second law of thermodynamics, conservation does not exist – unless you are talking about conservation of entropy, conservation of disorder, conservation of chaos, and conservation of physical matter.

Scientific Hypothesis: If the second law of thermodynamics is true, then we will observe that order, organization, and life do not exist. If the second law of thermodynamics is true, then we will observe that we do not exist and that everything is random chaos instead.

Scientific Observations: The order, organization, and syntropy have to exist somewhere within the fabric of this universe, or we wouldn't exist. We don't observe any of the things that the second law of thermodynamics predicts that we should be observing. We don't observe heat death. "Heat death" is steroids at the quantum level. Superconductors function best at or near absolute zero temperature. We don't observe proton decay. We don't observe an ever-encroaching gray goo coming in at us from all sides. We don't observe the non-existence of order and organization. We don't observe the conservation of disorder or entropy. We don't observe any of the things that the second law of thermodynamics predicts that we should be observing.

Scientific Conclusion: Therefore, the second law of thermodynamics is false and has been falsified by the Scientific Method and by negating the consequent.

We don't observe what the second law of thermodynamics predicts that we should be observing; therefore, the second law of thermodynamics is obviously false. We don't observe the non-existence of order and organization; therefore, the second law of thermodynamics can't possibly be true. We don't observe the non-existence of life; therefore, the second law of thermodynamics can't possibly be true.

The quantum fields are perfectly organized and perfectly conserved perpetual motion machines. The quantum fields are massless and entropyless. The quantum fields are syntropic, exergic, and conserved. Quantum Field Theory falsifies the second law of thermodynamics. The Perpetual Motion Cycle ($E = mc^2$) also falsifies the second law of thermodynamics. The equations for heat and entropy falsify the second law of thermodynamics. The second law of thermodynamics predicts that you shouldn't be here reading this right now. Your very existence is scientific proof that the second law of thermodynamics is false. The second law of thermodynamics will never be true as long as the quantum fields exist. You would have to destroy the quantum fields in order to make the second law of thermodynamics even remotely true.

Order and organization obviously exist, or we wouldn't exist.

Anything that is obviously made obviously has a Maker who made it. Anything that is obviously organized obviously has an Organizer who organized it. Anything that is obviously designed obviously has a Designer who designed it.

The quantum fields were obviously designed, organized, and made; therefore, the quantum fields obviously have a Designer, Organizer, and Maker who designed them, organized them, and made them. Physical atoms, proteins, genes, genomes, eyes, brains, and life forms were obviously designed, organized, and made; therefore, these things obviously have a Designer, Organizer, and Maker who designed them, organized them, and made them.

The truth is obvious and self-evident. We keep seeing and experiencing the truth all around us and all through us. In contrast, once we have successfully identified the hidden premises behind the denialistic philosophies of science, it is immediately obvious that those hidden premises are false and can be easily falsified with observational evidence.

This is the pinnacle of the Philosophy of Science. Identify the hidden premises within a denialistic philosophy, falsify them, and then move on to the next one. If you successfully identify and eliminate everything that is false, then only the truth will remain; and, the truth that remains will consist of everything that has ever been experienced and observed by someone, somewhere, sometime.

—

CONCLUSION

The denialistic philosophies of science are easy to falsify because they deny the existence of the evidence that is needed to verify them. In other words, the hidden premises of the denialistic philosophies cannot be verified with observational evidence. Instead, the denialistic philosophies have to be taken on blind faith alone, as being real and true.

This process of "negating the consequent" requires a bit of effort and thought, which explains why nobody has ever done it before now. They also haven't done it because nobody wants to prove that materialism, naturalism, and their derivatives are false. Instead, these people have convinced themselves that materialism, naturalism, nihilism, and atheism can't be falsified.

You only "negate the consequent" if you want to find and know the truth.

I might be the only person on this planet who knows what a denialistic philosophy of science is. As far as I can tell, I am the only person on this planet who knows how to identify the hidden premises within the denialistic philosophies of science, and then falsify them by negating the consequent. However, if you have been following along, you now know how to do it too, and these people will never be able to deceive you ever again.

I just set you free.

I just exposed the lie. The Materialists, Naturalists, Darwinists, Nihilists, Behaviorists, and Atheists deliberately define "science" as Materialism, Naturalism, Darwinism, Nihilism, Behaviorism, and Atheism so that their "science" will always be true and so that their "science" cannot be falsified by observational evidence. They cheat. Identify their hidden premises, falsify those hidden premises with observational evidence or experimental evidence, and that's the end of the denialistic philosophies of science.

By axiom, by fiat, by law, by peer review, and by decree, the denialistic philosophies of science deny the existence of the evidence that falsifies them. *Denying the existence of evidence* is a logic fallacy. *Deliberating rejecting the evidence* that you personally don't like or that falsifies your personal beliefs is a logic fallacy.

Begging the question, these people use their denials of evidence as the hidden premises or as the scientific evidence within their theories. *Using one's denials as scientific evidence* is a logic fallacy. You can't use the non-existence of something as scientific evidence, yet that's precisely what the creators of these denialistic philosophies do. They use their ignorance or their denial of evidence as scientific evidence and scientific proof that what they are telling us is real and true. These people have no observational evidence to support their beliefs, so they make up some "evidence" out of thin air and hope that we will fall for the ruse.

Science is observation, experience, evidence, truth, and knowledge; therefore, *assuming that that evidence does not exist* is a logic fallacy. *Assuming that nothing can falsify your beliefs* is a logic fallacy.

Rejecting scientific evidence, observational evidence, verified evidence, or experiential evidence because it falsifies your beliefs is a logic fallacy. You should adjust your beliefs and make them match with the evidence instead. Remember, *denying the existence of evidence* is a logic fallacy. *Deliberately rejecting evidence* because you don't like it is a logic fallacy.

Refusing to allow evidence into evidence is a logic fallacy. The denialistic philosophies refuse to allow the evidence that falsifies them into evidence. Science is observation, experience, evidence, truth, and knowledge. Therefore, all of the evidence should be allowed into evidence; and then, we should try to figure out what that evidence means and try to pursue a preponderance of that evidence. If a phenomenon is a common one, then it has meaning, because it is real and truly exists.

When it comes to the Materialists, Naturalists, and Atheists, the main flaw in their logic or reasoning is that they deny the existence of the evidence that falsifies their beliefs, and whenever they encounter that type of evidence, they refuse to allow it into evidence. Atheism of any kind is based upon a refusal to look at evidence, particularly the evidence that falsifies their beliefs. The denialistic philosophies are head-in-the-sand philosophies.

Whenever I encounter one of these denialistic philosophies of science, that is denying the existence of one thing or another and treating their denial as if it were science or scientific evidence, I automatically identify the hidden premises or the false axioms that these people are using to prove that they are right, and I immediately start falsifying these hidden premises or deliberate lies with observational evidence or experiential evidence to the contrary. Every denialistic philosophy has a hidden premise or primary axiom that has no supporting evidence and is obviously false. Identify it and falsify it, and then move on to the next one. If you successfully identify and eliminate everything that is false, then only the truth will remain; and, the truth that remains will consist of everything that has ever been experienced or observed. This is Philosophy of Science 101.

The Materialists, Naturalists, and Darwinists on this planet can't deceive me anymore, because I know that they are lying to me and know how to identify and falsify the hidden premises that they are using to deceive us. It's too late for me now. I can't go back to my materialism, naturalism, nihilism, and atheism because now I know why I was wrong to hold such beliefs. When I was finally ready for the truth, I wish I would have had an essay like this one to point me to the truth; but alas, I had to figure it out all on my own because I didn't know where to look at the time.

I'm seeing things differently now that I have overcome my materialism, naturalism, nihilism, and atheism.

I have observed that the false is falsified by the truth, and that the truth is repeatedly and constantly verified, experienced, and observed by Someone Psyche or Someone Intelligent somewhere, somehow, sometime. If a phenomenon has never been experienced nor observed by anyone, then it can't possibly be real or true. The non-existence of the non-physical has never been experienced or observed by anyone; therefore, it can't possibly be real or true, because the non-physical or your quantum or your energy or the light has always existed and will always exist. In other words, these concepts have always been true and will always be true so long as the energy, quantum waves, quantum fields, quantum information, and quanta or psyches exist.

Mark My Words

© January 2020
All Rights Reserved.

COMPARATIVE SCIENCE

The Real Scientists on our earth are thousands of years ahead of the materialists, naturalists, nihilists, and atheists because the Real Scientists don't deny the existence of the things that have been experienced, observed, and verified.

Our scientists have yet to discover the Primary Axiom of Science and Statistics. A few of our scientists and statisticians are starting to dance around it and getting close to it, but they have yet to declare it openly to our science community at-large. As far as I can tell, nobody on our planet has discovered the Primary Axiom of Science and Statistics. They aren't looking for it; and currently, they don't want it. Our scientists prefer the deceptions and the lies instead.

Collectively, our scientists as a whole are centuries away or even millennia away from actively embracing and accepting the Primary Axiom of Science and Statistics, because they have chosen to embrace and promote the obvious falsehoods in philosophy and "science" rather than finding and accepting the Truths in Science. Our culture and our scientists can't have the Truth in Science if they have chosen to accept and promote the deceptions and the lies instead. That's just logical common sense. Our scientists have no idea that their beliefs are wrong, or why they are wrong.

In this essay, I try to compare the Truths in Science with the obvious falsehoods that our scientists have designed and embraced.

Science is Knowledge of the Truth. If a concept is false or a concept has been falsified, then it isn't Science because it isn't true. Science is Observation and Experience. If a phenomenon or a theory has never been experienced nor observed by anyone in the universe, not even God, then it can't possibly be real or true. A phenomenon has to be experienced and observed by someone sometime somewhere, or it can't possibly be Real and True. This is logical common sense. In contrast, our scientists take things that have never been experienced nor observed by anyone, and they erroneously call these things "science", "knowledge", or "truth".

There's a big difference between Science and technology. A society like ours can be technologically advanced yet have no knowledge of the truth. There's a difference between the two. Having technology doesn't mean that you know how it works or know what makes it work. Our scientists are ever learning but never able to come to a knowledge of the truth, because they have embraced obvious falsehoods instead.

Our scientists as a group reject Transdimensional Physics and have chosen to believe that the transdimensional or the non-physical does not exist.

Transdimensional means non-local or non-physical. Transdimensional Physics is what our scientists call Quantum Mechanics – it is the scientific study of how energy or light works in the Non-Physical Realm or Transdimensional Realm. Quantum Mechanics, Quantum Field Theory, Conservation of Quantum Information, and the Conservation of Energy are the closest that our scientists have gotten to the Truths in Science; but, they still have a long way to go before they figure out how everything really works, because they only allow themselves to go so far and no further when it comes to Science, or a Knowledge of the Truth. Collectively, our scientists refuse to allow all of the evidence into evidence, so they can only go so far before they hit a dead-end and run into unresolvable problems with their theories and their falsified ideas and beliefs.

Real Scientists tend to define Science as "knowledge, consisting of the things that have actually been experienced and observed". That's one of the reasons why our Real Scientists are thousands of years ahead of the materialists, naturalists, nihilists, and atheists. The Materialists, Naturalists, Darwinists, Nihilists, and Atheists actively and deliberately reject the observations and the experiences of the human race as a whole. That's why they get things wrong.

WHY OUR SCIENTISTS GET THINGS WRONG

People on our earth used to define Science as "knowledge"; but in our modern era, our scientists now define "science" axiomatically as Materialism, Naturalism, Darwinism, Nihilism, Atheism, Creation by Chance, Creation Ex Nihilo, Spontaneous Generation, Abiogenesis, Chemical Evolution, the Theory of Evolution, or the Second Law of Thermodynamics. Our scientists literally define "science" as being everything that has been falsified by Science, or by Observation and Experience.

Defining "science" as Materialism and Naturalism and Darwinism is *circular reasoning*, which is a logic fallacy. It is also *jumping to conclusions* and *begging the question*. Our scientists start with the conclusion that "science" is Materialism, Naturalism, Darwinism, and Atheism; and then *begging the question*, they use Materialism, Naturalism, Darwinism, and Atheism as proof or as evidence that their "science" is true. It's illogical, but that's what they do.

Our scientists define Science as everything that has been falsified by Science, or Knowledge, or Observations. Our scientists define Science as everything that is known to be false.

Quantum Mechanics, Quantum Field Theory, Non-Locality, Action at a Distance, Quantum Waves or Photons or Tachyons, and Transdimensional Physics ONLY make sense from a non-physical spiritual perspective; and, our scientists don't even know it.

The Theory of Evolution and the Second Law of Thermodynamics are obviously false. They have never been experienced nor observed, which means that they can't possibly be real or true. None of the things which they predict have ever been experienced or observed, which means that they are false and have in fact been falsified by Negating the Consequent, the most powerful form of the Scientific Method.

WE KNOW that Quantum Field Theory and Quantum Mechanics are true because we have experienced and observed everything that they predict that we should be observing and experiencing. That's the way Science should work.

Negating the Consequent is philosophically valid and logically sound. That's why Negating the Consequent is the most powerful form of the Scientific Method.

None of the scientists on our planet know what Negating the Consequent is, that it exists, how it works, how to use it, or why it is the most powerful and the most reliable form of the Scientific Method. You can't use the Scientific Method to find and know the truth, if you don't know what Negating the Consequent is, how it works, and how to use it to find the truth.

That's why our scientists are always getting everything wrong and don't even know it.

CHANCE CANNOT DO CAUSATION

The theory of evolution is defined as Creation by Chance or Creation by Random Processes. This idea is obviously false, because chance cannot design or create anything.

Our leading scientists have erroneously chosen to believe that the Big Bang, the Prime Event, or the Origin of our Physical Universe was caused by random fluctuations in entropy, random quantum fluctuations, or by sheer chance alone. This idea is obviously false. Creation by Chance or Creation by Random Luck is obviously false, so there is no way in our universe that it could ever be Science or Knowledge. It has never been experienced nor observed. Chance does not do causation.

Chance by definition is the absence of causation. If there is no causation involved in a specific event, then you are looking at pure chance alone. Chance cannot do causation. Chance and causation are mutually exclusive. They falsify each other. They preclude each other. Therefore, any type of Creation by Chance or Chance Causation is obviously false, which means that the theory of evolution and the second law of thermodynamics are obviously false. Finding the truth is that simple – you just eliminate everything that was produced by chance alone, because CHANCE cannot do truth, choice, correlation, design, creation, or causation.

Only non-local non-physical intelligence, or psyche, or life force, or quantum consciousness is capable of making choices, processing quantum information, producing quantum waves, collapsing wave functions, and therefore capable of doing causation at every level of existence and reality including the quantum level or the transdimensional non-physical level. This is obviously true, because chance cannot do choice, creation, correlation, or causation. This is obviously true because it has been experienced and observed. Psyche or intelligence or consciousness has

actually been caught in the act, both at the quantum non-physical level as well as the physical level. Psyche has been experienced and observed.

Therefore, Real Scientists don't reject nor deny the existence of Quantum Consciousness, Psyche, Intelligence, Nature's Psyche, or Universal Consciousness and Intelligence as most of our scientists deliberately do. Intelligence obviously exists. Choice obviously exists. Psyches or Quantum Wave Processors obviously exist. Therefore, it is obvious that some type of Quantum Intelligence or Universal Consciousness or Massless Non-Physical Psyche exists, who is capable of making, transmitting, receiving, processing, analyzing, transforming, and storing Quantum Information or Quantum Waves. Such a thing has to exist, or Quantum Waves and Quantum Information wouldn't exist. There has to be something at the quantum level that is capable of collapsing wave functions thereby transforming omnipresent quantum waves or photons into some type of localized mass or heat instead. Psyche is it. Psyche is a conserved quantum or a conserved photon who is capable of processing, transforming, transceiving, and storing quantum waves or conserved quantum information.

It is obvious that some type of universal consciousness exists. There will always be a great deal of debate as to the true identity of this Universal Consciousness, Universal Psyche, Nature's Psyche, Servitor in the System, Ghost in the Machine, or Universal Intelligence; but, there is no doubt that it exists. The existence of Quantum Information or Quantum Waves proves that Psyche or Intelligence exists at the quantum level in the Non-Physical Transdimensional Realm. Some type of Conserved Quantum Information Processor or Conserved Quantum Wave Processor has to exist, or Conserved Quantum Information and Physical Laws and Quantum Waves wouldn't exist. This is Logic 101. The order and organization in this universe prove that Nature's Psyche, the Universal Chooser, the Ultimate Quantum Processor, or the Ultimate Causal Agent exists. You can debate for thousands of years as to its true identity, but there is no doubt that it exists. Its existence is obvious.

The Real Scientists and Open-Minded Scientists don't deny the existence of Conserved Quanta, Conserved Quantum Wave Processors, Conserved Quantum Information Processors, Conserved Quantum Intelligences, or Conserved Quantum Psyches as the materialists, naturalists, nihilists, and atheists currently do.

Some type of Universal Psyche or Conserved Quantum Consciousness has to exist; or, Quantum Information, Choice, Intelligence, and Quantum Waves (Thoughts) would not exist. We think and choose both in the Non-Physical Quantum Realm and in this Physical Realm; therefore, the non-physical transdimensional Psyche, or the Quantum Chooser, or the Quantum Information Processor obviously exists. This is what has actually been experienced and observed. Is it not? Something at the quantum level is choosing to make the quantum waves or the quantum information in the first place. Likewise, that same something at the quantum level is choosing to collapse the quantum wave function, thereby producing a physical particle instead. Something at the quantum level is running or operating this Perpetual Motion Cycle. Psyche is it.

This is obviously true. Quantum Intelligence, or Transdimensional Psyche, or a Quantum Information Processor, or a Quantum Wave Processor, or Universal Consciousness obviously exists. There's no sense denying it, as our Materialists and Naturalists do. Psyche or intelligence is identified by the choices that it makes, both at the non-physical quantum level and at the physical level. Psyche is the ultimate cause of everything. Psyche is the ultimate causal agent. Psyche is the thing that does selective attention, which means that from all of the sensory input that psyche is receiving from its physical body or spirit body, psyche chooses what it wants to pay attention to. Psyche is also identified by the effects that it has on physical matter. Psyche is the thing that influences physical atoms at the quantum level. Psyche is the thing that handles communication between physical atoms, and within physical atoms. Psyche affects physical matter.

WHO MAKES THE QUANTUM WAVES AND COLLAPSES THE WAVE FUNCTION?

Transduction, in respect to psychology, refers to the process of converting one form of energy into a different form of energy. Our scientists collectively reject transduction or energy transformation where physical matter is concerned.

Our scientists erroneously teach, preach, and believe that physical matter is conserved. They make this error because they don't realize that a physical atom is made up of many different forms of energy, and that the Controlling Psyches within Nature can transform physical matter or a physical atom into many different massless, entropyless, and non-physical forms of energy – anytime and anywhere that Nature's Psyche chooses to do so. Nature's Psyche is constantly transforming the physical matter, mass, entropy, heat, resistance to acceleration, and mass's heat storage capacity within our stars INTO massless, entropyless, chargeless, heatless, syntropic, exergic, infinite acceleration Quantum Waves or Photons or Light all the time. This phenomenon has been experienced and observed. Has it not?

Our scientists don't know that there is nothing sacred or conserved about physical matter. Physical matter and physical atoms can be transformed into massless, entropyless, heatless, omnipresent quantum waves or photons anytime that Nature's Psyche chooses to do so. Our scientists haven't yet discovered the Law of Psyche, so they have no scientific explanation for how quantum waves get made and collapsed. They think it happens by magic or by blind luck, and it doesn't. They are wrong. Our scientists are always wrong at a fundamental level, because their philosophy of science or interpretation of science is fundamentally wrong.

A couple of our scientists have correctly observed that it is Nature's Psyche or the "Controlling Psyches within the Energy" who are the ones who are making the quantum waves in the first place and then later collapsing those wave functions and thereby transforming non-local, massless, entropyless, non-physical, infinite acceleration, omnipresent quantum waves or photons INTO localized mass, heat, resistance to acceleration, or mass's heat storage capacity (entropy) instead.

This is the Perpetual Motion Cycle, and our scientists haven't yet discovered the Perpetual Motion Cycle – what it is and how it works. Our scientists have chosen and embraced the second law of thermodynamics instead, because in their minds it successfully falsifies Intelligence, Psyche, Quantum Consciousness, and the Perpetual Motion Cycle – the scientific truths that our scientists have formally and officially rejected.

It's not so obvious who designed the Quantum Fields, but it is obvious that it was Nature's Psyche or the "Controlling Intelligences within the Energy" who made the Quantum Fields and now keep the massless and entropyless and heatless Quantum Fields in existence.

According to the LAW of Psyche, every Psyche has a certain amount of energy that's under its control, and that Controlling Psyche forms or transforms the energy under its control into anything that it wants that energy to be, anytime and anywhere that it chooses to do so. Nature's Psyche makes the quantum fields and the quantum waves, and Nature's Psyche collapses the wave function thereby transforming energy or quantum waves INTO physical matter or heat instead. These realities and truths apply to all of us in this particular universe, so there's no sense denying it.

Psyche or Intelligence or Quantum Consciousness is the ultimate cause of everything that was ever designed and made. Psyche, Intelligence, Quantum Consciousness, the Spark of Life, or Life Force is the ultimate causal agent in this multiverse. Psyche is the only thing that we know of that can design and make things at the quantum level in the Transdimensional Realm or the Non-Physical Realm – including physical matter. A physical atom is made from a whole bunch of different non-physical forces, fields, and energy. Its physicality is simply an illusion. Anything that obviously made obviously has a Maker who made it. Physical atoms, quantum waves, and quantum fields are obviously made, which means that they obviously have some type of Psyche, Intelligence, or Maker who made them. This is obviously true, so there's no sense denying it.

Psyche or Intelligence is identified by the choices that it makes and the things that it makes. Anything that is obviously made obviously has a Maker who made it. Quantum fields and quantum waves are obviously made; therefore, they obviously have a Maker or a Psyche who made them. Physical atoms are obviously made from non-physical forces, fields, and energy; therefore, physical atoms obviously have some type of non-physical Psyche or Maker who makes them.

Our scientists have stated that the LAW of Quantum Information Conservation must be true in order for Quantum Mechanics and Quantum Field Theory to be true. In order for the LAW of Quantum Information

Conservation to be true, there must be some type of non-physical conserved psyche or conserved intelligence at the quantum level who is capable of making, transmitting, receiving, processing, analyzing, transforming, and storing Quantum Information or Quantum Waves. This is obviously true because the Quantum LAW of Information Conservation is obviously true and has to be true in order for us to exist in the first place. Psyche or Non-Local Consciousness is the ultimate Quantum Wave Processor, Quantum Information Processor, and Quantum Memory Storage Device. Conserved Psyches or Conserved Quantum Information Processors obviously have to exist, or the LAW of Quantum Information Conservation could never possibly be true. This is logical common sense.

Anything that is obviously made obviously has a Maker who made it. The Quantum Fields were obviously made. Therefore, the Quantum Fields obviously have a Maker who made them.

Nature's Psyche or the "Controlling Intelligences within the Energy" is the ONLY thing that we know of, who would be capable of making the massless, entropyless, heatless, and non-physical Quantum Fields out of exergy or the available energy within this universe.

According to Quantum Field Theory, the Gods or the Controlling Psyches in Nature had to design and make the Quantum Fields BEFORE they could make and sustain physical matter. No Quantum Fields, then no physical matter. The Quantum Fields are obviously massless, heatless, entropyless, conserved, syntropic, non-physical, immaterial, and intangible. The Quantum Fields are needed to make physical matter and to sustain the existence of physical matter. Therefore, there has to be something non-physical that is capable of designing and making Quantum Waves and Quantum Fields; otherwise, physical matter would not exist.

Now, it's fully possible that the Controlling Psyches in Nature have always known how to make and sustain the Quantum Fields. Therefore, it's possible that the Quantum Fields have always existed and that some type of unorganized matter has always existed. If that's the case, then it ends up being the Gods who organized the planets, stars, and galaxies out of all of that unorganized matter that existed BEFORE the Gods arrived in this part of the multiverse. It could have gone either way; but either way, something quantum and non-physical has to KNOW how to make and sustain Quantum Fields so that the physical atoms can actually exist in the first place.

Our scientists deny the existence of these obvious truths. When it comes to most of the scientists on this planet, their Materialism, Naturalism, Darwinism, and Atheism are holding them back and keeping them in the Dark Age in which they currently exist.

For fallen mortal physical beings, this is Pure Science rather than applied science. It is Pure Science because its explanatory power is millions of years ahead of anything that the Materialists, Naturalists, Nihilists, and Atheists will ever be able to produce. But unfortunately for fallen mortal beings, it's typically NOT an applied science. Mortal physical beings have no control over Transdimensional Physics or Quantum Mechanics; therefore, we spend all of our effort building barriers within our computer chips trying to prevent quantum tunneling from happening, because we can't figure out how to use quantum mechanics and quantum tunneling to do anything productive or useful. WE KNOW that quantum tunneling exists because we observe its effects and go out of our way to prevent it from happening; but, as physical fallen beings, we cannot quantum tunnel nor teleport at will like spiritual beings have been observed doing.

When it comes to the Quantum Fields, we can debate as to whether God the Father designed the system or not; but, it is obvious that the massless, entropyless, and non-physical Quantum Fields were purposefully designed and made to fulfill specific functions; therefore, it is obvious that there is something quantum or something non-physical that made them and currently conserves them. Nature's Psyche is it. Nature's Psyche makes the quantum waves in the first place; and then later, it is Nature's Psyche or the Intelligences within the Energy who collapse the wave function thereby transforming non-local, non-physical, omnipresent, infinite acceleration INTO localized mass, heat, resistance to acceleration, or mass's heat storage capacity (entropy) instead. This is the Perpetual Motion Cycle, and our scientists haven't yet discovered the Perpetual Motion Cycle, having chosen to believe in the second law of thermodynamics instead.

It's time to start embracing and accepting these obvious Truths in Science. But, our scientists do not. Instead, our scientists embrace and promote the things that are obviously false. Real Scientists wouldn't define the lies and the deceptions as "science" or "knowledge", like our scientists currently do. The materialists, naturalists, nihilists, and atheists promote the deceptions and the lies rather than the Truths in Science that have actually been experienced and observed.

Our people can't have the Truths in Science when they are constantly pursuing, accepting, and promoting the deceptions and the lies instead.

SYNTROPY MUST EXIST OR WE WOULDN'T EXIST

Some type of Syntropy must exist, or we wouldn't exist.

So, where is it?

This Syntropy should be all around us and all through us, since we actually exist. The fact that we exist is scientific proof that the second law of thermodynamics is false, because it predicts that we shouldn't exist. If there were no Syntropy, then the second law of thermodynamics would actually be true, and then all we would observe around us would be nothing more than random disorder and complete chaos, as the second law of thermodynamics predicts.

So, where is all this Syntropy or Order, and why haven't our scientists discovered Syntropy? Syntropy is the theoretical opposite of entropy. So, what is entropy and what is Syntropy?

In physics, the word "conserved" means entropyless or syntropic. Entropy means "not conserved". Syntropy is order and organization. Syntropy is the conservation of energy and conserved quantum information within conserved psyches or conserved packets of energy.

All of that missing Syntropy that our scientists haven't discovered yet is found within the Conservation of Energy and Psyche. Whether they know it or not, our scientists have formally rejected the Conservation of Energy and have replaced it with the second law of thermodynamics instead. The first law of thermodynamics falsifies the second law of thermodynamics, and vice versa. If one of them is demonstrably true, then the other one has been automatically falsified. So, which one is true, and which one is false? The one that has actually been experienced and observed is the one that's actually real and true. We don't observe any of the things that the second law of thermodynamics predicts that we should be observing; therefore, the second law of thermodynamics is false.

Some of that missing Syntropy is also found within the Conservation of Quantum Information within Psyche, which means that a lot of that missing Syntropy is found within Psyche, which is in fact a Quantum Information Processor and a Quantum Memory Storage Device, that is constantly conserved. Syntropy means conservation. Syntropy means entropyless. Conserved means entropyless. Entropy means "not conserved". Whereas, our second law of thermodynamics is "conservation of entropy", and it emphatically and erroneously states that the entropyless does not exist. The second law of thermodynamics is obviously false. Massless and entropyless quantum waves, photons, and quantum fields obviously exist; therefore, the second law of thermodynamics is obviously false.

Syntropy is found in ALL of the order and organization that we see around us, which our second law of thermodynamics says should not exist.

Ultimately, Syntropy is currently being conserved or stored within the perpetual motion machines that our scientists call the Quantum Fields. Quantum Field Theory and the massless, entropyless, heatless, syntropic, and conserved Quantum Fields falsify the second law of thermodynamics which erroneously states that Syntropy, the conserved, and the entropyless do not exist.

The second law of thermodynamics erroneously states that the total amount of disorder or entropy in the universe is constantly increasing and can never decrease or go to zero. This idea is obviously false. It has never been experienced nor observed. It is falsified by the Perpetual Motion Cycle ($E = mc^2$), and it is falsified by Quantum Field Theory and the Quantum Fields. It is also falsified by the Conservation of Energy and Psyche, as well as the Conservation of Quantum Information within Psyche.

Can you see now where all the hidden Syntropy is being stored? Can you see now why our "science" is mostly wrong? Our scientists haven't yet discovered Syntropy and the Perpetual Motion Cycle. They have no idea what it is, that it exists, or how it works. They have erroneously chosen to believe instead that entropy and physical matter are conserved.

The scientists on our planet erroneously define "entropy" or "mass's heat storage capacity" as random disorder or random chaos. The second law of thermodynamics is obviously false. The second law states that the total amount of entropy or disorder in our universe is constantly increasing and that it can never decrease and go to zero. The second law states that entropy or disorder is conserved. The second law is conservation of entropy. This idea is obviously false. It has never been experienced nor observed. The fact that we actually exist is scientific proof that the second law of thermodynamics is false.

Truth matters. Our scientists don't have the Truths in Science because they have deliberately embraced the falsehoods instead.

Our scientists make this egregious error because they have no idea what entropy is or how it works. They erroneously define entropy as "disorder", when in fact it is obvious that there is no correlation whatsoever between disorder and thermodynamics. Disorder doesn't cause heat or thermodynamics, and disorder doesn't eliminate it either. In fact, heat or thermodynamics represents a great deal of order and organization at the physical level, as does a physical atom that has no heat and is at absolute zero temperature. The order and organization remain, no matter what temperature or how much heat a physical atom contains. Even at absolute zero temperature, a lone physical atom remains perfectly organized and fully functional. Does it not?

Furthermore, the massless, entropyless, and conserved Quantum Fields represent perfect order and organization at the quantum level no matter what their temperature might happen to be. They function perfectly and flawlessly at absolute zero, as well as at billions of degrees. The Quantum Fields are the ultimate perpetual motion machines, and their proven and verified existence falsifies the second law of thermodynamics. The second law of thermodynamics will never be true as long as the Quantum Fields exist. Conserved means entropyless or syntropic; and, the syntropic or exergic Quantum Fields are both massless and entropyless, which means that they are being conserved. This is obviously true, or we wouldn't exist, and physical matter wouldn't exist either. Quantum Field Theory and the Quantum Fields falsify the second law of thermodynamics, and our scientists don't even know it.

Our scientists can't see any of this nor understand any of this because they have chosen an incorrect definition for entropy instead of pursuing the truth. You can't have the Truth in Science when you have the wrong definition for entropy. They have deliberately embraced the erroneous and falsified second law of thermodynamics, because for them the second law is just another version of Creation Ex Nihilo or Creation by Chance or Creation by Random Disorder, which is just another erroneous and falsified concept that they have chosen to believe in.

Our scientists believe in and promote Creation Ex Nihilo and Creation by Chance, which are obviously false. Our scientists define "science" as everything that is known to be false, which is why our scientists don't have the Truth in Science.

Transdimensional means non-physical, and the scientists on our earth collectively reject Transdimensional Physics or Non-Physical Physics or Quantum Mechanics because they don't know what it is and how it works.

As a group, our scientists claim that the non-physical does not exist, when in fact it is obvious that the quantum or the non-physical does exist. Photons or quantum waves are obviously massless, entropyless, chargeless, heatless, syntropic, exergic, and non-physical. The energy within them is constantly conserved. Conserved means entropyless. Photons or quantum waves are obviously massless and entropyless. Therefore, the energy and the intelligence within them are constantly conserved.

Our scientists reject these obvious truths, because they have convinced themselves that the non-physical does not exist and that entropy can never go to zero and cease to exist. Our scientists can't have the Truth in Science when they are constantly rejecting every truth in science.

SOME OF THE THINGS OUR SCIENTISTS HAVEN'T DISCOVERED YET

I find comparisons fascinating.

Our scientists define "science" by everything that is known to be false. They have deceived themselves, and they don't even know it. I find that fascinating. Self-deception works, and it works every time. That, too, is an interesting tidbit when it comes to human psychology or the scientific study of the human psyche. Self-deception works, and it works every time. Our greatest scientists and our greatest minds are not immune. In fact, they are the first ones to fall for all the deceptions and the lies, because they created them in the first place.

I, too, have a genius level intellect, but intelligence or a high IQ is absolutely worthless when a genius chooses to believe in everything that is false, as I did when I was a materialist, naturalist, nihilist, and atheist. I KNOW because I have been there and done that. At no other time in my life has my genius or intelligence been so wasted and worthless, as it was when I was a materialist, naturalist, nihilist, and atheist. Materialism, Naturalism, Darwinism, Nihilism, Atheism, and their derivatives were designed to keep people stupid, ignorant, and in the dark; and, it works even on geniuses who should know better. These falsified philosophies of science, falsified interpretations of science, or falsified religions were designed to prevent us from finding and knowing the Truths in Science. They work as intended. Genius or intelligence is worthless of one is constantly choosing and embracing the lies within his or her society.

I deliberately turned myself into an outsider; so now, it is obvious to me what's wrong with our earth's "science" because I'm not beholden to anyone and

I'm free to tell it as I see it; whereas, the indoctrinated who created the monstrosity in the first place are completely blind to its weaknesses, flaws, and falsehoods because they have become addicted to it or conditioned to it. Furthermore, I overcame my materialism, naturalism, nihilism, and atheism by studying the Truths in Science that our scientists don't know about or have officially rejected. I have certain advantages and a fresh new perspective that our scientists don't have.

Our scientists may never discover these Truths in Science because their blanket of self-deception is too strong, and they may never see through it. Human dispensations, or human plantings, can go for thousands of years and never discover these Truths in Science. Our earth may end up being one of them. Humans on our earth may never discover these Truths in Science because our scientists will most likely continue to reject them even when they finally know about them. The scientists on our earth, in general, are not looking for the Truths in Science because they have developed and accepted lies instead.

The scientists on our earth haven't yet discovered the Primary Axiom of Science and Statistics; and, they probably never will, because they aren't looking for it and don't want to discover it. Even when they stumble upon it, they reject it, because they have chosen to believe that it isn't true or that it doesn't exist. Consequently, our scientists will never know that the Primary Axiom of Science and Statistics falsifies the Null Hypothesis, Darwinism, the Theory of Evolution, Creation Ex Nihilo, Creation by Entropy, Creation by Random Disorder, or Creation by Chance. The false is always falsified by the truth, and our scientists haven't discovered that yet.

Our scientists haven't yet discovered that photons, tachyons, virtual particles, and quantum waves are entropyless, massless, heatless, chargeless, omnipresent, spiritual, non-local, and non-physical. Our scientists don't have what they need to discover, understand, accept, and develop an Ultimate Law of Thermodynamics because they have chosen to believe in their erroneous and falsified second law of thermodynamics instead. Their falsehoods, deceptions, and lies are preventing them from discovering the Truths in Science, including the Ultimate Law of Thermodynamics.

The Ultimate Law of Thermodynamics effectively replaces the second law of thermodynamics with Quantum Field Theory instead. Our scientists don't know yet that Quantum Field Theory, Transdimensional Physics, and the Quantum Fields falsify the second law of thermodynamics.

Our scientists don't have what they need to discover and develop a Quantum Law of Thermodynamics because they have erroneously chosen to believe that the quantum, the non-physical, or the transdimensional does not exist. Our scientists have no knowledge of Quantum Phase-Shifting – what it is, how it works, or why it is important. Our scientists have deliberately overlooked and rejected the Perpetual Motion Cycle. They have no idea what it is or how it works. Our scientists can't have the Truths in Science when they are constantly identifying and rejecting the Truths in Science.

The spiritual, transdimensional, or non-physical aspect is most evident when it comes to human beings. Human beings can override and break their addictions and their conditioning at will. Animals cannot. Behavioral extinction can be immediate when it comes to human beings; whereas, it can take a while and requires the removal of the conditioner, manipulator, or brainwasher before animal conditioning goes extinct. Our scientists haven't discovered this yet. They have convinced themselves that conditioning, addictions, or brainwashing are permanent. It's my contention that psyche should be put back into the physical sciences where it belongs. Our scientists have rejected psyche or quantum consciousness in all its different forms and manifestations. Our scientists will never discover it because they have officially rejected it.

Our scientists haven't yet discovered and implemented the most powerful form of the Scientific Method, which is called Negating the Consequent. Our scientists certainly haven't used Negating the Consequent to identify and eliminate the obvious falsehoods in their "science" or "knowledge".

Transdimensional physicists and out-of-body explorers have observed that when it comes to quantum mechanics or non-physical physics, there is a Servitor in the system or a God in the machinery who can read your mind and grant your requests. Whenever the human psyche enters into a non-consensus reality in the spirit world, the whole environment adjusts itself to the demands and expectations and requests of that human psyche.

Our scientists completely ignore all of this and pretend that it doesn't exist. *Denying evidence* or *rejecting evidence* is a logic fallacy. Collectively, our scientists deny the existence of the non-physical or the transdimensional. They have concluded in advance that it does not exist. However, it is impossible to establish through a preponderance of the evidence that the non-physical does not exist, especially since the vast majority of our universe is obviously non-physical, transdimensional, or quantum in nature and origin, being comprised of massless and entropyless space or quantum fields.

Our scientists haven't yet discovered that there is no such thing as "heat death". Absolute zero temperature is steroids at the quantum level or the perpetual motion level, where everything is conserved. It has been observed that superconductors, spirit matter, psyches, photons, quantum waves, and quantum fields function best and most efficiently at or near absolute zero in the state that our scientists erroneously call "heat death". At "heat death", everything comes alive at the quantum level. This is the Quantum Law of Thermodynamics. Heat or thermodynamics is purely a physical phenomenon. It doesn't exist at the quantum level in the syntropic realm, conserved realm, perpetual motion realm, or quantum realm. Heat or thermodynamics is a function of mass, and mass is purely a physical phenomenon that doesn't exist at the quantum level. Our scientists haven't discovered any of this yet, because they have chosen to go with their erroneous and falsified "second law of thermodynamics" instead.

Our scientists haven't yet discovered that the denialistic religions or philosophies of science – such as materialism, naturalism, nihilism, the theory of evolution, creation by chance, creation ex nihilo, the second law of thermodynamics, and atheism – are worthless and lame, because you can't make scientific discoveries by denying their existence. These are false and falsified religions with NO evidence to support them. You can't make scientific discoveries by denying their existence. Our scientists haven't figure this out yet.

It is super easy to assume that materialism and naturalism are true; but, it takes a huge amount of work to falsify them every way that they can be falsified.

When it comes to Quantum Mechanics and Quantum Field Theory, you have to learn how to take it into the transdimensional or the non-physical; otherwise, you will never figure out what's really going on in our universe.

There is a noticeable and detectable difference between autonomic or automatic behaviors and chosen behaviors. We, our psyche, knows the difference between the two. Even your dog can sense whether you deliberately kicked him or accidentally stepped on him. Collectively, our scientists haven't figured this out yet, because they aren't looking for it, and have already concluded that it doesn't exist. Nevertheless, organismic variables, personally chosen variables, or personally chosen behaviors do indeed make a showing in human behavior. In other words, the human psyche chooses and decides what it wants to do with the stimuli that are constantly coming its way.

You can grind the whole universe down, and you will never find a single atom of love, mercy, friendship, justice, kindness, or hate. It's the love (or the hate) that survives the death of one's physical brain, because it is the human psyche who survives the death of one's physical body. Love and friendship don't exist until the human psyche chooses them into existence. The human psyche is the creator of the non-physical. Our scientists haven't discovered this truth yet, and they never will, because they have already rejected it and concluded in advance that psyche does not exist.

Anything that is obviously made obviously has a Maker who made it.

It's all made from energy or light.

Who does the making at the quantum level? Who makes the quantum waves or the photons in the first place? Who collapses the wave functions thereby transforming them into some type of mass or heat instead? Who designed and made the quantum fields at the quantum level? Who runs the quantum fields and keeps them in existence?

Psyche, Quantum Consciousness, Intelligence, or Life Force. Psyche is a conserved quantum, or a conserved packet of energy, or a conserved photon.

What is made from energy at the quantum level?

Different types of quanta, particles, or quantum waves, which can then be assembled into spirit matter or physical matter.

This is cool stuff, but only if you are looking for it and want to find it. Our scientists haven't discovered any of this yet because they aren't looking for it and don't want to find it. No seeking, then no finding. Instead, our scientists have chosen to believe that the non-physical or the transdimensional does not exist; consequently, they aren't looking for the non-physical or the transdimensional, and therefore they will never find it. Self-deception works, and it works every time. Our scientists are not immune.

Our scientists haven't yet discovered Syntropy. They don't know what it is and how it works. Our scientists haven't yet discovered the Perpetual Motion Cycle. They don't know what it is or how it truly works. Our scientists haven't yet discovered the LAW of Psyche. Instead, our scientists have chosen to believe that Syntropy, Psyche, and the Perpetual Motion Cycle do not exist. Our scientists have chosen instead to believe in the second law of thermodynamics or creation by chance. The very fact that we exist is scientific proof that the second law of thermodynamics is false; yet, our scientists have chosen to believe in the second law of thermodynamics anyway, because they have erroneously chosen to believe that chance or random disorder can design and create anything that it sets its mind to, if given enough time to do so.

Our scientists have deliberately rejected the Primary Axiom of Science and Statistics because it successfully falsifies the second law of thermodynamics.

You have got to know about, understand, and accept the Perpetual Motion Cycle as well as Syntropy and Quantum Field Theory if you want to have a scientific explanation for why we exist. You have got to know about, understand, and accept the Law of Psyche if you want to have a scientific explanation for how all of this order and organization that we observe came to be in the first place.

Instead, our scientists have dreamed up the second law of thermodynamics, random fluctuations in entropy, and random disorder as their designer and creator, even though the second law predicts or says that we shouldn't exist. Where Science is concerned, the second law of thermodynamics is completely worthless because it predicts that we shouldn't exist, can't explain why we exist, and can't explain scientifically where all of this order and organization came from to begin with. You can't have the Truth in Science when everything you believe in is an obvious lie.

Our scientists haven't discovered and embraced Psyche, Intelligence, or Quantum Consciousness yet. They have no idea what it is or how it works. They don't know that Psyche is eternal, everlasting, syntropic, and conserved. Our scientists have never thought of Quantum Wave Processors or Quantum Information Processors, which is just another name for Psyche. Our scientists haven't yet discovered and accepted the Law of Quantum Information Conservation within Conserved Psyches or Conserved Quanta. They don't have the foundation nor the background that they need to discover it.

Our scientists don't know that the Quantum Fields and Quantum Waves and Photons are massless, entropyless, heatless, non-physical, conserved, exergic, syntropic, eternal, everlasting perpetual motion machines made from pure energy by Nature's Psyche, and are capable of infinite acceleration or omnipresence. Our scientists have no idea that light or photons are spiritual, transdimensional, or non-physical in nature and origin. The quantum fields, photons, tachyons, virtual particles, and quantum waves are obviously massless or non-physical or immaterial; yet, most of our scientists erroneously claim that the non-physical does not exist. Our scientists have no idea that Quantum Field Theory falsifies the second law of thermodynamics. Our scientists deny and reject the obvious Truths in Science.

Our scientists are just barely starting to discover non-locality or non-physicality. Most of our scientists still reject Action at a Distance, even though they have finally proven that it is true. Our scientists officially reject Transdimensional Physics or Non-Physical Physics – the thing that some of them call Quantum Mechanics. They don't accept it, and they don't understand it, and they don't want it.

Collectively, our scientists have rejected the existence of spirit matter or dark matter. Our scientists have no understanding of Quantum Phase-Shifting, what it is or how it works. They have no idea that spirit matter, physical matter, dark matter, exotic matter, quantum waves, photons, tachyons, quanta, and psyches ARE ALL different phases, dimensions, or levels of matter. They are all packets of energy. They are all quanta or made from quanta. They are all matter! They are all different forms of matter, each existing in its own unique phase, level, frequency, timeline, or dimension.

ONLY physical matter or entropic matter is subject to physical limitations, resistance to acceleration, locality, entropy, and spacetime – the other types of matter are massless, non-physical, transdimensional, quantum objects; and, they function according to the entropyless, ageless, syntropic, and massless rules of Quantum Mechanics, Action at a Distance, Omnipresence, Infinite Acceleration, and Quantum Field Theory.

Our scientists are confused about what is conserved and what is not conserved. Our scientists have the wrong definition for conservation. They don't know that conserved means entropyless. Therefore, our scientists erroneously believe that entropy and physical matter are conserved. They are wrong. They are wrong because they have the wrong definition for entropy and because they don't know what it means for a quantum object or a non-physical object to be conserved.

Consequently, our scientists don't have what they need to develop the Ultimate Law of Thermodynamics or to falsify the second law of thermodynamics. Furthermore, our scientists haven't yet discovered the Quantum Law of Thermodynamics. Our scientists don't have any clue what an Ultimate Law of Thermodynamics should look like, and they have no idea that the Quantum Law of Thermodynamics exists, what it is, or how it works.

Our scientists haven't yet discovered and fully embraced the Law of Quantum Information Conservation within Psyche, because our scientists have formally and officially rejected Psyche, Intelligence, Quantum Consciousness, Quantum Information Processors, Quantum Memory Storage Devices, or Quantum Wave Processors. Our scientists haven't yet discovered the LAW of Psyche. They don't know what it is or how it works. Our scientists haven't discovered Quantum Neuroscience yet. They haven't made the slightest attempt to apply

Transdimensional Physics or Quantum Mechanics to neuroscience. Quantum Neuroscience is a completely empty and virgin field where our scientists are concerned.

Our earth's scientists still think and believe that a physical brain is a computer – complete with wires, RAM, and software. Our scientists have no idea that there are no wires, no RAM, and no memory engrams within a physical brain. Our scientists haven't discovered yet that neurotransmitters are nothing more than a single hardware bit at the physical level. It's physically impossible to transmit a complex message or a thought or a memory through a neurotransmitter or a single hardware bit at the physical level. Our scientists don't realize that a synapse scrambles or randomizes everything that comes its way. It's physically impossible to transmit a message, thought, or memory through a synapse at the physical level.

So, how are the neurons in your brain communicating with each other since it's physically impossible for them to do so at the physical level? How are long-term memories being stored within a physical brain, since it's physically impossible to do so at the physical level within a physical brain? When your human psyche decides to raise its finger off the table, who or what collapses the necessary wave functions in order to fire or trigger the specific neuron in your brain that raises that specific finger off the table? Who MAPPED that specific function onto that specific neuron within your physical brain in the first place? Who designed and made your specific physical brain? Who designed and implemented the Quantum Map of Physical Functionality at the quantum level for your physical brain? In other words, who MAPPED your brain at the quantum level to perform specific functions at the physical level? Who or what is running and controlling your physical brain? Who is storing the memories that show up in your after-death life review after your physical brain is dead and gone? Who or what collapses the wave function? Who or what stores your after-death life review memories? Who or what fires the specific neuron that raises your finger off the table? Who or what is communicating between the neurons and the atoms at the quantum level within your physical brain? It isn't the human psyche because the human psyche isn't consciously aware of any of these things.

Collectively as a group, our scientists haven't discovered yet that it is Nature's Psyche or the Controlling Psyches within Nature who are making the quantum waves, running the quantum fields, and collapsing the wave functions at the quantum level.

Our scientists haven't thought about nor discovered any of these things, and they never will because they are looking for them and don't want to find them. No seeking, then no finding.

Some of our scientists are just barely starting to figure out what's wrong with Creation by Chance, Creation by the Null Hypothesis, Creation by Random Disorder, Chemical Evolution, Macro-Evolution, Abiogenesis, Spontaneous Generation, the Theory of Evolution, the Second Law of Thermodynamics, and all the other different forms of Creation by Chance or Creation by Random Luck. Until they can figure out why Creation by Chance is impossible and why the Theory of Evolution is physically impossible, they will never have the Truths in Science that scientists are supposed to seek and find.

Consequently, our scientists haven't yet discovered the Primary Axiom of Science and Statistics. They don't know what it is or why it is true.

Why?

Why have our scientists failed to discover all these different Truths in Science that are hiding in plain sight waiting for our scientists to discover them?

It's because our scientists have officially rejected these Truths in Science and have adopted and fully embraced Materialism, Naturalism, Darwinism, Nihilism, and Atheism instead. You can't have Truth in Science when you have deliberately rejected it and have chosen to believe in the lies instead.

Materialism, Naturalism, Darwinism, Nihilism, Atheism (Creation Ex Nihilo), Physical Reductionism, Atomism, Determinism, Behaviorism, and Creation by Chance have NO explanatory power because they are not an attempt to explain anything scientifically. Instead, these philosophies of science or interpretations of science are nothing more than catalogs or lists of the different things that these people have chosen to believe DOES NOT EXIST. They are NOT science. They are philosophies of science or falsified interpretations of science. They are nothing more than a listing of the things that these people have chosen to reject, chosen to deny, and chosen not to believe in. That's NOT Science or Knowledge. You can't have evidence for, proof of, or knowledge about the things that by definition DO NOT EXIST. It's philosophically and logically impossible to prove that something does not exist. That can only be done by axiom or by decree; and typically, these false axioms or false decrees are easily falsified by the things that have actually been experienced and observed.

Materialism and its derivatives are pseudo-sciences, because they are based upon faulty assumptions that are demonstrably false.

By definition and by axiom, Creation by Chance, Causation by Chance, Creation by Random Chaos, Creation by Entropy defined as Disorder, or Creation Ex Nihilo does not exist; therefore, by definition and by axiom you can never have evidence for, proof of, or true knowledge about these things because they do not exist. The ONLY reason we KNOW that Creation by Chance or Causation by Chance or Creation Ex Nihilo DOES NOT EXIST is because they have NEVER been experienced NOR observed by anyone. They have been falsified by Negating the Consequent, which effectively means that they have never been experienced nor observed by anyone, not even God. Philosophies of Science or Interpretations of Science, that have NEVER been experienced NOR observed by anyone, can't possibly be real or true.

Our scientists haven't yet discovered and implemented Negating the Consequent, which is the most powerful form of the Scientific Method. They don't know what it is or how it works; and therefore, our scientists currently are not using Negating the Consequent to falsify the things that are obviously false.

Our scientists need a new fresh pair of eyes – someone who sees things differently than the way they have chosen to see them and interpret them. Our scientists need someone to think outside of the box for them, because they are trapped in the box that they have made for themselves and can see no way out of it.

Our scientists haven't yet discovered the transdimensional, or the non-physical, or the spiritual. Our scientists haven't yet discovered massless and entropyless matter. Our scientists erroneously assume that everything is made from physical matter. They are wrong. Matter of any kind is made from quanta. A particle of matter or a quantum is simply an organized packed of energy. By this definition for matter, massless and entropyless photons and psyches are in fact quanta or organized packets of energy. Massless and entropyless photons, psyches, virtual particles, and quantum waves are all different forms of matter that exist at different phases, dimensions, timelines, frequencies, and levels. Psyches and photons and quantum waves are massless and entropyless particles, matter, or quanta. They are organized packets of energy or quantum waves; and, they are made from quantum waves or organized packets of energy. It's all just different forms of matter in different phases or dimensions; and, it's all made from energy.

Physical matter or entropic matter or baryonic matter is the ONLY type of matter that is subject to physical limitations such as resistance to acceleration. The other types of matter or quanta are capable of infinite acceleration or omnipresence. A physical atom is made from many different types of quanta or particles. An entropic physical atom has mass or resistance to acceleration, whereas, the other types of matter or the other types of quanta do not have mass, entropy, or resistance to acceleration. It's really simple to understand; but, our scientists haven't discovered any of this yet.

Our scientists haven't yet discovered that space is not made from physical matter. Remember, space is not made from matter. Space is the medium through which the matter, or the particles, or the quanta travel. Space is made from massless and entropyless quantum fields.

Not everything is made from matter. Not everything is made from physical matter. Most of our scientists erroneously believe that everything is made from physical matter. They are wrong. Space is NOT made from physical matter or any other type of matter. Space is made from massless and entropyless Quantum Fields. Space or the Quantum Fields are the medium through which particles travel. Space, or the Quantum Fields, or the Light of Creation, or the Light of the Anointed One, or the Light of Christ are NOT the particles themselves. Many of our scientists have erroneously chosen to believe that space is empty. They are wrong. Space is comprised of Quantum Fields, or the Light of Creation, or the Light of Christ.

Not everything is a quantum, or a quantum wave, or a particle. Space is made from energy, quantum fields, the zero-point field, the quantum sea of light, the Light of Creation, the Light of God, the Light of the Anointed One, or the Light of Christ. Space or the quantum fields transmit matter or quanta or particles. Space is not matter, quanta, or particles. Space, or the quantum fields, or the

Light of Christ is the medium of transmission at the quantum level. Different types of matter or different types of massless and entropyless particles or quanta are what gets transmitted through the quantum fields, or the Light of Christ, or the quantum sea of light.

Quanta or particles are organized packets of energy. There are lots of different types of massless and entropyless particles, quanta, or matter. According to those who have seen psyche at the quantum level, psyche appears to be the ultimate point particle – an infinite singularity. Psyche is a conserved quantum, or a conserved photon. Yet, it is observed floating in space or floating within the invisible and immaterial quantum fields. So, the space, the quantum fields, the Light of Creation, the Light of God, the Light of the Anointed One, or the Light of Christ is even more refined and ethereal than matter, psyches, quanta, quantum waves, virtual particles, or organized packets of energy.

Anything that is obviously made obviously has a Maker who made it. Anything that is obviously designed obviously has a Designer who designed it. Anything that is obviously organized obviously has an Organizer who organized it. Anything that is obviously fine-tuned obviously has a Fine-Tuner who fine-tuned it. Anything that is obviously standardized obviously has a Standardizer or a Lawgiver who standardized it.

The scientists on our earth haven't discovered these obvious Truths in Science, yet, because they aren't looking for them and don't want to find them. But, just because our scientists haven't discovered them doesn't mean that they don't exist. That's the flaw with the modern-day scientists on our earth – they erroneously believe that if they haven't discovered something yet, then it doesn't exist. It's lazy and it's sloppy. It's the very definition of Bad Science.

The Quantum Fields were obviously designed, organized, fine-tuned, and made – each one with a specific purpose and function in mind. They are all made from energy, and each one of them functions differently than the others. That level of purpose, design, order, organization, and fine-tuning at the quantum level or the non-physical level doesn't just happen accidentally. Each Quantum Field was carefully designed, fine-tuned, organized, and made from the very beginning with a specific function or purpose in mind – and that purpose was to make it possible to organize, make, and sustain physical atoms and entropic physical matter. This is obviously true because the physical matter obviously exists, and so do the Quantum Fields.

The Gods, husband and wife, mother and father, step into an unorganized and chaotic region of space; and then, they lovingly start to organize everything from that central point outwards in all directions. They guide everything upwards towards their level of existence, organization, order, and love. First, they organize the Quantum Fields. Then when the Quantum Fields have been organized and are functioning properly, the matter or quanta or particles are standardized or organized so that they can function together as a unit and eventually be formed into physical atoms or entropic physical matter. After these Gods are done with the standardization or law-giving process, then a physical atom on one side of their organized physical universe will be completely compatible with a physical atom on the opposite side of their newly organized physical universe.

This organization and standardization process continues outwards in all directions from their centrally chosen starting point; and in this manner, a new physical universe is organized and made from the chaotic and incompatible mass, matter, forces, fields, and energy that used to reign supreme in that part of the multiverse. It's the standardization process, the law-giving process, or the organization process that signifies that the Gods have stepped into that region of chaotic and unorganized space, and have successfully organized it so that a physical atom made on one side of their newly organized physical universe will be completely compatible with a physical atom that is made on the opposite side of their newly organized physical universe.

Imposing physical limitations brings order, organization, and standardization to chaos. The same thing happens at the quantum level with the Quantum Fields. God's influence and presence is in all things and through all things thanks to the Quantum Fields, the Light of Creation, or the Light of the Anointed One. The Quantum Fields are one of the best and most convincing Scientific Proofs of Gods Existence, and so is the existence of entropic physical matter, whether we realize it or not.

Anything that is obviously designed and made obviously had a Designer and a Maker who designed it and made it. Compatible Quantum Fields, compatible quanta or non-physical particles, and compatible physical atoms were obviously designed and made with that specific purpose in mind. These invisible, intangible, and non-physical forces, fields, quanta, and particles work together perfectly to form or to make this physical universe that we see around us today. It didn't happen by accident. It's too perfect. There's too much standardization or law or order within it, for it to have happened by accident or by chance. It's the exact opposite of disorder, and chaos, or chance. It was obviously designed, organized, fine-tuned, standardized, and made. Order doesn't magically spring into existence from nothing or from chaos. Order is deliberately organized and made. Spontaneous generation was falsified in 1859 by Louis Pasteur, and it has been false ever since.

Our earth's scientists don't want to find and discover these types of non-physical or transdimensional things, so they never will. Over 95% of our physical universe is in fact transdimensional or non-physical, which means that the scientists on our earth have collectively rejected over 95% of Science. They don't want to know about it, so they never will. They will continue to die in ignorance just as they have been doing for millennia on this earth, because they refuse to look, learn, see, and understand. Their ongoing fear of God and religion will prevent them from discovering transdimensional physics or non-physical physics. Our scientists will never discover any of this because they don't want to find it and learn about it.

That's a lot of Science to deliberately reject, but that's what the materialists and naturalists do. These people automatically delete, reject, and refuse to look at everything that falsifies their pre-chosen beliefs. That's how these people do "science" – by deleting the evidence that they personally don't like. *Rejecting evidence* is a logic fallacy. *Deleting and banning evidence* is a logic fallacy. *Refusing to look at evidence* is a logic fallacy. *Assuming that evidence* doesn't exist is a logic fallacy. These are some of the hundreds of logic fallacies upon which Materialism (the Second Law of Thermodynamics), Naturalism, Darwinism (Creation by Chance), Nihilism, Behaviorism, and Atheism (Creation Ex Nihilo) are based. Our scientists haven't yet discovered how to falsify Materialism, Naturalism, and their derivatives. It's simple to do, which explains in part why they haven't done it yet. They aren't trying, and they don't want to.

Our scientists haven't discovered any of this. It's foolishness to them, because they don't understand it, don't want it, and aren't looking for it. Instead, they have chosen to believe that space is completely empty and that the Quantum Fields, or the Light of Creation, or the Light of God, or the Light of the Anointed One, or the Light of Christ does not exist. Our scientists have received precisely what they have chosen to believe. The scientists on our earth have chosen to believe in "nothing", and that's precisely what their "science" is as a result of their choice. Isn't that amazing? Self-deception works, and it works every time. Even a genius level intellect is not immune, when it comes to self-deception. They don't understand their own psychology, because they have rejected the existence of psyche.

I have become fixated on the obvious lies in "science" – trying to identify them, overcome them, falsify them, and then using them to point me to the Truths in Science that are hidden from our view. The Materialists and Naturalists have a knack for identifying the Truths in Science, and then rejecting them. I have learned to look at and take seriously anything and everything that the Materialists and Naturalists are automatically rejecting and don't want us to see, because invariably it is the truth.

This is a lot that our scientists haven't discovered yet, and they never will discover most of it, because they have already rejected it and concluded that it doesn't exist. Most of our scientists have died without having even the slightest clue that these things exist. They aren't looking for this observed and verified Science; and therefore, they haven't found it because they don't want to find it. They don't want it to be true. They don't want it to exist.

These are just some of the things that our scientists haven't discovered yet, and it's holding them back, whether they realize it or not. A society can't have these Truths in Science if they haven't discovered them. We reap what we sow, and we find what we search for. Our scientists went searching for "nothing", and that's precisely what they found.

SOME OF THE FALSEHOODS OUR SCIENTISTS HAVE EMBRACED

Scientism is the philosophical belief that "science" or the scientific method is the ONLY way to find and know the truth. This belief is obviously false, especially once you realize and know that our scientists have axiomatically defined "science" as Materialism, Naturalism, Darwinism, Nihilism, Creation by Chance, Creation Ex Nihilo, and Atheism. These things can NEVER be used to find and know the truth, because they are demonstrably false and have been falsified by Science or the observations and the experiences of the human race.

The scientific method is NOT the best way to find and know the truth. Our scientists don't know that yet. Direct observation and experience, both at the physical level and the quantum non-physical level, is the best and the fastest way to find and know the truth. Observation and experience are a lot more cost effective, believable, and efficient than the scientific method. Science is observation, experience, knowledge, and truth after all – or it should be. It is observation and experience that make the scientific method effective or believable in the first place; and, it is also observation and experience that's used to falsify our false ideas and false hypotheses, even when we are running a science experiment or using the scientific method.

Our scientists have the wrong definition for Science. They have erroneously defined science as Materialism, Naturalism, Darwinism, Nihilism, Atheism, Creation Ex Nihilo, and Creation by Chance.

Materialism is the philosophical claim that the Non-Physical or the Transdimensional does not exist. Well over 95% of the "stuff" in our physical universe is demonstrably Non-Physical. Dark matter or spirit matter by definition is non-physical. Dark energy, the quantum fields, quantum waves, photons, tachyons, quantum information, energy, space, time, gravity, magnetism, radio waves, x-rays, and microwaves are obviously Non-Physical, yet WE KNOW that they exist. Action at a Distance is non-local or non-physical, yet our scientists have finally proven that it exists. Dark matter is technically transdimensional or non-physical because we can't get our hands on it. The proven and verified existence of any one of these things FALSIFIES Materialism, Naturalism, Darwinism, Nihilism, and even Atheism (Creation Ex Nihilo), which claim as their primary axiom that the non-physical does not exist.

The Naturalists claim that the non-physical, or the supernatural, or the transdimensional, or the quantum DOES NOT EXIST. Naturalism is a rejection of and a denial of the existence of anything supernatural or quantum or transdimensional such as the quantum fields, photons, quantum information processors (psyches), quantum waves, conserved quanta, phase-shifted matter or spirit matter or dark matter, energy, and light – anything that is non-physical yet obviously exists. All of these things are non-physical or supernatural, yet it is obvious that they exist. They have been proven to exist because they have actually been experienced and observed, even though they are non-physical and intangible. The verified and proven existence of something non-physical or supernatural such as a photon or a quantum wave FALSIFIES Naturalism which claims that the non-physical or the supernatural does not exist.

Nihilism is a denial of the existence of a pre-mortal life and a non-physical spiritual afterlife. Nihilism makes the claim that our psyche, spirit, and memories cease to exist when we die. Nihilism makes the claim that there is no such thing as psyche, quantum memories, quantum information, a spirit world, or a transdimensional realm. Nihilism makes the claim that our physical lives have no purpose and no meaning outside of eating, drinking, and procreating. Yet, according to Quantum Field Theory, it is obvious that something non-physical and pre-physical had to design and make the Quantum Fields, or physical matter still wouldn't exist anywhere in our universe. The LAW of Quantum Information Conservation within Psyche falsifies Nihilism. Our scientists haven't yet discovered psyche or quantum information processors, so they haven't yet discovered the things that they need to discover in order to be able to falsify Nihilism.

Atheism states that God does not exist, that Creators do not exist, that Intelligence or Psyche does not exist, and that everything in our universe was designed and created by something that does not exist. Atheism is Creation Ex Nihilo – the creation of something from nothing by nothing.

Atheism is falsified by the LAW of Quantum Information Conservation within Psyche. Quantum Information Conservation has to be true in order for Quantum Mechanics and Quantum Field Theory to be true. Therefore, some type of Conserved Quantum Mechanism or Conserved Quantum Machine must exist at the quantum level or subatomic level that is capable of making, transmitting, receiving, processing, analyzing, transforming, and storing Quantum Waves or Quantum Information. That machine or that mechanism is Psyche or Intelligence. It's the ONLY one that we have experienced and observed that is capable of making quantum waves and then later transforming those massless and entropyless quantum waves INTO some type of mass or heat instead. The proven and verified and necessary existence of Quantum Wave Processors, or Quantum Information Processors, Conserved Quanta, or Conserves Psyche FALSIFIES Atheism which claims that such things do not exist.

Darwinism is Naturalism and Atheism. Darwinism is fruit from the poisoned tree. Darwinism, too, teaches and preaches that the transdimensional or the non-physical does not exist. Consequently, Darwinism is obviously wrong. Natural selection doesn't touch our genes. Natural selection doesn't do anything except sit around and wait for us to die. Natural selection is Creation by Death or Creation by Chance, both of which are obviously false.

Our scientists have erroneously chosen to believe that Chance or Natural Selection can design and create anything that it sets its mind to, if given enough time to do so. They treat Natural Selection or Chance as if it were God. They are wrong. Natural Selection doesn't touch our genes. It is genetic recombination, during the production of our gametes, that produces the errors or the mutations in our genome, and not Natural Selection.

Natural Selection doesn't exist as a person, psyche, intelligence, or God. However, the psyches or the intelligences who perform the genetic recombination process or program are allowed to make mistakes, because it is a physical process after all. God obviously permits physical processes to go wrong, and that does indeed introduce genetic deterioration, genetic entropy, or random mutations into our genome and gene pool.

It has been observed that there are on average a million times more harmful mutations than there are beneficial mutations; therefore, the overall trend is towards genetic deterioration, genetic entropy, and extinction. Random mutations in our genes are never going to turn us into Gods, or beings of light; and, random mutations in our genes are never going to give us superpowers – or control over the transdimensional, the quantum, or the supernatural. It doesn't work that way. Random mutations are going to drive us quickly to extinction, which we can see happening all around us all the time, thus requiring the periodic replanting or reseeding of this earth by the Gods after each mass extinction that has taken place on this earth.

Once our scientists finally realize that Natural Selection or Chance cannot design and create anything, then they will find themselves looking for a logical and rational explanation for how the genomes and life forms got onto this planet in the first place. Anything that is obviously made obviously has a Maker who made it. Software and hardware are obviously made. Genomes are hardware and software. Genomes are obviously made; therefore, genomes obviously have a Maker who made them. The genome in each and every one of your cells is Scientific Proof of God's Existence. Your genome is God's signature, and God's signature is written on every cell in your body that has a nucleus.

Our scientists have chosen to believe in a false god that they call Natural Selection or Random Chance, rather than the God who designed and created and deployed their genome.

Likewise, our scientists have erroneously chosen to believe that entropy, disorder, or random chaos can design and create quantum fields, entropic physical matter, genomes, life forms, planets, stars, galaxies, and physical universes if given enough time to do so. They treat entropy, chance, disorder, or random chaos as if it were God. They are wrong. The Primary Axiom of Science and Statistics states that chance of any kind cannot do causation. That means that chance can never do choice, design, correlation, causation, or creation. Chance is not causation! Our scientists make these kinds of errors and develop these kinds of falsehoods because they haven't yet discovered the Primary Axiom of Science and Statistics which says that chance cannot do causation.

This essay doesn't list all of the falsehoods that our scientists have chosen to embrace and promote; but, it does list some of the worst of them. Self-deception works, and it works every time. Our scientists don't understand Science – what it is and how it works – because our scientists have the wrong definition for Science. *Begging the question, jumping to conclusions,* and *using circular reasoning*, our scientists axiomatically and erroneously define "science" as Materialism, Naturalism, Darwinism, Nihilism, Behaviorism, Creation by Chance, Creation Ex Nihilo, Atheism, and everything else that has been demonstrated to be false. Our scientists actively identify, censor, and ban the Truths in Science, replacing them with a bunch of lies instead. You can't have the truth when everything that you have chosen to believe in is demonstrably false.

Our scientists haven't yet discovered the Primary Axiom of Science and Statistics which states that chance cannot do causation. Chance cannot do choice, correlation, design, creation, or causation.

It has always been obvious to me that chance or random disorder can never design and create anything such as genomes, galaxies, quantum fields, brains, physical universes, physical atoms, or life forms. These things were obviously designed and made. Anything that is obviously designed and made, obviously has a Designer and a Maker who designed it and made it. This is logical common sense that the Materialists and Naturalists deny.

Chance produces disorder or it destroys. The dominant law in biology is extinction, not natural selection. Natural selection doesn't touch our genes. Natural selection doesn't do anything but sit around and wait for us to die. Natural selection is Creation by Death. It isn't even Creation by Chance. Natural selection always results in death. Genetic recombination during the production of our gametes produces our mutations and shuffles our genes – not natural selection.

By definition, chance cannot design, create, nor cause anything. Once it starts doing so, then it is no longer chance but has become some type of deliberate choice or intelligent causation instead. Only psyche or intelligence can do design and creation, both at the quantum level and at the physical level. This is what we have actually experienced and observed. Is it not?

TRUTH IS REPEATEDLY EXPERIENCED AND OBSERVED

The false is falsified by the truth; and, the truth is repeatedly and constantly experienced and observed.

Psyche, or Non-Local Consciousness, the Spark of Life, Quantum Consciousness, or Intelligence has been experienced and observed by near-death experiencers and by out-of-body explorers. At the quantum level and the psyche level, a Psyche is observed as being a pinprick of light, a point particle, an infinite singularity, a conserved quantum, or a conserved photon. While a person is outside of his spirit body looking at his spirit body in the Transdimensional Realm or Quantum Realm, he notices that when he goes looking for himself, there is nothing to be seen. In other words, one's Psyche or Intelligence is experienced first-hand as being an immaterial viewpoint in space, while out-of-body looking at one's spirit body.

A Dark Age consists of the times in a world's history when their "science" or philosophies of science (personal belief system or personal religion) is based upon a wide variety of deceptions and lies. Rather than being Knowledge and Truth, their "science" is actually false.

Every world has a Dark Age of ignorance, idolatry, and superstition that is produced by the Materialists, Naturalists, Nihilists, and Atheists. Our earth is still in its Dark Age thanks to the deceptions and the lies that are being treated as "science" or "knowledge" by our Materialists, Naturalists, Darwinists, and Atheists who claim to be scientists but don't know what they are talking about when it comes to Science – particularly when it comes to Transdimensional Physics or Non-Physical Physics. You need to get these things written down, because the Materialists, Naturalists, Nihilists, and Atheists will try to destroy them; and, every inhabited planet in this universe has people whose only purpose in life is to destroy.

If one is up for it and ready for it, the comparisons between the two worldviews, philosophies of science, or interpretations of science can be fascinating to study and learn about. That's how we humans find the truth – by identifying and eliminating everything that's a lie.

Science is Knowledge of the Truth. It can't be Science, and it can't be Knowledge, whenever it is obviously false or has been falsified. Materialism, Naturalism, Darwinism, Nihilism, Atheism, Creation by Chance, the Theory of Evolution, and the Second Law of Thermodynamics have been falsified by Science or by the observations and the experiences of the human race. Therefore, WE KNOW that they are false because Science or Knowledge has proven them false. In Science, we can indeed use a wide variety of different Scientific Methods or Observations TO PROVE that certain theories or ideas are false.

The Scientific Methods can be used to prove things false; and whether people realize it or not, the Scientific Methods have proven in a wide variety of different ways and millions of times that Materialism, Naturalism, Darwinism, Nihilism, Atheism, Creation by Chance, Causation by Chance, Creation by Random Disorder, Creation Ex Nihilo, Determinism, Physical Reductionism, Radical Behaviorism, Mechanism, Atomism, Chemical Evolution, Abiogenesis, Spontaneous Generation, the Theory of Evolution, and the Second Law of Thermodynamics are FALSE.

If you successfully eliminate everything that is false, then ONLY the truth will remain; and, the truth that remains always consists of the things that have actually been experienced and observed by someone, sometime, somewhere – either here in the physical realm or there in the non-physical quantum realm. If a phenomenon has NEVER been experienced NOR observed by anyone, then it can't possibly be real or true. That's just the way it is.

I'm a Real Scientist because I actively and deliberately pursue and promote the Truths in Science that have actually been experienced and observed, while at the same time trying to identify, falsify, and eliminate the things that are obviously false because they have never been experienced nor observed by anyone. Creation by Chance of any type has never been experienced nor observed; therefore, it should be identified as the falsehood that it is and eliminated from Science because it is obviously false. It can't be Knowledge, Truth, or Science if it is false. Therefore, it shouldn't be treated as if it were Science. It should be tossed out instead.

The Primary Axiom of Science and Statistics states that chance cannot do causation. Chance cannot do choice, correlation, design, creation, or causation. Once it starts doing so, then it is no longer chance but has become some type of intelligent choice instead. This is obviously true, which is why our scientists automatically reject it.

    Mark My Words

THE OBVIOUS LIES IN OUR SCIENCE

    Our scientists have yet to discover the Primary Axiom of Science and Statistics. They aren't looking for it; and currently, they don't want to find it. Our scientists prefer the deceptions and the lies instead. The way our scientists are going, they may never discover the Primary Axiom of Science and Statistics. Even if they were to know what it is, most of them would still reject it because they have accepted and embraced its opposite instead.

    Furthermore, our scientists haven't discovered and don't know how to use the Negating the Consequent version of the Scientific Method. They don't know what it is, how it works, or why it is the most powerful, most useful, and most reliable version of the Scientific Method. A society can't have Truth in Science if they don't know anything about the Primary Axiom of Science or the Negating the Consequent version of the Scientific Method.

    Some of our scientists are starting to believe that massless and entropyless quantum waves or transformable photons exist; but, these same people refuse to believe that conserved photons or conserved quanta exist, because psyches are conserved quanta or conserved photons. Our scientists haven't discovered the Conservation of Quantum Information within Conserved Quanta, Conserved Psyches, Conserved Photons, or Conserved Intelligences. They don't know what it is or how it works. Our scientists haven't yet discovered Conserved Quanta or Conserved Photons. Our scientists have convinced themselves that Quantum Wave Processors, or Quantum Information Processors, or Quantum Information Storage Devices do not exist because these types of things are obviously immaterial or non-physical; and, our scientists have convinced themselves that the non-physical does not exist. Our scientists have formally rejected psyche, intelligence, choice, or consciousness convincing themselves that it does not exist.

    Our scientists, psychiatrists, and medical doctors have erroneously convinced themselves that drugs or medications can do psychotherapy or psyche-therapy. They have no idea that they are wrong, because they have no idea why they are wrong. The drugs can't touch nor change one's psyche or one's soul. The drugs obviously can't do psyche-therapy. Currently, through medications or drugs, our psychiatrists give people additional mental illnesses in an attempt to treat and cure their mental illness. The Psychiatrists, or Materialists, or Naturalists literally drive people insane while trying to treat their depression, anxiety, or mental illness. That's how they make their money and their living. Their drugs produce suicidal ideation, desperation, addictions, hallucinations, delusions, paranoia, and psychosis which means that their drugs can kill you and will kill you if you let them. These people treat your mental illness by giving you additional mental illnesses. The net effect of sleeping pills is to make it so that you can't sleep naturally. Before long, you can't sleep at all; and when that happens, you quickly go psychotic and insane.

Our scientists have embraced creation by chance or chance causation, erroneously convincing themselves that everything in this universe was produced by chance alone. Collectively as a group, our scientists have yet to discover Transdimensional Physics or the Non-Physical Version of Physics. Our scientists don't have any idea what entropy is. They have dozens of different, contradictory, and self-defeating definitions for entropy. Most of our scientists have erroneously convinced themselves that entropy is disorder, chaos, chance, or death. The truth is that disorder and thermodynamics have no correlation with each other whatsoever. Disorder doesn't cause heat, and heat doesn't cause disorder. There's no correlation between the two. Our scientists have no idea that the second law of thermodynamics is false, because they have no idea why it is false, having convinced themselves that it is true instead.

Our scientists have huge gaping holes in their Science or their Knowledge.

Finally, our scientists haven't yet discovered Syntropy or the Perpetual Motion Cycle. They don't know what it is or how it works. Instead, they have convinced themselves that it doesn't exist. Most of our scientists don't know anything about Quantum Field Theory or the massless, heatless, ageless, and entropyless Quantum Fields, convincing themselves instead that the massless, the entropyless, or the non-physical does not exist. Our scientists have no idea that the Quantum Fields are in fact entropyless, massless, ageless, non-physical perpetual motion machines. Most of our scientists have convinced themselves that quantum waves and the quantum fields do not exist. Most of our scientists have convinced themselves that perpetual motion machines do not exist. They don't exist at the physical level, but perpetual motion machines or conserved machines obviously exist at the quantum level, or we wouldn't be here right now. Conserved order and organization has to exist at the quantum level; otherwise, physical matter would not exist at the physical level.

Our scientists may never discover these things, because they aren't looking for them and don't want them.

Rather than pursuing the Truth in Science, our scientists have chosen to accept, embrace, and promote the obvious lies instead. Our scientists can't have Truth in Science because they have formally and deliberately rejected it. They will never discover it because they don't want it.

The lies in "science" are rather obvious to see and understand – they consist of the things that have NEVER been experienced NOR observed by anyone, not even God.

Our people and our scientists are still in the dark ages of ignorance and superstition due to the fact that our scientists have embraced and are promoting Materialism, Naturalism, Scientism, Darwinism, Nihilism, Creation by Chance, and Creation Ex Nihilo or Atheism instead of the truth. It's holding them back. To believe that chance can design and create things, as our scientists do, is a serious logic error. It doesn't match with anything that we have ever experienced or observed. Chance is not causation.

You can't have the truth if everything you believe in is false. A society such as ours can't have the Truth in Science if they are constantly choosing to embrace and promote the lies instead. That's just the way it is.

Quantum Mechanics is Energy Mechanics, Non-Physical Mechanics, Spiritual Mechanics, or Transdimensional Physics. Transdimensional means non-physical. Transdimensional Physics is the scientific study of the Non-Physical Realm, the Non-Local Realm, the Energy Realm, the Psyche Realm, the Perpetual Motion Realm, the Transdimensional Realm, the Conserved Realm, the Exergic Realm, the Syntropic Realm, or what some of our people call the Spirit World. Transdimensional Physics is the study of energy or light – what it is and how it works when it is in its non-physical, massless and entropyless, chargeless, heatless, omnipresent, infinite acceleration, wave-like format where it has NO physical limitations.

Collectively as a group, our scientists have rejected Quantum Mechanics, Quantum Field Theory, the Perpetual Motion Cycle, Syntropy, the Perpetual Motion Realm, Conserved Quanta, Psyche, Intelligence, Consciousness, the Non-Physical Realm, and Transdimensional Physics. It's holding them back.

Quantum Mechanics, Quantum Field Theory, and the Perpetual Motion Cycle ($E = mc^2$) ARE the closest that our scientists have ever gotten to the truth; but, they have collectively chosen to reject these things and have chosen to embrace and promote Materialism, Naturalism, Darwinism, Nihilism, Atheism, Classical Realism, Creation Ex Nihilo, and Creation by Chance instead. Einstein discovered the equation behind the Perpetual Motion Cycle, but due to his Materialism and Naturalism and Classical Realism, he failed to recognize it for what it truly is.

Most of our scientists have no idea what the Perpetual Motion Cycle is and how it works. They have been blinded by their Physicalism, Naturalism, Darwinism, Atheism, and the thing that they call the second law of thermodynamics. It's holding them back and preventing them from finding the Truth in Science.

The scientists on our planet have gotten many things wrong, and it's holding them back. On our earth, our science community has adopted and is promoting something that they call the second law of thermodynamics.

The second law of thermodynamics is obviously false. The phenomena that it predicts haven't been experienced nor observed by anyone. In other words, the things that the second law predicts don't exist. The second law of thermodynamics is nothing but science fiction or pseudo-science.

The perfect disorder or chaos that the second law predicts has never been experienced nor observed, ever since the Quantum Fields were designed by God and made by Nature's Psyche. Our scientists haven't figured out, yet, that Quantum Field Theory falsifies the second law of thermodynamics.

As a result of their ignorance, our scientists now erroneously define entropy as "disorder" or "chaos". Entropy is in fact mass's heat storage capacity; and as such, entropy, heat, and thermodynamics represent a great deal of order and organization at the physical level. "Disorder" or "chaos" has absolutely NO correlation with entropy, heat, or thermodynamics; and, everything at the quantum level or the non-physical transdimensional level is perfectly organized and conserved. Everything that exists falsifies the second law of thermodynamics.

WE KNOW that the second law of thermodynamics is false because we don't observe any of the things that it predicts. It's that simple.

If you want the truth, it is important to allow ALL of the evidence into evidence, and that includes ALL of the times when entropy or mass has gone to zero and ceased to exist.

The second law of thermodynamics states that the total amount of entropy or disorder in our universe is constantly increasing and that it can never decrease and go to zero. This claim is obviously false. It has never been experienced nor observed by anyone. The second law predicts that we, this universe, and this planet should not exist; yet, that is not what we observe. Now is it?

The very fact that we exist is scientific proof that the second law of thermodynamics is false. We haven't observed anything that the second law of thermodynamics predicts. We don't observe proton decay. We don't observe an ever-encroaching gray goo coming in at us from all sides. We don't observe ever-increasing disorder or chaos. We don't even observe heat death. Heat death doesn't exist.

Heat death or absolute zero represents Maximum Efficiency when it comes to energy, photons, quantum waves, psyches, conserved quanta, quantum consciousness, quantum information, virtual particles (thoughts), quantum fields, quantum mechanics, and superconductors. Does it not? These quantum objects or transdimensional objects function perfectly and function best at absolute zero, or at the state that our scientists erroneously name as "heat death". We don't observe anything that the second law of thermodynamics predicts, which is why it is false.

Instead, we observe constantly conserved order and organization thanks to the massless and entropyless Quantum Fields, the Conservation of Energy and Psyche, the Conservation of Quantum Information within Psyche, and the Perpetual Motion Cycle $E = mc^2$. The observation and the experience of these things falsifies the second law of thermodynamics as well as creation by chance, or creation by entropy, or creation by random disorder. The Quantum Fields are massless, entropyless, ageless, and even heatless perpetual motion machines. They will never get old and they will never wear out because they have always been conserved, ever since they were made by Nature's Psyche or by the Controlling Intelligences within Nature. In order to make the second law of thermodynamics true, you would have to destroy the Quantum Fields. Then we really would have chaos and true disorder in this universe.

In True Statistics or statistics that are true, disorder or chaos cannot be used to produce population data or sample data, because by definition true disorder or true chaos doesn't correlate with anything! There is NO population data where disorder or chaos is concerned! Furthermore, in the real world, true disorder or true chaos doesn't exist as long as the Quantum Fields exist. It is erroneous, illogical, and false to base a whole science on chaos or disorder, as they do with the second law of thermodynamics, because true disorder or chaos cannot be measured, has no size of effect or significance, and doesn't correlate with anything. There is NO correlation between disorder and thermodynamics. Thermodynamics or heat or entropy (mass's heat storage capacity) in actuality represents a huge amount of order and organization. The very existence of thermodynamics falsifies the second law of thermodynamics, which says that it shouldn't exist. In other words, the second law of thermodynamics is self-defeating or self-falsifying.

The second law of thermodynamics is just another version of Creation by Chance or the Null Hypothesis, which makes it obviously false.

Mark My Words

DENYING THE PERPETUAL MOTION CYCLE

The second law of thermodynamics erroneously teaches that the total amount of disorder or entropy in our universe can never go to zero and cease to exist.

The second law of thermodynamics was designed to prevent our scientists from discovering Syntropy and the Perpetual Motion Cycle. The second law of thermodynamics denies the existence of order and organization at the quantum level, as well as at the physical level. The second law of thermodynamics is a type of materialism; and therefore, the second law denies the existence of the quantum or the conserved or the non-physical.

The second law of thermodynamics is based upon the assumption that only physical matter exists. The second law completely ignores and rejects the quantum, the transdimensional, the non-physical, the conserved, or the perpetual – anything that is eternal and everlasting.

Lies are designed to prevent us from finding and knowing the truth; and, that's how materialism and naturalism work. They prevent us from finding and knowing the truth. The materialists and naturalists deny the existence of the evidence that falsifies their beliefs.

Any form of materialism or naturalism really is the bane of science, knowledge, and truth. Materialism, naturalism, nihilism, and atheism – along with the second law of thermodynamics and theory of evolution – were designed to prevent our scientists from finding and knowing the Truths in Science that are waiting to be found. Each one of these deceptive philosophies claims as its primary axiom that the non-physical, or the transdimensional, or the indestructible does not exist. Therefore, the observed existence of anything non-physical automatically falsifies them.

Thanks to their erroneous and falsified second law of thermodynamics, our scientists incorrectly teach, preach, and believe that there is no such thing as

Syntropy or a Perpetual Motion Cycle. Their second law prevents them from finding and knowing the truth.

However, if there were no such thing as Syntropy, if there weren't any Perpetual Motion Cycle, and if the second law of thermodynamics were actually true, then none of this would exist and we wouldn't exist either. There has to be some type of Syntropy or Perpetual Motion Cycle, or we wouldn't exist. So where is it, since it obviously has to exist?

Let's start our search by looking at one of the obvious lies that our scientists have chosen to believe in.

The second law of thermodynamics erroneously states that the total amount of entropy or disorder in our universe is constantly increasing and that it can never decrease and go to zero. The second law of thermodynamics predicts that we shouldn't be here.

This is not what we experience and observe. Is it? We don't observe any of the things that the second law of thermodynamics predicts that we should be observing. We don't observe proton decay. We don't observe an ever-encroaching gray goo coming in at us from all sides. We don't observe random chaos all around us. We don't observe constantly conserved disorder. We don't observe that we don't exist. Instead, we observe order and organization all around us, all through us, and all throughout our universe. Why? It's because the second law of thermodynamics is false.

Thanks to the second law of thermodynamics, our scientists have successfully convinced themselves that Syntropy and the Perpetual Motion Cycle do not exist. Do you see the flaw in that? ALL of these faulty and falsified "sciences" that our scientists have embraced and teach are based exclusively on their unilateral claim or belief that something specific does not exist. That's NOT science or knowledge. That's wishful thinking and nothing more. That's philosophy, dogma, and religion – not science or knowledge. By definition, you can't have knowledge of things that do not exist, and you can't science things that do not exist. Yet, our scientists erroneously base whole "sciences" on things that they say do not exist. That's a logic fallacy, and they don't even know it. The only way to know that something does not exist is to observe that it does not exist. That can't be done. However, we can falsify a belief by observing that it is in fact false.

Our scientists haven't yet discovered that there is no correlation between disorder and thermodynamics. Furthermore, the second law of thermodynamics predicts that we shouldn't be here. Yet, here we are. The very fact that we exist is obvious scientific proof that the second law of thermodynamics is false. The very fact that we exist suggests that our scientists are overlooking or rejecting something extremely important in Science, and they don't even know it yet.

The second law of thermodynamics erroneously states that the total amount of entropy or disorder in our universe is constantly increasing and that it can never decrease and go to zero. Our earth's second law of thermodynamics erroneously teaches that entropy, physical matter, and disorder are conserved. Does it not?

The second law predicts that we shouldn't be here and that none of this should exist. This is obviously false.

Our scientists haven't yet discovered an Ultimate Law of Thermodynamics that differentiates between what is conserved and what is not conserved. There are also gradations of conservation that our scientists haven't discovered yet. Our scientists erroneously believe that information is conserved, when in fact, information or energy is infinitely malleable. Information is a form of energy. The form of energy is NEVER conserved, because energy and information are infinitely malleable or infinitely transformable. It's the energy and the psyche that are in fact conserved, not the form of that energy. Our scientists would know this if they had anything like the Ultimate Law of Thermodynamics.

https://ultimate-law-of-thermodynamics.com/

Psyche and energy are conserved. Psyche is a conserved quantum or a conserved photon. Psyche is the ultimate point particle. Once the quantum fields were designed and made, they have been conserved by Nature's Psyche and the Gods. In contrast, entropy (mass's heat storage capacity), mass, heat, physical matter, quantum waves, resistance to acceleration, the aging process, energy's form, and information are not conserved. Energy is information. Energy is reusable and information can be transformed. Think of radio waves. The energy is always conserved, and the psyche that puts the information into the radio waves is always conserved, but the information transmitted through the radio waves or quantum waves is constantly being transformed as needed. Quantum waves are transformable energy.

Psyches communicate with each other at a distance through quantum waves, using the quantum fields as the medium of transmission. Quantum waves are WiFi at the quantum level. Radio waves produce WiFi at the physical level. Information is transferred from psyche to psyche through these various different types of waves, both at the physical level and the non-physical quantum level. This is what has actually been experienced and observed. The form is never conserved. Only the underlying energy and psyche are conserved. Psyche is a conserved quantum or a conserved photon. Psyche and energy are conserved; whereas, everything else is not conserved. This is the Ultimate Law of Thermodynamics. It's an attempt to explain everything that has ever been experienced and observed.

According to the Law of Psyche, every psyche or intelligence has a certain amount of energy or information that is under its control, and that controlling psyche can form or transform the energy and information that's under its control into anything that it wants that energy or information to be, anytime and anywhere that it chooses to do so.

Nature's Psyche is the thing that made and is now running the quantum fields. The quantum fields are the medium through which quantum waves travel and psyches interact. Nature's Psyche makes quantum waves or transformable photons at will from mass, heat, resistance to acceleration, or entropy; and then later, when that same psyche chooses to stop the quantum wave or photon, that psyche transforms the infinite acceleration omnipresent quantum wave it has made

INTO some type of localized mass, heat, resistance to acceleration, or entropy (mass's heat storage capacity) instead. This is the Perpetual Motion Cycle ($E = mc^2$). It has always existed, and it will always exist, because the underlying energy and psyche are always conserved. The quantum fields are syntropic and conserved by Nature's Psyche; and, the Perpetual Motion Cycle is Syntropy, or the conservation of energy and psyche.

Our scientists deny the existence of the Perpetual Motion Cycle. Our scientists deny the existence of Syntropy.

Thanks to their erroneous and falsified second law of thermodynamics, our scientists haven't yet discovered Syntropy or the Perpetual Motion Cycle. They don't know what this is, how it works, why it is important, or that it even exists. Instead, our scientists have erroneously concluded that these things do not exist. Our scientists will never discover Syntropy or the Perpetual Motion Cycle by denying its existence.

It is a logic fallacy to conclude a priori that Syntropy does not exist. It is a logic fallacy to conclude a priori that the Perpetual Motion Cycle does not exist. Since we exist, these things have to exist. It is obvious that there has to be some type of order and organization in this universe, or we wouldn't exist. There has to be some type of organizing force in this universe, or we wouldn't exist. This is self-evident and obviously true.

Just as it is obvious that it is impossible to make perpetual motion machines at the physical level due to physical limitations such as mass's heat storage capacity (entropy), friction, heat, mass, thermodynamics, and resistance to acceleration; it is equally obvious that thermodynamics, mass, mass's heat storage capacity, friction, and resistance to acceleration do not exist at the quantum level in the ageless realms, the conserved realms, the syntropic realms, the massless realms, the entropyless realms, the transdimensional realms, the non-physical realms, the heatless realms, or the perpetual motion realms. This is the Quantum Law of Thermodynamics.

https://quantum-law-of-thermodynamics.com/

Scientists, out-of-body explorers, and near-death experiencers have observed that there is no aging process, no resistance to acceleration, no mass, no heat-storage capacity, no friction, no heat, and no thermodynamics at absolute zero temperature in the quantum realm, spirit world, syntropic realm, conserved realm, or perpetual motion realm. Absolute zero temperature represents maximum efficiency in the quantum realm or conserved realm. There is no such thing as heat death, death, thermodynamics, or entropy in the entropyless quantum realm or massless psychic realm. This is what has actually been experienced and observed. Is it not?

In contrast, the second law of thermodynamics erroneously teaches that entropy, physical matter, and disorder are conserved. This is obviously false. Entropy means "not conserved"; therefore, it is impossible for entropy to be conserved. Disorder is not conserved. In fact, disorder doesn't exist at the quantum level thanks to the perpetual motion machines that we call the quantum fields. Entropic physical matter obviously is not conserved! Entropic physical matter can be transformed or transmuted anytime that Nature's Psyche chooses to do so.

Psyches are conserved quanta or conserved photons. Energy or information is infinitely malleable and can be transformed from one form to another by Psyche or Intelligence, but energy and psyche are always conserved. Energy and psyche cannot be made, and they cannot be destroyed. They have always existed, and they will always exist. They are eternal and everlasting. They are the foundation of the Perpetual Motion Cycle at the quantum level in the transdimensional realms or massless realms, where entropy, mass, resistance to acceleration, and thermodynamics have not been observed and apparently do not exist.

This is really simple to understand but proves impossible for a materialist, naturalist, nihilist, behaviorist, or atheist to accept. These people prefer the deceptions and the lies instead of the obvious truths that exist in Science. The materialists, naturalists, nihilists, and atheists are unable to understand and accept any of this because they have brainwashed or conditioned themselves not to believe it, as part of the dumbing-down of their society.

Because of the second law of thermodynamics, our scientists haven't truly studied and discovered Quantum Field Theory. These people don't know yet that Quantum Field Theory falsifies their second law of thermodynamics. The massless, entropyless, heatless, intangible, non-physical quantum fields are the ultimate perpetual motion machines. The quantum fields are the best and most reliable perpetual motion machines that have ever been made. The quantum fields permeate and fill the immensity of space, including the space within each physical atom. The quantum fields are the medium through which the different types of quantum waves travel, and the different psyches or quanta or energy packets interact. The quantum fields are perfect order and organization at the quantum level in the non-physical transdimensional realm. The quantum fields are Pure Syntropy. Their very existence falsifies the second law of thermodynamics which predicts that the quantum fields shouldn't exist.

Ever since the quantum fields were designed by the Gods and made by Nature's Psyche or the Controlling Psyches within Nature, the quantum fields have been conserved because they are made from pure energy or exergy. The quantum fields will never get old and never wear out. The second law of thermodynamics will never be true as long as the quantum fields exist. You would have to destroy the perfectly ordered and organized quantum fields at the quantum level in order to make the second law of thermodynamics true. If you can do that, then you are a God.

Our scientists haven't discovered any of this yet.

On our earth, Darwinism, Materialism, Naturalism, Nihilism, Behaviorism, Hard Determinism, Atheism, and their derivatives are united in their ongoing claim that the non-physical does not exist, that psyche or consciousness or mind doesn't exist, that choice or free will does not exist, and that God does not exist.

Therefore, observation and verification of any one of these things falsifies materialism, naturalism, behaviorism, and their derivatives which claim that these things do not exist.

The observation of choices being made falsifies the primary axiom of radical behaviorism and hard determinism which states that choice, free will, or agency does not exist. The obvious fact that we can identify the difference between a chosen behavior and an involuntary behavior tells us that radical behaviorism and hard determinism are false.

It has been observed that choice is a function of psyche, consciousness, or intelligence – both here in this physical world, and in the afterlife in the spirit world. You will either choose to study this observational and experiential evidence, or you will not. It's a choice; and, it's easily identified as a choice. Even a dog can discern whether you deliberately kicked him or accidentally stepped on him. Collectively, our scientists erroneously deny the existence of choice. They have chosen to believe that choice does not exist.

Our scientists haven't yet discovered the Perpetual Motion Cycle, choosing instead to believe that perpetual motion or syntropy does not exist. Nevertheless, it has to exist, or we wouldn't exist. This is obviously true, because we obviously exist.

Quantum mechanics literally feeds off of our desires and our choices. It seems like magic to us, but it was designed by the Gods to be that way.

When the psyche or intelligence within a quantum wave chooses to stop, that psyche literally transforms that omnipresent infinite acceleration quantum wave INTO some type of mass, heat, resistance to acceleration, information, or entropy (mass's heat storage capacity) instead. Nature's Psyche is the thing that chooses to collapse the wave function or chooses to stop the quantum wave or the photon. The collapse of the wave function or photon INTO mass or physical matter is synonymous with infinite deceleration and results in resistance to acceleration. It's a transformation of energy from one form to another. All the while, the energy is conserved. The form or the information changes, but the energy and psyche are always conserved. Psyches are conserved quanta, after all.

Likewise, Nature's Psyche also makes the photon or the quantum wave in the first place by transforming some type of physical matter, mass, heat, or entropy (mass's heat storage capacity) INTO massless, heatless, chargeless, entropyless, non-physical quantum waves or photons. Round and around it goes; and, it never stops because energy and psyche are always conserved. This is the Perpetual Motion Cycle ($E = mc^2$). This is Syntropy. It will never get old, and it will never wear out. It has always existed, and it will always exist. That is why it is the Perpetual Motion Cycle.

Our scientists will NEVER discover Syntropy or the Perpetual Motion Cycle, because they have already rejected it and concluded that it does not exist. They aren't looking for it. They don't want it. Therefore, they will NEVER find it.

The quantum fields are also massless, entropyless, and non-physical perpetual motion machines. Now that the quantum fields have been designed and made, they will exist forever, unless some Psyche or Intelligence chooses to disassemble them, because energy and psyche are always conserved. Forms can be conserved only if some Conserved Psyche or Conserved Quanta chooses to conserve them. Do you see how that works? It explains everything that has ever been experienced and observed.

It has been observed that in the spirit world or the transdimensional realm, our psyche's desires and choices literally shape, or form, or organize our surroundings. A non-consensus reality in the quantum realm literally organizes itself according to the demands and expectations of the human psyche. That's what makes a consensus reality, such as a physical reality, unique. A physical reality doesn't spontaneously readjust itself to the whims of every psyche that passes by, as a non-physical spiritual quantum non-consensus reality does. In a non-consensus reality, the human psyche chooses what it wants its environment to be. In a non-consensus reality in the spirit world, the human psyche is God.

Hard determinists and radical behaviorists are united in their claim that choice, free will, and agency do not exist; therefore, any observed and verified example of a choice being made falsifies or puts the lie to this specific primary axiom or truth claim from hard determinism and radical behaviorism.

Likewise, observation of massless, entropyless, chargeless, heatless, and non-physical quantum fields or quantum waves or photons puts lie to the materialistic and naturalistic claim that the non-physical or the transdimensional does not exist.

Our scientists haven't learned how to falsify materialism and naturalism with observational evidence. Until they do, they will never discover the truths in science that are waiting to be found.

Instead, our scientists have convinced themselves that Materialism, Naturalist, Darwinism, Nihilism, Behaviorism, Hard Determinism, Atomism, and Atheism cannot be falsified. These people erroneously teach, preach, and believe that their denialistic philosophies cannot be falsified by scientific evidence, observational evidence, or experiential evidence. These people erroneously teach, preach, and believe that their denialistic philosophies cannot be falsified by the Scientific Method. They are wrong, and they don't know it, because they have convinced themselves that they are always right. They don't want to be party to the Truths in Science that falsify their beliefs.

CHANCE CANNOT DO CAUSATION

Our scientists erroneously believe that chance can design and create anything that it sets its mind to, if given enough time to do so. Our scientists haven't yet discovered the Primary Axiom of Science and Statistics which states that chance cannot do causation. Chance is not causation. By definition, chance cannot do choice, correlation, design, creation, or causation. Once it starts doing so, then it is no longer chance, but has become some type of intelligent choice or deliberate causation instead.

The flaw in the second law of thermodynamics is that it defines entropy (mass's heat storage capacity) as "disorder". Disorder is the wrong definition for entropy, or heat, or thermodynamics. Disorder doesn't produce heat, and heat doesn't produce disorder. There's no correlation between the two.

Disorder or chance cannot do causation. They cannot design and create things either.

"Difference" and "effect" mean the same thing. There's NO difference between chaos, disorder, and random chance, which means that NONE of these can produce any kind of an effect. No effect, then no causation. Disorder or chance cannot do causation. Chaos by definition is a complete lack of coordination within the system – a complete lack of causation within the system. There is NO correlation between heat and disorder. There can't be. The one cannot cause the other, because there is no correlation between the two. Random disorder or chance cannot do causation, creation, or choice. Disorder doesn't produce heat, and disorder doesn't eliminate heat. There is NO correlation between thermodynamics and disorder!

In ANY science experiment or scientific observation, if the correlation between the independent variable and the dependent variable is insignificant or non-existent, then we don't have any confidence that there is a causal relationship between the variables. In fact, we are supposed to conclude that there is NO causal relationship between the variables. There can be no correlation whatsoever between true disorder and anything else. Once we start observing some type of correlation or relationship, then we are no longer looking at pure disorder or pure chaos.

It is a scientific error and a logic fallacy to USE random disorder, chaos, or chance as the independent variable or as the dependent variable – as the Materialists, Naturalists, Darwinists, Nihilists, and Atheists do – because random disorder, pure chaos, or chance cannot do causation of any kind despite what the Naturalists and Darwinists might erroneously claim. Chance of any kind cannot produce a real effect. If it is producing a measurable effect, then it is no longer chance. Chance and causation are mutually exclusive. They falsify each other.

These people don't know what they are talking about whenever they make the claim that chance can do causation. You automatically KNOW that these people are lying to you and trying to trick you and deceive you whenever they try to convince you that random fluctuations in entropy, random disorder, chaos, or chance designed and created everything that exists in this universe. You automatically KNOW that they are wrong, because chance of any kind cannot do design and creation – or choice and causation.

Chance, disorder, and chaos are the same thing. NONE of them can produce any kind of measurable effect, which means that NONE of them can do choice or causation. Once you understand this truth, then you automatically KNOW why the second law of thermodynamics and the theory of evolution are false. Chance of any kind cannot do design and creation. It's physically impossible. It can't be done. Therefore, the theory of evolution and the second law of thermodynamics are false because they are nothing more than Creation by Chance or Chance Causation, which is impossible.

THE ULTIMATE TEST OF TRUTH IN SCIENCE AND STATISTICS

Chance of any kind cannot do choice, correlation, design, creation, or causation. If someone says that it can, then you automatically KNOW that they are lying to you.

This is an extremely powerful Truth Test or Test of Truth. The whole of Science, Statistics, the Scientific Methods, and Science Experimentation is based upon it.

You automatically KNOW that they are lying to you and trying to deceive you if they are trying to convince you that chance can design and create things if given enough time to do so.

You can give chance an eternity to design and create things and it will NEVER do so, because once it starts designing and creating things, then it is no longer chance but has become some type of choice, causation, psyche, intelligence, or creation instead. Chance and causation are mutually exclusive. Chance and choice are mutually exclusive. Chance and psyche (intelligence) are mutually exclusive. Chaos and creation are mutually exclusive. Chance and intelligent creation are mutually exclusive. Random disorder and intelligent design are mutually exclusive. Chance cannot do causation, nor choice, nor creation. Once it starts doing so, then it is no longer chance. If your favorite scientists or your college teachers are trying to convince you that chance can do design and creation, then they are lying to you, whether they know it or not.

This powerful test of Truth Assertions in Science successfully and completely separates the truths in science from the lies in science.

If you are perceptive and paying attention, you will eventually notice that ALL of the lies in science are some form of Creation by Chance, Chance Causation, Design and Creation by Random Disorder, Creation by Entropy, or Creation by Death. These people really truly believe that Chance can design and create anything if given enough time to do so; and, they are wrong.

These people literally make chance or disorder their independent variable, and then without running any kind of science experiment, they *jump straight to the conclusion* that chance or disorder CAUSED the dependent variable (proteins, genes, genomes, life forms, planets, stars, galaxies, physical matter, and universes). In a True Science Experiment, the Null Hypothesis means that there is ZERO correlation between the independent variable and the dependent variable. These people literally use the Null Hypothesis, Chance Causation, or Creation by

Chance as their independent variable, and then *jump straight to the conclusion* that chance caused everything that exists in this universe. This is the very definition of Bad Science, whether they realize it or not. This is also Bad Statistics!

The Primary Axiom of Statistics states that we cannot prove the Null Hypothesis true from the data of any science experiment. In other words, we cannot prove Creation by Chance, Chance Causality, Creation by Random Disorder, Creation by Blind Luck, or Chance Causation true from the data of any science experiment. These things are automatically false, according to the Primary Axiom of Statistics.

Therefore, Materialism, Naturalism, Darwinism, Nihilism, Atheism (Creation Ex Nihilo), the Theory of Evolution (Creation by Chance), and even the Second Law of Thermodynamics (Creation by Random Chaos) ARE the Null Hypothesis. They are ALL Creation by Chance or Causation by Chance, which means that we will NEVER be able to prove them true from the data of any science experiment, because the Null Hypothesis or Creation by Chance is automatically false in every case, because chance of any kind by axiom or by definition cannot do design and creation. By the Primary Axiom of Statistics, Materialism, Naturalism, Darwinism, Nihilism, Atheism, the Theory of Evolution, the Second Law of Thermodynamics, and every other form of Creation by Chance ARE automatically false and have automatically been falsified because they are Causation by Chance, or Creation by Chance, or the Null Hypothesis to begin with.

Our scientists have defined the second law of thermodynamics as Creation by Random Disorder or Creation by Random Fluctuations in Entropy. Look up Boltzmann Brains, Random Fluctuations in Entropy, Chemical Evolution, Spontaneous Generation, Abiogenesis, Macro-Evolution, Quantum Fluctuations and the Big Bang Theory, or Natural Selection if you don't believe me. These people have random disorder, chaos, entropy, blind luck, or chance designing and creating genomes, proteins, planets, stars, galaxies, physical matter, and universes from scratch or from nothing. These people are lying to you and trying to trick you and deceive you, because chance of any kind cannot do design and creation no matter how much time it is given to do so. You automatically KNOW that these people are lying to you whenever they try to convince you that chance or random disorder can design and create things.

This is not hard to understand. It may be hard for some people to accept, but it's not hard to understand. Even somebody in Kindergarten or Elementary School should be able to understand this concept. Chance of any kind cannot do causation, choice, design, or creation no matter how much time it is given to do so. An infinite amount of time isn't long enough to get chance to do causation. Chance cannot do causation. This primary axiom in Science and Statistics is simple enough for a child to understand, yet apparently complex enough or hard enough for a PhD scientist to completely overlook it and reject it.

Our very best and brightest have deliberately and actively rejected this truth, because they desperately want Chance, Disorder, Entropy, Natural Selection (Death), Chaos, or the Null Hypothesis to be their Designer, Creator, and God.

Their motto is: Anything but God. They are willing to believe in anything but God. They are willing to believe in even the most stupid and obviously false ideas, so long as they don't have to believe in God. Their prejudice or bias blinds them to the truth. They literally can't handle the truth.

The other Primary Axiom of Statistics states that if we observe NO correlation between the independent variable and the dependent variable, then we automatically KNOW that there is NO causal relationship between the two variables.

True disorder or true chaos by definition can have NO correlation with anything else in this universe. If it starts to correlate with something, then it is no longer disorder or chaos. If we are observing true disorder or true chaos, then we are by definition observing NO correlation. NO correlation means NO causation. Pure disorder means no causal relationship with anything else.

The Null Hypothesis is synonymous with No Correlation. The Null Hypothesis and No Correlation are synonymous with No Causation. It's the same idea. It's the same Primary Axiom in Statistics. The Null Hypothesis means "no correlation"; and, "no correlation" means "no causation", which means that the Null Hypothesis cannot do causation. The Null Hypothesis and Creation by Chance are the same thing. The Null Hypothesis or Creation by Chance are synonymous with "no causation" or "no correlation". Once again, we see that the Null Hypothesis or Creation by Chance cannot do causation, correlation, choice, or creation – according to the primary axioms of Statistics and Science.

Any way we choose to look at it, the Primary Axioms of Statistics and Science repeatedly state that chance cannot do choice, correlation, creation, or causation. Therefore, it is unscientific and false to claim that chance can do design and creation if given enough time to do so. In Science and Statistics, the false is falsified by the truth. The truth is that chance of any kind cannot do design and creation no matter how much time it is given to do so. An eternity is not enough time for chance to do design and creation, because chance cannot do design and creation, according to the Primary Axioms of Science and Statistics.

By definition, true disorder or true chaos has NO correlation with anything else. Since there is NO correlation between "disorder" and "heat", and since there is NO correlation between "thermodynamics" and "entropy defined as disorder", there is NO way to do significance testing or statistics ON "disorder", a "correlation coefficient of zero", the "null hypothesis" or "creation by chance", or "entropy defined as disorder".

It's unscientific to base a "science" on chance or disorder as the Materialists and Naturalists do. "Entropy defined as disorder" doesn't correlate with anything, therefore, it has NO statistical significance and NO scientific significance. Disorder or chance is clearly the WRONG definition for entropy, heat, and thermodynamics. Disorder has NO correlation with heat or thermodynamics. Heat or thermodynamics represents a huge amount of order and organization at the quantum level, as well as the physical level, whether we realize it or not. Defining entropy as "disorder" or "creation by chance" falsifies the second law of thermodynamics, whether the realize it or not. It violates the primary axiom of Science and Statistics which states that chance cannot do causation.

The ONLY part of the second law of thermodynamics that is demonstrably true is the obvious fact that physical matter cannot be used to design and run perpetual motion machines. Physical matter wasn't designed to work that way. Physical matter isn't conserved; therefore, it can't function as a perpetual motion machine. It's that simple! Physical matter has physical limitations, resistance to acceleration, and locality which prevent it from functioning as a usable and dependable perpetual motion machine in the real world.

The ONLY part of the second law that is demonstrably true is the "thermodynamics"; and, thermodynamics or heat has absolutely nothing to do with "disorder" or "chaos".

Isn't that ironic? "Disorder" and the "disorder definition for entropy" have absolutely nothing to do with thermodynamics or heat. Disorder doesn't produce heat, and disorder doesn't eliminate heat. There's NO correlation between the two, which means that "disorder" is the wrong definition for thermodynamics or entropy. The second law of thermodynamics is internally inconsistent. It falsifies itself. The second law also doesn't match with what has actually been experienced and observed. It's obvious that the second law of thermodynamics is false.

Whether people realize it or not, there is a huge amount of order and organization at the quantum level within heat, thermodynamics, and entropy. Our scientists don't realize and know that entropy (mass's heat storage capacity) represents a huge amount of order and organization at the quantum level, even at "heat death" or absolute zero. The massless and entropyless Quantum Fields are perfect order and organization at the quantum level, especially at "heat death" or absolute zero. Over and over again, the Science reveals that you would have to destroy the Quantum Fields in order to make the second law of thermodynamics true. The Quantum Fields are the ultimate perpetual motion machines, because they are being conserved. Their very existence falsifies the second law of thermodynamics which says that they do not exist.

Remember, there is NO correlation between disorder and thermodynamics. Disorder does not produce heat, and heat does not produce disorder. By definition, there's NO correlation between disorder, or chaos, or chance AND anything else that we might experience and observe. Once chance starts to do something, then it is no longer chance.

The sun is often portrayed as maximum chaos or maximum disorder, and there's tons of mass and heat and therefore "entropy" or "heat storage capacity" that's associated with a sun. The absence of Heat or "heat death" doesn't maximize disorder as our second law of thermodynamics claims. Heat doesn't guarantee order, either. There's NO correlation between heat and disorder!

For contrast, notice that there is NO mass, NO heat, and NO entropy associated with photons, quantum waves, and the quantum fields, which is why these massless and entropyless quantum objects represent perfectly conserved syntropic Order and Organization at the quantum level in the Spirit World or the Quantum Realm. Our second law of thermodynamics erroneously states that massless and entropyless, perfectly organized, quantum objects or quantum waves or quantum fields DO NOT EXIST.

The Quantum Fields are massless, entropyless, syntropic, exergic, and conserved perpetual motion machines. The Perpetual Motion Cycle $E = mc^2$ runs on energy, which is why it is constantly and forever conserved. The Conservation of Energy and Psyche, as well as the Conservation of Quantum Information within Psyche or Intelligence, is what makes these quantum perpetual motion machines possible at the quantum level in the Transdimensional Non-Physical Realm. Quantum objects (and even Superconductors) don't need heat as a lubricant to make them run – in fact, these perpetual motion machines run best and most efficiently at or near "heat death" or absolute zero.

This is what we have experienced and observed, is it not?

These truths falsify the second law of thermodynamics.

The second law of thermodynamics will never be true as long as the Quantum Fields exist. You would have to destroy the Quantum Fields in order to return our universe to the Chaos Realm from whence it was originally organized. You would have to destroy the Quantum Fields in order to make the second law of thermodynamics even remotely true. Even then, the second law of thermodynamics would still be false thanks to the Conservation of Energy and Psyche. Photons, Quantum Waves, Quantum Fields, Conservation of Energy and Psyche, Conservation of Quantum Information within Psyche, Non-Physical Perpetual Motion Machines, and Syntropy ARE scientific proof that the second law of thermodynamics is false.

When our scientists finally figure out how things really work in the real world, they will vote to replace the falsified second law of thermodynamics with Quantum Field Theory, Conservation of Energy and Psyche, the Conservation of Quantum Information within Psyche, and the Perpetual Motion Cycle instead. They will replace the second law of thermodynamics with the things that have actually been experienced and observed.

THE PRIMARY AXIOM OF SCIENCE AND STATISTICS

So, what is the Primary Axiom of Science and Statistics?

In case you haven't guessed it yet, the Primary Axiom of Science and Statistics states that CHANCE cannot do choice, correlation, design, creation, or causation. The whole of Science and Statistics exists to demonstrate to us and prove to us that CHANCE cannot do causation.

Yet, what do we observe from the scientists on our earth? Collectively, in complete violation of the Primary Axiom of Science and Statistics, they have chosen to believe that CHANCE caused or produced everything that exists in this universe.

Our scientists have deliberately chosen to reject the Primary Axiom of Science and Statistics and have chosen to go with a falsified primary axiom instead. Our scientists reject this obvious truth that chance of any kind cannot do choice, creation, or causation. Therefore, our scientists will never have the truth until after they have changed their minds. It's that simple.

When it comes to our scientists, their falsified primary axiom of science states that CHANCE caused or produced everything that exists. This is the primary axiom of materialism, naturalism, nihilism, and atheism. For the few of us who know about this, it's simply called the Primary Axiom. Their Primary Axiom is obviously false, but that will never convince them to look for and accept something better instead, because they prefer the deceptions and the lies over the truth.

Our scientists start with the pre-chosen conclusion that CHANCE designed and created everything that exists, and then they use that pre-chosen conclusion as "scientific evidence" to prove that their pre-chosen conclusion is correct. Furthermore, our scientists force their philosophy of science, their interpretations of statistics, and their interpretations of science to match with this pre-chosen conclusion that CHANCE made everything that exists in this universe. That's why our scientists will never find and know the truth. They aren't looking for it, and they don't want it. Our scientists have already rejected the Primary Axiom of Science and Statistics which states that chance of any kind cannot do choice, correlation, creation, or causation.

The other "science" where our scientists have gotten everything wrong deals with Darwinism or what they call the theory of evolution. Like the second law of thermodynamics, the theory of evolution is obviously false. Both of these are in fact used as and presented to us as Creation by Chance or Causation by Chance, which means that they are obviously false, because chance of any kind cannot do design and creation.

WE KNOW that the second law of thermodynamics is false because we don't observe any of the things that it predicts. Likewise, WE KNOW that the theory of evolution is false because we don't observe any of the things that it predicts. It really is that simple! Evolution of any kind has never been experienced nor observed nor replicated anywhere, in the lab or in the wild. That's because the theory of evolution is false.

The thing that our scientists call "natural selection" doesn't touch our genes. Natural selection just sits around and waits for us to die. Natural selection isn't a person, place, or thing. Natural selection is death. In our science community, natural selection has become Creation by Death. Death cannot produce life. Our scientists truly believe that death or natural selection can design and create things. These people are obviously wrong. Natural selection is just another form of Creation by Chance or Chance Causality, which means that it's obviously false.

Likewise, the different forms of the theory of evolution such as chemical evolution, abiogenesis, spontaneous generation, macro-evolution, natural selection, and creation ex nihilo ARE design and creation by chance, which is why they are automatically false, because chance of any kind cannot design and create anything.

It has never been experienced nor observed doing so, because it can't. Chance cannot do causation. Once chance starts doing causation, then it is no longer chance, but has become some type of choice, creation, or intelligent causation instead. ONLY psyche or intelligence or consciousness can do causation or choice or creation at all levels of existence, including both the Non-Physical Quantum Level and the Physical Level. It's the ONLY thing that has been observed doing so.

Chance and causation are mutually exclusive. Once chance actually starts to cause something to happen, then it is no longer chance. Creation by Chance or Causation by Chance are impossible, which means that Materialism, Naturalism, Darwinism, Nihilism, Atheism, the Theory of Evolution, Creation by Random Disorder, and the Second Law of Thermodynamics are impossible because by definition they are produced by chance alone – they ARE Chance Causality, or Creation by Chance, or the Null Hypothesis. Chance cannot do causation, which is why the theory of evolution is false, because the theory of evolution is by definition Creation by Chance.

This must be repeatedly emphasized and eventually accepted if we truly want to find and know the truth in Science, and Statistics, and Science Experiments.

The Primary Axiom of Statistics states that we cannot prove the Null Hypothesis true from the data of any science experiment. In other words, we cannot prove Creation by Chance, Chance Causality, Creation by Random Disorder, Creation by Blind Luck, or Chance Causation true from the data of any science experiment. These things are automatically false, according to the Primary Axiom of Statistics. The Primary Axiom of Science and Statistics correctly states that CHANCE of any kind cannot do choice, correlation, creation, or causation. Anything that violates the Primary Axiom of Science and Statistics is automatically false.

Therefore, Materialism, Naturalism, Darwinism, Nihilism, Atheism (Creation Ex Nihilo), the Theory of Evolution (Creation by Chance), and even the Second Law of Thermodynamics (Creation by Random Chaos) ARE the Null Hypothesis. They are ALL Creation by Chance or Causation by Chance, which means that we will NEVER be able to prove them true from the data of any science experiment, because the Null Hypothesis or Creation by Chance is automatically false in every case, because chance of any kind by axiom and by definition cannot do design and creation. By the Primary Axiom of Statistics, Materialism, Naturalism, Darwinism, Nihilism, Atheism, the Theory of Evolution, the Second Law of Thermodynamics, and every other form of Creation by Chance ARE automatically false and have automatically been falsified because they are Causation by Chance, or Creation by Chance, or the Null Hypothesis to begin with.

This is how we separate the truths from the lies in Science and Statistics.

Remember, we cannot prove the Null Hypothesis or Creation by Chance true from the data of any science experiment, because according to the Primary Axiom of Statistics, Creation by Chance, or Chance Causality, or Causation by Chance is automatically false. Chance and causation are

mutually exclusive. Chance and choice are mutually exclusive. Chance and creation are mutually exclusive. Chance cannot do choice, causation, nor creation. Once chance starts doing something, then it is no longer chance but has become some kind of intelligent choice or intelligent causation instead. This reality and truth applies at every level of existence, whether we are talking about the quantum non-physical nonlocal level or the localized physical level that we are familiar with.

Whenever the correlation between the independent variable and the dependent variable is insignificant, or whenever our scientific data is near the null hypothesis (which is produced by chance alone), then we don't have any confidence that there is a causal relationship between the variables. We are therefore to end our science experiment by concluding that there is no causal relationship between our independent variable and our dependent variable. That's the way Science and Statistics are supposed to work, because chance of any kind cannot do causation.

There is NO correlation between true disorder and anything else in this universe, therefore, disorder or chance cannot be the true cause of anything in this universe, because chance or disorder by definition cannot produce a real effect. Furthermore, in True Science and True Statistics, chance or disorder cannot be used as our independent variable or causal variable as the Materialists, Darwinists, Atheists, and Naturalists do. These people have erroneously chosen to believe that everything in our universe was produced or caused by chance alone; and, these people are obviously wrong, according to the primary axioms of Statistics and Science.

This is the most powerful test of Truth Assertions ever designed and made in Statistics and Science. It automatically separates the Truths in Science from the lies in science. Creation by Chance or Causation by Chance is automatically false. It will never be true. If someone is trying to convince you that chance can do design and creation if given enough time to do so, then you just automatically KNOW that they are lying to you and trying to trick you and deceive you. The whole of Science and Statistics tells us that they are wrong. That's why we have Science and Statistics in the first place – to identify and eliminate anything that was produced by chance alone, starting with Materialism, Naturalism, Darwinism (Creation by Chance), Nihilism, Atheism (Creation Ex Nihilo), the Theory of Evolution (Creation by Natural Selection or Rocks or Death), and the Second Law of Thermodynamics (Creation by Random Disorder or Creation by Chaos).

The Primary Axiom of Science and Statistics, as well as its corollaries, states that chance of any kind cannot do choice, correlation, creation, or causation. If a theory violates the Primary Axiom of Science and Statistics, then that theory is automatically false and has automatically been falsified, whether we realize it or not. That's the power of the Primary Axiom of Science and Statistics. It falsifies the things that are actually false.

Materialism, Naturalism, Darwinism, Nihilism, Atheism, the Theory of Evolution, the Second Law of Thermodynamics, and every other type of Creation by Chance violate the Primary Axiom of Science and Statistics; therefore, they are automatically false and have automatically been falsified. Separating the Truths in Science from the lies in science really is this simple and obvious. The truth of it is hiding in plain sight where nobody can see it because nobody is looking for it and nobody wants it to be true. Self-deception works, and it works every time.

The sooner we discover the primary axiom in Science and Statistics, the better.

It is obvious that some type of Quantum Information Processor, or Quantum Wave Processor, or Intelligent Psyche has to exist at the quantum level; or, we wouldn't exist; and, we wouldn't have quantum waves, quantum fields, quantum information, photons, and collapsed wave functions (physical matter). We are still arguing as to the true identity of this Universal Intelligence or Universal Psyche who designed and organized and made everything; but, there is no doubt that it exists. You can't have quantum information and quantum waves without some type of Quantum Information Processor or Quantum Wave Processor that is capable of making, transmitting, receiving, processing, analyzing, and storing quantum waves or quantum information. You can't have conserved quantum information (memories, blueprints, physical laws, standardization) without some type of conserved quantum or conserved psyche to process it, analyze it, and store it.

Psyches or conserved quanta make and transmit quantum waves or quantum information; and, psyches or conserved quanta collapse the wave function thereby transforming a photon or a quantum wave into some type of mass or heat instead.

It's obvious that Nature's Psyche or the Controlling Psyches within Nature made the Quantum Fields, and this same Psyche or Intelligence has been conserving or maintaining the Quantum Fields ever since.

The question is whether there is a God who taught Nature's Psyche how to make the Quantum Fields and the Physical Matter, or whether the Controlling Psyches in Nature figured that out all on their own. Somewhere sometime there had to be The Person or The Intelligence who figured out how to make Quantum Fields and Physical Matter from the Quantum Fields. It is obvious from Quantum Field Theory that the Gods or the Controlling Psyches had to design and make the Quantum Fields BEFORE they could make and sustain physical matter. But, the identity of these Gods or Controlling Psyches isn't so obvious, unless they actually choose to introduce themselves to you in person.

Now, it's fully possible that the Controlling Psyches in Nature have always known how to make and sustain the Quantum Fields. Therefore, it's possible that the Quantum Fields have always existed and that some type of unorganized matter has always existed. If that's the case, then it ends up being the Gods who organized the planets, stars, and galaxies out of all of that unorganized matter that existed BEFORE the Gods arrived in this part of the multiverse. It could have gone either way; but either way, something quantum and non-physical has to KNOW how to make and sustain Quantum Fields so that the physical atoms can actually exist in the first place.

According to the Philosophy of Science, there are many different Scientific Methods or scientific methodologies that we can use in our pursuit of the truth.

ALL of the different Scientific Methods can be used to FALSIFY the Theory of Evolution, the Second Law of Thermodynamics, Materialism, Naturalism, Darwinism, Nihilism, Atheism, Creation by Chance, Creation Ex Nihilo, or anything else that is by definition produced by chance alone because chance by definition cannot do causation. When we scientists FALSIFY a theory or an idea, we do indeed PROVE that it is false. Negating the Consequent or Falsifying a Theory is a philosophically valid and logically sound Scientific Method than can be used to PROVE that our ideas and theories are false.

Therefore, it is NO exaggeration and it is NO joke whenever I state the obvious fact that the Scientific Methods PROVE that the Theory of Evolution and the Second Law of Thermodynamics are false, because that's precisely what they were designed to do.

NEGATING THE CONSEQUENT

Our scientists have yet to discover and implement the most powerful version of the Scientific Method – Negating the Consequent. They don't even know what it is or that it exists. They don't use it because they don't know about it, and they aren't looking for it and don't want it, because it falsifies the denialistic philosophies of science including materialism, naturalism, and the theory of evolution.

Here is the template and the syllogism for a generic version of Negating the Consequent. This is the most powerful form of the Scientific Method:

Scientific Hypothesis: If Theory X is true, then we will observe Y.

Scientific Observations: We don't observe Y.

Scientific Conclusion: Therefore, Theory X is false and has been falsified by the Scientific Method because we don't observe any of the things that Theory X predicts.

This IS the Scientific Method in action. The fact that we don't observe Y is scientific proof that Theory X is false. This is called "falsifying a theory" or "negating the consequent". We have indeed proven that Theory X is false. It's that simple, which is why Negating the Consequent is the most powerful and the most useful form of the Scientific Method. When you falsify a theory, you have indeed proven that it is false.

—

Here's how the Negating the Consequent version of the Scientific Method is used in real life:

> **Scientific Definition:** The Theory of Evolution makes the claim that every genome and every life form on this planet was produced by chemical evolution, macro-evolution, abiogenesis, spontaneous generation, and random chance.
>
> **Scientific Hypothesis:** If the Theory of Evolution is true, then we will observe chemical evolution, macro-evolution, abiogenesis, spontaneous generation, and creation by random chance.
>
> **Scientific Observations:** We don't observe chemical evolution, macro-evolution, abiogenesis, spontaneous generation, or creation by random chance in the lab or in the wild. We don't observe any of the things that the Theory of Evolution predicts that we should be observing. Instead, we observe that anything that is made obviously needed a Maker to make it.
>
> **Scientific Conclusion:** Therefore, the Theory of Evolution is false and has been falsified by the Scientific Method because we don't observe any of the things that the Theory of Evolution predicts.

By Negating the Consequent and through the Scientific Method, we have indeed falsified the Theory of Evolution or proven that it is false. It's that simple. We don't observe any of the things that the Theory of Evolution predicts, therefore we KNOW that the Theory of Evolution is false.

—

Let's apply the Negating the Consequent version of the Scientific Method to Radical Behaviorism or Hard Determinism:

> **Scientific Definition:** Radical Behaviorism or Hard Determinism makes the claim that there is no such thing as choice, free will, free agency, or volition.
>
> **Scientific Hypothesis:** If Radical Behaviorism or Hard Determinism is true, then we will observe nothing on this planet making choices of any kind. Human beings and the animals would be exactly like the rocks – completely reactionary – if Radical Behaviorism or Hard Determinism were true. If Radical Behaviorism were true, then you would have never chosen to find and read this essay.
>
> **Scientific Observations:** We don't observe the complete absence of choice on this planet. We don't observe any of the things that Radical Behaviorism or Hard Determinism predicts that we should be observing. Instead, we observe humans and animals making choices every single second of their lives. We have a definition for choice because it has actually been experienced and observed, and we recognize it when we see it.
>
> **Scientific Conclusion:** Therefore, Radical Behaviorism or Hard Determinism is false and has been falsified by the Scientific Method

because we don't observe any of the things that Radical Behaviorism or Hard Determinism predicts.

By Negating the Consequent and through the Scientific Method, we have indeed falsified Radical Behaviorism or proven that it is false. It's that simple. We don't observe any of the things that the Radical Behaviorism or Hard Determinism predicts, therefore we KNOW that the Radical Behaviorism is false.

We don't observe the inability to choose in any living organism, animal, or human being. Even the ones that are crippled or mentally handicapped are still making choices.

—

Atheism or Creation Ex Nihilo is falsified by the Negating the Consequent version of the Scientific Method due to a complete lack of observational evidence supporting the concept:

Scientific Definition: Atheism of any kind is Creation Ex Nihilo – the creation of something by nothing from nothing. If you observe carefully, even the majority of the Christians on our planet are Atheists and have chosen to believe in Creation Ex Nihilo or Magic.

Scientific Hypothesis: If Atheism true, then we will observe creation ex nihilo, spontaneous generation, and abiogenesis in the lab, in a vacuum, and in the wild. We will observe life forms, rocks, and water springing into existence out of thin air and within a vacuum all around us all the time in real time. An empty vacuum will be constantly producing rocks, minerals, and life forms, if Creation Ex Nihilo or Atheism is true.

Scientific Observations: We don't observe chemical evolution, macro-evolution, abiogenesis, spontaneous generation, creation by random chance, or creation ex nihilo in the lab, in a vacuum, or in the wild. We don't observe any of the things that Creation Ex Nihilo or Atheism predicts that we should be observing. In fact, Louis Pasteur falsified Spontaneous Generation, Abiogenesis, Chemical Evolution, Creation Ex Nihilo, Materialism, Naturalism, Atheism, and the Theory of Evolution in 1859.

Scientific Conclusion: Therefore, Atheism or Creation Ex Nihilo is false and has been falsified by the Scientific Method or by the Observations of the human race because we don't observe any of the things that Creation Ex Nihilo or Atheism predicts.

By Negating the Consequent and through the Scientific Method, we have indeed falsified Creation Ex Nihilo or Atheism and have proven that it is false. It's that simple. We don't observe any of the things that Atheism predicts, therefore we KNOW that Atheism or Creation Ex Nihilo is false.

The truth consists of the things that have actually been experienced and observed. The Materialists, Naturalists, Darwinists, Nihilists, Behaviorists,

Determinists, and Atheists predict and claim that the non-physical does not exist. Therefore, the proven and verified existence of anything non-physical – such as quantum waves or thoughts, quantum fields, gravity, magnetism, radio waves, microwaves, dark energy, quantum information or non-physical information, psyche or intelligence or consciousness, photons, or light – falsifies Materialism, Naturalism, Darwinism, Nihilism, Atheism, and their derivatives which claim that these non-physical things do not exist. Any one of these is sufficient to falsify Materialism, Naturalism, Darwinism, Nihilism, and Atheism; and collectively, they make it obvious or should make it obvious that Materialism, Naturalism, Darwinism, Nihilism, Behaviorism, Atheism, and their derivatives are false. We don't observe the non-existence of these things as Naturalism and Atheism predict that we should be observing.

We don't observe the non-existence of psyche, intelligence, or consciousness; therefore, WE KNOW that Materialism, Naturalism, and Atheism are false because they predict the non-existence of these things. We don't observe the non-existence of quantum waves or thoughts, quantum fields or organized energy, quantum information or non-physical information, and quantum waves or photons; therefore, WE KNOW that Materialism, Naturalism, and Atheism are false because they predict the non-existence of these things.

We don't observe any of the things that the different forms of Atheism predict, therefore WE KNOW that the different forms of Atheism are false. It's that simple; and, we really have used the Scientific Method TO PROVE that Materialism, Naturalism, Darwinism, Nihilism, Atheism and their derivatives are false.

If someone is trying to convince you that the Scientific Method can't be used to prove anything, then you automatically KNOW that they are lying to you, because the Scientific Methods can be used TO PROVE that Materialism, Naturalism, Darwinism, Nihilism, Atheism, and their derivatives are false.

—

Here's another real-life theory that is falsified by the Negating the Consequent version of the Scientific Method:

> **Scientific Definition: Almost every scientist on our planet has chosen to believe that Chance can design and create things if given enough time to do so. They have chosen to believe that random chance designed and created everything that exists in this universe. This belief system is called Chance Causation, Creation by Chance, the Null Hypothesis, or Chance Causality.**
>
> **Scientific Hypothesis: If Creation by Chance is true, then we will observe Chance designing and creating things in the lab and in the wild. We will observe life forms spontaneously generate out of the swamps in real time. We will observe the metal and the rocks come alive all around us all the time, if Creation by Chance is true. Computers and aircraft will suddenly come into existence by chance alone if Chance Causation is true.**

Scientific Observations: We don't observe Chance designing and creating anything. We don't observe chemical evolution, macro-evolution, abiogenesis, spontaneous generation, or creation by random chance in the lab or in the wild. We don't observe any of the things that Creation by Chance predicts that we should be observing. Instead, we have observed that Chance of any kind cannot do choice, correlation, design, creation, or causation no matter how much time it is given to do so.

Scientific Conclusion: Therefore, Creation by Chance is false and has been falsified by the Scientific Method because we don't observe any of the things that Creation by Chance predicts.

By Negating the Consequent and through the Scientific Method, we have indeed falsified Creation by Chance or proven that it is false. It's that simple. We don't observe any of the things that Creation by Chance predicts, therefore we KNOW that Creation by Chance is false.

Chance cannot do choice, correlation, creation, or causation. It has NEVER been observed doing so because it can't. The whole purpose of Science and Statistics is to find the True Cause of an event, and chance can never be the True Cause of anything because chance cannot produce a real effect. You don't want any of your scientific conclusions or scientific interpretations to be based upon chance factors, because chance will falsify any scientific interpretation that you happen to make, because chance of any kind cannot produce causation or a real effect.

Creation by Chance or Chance Causation is the Null Hypothesis, and once you have chosen to retain the Null Hypothesis in a science experiment, then you automatically KNOW that there is NO causation between your independent variable and your dependent variable, and you automatically KNOW that your Alternative Hypothesis is false, because you automatically KNOW that chance of any kind cannot do causation. Chance and causation are mutually exclusive. Chance and choice are mutually exclusive. Chance and intelligent design are mutually exclusive. Once "chance" starts doing causation, then it is no longer chance but has become some type of order, organization, intelligent choice, correlation, deliberate design, creation, or causation instead.

Chance cannot do choice, correlation, creation, or causation. This is the Primary Axiom of Science and Statistics, whether we realize it or not.

ALL of the falsehoods and lies in our "science" would begin to disappear and over time hopefully completely disappear if our scientists were to accept the Primary Axiom of Science and Statistics and were to embrace the obvious truth that chance cannot design and create anything, no matter how much time it is given to do so.

—

Here's another example of the Negating the Consequent version of the Scientific Method in action:

Scientific Definition: The Second Law of Thermodynamics states that the total amount of entropy or disorder in our universe is constantly increasing and that it can never decrease and go to zero.

Scientific Hypothesis: If the Second Law of Thermodynamics is true, then we will observe proton decay, an ever-encroaching gray goo coming in at us from all sides, and ever-increasing amounts of disorder and chaos all around us. We will observe the whole universe falling into chaos all around us.

Scientific Observations: We don't observe proton decay, an ever-spreading gray goo, or ever-increasing disorder and chaos in the lab or in the wild. We don't observe any of the things that the Second Law of Thermodynamics predicts that we should be observing. Instead, we observe constantly conserved order and organization all throughout our universe. The very fact that we exist is scientific proof that the second law of thermodynamics is false, because the second law predicts that we shouldn't exist.

Scientific Conclusion: Therefore, the Second Law of Thermodynamics is obviously false and has been falsified by the Scientific Method because we don't observe any of the things that the Second Law predicts.

By Negating the Consequent and through the Scientific Method, we have indeed falsified the Second Law of Thermodynamics or proven that it is false. It's that simple. We don't observe any of the things that the Second Law predicts, therefore WE KNOW that the Second Law of Thermodynamics is false.

You can't have the Truth in Science if you don't know anything about the Perpetual Motion Cycle – what it is and how it works. Instead of the Perpetual Motion Cycle, we have the second law of thermodynamics, which is obviously false. The second law of thermodynamics predicts that nothing should exist, and that everything should be random chaos or complete disorder. But, that's not what we observe. Is it?

Instead of complete disorder and chaos, we observe Syntropy, Conserved Order and Organization, or the Perpetual Motion Cycle all around us all the time.

The one that is observed is the one that's actually real and true.

WE KNOW that the second law of thermodynamics is false because we don't observe any of the things that it predicts that we should be observing. We observe and experience the exact opposite instead. Therefore, WE KNOW that something within the second law of thermodynamics is fundamentally flawed, and we should find ourselves looking for the Perpetual Motion Cycle, Exergy, or Syntropy instead.

As scientists, we should look for and study the things that are actually being experienced and observed, rather than accepting and embracing the things that are obviously false and have never been experienced nor observed.

—

That's Negating the Consequent or Falsifying a Theory in action. This is extremely powerful Science and a supremely powerful and useful version of the Scientific Method because it proves things false. We really have proven that these things are false. If you successfully eliminate everything that is false, then only the truth will remain. This is Logic 101. We find the truth through a process of elimination. In Science, we find the truth by identifying and eliminating everything that is false. The false is falsified by the truth, and the truth ends up being everything that has ever been experienced or observed by someone sometime somewhere.

This is the pinnacle of the Philosophy of Science; and, you KNOW that it is true because the materialists and naturalists have already rejected it without having seen it and without knowing that it exists. The materialists, naturalists, nihilists, and atheists automatically deny the existence of the evidence that falsifies their beliefs, therefore, it quickly becomes obvious which observational evidence or scientific evidence falsifies their beliefs, because these people pretend that it does not exist. The evidence that they don't want you to see is the evidence that falsifies their beliefs.

It's obvious that our scientists haven't yet discovered the Negating the Consequent version of the Scientific Method, or they would already be using it to falsify the things that are obviously false. One cannot use something to find and know the truth, if they don't know that it exists or don't know how to use it.

Remember, if we don't observe any of the things that a theory predicts, then we automatically KNOW that it is false. A phenomenon as well as a theory has to be experienced and observed in order for it to be real and true. If a phenomenon or a theory has never been observed nor experienced by anyone, then WE KNOW that it is false and does not exist. WE KNOW that the Theory of Evolution and the Second Law of Thermodynamics are false because they don't match with anything that we have experienced and observed. In other words, they have NO observational support, so they are obviously false.

Falsifying the lies in science really is that simple; yet, our scientists will continue to reject this concept because they don't want it to be true. They prefer the deceptions and the lies instead. That's just the way it currently is on our planet. Knowing about the Negating the Consequent version of the Scientific Method doesn't mean that they will actually adopt it. Knowing the truth and accepting the truth are two completely different things.

A planet can go for thousands of years or even millions of years without ever "discovering" the Theory of Evolution or the Second Law of Thermodynamics, because those two are obviously false. They shouldn't have been "discovered" in the first place, because they are demonstrably wrong and because they have been falsified by Science, Knowledge, Observations, Experiences, and the Scientific Methods. One cannot have the truth if he or she chooses to embrace the lies instead. Some lies are better left undiscovered; and, some lies are avoidable, simply by never creating them in the first place. The Theory of Evolution and the Second Law of Thermodynamics are two such lies. Our scientists would be better off if these lies had never been created to begin with. These lies were avoidable because they are obviously false.

However, every human-inhabited planet in the universe has Materialists, Naturalists, Mechanists, Physicalists, Determinists, and Atheists who have erroneously chosen to believe that the non-physical does not exist, that the transdimensional realm does not exist, that the conservation of energy does not exist, that quantum mechanisms or spiritual mechanisms do not exist, that quantum waves and quantum fields do not exist, that psyche or intelligence or consciousness does not exist, that quantum information conservation does not exist, that quantum information or non-physical information does not exist, and that God does not exist because they personally have never experienced these things or think that they have never experienced these things.

Every photon, quantum field, and quantum wave is scientific proof that the non-physical exists, which means that every photon, quantum field, and quantum wave is scientific proof that Materialism, Naturalism, Darwinism, Nihilism, Atheism, and their derivatives are false.

Just because WE KNOW that Naturalism and Atheism are false, that doesn't prevent a human society from having Naturalists and Atheists. The existence of these types of atheistic and naturalistic humans is unavoidable because each person chooses for himself or herself what they want to believe, what they want to promote, and what they want to reject – irrespective of what is actually real and true. Proving to these people that they are wrong won't get them to change their minds. Changing their minds is something that they have to do for themselves, or it's not going to happen. It's not something that you can do for them. It wouldn't be real, and it wouldn't be true, if you did it for them or forced it upon them. Conversion to the Truth is something that has to be done from within the individual at the psyche level or the spiritual level, or it isn't real, and it doesn't last.

Truth is something that has to be found and owned by the individual. Each person has to choose to find the truth, choose to accept the truth, choose to own the truth, and choose to believe the truth; or, it isn't real, and it doesn't take. The truth can't be forced upon us, especially if we don't want it to begin with. Most of the time the truth hurts, especially if it is falsifying one's cherished lies or pre-chosen beliefs; but, the truth also sets us free once we finally choose to accept it. Truth is like a two-edged sword – it cuts both ways – it can hurt, but it can also be liberating and set us free. It all depends upon what one it trying to accomplish with one's life and one's choices.

OBSERVATION AS A SCIENTIFIC METHOD

OBSERVATION is one of the Scientific Methods; and, what do we observe?

We observe that we exist.

The very fact that we exist is scientific proof that the second law of thermodynamics is false, because the second law of thermodynamics predicts or says that we shouldn't exist. It's obvious that the second law is false.

If disorder or entropy were truly a conserved substance as the second law of thermodynamics claims, then we wouldn't exist. There has to be some type of Syntropy or Exergy within the system in order for us to exist in the first place. That massless entropyless Syntropy or Conservation exists and is stored within the Energy, the Quantum Waves, the Quantum Information, the Psyches or Intelligences, the Photons, and the Quantum Fields. These things are different types of non-physical, immaterial, perpetual motion machines. Psyche is a conserved quantum. The Quantum Realm or Psyche Realm is the Perpetual Motion Realm, or the Energy Realm, or the Conserved Realm, or the Non-Physical Realm, or the Syntropic Realm. It will never get old, never wear out, and never die because it is constantly conserved. It is eternal and everlasting. It has always existed, and it will always exist. Energy cannot be made, and it cannot be destroyed. It's that simple. The Truth in Science always is.

You would have to destroy the Energy, the Psyches or Quantum Consciousness, the Quantum Information within Psyches or Intelligences, the Quantum Realm or Transdimensional Realm, the Perpetual Motion Cycle $E = mc^2$, the Quantum Fields, and God in order to make the second law of thermodynamics true.

That's never going to happen, so the second law of thermodynamics will never be true. Our scientists have blinded themselves to these truths, and they will never see it nor understand it, until after they have figured out for themselves why the second law of thermodynamics and the theory of evolution are false.

Finding the Truth in Science is really simple. It starts by identifying and eliminating the obvious lies; and then, you systematically replace the obvious lies with the obvious truths instead. The obvious truths end up being the things that have actually been experienced and observed, either in the flesh or out-of-body in the spirit world.

It's obviously true that non-physical, invisible, entropyless, massless Quantum Information, Quantum Fields, Quantum Waves, Photons, and Quantum Processors (Psyches) exist. Therefore, it is obvious that Materialism and Naturalism and Physical Reductionism are FALSE because these things make the claim as one of their primary axioms that the non-physical does not exist and that Psyches or Quantum Processors do not exist.

Quantum Information Processors have to exist in order for Quantum Information to exist; and, the Conservation of Quantum Information within Conserved Quantum Processors (Psyches) has to be true in order for Quantum Mechanics and Quantum Field Theory to be true. Something quantum or non-physical has to be designing, building, and then running the massless, entropyless, heatless, syntropic, exergic, non-physical Quantum Fields. Something quantum or non-physical has to be making or instantiating the quantum waves or the photons, and then later collapsing the wave function thereby transforming that omnipresent quantum wave into some type of localized mass or heat instead. It is obvious that something non-physical or something quantum has to be running or operating the massless, entropyless, non-physical quantum waves and quantum fields. It is obvious that something at the quantum level or non-physical level has to be functioning as a Quantum Information Processor and Quantum Wave Processor. Psyche, or intelligence, or life force, or consciousness is that thing. It's the only one that has been experienced and observed.

Materialism, Naturalism, and their derivatives axiomatically state that the non-physical or the massless DOES NOT EXIST. These denialistic philosophies claim that the Quantum Waves, Photons, Psyches, and Quantum Fields DO NOT EXIST. These denialistic philosophies claim that Quantum Information and Quantum Information Processors (Psyches) DO NOT EXIST. These claims are obviously false. If they were true, then we wouldn't exist.

By definition, physical matter is a collapsed wave function. Therefore, something non-physical and pre-physical HAS TO EXIST in order to make or instantiate the quantum waves or the quantum information in the first place, and then later to collapse that wave function and TRANSFORM that energy into the components for the first physical atom that was designed and made. The primary axiom of Quantum Field Theory states that particles are born, and particles die; and, a physical atom is comprised of quanta, or particles, or bundles of energy, or quantum waves. Particles are MADE, and particles can be unmade or transformed into some other type of particle or form of energy. Particles, including physical atoms, are MADE or instantiated by Quantum Wave Processors or Quantum Information Processors, which means that those same Psyches can UNMAKE or TRANSFORM that physical matter back into quantum waves or photons, or even absorb them back into the quantum fields, anytime and anywhere those Psyches or Quantum Wave Processors CHOOSE to do so.

This is obviously true. Some type of non-physical and pre-physical Quantum Wave Processor (Psyche or Intelligence) had to design and make and instantiate the first physical atom or the first particle of physical matter; otherwise, physical matter wouldn't exist. First things first. You have got to have organized and functional quantum waves or quanta BEFORE they can be transformed into physical atoms or physical matter; and, you have got to have some type of Quantum Wave Processor or Quantum Information Processor or Conserved Quantum BEFORE you can have usable and functional and meaningful quantum waves. This is logical common sense.

Materialism, Naturalism, and their derivatives such as Darwinism and Atheism are obviously false. So, toss them out and get rid of them. Then replace them with Quantum Field Theory, Conservation of Quantum Information within Psyches or Quantum Processors, Quantum Waves or Photons or Intelligence or Consciousness or Light, Conservation of Energy and Psyche, Quantum Mechanics, and the Perpetual Motion Cycle $E = mc^2$ instead. If you want the Truth in Science, then trash all of the lies, and then replace those lies with observed truths instead.

Simple. Powerful. Parsimonious. Logical. True.

Materialism, Naturalism, Darwinism, Nihilism, and Atheism are NOT science or knowledge because they are obviously false. Something cannot be science or knowledge when it is false. These things are philosophies of science or interpretations of science and nothing more. So, get rid of them, because they are nothing but falsified philosophies of science or falsified interpretations of science.

Likewise, it is obvious that the theory of evolution and the second law of thermodynamics are false. So, toss them out, and then replace them with obvious truths instead.

Both of these are Creation by Chance, Creation by Random Processes, Creation by Blind Luck, Creation by Disorder, or Causation by Chance. That's why both the theory of evolution and the second law of thermodynamics are obviously FALSE. They are the Null Hypothesis and were produced by chance alone. It's obvious that chance of any kind cannot design and create anything. Once it starts doing so, then it is no longer chance, but has become some sort of intelligent choice or causation instead. Causation and chance are mutually exclusive. Choice and chance are mutually exclusive. They falsify each other or eliminate each other. Chance cannot do causation, so eliminate chance as one of your causes. Chance cannot do choice, so eliminate chance as one of your choosers.

Everything that exists falsifies the second law of thermodynamics which says that it shouldn't exist.

Everything that was deliberately caused falsifies the theory of evolution which says that there is no such thing as intelligent causation.

The false is falsified by the truth; and, the truth is constantly and repeatedly experienced and observed. It's that simple.

The Theory of Evolution, the Second of Thermodynamics, Creation Ex Nihilo, and Creation by Chance are obviously false. So, toss them out and get rid of them. Then replace them with Quantum Field Theory, Conservation of Quantum Information within Psyches or Quantum Processors, Quantum Waves or Photons or Intelligence or Consciousness or Light, Conservation of Energy and Psyche, Quantum Mechanics, and the Perpetual Motion Cycle $E = mc^2$ instead. If you want the Truth in Science, then trash all of the lies, and then replace those lies with observed truths instead.

Simple. Powerful. Parsimonious. Logical. True.

If you do as I suggest, then you will start to have the Truth in Science, instead of all the deceptions and lies that we currently have embedded in our "science". Of course, most of our scientists won't do as I suggest, because they actually prefer the deceptions and the lies, because that's how they have convinced themselves that they are right, and that's how they make their money. Self-deception works, and it works every time. These people prefer the deceptions and the lies in our science because they are the people who created these falsehoods and lies in the first place. Every scientist loves his brainchild, or the creation of his hands, no matter how ugly or false it might be.

Science is Observation and Experience. If a phenomenon or theory has never been experienced nor observed by anyone in this multidimensional universe, then it can't possibly be real and true. Toss it out and replace it with something that has actually been experienced and observed. If you do, then you will finally start to find the Truths in Science that exist to be found.

EXPANDING CLOUDS OF HYDROGEN GAS

The scientists on our earth have chosen to believe in the Big Bang Theory, which says that after a major explosion, hydrogen and helium gas spread out in all directions throughout the whole of our observable universe.

According to our scientists, all there was after the Big Bang was expanding clouds of hydrogen and helium gas. Then these scientists take a monumental leap of blind faith into the physically impossible and conclude that those expanding clouds of hydrogen and helium gas magically coalesced into rocks, planets, stars, and galaxies under the effects of gravity. Such a thing is physically impossible; but, our scientists don't know that. They believe it anyway.

Isn't that amazing? Our scientists erroneously believe that expanding clouds of hydrogen and helium gas can coalesce into rocks, planets, stars, and galaxies. Such a thing is physically impossible, but our scientists believe that it happens anyway.

Hydrogen atoms will combine together into hydrogen molecules, but that's where their natural coalescence ends. Hydrogen molecules naturally repel each other. There's no way in the universe that an expanding cloud of hydrogen gas is ever going to coalesce into a planet, or a star, or a galaxy under the effects of gravity.

Look at what we observe!

What happens to a tank full of hydrogen on our earth?

The stuff is under great pressure and the stuff is also under the influence of gravity. Does that tank full of hydrogen coalesce into a rock, a planet, or a star? Of course not! If it's not going to happen on our earth, then it's definitely not going to happen in the vacuum of space. Expanding clouds of hydrogen gas are NEVER going to coalesce into a rock, planet, star, or galaxy. That's physically impossible.

Helium gas is the closest thing that we have to an ideal gas. Helium is inert and doesn't really interact with anything. There seems to be a few ways to force the stuff to interact; but, it doesn't happen naturally. If it happened naturally, then we would observe pebbles or liquid forming inside of our helium balloons. The fact that tanks and balloons full of helium don't coalesce under pressure and under the effects of gravity on our earth into rocks and planets and stars is scientific proof that it's NEVER going to happen out in the vacuum of space. It can't happen because it's physically impossible.

The fact that contained and pressurized clouds of hydrogen and helium gas do not coalesce naturally into rocks, planets, and stars under the influence of gravity on the surface of our earth is scientific proof that it's never going to happen naturally out in the vacuum of space. It can't happen. It's physically impossible.

According to the Big Bang Theory, after the Big Bang, every single physical atom in this universe was moving away from every other physical atom in this universe; and, that's precisely what we should be observing now, all around us, if the Big Bang Theory were actually true.

We don't observe what the Big Bang Theory predicts; therefore, WE KNOW that there is something fundamentally flawed and wrong with the Big Bang Theory and that our planet's theories of origins need to be rethought and reworked.

Looking through our telescopes, we observe that planets, stars, and galaxies are made all at once or born whole; and then, they evolve from there or change from there, naturally, in the ways that we currently observe.

If tanks full of pressurized hydrogen and helium gas refuse to coalesce under the effects of gravity into rocks, planets, and stars on our earth, then it's certainly not going to happen out in the vacuum of expanding space.

Who or what do we know of at the quantum level in the transdimensional realm that would be capable of making planets, stars, and galaxies at the quantum level or in the spirit world; and then be capable of phase-shifting those planets, galaxies, and stars all at once into existence at the physical level where we currently live? Physical brains aren't capable of doing such a thing, so what is? What would be capable of organizing spirit matter and energy at the quantum level into planets, stars, and galaxies; and then be capable of phase-shifting those planets, stars, and galaxies all at once into our physical localized spacetime dimension where we currently exist?

Quantum phase-shifting has been experienced and observed, so who or what is the Master of phase-shifting?

Well, we fallen mortal beings don't seem to have any mastery or control over phase-shifting, so who does? If you can answer that question, then you know who designed and made our physical universe including its planets, stars, gas clouds, and galaxies because you know that it didn't happen naturally under the effects of gravity within an expanding vacuum.

Transdimensional means non-physical. Transdimensional physics is the study of non-physical physics, and it is obvious that the vast majority of physics is in fact transdimensional physics, or quantum mechanics, or quantum field physics. Even a physical atom is totally comprised of the non-physical, if you get right down to it. Everything is non-physical. Even a physical atom is completely non-physical or spiritual, being made completely from non-physical energy, forces, fields, and quantum waves.

Planets, stars, and galaxies are built or made in the Transdimensional Realm from spirit matter or dark matter, where there are NO physical limitations; and then, when they are done being built, they are phase-shifted whole INTO our physical spacetime localized realm, where all of the physical limitations instantly come into play. That's the way it is done at the micro-scale with something the size of a physical atom, through the Perpetual Motion Cycle $E = mc^2$, and with the collapsing of the wave function or phase-shifting; and, that's the way it is done at the macro-scale as well with something the size of a galaxy. The wave function of the whole spiritual galaxy is collapsed, and the whole thing becomes physical all at once. That's the only way it can be done, since it is physically impossible for expanding clouds of hydrogen and helium gas to coalesce into planets, stars, and galaxies.

Collapsing the wave function is the way that phase-shifting or the transition from the spiritual to the physical works; and, it works the other way around as well, according to the Perpetual Motion Cycle $E = mc^2$. Within a sun or a star, mass, heat, resistance to acceleration, and mass's heat storage capacity (entropy) are constantly being transformed by Nature's Psyche INTO massless, heatless, chargeless, and entropyless, omnipresent, infinite acceleration quantum waves or photons. The Perpetual Motion Cycle works perfectly both coming and going. It will never get old, and it will never wear out, because it is constantly conserved.

INFINITE MEMORY STORAGE CAPACITY

Quantum Field Theory is the best proven and most verified Science that our scientists have; yet, what do we observe from our scientists? They reject it because it falsifies their pre-chosen beliefs that the non-physical does not exist.

Furthermore, they refuse to apply Quantum Field Theory, Transdimensional Physics, Non-Physical Physics, or Quantum Mechanics to something as simple and obvious as Neuroscience. The Neuroscience on our world is worthless because it is nothing more than Materialism, Naturalism, Darwinism, Nihilism, and Atheism masquerading as "science". NONE of our scientists have ever thought to apply Quantum Mechanics and Quantum Field Theory to help them to interpret and understand Neuroscience. Our scientists refuse to allow this evidence into evidence because it falsifies their belief system or their personal religion.

Millions of us if not billions of us have died and experienced an After-Death Life Review in which they re-lived or re-experienced their whole life, with commentary, in a matter of seconds. Many have stated that God or the Being of Light and Love is there with them, watching their life with them, and commenting on their life while they are both watching their life unfold.

It is obvious that the memories, that survive the death of our physical brain and end up in our After-Death Life Review, are not stored within our physical brain.

Some people have reported that their After-Death Life Review was in 360-degree spherical vision. The human brain doesn't "see" and process video in 360 degrees. Therefore, it is obvious to the people who have experienced a 360-degree Life Review that it isn't the human brain that creates and stores our After-Death Life Review.

Once again, it is obvious that the memories, that survive the death of our physical brain and end up in our After-Death Life Review, are not stored within our physical brain.

Each human being has direct access to and uses an infinite amount of memory storage capacity; and, that's physically impossible. At the physical level, memory storage capacity can only be so dense and so compacted before you completely run out of memory storage capacity and have to start dumping stuff. Yet, NO human being completely runs out of memory storage capacity anywhere or at any time during their mortal life. That reality is physically impossible.

I have been studying and learning for decades, yet I have never hit a wall where I'm no longer able to learn something new that I have never thought about before. There is NO limit to our memory storage capacity. We each have access to and use an infinite amount of memory storage capacity. That's physically impossible, but it is true.

Psyches are conserved quanta or conserved photons. Psyches also appear to be infinite singularities with possibly an infinite amount of memory storage capacity. Furthermore, a spirit body appears to be a galaxy of stars or a galaxy of psyches held together by forces and fields. Should your spirit body ever run out of memory storage capacity, then just add another psyche or conserved quantum into the mix. Furthermore, energy is information. If you want more information or knowledge in your psyche, then just add more energy to your psyche, and then your psyche will become brighter and more intelligent. There's really no limit to the amount of information that can be stored within a psyche or an infinite singularity. If you want your psyche to become more intelligent, then add more energy, information, or light to it.

Despite the fact that they have proven that there are NO memory engrams within a physical brain and that there is NO physical RAM within a physical brain, most of our neuroscientists have erroneously chosen to believe that all of a person's memories are stored within a person's physical brain and nowhere else. Many of our top scientists have erroneously claimed that all of our memories are stored in synapses. They have no idea what they are talking about, and they claim to be neuroscientists.

Synapses are gaps in space. If our memories are being stored within synapses, then they are being stored within empty space, according to our neuroscientists; or, they are being stored within our cerebral spinal fluid according to our neuroscientists. Such an idea is falsified by the Materialism and the Naturalism that these neuroscientists have chosen to believe in. According to these people, there is nothing but empty unusable space within a synapse, so there's no way in this universe to store any information within a synapse at the physical level; and therefore, their materialistic, naturalistic, and atheistic claims that our memories are stored within our synapses is self-defeating or self-falsifying because their claim is physically impossible.

If our memories are indeed being stored within our synapses, then our memories are in fact being stored as quantum waves within the quantum fields that exist within all of that "empty space" that exists between the nucleus and the electrons of every atom within your physical brain. The only way to store an infinite amount of memory or an infinite amount of quantum information is at the quantum level either as quantum waves within the quantum fields or as quantum information within the infinite singularities that we call Psyche or Intelligence.

Furthermore, neurotransmitters are dumped randomly into a synapse or a synaptic cleft; therefore, there is NO way to store a memory within a synapse or to transfer a memory through a synapse from one neuron to the next. A synapse literally scrambles everything that comes its way! A neurotransmitter represents one single physical bit of information, because all that a neurotransmitter does is act as a key to open an ion-gate for a specified amount of time. There's NO way in this universe to transfer a message or to store a message within a neurotransmitter at the physical level. You can't store gigabytes of information within a single physical hardware bit at the physical level. It's physically impossible.

There are no wires within our brains; therefore, it is impossible for the neurons in our brains to communicate with each other at the physical level. Consequently, it is obvious that the neurons must be doing all of their communication and interaction at the quantum level, because it's certainly not happening at the physical level. It's impossible at the physical level.

It is obvious that our memories are NOT stored within our physical brain.

The only type of memory that actually seems to be stored within our physical brain is the stuff that they call "working memory" or "short term memory" – our memory of sequential lists or honey-do lists; and, it is clear that our working memory is indeed limited and has physical limitations. It has been observed that our brain can successfully store seven characters or numbers in a row before it starts dumping stuff in order to make room for more. The true memory storage capacity of a physical brain is pitiful and paltry.

Even when it comes to video which represents a huge amount of information, the ONLY thing that can be stored within our physical brain is our current video image or visual image, and it is immediately dumped in order to make room for the next video image that is incoming. Our physical brain does not store video and cannot store video. A physical brain doesn't have instant replay. ALL of our visual memories or video memories are being stored someplace else besides our physical brain. This has to be true, otherwise they wouldn't be able to show up in our After-Death Life Review.

In order to remember more than seven or eight characters in a row and in order to remember or store video, we humans actually have to train ourselves to tap into and use our infinite non-physical memory storage capacity that is being stored and processed at the quantum level as Quantum Waves (Thoughts or Memories) by Quantum Information Processors (Psyches). At the quantum level or the psyche level, there are NO physical limitations, which means that there are NO limits to the amount of memory that a human being can use, process, access, and store at the quantum level within the Light, or the Quantum Waves, or the

Quantum Fields. Through holography, a massive amount of information can be stored within the light.

Each one of us experiences an infinite amount of memory storage capacity, and some of us actually learn how to tap into it while we are still mortal and limited. This is what we actually experience and observe, and it completely falsifies the claims of our neuroscientists who state that ALL of our memories are being stored as bits and bytes within our physical brain or within the synapses (empty space) of our physical brain, and that ALL of our memories cease to exist when we die or our physical brain dies.

The memories within conserved psyches or conserved quanta can't cease to exist, because conserved psyches have the ability to process, transform, and store conserved quantum information. Conserved psyches are energy. Conserved quantum information is energy. Energy cannot be destroyed.

The truth of the situation is completely different than the falsehoods and the lies that our scientists and neuroscientists are currently promoting and disseminating.

There are NO physical wires within a physical brain, there are NO memory engrams, and there is NO physical RAM within a physical brain; yet, each human being has access to and uses an infinite amount of memory storage capacity, whether they realize it or not. An infinite amount of memory storage capacity is physically impossible at the physical level, yet easily possible and obviously possible within the Infinite Singularity, the Pinpoint of Light, or the Spark of Light that we have identified and called a Psyche, an Intelligence, a Quantum Consciousness, a Life Force, a Spark, a Conserved Quantum, a Quantum Wave Processor, or a Quantum Information Processor.

The false is falsified by the things that we have actually experienced and observed. The fact that each one of us has experienced or will experience an infinite amount of memory storage capacity is scientific proof that our memories are NOT being stored within nor processed by our physical brain. A physical brain just doesn't have what it needs to process and store an infinite amount of memory. It's obvious that our memories are being stored someplace else besides our physical brain, and that's physically impossible, which means that our memories and thoughts are being processed, transmitted, received, and stored as Quantum Waves by Quantum Wave Processors or Quantum Information Processors which have NO physical limitations.

The proven and verified existence of the physically impossible falsifies Materialism, Naturalism, Darwinism, Nihilism, Atheism, the Theory of Evolution, and the Second Law of Thermodynamics which claim that the non-physical does not exist, quantum waves do not exist, quantum fields do not exist, and that the human psyche does not exist. Scientists should go with what has been experienced and observed, rather than scientific inferences, philosophical speculation, and wishful thinking as our scientists do.

These competing theories or ideas falsify each other and are mutually exclusive. So, which one of them is false, and which one of them is true? The one that is true ends up being the one that has actually been experienced and observed.

The non-existence of the non-physical, as well as the non-existence of psyche or intelligence or consciousness, has NEVER been experienced NOR observed by anyone who is conscious, intelligent, and alive; therefore, WE KNOW that the non-existence of the non-physical or the non-existence of consciousness is not real and is not true.

That leaves infinite memory storage capacity and Quantum Wave Processors (Psyches) as the only thing that could possibly be real and true, and as the only thing that can explain our infinite memory storage capacity and our after-death life review successfully and scientifically in a way that actually makes logical sense.

Go with the one that has actually been experienced and observed, and then dump all the rest. That's what I chose to do, and you can too. That's what True Scientists are supposed to do; and, I'm a True Scientist whether you believe it or not.

QUANTUM FIELDS WERE OBVIOUSLY MADE

The primary axiom of Quantum Field Theory states that particles are born, and particles die. In other words, particles are made, and particles are absorbed back into the Quantum Fields from which they were made. The Quantum Fields are organized energy or light.

Organized by whom?

Anything that is obviously organized obviously has an Organizer who organized it.

The Quantum Fields are conserved perpetual motion machines. They will never get old, and never wear out because they are being conserved.

Being conserved by whom?

Energy can be transformed into anything at any time, so who is keeping the Quantum Fields in existence? It, too, has to be something that is being conserved. We can't have Conservation of Quantum Information without some type of conserved Quantum Wave Processor, or Quantum Information Processor, or Quantum Memory Storage Device to conserve it.

Who or what is capable of making and conserving Quantum Fields at the quantum level or the non-physical level? Physical matter can't touch nor control the quantum level, or the psyche level, or the non-physical level. According to the Obvious Law of Physics, when it comes to physics of any kind, the smaller dwells within and controls the larger.

Something smaller than the quanta, or the energy packets, or the particles is dwelling within them and controlling them.

This is obviously true. We wouldn't exist if it weren't true.

So, what are particles?

Particles are quanta or wave packets of energy. Particles are quantum waves. Particles are also collapsed wave functions or physical atoms. Particles are quantum waves, photons, tachyons, and physical atoms.

Collectively and unitedly, our scientists deny the existence of the things that were obviously made. Our scientists are some of the most ignorant people who have ever existed in this universe because they choose to be.

Anything that is obviously made obviously has a Maker who made it.

The quantum fields, quantum waves, quantum information, and collapsed wave functions (mass or physical matter) were obviously made.

Therefore, quantum fields, quantum waves, quantum information, and physical atoms obviously have a Maker who made them.

This Maker obviously exists. So, who or what is this Maker?

What do we know of that has actually been experienced and observed that is capable of making quantum waves and capable of collapsing quantum waves thereby producing mass or physical matter?

For this, we are looking at the Ultimate Point Particle, an Infinite Singularity, a Conserved Quantum, or a Quantum Pinpoint of Light that we call psyche, intelligence, the spark, life force, or consciousness. We are looking for some kind of Quantum Wave Processor or Quantum Information Processor. We are looking for Psyche or Intelligence.

Our scientists deny the existence of Psyche or Intelligence or Consciousness; therefore, they have NO explanation for how quantum waves, quantum fields, and collapsed wave functions are made or done. They believe that it happens by magic spontaneously out of thin air or in a vacuum from nothing. They are obviously wrong, but they don't know it. Self-deception works, and it works every time.

Physical matter doesn't just spring into existence out of thin air from nothing as the Naturalists and Atheists claim. Physical matter is purposefully and deliberately made by something that is smaller than it and by something intelligent that dwells within it.

Because they are Materialists, Naturalists, Darwinists, Nihilists, and Atheists, our scientists have NO way of discerning or separating the things that were obviously made from the things that have always existed, because they have never thought about any of these things and refuse to think about these things.

Anything that is obviously made obviously has a Maker who made it.

What isn't so obvious in this process is that heat or mass's heat storage capacity (entropy) is also made, which means that it too has a Maker who makes it and then later transforms it back into massless, entropyless, chargeless, and heatless quantum waves or photons.

The natural state of everything is absolute zero.

Everything in this universe moves towards absolute zero, because that is its ground state or native state. The Quantum Realm, Transdimensional Realm, Psyche Realm, or Spirit World is entropyless and heatless, and it functions BEST and most efficiently at absolute zero. There is NO heat, thermodynamics, or heat storage capacity in the Quantum Realm. It isn't needed. Heat, thermodynamics, or mass's heat storage capacity (entropy) represents inefficiency or physical limitations; and, they aren't needed nor wanted at the quantum level in the Syntropic Realm, the Ageless Realm, or the Conserved Realm of energy and light. Mass, resistance to acceleration, heat, and mass's heat storage capacity are purely physical phenomena. They don't exist as such at the quantum level in the Transdimensional Realm, or the Spirit World, or the Psyche Realm.

Our scientists have discovered the Perpetual Motion Cycle $E = mc^2$, but they don't know what it is or how it works.

Currently, the mass, heat, and entropy in our sun and in every star in this universe are being transformed by Nature's Psyche (or the Collective Intelligences in Nature) INTO massless, heatless, chargeless, entropyless, infinite acceleration, omnipresent quantum waves or photons. It's their job. That's what these psyches do.

Later, when a quantum wave or a photon CHOOSES TO STOP, then the Psyche or Intelligence within that quantum wave or photon TRANSFORMS its infinite acceleration INTO some type of mass, heat, resistance to acceleration, or mass's heat storage capacity (entropy) instead. The psyche or intelligence within the photon or the quantum wave CHOOSES what it wants its energy to become. It can transform the energy under its control into anything at any time.

Most of the time, a quantum wave's infinite acceleration is transformed into heat rather than mass; and, that's how heat gets made.

However, Nature's Psyche can make brand new mass, or physical matter, or physical atoms anytime and anywhere that it chooses to do so. Nature's Psyche can make whole galaxies all at once out of the energy within the Quantum Fields anytime and anywhere that it chooses to do so. Looking into our telescopes, it soon becomes obvious that galaxies are made whole, all at once, and then they start to wind up from there. Galaxies are born. Galaxies are made. Then over time, the mass, heat, resistance to acceleration, and mass's heat storage capacity (entropy) within the suns are transformed back into massless, heatless, chargeless, and entropyless quantum waves, photons, or quantum fields instead.

This is the Perpetual Motion Cycle. It has gone on for eternity, and it will go on for eternity. It is syntropic, and it is conserved. It works according to the LAW of Psyche, as well as the Conservation of Quantum Information within Psyche.

According to the LAW of Psyche, every psyche or intelligence in this multiverse has a certain amount of energy that's under its control; and, that controlling psyche can form or transform the energy under its control

into anything that it wants that energy to be, anytime and anywhere that it chooses to do so.

Psyche or Intelligence cannot be made, and it cannot be destroyed. It has always existed, and it will always exist. Sometime along the way, the Gods pulled your psyche or intelligence out of the infinite darkness and chaos; brought it into their glory, light, order, and love; and gave it a spirit body and made it a child of God. Psyche, intelligence, or consciousness has to exist, or you wouldn't exist.

There has to be something conserved at the quantum level that is capable of making, transmitting, receiving, processing, analyzing, and storing Quantum Waves or Quantum Information, or the LAW of Quantum Information Conservation would be impossible. Psyche is that Quantum Wave Processor or Quantum Information Processor. It's the only one that we have experienced and observed. Psyche is a conserved quantum or a conserved photon. If transformable photons exist, then conserved photons can exist as well.

Psyche is the thing that makes massless, heatless, chargeless, and entropyless quantum waves, thoughts, memories, or quantum information from mass, heat, entropy, energy, exergy, or the quantum fields; and, Psyche is the thing that collapses those quantum waves or photons thereby TRANSFORMING them into mass, heat, resistance to acceleration, or mass's heat storage capacity (entropy) instead. Heat, thermodynamics, mass's heat storage capacity, or entropy is made by Nature's Psyche. Mass or resistance to acceleration is also made by Nature's Psyche. Then inside our suns or stars, that mass, heat, resistance to acceleration, or entropy (mass's heat storage capacity) is TRANSFORMED by Nature's Psyche back into massless, heatless, chargeless, and entropyless quantum waves, infinite acceleration, omnipresence, or photons.

Round and around and around it goes for all eternity. It's one eternal round. This is why it is called the Perpetual Motion Cycle. It will never wear out, never get old, and never stop working because it is conserved. Our scientists have yet to discover the Perpetual Motion Cycle – what it is, and how it works. They don't know anything about the Perpetual Motion Cycle and have convinced themselves that it doesn't exist. They aren't looking for it, so they will never find it.

Mark My Words

HOW THINGS REALLY WORK

Have you ever studied the interference pattern produced by individual photons or electrons launched towards a detector screen? The electrons and photons don't land as a pinpoint dot on the screen. They splash down on the screen across the whole screen.

Most of the energy splashes down within one standard deviation of the electron's chosen landing spot, making it look somewhat like a dot or a point. Yet, you can see with your eyes the fuzzy haze of the energy that lands within two standard deviations away from the central chosen landing spot. However, just like a bell curve that goes out to infinity at the ends, a small part of that electron or photon lands all across the whole screen, and an even smaller portion of that omnipresent electron or photon lands all across our universe where it was in the first place.

Photons and electrons and tachyons are transformable quanta or traveling quanta. They travel as quantum waves through the quantum fields. Photons are also omnipresent while they "travel", because they are capable of infinite acceleration and infinite velocity, because they have NO entropy or NO resistance to acceleration, because they have NO mass while they travel.

From our perspective, photons choose to "travel" at the speed-of-light. That's how they reconcile their timelessness with our spacetime realm. There has to be some kind of conversion factor, and the speed-of-light is that standardized conversion factor that the Gods have made for it to follow. Tachyons choose to "travel" faster than the speed-of-light. Quantum waves or traveling quanta choose the speed at which they travel. They can quantum tunnel instantly to their destination if they choose to do so.

An electron is unique in that it is half-in and half-out while it travels. Half of the time an electron has mass or resistance to acceleration while it is traveling, and half of the time an electron is completely spiritual or non-physical while it travels at the speed-of-light from our perspective. An electron oscillates back and forth between physicality and non-physicality while it travels. Half the time an electron is local, and the other half of the time it is non-local. That's how a single electron is able to create a shell or a forcefield around the nucleus of a single atom. Half the time, an electron is omnipresent and spread throughout the whole universe all at once.

This is what we have actually experienced and observed; therefore, we KNOW that it is real and true.

In contrast, psyches or intelligences are conserved quanta or conserved photons. Psyches are also infinite singularities, conserved quantum computers, conserved quantum wave processors, conserved quantum information processors, and conserved quantum information storage devices. Psyches are the ultimate point particle. Psyche is the fundamental unit of reality and existence. This is a psyche ontology and the Ultimate Model of Reality.

When seen at the quantum level in the spirit world, a psyche or a conserved quantum seems to be a pinpoint of light or a living spark. However, psyches too function along a bell curve. Most of their light and energy is found within one standard deviation of their current central point of existence. However, parts of that psyche are omnipresent and exist simultaneously throughout our whole universe all at once. A psyche is an infinite singularity that is omnipresent in range or scope. Psyche is the best of both worlds.

This is how things really work at the quantum level in the spirit world or the transdimensional realm.

The only way to have the Truth in Science is to identify and eliminate the lies. We have to be willing to study and accept the things that have actually been experienced and observed. As you can tell, I concentrate a lot on the lies, because I use the lies in science to point me to the truths in science.

During my rebellious phase, I used to be a materialist, naturalist, nihilist, and atheist until Science or Knowledge convinced me that I was wrong. I guess you can say that I'm a convert to The Truth. I'm one of those who had to experience nihilism and atheism in order to know what's wrong with it. I have always been that way. I never trusted anyone. I was the kid who had to stick my hand on the stove to see if it really was hot. My only redeeming grace was that I considered myself to be an open-minded scientist; therefore, I was willing to examine and study evidence of any kind, including the evidence that falsified my beliefs. As a result, I mastered the art of critical thinking.

I wasn't a Darwinist and never fully believed in the theory of evolution because that one is obviously false. There was a time when I believed that natural selection exists and that it could do the things that they say that it does; but, I've always known that spontaneous generation, abiogenesis, chemical evolution, or macro-evolution is false. The second law of thermodynamics is obviously false and never sat well with me either. I used to be a materialist, naturalist, nihilist, and atheist. It's possible to believe in certain lies while at the same time rejecting others.

In this part of this book, I tried to help you identify some of the things that are wrong with our "science" as well as trying to suggest a few possible fixes or replacements that have actually been experienced and observed. Obviously, I haven't identified everything that is wrong with the "science" on our planet, just some of the most egregious flaws.

Creation by Chance or Chance Causation seems to catch or include the most severe falsehoods in our "science" – the things that WE KNOW are wrong, because they are obviously false and because they have never been experienced nor observed by anyone in this universe. The fact that Creation by Chance has never been experienced nor observed by anyone is scientific proof that the concept isn't real and isn't true. There's no such thing as spontaneous generation or creation ex nihilo. It has never been experienced or observed; therefore, it has to be false.

The Primary Axiom of Science and Statistics states that chance is not causation. Chance is the opposite of causation. Chance and causation are mutually exclusive. They falsify each other. Chance of any kind cannot do choice, correlation, design, creation, or causation. Once chance starts doing something, then it is no longer chance but has become some type of intelligent choice or deliberate causation instead.

Our scientists have erroneously concluded that chance alone caused everything that exists in this universe, in complete violation of the Primary Axiom of Science and Statistics. That's why their "science" is demonstrably false. Therefore,

Creation by Chance or Chance Causation ends up being the most obvious thing that's wrong with the "science" on our planet.

Once a scientist decides that chance cannot do causation, then instantly he finds himself looking for the True Cause of each event because he knows that chance cannot be the true cause of anything. According to the Primary Axiom of Statistics, chance cannot produce a real effect. Chance cannot do causation. Once it starts doing causation, then it is no longer chance.

You automatically KNOW that scientists are lying to you whenever they try to convince you that chance can design and create things if given enough time to do so. Don't believe them, because it isn't true. If they are making that kind of claim, then they don't know what they are talking about.

Our scientists haven't yet discovered and implemented the Primary Axiom of Science and Statistics. We will always have Militaristic Atheists, Physicalists, Materialists, Naturalists, Mechanists, Determinists, and Nihilists who are trying to destroy our way of life as well as the truths that we hold dear. I used to be one of them, until Science or The Truth convinced me that I was wrong.

The falsehoods in "science" are dangerous. Their lies can kill you.

If you live with the lies long enough, you start to believe that they are true. Well, I have finally lived with the truth long enough that I'm starting to believe that it is true. I can actually explain it to others, because I KNOW why it is true and because, for the sake of contrast and comparison, I have actually identified many of the things that are obviously false.

When it comes to the Materialists, Naturalists, Nihilists, and Atheists, their philosophy of science or interpretation of science is fundamentally flawed. These people have erroneously chosen to teach and believe that only physical matter exists and that the non-physical does not exist. They are obviously wrong. At times, it makes me wonder how I was ever able to be a materialist, naturalist, nihilist, and atheist because it is so obvious to me now that they are wrong. These people don't know what they are talking about because they don't understand Science. These people have completely rejected transdimensional physics or non-physical physics. They can't explain it, and they don't know what it is or how it works.

Light, energy, exergy, quantum waves, photons, and quantum fields are obviously massless, entropyless, and non-physical. In fact, by definition, everything in this universe is non-physical, except for a physical atom; and, everything within a physical atom is non-physical or spiritual as well, so even an atom is technically non-physical. Space and time are obviously non-physical, and so are gravity, magnetism, radio waves, x-rays, dark energy, and microwaves.

A physical atom really isn't physical or solid. The electrons orbiting an atom move at an infinite velocity thereby producing the illusion that they have formed a solid shell or a force field around the nucleus of the atom; but, that is ONLY an illusion. It's all made from energy, which means that it's all non-physical. A physical atom is also non-physical. Its physicality is simply an illusion, which means that its physicality really doesn't exist.

Therefore, whenever a scientist or college professor is trying to convince you that the massless, the entropyless, or the non-physical does not exist, they are lying to you and trying to trick you and deceive you, whether they realize it or not. These people don't have the Truth in Science because what they are teaching and what they believe has been falsified by Science or by the Observations of the human race. These people don't have Science or Knowledge of how things really work in the non-physical realm, the transdimensional realm, the spiritual realm, the energy realm, the psychic realm, the conserved realm, or the quantum realm. Instead, they pretend that these things do not exist, and they refuse to look at evidence. I KNOW because I used to be one of them.

—

Electrons are half-in and half-out.

When electrons are quantum or non-physical or immaterial or massless, they travel at an infinite velocity or quantum tunnel, which is how they are able to create a "solid" shell around the nucleus of an atom. The electrons create a force field or a shield around the atom while they are in their quantum, non-physical, infinite velocity format.

The other half of the time, electrons are localized here in our physical spacetime realm, and they actually have mass or resistance to acceleration while they are localized or have a presence in spacetime.

Electrons are the best of both worlds; whereas, photons are completely massless, entropyless, chargeless, and even heatless, non-physical, infinite acceleration, omnipresent quantum waves. It's ONLY when photons CHOOSE to stop that they transform their massless and entropyless infinite acceleration INTO some type of localized mass, or heat, or resistance to acceleration, or entropy (mass's heat storage capacity) instead.

According to the Law of Psyche, each psyche has a certain amount of energy that's under its control, and that controlling psyche can form or transform the energy under its control into anything that it wants that energy to be, anytime and anywhere that it chooses to do so.

The Perpetual Motion Cycle $E = mc^2$ makes this reality and truth obvious for anyone who is willing to look and see.

Every time that a massless and entropyless photon or quantum wave CHOOSES to stop, it typically transforms itself into mass, heat, resistance to acceleration, information, or entropy (mass's heat storage capacity) instead. It CHOOSES what it wants to become.

Most of the time, instead of transforming itself into mass or physical matter, a photon typically transforms itself into heat, and we actually feel the heat on our skin. Every photon has a psyche or intelligence within it, and that controlling psyche or intelligence can transform the energy under its control into anything that it wants that energy to be, anytime and anywhere that it chooses to do so. Photons or quantum waves are constantly transforming themselves into localized heat, all around us all the time. This phenomenon has actually been experienced and observed; therefore, WE KNOW that it is real and true.

The Perpetual Motion Cycle E = mc² works flawlessly in the other direction as well; and throughout it all, the energy is always conserved, which means that the total amount of energy never increases, and it never decreases either. The energy has always existed, and it will always exist, because it is conserved. That's why this is the Perpetual Motion Cycle.

The second law of thermodynamics is the reason why our scientists haven't yet discovered the Perpetual Motion Cycle, because the second law erroneously tells them that the Perpetual Motion Cycle does not exist. The lies prevent us from discovering and accepting the Truths in Science. You can't have the truth if you constantly choose to embrace and promote the lies instead. That's just the way it is.

In our sun and within every star, Nature's Psyche or Nature's Intelligence is currently TRANSFORMING mass, heat, resistance to acceleration, and entropy (mass's heat storage capacity) INTO massless, entropyless, chargeless, and heatless, infinite acceleration, non-physical, omnipresent photons or quantum waves instead. It's happening right now in every star that we observe, and all the while, the energy is conserved. This phenomenon has actually been experienced and observed; therefore, WE KNOW that it is real and true.

The Perpetual Motion Cycle is eternal and everlasting. It will never wear out and never stop working, because it is constantly conserved. Energy of any kind and in any form is constantly conserved. Energy is entropyless. Energy is syntropic, which means that it is timeless, ageless, will never get old, and never die or cease to exist. Energy has always existed, and it will always exist. Energy cannot be made, and it cannot be destroyed. This phenomenon has actually been experienced and observed; therefore, WE KNOW that it is real and true.

The Perpetual Motion Cycle $E = mc^2$ is the ultimate perpetual motion machine. The Quantum Fields are also perpetual motion machines. The massless, entropyless, and non-physical Quantum Fields have always existed and have always been conserved ever since they were designed and made. Their very existence falsifies the "second law of thermodynamics" which states and claims that the Perpetual Motion Cycle, and the massless and entropyless Conserved Quantum Fields, DO NOT EXIST. Everything that exists falsifies the second law of thermodynamics, which predicts that it shouldn't exist.

The false is falsified by the truth or by the things that have actually been experienced and observed. The "second law of thermodynamics" is falsified by the Conservation of Energy and Psyche, the Conservation of Quantum Information within Psyche, the Perpetual Motion Cycle $E = mc^2$, as well as the massless and entropyless Quantum Waves, Photons, and Quantum Fields. The "second law of thermodynamics" says that these things do not exist, and it's obvious that they do.

—

On this planet for billions of years, we didn't have Darwinism, the theory of evolution, nor "creation by death or natural selection" because these things are obviously false. I find it fascinating to observe how easily our scientists have fallen for these deceptions and lies because they desperately want them to be true. Self-deception works, and it works every time.

Spontaneous generation, abiogenesis, chemical evolution, macro-evolution, creation ex nihilo, creation by chance, causation by chance, creation by random disorder, and creation by blind luck are obviously false. They have NEVER been experienced NOR observed because they are physically impossible.

Creation Ex Nihilo or Atheism is the theory that was created out of nothing. It has NO substance. Nothing is nothing; but, the Naturalists and Atheists try to turn nothing into something. It's illogical, and it's irrational. But that's what they do. Creation Ex Nihilo is the theory that was created out of nothing; and, the Theory of Evolution or Creation by Chance is right up there with it. It too is nothing. There's NO substance there. There's NO truth there because chance, chaos, or disorder by definition and by axiom cannot design and create anything. Creation by Chance is physically impossible. Once you start seeing Creation, Choice, or Causation, then you are no longer looking at chance; and, ONLY psyche, intelligence, or consciousness can do creation, choice, and causation at every level of existence including the non-physical quantum level.

Whenever a scientist or college professor tries to convince you that chance of any kind can design and create things if given enough time to do so, you just automatically KNOW that he or she is lying to you and trying to trick you and deceive you. Chance cannot design or create anything; and, once it starts doing so, then it is NO LONGER chance, but has become some type of intelligent choice or deliberate causation instead.

That's the way things really work, whether people realize it or not.

These obvious Truths in Science are deliberately rejected and denied by the Materialists, Naturalists, Darwinists, Nihilists, and Atheists which is why their theories and ideas are false.

Abiogenesis is spontaneous generation – life from lifelessness, or life spontaneously generating from inert lifeless physical matter. Abiogenesis or spontaneous generation or chemical evolution or creation by chance is what's called Macro-Evolution, Darwinism, and the Theory of Evolution on our earth.

Abiogenesis, Naturalism, Materialism, Darwinism, Chemical Evolution, the Theory of Evolution, and Spontaneous Generation were falsified or proven false in 1859 by Louis Pasteur. Ironically, that's the same year that Charles Darwin published his book, "On the Origin of Species". The Theory of Evolution has always been false, and it was officially falsified the very same year that it was presented to our world. The reason that our scientists embraced the theory of evolution is because it is a deceptive lie, and they desperately wanted that lie to be true; but, wanting it to be true doesn't make it true. Creation by Chance of any kind is obviously false.

Anything that can be attributed to Evolution or Abiogenesis can also be attributed to an Intelligent Designer; but, the Intelligent Designer is more plausible, parsimonious, logical, possible, credible, sustainable by evidence and observation, and believable than Creation by Rocks or design and creation by Random Mutations and Natural Selection, which cannot design and create, and have never been observed doing so.

Remember: Evolution, Genetic Drift, Genetic Entropy, Genetic Deterioration, Natural Selection, and Random Mutations didn't even exist until AFTER God designed and created the genomes, the physical brains, and the physical bodies in the first place.

—

Chance and games of chance have to be given an intelligent assist or an intelligent start, or they will never exist to begin with. You have to have an intelligent psyche or intelligent consciousness who decides to make the coins, toss the coins, count the coins, and observe or verify and remember and recognize that there were all heads or all tails; otherwise, nothing would ever happen in the first place. In other words, games of chance don't design and run themselves. Someone Psyche or Someone Intelligent has to be there to keep score. It requires psyche, consciousness, or intelligence in order to design and run a successful game of chance. Evolution, Genetic Drift, Natural Selection, or Random Mutation is a game of chance, which means that someone intelligent had to design and create the genomes and the physical bodies in the first place; otherwise, that particular game of chance would not exist.

This is logical common sense which the Materialists, Naturalists, Darwinists, and Atheists deliberately and purposefully ignore and reject and delete, which is why these people seldom make any significant discoveries in the fields of Non-Local Consciousness and Transdimensional Physics – Psyche and Quantum Mechanics respectively. These people purposefully and deliberately ban, block, censor, burn, ridicule, delete, and destroy anything to do with the Non-Local Sciences or the Spiritual Sciences or the Quantum Sciences.

Some of the Darwinists and Naturalists literally lose their mind and turn into rabid dogs when you try to explain to them why the theory of evolution is false. They come unglued. They can't accept it. They can't handle it. They won't accept it, because they have turned evolution into their religion and their God. But, Creation by Chance of any kind is obviously false. Chance and causation are mutually exclusive. They falsify each other, which means that chance can NEVER do causation. Once it starts doing causation or choice or creation, then it is no longer chance but has become some kind of intelligent intervention, or intelligent causation, or intelligent choice instead. ONLY intelligent scientists can do Science, and the Gods are the most intelligent scientists that exist in this universe.

Random chance, blind luck, chaos, or disorder cannot do Science.

Natural selection is Creation by Death. Natural selection doesn't do anything except sit around and wait for you to die. Natural selection doesn't touch our genes. Creation by Death or "creation by natural selection" is also obviously false.

Death is effectively the same thing as chance – it cannot design and create, nor can it do causation or choice. There isn't a person or a psyche there called Death or Natural Selection who is going around the universe doing things for us or to us. Death or Natural Selection isn't some type of God, but the Darwinists treat it as if it were.

Chance of any kind cannot do design, science, and creation. That's another one of the reasons why the second law of thermodynamics is false. The second law is also Creation by Chance or Creation by Random Disorder. Don't believe me? Look up Boltzmann Brains or "Creation by Random Fluctuations in Entropy" to see for yourself what the second law of thermodynamics was intended to be. It's Creation by Chance or Causation by Chance – the very same falsehood, deception, and lie as the theory of evolution. It's all part of the same erroneous package – Materialism, Naturalism, Darwinism, Nihilism, Behaviorism, Determinism, Physical Reductionism, and Atheism. ALL of these are Creation by Chance, Chance Causation, the Null Hypothesis, or Creation by Random Disorder and Blind Luck, which is why all of them are false.

—

The second law of thermodynamics is obviously false. The very fact that we exist is scientific proof that the second law is false. The second law of thermodynamics claims that the total amount of entropy or disorder in our universe is constantly increasing and that it can never decrease and go to zero. This claim is obviously false. The second law predicts that we shouldn't be here, yet here we are.

The second law of thermodynamics predicts proton decay. Do we observe proton decay? No, we do not. Why not? Because the second law of thermodynamics is false!

Here we just used the Scientific Method and Negating the Consequent to falsify the second law of thermodynamics, and we did indeed PROVE that the second law of thermodynamics is false. It's that simple. We don't observe any of the things that the second law of thermodynamics predicts, which means that the second law of thermodynamics is false. It really is that simple to falsify the second law of thermodynamics and the theory of evolution.

What else does the second law predict? The second law of thermodynamics predicts that we should observe an ever-encroaching gray goo coming in at us from all sides. We should be able to see it because the second law predicts it. The second law predicts that the total amount of chaos or disorder that we observe all around us should be ever-increasing. We should be able to see it because the second law predicts it. But, is that what we actually experience and observe? NO, it is not. Instead, we OBSERVE constantly conserved order and organization all throughout our universe and all around us. Why? Because the second law of thermodynamics is false! Why? Because the order and the organization are being conserved, not the disorder and chaos! The second law of thermodynamics is false because it is by definition the Conservation of Disorder, or the Conservation of Chaos, or the Conservation of Entropy which obviously does not exist.

Even at what the scientists call "heat death" or absolute zero, the physical matter continues to exist, and the black holes continue to exist, and the quantum fields or space continues to exist. Why? Why are they being conserved? Well, they are NOT being conserved by their disorder, or their chaos, or their heat death. They are being conserved by Nature's Psyche, by the Conservation of Energy and Psyche, by the Conservation of Quantum Information within Psyche, by the Perpetual Motion Cycle $E = mc^2$, and by the Quantum Fields. The Quantum Fields are perpetual motion machines, and they function at optimal efficiency at "heat death" when the temperature is absolute zero. This IS what we have actually experienced and observed; therefore, WE KNOW that it is real and true.

Superconductors, quantum waves, photons, and quantum fields function best and function most efficiently at "heat death" or near absolute zero. Heat death is steroids and adrenaline for quantum objects and Quantum Mechanics. Heat death isn't "death". Heat death is Maximum Efficiency at the quantum level in the spirit world, the non-physical realm, the quantum realm, or the transdimensional realm where everything is conserved. This is what we have actually experienced and observed. Consequently, WE KNOW that it is real and true; and, it successfully falsifies the second law of thermodynamics.

These scientific observations falsify the second law of thermodynamics. Everything that exists falsifies the second law of thermodynamics, which says that it shouldn't exist. The false is falsified by the truth, and the truth consists of all the different things that have actually been experienced and observed. The second law of thermodynamics and the theory of evolution are obviously false because NONE of the things that they predict have ever been experienced or observed. It's that simple. One of the most efficient and effective ways to find and know the Truth in Science is to identify and then eliminate the obvious lies from "science". It works, because it is true. We find and know the truth through the process of identifying and eliminating everything that is false – everything that has NEVER been experienced NOR observed.

It's even more obvious that the second law of thermodynamics is false than it is that the theory of evolution is false, yet there are many scientists on our planet who have chosen to believe in the second law religiously; and, many of them have also chosen to make disorder or what they call "entropy" their Designer, and their Creator, and their God. Yet, it is obvious that disorder, chaos, or random chance cannot design and create anything, not even if it is given an eternity to do so. Creation by Chance or Creation by Random Disorder is physically impossible. It cannot be done, which means that it wasn't done that way. We need a better explanation for Origins besides creation ex nihilo or creation by chance, because chance causation of any kind is obviously false and should be eliminated from "science". The false is falsified by the truth, and the truth consists of the things that have actually been experienced and observed.

—

The Primary Axiom of Science and Statistics states that chance of any kind cannot do choice, correlation, creation, or causation. This simple axiom identifies and eliminates most everything that is false in "science". Chance cannot do causation. Chance is not causation. Anything that violates the Primary Axiom of Science and Statistics is automatically false.

Simple. Logical. Parsimonious. Rational. True.

This simple axiom identifies and eliminates most everything that is false or wrong with our "science".

—

Quantum Mechanics or Transdimensional Physics is the study of spiritual mechanisms or transdimensional mechanisms, which means that Quantum Mechanics or Transdimensional Physics is the study of how energy, light, spirit matter, quantum waves, and psyche function and work in the Non-Local Realm, Transdimensional Realm, Non-Physical Realm, or Spirit World.

Psyche goes by many different names – Non-Local Consciousness, Quantum Consciousness, the Spark of Life, an Infinite Singularity, Intelligence, Light and Truth, Life, the Breath of Life, Life Force, and even the Soul. Psyche or Non-Local Consciousness or Intelligence is something completely different than spirit matter, spirit bodies, physical matter, and physical bodies. When the out-of-body travelers go looking for their Psyche, there's nothing there to be seen or perceived. Psyche is a disembodied life-force, a viewpoint in space, and an infinite singularity or pinpoint of light which commands and controls both spirit matter and physical matter telepathically and telekinetically. Psyche or intelligence or consciousness is pure energy and conserved energy.

Psyche is a conserved quantum or a conserved photon. Psyche is the ultimate Quantum Information Processor. In order for the Quantum Law of Information Conservation to be true, some type of massless, non-physical, immaterial, transdimensional, entropyless, and conserved quantum machine, quantum computer, or quantum entity MUST EXIST in order to make, transmit, receive, process, analyze, transform, and store Quantum Information and Quantum Waves. Psyche IS that conserved quantum computer or conserved quantum information processor. Psyche is life. Psyche is the Ultimate Chooser and the Ultimate Causal Agent.

Psyche, Intelligence, or Consciousness is the thing that makes or instantiates the quantum waves in the first place; and then later, Psyche is the thing that stops the photon and then TRANSFORMS the energy and infinite acceleration within that quantum energy packet or quantum wave or photon INTO some type of mass, heat, resistance to acceleration, or entropy (mass's heat storage capacity) instead. All the while, the Conserved Quantum Information within the Conserved Psyche and used by the Psyche is constantly conserved. Our memories are made by Psyche and stored within Psyche, because Psyche makes Quantum Information, transmits Quantum Information as Quantum Waves, and then receives and stores Quantum Information as Memories. This is how things really work. This is Real Science.

This is True Science. This is a True Knowledge of how things really work at all levels of existence and reality.

—

By definition, physical matter is a collapsed wave function. Therefore, something non-physical and pre-physical HAS TO EXIST in order to make or instantiate the quantum waves or the quantum information in the first place, and then later to collapse that wave function and TRANSFORM that energy into the components for the first physical atom that was designed and made. The primary axiom of Quantum Field Theory states that particles are born, and particles die; and, a physical atom is comprised of quanta, or particles, or bundles of energy, or quantum waves. Particles are MADE, and particles can be unmade or transformed into some other type of particle or form of energy. Particles, including physical atoms, are MADE or instantiated by Quantum Wave Processors or Quantum Information Processors, which means that those same Psyches can UNMAKE or TRANSFORM that physical matter back into quantum waves or photons, or even absorb them back into the quantum fields, anytime and anywhere those Psyches or Quantum Wave Processors CHOOSE to do so.

This is obviously true. Some type of non-physical and pre-physical Quantum Wave Processor (Psyche or Intelligence) had to design and make and instantiate the first physical atom or the first particle of physical matter; otherwise, physical matter wouldn't exist. First things first. You have got to have organized and functional quantum waves or quanta BEFORE they can be transformed or organized into physical atoms or physical matter; and, you have got to have some type of Quantum Wave Processor or Quantum Information Processor BEFORE you can have usable and functional and meaningful quantum waves. This is logical common sense.

Quantum Information Processors, Quantum Wave Processors, Quantum Wave Instantiators, Quantum Wave Collapsers, Psyches, Intelligences, Consciousness, Choosers, or Creators have to exist; otherwise, we wouldn't exist, everything would be nothing but random chaos, and the second law of thermodynamics would actually be true. This is obviously true because we obviously exist; and, the fact that we exist is scientific proof that the second law of thermodynamics is false.

Everything that exists and everything that has ever been experienced and observed is scientific proof that the second law of thermodynamics is false, because the second law predicts and claims that nothing should exist. The second law predicts complete and total chaos and disorder; yet, we observe the exact opposite thanks to the Quantum Fields, the Perpetual Motion Cycle, and the Quantum Wave Processors which are ALL conserved, syntropic, eternal, and everlasting perpetual motion machines at the quantum level.

In order for Quantum Information and Quantum Waves to exist, there has to be some type of non-physical, syntropic, conserved quantum machine capable of making, broadcasting, transmitting, receiving, processing, analyzing, transforming, and storing Quantum Waves and Quantum Information. Psyche IS that quantum machine, quantum information processor, or quantum wave processor. Psyche has to exist at the quantum level, or quantum waves and quantum information storage wouldn't be possible at the quantum level. This is logical common sense. WE KNOW that Psyches or Quantum Information Processors exist, or quantum waves and quantum information and memories wouldn't exist. Psyche is that Quantum Wave Processor. It is the ONE that has actually been experienced and observed.

Out-of-body explorers have observed that – while they are separated from their spirit body and looking at their spirit body – when they go looking for themselves there is nothing to be seen. Why? It's because a Psyche has been observed or seen as being a pinpoint of light, a spark of light, or a photon at the quantum level as well as the psyche level. Psyche is the ultimate point particle. Psyche is an infinite singularity. Psyche is a conserved quantum. Psyche is the thing from which and by which universes are made. Psyche is a photon or an infinite singularity or the ultimate point particle even at the quantum level and the psyche level, which explains why psyche, or intelligence, or consciousness, or a quantum is always experienced first-hand as being an immaterial viewpoint in space. When out-of-body explorers are separated from their spirit body looking at their spirit body, they report themselves or their Psyche or their Intelligence as being an immaterial viewpoint in space, which it is. It is a pinpoint of light, even at the quantum level, that is capable of omnipresence and omniscience. Psyche goes all the way down the rabbit hole while simultaneously going to infinity and beyond. Psyche is an infinite singularity that has infinite range and scope.

Human beings are comprised of a disembodied Psyche, a Spirit Body, and a Physical Body. Spirit Bodies and Physical Bodies have a beginning, a time that they first get organized by God's Psyche or by God's Spirit or by our Heavenly Parents; but, Intelligence or Psyche or Non-Local Consciousness is eternal, immortal, and indestructible. Raw unorganized spirit matter is also eternal and uncreated and indestructible in that it's pure energy or pure exergy; however, each particle of physical matter has a beginning and can therefore theoretically have an end by being converted back into spirit matter or usable energy once again by God's Psyche or by Nature's Psyche.

This is how things really work at the quantum level, whether we know it or not. There has to be a logical explanation for how quantum waves and quantum information are made, transmitted, received, processed, transformed, and stored. Psyche, Intelligence, Quantum Consciousness, Quanta, Life Force, the Spark of Life, Quantum Information Processors, or Quantum Wave Processors ARE the scientific explanation that we have been searching for. This thing has to exist, or quantum information, quantum waves, and collapsed quantum waves (physical particles) would not exist.

I know that this will be too much Science or Knowledge for some people to be able to understand and accept, but we have to start somewhere if we are ever going to figure out how things really work at the quantum level in the Non-Physical Realm, Quantum Realm, Psychic Realm, or Transdimensional Realm. We have to start with a common reference frame that we KNOW exists, and then build our knowledge and understanding from there. We KNOW that quantum waves and quantum information exist, so now it's time to figure out how and why they exist.

It's time for our scientists to discover non-physical and conserved Quantum Wave Processors or Quantum Information Processors, because they have to exist in order for quantum waves, collapsed quantum waves (physical matter), and quantum information to exist.

—

Quantum Mechanics or Transdimensional Physics measures and tests the interplay between the Human Psyche and Nature's Psyche. The Human Psyche commands and controls its assigned physical brain; and, Nature's Psyche commands and controls each and every particle or atom of physical matter within that physical brain. It's Nature's Psyche who collapses the wave functions and fires the neurons in our brains, because it's Nature's Psyche who made our physical brains and MAPPED the meaning onto our physical brains at the quantum level in the first place. The Human Psyche isn't consciously aware of any of these things. It's all being handled and run by Nature's Psyche.

The Human Psyche interacts with Nature's Psyche telepathically and even at times telekinetically, through a process that's called the Quantum Zeno Effect. Telepathy is like radio waves, but much more refined and functioning at a much higher velocity and frequency than radio waves. Telepathy is WiFi at the quantum level. Transpersonal Psychology is the scientific study of psyche to psyche interaction, or person to person interaction, both at the quantum level and the physical level.

Spirit Matter, and Psyche and Telepathy, exist and function at velocities greater than the speed-of-light in the Non-Local Realm or Quantum Realm. There are theoretically no speed limits or distance limitations or physical limitations or time restrictions in the Spirit World, which means that Quantum Tunneling or Teleportation is a natural part of the Non-Local Realm or the Quantum Realm where psyche and spirit matter are concerned. The only limitations experienced in the Spirit World are those put there by God.

There is no heat, mass, or entropy in the Transdimensional Realm. It's not needed, because it slows things down. There's NO aging process in the Non-Physical Realm or the Conserved Realm. Everything is eternal and everlasting at the quantum level, because the energy is always conserved. Energy can be formed or transformed into anything, including physical matter and spirit matter; but, the energy itself is always conserved. The form of that energy isn't conserved, but the energy is conserved.

This is what has been observed and experienced by Human Psyches while they were exploring the Non-Local Realm, or Transdimensional Realm, or Spirit Realm, or Quantum Realm.

—

When it comes to physical matter and physical bodies, God deliberately slowed everything down to sub-light speeds so that we human psyches can live it, experience it, and remember having done so. A Physical Realm is the ultimate Consensus Reality. In a Physical Reality, I can depend upon this book being there on my computer tomorrow morning when I go looking for it. Here in this physical reality, I can depend upon my house being there, my job being there, my car being there, my food being there, my wife being there, my computer being there, my books being there, and my dog being there when I go looking for them tomorrow. Such a thing isn't possible in the unorganized non-consensus realities of the Spirit Realm or Transdimensional Realm. Remember, a Physical Reality is the Ultimate Consensus Reality, and in a Physical Reality God deliberately slowed everything down for us and deliberately organized everything so that we can live it, learn from it, experience it, and remember having done so.

A photon traveling at the speed-of-light has NO experiences and learns nothing from the event. From its perspective, it simply teleports directly to its destination, with no entropy or no passage of time taking place for it during the process because for it the passage of time has stopped. There is NO entropy or no aging process in the Spirit Realm or Transdimensional Realm. The Quantum Realm is an Ageless Realm because the whole thing is being conserved. Time flows differently in a completely different timeline in the Quantum Realm; and, nothing ages because everything is being conserved.

From our slowed-down perspective, it might have taken that photon 13 billion years to reach us; but, from its infinite acceleration quantum perspective, a photon arrives at its destination instantaneously the very moment that it launches. This is the time dilation effect that's the result of the Theory of Relativity or Special Relativity. At the speed-of-light or faster, TIME STOPS, distance contracts to zero, resistance to acceleration drops to zero, the aging process stops, and entropy ceases to exist; and, from the perspective of that quantum object or quantum wave or photon, it simply quantum tunnels to its destination experiencing nothing during its journey. When TIME STOPS at the speed-of-light or faster, there is nothing to experience and nothing to remember because the passage of time has stopped. The object seems to quantum tunnel instantly to its destination from its perspective.

That's how the quantum is reconciled with the physical. It's all handled through time. It's all reconciled through time. From our perspective trapped in time, it took 13 billion years for that photon to get to us. From its timeless and ageless perspective, that same photon quantum tunneled instantly to its destination because from its perspective the passage of time had completely stopped during its journey.

Time flows relative to one's perspective. From the perspective of a massless and entropyless Photon traveling at the speed-of-light, it experiences no entropy, no passage of time, and no aging whatsoever which means that it experiences and learns and remembers nothing during the journey. In contrast, God deliberately slows everything down for us in this physical realm by introducing physical limitations, speed limitations, velocity and acceleration limitations, resistance to acceleration, space, time, distance, and entropy so that we can live this Physical Realm, experience this Physical Realm, learn from this Physical Realm, and remember having done so.

—

The proven and verified existence of the massless, entropyless, non-physical, omnipresent photons, quantum waves, and quantum fields FALSIFIES Materialism, Naturalism, Darwinism, Nihilism, Atheism, the Theory of Evolution, and the Second Law of Thermodynamics which claim that these non-physical things do not exist. The false is falsified by the truth, and the truth consists of the things that have been constantly and repeatedly experienced and observed.

This is what I have discovered, and this is how I explain it. I sometimes wonder how close I have gotten to the truth, and how much more I have yet to discover and learn. But, this is infinitely better than anything that I have ever gotten from the materialists, naturalists, nihilists, and atheists. All that these people have ever done is to claim that NONE of this quantum or transdimensional stuff exists. That's completely worthless, because it doesn't explain what the quantum is or how it works! They simply say that it doesn't exist and leave it at that. It's garbage!

Everything in this universe is non-physical. Even a physical atom is non-physical. A physical atom is made from a bunch of non-physical quanta, psyches, forces, and fields. The Materialists, Naturalists, Darwinists, Nihilists, and Atheists deny the existence of everything that exists, except for physical matter. It's stupid and idiotic, but that's what they do. The way that they get away with it is by hiding what they are doing, so that we can't see it nor discover it. They work in the dark and have become experts at keeping us away from the truth or the light.

If something is true, then you should be able to demonstrate that it is true. Materialism and naturalism claim that the non-physical does not exist. If that is true, then they should be able to demonstrate that it is true. They can't! It's obviously false. The massless, entropyless, non-physical quantum fields, quantum waves, quantum information, and photons obviously exist; therefore, materialism and naturalism are obviously false. Materialism or physicalism is a lie. The verified and proven existence of light, photons, quantum waves, radio waves, microwaves, x-rays, gravity, and dark energy FALSIFIES materialism and naturalism which claim that these non-physical things do not exist. The false is falsified by the truth! The false is falsified by the things that have actually been experienced and observed.

In summary, there is NO WAY to salvage Materialism, Naturalism, Darwinism (Creation by Chance), Creation by Death (Natural Selection), Nihilism, Atheism (Creation Ex Nihilo), the Theory of Evolution (Spontaneous Generation), or the Second Law of Thermodynamics (Creation by Disorder or Chaos). They are obviously false. If you want the Truth in Science, if you want True Knowledge, then you have no choice but to toss them out and to start over with something that has actually been experienced and observed – something like the Conservation of Energy and Psyche, the Conservation of Quantum Information within Psyche, the Perpetual Motion Cycle ($E = mc^2$), as well as the massless and entropyless Quantum Waves, Photons, and Quantum Fields.

Even space, time, gravity, magnetism, x-rays, radio waves, and microwaves falsify Materialism, Naturalism, and their derivatives such as Darwinism (Creation by Pure Chance) and Atheism (Creation Ex Nihilo) which claim that these non-physical things do not exist. The false is falsified by the truth or by the things that have actually been experienced and observed.

This is how things really work, and these were the things that I wanted to know most. I wanted to find the Truth in Science, and the only way to do that is to identify and eliminate all the lies. Over time, all of these things became clear to me; and over time, I came to KNOW how things really work in the Conserved Realms, or Transdimensional Realms, or Quantum Realms, or Spirit Realms – at least as well as any physical being could ever possibly know.

Now I can rest in peace knowing that I have finally found the truth. Now I can explain everything that has ever been experienced or observed.

Mark My Words

APPENDIX

The purpose of the Scientific Method is to help us to find THE TRUTH, through a preponderance of the evidence.

the Scientific Method has no value to us if we use it to convince ourselves that a LIE is TRUE, as the Materialists and Darwinists always seem to do.

That's what I discovered during my Pursuit of the True Reality of All Things, and during my usage of the Scientific Method.

It is July 2016 as I write this.

I AM NEW to all of this that I have written in this book — it having come together for me during the past year or so. But, I KNOW that I have finally found THE TRUTH.

I JUST KNOW IT. I CAN FEEL IT!

THE TRUTH tastes good. THE TRUTH feels good. THE TRUTH is exhilarating, refreshing, stimulating, and exciting! I have NO DOUBTS whatsoever that I have finally found THE TRUTH.

In comparison, during the times in my life when I thought that the Theory of Evolution might be true, I had lots of little niggling doubts in the back of my mind making me wonder if I might be wrong.

And, for me personally, I was emotionally and psychologically miserable during the time when I was an Atheist and a Materialist. Materialism, Naturalism, Darwinism, and Atheism are the types of religion that bring no happiness, no comfort, no peace, no love, no charity or mercy, no spirituality, no enduring friendships, no surety, no excitement, no inspiration, and no joy. There was always that constant nagging feeling that something was wrong or that I had gotten something seriously wrong, while I was in an Atheistic and Materialistic frame of mind. The doubts and confusion were a multiplying.

The grass really IS greener on the Theistic side of the fence, but ONLY when you finally want to be there.

Replacing the Lies with the Truth

I used to be a materialist, naturalist, nihilist, and atheist back in 2012. Back then, I knew what an Atheist and a Darwinist is; but, I didn't have a definition for materialism, naturalism, or nihilism. You can be one of these without actually knowing what it is or what they believe.

When someone finally explained physicalism, atomism, physical reductionism, or materialism to me, I knew that I was one and had been one by default and by training. The purpose of our public schools is to indoctrinate you into Materialism, Naturalism, Darwinism, Nihilism, and every other form of Atheism. If you are not an atheist by the time you graduate from college, then your college professors have failed in their primary mission. After fifty years of life on this planet, I had this stuff ingrained into me so deeply that it had become an integral part of me; and, I couldn't see anything wrong with it.

When I went searching for the truth, my materialism or physicalism was the first lie to be abandoned, because it is obviously false once you know what it is.

Physical matter obviously exists. Physical matter has been experienced, observed, and verified. It's stupid to claim that physical matter does not exist. And, that's how they get you. They get you by *affirming the consequent*!

If materialism or physicalism is true, then we will observe the existence of physical matter. We observe the existence of physical matter; therefore, materialism or physicalism is true.

Materialism or physicalism defined as the verified and proven existence of physical matter is in fact true. Physical matter has indeed been experienced, observed, verified, and proven to exist. That much of physicalism or materialism is in fact true. It's demonstrably true.

If they left it at that, then everything would be fine, and materialism would in fact be true. But, what these people do is start tacking on a bunch of hidden premises and hidden conclusions by surreptitiously changing the definition of materialism or physicalism.

Materialism or physicalism is actually defined as the philosophy of science or the interpretation of science that states that ONLY physical matter exists and that EVERYTHING in this universe is made from physical matter. The Materialists, Naturalists, Darwinists, Nihilists, and Atheists claim that physical matter is the ONLY thing that exists.

They get you with the truth or the obvious fact that physical matter does indeed exist, then then they start attaching a bunch of different lies and want you to swallow them whole as well.

Materialism, or physicalism, or atomism, or physical reductionism makes the scientific truth claim that the non-physical, the quantum, or the transdimensional DOES NOT EXIST. That's the LIE which gets attached to the obvious truth that physical matter exists. These people want you to believe that ONLY physical matter exists, so that there is no room for the possibility that God, Conserved Quanta (Psyches), or Creators might exist.

When it comes to the denialistic philosophies of science which claim that one thing or another DOES NOT EXIST, the way you falsify them with the Scientific Method is to identify their hidden assumptions or their hidden conclusions, because their hidden premises are obviously false.

Materialism or physicalism starts with the hidden assumption and ends with the hidden conclusion which states that the non-physical, the transdimensional, or the quantum DOES NOT EXIST. That's its hidden assumption, as well as its hidden conclusion; and, it's obviously false. Even physical atoms are made from a bunch of non-physical quanta, psyches, forces, and fields. Everything in this universe is non-physical, except for a physical atom. Space, time, magnetism, radio waves, gravity, dark energy, microwaves, x-rays, photons, quantum waves, and quantum fields are obviously massless, entropyless, and non-physical. Therefore, the hidden axioms of materialism or physicalism are obviously false.

The LIES, that they tack onto the obvious truth that physical matter exists, are obviously false. Falsify the hidden premises, and you falsify the theory.

Affirming the Consequent

It is important to consider the logic used by the Materialists, Naturalists, Darwinists, Nihilists, and Atheists to convince us that what they are telling us is true.

A prediction based on a theory usually takes this form: "If Theory X is true [antecedent], then we will observe Y [consequent]."

The following logical argument outlines the typical fallacious approach that has been taken by scientists, called *affirming the consequent*:

If Theory X is true, then we will observe Y.

We observe Y.

Therefore, Theory X is true.

This sort of thinking, however, reflects a logical fallacy called *affirming the consequent*. Here is a comparable example to demonstrate:

If Sally's pet is a cat, it will have a tail.

Sally's pet has a tail.

Therefore, Sally's pet is a cat.

We can easily see that this logic is fallacious. Just because we observe Y (a pet with a tail) that does not mean that our theory X (the pet is a cat) is true. After all, dogs and lizards have tails too.

How They Trick Us and Deceive Us

Materialism, atomism, physicalism, or physical reductionism starts with the obvious truth that physical matter exists, and then they start tacking on a bunch of different lies in the hope that you will swallow those as well. Their ultimate goal is to convince you that the non-physical, the quantum, or the transdimensional does not exist.

Naturalism starts with the obvious truth that nature exists, and then they start attaching a bunch of lies trying to convince you that the supernatural or the quantum does not exist.

Darwinism, or Creation by Chance, or Creation by Death (Natural Selection) starts with the obvious truth that the fossil record does indeed exist; and then, they start attaching a bunch of lies trying to convince you that Creators and Intelligent Designers do not exist.

Atheism starts with the obvious truth that we exist; and then, the atheists start attaching a bunch of lies trying to convince us that "nothing" made us exist or that we were designed, engineered, fine-tuned, and field-tested by random chance or by nothing at all. Tricky, huh?

The Materialists, Naturalists, Darwinists, Nihilists, and Atheists trick us and deceive us by *affirming the consequent*. They start with an obvious truth as the bait or the hook, and then they switch to or attach a bunch of different lies and try to get you to swallow those as well. It's devious, and it's evil; and, it works.

Identify the hidden lies or the hidden premises, falsify them with observational evidence to the contrary, and you have indeed successfully falsified materialism and naturalism in its many different forms, using the scientific method to do so. Observation is a scientific method! The denialistic philosophies of science are falsified with observational evidence to the contrary. Observe anything non-physical such as time, space, quantum fields, photons, radio waves, x-rays, dark energy, magnetism, gravity, or quantum waves; and, you have in fact successfully falsified materialism, naturalism, and their derivatives which claim that the non-physical does not exist.

When materialism or physicalism goes down, the rest of the denialistic philosophies of science go down with it as fruit from the poisoned tree, because each one of the denialistic or atheistic philosophies of science surreptitiously teaches and preaches that the non-physical, the quantum, the supernatural, or the transdimensional DOES NOT EXIST.

If you successfully identify and eliminate everything that is false, then only the truth will remain; and, the truth that remains will consist of everything that has ever been experienced or observed.

Mark My Words

Einstein's Biggest Blunders

https://markme.website/einsteins-biggest-blunders/

My purpose is to set everything straight; and, it began with the long and painful process of setting myself straight, first. Back in 2012, I was a Materialist, Naturalist, Nihilist, and Atheist. I had a long ways to go before I finally saw the light and figured out how things really work.

Einstein's biggest blunder or greatest error came when he rejected Action at a Distance and concluded that the speed-of-light limitation applies at the quantum level in the Quantum Realm, Energy Realm, Psychic Realm, or Spirit World. This error is based upon the conclusion that physical matter is the only thing that exists; and, it has been proven false.

https://ultimate-model-of-reality.com/falsifying-einstein-and-classical-realism/

Instantaneous Action at a Distance is the rule or the law in the Quantum Realm. My best friend (and thousands of other people) has been there in the Quantum Realm or the Spirit World, and he told me that he can testify as an eye-witness that the speed-of-light limitation does NOT apply in the Spirit World or the Quantum Realm. He was participating and communicating instantaneously with humans who were on opposite ends of the universe. Even though these people reside on different planets in vastly separated parts of this universe, their consciousness was there witnessing and experiencing the same event that my friend was witnessing and experiencing. There are NO physical limitations in the Quantum Realm! In the Spirit World, they communicate telepathically, at a distance, instantaneously.

It's obvious that it has to be this way. There is no entropy and can be no entropy in the Quantum Realm; otherwise, something like the Big Bang would never have been possible. If the Big Bang were simply a random physical event as the Materialists and Naturalists claim, then there is nothing to stop it from happening randomly once again in your back yard, wiping us out in the process.

Einstein's other really great blunder was to assume that entropy applies to the Quantum Realm or Spirit World. It doesn't. It can't. The Quantum Law of Thermodynamics states that there is no entropy and no thermodynamics at the quantum level in the Light Realm or the Quantum Realm. It has to be this way, or we wouldn't exist. There could never have been something like a Big Bang if entropy were in fact building up in the Quantum Realm. This is a serious blunder that we all have made. We deny the Conservation of Energy (or the Conservation of Psyche) at the most fundamental level by insisting that entropy or the second law of thermodynamics applies at the quantum level in the Psychic Realm or Quantum Realm.

There has to be some type of Syntropy or Organizing Force or Life Force at the quantum level, or the Big Bang could never have happened, the quantum fields could never have been made, and physical matter would have been impossible to build and sustain. You see, the Gods or the Controlling Psyches had to design and make the quantum fields BEFORE it became possible for them to make and sustain physical matter. Quantum fields are obviously made from energy by Someone Psyche, which means that the same has to be said of physical matter. There has to be Syntropy or Conservation of Energy at the quantum level, or our physical level would not have been possible to make.

Syntropy: The Answer to Life, the Universe, and Everything

https://www.amazon.com/gp/product/B07BPT3W8R/

Einstein's biggest blunder came when he assumed that physical laws, physical limitations, and physical restrictions apply at the quantum level in the Quantum Realm; and, our biggest blunder comes when we choose to believe him. Whenever we choose to give physical limitations and physical laws priority over Syntropy or the Conservation of Energy, then we make the most serious blunder that we can possibly make in physics and effectively destroy everything that we are trying to explain, sustain, and understand. Materialism and Naturalism destroy science, especially Quantum Field Theory and Quantum Mechanics. They both can't be true simultaneously. They are mutually exclusive. Something has to give.

During the past few years, I have repeatedly observed that Quantum Mechanics and Quantum Field theory are constantly verified; whereas, Materialism and Naturalism are constantly falsified. Therefore, we KNOW which one of them is true, and which one of them is false. When it comes to Science, the false is falsified by the truth, and the truth is repeatedly experienced and observed. We have repeatedly observed that physical limitations and physical laws DO NOT APPLY at the quantum level in the Psyche Realm, Syntropy Realm, or Quantum Realm. The Science – the Real Science – changed the way I look at things; and, it helped me to overcome some of Einstein's greatest blunders.

One of my greatest conceptual breakthroughs came when I first understood what really happens when an object or particle reaches the speed-of-light. According to the theory of relativity, at the speed-of-light, time stops, length contracts to zero, and distance goes to zero. Do you understand what this really means? It means that at the speed-of-light, in the quantum realm, velocity or speed goes to infinity. Speed is instantaneous or infinite in the Quantum Realm at the quantum level. Objects in the quantum realm can be stationary, of course; but, whenever they choose to move, they can quantum tunnel or teleport instantly to their destination at an infinite velocity.

One of my greatest conceptual breakthroughs came when I finally reconciled the Theory of Relativity with Quantum Mechanics. It came with the realization of what happens to a photon traveling at the speed-of-light.

You see, from our perspective trapped in space-time and this physical reality, that photon took 13.2 billion years to reach us as it traveled through space. However, at the speed-of-light from the perspective of the photon, time stopped, entropy stopped, there was no passage of time, distance collapsed to zero, length contracted to zero, and that photon from its perspective literally quantum tunneled or teleported instantaneously to its destination. In other words, that photon from its perspective at the speed-of-light experienced nothing from start to finish. It simply teleported or quantum tunneled instantaneously to its destination. This leads to the quantum realization that at the quantum level in the Quantum Realm from the perspective of the photon, a photon knows where it is going to land before it launches, or it doesn't launch.

This "infinite velocity" or "quantum tunneling" at the quantum level also lead to my realization that the Gods or Controlling Psyches created this physical realm for us, in order to slow things down for us, so that we can actually experience them, learn from them, experiment with them, and remember having done so. You see, at the speed-of-light, space-time goes to zero, distance goes to zero, time stops or goes to zero, and length contracts to zero. When time stops, there's nothing to experience. When time stops, there is no entropy and no aging process. Isn't that so? Space-time as we know it ceases to exist at the quantum level in the Quantum Realm or the Spirit World. They have to switch over to some other time dimension or timeless dimension in the Spirit World because the space-time dimension that we are familiar with here in this physical realm ceases to exist at the quantum level in the Energy Realm or the Quantum Realm.

This perspective-shift successfully reconciles the Theory of Relativity with Quantum Mechanics. From the perspective of the photon, it simply quantum tunnels to its destination experiencing NO passage of time while it is "traveling" to its destination. From our perspective mired in space-time, it took that same photon 13.2 billion years to reach us. This perspective-shift from the quantum realm to the physical realm successfully reconciles quantum mechanics with the theory of relativity. Does it not? We are talking about the same photon but from different perspectives. From its timeless perspective, it quantum tunnels or teleports to its destination. From our space-time perspective, it is limited by the speed-of-light. It's the same photon but different perspectives; and, it works! It matches with reality.

These truths and observations led to my discovery of the Quantum Law of Thermodynamics. So, what is the Quantum Law of Thermodynamics? Well, at the speed-of-light, time stops according to the Theory of Relativity which means that any aging process stops, which also means that entropy stops. In other words, there is NO entropy, or thermodynamics, or aging process at the quantum level in the Quantum Realm. Instead, energy or psyche is syntropic in the Quantum Realm, which means that energy or psyche is always conserved at the quantum level in the Quantum Realm. Energy or psyche is eternal and everlasting in the Quantum Realm or Psyche Realm – without a beginning of days or an end of years.

https://quantum-neuroscience.com/the-quantum-law-of-thermodynamics/

https://syntropy.site/syntropy-must-exist/

The greatest blunder of any Materialist, Naturalist, Nihilist, Classical Physicist, Darwinist, or Atheist is to assume that the Second Law of Thermodynamics applies at the quantum level in the Energy Realm, Spirit World, or Quantum Realm. They are wrong. There is NO entropy at the quantum level in the Transdimensional Quantum Realm. Entropy ceases to exist at the speed-of-light, because time stops at the speed-of-light. Entropy or the second law of thermodynamics is exclusively a physical phenomenon. It does not apply and does not exist in the Quantum Realm.

The Quantum Law of Thermodynamics states that ALL of the energy is available all of the time at the quantum level and the psyche level. In other words, anergy, entropy, unavailable energy, thermodynamics, heat death, death, and the second law of thermodynamics DO NOT EXIST in the Quantum Realm. This is what has been experienced and observed by Out-of-Body Travelers and Near-Death

Experiencers. The BEST and most convincing Proof of Heaven is to go there and see it for yourself, or to choose to trust someone who has.

Well, that's a major discovery, isn't it? But, you'll never find out about it from the Materialists, Naturalists, Darwinists, Nihilists, Behaviorists, and Atheists because they have rejected it. According to these people, the non-physical Psychic Realm or Quantum Realm does not exist. According to these people, Psyche, Energy, Life Force, or Light does not exist. According to these people, only physical matter exists, which means that only entropy exists. This also means that these people really truly believe that entropy and entropic physical matter are conserved. The Materialists and Naturalists erroneously teach and believe that entropy is conserved or that entropy is eternal. These people teach and believe that only entropic physical matter exists; therefore, these people erroneously teach and believe that physical matter and entropy are conserved. Entropy is death. These people literally teach and believe that entropy or death is conserved. Do they not? That's what they believe!

The fact that these people have been proven wrong by Science led me to the Ultimate Law of Thermodynamics. So, what is the Ultimate Law of Thermodynamics?

The Ultimate Law of Thermodynamics differentiates between what is being conserved and what is not conserved. Psyche, Life Force, Energy, Intelligence, Information, and Syntropy are conserved; whereas, physical matter, death, and entropy are not conserved. The First Law of Thermodynamics tells us that the energy is constant, and that the energy is conserved – not the FORM of that energy. Energy can be transformed from one form to another, which means that the FORM is never conserved. The Ultimate Law of Thermodynamics teaches that the FORM of the energy is never conserved. Only the energy or the psyche is conserved. Consequently, physical matter, death, and entropy are NOT conserved at the quantum level. In fact, they don't really exist at the quantum level. There is no entropy or death in the Quantum Realm or Energy Realm because everything is conserved in the Syntropic Realm.

http://ultimate-law-of-thermodynamics.com/defined/

I used to be a Physicalist, Naturalist, Nihilist, and Atheist; but, I'm no longer satisfied being dumb and blind without a soul or a mind. Nowadays, I have to know what things are and how they work. Like I said, my purpose is to set everything straight; and, that process began by setting myself straight.

Psyche is the innate intelligence within all the different forms of energy or quanta. Quanta are organized packets of energy. Quanta or "particles" are waves of energy moving through a quantum field. Someone Psyche has to start the quanta moving in the first place or nothing would ever happen. Syntropy is the Conservation of Energy. Syntropy is the First Law of Thermodynamics. Syntropy is the opposite of entropy. Psyche or Energy is syntropic, which means that Psyche or Energy is conserved.

In contrast, entropic physical matter is made from different forms of energy. The form is NEVER conserved. The Ultimate Law of Thermodynamics states that the form is never conserved. Energy is infinitely malleable or infinitely transformable. Each psyche has a certain amount of energy that's under its control; and, that controlling psyche can form the energy under its control into anything that it wants that energy to be. At some point in the future – if not before – when all of the physical matter in our physical universe has reached thermal equilibrium or heat death, the Controlling Psyches who made the physical matter in the first place can simply convert that physical matter back into raw reusable energy, and then use that energy again to make quantum fields and another physical universe. There's no such thing as heat death in the Quantum Realm. Heat or thermal disequilibrium is irrelevant at the quantum level, where everything is made and sustained.

Physical matter, entropy, and time are made or caused to begin, which means that they are NOT conserved. The Gods can transform entropic physical matter into a different form of energy anytime they choose to do so. This is the Ultimate Law of Thermodynamics. It states that Psyche, Energy, Life Force, and Syntropy are conserved; whereas, entropic physical matter is NOT conserved. The Ultimate Law of Thermodynamics simply differentiates between what is being conserved and what is NOT conserved.

The Ultimate Law of Thermodynamics is a necessary adjustment or refinement to the classical laws of thermodynamics because the Materialists, Naturalists, Darwinists, Nihilists, and Atheists erroneously teach and believe that physical matter, death, and entropy are conserved because these people erroneously teach and believe that entropic physical matter is the ONLY thing that exists. The Ultimate Law of Thermodynamics corrects that flaw by stating that entropy, death, and physical matter are NOT conserved. Therefore, if there is any conflict between the first law of thermodynamics and the second law of thermodynamics, the Ultimate Law of Thermodynamics declares the first law of thermodynamics the winner every time.

The Ultimate Law of Thermodynamics states that Psyche, Energy, Life Force, and Syntropy are always conserved; whereas, entropy, death, and physical matter are NOT conserved. This is what has actually been experienced and observed by Out-of-Body Travelers and Near-Death Experiencers. Science is observation and experience, not wishful thinking and philosophical speculation. The Ultimate Law of Thermodynamics gives precedence to whatever has been experienced and observed.

https://syntropy.website/the-law-of-psyche/

https://psyche-ontology.com/psyche-and-energy-are-essentially-synonymous/

Truth sustains truth. Within Science, the false is falsified by the truth; and, the truth is repeatedly experienced and observed.

According to the Law of Psyche, every Psyche, Intelligence, Life Force, or Consciousness has a certain amount of energy that's under its control. The controlling psyche can form the energy under its control into anything that it wants that energy to be. Energy is infinitely malleable or infinitely transformable, because energy or psyche is always conserved. Energy is the ultimate perpetual motion machine. It never wears out because it is always conserved.

The Law of Psyche states that Psyche, Consciousness, Life Force, or Intelligence has been experienced and observed; therefore, it is real, truly exists, and must be explained and explainable by Science, when our Science finally has enough depth to do so.

Psyche or Consciousness is the innate intelligence within all the different forms of energy. The Ultimate Law of Thermodynamics states that Psyche, Intelligence, Life Force, Energy, the Quantum Realm, and Syntropy are conserved; whereas, physical matter, death, and entropy are NOT conserved. The Ultimate Law of Thermodynamics simply differentiates between what is being conserved and what is NOT conserved. It all fits together perfectly because it is true. It is true because it has actually been experienced and observed.

https://ultimate-law-of-thermodynamics.com/

The Quantum Law of Thermodynamics states that heat death is impossible at the quantum level because the second law of thermodynamics or entropy doesn't exist at the quantum level. The Quantum Realm is an Isothermal Realm which means that in the Quantum Realm or Spirit World all of the energy is available for use all of the time, and that energy is always conserved. There is NO anergy or entropy in the Quantum Realm. Heat death is impossible at the quantum level. Consequently, the Quantum Law of Thermodynamics states that death is impossible at the psyche level. Your psyche, life force, or intelligence was not made, and it cannot be destroyed. Your Psyche cannot die. It is always conserved. This is what has been experienced and observed.

https://quantum-law-of-thermodynamics.com/

You see, the fermions comprise matter and the bosons communicate or transfer the fundamental forces. Alas, both fermions and bosons are made from energy by some controlling psyche. The controlling psyche or controlling intelligence chooses whether the energy under its control will be fermions or bosons. David Bohm stated that physical matter is frozen light or frozen energy. It's energy that's slowed down or brought to a standstill within locality or space-time. In contrast, bosons are quantum waves or thoughts. They travel from one psyche to the next instantaneously communicating information or forces. Massless bosons have NO speed limit, which means that they are capable of Instantaneous Action at a Distance at the quantum level. It's the fermions that have been slowed down and made subject to physical limitations, by having some type of clock or entropy or aging process placed within them.

Energy is the ultimate perpetual motion machine – you get the same amount out as what you put in – because energy or psyche is always conserved. But these "perpetual motion machines" are only realistically functional at the quantum level. They are impossible at the physical level thanks to the Laws of Thermodynamics that exist at the physical level. The physical has physical limitations. The quantum doesn't. That's something that Einstein and the other materialistic and naturalistic geniuses never realized. These people truly believed that the speed-of-light limitation applied just as much to the quantum realm as it does to the physical realm; and, they were wrong, and have subsequently been proven wrong.

PBS NOVA "Einstein's Quantum Riddle – Instantaneous Action at a Distance Proven True Once Again" - 1/9/19

https://www.youtube.com/watch?v=ZEhoR-LCDlo
https://www.amazon.com/gp/video/detail/B07MJK3GMV/

Einstein's biggest blunders became our biggest blunders.

Every time that Instantaneous Action at a Distance, Non-Locality, Non-Physicality, or Quantum Entanglement is proven true or verified during our science experiments, Materialism, Naturalism, Darwinism, Nihilism, Behaviorism, Determinism, and Atheism are FALSIFIED. The false is falsified by the truth; and, the truth is repeatedly experienced, observed, witnessed, and verified. Instantaneous Action at a Distance, Telepathy, Quantum Tunneling, Quantum Entanglement, Non-Locality, Non-Physicality, Psychic Interaction, Instantaneous Transpersonal Communication, or Infinite Velocity at the Quantum Level is repeatedly verified during our science experiments; and, its verification falsifies the claims of Materialism, Naturalism, and their derivatives which state that these things do not exist. The repeated verification of Quantum Phenomena or Psychic Phenomena or Action at a Distance FALSIFIES Materialism, Naturalism, Darwinism, Nihilism, Behaviorism, Determinism, and Atheism which claim that these things do not exist.

Einstein's biggest blunder, and the one that most of our scientists continue to make, is the chosen belief that physical limitations such as entropy and the speed-of-light apply at the quantum level in the Quantum Realm or Spirit World. They don't! There is NO entropy or thermodynamics at the quantum level where energy or psyche is concerned. Instead, psyche or energy is always conserved at the quantum level.

Furthermore, there are no speed limits enforced at the quantum level. At the quantum level in the Spirit World, it has been observed that spirit bodies and quantum objects are capable of quantum tunneling or teleporting across the universe instantaneously if they choose to do so. There are no speed limits in the Quantum Realm, which means that physical limitations do not apply in the Spirit World. Things don't age, and things don't wear out and die in the Spirit World. Psyche or energy is conserved in the Quantum Realm, which means that it doesn't age or wear out. Entropy or death, as we know it, doesn't exist at the quantum level in the Quantum Realm.

This is it. This is what I have been searching for, for all of my life.

As a scientist, one of the biggest blunders I ever made was to assume that Natural Selection or Evolution can design and create. I was wrong. Natural Selection or Evolution is entropy or death. Entropy and death cannot design and create. They can only destroy. Natural Selection doesn't do anything except sit around and wait for you to die. In the case of Natural Selection, "selection" is death. It is natural for us to die or get selected against. Natural Selection doesn't touch our genes. It can't, because it isn't a physical objective being. It also doesn't have a soul, intelligence, consciousness, life force, or a mind. Only psyche or intelligence can design and create. This is what we have actually experienced and observed. Natural Selection doesn't have the power nor the ability to do anything. It's simply a philosophical concept and nothing more. Go figure! I don't know if Einstein fell for this particular blunder, but most of our modern-day scientists have.

https://evolution-is-entropy.com

The original creative powers – the ones that existed BEFORE the first particle of physical matter was designed and made – were found at the quantum level within Consciousness, Intelligence, Life Force, or Psyche. This is obviously true. In fact, my research into Consciousness, Quantum Field Theory, and Quantum Mechanics eventually led to my discover of what I call The Obvious Law of Physics, which states that the smaller dwells within and controls the larger. Energy is infinitely malleable. You can form energy into anything that you want it to be. Who or what within the energy itself decides whether to form that energy into an electron or a quark? Someone conscious and intelligent is making that choice. It doesn't just happen randomly. There is a choice being made. Someone intelligent and psychic within the energy actually knows how to form the energy under its control into either an electron or a quark. It knows how an electron or a quark is supposed to act, and how they are supposed to interact. There's intelligence there and consciousness there. Energy carries within it intelligence, consciousness, life force, light, intention, perception, attention, and information.

According to the Law of Psyche, every psyche has a certain amount of energy that's under its control; and, that controlling psyche decides what form the energy under its control will assume. That controlling psyche KNOWS what an electron is and what a quark is, and that controlling psyche DECIDES whether the energy under its control will be an electron, or a quark, or part of a quantum field, or a quantum wave, or a photon. Remember, according to the Obvious Law of Physics, the smaller dwells within and controls the larger. There is some type of intelligence, life force, consciousness, or psyche within each electron, photon, field, force, and photon who actually decided that the energy under its control would function as one of these things and be one of these things.

There are different levels of reality, or different phases or dimensions in the Spirit World. It's layered, and it has depth. Objects out of phase with each other can occupy the same space at the same time without colliding, interacting, or interfering with each other. This is what has actually been experienced and observed. The Quantum Realm, Non-Physical Energy Realm, Psychic Realm, or Spirit World is right here within us; and, it is completely different than our physical realm with its obvious physical limitations and speed-limits. Instantaneous Action at a Distance, Conservation of Psyche or Energy, a complete lack of Entropy or Thermodynamics, and Potential Infinite Velocities or Quantum Tunneling are an integral part of the Quantum Realm. Einstein's biggest blunder was to assume that physical limitations such as entropy, the aging process, death, the theory of relativity, and his speed-of-light limitation apply to the Quantum Realm or the Spirit World. It's a blunder or a mistake that our modern-day physicists continue to make because they don't want Action at a Distance and the Immortality of their Psyche to be true.

Remember, Einstein's biggest blunders became our biggest blunders. I find all of this fascinating – figuring out what I and Einstein got wrong, and why we got it wrong. After a hundred years of verifying Quantum Mechanics and falsifying Naturalism or Classical Physics, we are now in a much better position to see what Einstein got wrong than they were back then. I like Quantum Mechanics and Quantum Field Theory because they make me think and learn.

Space Is Not Empty

https://www.youtube.com/playlist?list=PLsPUh22kYmNAHB1W2_Ka2F83sOb dczwKr

Quantum Mechanics

https://www.youtube.com/playlist?list=PLsPUh22kYmNCGaVGuGfKfJl-6RdHiCjo1

Quantum Field Theory

https://www.youtube.com/playlist?list=PLsPUh22kYmNBpDZPejCHGzxyfgitj2 6w9

Quantum Tunneling Is Faster than Light

https://www.youtube.com/watch?v=-IfmgyXs7z8

Quantum tunneling is faster than light, except at the physical level, where everything was designed to be slower than the speed-of-light. It's a good thing that God gave physical matter a short De Broglie Wavelength; otherwise, the atoms within your physical body could quantum tunnel away from you at will, and you would literally dissolve into thin air. It's a good thing that the atoms in your physical body can only quantum tunnel short distances and that they are subject to physical laws and physical restrictions. If the atoms within your physical body had a long De Broglie Wavelength or an infinite De Broglie Wavelength, then they could quantum tunnel to the other side of the universe instantaneously at will, leaving your spirit with no physical body or physical world to live within. God put the physical laws and physical restrictions into place for a good reason.

The Impossibility of Perpetual Motion Machines

https://www.youtube.com/watch?v=rckrnYw5sOA

Perpetual motion machines are physically impossible at the physical level because the conservation of energy or the first law of thermodynamics states that the best we can do is to get out precisely what we put in. In other words, it's impossible to get more energy out than what we put in at the physical level, without tapping into some kind of hidden resource at the quantum level – think of a nuclear explosion which is a quantum phenomenon. When it comes to a nuclear explosion, we tend to get more energy out than what we put in, and that's thanks to quantum mechanics or energy mechanics. $E = mc^2$. Typically, though, when it comes to machines at the physical level, the combined influence of friction, gravity, and entropy interfere, which means that we always get out less energy than what we put in when it comes to our physical machines.

The only way to get true perpetual motion is at the quantum level where entropy and thermodynamics don't exist and where energy or psyche is always conserved. If we could tap into the Spirit World or the Quantum Realm, we would never run out of energy because energy or psyche is always conserved at the quantum level and there are no thermodynamics to have to deal with at the quantum level.

It's pretty obvious that this is the way things really work; but, it wasn't the least bit obvious when I was a Materialist, Naturalist, Nihilist, and Atheist. I couldn't see it, until I saw it. There are a lot of brilliant physicists who still can't see it because they don't want it to be true. They are banking on the heat death of the universe instead.

Back in 2012, I was a Materialist, Naturalist, Nihilist, and Atheist. Then I started interacting with these people online; and as the months passed by, I started to notice that these people were lying to us and many of them were deliberately trying to trick us and deceive us. It was obvious to me; and, it was also obvious that if their philosophies, theories, and ideas were in fact correct, true, and right as they claimed they were, then there should be no need for them to lie to us, trick us, or deceive us in order to get us to believe what they wanted us to believe.

When I realized that the Scientific Naturalists, Darwinists, Nihilists, Behaviorists, Determinists, and Atheists are lying to us and trying to trick us and deceive us, that's when they began to lose me. Eventually, I decided to go in search of the truth instead. The truth ended up being noticeably different than what the professional Darwinists and Naturalists were claiming that it is. The more I studied, the more obvious it became that these people were lying to us and were deliberately trying to trick us and deceive us. I was more interested in the truth than in having these people like me. I wanted to know how everything really works.

Eventually I realized that the Materialists, Naturalists, Darwinists, Nihilists, and Atheists have no observations or experiences to support their claims that the non-physical does not exist, the supernatural or transdimensional does not exist, intelligence or psyche does not exist, the quantum or the supernatural or the spiritual does not exist, a pre-mortal life and an after-life do not exist, and God or the Being of Light and Love do not exist.

Instead, ALL of the observations and experiences were on the side of Quantum Mechanics, Energy Mechanics, Intelligence or Psyche, Choice or Psyche Determinism, Action at a Distance, Non-Locality or Non-Physicality, Quantum Field Theory, and the observed and experienced existence of the resurrected Biblical God Jesus Christ. It is obvious to me that Quantum Fields are non-physical and pre-physical in nature and origin, which means that God or the Controlling Psyches had to design and make the Quantum Fields BEFORE they could make and sustain physical matter.

It also became obvious to me that our resurrected Lord Jesus Christ, or that Being of Light and Love experienced while out-of-body during Near-Death Experiences, has in fact been experienced and observed. Furthermore, it is obvious that Intelligence, Consciousness, Life Force, Psyche, or Soul has in fact been experienced and observed. If you can read and understand this, then it is obvious that you are Intelligent, Conscious, Alive, and have some type of Psyche or Soul within you somewhere at some level.

It was then that I developed a new Philosophy of Science that is based directly upon observation and experience and verification. I called it Science 2.0. Science 2.0 is based upon everything that has been experienced and observed. If it has been experienced, observed, or witnessed, then it qualifies as a part of Science 2.0. Soon I noticed that the false is always falsified by the truth; and, the truth is repeatedly experienced and observed.

Science 2.0: I Upgraded My Science

https://www.amazon.com/gp/product/B0771K6WTX/

Then, I was well on my way to developing a new model of reality, the Ultimate Model of Reality, which is a model of reality that is based upon what has actually been experienced and observed. If it hasn't been repeatedly witnessed, experienced, and observed, then it is worthless and isn't a part of reality as we know it and have experienced it. Materialism, Naturalism, Darwinism, Chemical Evolution, Design and Creation by Natural Selection, Spontaneous Generation, Abiogenesis, Macro-Evolution or Cats Giving Birth to Dogs, Non-Existence, and Creation Ex Nihilo or Atheism are NOT a part of reality as we have experienced it to be. These falsified philosophies are wishful thinking or a denial of reality rather than actual observations and experiences. That's what I experienced and observed. It's time for a new model of reality that actually matches with reality.

David Bohm is probably my most favorite theoretical physicist; but, it takes a bit of effort to translate what he is talking about into ordinary everyday English.

https://epdf.tips/the-essential-david-bohm.html

https://philosophy-of-science.com/The-Essential-David-Bohm

With his implicate order, David Bohm opened the door to the Quantum Realm and started to explain how it really works.

What we are suggesting here is that the implicate order, the non-physical order, the energy order, the psyche order, the spiritual order, or the quantum order be taken as fundamental. What Bohm is repeatedly suggesting is that we start with the truth; and, the truth is that the physical is made from the non-physical by Consciousness or by Someone Psyche. In contrast, all the Materialists, Naturalists, Nihilists, Darwinists, Behaviorists, and Atheists can tell us is that the non-physical does not exist. That's worthless! Claiming that the non-physical doesn't exist doesn't tell us what it is or how it works.

According to Bohm, the essence of consciousness resides within the Holomovement or the Implicate Order, which is the quantum level or the spirit world. It has to reside someplace, because consciousness has been experienced and observed. It's real. It truly exists. Furthermore, the Explicate Order or our physical reality unfolds or explicates from this much deeper Holomovement or Implicate Order. The physical is built from the quantum, or the spiritual, or the non-physical, just like Quantum Field Theory states and just like the Biblical God Jesus Christ tries to teach us. It would take quite a bit of Intelligence, Information, or Psyche to design and create a quantum field or a holomovement and then convince the energy within it to act that way.

Bohm's Holomovement is synonymous with his Implicate Order, which are synonymous with quantum fields and quantum mechanisms. Quantum Mechanics explains how energy, or psyche, or light acts at the quantum level or the implicate level.

Scientific Axiom: Energy or light can take on any form that it desires to assume, including spirit matter and physical matter. Energy or psyche is always conserved, but it can change form at will. This is what has been experienced and observed.

Scientific Observation: Anything that is obviously organized obviously has an Organizer who organized it.

Scientific Observation: Quantum fields were obviously organized from raw chaotic energy.

Scientific Conclusion: Therefore, quantum fields obviously have an Organizer who organized them and taught the energy within them to act that way or to form that way.

David Bohm was teaching these very same principles back in 1980. He wrote:

> Indeed, if one applies the rules of quantum theory to the currently accepted general theory of relativity, one finds that the gravitational field is also constituted of such "wave-particle" modes, each having a minimum "zero-point" energy. As a result, the gravitational field, and therefore the definition of what is to be meant by distance, cease to be completely defined. As we keep on adding excitations corresponding to shorter and shorter wavelengths to the gravitational field, we come to a certain length at which the measurement of space and time becomes totally undefinable. Beyond this, the whole notion of space and time as we know it would fade out, into something that is at present unspecifiable. So, it would be reasonable to suppose, at least provisionally, that this is the shortest wavelength that should be considered as contributing to the "zero-point" energy of space.
>
> When this length is estimated it turns out to be about 10^{-33} cm. This is much shorter than anything thus far probed in physical experiments (which have got down to about 10^{-17} cm or so). If one computes the amount of energy that would be in one cubic

centimeter of space, with this shortest possible wavelength, it turns out to be very far beyond the total energy of all the matter in the known universe.

What is implied by this proposal is that what we call empty space contains an immense background of energy, and that matter as we know it is a small, "quantized" wavelike excitation on top of this background, rather like a tiny ripple on a vast sea. (David Bohm, *The Essential David Bohm*, p. 98.)

Imagine it! According to Zero-Point Theory, ALL of the energy needed to make ALL of the physical matter in our observable physical universe can be found or contained within a single cubic centimeter of space at the quantum level.

We have the truth on the one hand and the lie on the other hand. Most scientists have chosen to believe in the lie. However, the truth is that it's all made from energy or light – what the new agers call the Quantum Sea of Light. We'll never run out energy that can be converted into physical matter. And, whenever physical matter reaches thermal equilibrium or maximum entropy, the psyches controlling it can turn it back into raw usable energy anytime they choose to do so. In other words, entropy is a non-starter. Entropy is an illusion. There is more energy in one cubic centimeter of space at the quantum level than in all of the physical matter in the whole universe combined. All it takes is the energy of one cubic centimeter of space at the quantum level to make the whole physical universe that we see around us today. Are you starting to see it yet?

There is no such thing as heat death, or entropy, or running out of energy where the Zero-Point Field of Light is concerned. At the quantum level, there's enough energy in one cubic centimeter of space to make a whole other physical universe just like this one, at will. In other words, if you are a Psyche with a cubic centimeter of space under your complete control, you have enough energy under your control to make a whole other physical universe like this one out of the energy that is under your control. That's completely different than what the Materialists, Naturalists, Nihilists, and Atheists are teaching us when they tell us that we and the universe are going to end in entropy or heat death, and that we cease to exist when we die.

So, who is right and who is wrong? Which model has actually been seen, observed, experienced, and experimentally verified? Do we observe the existence of Psyche or Intelligence; or, do we observe the non-existence of Psyche, Consciousness, or Intelligence? Answer that, and you figure out how everything really works! As I see it, in order to be an Observer or a Scientist, you actually have to be conscious or intelligent, which means that you have to have some kind of Psyche or Life Force in there somewhere, or you couldn't be an Observer nor a Scientist.

Back in 2012, I was a Materialist, Naturalist, Nihilist, and Atheist; but, I was also an open-minded scientist who was looking for the truth. All I wanted was the truth, and I eventually got tired of these people lying to me. After interacting with these people online for two or three years, I started to realize that they were lying to us and trying to trick us and deceive us. I started to see through the charade. Eventually, I got to the point where I wanted to find out what they were getting wrong. So, what is the thing that Einstein, the Materialists, Naturalists, Atheists, Darwinists, Nihilists, Classical Physicists, and Mechanists are getting wrong?

As I see it at this point in time (2019), the main problem that these people have is that they don't realize that space is Multi-Phasic. The Multiphasic nature of space is what they are constantly getting wrong. Only a few of the Quantum Physicists like David Bohm, Henry P. Stapp, and Pim Van Lommel are in fact getting it right. Everyone else has been getting it wrong.

Again, the answer and the truth are hidden in what David Bohm states here:

> **When this length is estimated it turns out to be about $10-33$ cm. This is much shorter than anything thus far probed in physical experiments (which have got down to about $10-17$ cm or so). If one computes the amount of energy that would be in one cubic centimeter of space, with this shortest possible wavelength, it turns out to be very far beyond the total energy of all the matter in the known universe.**
>
> **What is implied by this proposal is that what we call empty space contains an immense background of energy, and that matter as we know it is a small, "quantized" wavelike excitation on top of this background, rather like a tiny ripple on a vast sea.** (David Bohm, *The Essential David Bohm*, p. 98.)

How is it possible for one cubic centimeter of space to contain enough energy to make ALL of the physical matter in our known universe? Isn't that an oxymoron or a contradiction in terms? It is, if physical matter and our physical universe are the only thing that exists, as the Materialists, Naturalists, Darwinists, Nihilists, Mechanists, and Atheists claim. However, these people are always wrong, and always proven wrong.

So, what gives? What are they always getting wrong?

The answer is that these people don't realize and accept the fact that Space is multi-use or multi-phasic. Space has depth, or levels, or dimensions, or phases. There's enough energy in a centimeter of space at the quantum level or the zero-point level to make all of the physical matter that we see around us here at the physical level. Space is comprised of different levels of reality, different phases, or different dimensions. Space is not all one thing, and it's not completely empty as the Materialists and Naturalists claim. At the quantum level in the Spirit World, space is chock full of energy. The Spirit World is right here within us. The dark matter and dark energy are right here within us. The gravity waves and the quantum waves and the psychic telepathic waves are right here within us. It's all taking place within the same space at the same time, just at different levels of existence. There's a psyche level, there's a quantum level, and there's a physical level.

The FLAW and WEAKNESS of the Materialists, Naturalists, Darwinists, Nihilists, and Atheists is their claim that physical matter or the physical level is the only thing that exists. These people really truly teach and believe that entropy, death, and therefore entropic physical matter are conserved. They are demonstrably wrong. Space is multiphasic. It contains multiple levels within it. Einstein's biggest blunders became our biggest blunders. Remember, Quantum Field Theory and Quantum Mechanics FALSIFY Materialism, Naturalism, and their derivatives such as Atheism (Creation Ex Nihilo By Nihilo), Nihilism (Space Is Empty), and Darwinism (Spontaneous Generation or Abiogenesis).

Within his theoretical physics, David Bohm talks about an explicate order or a physical level of existence, an implicate order or a quantum level of existence, and a super-implicate order or an organizing principle which would in fact be a psyche level of existence. These are different levels or phases of reality that exist within the same space at the same time. The physical level or explicate order appears to be, from our perspective at the physical level, rarefied or spread-out or limited in energy. Almost all of the energy is hidden at the lower levels where we can't get at it. In contrast, Psyche and Spirits can get at that hidden energy all the time, because at their level the energy is available all the time. Energy is exergy (fully available) at the quantum level and the psyche level. There is NO entropy or anergy at the quantum level. This is the thing that the Materialists and Naturalists don't realize. There are NO physical limitations at the quantum level within Bohm's implicate order.

At the quantum level and the psyche level, energy is isothermal and non-thermal and exergic, which means that all of the energy is available all the time and also means that the energy is always conserved or preserved. At the quantum level there is no thermodynamics, which means that the second law of thermodynamics or entropy does not exist at the quantum level. Only the conservation of energy or the conservation of psyche exists at the quantum level. At the quantum level or the psyche level, energy is exergy, which means that there is NO entropy at the quantum level or the psyche level. This is the thing that Einstein, and the Materialists and Naturalists are getting wrong. They have erroneously chosen to believe that the physical limitations such as the speed-of-light and entropy apply at the quantum level and the psyche level, and they don't.

> **Thus, the super-quantum potential expresses the activity of a new kind of implicate order. This implicate order is immensely more subtle than that of the original field, as well as more inclusive, in the sense that not only is the actual activity of the whole field enfolded in it, but also all its potentialities, along with the principles determining which of these shall become actual.**
>
> **I was in this way led to call the original field the first implicate order, while the super-quantum potential was called the second implicate order (or the super-implicate order). In principle, of course, there could be a third, fourth, fifth implicate order, going on to infinity, and these would correspond to extensions of the laws of physics going beyond those of the current quantum theory, in a fundamental way. But for the present I want to consider only the**

second implicate order, and to emphasize that this stands in relationship to the first as a source of formative, organizing, and creative activity.

It should be clear that this notion now incorporates both of my earlier perceptions – the implicate order as a movement of outgoing and incoming waves, and of the causal interpretation of the quantum theory. So, although these two ideas seemed initially very different, they proved to be two aspects of one more comprehensive notion. This can be described as an overall implicate order, which may extend to an infinite number of levels and which objectively and self-actively differentiates and organizes itself into independent sub-wholes, while determining how these are interrelated to make up the whole.

Moreover, the principles of organization of such an implicate order can even define a unique explicate order, as a particular and distinguished sub-order, in which all the elements are relatively independent and externally related. To put it differently, the explicate order itself may be obtainable from the implicate order as a special and determinate sub-order that is contained within it.

All that has been discussed here opens up the possibility of considering the cosmos as an unbroken whole through an overall implicate order. Of course, this possibility has been studied thus far in only a preliminary way, and a great deal more work is required to clarify and extend the notions that have been discussed in this paper. (David Bohm, *The Essential David Bohm*, pp. 196-197.)

Space had depth. Space has layers. Space is Multi-Phasic. Space at the physical level contains tons of non-physical stuff within it! Space at the physical level contains lots of non-physical layers within it. Energy is mostly non-physical. The only time energy is physical is when the Gods or the Controlling Psyches have formed that energy into entropic physical matter. The rest of the time, energy is non-physical. The verified and proven existence of the non-physical FALSIFIES Materialism, Mechanism, Classic Physics, Newtonian Physics, Naturalism, Physicalism, Atomism, Darwinism, Atheism, and their derivatives such as Behaviorism and Determinism.

https://psyche-ontology.com/buhlman/

https://psyche-ontology.com/psyche-experienced-and-observed/

https://psyche-ontology.com/psyche-observed/

This is it. This explains what the Materialists, Naturalists, Darwinists, Nihilists, Behaviorists, Determinists, and Atheists are getting wrong. Physical space contains within it some type of quantum space or spiritual space, which contains within it some kind of psyche space or organizing space or conscious space, which contains within it all of the energy or exergy. It's nested, or layered, or multiphasic; and for all we know, there could be an infinite number of levels at the same time within the same cubic centimeter of space. That's how it's possible for one cubic centimeter of space at the quantum level or zero-point level to contain within it enough energy to make all of the physical matter that we see around us here at the physical level within our physical universe.

This means that there is NO limit to the number of physical universes that the Gods or the Controlling Psyches can make from all of the exergy or energy that exists at the quantum level, and the psyche level, and the infinite number of levels that exist at the zero-point level or the exergy level – where entropy, thermodynamics, and anergy do not exist and cannot exist thanks to the Conservation of Energy that reigns supreme at the super-implicate level or lowest levels of existence.

The destroyed exergy, or expended energy, or used exergy has been called anergy or entropy. Exergy is the energy that is available to be used. We see and experience a lot of entropy or anergy at the physical level that simply doesn't exist at the quantum level. Entropy, or thermodynamics, or the aging process is exclusively a physical phenomenon. It doesn't exist at the quantum level. It's ALL exergy or syntropy at the quantum level. It has to be, or the quantum level would have filled up with entropy eons ago, and we wouldn't and couldn't exist today. Someone Psyche at the quantum level has to KNOW how to do syntropy, or the reversal of entropy, or the conservation of energy; otherwise, quantum fields and physical matter would be impossible to make and sustain today.

Whenever I choose to let Psyche and Quantum Mechanics in to play, I can literally explain everything that comes my way. The explanatory power of Intelligence and Quantum Mechanics is vastly superior to anything that we can get from the Materialists, Naturalists, Darwinists, Nihilists, and Atheists who assure us that Psyche, Intelligence, Consciousness, Quantum Fields, Life Force, Syntropy, and Quantum Mechanics DO NOT EXIST. How does claiming that something does not exist explain what it is and how it works? It doesn't! And, that's why Materialism and Naturalism fail in the end. They lack explanatory power!

Remember, exergy is the energy that is available to be used; and at the quantum level, ALL of the energy is exergy thanks to the Conservation of Psyche or the Conservation of Energy that predominates at the quantum level and the psyche level. There is NO thermodynamics, anergy, or entropy at the quantum level. It doesn't exist at the quantum level. Only the First Law, or the Conservation of Energy, or Syntropy, or Exergy exists at the quantum level. This is the Quantum Law of Thermodynamics.

This is what the Materialists, Naturalists, Darwinists, Nihilists, Mechanists, and Atheists are constantly getting wrong, and this is why their theories, philosophies, and beliefs are constantly being FALISFIED by Quantum Mechanics, Action at a Distance, the Quantum Zeno Effect, Quantum Tunneling, and Quantum Field Theory. The proven and verified existence of Quantum Field Theory, Action at a Distance, Quantum Mechanics, the Quantum Zeno Effect or Telepathy, Quantum Tunneling or Teleportation, Quantum Phase-Shifting, and the Conservation of Psyche or Energy FALSIFIES Materialism, Naturalism, and their derivatives such as Darwinism and Atheism.

The false is falsified by the truth; and, the truth is repeatedly experienced and observed. Psyche, Intelligence, or Consciousness is constantly being experienced and observed; yet, the Materialists and Naturalists are constantly telling us that it doesn't exist. It's obvious that they are wrong. I'm seeing or observing Intelligence, Consciousness, Self-Awareness, Life Force, or Psyche all around me all the time. It's unavoidable.

My dog is just smart enough to get into trouble all the time, but not smart enough to get out of it every time. The same reality applies to the Materialists, Naturalists, Darwinists, Classical Physicists, Mechanists, and Atheists. They are just smart enough to create a whole host of unsolvable problems, but they aren't smart enough to solve them. Psyche, Intelligence, Life Force, Consciousness, Quantum Mechanics, Quantum Field Theory, the Quantum Zeno Effect, Quantum Tunneling, Quantum Multi-Phasing, Quantum Superposition, and Quantum Field Theory SOLVE everything that the Materialists and Naturalists can't solve with their falsified philosophies and beliefs. This is what has been experienced and observed.

You are going to have to discover these truths for yourself, or go without. The best I can do for you is to show you where to look; but, you are going to have to do the work of studying and learning for yourself. You won't find these truths in our public school system, though, because in the Western World our public education systems are run by the Materialists and Naturalists and the purpose of a public education is to indoctrinate you in the religious dogma of Materialism, Naturalism, Darwinism, Nihilism, and Atheism. If you are not an Atheist by the time you graduate from college, then your college professors have failed in their primary mission where you are concerned. Their ultimate goal is to have you dumb and blind without thoughts or a mind. It's what they do.

Ultimately, the Materialists, Naturalists, Darwinists, Nihilists, Behaviorists, Mechanists, and Atheists weren't able to explain to me what I needed to know and wanted to know, so it was time for me to upgrade my Science to something better. And, I did. With Psyche and Quantum Mechanics under my sway, I can literally explain everything that comes my way. It looks like to me that Quantum Mechanics, Action at a Distance, Psyche, and Quantum Field Theory are here to stay. Materialism and Naturalism are dead. Long reign the new paradigm – Quantum Field Theory and Quantum Mechanics!

It's good to finally have the truth and to know how everything really works.

Mark My Words

ACKNOWLEDGEMENT OF MENTORS AND TEACHERS

Obviously, I didn't write this in a vacuum. I took inspiration and insight and SCIENCE from people, whom I chose to be my teachers and mentors in this journey of exploration and discovery. I thank them sincerely for taking the time to study and prepare themselves to teach.

I want my readers to be able to get access to the following FREE information, which I have listed and archived at the following link:

http://www.markme.us/id/

I thank the original contributors for it!

—

I express gratitude to John C. Sanford for his written permission to quote from any of his books as much as I want, so long as I cite the source. His books are in fact the top three that I recommend to my readers and critics as the first books about Evolution that they should buy and read. It also helps that a couple of his books can be found for free online. John C. Sanford is a wonderful individual, which I can't say about many if not most of the Darwinists whom I have encountered online.

When it comes to the Theory of Evolution, the Darwinists themselves are often its worst enemy. The Darwinists and Materialists and Atheists can be very unpleasant people to interact with.

If you are trying to take down the Theory of Evolution, all you really need is John C. Sanford's scientific research and his books.

I archived his Free Book which he permitted me to quote at the following link:

http://www.markme.us/wp-content/uploads/2016/06/Biological-Information-Synopsis.pdf

I downloaded and zipped up their much larger FREE 584 page science book For Personal Use at this link:

http://www.markme.us/wp-content/uploads/2016/06/Biological-Information-New-Perspectives.zip

The original sources are listed at this link:

http://www.markme.us/id/

You should have noticed that I quoted John Sanford's science book "Genetic Entropy" 4th Edition THE MOST in this book, because his book was the book which did me the most good when it came to the Theory of Evolution. Go with the best! "Genetic Entropy" became my "bible" when it comes to the Theory of Evolution or Creation by Evolution.

—

The people at Discovery Institute never seem to look at their email; and, the email contact links on their websites are broken or dead. Apparently, people are writing them email that they don't want to read. It took EIGHT months to get my first email response from them to a question that I had asked them. My inability to correspond with them hasn't been for a lack of trying. However, I do thank them for their FREE ARTICLES which I have tried to promote when appropriate, and point my readers to; and, I do thank them for their Books and Movies which have explained to me many different reasons why the Theory of Evolution is FALSE.

When I finally got that response from the Discovery Institute, they gave me a link to a FREE "book" online that answered my question, with the implication that they want me sharing that FREE book with all of my Darwinist and Atheist friends who are willing to look at it and read it. It was 92 pages, so I simply linked to it here in this book.

http://www.discovery.org/scripts/viewDB/filesDB-download.php?command=download&id=10141

I also dedicated a page to it at this book's associated website:

http://www.markme.us/id/

Instead, since they told me to share their free material with my Atheistic and Darwinist friends, I downloaded and linked to their much shorter review:

http://www.signatureinthecell.com/responses/response-to-darrel-falk.php

Good enough.

I have indeed made the attempt to quote and PROMOTE their message within the limits of the FAIR USE copyright laws — meaning that I go out of my way to point my readers to them without going overboard quoting them. I particularly PROMOTE the writings of Stephen C. Meyer and Jonathan Wells, because their books have given the Darwinists the most grief — here in America at least.

—

I send out a hello and a salute to Harun Yahya, another person that the Darwinists definitely do not want us seeing and reading, for good reason. In every one of his books, he has a chapter explaining why Darwinism is EVIL and how the Theory of Evolution is FALSE. I have quoted Harun Yahya's material sparingly because it can ALL be found for free online at his websites.

Harun Yahya's websites and 238 free books are the things that I chose to highlight and promote in this book. I'll leave the rest of it up to you.

There are 238 English books from Harun Yahya which you can download for free in their pristine and beautiful PDF format!

http://www.harunyahya.com/list/type/1/name/Books

As a way of introducing and promoting Harun Yahya, I also link to his FREE controversial introductory article about Evolution and Walcott, which Yahya was using to promote his four volume "Atlas of Creation":

http://www.harunyahya.com/en/Articles/9149/the-evolutionist-tradition-of-concealing

Archived at:

http://www.markme.us/id/

I chose this article among the hundreds that Yahya has available, because it is the article which the Darwinists wanted to debate about most online, and because it seemed to get the Darwinists all stirred up causing them the most consternation and grief.

Good enough!

—

My thanks goes out to Hugh Ross. I have purchased all of his books and have been reading them. Hugh Ross provides the very best Scientific Evidence demonstrating that the Bible is true and scientific.

I think Hugh Ross alone has pointed me to a thousand different reasons why the Theory of Evolution is worthless and false.

I have tried to contact Hugh Ross multiple times and finally got a response from his secretary. I was told that I could quote from "The Genesis Question" based upon FAIR USE, if I kept the number of quoted words from Hugh Ross under 200 words. So, whatever I might quote from Hugh Ross will be quoted sparingly.

However, I can send you to his FREE website for lots of FREE and interesting articles which he and his associates have written about Science, Evolution, and the Biblical God.

http://www.reasons.org/

Hugh Ross seems to be one of the first scientists to figure out why the Theory of Evolution is FALSE. As the book "The Genesis Question" demonstrates, Hugh Ross knew what was wrong with the Theory of Evolution back in 1998 when he first published "The Genesis Question". Hugh Ross knew THE TRUTH about Evolution TWENTY years before I started to figure out what's wrong with Darwinian Chance and Evolution. Cool, huh?

I quote Hugh Ross sparingly within the limits of the FAIR USE copyright laws, but refer to him generously whenever it feels right to do so. I'm currently working on a book entitled, "Scientific Proof of God's Existence"; and, I intend to reference Hugh Ross a lot in that book, which is one of the reasons why I refer to him infrequently in this book. I just want people to know that Hugh Ross and his books can explain to you why the Theory of Evolution, and "Creation by Darwinian Chance", and Creation by Evolution must be false.

One of the purposes for this book is to use the Scientific Method to explain to people why I am no longer an Atheist and why I no longer believe in the Theory of Evolution. Hugh Ross definitely had a hand in that process.

—

I thank Gerald L. Schroeder for his book, "The Science of God: The Convergence of Scientific and Biblical Wisdom". That book is a Top Ten favorite of mine. It gave me a working model of the Big Bang and the 15 billion year age of the universe, which I had never considered before.

As a way of introducing Dr. Schroeder, with his written permission, I included a link to his FREE article from his website, where he explains what's wrong with the Theory of Evolution. I thank Dr. Schroeder for giving me written permission to include his essay in my books, even though I didn't include it in this particular book.

http://www.geraldschroeder.com/Evolution.aspx

Archived at:

http://www.markme.us/id/

—

I thank William Buhlman for writing "Adventures Beyond the Body: How to Experience Out-of-Body Travel" and sharing his personal experiences with us. That book is also a Top Ten favorite of mine because it single-handedly put an end to Materialism, for me.

I never contacted William Buhlman for permission to quote from his book, because the copyright on that book permits limited quoting for review and promotional purposes. My quotes were indeed a small fraction of that book.

—

I believe that all the other quotes in this book are within the FAIR USE copyright laws and are used for review and promotional purposes in an attempt to get my readers to look at and read these people's books. These were the books and these were the people who helped me to see and understand why Materialism and the Theory of Evolution are WRONG.

I thank these individuals sincerely for all that they have done for me over the past year or two.

Obviously, if a particular author doesn't want me quoting and promoting his or her book in the book that I have created here, then I will be more than happy to remove their quotes from a future edition of this book. But, I think I have covered all of the bases and gotten all of the permissions that I needed to acquire.

I thank these individuals for being my teachers in this journey of exploration.

—

The purpose of the Scientific Method is to help us to find THE TRUTH, through a preponderance of the evidence.

the Scientific Method has no value to us if we use it to convince ourselves that a LIE is TRUE, as the Materialists and Darwinists always seem to do.

That's what I discovered during my Pursuit of the True Reality of All Things, and during my usage of the Scientific Method.

Evolution is dumb and blind without hands and a mind!

Each failure of Evolution to do what that Darwinists say that it does becomes yet another miniature proof of God's existence.

Peer-Reviewed Science Articles Supporting Intelligent Design!

When it comes to Intelligent Design, the biggest criticism which I always hear from the Atheistic Darwinists is that Intelligent Design Theory has NO peer-reviewed Science Articles supporting it.

It's a dodge!

That's just another one of their Darwinian myths or lies, which the Materialists are desperate to believe in and try to convince us is true. THE FACT IS that the Darwinists' own Scientific Research supports the Intelligent Design Theory BETTER than it supports the Theory of Evolution.

Nevertheless, I have Atheists and Darwinists pestering me for peer-reviewed science articles that cast doubt on the Theory of Evolution, because they say that there are none.

EIGHT months after asking Discovery Institute for a list of peer-reviewed articles supporting ID, someone at the Discovery Institute finally sent me a link to peer-reviewed research supporting the Theory of Intelligent Design. Thank you!

Here is a part of the message that I received from the Discovery Institute:

> Here's a link to a regularly updated, annotated bibliography of ID-friendly and ID-supporting peer-reviewed research:
>
> http://www.discovery.org/scripts/viewDB/filesDB-download.php?command=download&id=10141
>
> Food for thought for your Atheist and Darwinist friends . . .

I took this reply to mean that they wanted me sharing this 92 page book with all of my Atheist and Darwinist friends. I had indeed asked for something that I could share with my friends.

This "book" from the Discovery Institute is entitled, "BIBLIOGRAPHIC AND ANNOTATED LIST OF PEER-REVIEWED PUBLICATIONS SUPPORTING INTELLIGENT DESIGN".

I also dedicated a page to this FREE book on my associated website, in case their link should ever go dead.

http://www.markme.us/id/

Peer-reviewed science articles supporting Intelligent Design!

This is another one of those things that the Darwinists and Materialists don't want us seeing and reading, because they don't want us finding out THE TRUTH about Evolution. I have had an endless number of Materialists tell me that there is no Science supporting Intelligent Design Theory, which is a lie, especially given the fact there is NO SCIENCE supporting goo-to-you evolution and Creation by Darwinian Chance or Creation by Evolution.

The Materialistic Darwinists are losing this war, which is why they have to lie and attempt to discredit their opponents in order to make a case for the Theory of Evolution, because the Scientific Evidence has turned against them.

—

I link to the following FREE Article with IMPLIED PERMISSION from the Discovery Institute, because I saw this article as the best introduction to Intelligent Design Theory and the book "Signature in the Cell" by Stephen C. Meyer; and, it's a lot shorter than their 92 page book which they encouraged me to share with my friends.

Source: http://www.signatureinthecell.com/responses/response-to-darrel-falk.php

I also have a backup up this article at this book's associated website, in case their link should ever go dead:

http://www.markme.us/wp-content/uploads/2016/06/Signature-in-the-Cell-Responses.pdf

I was told in writing by the Discovery Institute to share their free material with my Atheistic and Darwinistic friends. This is the article of theirs which I chose to share and promote in its entirety within this book. I wasn't going to include their 92 page book into this thing, especially since their free "book" is considered to be a work in progress. Even though I requested direct permission to quote this article in my books, if history is any indication, I probably won't get a response from them until a year from now, if at all.

—

I thank these individuals for being my teachers in this journey of exploration.

Evolution is dumb and blind without hands and a mind!

Each failure of Evolution to do what that Darwinists say that it does becomes another miniature proof of God's existence.

NOT A CHANCE

While writing the book "The Theory of Evolution Proves that God Exists: Why I Am No Longer an Atheist and Why I No Longer Believe in the Theory of Evolution", I came late to "Not a Chance: God, Science, and the Revolt against Reason" by R. C. Sproul and Keith Mathison.

Their book provided the final piece to the puzzle; and, I thank them for it.

Their book ended up putting a capstone on my research about the Theory of Evolution, explaining to me that CHANCE does not exist; therefore, it couldn't have designed and created anything, let alone designing and creating everything as the Materialists claim that it did.

If you read that book, you will see that they explain why CHANCE couldn't have designed and created this universe; but, whatever they say about CHANCE in their book applies equally as well to Darwinian Chance. You find out that the dude doesn't exist, whenever you try to look him up.

I quoted very sparingly from their book in my review per FAIR USE copyright laws, since my primary purpose was to point you to their book and their claim that CHANCE cannot design and create anything, because CHANCE doesn't exist. I don't need to quote the whole book in order to get that point across.

—

My motto and deductive reasoning for this book came from Sherlock Holmes:

How often have I said to you that when you have eliminated the impossible, whatever remains, *however improbable*, must be THE TRUTH? — Sherlock Holmes

The purpose and theme of this book has been to eliminate the IMPOSSIBLE, so that we are left staring at THE TRUTH.

Creation by Darwinian Chance or Creation by Evolution is IMPOSSIBLE, so we have to find another explanation for the origin of all the life on this planet. It's elementary my dear Watson!

Abductive Reasoning seeks for the best and most parsimonious interpretation or explanation for the available evidence. I also employ deductive reasoning. By eliminating ALL of the FALSE PREMISES, we can deduce the CORRECT CONCLUSION. It ALL points to God.

—

CHANCE is a non-person or a non-entity.

CHANCE DOES NOT EXIST.

CHANCE IS NOTHING.

When was the last time that you got NOTHING to do something for you?

The Darwinists have placed ALL of their faith, hope, and trust into Darwinian Chance or Macro-evolution, which doesn't exist.

The Darwinists have placed ALL of their faith, hope, and trust into NOTHING.

The Darwinists literally believe that NOTHING designed and created everything.

What are the odds that NOTHING could have designed and created everything in this universe? NOT A CHANCE! NOTHING by definition in principle does not exist; and, something that does not exist cannot design and create anything.

Goo-to-you evolution IS NOTHING. It does not exist, never existed, and will never exist. It would take a God to bring such a thing into existence.

The Theory of Evolution is NOTHING, because evolution of any kind cannot design and create anything at all. Evolution doesn't exist as a person or an entity. Mutation and Selection working together cannot design and create anything. Knowing that this is true is EVERYTHING that you really need to know about the Theory of Evolution. The Theory of Evolution is NOTHING, meaning that the Theory of Evolution or Creation by Evolution is FALSE. What are the odds that Evolution could have designed and created it all? NOT A CHANCE!

If you are a Darwinist, or a Materialist, or an Atheist, you have literally given your life away for NOTHING and turned your life over to NOTHING, receiving NOTHING in the return. What will you have accomplished from it all after your life is over? NOTHING! What do I have to show for My Atheism? NOTHING!

When you finally remove Darwinian Chance or Macro-evolution from the picture, you are left staring into the face of SOMEONE real and true, the Biblical God Jesus Christ, who actually says repeatedly to us that He is the One who designed and created everything in this universe.

So, which idea is more rational, logical, scientific, and believable — the idea that NOTHING designed and created it all or the idea that the Biblical God Jesus Christ designed and created it all? Which idea is the most parsimonious and fits best with Occam's razor? Which one is actually SUFFICIENT to the task of being a designer, creator, and causal agent? Which one is alive and which one doesn't exist? If your life actually depended upon it which one would you choose to help you and save you?

If you are honest with yourself, then you just simply KNOW that NOTHING has nothing to offer you, because NOTHING doesn't exist and NOTHING can't do anything at all.

This little thought experiment is really SOMETHING, isn't it? Only an Atheistic Materialist will get NOTHING out of it.

—

The Darwinists employ Chance Mutations as their Designer and Creator. That's silly and illogical, because Chance cannot design and create anything at all! Employing Chance as a Designer and Creator IS a Category Error, which is a logic fallacy. The whole Theory of Evolution is based exclusively on these kinds of logic fallacies! Garbage in, then garbage out!

It's too late for me now. I can't go back to Atheism, Materialism, and Darwinism because I can't think of any way to defend them and promote them anymore.

—

Once again, I thank these individuals for being my teachers in this journey of exploration.

The purpose of the Scientific Method is to help us to find THE TRUTH, through a preponderance of the evidence.

the Scientific Method has no value to us if we use it to convince ourselves that a LIE is TRUE, as the Materialists and Darwinists always seem to do.

That's what I discovered during my Pursuit of the True Reality of All Things, and during my usage of the Scientific Method.

Evolution is dumb and blind without hands and a mind!

Each failure of Evolution to do what that Darwinists say that it does becomes yet another miniature proof of God's existence, because God MUST EXIST in order to do all of the Science and Creation that evolution or chance could NEVER have done.

Cool, huh?

The Associated Websites

http://www.markme.us/

http://www.markme.website/

The Facebook Page for Mark My Words:
https://www.facebook.com/MarkMyScience/

Twitter for Mark My Words:
https://twitter.com/Mark_Me_Words

—

Documenting my understanding of the Scientific Method.

Demonstrating how I use the Scientific Method in my pursuit of THE TRUTH.

The Associated Website: http://www.markme.website/scientific-method/

The Associated Forum: http://www.markme.us/forums/forum/scientific-method/

—

If you liked this book and want to spread the word, then be sure to send your friends to one of the following links where they can get their own free copy of another book's introductory summary entitled, "The Theory of Evolution Proves that God Exists — A Summary":

http://www.markme.website/free-book/

http://www.markme.us/free-book/

—

If you find grammatical and spelling errors in any of my books, please sign up at the associated forum and report them.

http://www.markme.us/forums/forum/report-errors-in-mark-my-words-books/

Please don't sign up just to tell me that I'm "effin knutts" because I know that already and other Materialists and Darwinists have beaten you to it. Please tell me something that I don't already know.

—

The purpose of the Scientific Method is to help us to find THE TRUTH, through a preponderance of the evidence.

the Scientific Method has no value to us if we use it to convince ourselves that a LIE is TRUE, as the Materialists and Darwinists always seem to do.

That's what I discovered during my Pursuit of the True Reality of All Things, and during my usage of the Scientific Method.

—

I wish that I would have had access to the information in this book fifty or a hundred years ago; but, most of what I used to convince myself that the Theory of Evolution or Creation by Evolution is FALSE was not available a decade ago. During the past decade, there has been a veritable explosion of knowledge and scientific information, all of which demonstrates clearly and conclusively that Creation by Evolution is IMPOSSIBLE.

Each failure of Evolution to do what that Darwinists say that it does becomes yet another miniature proof of God's existence. God MUST EXIST in order to have done all of the design, creation, manufacturing, and SCIENCE which Evolution and Chance could never have done.

Creation by Evolution is IMPOSSIBLE and FALSE.

Creation by Intelligent Beings is REAL and TRUE.

Keep the best and get rid of all the rest. That's what I try to do.

Mark My Words!

Other Books by This Author, Mark My Words!

My Author Page on Amazon:

https://www.amazon.com/-/e/B01IAEF2Y6

https://amazon.com/author/science

My Facebook Page:

https://www.facebook.com/MarkMyScience/

My Twitter Page:

https://twitter.com/Mark_Me_Words

—

1. "Summary Of: The Theory of Evolution Proves that God Exists"

https://www.amazon.com/dp/B01GQCWED6

https://www.amazon.com/dp/1521130485

I used to be a Materialist, Naturalist, Nihilist, and Atheist. I was a practitioner and promoter of Scientism; but, I naturally resisted Darwinism due to all the obvious holes in the theory. I had to fully understand what these philosophies are and whether they had been verified or falsified before I could move on to other types of science. This book documents part of that journey.

An introductory summary to my much larger book, "The Theory of Evolution Proved to Me that God Exists: Why I Am No Longer an Atheist and Why I No Longer Believe in the Theory of Evolution".

The second edition of this book contains an introduction to my much larger book, "The Ultimate Model of Reality: Psyche Is the Ultimate Cause", which is my core message to the world.

07JUN2016

—

2. "The Theory of Evolution Proved to Me that God Exists: Why I Am No Longer an Atheist and Why I No Longer Believe in the Theory of Evolution"

https://www.amazon.com/dp/B01HZYBZ7K

https://www.amazon.com/dp/1521131228

I used to be a Materialist, Naturalist, Nihilist, and Atheist. I was a practitioner and promoter of Scientism; but, I naturally resisted Darwinism due to all the obvious holes in the theory. I had to fully understand what these philosophies are and whether they had been verified or falsified before I could move on to other types of science. This book documents part of that journey.

Over the period of a few months, I had an epiphany or a major breakthrough; and in the process, the Theory of Evolution proved to me clearly and conclusively that God Exists. I have had no doubt about God's existence, after the Theory of Evolution proved to me that God Exists. By that point in time, I was officially done with My Materialism and My Atheism and firmly in a different camp. I'm now a part of God's army. This huge book, "The Theory of Evolution Proved to Me that God Exists", explains in great detail why I am no longer an Atheist and why I no longer believe in the Theory of Evolution.

This book documents my journey of discovery and the defense of my thesis.

04JUL2016

—

3. "The Scientific Method Proves That the Theory of Evolution Is False"

https://www.amazon.com/dp/B01IAAIRT2

https://www.amazon.com/dp/1521133611

I used to be a Materialist, Naturalist, Nihilist, and Atheist. I was a practitioner and promoter of Scientism; but, I naturally resisted Darwinism due to all the obvious holes in the theory. I had to fully understand what these philosophies are and whether they had been verified or falsified before I could move on to other types of science. This book documents part of that journey.

In this book, I document what I know about the Scientific Method, and I demonstrate how I use The Scientific Method in my pursuit of The Truth.

I have learned to trust The Scientific Method, The Rules of Science, Abductive Reasoning, and Deductive Reasoning to give me a clear and accurate assessment of the evidence at hand. In this book, I use Science, The Scientific Method, the Rules of Science and Scientific Research, Abductive Reasoning, and Deductive Reasoning to prove that the Theory of Evolution is false and why it is false. You will have to judge for yourself if I meet my Burden of Proof through a preponderance of the evidence. I don't hold anything back!

Remember, we can indeed use the Scientific Methods to falsify theories or to prove theories false. Most people don't realize that, but it's true. That's what I did to the

Theory of Evolution – I used various different scientific methods to falsify Darwinism, Naturalism, Materialism, Creation by Rocks, and the Theory of Evolution. Once you have successfully eliminated ALL of the falsehoods such as Materialism, Atheism, Naturalism, and Darwinism, you are left staring at THE TRUTH. It's elementary my dear friend.

The second edition of this book contains an introduction to my much larger book, "The Ultimate Model of Reality: Psyche Is the Ultimate Cause", which is my core message to the world.

11JUL2016

—

4. "The Second Comforter: Supping with Our Resurrected Lord Jesus Christ"

https://www.amazon.com/dp/B01IAKHTY6

https://www.amazon.com/dp/152113281X

I used to be a Materialist, Naturalist, Nihilist, and Atheist. I had to understand Christianity and be able to defend it before I could become a dedicated part of it. This book documents a small part of my efforts in that regard.

1) – 3) The first three parts of this book, are my standard defense package against Materialism. I found that the best defense is a good offense. I first had to get rid of my residual Atheism, Materialism, and Darwinism, BEFORE I was free to pursue knowledge and information about Spirit, Spirituality, Quantum Mechanics or Spiritual Mechanisms, Revelations from God, Revelations of God, the Nature of God, the Scriptures of God, the Holy Ghost, and THE SECOND COMFORTER.
4) Exploring THE SECOND COMFORTER Experience and what it really means to sup with the Lord. The Focal Point of this book.
5) What is an "Anointed Savior", a "Jesus Christ"? Who can be called to fill that Position or Role?

The second edition of this book contains an introduction to my much larger book, "The Ultimate Model of Reality: Psyche Is the Ultimate Cause", which is my core message to the world.

I also included my essays on Lived Experience and My Scientific Discoveries, at the end of this book.

11JUL2016

—

5. "Quantum Mechanics from a Non-Physical Spiritual Perspective"

https://www.amazon.com/dp/B01J023TGU

https://www.amazon.com/dp/1521132380

This was my first venture into Quantum Mechanics. This book was one of my first positive contributions to science. After doing research for this book and after writing this book, I felt like for the first time in my life that I actually understood Quantum Mechanics. Quantum Mechanics makes absolutely NO sense whatsoever from the perspective of Materialism, Naturalism, and Classical Physics. In order to write this book, I had to develop a new and better perspective on the subject of Quantum Mechanics.

I realized that Quantum Mechanics makes perfect logical sense from a Non-Physical Spiritual Perspective. Quantum Mechanics makes no sense whatsoever from a Materialistic Perspective. Materialism is the worst way and the most-limited way to interpret anything, especially Quantum Mechanics. The purpose of this treatise is to demonstrate clearly, logically, scientifically, and conclusively that the Trans-Dimensional Realm, or Non-Local Realm, or Non-Physical Realm, or Spirit Realm, or Quantum Realm truly exists, and that Materialism is FALSE. You will have to decide for yourself if I meet my Burden of Proof through a preponderance of the evidence. What I wrote in this book ended up becoming core foundational material for many of my other books.

The second edition of this book contains an introduction to my much larger book, "The Ultimate Model of Reality: Psyche Is the Ultimate Cause", which is my core message to the world.

24JUL2016

—

6. "Using the Scientific Method to Eliminate the Usual Suspects and to Prove the Truth"

https://www.amazon.com/dp/B01J6STHP0

https://www.amazon.com/dp/1521133581

I used to be a Materialist, Naturalist, Nihilist, and Atheist. I was a practitioner and promoter of Scientism; but, I naturally resisted Darwinism due to all the obvious holes in the theory. I had to fully understand what these philosophies are and whether they had been verified or falsified before I could move on to other types of science. This book documents part of that journey.

Using Science and The Scientific Method to eliminate the Usual Suspects, in an attempt to demonstrate and prove THE TRUTH! In this book, I line up the Usual Suspects, gather and define the Evidence; and then, I apply The Scientific Method, common sense logic, abductive reasoning, and deductive reasoning to them, in order to see if any one of them could have done the job of designing and creating this Physical Universe, this Earth, and ALL of the Genomes and Life Forms on this Earth. The goal is to find the Correct Suspect among all of the Usual Suspects, who are typically presented to us for our consideration. This was a fun book to write, because it made logical sense to me from beginning to end.

Remember, we can indeed use the Scientific Methods to falsify theories or to prove theories false. Most people don't realize that, but it's true. That's what I did to the Theory of Evolution – I used various different scientific methods to falsify Darwinism, Naturalism, Materialism, Creation by Rocks, and the Theory of Evolution. Once you have successfully eliminated ALL of the falsehoods such as Materialism, Atheism, Naturalism, and Darwinism, you are left staring at THE TRUTH. It's elementary my dear friend.

The second edition of this book contains an introduction to my much larger book, "The Ultimate Model of Reality: Psyche Is the Ultimate Cause", which is my core message to the world.

27JUL2016

—

7. "NATURE vs. NURTURE vs. NIRVANA: An Introduction to Reality"

https://www.amazon.com/dp/B01JWRCSVA

https://www.amazon.com/dp/1521132615

I used to be a Materialist, Naturalist, Nihilist, and Atheist. As I began my in-depth study of science and reality, I slowly realized that Materialism, Naturalism, and Darwinism are a horrible and devastating platform from which to do science.

The philosophical dogmatic religions associated with Materialism, Naturalism, and their derivatives are predatory and exclusionary. Materialism and Naturalism were designed to exclude, ban, censor, and block whole branches of science from consideration – things like Quantum Mechanics and Quantum Non-Local Consciousness. The proven and verified existence of quantum mechanisms falsifies Materialism, Naturalism, and their derivatives, which is why the Materialists and Naturalists ban, censor, hide, and block these types of science.

In this book, I take the greatest and most controversial philosophical issues in human history, and I solve them all quickly and easily through a simple change in one's worldview or point-of-view. I introduce a bit of REALITY into the equation, which solves all of these different problems quite nicely.

This book is based upon one of my epiphanies; and thus, the contents of this book form part of my core foundational message to the world.

The second edition of this book contains an introduction to my much larger book, "The Ultimate Model of Reality: Psyche Is the Ultimate Cause", which is my core message to the world.

07AUG2016

—

8. "I Am Not a Creationist: So What Am I?"

https://www.amazon.com/dp/B071XTM8XY

Creationism is always presented by the Materialists, Naturalists, Behaviorists, and Atheists as some type of magic or voodoo.

My scientific observation is that Materialism, Naturalism, and Atheism are in fact THE MAGIC in all of this because Materialism, Naturalism, and Darwinism reduce to "design and creation by rocks", which is synonymous with MAGIC, because the rocks have NEVER been caught in the act of design and creation and never will be.

ONLY intelligent beings or intelligent psyches have the capability to design, create, program, organize, engineer, field-test, manufacture, deploy, and do science.

Quantum Mechanics tells us that Psyche or Non-Local Consciousness and its associated Conscious Observation or Word of Command is required to convert spirit matter into physical matter. Furthermore, scientific observation tells us that ONLY intelligent beings or intelligent psyches have the ability to design, program, engineer, manufacture, field-test, organize, and create new and unique physical objects, including genomes and life forms.

Materialists, Naturalists, Darwinists, and Atheists ridicule and mock Theistic Creation while at the same time choosing to believe that the rocks or raw physical matter designed and created ALL of the genomes and life forms on this planet. As a result, these Materialists and Atheists present creation and creationism as some sort of magical Creation Ex Nihilo. I'm NOT that type of creationist. I do NOT believe in Creation Ex Nihilo or magic. And, I'm not the type of creationist who believes that the rocks or physical matter can design and create genomes and life forms. In other words, I'm NOT a Materialistic, Naturalistic, and Darwinistic Creationist either!

Instead, I choose to believe in the Truths which are presented to us by the Lived Experiences of the human race. We KNOW from the Lived Experiences of the human race that Psyche, Consciousness, Intelligence, or Life exists. We KNOW from the Lived Experiences of the human race that spirit matter is something completely different than Psyche. We KNOW from Quantum Mechanics that matter or Quantum Objects can be in either a spirit matter state of existence or a physical matter state of existence but NOT in both states simultaneously. We KNOW from Quantum Mechanics that Psyche or Conscious Observation is required to convert spirit matter into physical matter, which makes it kind of clear and obvious that Psyche or Non-Local Consciousness is immaterial and doesn't consist of spirit matter nor physical matter.

According to these Truths and KNOWLEDGE which we have gleaned from the Lived Experiences of the human race, we KNOW that Psyche or Non-Local Consciousness has always existed and will always exist. Psyche is immortal, immaterial, infinite, transcendent, and co-eternal with God. We also KNOW that spirit matter has always existed and will always exist; and, we KNOW that psyche is something completely different than spirit matter. Psyche acts. Matter, including spirit matter and physical matter, are acted upon by Psyche.

Again, we KNOW from Quantum Mechanics that ONLY Psyche can convert spirit matter into physical matter. Finally, we KNOW from the observations and lived experiences of the human race that ONLY Psyche or Intelligent Beings can organize physical matter into new, unique, and useful forms such as genomes, physical life forms, computers, cars, skyscrapers, bacteria, dinosaurs, planets, stars, galaxies, universes, and anything else that you can imagine. This scientific observation has been verified trillions of times with an infinite number of more times to go. ONLY intelligent beings or intelligent psyches can design and create and work with physical matter. Scientific methods or scientific observations will ALWAYS verify the truth. In contrast, Materialism and Naturalism and Atheism have been falsified by the Scientific Methods trillions of different times in thousands of different ways, because the rocks can't design and create.

So, technically, I am not a Creationist, because there really is no such thing as Creation Ex Nihilo; and, there's definitely NO such thing as Creation by Rocks or Creation by Physical Matter, because the rocks can't design and create.

So, what am I?

I'm more of an Organizationist and Scientist.

We KNOW from the Lived Experiences of the human race that Psyche and Spirit Matter have always existed and will always exist. We KNOW from Science, Logic, and Quantum Mechanics that ALL physical matter has a beginning when Psyche or Non-Local Consciousness commands spirit matter to convert into physical matter. But, converting spirit matter into physical matter is NOT creation ex nihilo – it is instead transmutation or transformation from one state of existence to a different state of existence.

Finally, when God's Psyche organizes physical matter into genomes and life forms, that process is NOT creation either, although most people would tend to classify it as a creative process. However, technically, God is organizing that physical matter into new and useful types and forms. God is NOT a Creationist either, because God doesn't do magic or creation ex nihilo. God is a Scientist, Engineer, and Manufacturer.

This book explains in part how I finally came to this Ultimate Conclusion. I am NOT a Creationist. I don't believe in Creation by Rocks or creation by physical matter, as the Darwinists, Materialists, Naturalists, and Atheists do. And, I don't believe in Creation ex Nihilo or Creation by Magic as many of the Theists and Christians do.

I'm an Organizationist or Scientist; and, I KNOW from observation and the Lived Experiences of the human race that the Biblical God Jesus Christ and His Father are the Ultimate Scientists and the Ultimate Organizationists. I'm NOT a Creationist, and God and Christ are NOT Creationists either. They are Scientists and Organizationists. They are in the Construction Business.

God and Christ took what already existed and lifted it to a greater and higher form or state of existence and organization. That's technically NOT creation. That's applied Science and Organization and Construction. God and Christ commanded physical matter and our physical space-time into existence; and, then they proceeded to organize some of that physical matter into planets, stars, galaxies, genomes, and physical life forms. That's the work of Scientists, not Creationists.

It's elementary my dear friend!

This book documents a lot of what I went through during 2015 and 2016 while I was trying to discover this information and gain this knowledge.

18APR2017

—

9. "The Ultimate Model of Reality: Psyche Is the Ultimate Cause"

https://www.amazon.com/dp/B071NC9JK6

Eventually, it became clear to me that I had a serious hole in my knowledge, due to five decades of exposure to Materialism, Naturalism, Scientism, Skepticism, and the Theory of Evolution. It was time for a paradigm shift in my own way of thinking about science and doing science.

I dedicated six months of my life to studying the Philosophy of Science and developing a new ontology. I found out what's wrong with the Scientific Method as the Darwinists, Materialists, Naturalists, and Atheists tend to use it. I developed my own Philosophy of Science that attempts to include ALL of the evidence that we have on hand as a race. This knowledge was the KEY to everything else that I was about to develop.

This book introduces my Psyche Ontology or my Ultimate Cause Model of Reality, wherein Psyche is the fundamental unit of reality.

In this book, I present a new and unique Philosophy of Science which demonstrates the existence of Psyche and then proceeds to employ Psyche in ALL aspects of Philosophy, Science, Application, Spirituality, Lived Experience, and Knowledge.

This becomes a Psyche Paradigm or the Ultimate Model of Reality.

22APR2017

—

10. "Putting Psyche Back into Psychology: Restoring Science to Consciousness"

https://www.amazon.com/dp/B071NC987S

Eventually, it became clear to me that I had a serious hole in my knowledge, due to five decades of exposure to Materialism, Naturalism, Scientism, Skepticism, and the Theory of Evolution. It was time for a paradigm shift in my own way of thinking about science and doing science.

I dedicated six months of my life to studying the Philosophy of Science and developing a new ontology. I found out what's wrong with the Scientific Method as the Darwinists, Materialists, Naturalists, and Atheists tend to use it. I developed my own Philosophy of Science that attempts to include ALL of the evidence that we have on hand as a race. This knowledge was the KEY to everything else that I was about to develop.

This book further develops my Psyche Ontology or Ultimate Cause Model of Reality, wherein Psyche is the fundamental unit of reality.

In this book, I document how Psyche was originally an integral part of Psychology, how Behaviorism caused Psychology to lose its mind, and the different types of Personality Theory or Psyche Theory which have attempted to put Psyche back into Psychology.

It is my intention and proposition to include Near-Death Experiences (NDEs), Shared-Death Experiences (SDEs), Out-of-Body Experiences (OBEs), and other types of Lived Experience into Psychology or the "Study of Psyche" as an integral and essential part of Psychology.

NDEs, OBEs, SDEs, Spiritual Experiences, and other types of Lived Experience ARE scientific proof of Psyche's existence. These things should be a part of the Science of Psychology or the Science of Psyche.

22APR2017

—

11. "BioPsychoSocial: Including Psyche or Light into our Theoretical Models"

https://www.amazon.com/dp/B0713NDHVW

Eventually, it became clear to me that I had a serious hole in my knowledge, due to five decades of exposure to Materialism, Naturalism, Scientism, Skepticism, and the Theory of Evolution. It was time for a paradigm shift in my own way of thinking about science and doing science.

I dedicated six months of my life to studying the Philosophy of Science and developing a new ontology. I found out what's wrong with the Scientific Method as the Darwinists, Materialists, Naturalists, and Atheists tend to use it. I developed my own Philosophy of Science that attempts to include ALL of the evidence that we have on hand as a race. This knowledge was the KEY to everything else that I was about to develop.

This book further develops my Psyche Ontology or Ultimate Cause Model of Reality, wherein Psyche is the fundamental unit of reality.

In this book, my primary focus is on Mental Illness and various different forms of Psychotherapy. This book chronicles my search for the Ultimate Psychotherapy, and it will also chronicle in part my brush with mental illness and addiction. Addiction causes psychosis and mental illness. Everyone who goes through withdrawal will experience some kind of psychosis or mental illness.

I promote the BioPsychoSocial Model, compare it to other models for psychology, and then try to explain why the BioPsychoSocial Model is better and has greater explanatory power.

I end this book by discussing what I believe to be the BEST forms of psychetherapy.

22APR2017

—

12. "God Is in the Light: God is light, and in Him is no darkness at all."

https://www.amazon.com/dp/B07168S37N

I used to be a Materialist, Naturalist, Nihilist, and Atheist.

This book was my second solid venture into religion. I felt like I had to understand Psyche and God better than I did before I could move on to other types of research. I wrote most of this book a year and a half before I released it.

In this book, I explore Light and God's Psyche. Psyche is Living Light, or light which is conscious, alive, intelligent, self-aware, and universally connected with its surroundings. While writing this book, I came across a lot of interesting things to think about and study.

23APR2017

—

13. "Tripping the Light Fantastic: How Prescription Drugs Almost Killed Me"

https://www.amazon.com/dp/B071RJP9T8

I got ill.

I willingly let The Doctors get me addicted to half a dozen different prescription drugs; and at the peak of my misery, I was taking over a dozen different prescription drugs on any given day.

I went insane, and I tried to kill myself.

This book chronicles my addiction, and some of the things I went through trying to get sober. I discuss some of the strange things that I experienced during the withdrawal process while I was out of my mind. And, I discuss many of the interesting things that I learned along the way.

This is the book that my friends told me that I must write, because the world needs to hear my story.

23APR2017

—

14. "Scientific Proof of God's Existence: A Primer"

https://www.amazon.com/dp/B071713NNL

https://www.amazon.com/dp/1521325170

This book is a basic Primer explaining how I developed a Scientific Proof of God's Existence. The concepts and ideas were three or four years in the making. It wasn't obvious at first how this should be done or if it could be done.

This book and these essays were the foundational core for everything that was to come next during my scientific research.

18MAY2017

—

15. "Science 2.0: I Upgraded My Science"

https://www.amazon.com/dp/B0771K6WTX

Based upon what I had learned from the Philosophy of Science, I felt that it was finally time for the world to develop a new and better way of doing science – an all-inclusive paradigm, approach, or model that actually allows ALL of the evidence into evidence instead of banning, blocking, censoring, ridiculing, mocking, hiding, and destroying huge chunks of the evidence as the Materialists, Naturalists, Darwinists, and Atheists do.

I upgraded my science to Science 2.0. Under Science 2.0, ALL of the evidence is allowed into evidence; and, Science 2.0 pursues a preponderance of the evidence. *Refusal to look at evidence* is a logic fallacy. Science 2.0 defines science as first-hand observation and experience. Science 2.0 gives preference and priority to evidence, experience, observation, phenomenology, and experimentation OVER personal philosophical speculation, official dogmatic interpretations of scientific research, philosophical sophistry, and scientific inferences.

Science 2.0 makes the claim that the BEST and fastest way to find and know the truth is to live it and experience it for yourself, or to choose to trust someone who has. The second-best way to find and know the truth is to use various different scientific methods to falsify and eliminate everything that is false so that only the truth remains. If you successfully eliminate everything that is false – such as Materialism, Naturalism, Darwinism, Nihilism, Scientism, Behaviorism, Determinism, and Atheism – then ONLY the truth will remain. This is Logic 101.

It's fascinating to observe and study what remains, after all of the falsehoods have been eliminated.

31OCT2017

—

16. "Scientific Proof of God's Existence: Finding God Where the Atheists Refuse to Look for Him"

https://www.amazon.com/dp/B07B26CRHX

I have been working on this book for half a decade now. Anytime I have come across a Scientific Proof of God's Existence, I have tried to get a copy of it into this book.

The Materialists, Naturalists, Darwinists, and Atheists told me that there is NO scientific proof of God's existence and that there will NEVER be any scientific proof of God's existence. I believed them. I believed them for over fifty years of my life, until I discovered that they are wrong. These poor people are almost always wrong. Once I found my first convincing Scientific Proof of God's Existence, dozens more were forthcoming over the years, as I kept my eyes open and my mind open to the possibilities. I like to think that there are always possibilities.

25FEB2018

—

17. "Quantum Neuroscience: The Answer to Life, the Universe, and Everything"

https://www.amazon.com/dp/B079Z6QQQB

This book is my Magnum Opus. All of my research and all of the evidence led me to write this book. Everything in my life has led me to write this book. I made some unbelievable discoveries while researching and writing this book. I'm still having a hard time believing some of the things that I discovered about Quantum Mechanics and the Physical Brain. This book is guaranteed to shock you and surprise you. You are going to resist this one because you have been trained, conditioned, and brainwashed to do so. I KNOW, because I resisted it as well, until the preponderance of the evidence got through to me and convinced me that I was wrong. Enjoy the ultimate Mind Game!

25FEB2018

—

18. "Syntropy in Defense of Quantum Mechanics: The Answer to Life, the Universe, and Everything"

https://www.amazon.com/dp/B07BPT3W8R/

This book is my summation, which I will use to summarize and promote everything that I have discovered so far about science, nature, and reality. Parts of it will be incorporated into my other books as time allows.

25MAR2018

19. "Stealth Ascendancy: The Timeline Chronicles"

https://www.amazon.com/dp/B07BPT3W8R/

This book is the first book on my science fiction series. All of the science fiction and fantasy in this world is based upon Materialism, Naturalism, Darwinism, Nihilism, and Atheism – or magic. This is the first science fiction book on the planet that attempts to use what has actually been experienced and observed as its basis or its "Bible". This book attempts to stay with science that is actually true.

31OCT2019

---

20. "The Denialistic Philosophies of Science: The Rise and Fall of Behaviorism"

https://www.amazon.com/dp/B084LDX7RM/

The Denialistic Philosophies of Science are my own unique discovery; so, I decided to write a book on this topic instead of just short essays. I'm the premier expert in the Philosophy Science, so you will only get this kind of information from me during this century. Our scientists have yet to discover the denialistic philosophies, so they have nothing to offer us when it comes to this topic. By knowing about the denialistic philosophies and what they represent, you can easily separate the truths from the lies in Science and Statistics. The Materialists, Naturalists, Darwinists, Nihilists, Behaviorists, and Atheists have already rejected the Truths in Science that are waiting to be found, so you won't find anything like this in our public schools as long as the atheists continue to control the curriculum in our public schools.

25FEB2020

If your Science isn't proving stuff to you, then you ain't doing it right or your theories are false. Remember, a false theory can be used to point us to the truth, because the truth is often the opposite of any theory which has been demonstrated to be false. Therefore, false theories can be used to point us to the truth and thereby prove the truth.

THE TRUTH, whenever I find it and wherever I find it, IS parsimonious and makes logical sense. THE TRUTH tastes good and feels good. There is NO confusion or doubt associated with THE TRUTH. You just know that it is TRUE.

The primary goal of my books is to set people free from the chains of Materialism, as I have been set free. I hope to do for my readers what other authors have so kindly done for me. To accomplish that goal, I try to explain in every way possible why I am no longer an Atheist and why I no longer believe in Darwinism or Materialism. I have changed a great deal in the past few of years, and I am no longer the person that I used to be. My blinders have been taken off, and I have been set free to pursue my full potential. Come join me!

Salutations!

The purpose of the Scientific Method is to help us to find THE TRUTH, through a preponderance of the evidence.

the Scientific Method has no value to us if we use it to convince ourselves that a LIE is TRUE, as the Materialists and Darwinists always seem to do.

That's what I discovered during my Pursuit of the True Reality of All Things, and during my usage of the Scientific Method.

Have a nice day, and a wonderful life!

Mark My Words!

Original Release: 12JUL2016
Final Version: 25FEB2020
Current Version: 24FEB2020

www.ingramcontent.com/pod-product-compliance
Lightning Source LLC
Chambersburg PA
CBHW071347210526
45465CB00001B/1